Microwave Superconductivity

NATO Science Series

A Series presenting the results of activities sponsored by the NATO Science Committee. The Series is published by IOS Press and Kluwer Academic Publishers, in conjunction with the NATO Scientific Affairs Division.

A. Life Sciences	IOS Press
B. Physics	Kluwer Academic Publishers
C. Mathematical and Physical Sciences	Kluwer Academic Publishers
D. Behavioural and Social Sciences	Kluwer Academic Publishers
E. Applied Sciences	Kluwer Academic Publishers
F. Computer and Systems Sciences	IOS Press

1. Disarmament Technologies	Kluwer Academic Publishers
2. Environmental Security	Kluwer Academic Publishers
3. High Technology	Kluwer Academic Publishers
4. Science and Technology Policy	IOS Press
5. Computer Networking	IOS Press

As a consequence of the restructuring of the NATO Science Programme in 1999, the NATO Science Series has been re-organized and new volumes will be incorporated into the following revised sub-series structure:

I. Life and Behavioural Sciences	IOS Press
II. Mathematics, Physics and Chemistry	Kluwer Academic Publishers
III. Computer and Systems Sciences	IOS Press
IV. Earth and Environmental Sciences	Kluwer Academic Publishers
V. Science and Technology Policy	IOS Press

NATO-PCO-DATA BASE

The NATO Science Series continues the series of books published formerly in the NATO ASI Series. An electronic index to the NATO ASI Series provides full bibliographical references (with keywords and/or abstracts) to more than 50000 contributions from international scientists published in all sections of the NATO ASI Series.
Access to the NATO-PCO-DATA BASE is possible via CD-ROM "NATO-PCO-DATA BASE" with user-friendly retrieval software in English, French and German (WTV GmbH and DATAWARE Technologies Inc. 1989).

The CD-ROM of the NATO ASI Series can be ordered from: PCO, Overijse, Belgium

Series E: Applied Sciences – Vol. 375

Microwave Superconductivity

edited by

Harold Weinstock
Air Force Office of Scientific Research,
Arlington, VA, U.S.A.

and

Martin Nisenoff
M. Nisenoff Associates,
Kensington, MD, U.S.A.

Kluwer Academic Publishers

Dordrecht / Boston / London

Published in cooperation with NATO Scientific Affairs Division

Proceedings of the NATO Advanced Study Institute on
Microwave Superconductivity
Millau, France
29 August–10 September 1999

A C.I.P. Catalogue record for this book is available from the Library of Congress.

ISBN 1-4020-0445-1

Published by Kluwer Academic Publishers,
P.O. Box 17, 3300 AA Dordrecht, The Netherlands.

Sold and distributed in North, Central and South America
by Kluwer Academic Publishers,
101 Philip Drive, Norwell, MA 02061, U.S.A.

In all other countries, sold and distributed
by Kluwer Academic Publishers,
P.O. Box 322, 3300 AH Dordrecht, The Netherlands.

Printed on acid-free paper

dedicated to the memory of
James E. Zimmerman *and* Christoph Heiden

TABLE OF CONTENTS

INTRODUCTION

During the first two years following the discovery of superconductors with transition temperatures greater than 77 K (the boiling point of liquid nitrogen at atmospheric pressure), many fanciful predictions were made about the immediate applications of these ceramic materials. After more than a decade of often-feverish worldwide research activity, most of those euphoric hopes have yet to be realized. This is true despite considerable progress in understanding the fundamental mechanisms responsible for superconducting behavior in perovskite ceramics and in significant advances in processing technology to produce thin-film, thick-film and tape structures. However, if there is one application area that already has had a significant impact in the commercial and military markets, it is the area of passive microwave components. Specifically, thin-film and thick-film HTS microwave resonators with extremely high values of the quality factor, Q, have been used to fabricate filters that provide very low insertion loss and very sharp frequency roll-off characteristics ("skirts") that only can be realized using conventional filters that are about two orders of magnitude larger in volume and one order of magnitude larger in mass. A complete filter system, containing either six or twelve filters, *with its associated closed-cycle cooling system*, can be packaged to fit into a conventional electronic instrument rack, with the height of the system only 10 to 20 cm. The operator need not know that the key components are operating at very cold temperatures, typically, close to the boiling point of liquid nitrogen (77 K).

Currently, there are many - possibly eight to ten – (mostly) small companies throughout the world manufacturing such filter systems. It is estimated that the total number of such systems manufactured to date is about 1,000 units. Many units are undergoing beta-site testing in the base stations of numerous cellular communications systems. There have been published reports from the HTS filter vendors that the use of HTS filter systems reduces the number of "dropped calls" and, correspondingly, this should produce increased revenue as a result of this more efficient operation – see the chapter in this volume by Balam Willemsen for details. Despite the obvious improvement in system performance, the service providers are still performing cost-benefit trade-off studies before committing to large-scale introduction of HTS filter systems to their operating systems.

It is clear also that the world's military forces are considering the insertion of HTS filters into their surveillance and communications systems. Although it is difficult to predict the degree to which military usage will drive this industry, it is probably correct to say that military adaptation of the technology is a certainty; only the magnitude is uncertain.

To summarize, there are numerous emerging markets for passive HTS microwave filters and related components. While it is difficult to predict the ultimate size of these markets, under the most optimistic scenarios, the market could exceed 100,000 units per year globally. Within the next two to three years, it should be far more certain what the actual size of the market will be for these products. At this time, however, there is considerable interest in learning how to improve the microwave properties of HTS materials, how to characterize the materials and evaluate circuit parameters, how to improve the design of HTS microwave and radio-frequency circuits, and how to provide efficient, compact cryogenic packaging.

This volume, based upon lectures presented at the NATO Advanced Study Institute (ASI) held August 29 to September 10, 1999 in Millau, France, has been specifically planned to provide the basis for a graduate-level course on microwave superconductivity. Additionally, it is designed to serve as a convenient reference for people conducting research in this field. The inspiration for the NATO ASI itself was a desire on the part of university researchers to teach a class on this subject, but who found there wasn't a suitable textbook upon which to base such a course. Although the discussion above speaks only of the major application of HTS microwave filters, a course in microwave superconductivity must be comprehensive. It should cover the relevant fundamentals of electromagnetism, fabrication and characterization of materials, a variety of applications, and cryocooler technology. Although the focus is primarily on HTS superconductors, some applications involving liquid-helium-based (or LTS) superconductivity have been included. Applications or structures covered (either HTS or LTS) involve antennas, lumped elements, RF accelerators, RF SQUIDs, RF SQUID-based amplifiers, long-Josephson-junction circuits, and (of course) filters. While it is impossible to claim all applications have been covered, those that have been covered are the ones that already have been found to be (or show promise to be) significant with regard to the commercial, military and research communities. Additionally, measurement techniques and cryogenic systems have been considered.

Arlington, Virginia — Martin Nisenoff
October 2001 — Harold Weinstock

ACKNOWLEDGMENTS

This book comprises the official proceedings of the NATO Advanced Study Institute (ASI) held 29 August to 10 September 1999 at the International Hotel in Millau, France. The Institute would not have been possible without generous financial support from the Scientific and Environmental Affairs Division of NATO. Hence, we first must express our gratitude to NATO for making the Institute and these proceedings possible. Its support of six sequential ASIs that were proposed and directed by one of us (HW) in the field of superconductivity since 1988 is greatly appreciated. It is worth noting that the other co-director (MN) of the 1999 ASI also was a co-director of the first ASI in 1988; thus completing the cycle as it had begun. The (United States) Air Force Office of Scientific Research (AFOSR) through its European Office of Aerospace Research and Development provided substantial additional financial support. Furthermore, HW must thank AFOSR for allowing him to work on activities related to the ASI during the course of his normal duties and for providing other institutional resources. It is a pleasure, as well, to acknowledge financial support from the (United States) National Science Foundation, which provided travel support to US researchers, primarily graduate students.

The ASI would not have been possible without the dedicated and superior efforts of Mrs. Sandra Ronayne, a Management Assistant at AFOSR. She handled all of the ASI administrative tasks, created the ASI Web site and handled most correspondence with participants. It is regretful that Mrs. Ronayne was unable to attend the ASI itself due to health problems that fortunately have had a happy ending. She was sorely missed, although we would like to thank Mrs. Linda Weinstock for her service at the ASI registration desk and library, and for doing whatever else helped ease the burden of two rather busy co-directors.

We reserve our greatest thanks for the fifteen distinguished scientists who were ASI lecturers and who (in some cases together with their collaborators) subsequently became the authors of the chapters in these proceedings. During the ASI all of them devoted from one to two weeks of their valuable time to lectures, workshops and informal discussions with about eighty mostly young researchers from fifteen countries. Surely their most difficult task was to prepare a manuscript - two manuscripts for six of the lecturers - for these proceedings and to deal with two rather fussy editors (MN and HW). For all of these reasons, we are very much in their debt. Thanks are due also to the aforementioned ASI participants who provided the enthusiasm for microwave superconductivity that helped motivate the efforts of the lecturers.

As always, we must acknowledge the patience and good will of our families for excusing the many distractions caused by the ASI and the editing of these proceedings.

Arlington, Virginia — Martin Nisenoff
October, 2001 — Harold Weinstock

FUNDAMENTAL CONSIDERATIONS OF SUPERCONDUCTORS AT MICROWAVE FREQUENCIES

M. J. Lancaster
The University of Birmingham
School of Electronic and Electrical Engineering
Edgbaston
Birmingham
UK
B15 2TT

1. Introduction

This chapter presents a fundamental analysis of the interaction of electromagnetic waves with superconducting materials. This presentation is based on simple electromagnetic theory and provides a basic understanding of the physical phenomenon involved. The model developed is of considerable practical use and is the basis for many calculation of the performance of superconducting devices at high frequency.

The chapter first introduces the concept of surface impedance followed by the two fluid model. Once these have been understood, basic electromagnetic theory is used to examine the propagation of plane electromagnetic waves in superconductors. Examination of this process leads to an understanding of the basic properties of superconductors at high frequencies and develops expressions for use in practical situations. Power and energy considerations are examined through a discussion of Poynting's theorem in relation to superconducting materials. This is followed by examples of experimental measurements of the important material parameters.

2. Surface impedance

Surface impedance is probably the most important concept when considering the interaction of electromagnetic waves with almost any material and is particularly important when dealing with superconductors. Consider the scenario in Figure 1, where an electromagnetic wave with electric field, represented by E_i, is incident on a boundary between two materials. The material on the left could be free space and the material on the right could be a block of superconductor. The incident wave produces a reflected wave, E_r and a transmitted wave E_t, the amplitude and phase of which are determined by the material properties of the two media.

H. Weinstock and M. Nisenoff (eds.), Microwave Superconductivity, 1–20.

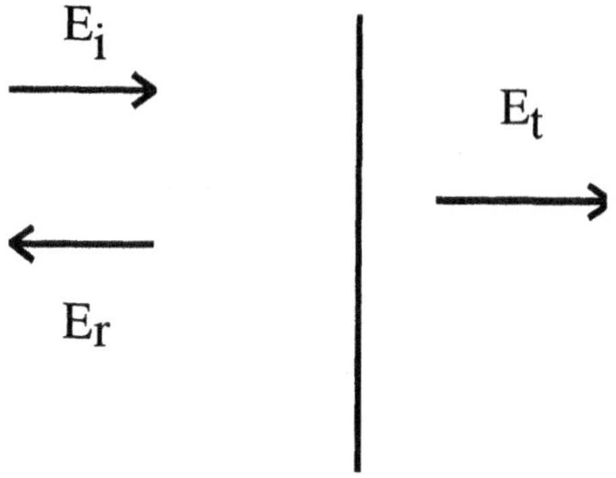

Figure 1 Electromagnetic waves at a boundary.

Assuming the waves are sinusoidal, they can be written as,

$$E_i = E_{oi} e^{-\gamma_1 z} \qquad H_i = \frac{E_{oi}}{Z_1} e^{-\gamma_1 z}$$

$$E_t = E_{ot} e^{-\gamma_2 z} \qquad H_t = \frac{E_{ot}}{Z_2} e^{-\gamma_2 z} \tag{1}$$

$$E_r = E_{or} e^{+\gamma_1 z} \qquad H_r = -\frac{E_{or}}{Z_1} e^{+\gamma_1 z}$$

E_{0i}, E_{0t}, and E_{0r} are the amplitudes of the waves, z is the usual Cartesian co-ordinate and is in the same direction of the propagated wave. Z_1 and Z_2 are the intrinsic impedance of the two materials and, γ_1 and γ_2 are the propagation constants of the two materials.

The tangential electric and magnetic fields at the boundary must add as follows

$$E_i + E_r = E_t$$

$$H_i + H_r = H_t$$

We define the surface impedance of a material as the ratio of the tangential electric to magnetic fields, hence,

$$Z_s = \frac{E_t}{H_t} = \frac{E_i + E_r}{H_i + H_r} \tag{2}$$

If one substitutes values for the fields from (2) and take $z = 0$ then

$$Z_s = Z_2 = Z_1\left(\frac{E_{oi} + E_{or}}{E_{oi} - E_{or}}\right)$$

The importance of this relationship is that the intrinsic impedance of a material Z_2 is equal to the surface impedance of the material Z_s. Of course, a transmission and reflection coefficient can be defined as the ratio of the transmitted and reflected fields to the incident field respectively are simply shown to be

$$\frac{E_{0t}}{E_{0i}} = \frac{2Z_2}{Z_1 + Z_2} = T\ ,$$

$$\frac{E_{0r}}{E_{0i}} = \frac{Z_2 - Z_1}{Z_2 + Z_1} = R\ .$$

Both reflection and transmission coefficients are strongly dependant upon the surface impedance of medium 2. Knowing the surface impedance allows simple calculations of the reflected and transmitted waves to be made.

The concept of surface impedance can be generalised to non-planar boundaries. Figure 2 shows an enclosed cavity, having either metallic or superconducting walls. From the point of view of analysing the structure, it can be assumed that the boundary has a surface impedance Z_s. Analytical calculations can be made of the resonant frequencies and quality factor of the cavity, and then the value of Z_s can be substituted at the end. Thus, metallic, superconducting or more complex boundaries can be considered after the main calculation.

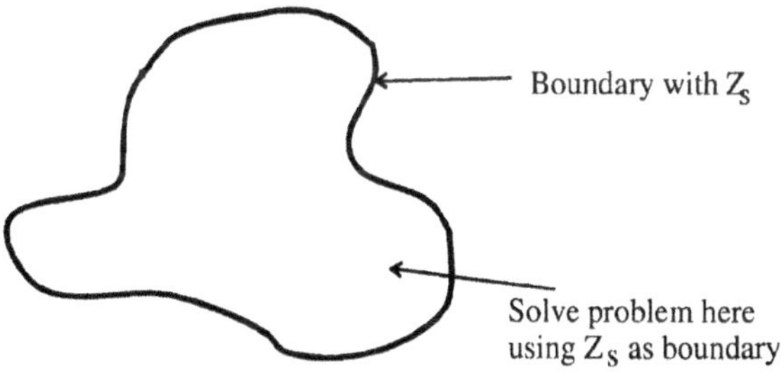

Figure 2 Generalised enclosed structure with a boundry with surface impedance Z_s.

For example, the walls of the cavity may be made up of the multi-layer structure shown in Figure 3. By simple calculations, the surface impedance of the structure can be calculated with the effective surface impedance being the combined surface impedances of the HTS, dielectric and air. Of course this impedance can then be used for the calculation of the properties of the cavity of Figure 2[1].

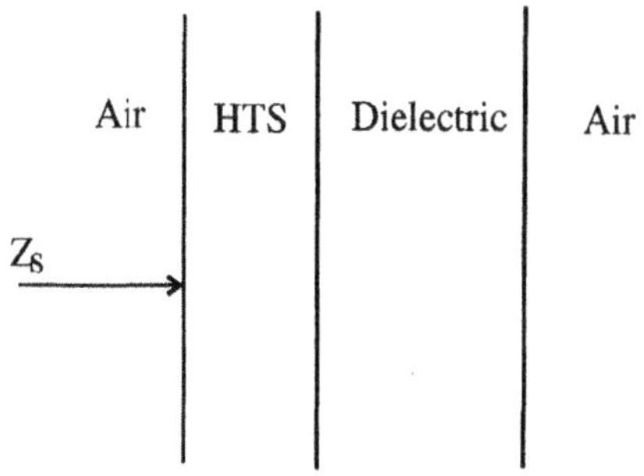

Figure 3 Multi-layer structure

The main point here is that the concept of surface impedance is extremely useful and this point will be enlarged upon later in this text.

The surface impedance can also be considered as the impedance of a square piece of material. Consider Figure 4 where a piece of material length *l* and width *w* has a current *I* flowing through it generating a potential *V* across the length *l*.

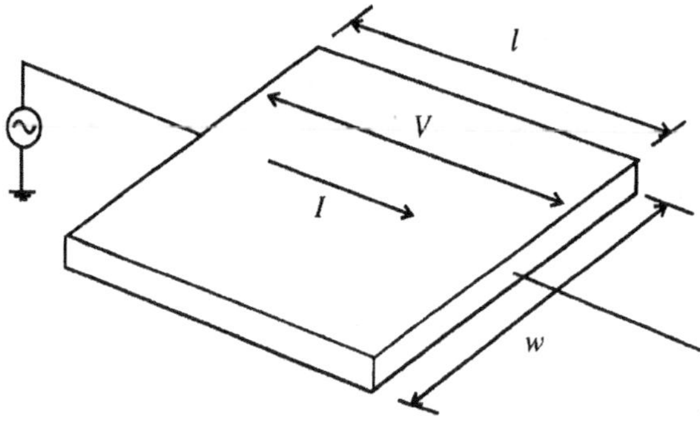

Figure 4 Applied voltage and induced current across the surface of a piece of material.

The surface impedance is given as the ratio of the voltage to current. This can be written as the ratio of the electric field *E* multiplied by the length *l* to the surface current density *K* multiplied by the width *w*. If the width is equals to the length, then this results in the ratio of the surface electric to magnetic fields since the surface current density is equal to the magnetic field at the surface.

$$Z_s = \frac{V}{I} = \frac{El}{Kw} = \frac{E_t}{H_t}.$$

1 It is acceptable to use this concept provided the skin depth or penetration depth are small, compared with the curvature of the object under investigation.

A second concept will now be considered before we can proceed with the basic calculation of electromagnetic field with superconductors.

3. Two fluid model

In normal metals, current is carried by electrons, while in superconductors current is carried by a combination of normal electrons and electron pairs. This is depicted in Figure 5. This is a well-known phenomena and examined thoroughly in theoretical and experimental studies. The electron pairs are bound together by lattice interactions and travel without loss.

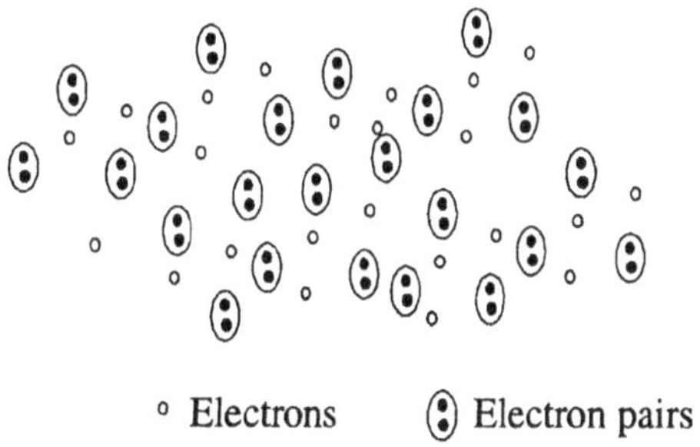

Figure 5 Electron and Electron pairs flowing inside a superconductor

The ratio of normal electrons to electron pairs depend upon the temperature, with the concentration of paired electrons decreasing as the temperature increases and going to zero at the transition temperature. We will now use this picture to form some fundamental expression and ideas.

For an electron pair, the force acting on the paired electrons is given by

$$2m\frac{d\underline{v}_s}{dt} = -2e\underline{E} \tag{3}$$

Here, m is the mass, $\underline{v}_s$ the velocity of the paired electrons, e the electron charge and $\underline{E}$ applied electric field.

For a normal electron there is also a force due the electrons scattering off the lattice.

$$m\frac{d\underline{v}_n}{dt} + m\frac{\underline{v}_n}{\tau} = -e\underline{E} \tag{4}$$

It is interesting to note that the term $m\frac{\underline{v}_n}{\tau}$ provides for the decay of current. If the electric field is removed, putting $\underline{E}$ to zero, the equation becomes

$$m\frac{d\underline{v}_n}{dt} = -m\frac{\underline{v}_n}{\tau} .$$

This has a solution

$$v_n = ke^{-\tau t} \qquad \underline{v}_n \propto e^{-t/\tau} .$$

Here τ is the momentum relaxation time and k a constant. It can be seen that the velocity decays exponentially once the field is removed.

The current density due to the electron pairs can be written as

$$\underline{J}_s = -n_s e\underline{v}_s \tag{5}$$

While the current density due to the normal electrons is

$$\underline{J}_n = -n_n e\underline{v}_n \tag{6}$$

Here n_s and n_n are the paired and normal electron densities. (3) to (6) now define current density in the superconductor. We will now look at two particular cases.

In the first case, it will be assumed that the normal, unpaired electron density is very small. This is the case at low temperatures where almost all the electrons are paired. This is also an approximation for a low frequency analysis as will be seen below.

Substituting (5) into (3) gives:

$$\Lambda\frac{\partial \underline{J}_s}{\partial t} = \underline{E} \tag{7}$$

Where

$$\Lambda = \frac{m}{n_s e^2} .$$

(7) is called the first London equation [1]. Now take the curl of (7), giving

$$\Lambda\frac{\partial}{\partial t}(\underline{\nabla} x \underline{J}_s) = \underline{\nabla} x \underline{E} \tag{8}$$

Faradays Law is given by

$$\underline{\nabla} x \underline{E} = -\frac{\partial \underline{B}}{\partial t} .$$

Making a substitution for the curl of E into (8), and integrating with respect to time gives

$$\Lambda \underline{\nabla} x \underline{J}_s = -\underline{B}.$$

This is the second London equation. It should be noted that there should be a constant of integration in the above step. This has been arbitrarily set to zero. The main reasoning for this is that the second London equation has been verified experimentally and no constant has been deemed necessary for agreement. Equation (7) together with Maxwell's equations gives a simple method for making calculations on superconducting materials.

Now let us look at a second case. Here it will be assumed that all the electromagnetic fields vary in a sinusoidal manner. Hence

$$\underline{J}_s = \underline{J}_{so} e^{j\omega t} \qquad \underline{J}_n = \underline{J}_{no} e^{j\omega t} \qquad \underline{E} = \underline{E}_o e^{j\omega t}.$$

Using these sinusoidal time dependencies and substituting (5) and (6) into (4) and (3) results in expressions for the super and normal current densities. By summing these, a total current can be deduced.

$$\underline{J}_o = \underline{J}_{no} + \underline{J}_{so} = \left\{ \frac{n_n e^2 \tau}{m(1+\omega^2\tau^2)} - j\left(\frac{n_s e^2}{\omega m} + \frac{n_n e^2 \tau^2 \omega^2}{\omega m(1+\omega^2\tau^2)} \right) \right\} \underline{E} \tag{9}$$

This expression for the total current can be written

$$\underline{J}_o = (\sigma_1 - j\sigma_2)\underline{E}_o, \tag{10}$$

where

$$\sigma_1 = \frac{n_n e^2 \tau}{m(1+\omega^2\tau^2)} \qquad \text{and} \qquad \sigma_2 = \frac{n_s e^2}{\omega m} + \frac{n_n e^2 \tau^2 \omega^2}{\omega m(1+\omega^2\tau^2)}. \tag{11}$$

Equation (10) defines a complex conductivity. We can now use Maxwell's equations with sinusoidal time dependencies and the complex conductivity for calculations. That is, replace $\underline{J} = \sigma \underline{E} by J = (\sigma_1 - j\sigma_2)\underline{E}$.

Figure 6 depicts the current flow in a superconductor which is represented as a simple circuit. There is both an inductive term and a resistive term, as we would expect from the complex conductivity. At low frequencies, the inductive part provides a short and there is little effect from the resistive term. In fact, when a non-time varying signal is applied, the resistor has no effect and the superconductor has no loss as is well known. The inductive part of the circuit becomes highly reactive at high frequencies and then losses occur due to current flowing through the resistor.

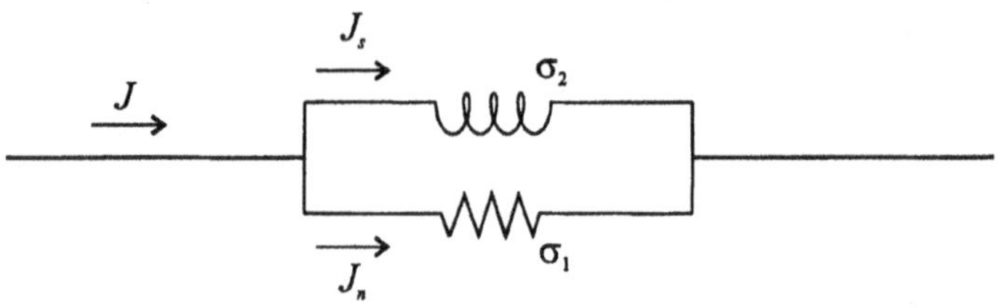

Figure 6 Circuit reresentation of current flow in superconductors.

The calculation resulting in (9) and (10) is very basic. However the main aim here has been to show that a superconductor can be represented by a complex conductivity. There are much more accurate methods based on the BCS theory to calculate this complex conductivity in terms of fundamental material parameters [2,3,4,5,6]. However, for practical use of these materials, these expressions are seldom required.

As stated above, as the temperature increases, the number of electron pairs decrease resulting in more electrons. The commonly used expressions for the number of carriers are given by [7].

$$n_s = n_o(1-t^4) \qquad n_n = n_o t^4 \tag{12}$$

Here $t=T/T_c$, where T is the actual temperature and T_c is the superconducting transition temperature.

4. Maxwell's equations, the wave equation and superconductors

Now that we understand surface impedance and complex conductivity, it is possible to examine fundamental electromagnetic properties of superconductors. This will be done by looking firstly at Maxwell's equations for sinusoidal time dependence and seeing that a wave equation can be developed from them. Using the wave equation, a propagation constant and impedance can be extracted and then these parameters will be examined using complex conductivity.

For sinusoidal time dependencies, Maxwell's equations are:

$$\underline{\nabla} x \underline{H} = \underline{J} + j\omega \underline{D}\,, \tag{13}$$

$$\underline{\nabla} x \underline{E} = - j\omega \underline{B}\,, \tag{14}$$

$$\underline{\nabla} \cdot \underline{B} = 0\,, \tag{15}$$

$$\underline{\nabla} \cdot \underline{E} = 0\,. \tag{16}$$

Here we have assumed a region free of static charges. Now take the curl of (14) and put $\underline{B}=\mu\underline{H}$

$$\underline{\nabla}x\underline{\nabla}x\underline{E}=-j\omega\mu\underline{\nabla}x\underline{H}\,.$$

Dividing by -jωμ and substitute for $\underline{\nabla}x\underline{H}$ from (13) gives

$$\frac{\underline{\nabla}x\underline{\nabla}x\underline{E}}{-j\omega\mu}=\underline{J}+j\omega\underline{D}\,.$$

Now put $\underline{J}=\sigma\underline{E}$,.

$$\underline{\nabla}x\underline{\nabla}x\underline{E}=-j\omega\mu(\sigma+j\omega\varepsilon)\underline{E}\,. \tag{17}$$

The conductivity σ is complex for a superconductor and real for a normal metal but will be left just as σ for the moment.

Using the vector identity

$$\underline{\nabla}x\underline{\nabla}x\underline{E}=\underline{\nabla}(\underline{\nabla}\cdot\underline{E})-\nabla^2\underline{E}\,,$$

but with $\underline{\nabla}(\underline{\nabla}\cdot\underline{E})=0$ from (16), gives

$$\underline{\nabla}x\underline{\nabla}x\underline{E}=-\nabla^2\underline{E}\,.$$

Substitution into (17) gives

$$\nabla^2\underline{E}=j\omega\mu(\sigma+j\omega\varepsilon)\underline{E}\,.$$

This is the wave equation for the electric field. A similar one for $\underline{H}$ exists and is calculated in a similar way. This equation can be rewritten as

$$\nabla^2\underline{E}=\gamma^2\underline{E} \tag{18}$$

Where γ is the propagation constant. This equation is very general, and will be used to get a better understanding of the propagation of electromagnetic waves in superconductors that will now be simplified by assuming plane waves.

5. Plane waves in superconductors

For a plane wave travelling in the z direction the simplest expressions for the electromagnetic fields are.

$$E_x = E_o e^{-\gamma z}$$
$$H_y = H_o e^{-\gamma z} \tag{19}$$

Here the electric field is directed in the x direction and the magnetic field at right angles to this in the y direction. The variation with time is $e^{j\omega t}$ but this has been dropped to simplify the expressions. To verify that these equations are solutions to the wave equation, they can be substituted into (18). If this is done, it can be easily shown that the propagation constant is given by

$$\gamma = \sqrt{j\omega\mu(\sigma + j\omega\varepsilon)}\ . \tag{20}$$

This expression can be written as

$$\gamma = \alpha + j\beta \tag{21}$$

Defining the propagation constant in this manner allows separate consideration of the losses (real part) and the phase (imaginary part) of the electromagnetic wave.

It is interesting to note that $\underline{E}$ and $\underline{H}$ are not independent for the plane waves under consideration, and are connected through Faradays law (14). For plane waves, Faradays law becomes

$$-j\omega\mu H_y = \frac{\partial E_x}{\partial z}\ .$$

Substituting for E_x and H_y from (19) above gives

$$Z_s = \frac{E_x}{H_y} = \frac{j\omega\mu}{\gamma} = \sqrt{\frac{j\omega\mu}{\sigma + j\omega\varepsilon}}\ . \tag{22}$$

This is the intrinsic impedance or surface impedance of the material, which is very general and can be considered to be any material at the moment. Intrinsic impedance is used for any materials, while the surface impedance is usually used for good conductors.

6. Plane waves in good superconductors

We will now calculate the propagation constant and surface impedance of a superconductor. This can be done by the substitution of a complex conductivity into (20) which yields

$$\gamma = \sqrt{j\omega\mu(\sigma_1 - j\sigma_2 + j\omega\varepsilon)}\ .$$

This can be written as

$$\gamma = \sqrt{j\omega\mu(-j\sigma_2)(\frac{\sigma_1}{-j\sigma_2} + 1 + \frac{j\omega\varepsilon}{-j\sigma_2})}\ .$$

We will now assume $\frac{j\omega\varepsilon}{-j\sigma_2}$ is very small which is the situation when the displacement current is much less than the supercurrent in a superconductor. This is always true at microwave frequencies because of the high conductivity. The expression for the propagation constant now becomes

$$\gamma = \sqrt{\omega\mu\sigma_2}\sqrt{1 + j\frac{\sigma_1}{\sigma_2}}\ .$$

Now assume that the temperature is much lower than the transition temperature and the supercurrent dominates the normal current. Then $\sigma_2 >> \sigma_1$ and a binomial expansion of the above expansion gives

$$\gamma = \sqrt{\omega\mu\sigma_2}(1 + j\frac{\sigma_1}{2\sigma_2})\ .$$

Using (21) the real part of γ is the attenuation coefficient which can be written as

$$\alpha = \sqrt{\omega\mu\sigma_2}\ .$$

This represents the decay of the electromagnetic wave as it travels in a superconductor. The decay is exponential i.e. $e^{\alpha z}$. It is possible to define a characteristic depth which is the depth where the waves decays by e^{-1}. This characteristic depth is called the penetration depth for superconductors and is given by

$$\lambda = \frac{1}{\sqrt{\omega\mu\sigma_2}}\ . \tag{23}$$

For high temperature superconductors the penetration depth is of the order of 100 nm and obviously varies with temperature and the particular superconductor. For a perfect superconductor well below the transition temperature, it is independent of frequency

($\sigma_2 \alpha \frac{1}{\omega}$ see later). For a normal metal the corresponding depth is called the skin depth δ, which is dependent upon frequency.

The imaginary part of the propagation constant is:

$$\beta = \frac{\sigma_1}{2\sigma_2}\sqrt{\omega\mu\sigma_2}\,,$$

$$\beta = \sqrt{\frac{\omega\mu\sigma_1^2}{4\sigma_2}}\,.$$

Hence, the velocity of an electromagnetic wave can be calculated.

$$c = \frac{\omega}{\beta} = \omega\sqrt{\frac{4\sigma_2}{\omega\mu\sigma_1^2}} = \sqrt{\frac{4\omega\sigma_2}{\mu\sigma_1^2}}\,.$$

Now using the expression for the surface impedance of the superconductor that was given in (22), which is

$$Z_s = \sqrt{\frac{j\omega\mu}{\sigma + j\omega\varepsilon}}\,,$$

and substituting in the complex conductivity gives

$$Z_s = \sqrt{j\omega\mu}\,(\sigma_1 - j\sigma_2 + j\omega\varepsilon)^{-1/2}\,.$$

Re-arranging terms gives

$$Z_s = \sqrt{\frac{j\omega\mu}{-j\sigma_2}}\left(\frac{\sigma_1}{-j\sigma_2} + 1 + \frac{j\omega\varepsilon}{-j\sigma_2}\right)^{-1/2}\,.$$

As described above the displacement current term can be neglected, i.e. $\frac{j\omega\varepsilon}{-j\sigma_2}$ is small.

Also, using $\sigma_2 >> \sigma_1$ implying a good superconductor and expanding gives

$$Z_s = \sqrt{-\frac{\omega\mu}{\sigma_2}}\left\{1 - \frac{j\sigma_1}{2\sigma_2}\right\},$$

$$Z_s = \sqrt{\frac{\omega\mu}{\sigma_2}}\left\{\frac{\sigma_1}{2\sigma_2} + j\right\}.$$

By looking at the real and imaginary parts of this expression, the resistance and reactive parts of the surface impedance can be extracted.

$$R_s = \frac{\sigma_1}{2\sigma_2}\sqrt{\frac{\omega\mu}{\sigma_2}} \quad X_s = \sqrt{\frac{\omega\mu}{\sigma_2}}.$$

The value of σ_2 can now be substituted from (23) to get the equations in terms of λ

$$R_s = \frac{\omega^2\mu^2\sigma_1\lambda^3}{2} \quad X_s = \omega\mu\lambda.$$

From the reactance, an inductance can be defined

$$L_s = \mu\lambda.$$

This is called the kinetic inductance

7. Comparison of good superconductors with normal conductors

It is important to compare the electromagnetic response of good superconductors with that of normal conductors.

As was done with superconductors, the displacement current is assumed to be much less than the conduction current, so for good conductors

$$\frac{\sigma_n}{\omega\varepsilon} >> 1.$$

Here we have written σ_n for the purely real conductivity of a normal conductor. Knowing this, expressions for the propagation constant and impedance for a normal conductor can be developed using similar arguments to the above.

With this approach the following comparisons are made. Here expressions for superconductors are on the left side of the page and normal conductors on the right. The attenuation constant is given by

$$\alpha = \sqrt{\omega\mu\sigma_2} \qquad \alpha_{\text{normal}} = \sqrt{\frac{\omega\mu\sigma_n}{2}}$$

Inverting these expressions gives the penetration depth and skin depth

$$\lambda = \frac{1}{\sqrt{\omega\mu\sigma_2}} \qquad \delta_{\text{normal}} = \sqrt{\frac{2}{\omega\mu\sigma_n}}$$

As mentioned above, for a superconductor, $\sigma_2 \alpha \frac{1}{\omega}$ hence λ is independent of ω. While for a normal conductor, σ_n is independent of ω, hence the skin depth is inversely proportional to the square root of frequency.

A comparison of the propagation velocity can be made

$$c = \sqrt{\frac{4\omega\sigma_2}{\mu\sigma_1^2}} \qquad c_{\text{normal}} = \sqrt{\frac{2\omega}{\mu\sigma_n}}$$

Also expressions for surface resistance can be deduced

$$R_s = \frac{\sigma_1}{2\sigma_2}\sqrt{\frac{\omega\mu}{\sigma_2}} \qquad R_{s\,\text{normal}} = \sqrt{\frac{\omega\mu}{2\sigma_n}}$$

For a superconductor with σ_2 inversely proportional to frequency, it can be seen that the surface resistance is proportional to frequency squared. However for a normal conductor, σ_n is independent of frequency so the surface resistance is proportional to the square root of frequency. This means that the superconductor surface resistance increases more rapidly than the normal conductor surface resistance. At lower frequencies, the superconductor surface resistance is much smaller than for a normal metal but as the frequency increases it eventually become larger. The frequency where the superconductor surface resistance is equal to the normal metal surface resistance is called the 'crossover frequency'. For copper and thin film high temperature superconductor this is normally several hundred gigahertz. Figure 7 shows a graph of some measurements of surface resistance of the high temperature superconductor $YBa_2Cu_3O_7$, which is compared to copper surface resistance at 77K. Niobium at the same reduced temperature is included for comparison.

Finally a comparison can be made of the surface reactance of a superconductor and normal metal,

$$X_s = \sqrt{\frac{\omega\mu}{\sigma_2}} \qquad X_{s\,\text{normal}} = \sqrt{\frac{\omega\mu}{2\sigma_n}}.$$

It should be noted that for a normal metal R_s=X_s which is not true for a superconductor.

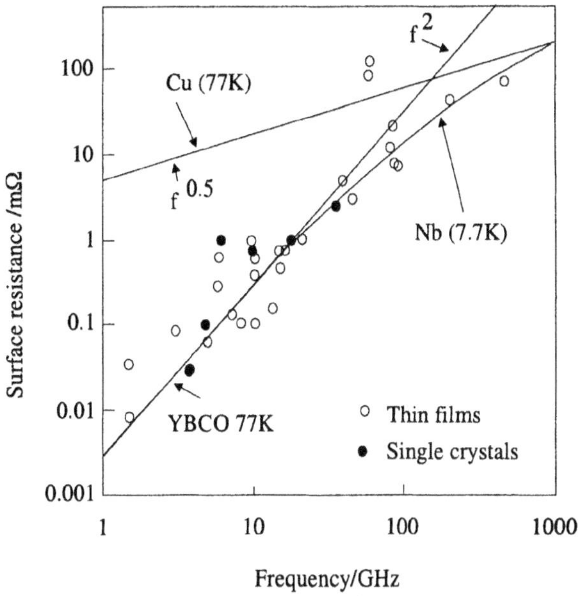

Figure 7 The surface resistance of superconductors and copper as a function of frequency.

8. Poynting's theorem for superconductors

The well known Poynting's theorem is

$$-\int_s \underline{ExH}\, ds = \frac{\partial}{\partial t}\int_v \frac{1}{2}\left(\varepsilon E^2 + \mu H^2\right) dv + \int_v \underline{E}.\underline{J}\, dv \,. \tag{24}$$

This theorem describes power flow in a volume v enclosed in surface s. The term on the left hand side represent the power flowing into the volume through the surface. The first term on the right hand side represent the rate of change if stored electromagnetic energy within the volume and the last term on the right hand side represents the power dissipated in the volume.

For a superconductor, the current density and electric field are related by the complex conductivity,

$$J = \left(\sigma_1 - j\sigma_2\right)E \,.$$

However, this is only true for sinusoidal time dependencies, whereas (24) is much more general. A more general expression for the relationship between current density and electric field is given by

$$\frac{\partial \underline{J}}{\partial t} = \frac{\partial \underline{J}_n}{\partial t} + \frac{\partial \underline{J}_s}{\partial t} = \frac{\partial \underline{J}_n}{\partial t} + \frac{1}{\mu\lambda^2}\underline{E} .$$

The last term is from London's 1stequation, (7), and equations (11) and (22). Also,

$$\frac{\mu\lambda^2}{2}\frac{\partial J_s^2}{\partial t} = \frac{\mu\lambda^2}{2}J_s 2\frac{\partial J_s}{\partial t} = \mu\lambda^2 J_s \frac{1}{\mu\lambda^2}\underline{E} = \underline{J}_s.\underline{E} ,$$

and

$$\sigma_1 E^2 = \underline{J}_n.\underline{E} .$$

With substitutions, Poynting's theorem becomes:

$$-\int \underline{E}x\underline{H}ds = \frac{\partial}{\partial t}\int_v \left(\frac{\varepsilon E^2}{2} + \frac{\mu H^2}{2} + \frac{\mu\lambda^2}{2}J_s^2\right)dv + \int_v \sigma_1 \underline{E}^2 dv .$$

The first three terms on the right hand side represent energy stored, last term is energy dissipated. The extra term for storage $\frac{\mu\lambda^2}{2}J_s$ is the energy stored in the kinetic energy of the travelling electron pairs.

9. Measurements of surface impedance and penetration depth

There are many methods of measuring the surface impedance of superconductors [8] and some are discussed elsewhere in this book. Here we briefly discuss a measurement using the coplanar resonator [9]. We are mainly concerned in putting some numbers to the parameters discussed above to get a feel for their magnitudes. The coplanar resonator, shown in Figure 8, consists of a patterned thin film on the surface of a single crystal substrate. The resonator consists of two large ground planes and a central conductor. This central conductor has resonant frequencies when its length is an integer number of half wavelengths.

Energy is coupled into and out of the device by very weak capactive coupling at each end of the resonant section. The response measured on a network analyser is shown in Figure 9. The quality factor and the central frequency of the resonance are measured, and the quality factor as a function of temperature is shown in Figure 10. The quality factor is defined in the usual manner as the ratio of the centre frequency to the half power bandwidth.

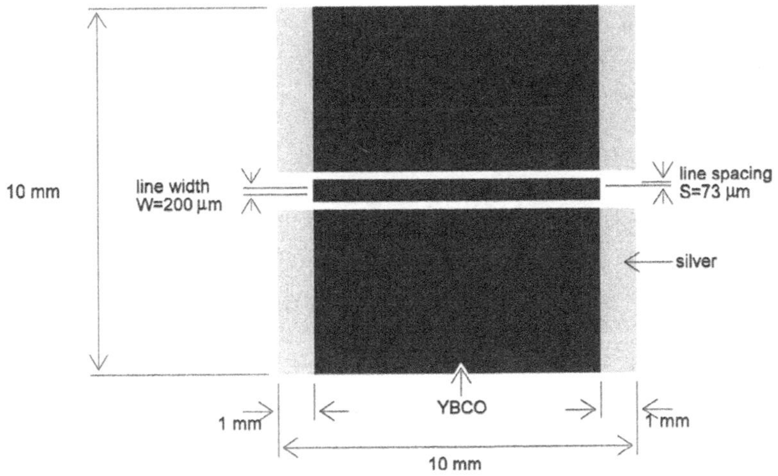

Figure 8 The coplanar resonator

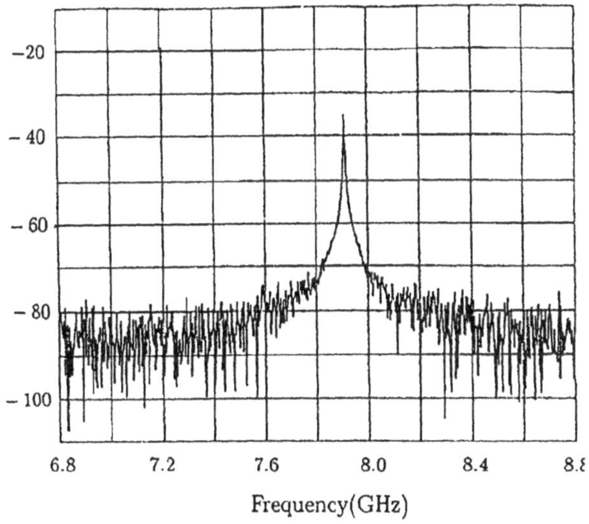

Figure 9 Frequency response of the coplanar resonator.

In order to translate this quality factor and central frequency into surface impedance a calculation of the volume current distribution in the superconductor is required. This can be done numerically [9] and is shown in Figure 11.

Knowing this current distribution, the surface impedance of the superconductor can be calculated. Two different coplanar resonators are required to deduce the full impedance. Figure 12 shows the calculated surface resistance for a $YBa_2Cu_3O_7$ thin film. Figure 13 shows the inverse square of the penetration depth that is proportional to σ_2. Also shown on this figure are a number of theoretical temperature dependencies, it can be seen the variation described by (12) is not a good fit, but that a better one is a squared dependence.

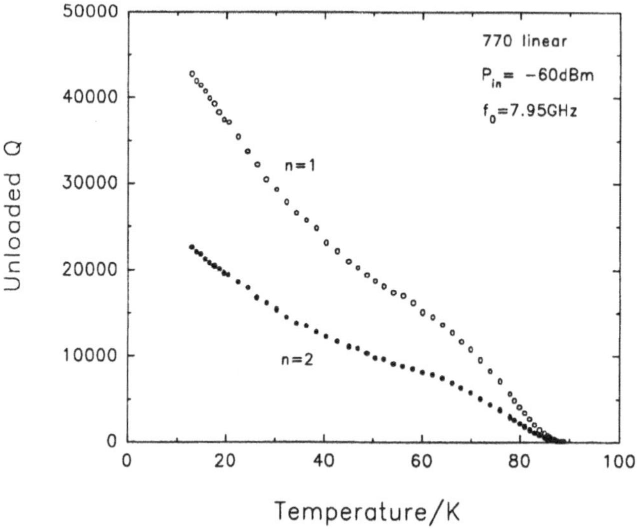

Figure 10 Quality factors of the fundamental and first harmonic of a HTS coplanar resonator

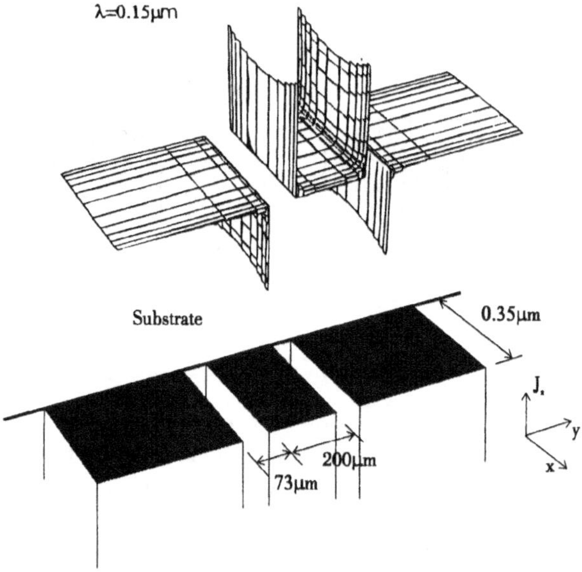

Figure 11 Volume current distribution in the coplanar resonator.

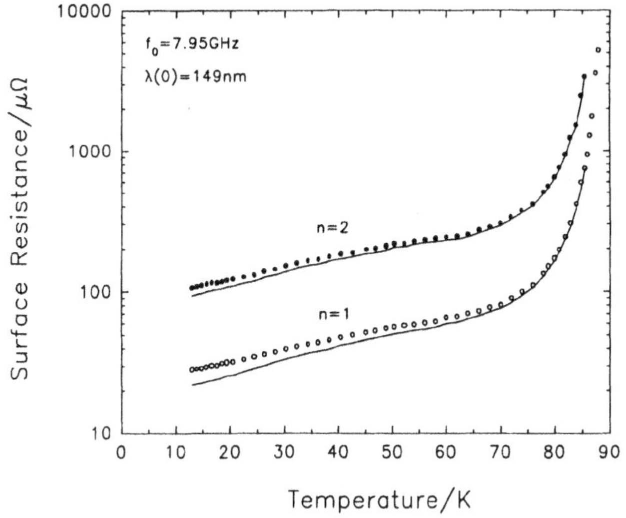

Figure 12 Surface resistance of HTS

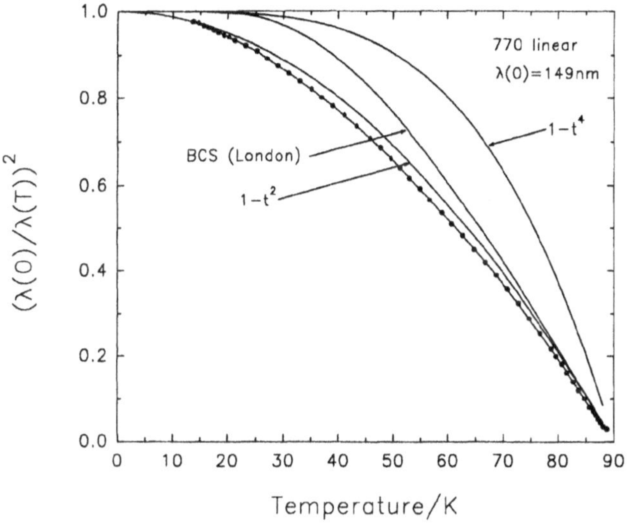

Figure 13 Penetration depth of a HTS thin film

10. Conclusion and summary

This article has described the fundamental nature of the interaction of electromagnetic waves with superconductors. The two fluid model has been introduced and through this a complex conductivity defined. The complex conductivity has been used with Maxwell's equations to derive expression for the superconducting penetration depth, propagation constant, wave velocity and surface impedance. These quantities have been deduced by looking at plane electromagnetic wave propagation inside and at the boundary of superconducting material. The concepts developed have been discussed with reference to typical measurements made on high temperature superconductors.

11. References

1 London, F., and London, H. (1935) The electromagnetic equations of the superconductor, Proc. Roy. Soc. (London) **A149,** 71-79.
2 Waldram, J. R. (1996) *Superconductivity of metals and cuprates,* Institute of Physics, London.
3 Tilley, D. R., and Tilley J. (1990) *Superfluidity and superconductivity,* Adam Hilger, New York.
4 Mattis, D. C., and Bardeen J. (1958) Theory of the anomalous skin effect in normal and superconducting metals, *Physical Review* **111,** 412-417.
5 Bardeen, J., Cooper, L. N., and Schrieffer, J. R. (1957) Theory of superconductivity, *Phys. Rev.* **10,** 1175-1204.
6 Tinkham, M. (1996) *Introduction to superconductivity,* McGraw Hill, Singapore.
7 Gorter, C. J. and Casimir H. B. G. (1934) On superconductivity, Physica **1,** 305-320
8 Lancaster, M. J. (1997) *Passive microwave device applications of high temperature superconductors,* Cambridge University Press, Cambridge, UK.
9 Porch, A., Lancaster, M. J., Humphreys, R. (1995) Coplanar resonator technique for the determination of the Surface Impedance of Patterned thin films, *IEEE Trans. on Microwave Theory and Techniques* **43,** 306-314.

MICROWAVE PROPERTIES OF SUPERCONDUCTORS

MATTHIAS A. HEIN
Superconductivity Research Group, University of Birmingham, UK
on leave from Dept. of Physics, University of Wuppertal, Germany.

1. Introduction

This contribution describes the microwave properties of superconductors with emphasis on experimental data. Typical results for the surface impedance Z_s are illustrated for Nb_3Sn films on sapphire as a classical superconductor, and compared with those of high-temperature superconducting (HTS) $YBa_2Cu_3O_{7-x}$ films on $LaAlO_3$, MgO, and CeO_2-buffered sapphire substrates. Section 2 reviews the relation of Z_s-measurements to the microscopic features of superconductors and describes systems suited to characterize unpatterned films. Section 3 provides data on the temperature dependence of Z_s for both types of superconductors at low microwave field amplitudes (*linear* response). Analyzing features that depend on sample quality is crucial to distinguish between conventional and nonconventional pairing mechanisms. Section 4 describes the *nonlinear* microwave response of superconducting films. Models and analytical estimates of critical field levels lead to a distinction between different nonlinear mechanisms, and to approaches for material optimization. The implications for a consistent theoretical understanding and for microwave applications of HTS are summarized in Section 5.

2. The Surface Impedance of Superconductors

2.1 RELEVANCE FOR FUNDAMENTAL ASPECTS AND APPLICATIONS

The surface impedance of superconductors is determined by the number densities n_S and n_N of paired and unpaired charge carriers (quasiparticles), and by the scattering mechanisms acting on them. These *microscopic parameters* are closely related to the density

H. Weinstock and M. Nisenoff (eds.), Microwave Superconductivity, 21–53.

of quasiparticle states [1-4]. Beside tunneling spectroscopy, Z_s-measurements are therefore a key technique to assess the pair state. This is best illustrated for the local limit which applies to Nb_3Sn (T_c=18K, λ_L=60nm) and $YBa_2Cu_3O_{7-x}$, (YBCO, $T_c \approx$ 89K for x<0.2, λ_L=140nm), since their London penetration depths λ_L are large compared to the coherence lengths (ξ_0=2.5 and 7nm, respectively [4]) and quasiparticle mean free paths ($l \leq 100$ and $\approx$10nm, respectively). The surface impedance in the local limit is given by $Z_s=R_s+iX_s=(i\omega\mu_0/\sigma)^{1/2}$, with $\sigma=\sigma_1-i\sigma_2$ the complex conductivity, $\omega=2\pi f$ the circular frequency, R_s the surface resistance, and $X_s=\omega\mu_0\lambda$ the surface reactance. In the Drude model of metals, σ_1 depends on $n_N(T)$, the effective quasiparticle mass m*, and the scattering time $\tau(T)$ which is often assumed to be frequency independent:

$$\sigma_1(T) = \frac{e^2 n_N(T)}{m^*(T)} \times \frac{\tau(T)}{1+[\omega\tau(T)]^2} . \tag{1}$$

The renormalized quasiparticle mass $m^*(T)=m_0[1+\Lambda(T)]$ is related to the strength $\Lambda(T)$ of the electron-phonon interaction and can be temperature dependent, especially in conventional strong-coupling superconductors [5,6]. The imaginary part of σ defines the effective penetration depth λ:

$$\sigma_2(T) = \frac{e^2}{\omega m^*(T)} \times \left[n_S + n_N \frac{(\omega\tau)^2}{1+(\omega\tau)^2} \right] = \frac{1}{\mu_0 \omega \lambda^2(T)} . \tag{2}$$

At low frequencies, $\omega\tau \ll 1$, λ approaches the London value $\lambda_L=[m^*(T)/\mu_0 e^2 n_S(T)]^{1/2}$. In BCS superconductors at low temperatures, $T \ll T_c$, m* is constant and the pairing is complete $[n_N(T)+n_S(T)=1]$. The pair density $n_S(T)$ is related to the quasiparticle density of states N(E) [1,7]:

$$\frac{n_S(T)}{n_S(0)} = 1 + 2 \int_{\Delta(T)}^{\infty} dE \frac{N(E,T)}{N_0} \times \frac{\partial f(E,T)}{\partial E} \tag{3}$$

with Δ the energy gap, N_0 the density of states in the normal-state, and f(E,T) the Fermi distribution function. In BCS-theory, $N(E)/N_0=E/(E^2-\Delta^2)^{1/2}$, and Eq. (3) leads at $T<T_c/2$ to the well-known exponential temperature dependences of λ and R_s [8]:

$$\frac{\lambda(T)-\lambda(0)}{\lambda(0)} = \left(\frac{\pi\delta}{2t}\right)^{1/2} \times e^{-\delta/t} \tag{4a}$$

$$R_s(T) = A \times \frac{1}{t} \times \ln\left(\frac{4t}{\nu\delta}\right) \times e^{-\delta/t} + R_{res} , \tag{4b}$$

where A is a constant and $t=T/T_c$, $\nu=hf/\Delta$, and $\delta=\Delta(0)/k_BT_c$ are the reduced temperature, frequency, and energy gap. The experimental observation of non-exponential temperature dependences and high residual values $R_s(0)=R_{res}$ indicate that the density of states of HTS deviates from the singular BCS shape (Sec. 3.3). Detailed descriptions of the fundamental microwave properties of superconductors including HTS are available in [1-4,7-11].

The benefit of superconductors for *microwave applications* is their extremely low power dissipation, leading to high unloaded quality factors Q_0 of resonators. The advantages of high Q_0-values, e. g., for microwave filters, are smaller size, steeper frequency roll-off characteristics ("skirts"), lower insertion loss, or higher functionality [12-15]. The Q_0 of a single resonator is limited by the largest of three dissipation mechanisms:

$$\frac{1}{Q_0(T)} = \gamma_{cond} R_s(T) + \frac{1}{Q_{para}} + \gamma_{diel} \tan\delta . \tag{5}$$

The first term is the surface resistance multiplied by the geometric factor γ_{cond}. Except for lumped-element devices (large γ_{cond}), this term, at f=1 GHz and T=50–70 K, is of order 10^{-7}–10^{-6}. The second term measures the ratio of parasitic power dissipation in the housing to the circulating power $P_{circ}=\omega W$. For optimized designs of circuit and housing, values below 10^{-6} could be achieved with microstrip devices [16-18]. The last term denotes losses in the dielectric volume V ($\gamma_{diel}\approx 1$ for most devices) which, except for dielectric resonators, is determined by the substrates of the SC films. Compared to typical $\tan\delta$-values of 10^{-6} for sapphire at 70 K [19,20], the loss tangent of $LaAlO_3$ might exceed 10^{-5} in this interesting temperature range [20,21], thus limiting Q_0 [22].

The sensitivity of a filter to nonlinear power handling is determined by the field dependent surface impedance $Z_s(B_s)$. The circulating power of a resonator is related to the microwave field amplitude B_s acting on the superconducting surface [4,23]:

$$B_{s,max}^2 = \frac{\mu_0}{\pi f_0 V} \times \frac{\max\{B_s^2\}_A}{\text{ave}\{B_s^2\}_V} \times P_{circ} . \tag{6}$$

The conversion of P_{circ} into field amplitudes is given by V and by the microwave field distribution. Field enhancement at the contours of quasi-one dimensional circuits leads to large $\max\{B_s^2\}_A/\text{ave}\{B_s^2\}_V$-ratios which are deleterious to power handling. In contrast, 2- or even 3-dimensional resonators are appropriate for handling high power.

Many groups have been successful in exploiting the benefit of epitaxial HTS films for microwave applications between 0.01 and 10 GHz [24-30], e. g., for ground- and satellite-based communications [14,15] or for nuclear magnetic resonance [31].

2.2 INTEGRAL RESONANT MEASUREMENT TECHNIQUES

Numerous methods to investigate absolute level, linearity and homogeneity of Z_s have been developed [4,30-34], with compromises between *simplicity* and *high resolution*. In cavity resonators, the sample forms one endplate of a cylindrical cavity (e. g., [35]). Other systems employ dielectrics inside a metallic housing with minimized radiation losses to store the field energy [36]. A strongly miniaturized resonator type employs two films on both sides of a dielectric (parallel plate resonator), e. g., by use of a double-sided film. If operated in rotationally symmetric edge-current free modes, such a *disk resonator* constitutes a promising approach for compact high-power filters [16].

The surface impedance can be deduced from experiment by measuring the center frequency f and the full width at half maximum $\Delta f_{1/2}$ of the resonant curve in the frequency domain. To obtain Q_0 and R_s [Eq. (5)], the loaded quality factor $Q_L = f/\Delta f_{1/2}$ is corrected for the power radiated through the couplings into and out of the resonator [4,30]. Much faster measurements can be performed in the time domain by measuring the insertion loss $P_{out}/P_{in} \propto Q_L^2$ of the resonator when excited by short periodic square-pulses of microwave power. The duty cycle, defined as pulse duration times repititon rate, can be as low as 10^{-6} to minimize microwave heating. Changes of the penetration depth are deduced from changes Δf of the resonant frequency according to $-\omega\mu_0\Delta\lambda = 2/\gamma_{cond} \times \Delta f/f$. The microwave field amplitude B_s is given by the stored energy W [Eq. (6)] which can be derived from the transmitted power at resonance.

3. Linear Surface Impedance of Superconducting Films

3.1 POLYCRYSTALLINE Nb_3Sn FILMS ON SAPPHIRE SUBSTRATES

The intermetallic compound Nb_3Sn is a valuable reference superconductor for YBCO [4] because it has a high T_c (the highest of all metallic SCs except for the unstable Nb_3Ge), it displays strong electron-phonon coupling, a short coherence length, a pronounced dependence of the electronic properties on the crystalline order, and it is (weakly) anisotropic. The Nb atoms form three perpendicular chains which lead to peaks in the density of states near the Fermi energy, and which are essential for the high transition temperature [37]. The pairing in Nb_3Sn results from electron-phonon coupling with an s-wave symmetry of the order parameter. All features, except for the nature and the symmetry of the pair state, are known to apply also for YBCO. Therefore, careful and comparative analysis of the surface impedance of both superconductors is valuable to distinguish between intrinsic and extrinsic and between conventional and unconventional microwave properties.

Figure 1 displays typical results at 87 GHz for the temperature induced increase of the surface resistance $\Delta R_s(T) = R_s(T) - R_{res}$ and the penetration depth $\Delta\lambda(T)$ [38,39]. Exponential behavior with a reduced energy gap $\delta = 1.8$-2.3 [see Fig. 3 (d)] is observed for both quantities as expected from Eq. (4). The slightly different slopes in Fig. 1 reflect the temperature dependent prefactors in Eqs. (4a) and (4b), as well as impurity effects

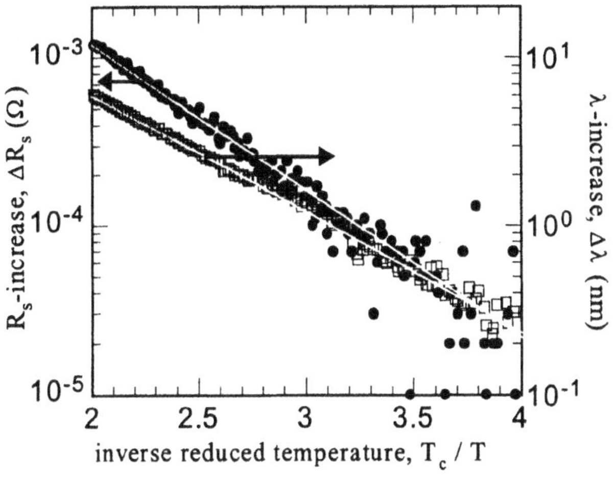

Figure 1. $\Delta R_s(T)$ (filled symbols, left ordinate) and $\Delta\lambda(T)$ (open symbols, right ordinate) at 87 GHz of polycrystalline Nb_3Sn films on sapphire [38]. The lines emphasize the exponential behavior.

on $\Delta\lambda(T)$ (see below). The R_s-data were corrected for a residual value, which corresponded to the resolution of the measurement system [35]. In general, normalconducting inclusions due to magnetic flux or imperfections of the film structure are known to contribute to finite values of R_{res}, leaving it an extrinsic property of a superconductor.

In contrast to the absolute values of R_s, only changes of the penetration depth can be deduced from cavity and most dielectric resonator measurements. Absolute λ- and, equivalently, X_s-values were obtained by fitting $\Delta\lambda(T)$ to the temperature variation expected from the scaled BCS theory [8,35]. Such a procedure makes it possible to reconstruct $\sigma_1(T)$ and $\sigma_2(T)$ from the surface impedance $Z_s(T)$ according to [11]:

$$\frac{\sigma_1 - i\sigma_2}{\omega\mu_0} = \frac{2R_sX_s - i(X_s^2 - R_s^2)}{X_s^2 + R_s^2} \tag{7}$$

Figure 2 shows as an example $\sigma_1(T)$ for a typical Nb_3Sn film in comparison with two epitaxial YBCO films (Sec. 3.2) and the theoretical result for a BCS superconductor in the low-frequency limit (curve, [4,7]). The agreement between experiment and theory leads to the identification of the maximum of $\sigma_1(T)$ with the coherence peak, which is suppressed in clean or strongly coupled superconductors and absent in HTS [40].

The Nb_3Sn films presented here were prepared by tin vapor diffusion into Nb precursor films [38]. Due to the distributed nucleation of Nb_3Sn sites and the diffusive grain growth, the resulting films were polycrystalline. The average grain size was found to increase with the square root of the Nb film thickness in the range from 0.1 to 3 μm [39,

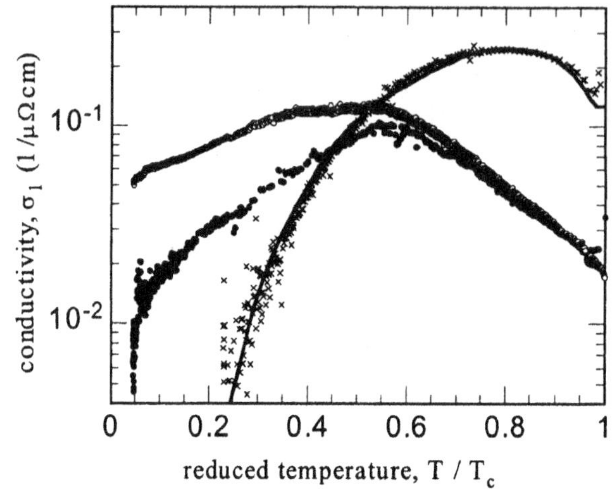

Figure 2. Quasiparticle conductivity $\sigma_1(T)$ at 87 GHz for the Nb_3Sn film of Fig. 1 (crosses) and for two epitaxial YBCO films on LAO (circles). The curve denotes the BCS result in the low-frequency limit [4,38].

41]. Given the short coherence length of Nb_3Sn, it is crucial to consider granular effects on Z_s and related transport properties. The main results of a study of films of different

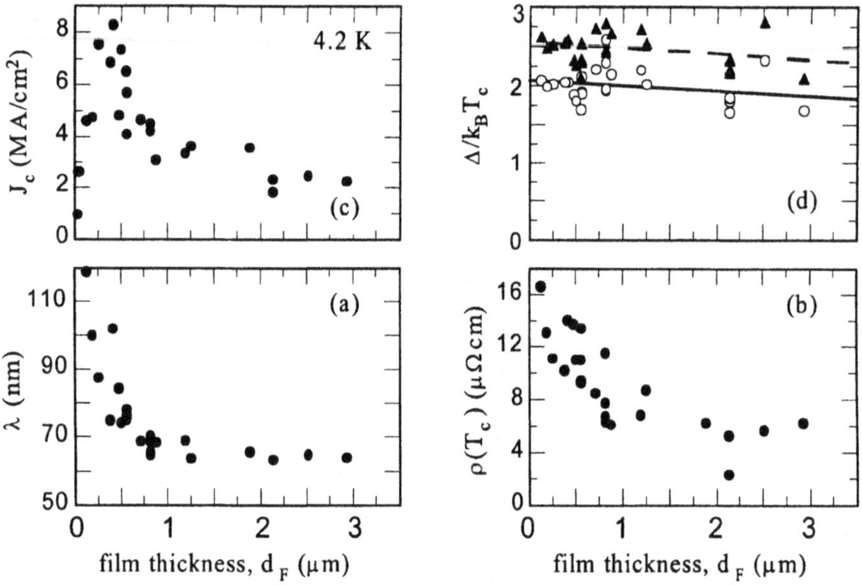

Figure 3. Effect of film thickness d_F on the penetration depth λ (lower left), resistivity $\rho(T_c)$ (lower right), critical current density $J_c(4.2K)$ (upper left) and reduced energy gap $\delta = \Delta/k_BT_c$ (upper right) of polycrystalline Nb_3Sn films on sapphire prepared by tin vapor diffusion. Open and filled symbols in panel (d) refer to a pure exponential and to Eq. (4b), respectively, with the lines indicating the average trend [4,38,39,41].

thickness are summarized in Fig. 3 in terms of the penetration depth (lower left panel a,), resistivity $\rho(T_c)$ at $T=T_c$ (lower right panel b, deduced from R_s according to the normal skin effect), critical AC shielding current density J_c at 4.2K (panel c), and reduced energy gap deduced from $R_s(T)$ according to Eq. (4b) (panel d). Overall, λ and $\rho(T_c)$ decrease with increasing film thickness, and thus with increasing grain size, indicating grain boundary scattering as the source reducing the quasiparticle mean free path. The λ- and ρ-data saturate at d_F>1 μm at the London value λ_L=60 nm and at the intrinsic resistivity $\rho(T_c)\approx5$ μΩcm of clean Nb_3Sn of high crystalline order [42]. Figure 3 (c) reveals a non-monotonic effect of the film structure on J_c. The decrease of J_c above $d_F\approx0.5$ μm is attributed to grain boundary pinning, which becomes weaker for larger grains. The decrease of J_c below $d_F\approx0.5$ μm results from weak grain boundary coupling [39,41]. Finally, the trend of the reduced energy gap to decrease with increasing grain size [Fig. 3 (d)] indicates a slight anisotropy which becomes more pronounced as the texture of the films improves.

While the Z_s-data of Nb_3Sn display, at first glance, textbook-like BCS behavior, the results of Fig. 3 emphasize the necessity of a careful quantitative analysis. They also shed doubt on simple definitions of sample quality as, e. g., high J_c-values do not coincide with low λ-values. Finally, subtle effects depending on the details of sample preparation can imitate unconventional behavior. This is illustrated in Fig. 4 for one Nb_3Sn film (open triangles) where $\Delta\lambda(T)$ increased linearly up to at least $0.4T_c$, in contrast to the BCS-like behavior observed for another, Ti-doped, film (open inverted triangles). Possible explanations of this difference are associated with the coupling strength, a temperature dependent effective mass (Sec. 3.3), or granularity. It is worth anticipating that similar $\Delta\lambda(T)$-dependences were also observed for epitaxial YBCO films (circles in Fig. 4, Secs. 3.2 and 3.3).

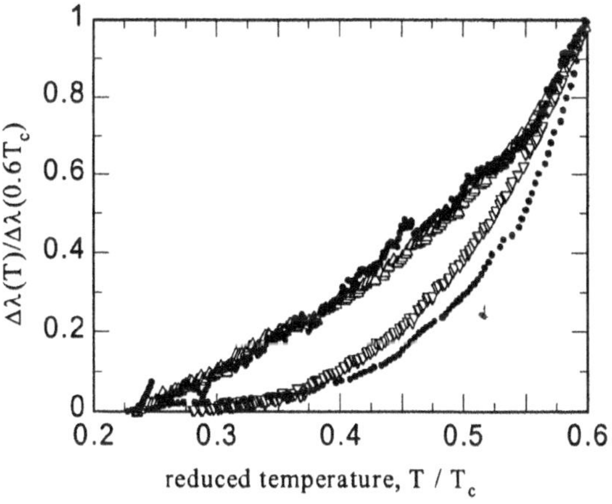

Figure 4. Temperature dependent change of the penetration depth, $\Delta\lambda(T)/\Delta\lambda(0.6T_c)$, at 87 GHz for two Nb_3Sn films (open triangles) and two epitaxial YBCO films (filled circles) [39]. Note broken x-scale.

3.2 EPITAXIAL $YBa_2Cu_3O_{7-x}$ FILMS ON VARIOUS SUBSTRATES

It is impractical to review the entire body of $Z_s(T)$-data of HTS crystals and epitaxial films of different compounds [2-4,9-11,43,44]. Instead, some features discovered recently with YBCO films on different substrates are described in the following [45-47], which appear essential to understand the quasiparticle and Cooper pair transport.

Figure 5 compares the temperature dependent surface resistance of eight epitaxial YBCO films, on sapphire (group 1), MgO (group 2), and $LaAlO_3$ (group 3) at 19 GHz in the range 4-40 K [46-48]. While the steep drop of $R_s(T)/R_s(T_c)$ near T_c due to the onset of Meissner screening was similar to that of Nb_3Sn and comparable for all films, the YBCO films could be distinguished below $t \leq 0.7$ in terms of their slopes $\partial R_s/\partial T$ [45]. The films on $LaAlO_3$ displayed comparable curvature on an absolute scale. They also showed a shallow shoulder of $R_s(T)$ around $t \approx 0.6$, which was attributed to strongly temperature dependent quasiparticle scattering (see Figs. 2 and 6) [35,49,50]. In contrast to YBCO/$LaAlO_3$, $R_s(T)$ decreased less steeply below 40 K for YBCO/MgO and even less for YBCO/sapphire. This behavior was attributed to an increasing competition between pair condensation and quasiparticle scattering at impurities (Sec. 3.3, [46,48]).

All YBCO films in Fig. 5 approached a comparable residual surface resistance $R_{res} \approx 130$ μΩ, well above the measurement accuracy, which was independent of the slope $\partial R_s/\partial T$. It was thus tempting to subtract the R_{res}-values, according to the usual procedure with conventional superconductors (Sec. 3.1). Figure 6 summarizes the resulting $\Delta R_s(T)$-

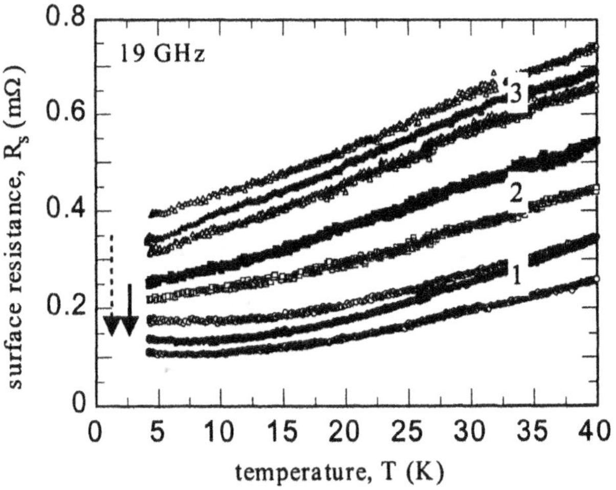

Figure 5. $R_s(T)$ at 19 GHz for epitaxial YBCO films on sapphire (group 1, circles, three different samples [51,52]), MgO (group 2, squares, two samples [53]) and on $LaAlO_3$ (group 3, triangles, three samples [52,54]). The arrows indicate offsets of 0.1 mΩ (solid line, MgO) and 0.2 mΩ (dashed line, $LaAlO_3$) [46].

data. The films on $LaAlO_3$ and MgO displayed power-law behavior of $\Delta R_s(T) \propto t^{\alpha}$, with $1 \leq \alpha \leq 2$ as reported by others [43,44,55]. In contrast, YBCO on sapphire showed clear exponential behavior, $\Delta R_s(T) \propto \exp(-\delta/t)$ with an apparent reduced energy gap $\delta = \Delta/k_B T_c$ between 0.7 and 1.0 [46], which is small compared to the BCS value of 1.76. However, the exponential behavior proves the existence of a finite energy gap, remaining after averaging over all quasiparticle momentum directions across the Fermi surface. It confirms earlier observations of activated behavior in well-oxygenated YBCO films and single-crystals [56-58]. Given the systematic influence of the substrate material, the data provide furthermore insight into the mechanisms leading to the formation of an energy gap (Sec. 3.3).

In contrast to the results for epitaxial YBCO *films* on sapphire substrates, most Z_s data on HTS *single-crystals* did not reveal exponential temperature dependences. Instead, the variations of R_s and especially of $\Delta\lambda$ at low temperatures often followed power-laws with a power of 1-2. This was considered an important indication of unconventional pairing with a resulting d-wave symmetry of the order parameter in the cuprates [2,43,55]. A similar behavior was sometimes observed for epitaxial films (see Fig. 4 for an example). However, as emphasized by the comparison with ΔZ_s-data on Nb_3Sn, neither is a power-law behavior unique for all types of YBCO samples, nor is it a proof of an unconventional nature of the pair state [55].

Another specific feature that distinguishes YBCO ($R_{res} \approx 130\mu\Omega$ at 20 GHz) from Nb_3Sn ($R_{res} \approx 170 n\Omega$ at 20 GHz [59]) is the much higher residual resistance R_{res} which indicates a sizeable number of unpaired charge carriers [60]. There have been much speculation about the sources of R_{res}, such as normalconducting impurities [61], weak

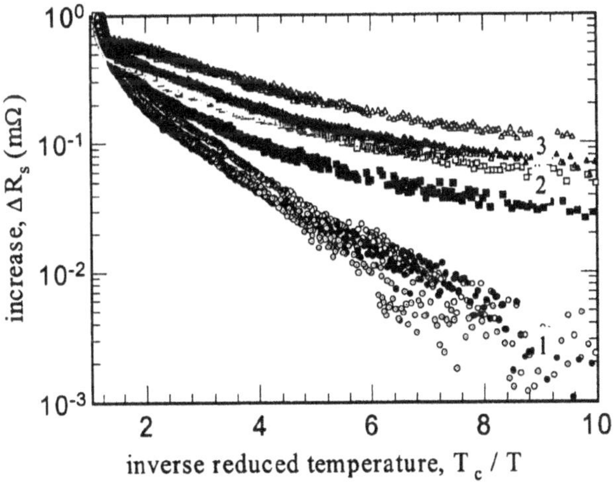

Figure 6. Temperature dependence of $\Delta R_s(T) = R_s(T) - R_{res}$ corresponding to the data displayed in Fig. 5.

links [62], intrinsically unpaired charge carriers of d-wave superconductors [50,63,64], or magnetic pairbreaking [10,46]. Yet in the absence of a conclusive answer, empirical analysis of R_{res}-data of YBCO films is helpful for an improved understanding. Figure 7 correlates the temperature-induced increase $\Delta R_s = R_s(0.85T_c) - R_s(4.2K)$ with the 4.2K-level of R_s for polycrystalline and epitaxial YBCO films [4]. The data for granular (triangles) and epitaxial samples (circles) are clearly distinguishable. The overall trend, indicated by the dashed line, indicates Josephson coupling between the grains [62,65]. In contrast, the ΔR_s-data of good epitaxial films tend to saturate around 10 mΩ. This value comes close to the intrinsic R_s value expected for YBCO at 87 GHz and $0.85T_c$. The influence of granularity on $R_s(T)$ thus seems to disappear in high-quality epitaxial films. Even so, the $R_s(4.2K)$- and hence the R_{res}-values extend over almost two orders of magnitude. The scatter of R_{res} at constant ΔR_s indicates that absolute level and temperature dependence of R_s are affected by several, probably independent, mechanisms. Further studies of R_{res} require improved R_s-resolution, better than 10 μΩ at 10 GHz.

3.3 HINTS FOR (UN)CONVENTIONAL PAIR INTERACTIONS

The most relevant indications of unconventional pairing in the hole-doped cuprate superconductors came from experiments with Josephson and tunnel junctions (e.g., [66,67]), which sense the phase of the order parameter directly, and from microwave spectroscopy. However, conclusions from Z_s data about the nature and symmetry of the pair state and about the corresponding density of quasiparticle states are indirect (e. g.,

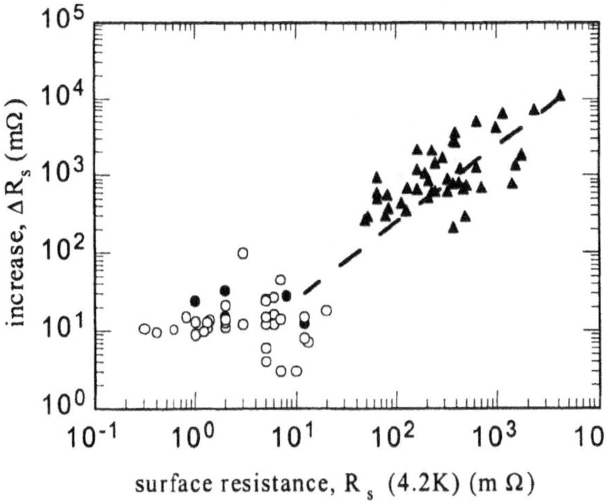

Figure 7. Empirical correlation between the increase of the surface resistance $\Delta R_s = R_s(0.85T_c) - R_s(4.2K)$ and $R_s(4.2K)$ for granular YBCO layers (triangles) and epitaxial films (filled circles) at 21.5 GHz and at 87 GHz (open circles) [4]. The 21GHz data were f^2-scaled to 87 GHz. The dashed line indicates the overall trend.

[35]) and must hence be considered with care. Three specific microwave properties of the cuprates (with emphasis on YBCO) that could be related to unconventional pairing are briefly analyzed in the following, in relation to the features mentioned in the preceding section.

The initial reduction of the residual resistance of epitaxial HTS films was related to improved preparation (Fig. 7), but there has been very little further progress over the past five years. The reproducibly achieved level around 20 $\mu\Omega$ at 10 GHz is still above the minimum level of about 5 $\mu\Omega$ anticipated for clean d-wave superconductors in the zero-frequency strong-scattering limit [35,63,64]. Any explanation of enhanced R_{res}-values should be decoupled from the temperature dependence of R_s, as indicated by experiment (Fig. 5). This constraint implies, within the framework of a two-fluid model (Sec. 2.1), that the residual fraction $n_N(0)$ of unpaired charge carriers should increase in proportion to the residual scattering rate $\tau^{-1}(0)$, thus keeping $R_{res} \propto n_N(0)\tau(0)$ constant [46]. Quasiparticle scattering and pairbreaking should accordingly have the same basic origin, similar to spin flip scattering at magnetic impurities in conventional superconductors or ordinary impurity scattering in d-wave superconductors. To keep the temperature dependence of $R_s(T)$ unaffected in the latter case would demand low impurity concentrations. Further distinction between the different types of quasiparticle scattering, and between intrinsic or extrinsic sources of the residual resistance, requires high-resolution measurements of $Z_s(T)$ at very low temperatures $T<4$ K.

The second specific microwave property of HTS is the low-temperature variation of $Z_s(T)$. While power-law behavior is not sufficient to prove a d-wave symmetry of the order parameter (Fig. 4 and [3,35,55]), "untypical" conventional mechanisms like magnetic pairbreaking [68] or strong pair coupling [69,70] should be considered as well. The latter effect was recently studied in terms of the temperature dependent pair coupling strength $\Lambda(T)$, which affects the renormalization of the effective quasiparticle mass $m^*(T)$ and hence the temperature dependences of related quantities like the penetration depth or the quasiparticle conductivity and hence the surface resistance (Sec. 2.1) [6]. Figure 8 compares $Z_s(T)$-data obtained with an epitaxial YBCO film at 87 GHz (sample "A" in [35]) with an extended two-fluid model incorporating the effects of strong electron-phonon coupling. For simplicity, an Einstein spectrum located at the phonon frequency Ω_E was assumed for the phonon density of states, yielding $\Lambda(0)=4$ in accordance with literature [10,68,69]. The ratio $\Psi \equiv k_B T_c/h\Omega_E$ was the only free parameter to fit the penetration depth data. Fitting the $R_s(T)$-data involved two additional parameters, namely the residual resistivity $\rho_r=1/\sigma_1(0)$ and the fraction $\varphi=n_N(0)/n_N(T_c)$ of unpaired charge carriers at zero temperature. The results in Fig. 8 were self-consistent for $\Psi=0.25$ [71], $\rho_r=1.8$ $\mu\Omega$cm and $\varphi=0.04$ [6]. Given the obvious oversimplification of the model, the agreement between data and model is surprisingly good.

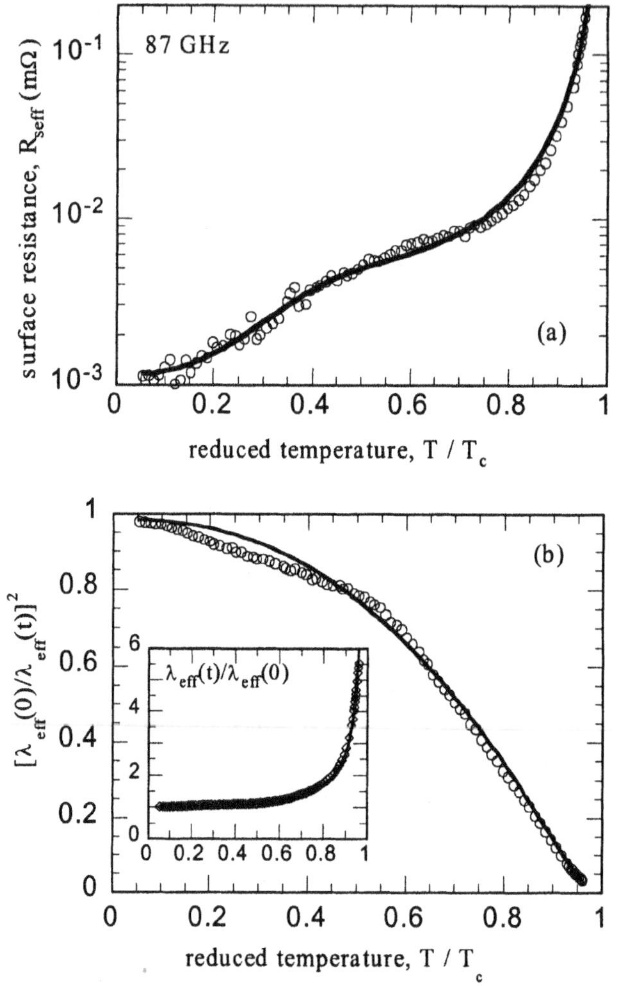

Figure 8. Surface resistance (a) and reduced pair density (b) versus reduced temperature $t=T/T_c$ derived from experiment (symbols, [35]) at 87 GHz (without corrections for finite film thickness). The inset to Fig. 3(b) shows λ(T). The curves represent the best fits to the two-fluid model assuming a temperature dependent effective mass, based on Λ(0)=4 and a simple Einstein phonon spectrum [6].

A strictly linear T-dependence of $\lambda^{-2}(t)$ at $t \ll 1$ could not be obtained with the simple Einstein spectrum. However, as pointed out by Eliashberg [70], such a behavior can be simulated with more realistic phonon spectra, where the density of phonon states extends to low frequencies. Along this line, the subtle differences in λ(T) between single-crystals [where in many cases $\lambda(T) \propto T$ has been observed] and epitaxial films [where predominantly $\lambda(T) \propto T^2$ is found] may find a natural and consistent explanation in terms of different phonon spectra, given that strain and defects are present in films but absent in high-purity crystals (see also next paragraph).

The third specific microwave property of HTS is the effect of strain on the superconducting state (Figs. 5 and 6), which is potentially related to the appearance of an exponential temperature dependence of R_s. In order to evaluate this substrate dependent behavior quantitatively, $R_s(T)$ was approximated in the low-temperature range (4–20)K by a polynomial of degree two, with $\partial R_s/\partial T$ representing the linear slope. The T_c-values were used to describe the crystalline quality of the samples which was affected by strain [46]. As shown in Fig. 9 for a variety of films (filled symbols), $\partial R_s/\partial T$ is obviously correlated with T_c. Also included in the figure are published data for YBCO crystals [72,73] (open diamonds), which fit to the overall trend. The temperature dependence of R_s becomes steeper with increasing crystalline perfection, with no general difference between films and crystals.

The origin of the correlation of $\partial R_s/\partial T$ with T_c and the related change from a power-law to an exponential temperature dependence of R_s as a function of the substrate bears important information about the pair coupling, at least in YBCO [46]. As discussed in [74,75], lattice strain can affect the pair state significantly. For YBCO films, the separation between two adjacent CuO_2 planes due to in-plane compressive strain is expected to reduce T_c. This could explain the lower critical temperatures of the films on sapphire compared to those on $LaAlO_3$. Another effect of an elongation of the lattice along the crystal c-direction is to modify the electronic interactions between the CuO_2-planes and the charge reservoir in the CuO chains, which are suspected to induce superconductivity in the chains [10,76,77]. Whether the chains display no or a small energy

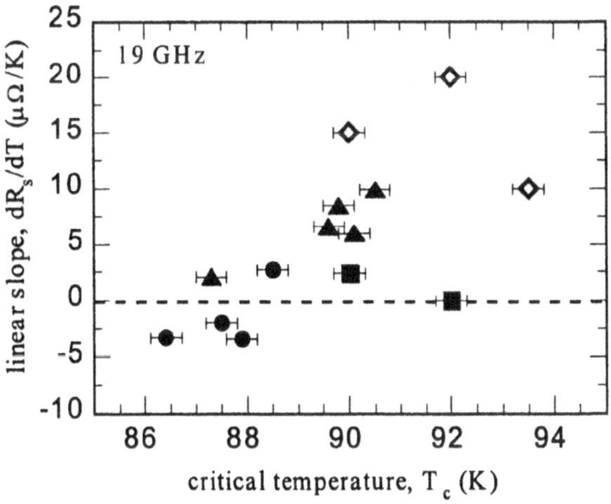

Figure 9. Slope $\partial R_s/\partial T$ versus T_c for various YBCO films on sapphire (circles), MgO (squares), and $LaAlO_3$ (triangles), and for single-crystals (diamonds, [72,73]). The horizontal line marks constant $R_s(T)$ [46].

gap depends not only on the oxygen deficiency, but also on magnetic interactions between the out-of-plane ("apex") oxygen sites and the chains. Increasing the plane-chain distances could therefore convert the chains from a gapless into a state with a small but finite energy gap. Such a mechanism would also explain the activated behavior to occur predominantly in strained films rather than in high-purity crystals.

An alternative explanation of superconductivity with a distinct energy gap in YBCO could be scattering at non-magnetic impurities, which are likely to be present in strained films, and are known to reduce the critical temperature of d-wave superconductors [66,78]. It is speculated that defects and interfaces could frustrate the d-wave symmetry, and induce a subdominant s-wave component extending over a few coherence lengths [79-81]. However, it is questionable if such a scenario can account for the observed exponential temperature dependence of R_s, given that this quantity integrates over the entire sample area, and over all quasiparticle energies and momentum directions. Finally, the small energy gap apparent in Fig. 6 is not sufficient to describe the steep drop of R_s near T_c, but requires the assumption of a second, much larger, energy gap (approximately $2\Delta/k_BT_c\approx6$) as discussed in [35]. It is therefore tempting to assume that the appearance of an s-wave component indicates the existence of two electronic bands which reflect the electronic interactions between chains and planes.

4. Nonlinear Surface Impedance of Unpatterned Films

While the temperature dependence of the surface impedance of HTS is well established empirically, and common trends are emerging for different types of samples (Sec. 3), convergence is still lacking for the dependence of R_s and λ on the microwave field amplitude B_s. This is due to a variety of features involved in the nonlinear response of the cuprates which complicates technological improvements as well as theoretical understanding. Investigation of $Z_s(B_s)$ at elevated microwave power levels also requires extended instrumentation. Because of available power sources, it is limited to frequencies below about 20 GHz and hence, in case of unpatterned samples, to large films [4, 9,30]. Since analysis of the nonlinear response is crucial for improving the microwave performance of HTS films and devices built of them, numerous groups have focused on the investigation of *patterned circuits* [23]. However, the following sections describe the phenomenology of nonlinear effects in *unpatterned films* in terms of the microwave field dependent surface impedance. Dynamic nonlinearities are beyond the scope of this paper but are covered, e. g., in [4,23,82,83]. The emphasis of the analysis presented here is on $R_s(B_s)$, while similar arguments hold for the nonlinear penetration depth $\lambda(B_s)$. This similarity does not hold for dynamic nonlinearities where the microwave field (or current) dependence of λ dominates the generation of harmonics and two-tone frequency intermodulation, especially in transmission line resonators [82,84].

Typical $R_s(B_s)$ curves measured with Nb_3Sn and YBCO films are displayed in figures 11 and 14 [39,47] (see sections 4.2 and 4.3 for more detailed descriptions). Both sets of curves display similarities in the sense that R_s stays constant at comparably low fields and then increases at higher levels. This increase can be magnetic and/or thermal in nature and is induced predominantly at defects [4]. The possible limitations of the linear microwave response, i. e., the range of constant $R_s(B_s)$, can be analyzed for both superconductors on the same basis (Sec. 4.1) [4]. As far as experimental data are concerned, a critical microwave field B_s* is used in the following as that field level at which R_s increased by 20% above its low-field value (unless otherwise stated). While this choice is arbitrary, it provides a common and practical basis to compare published $R_s(B_s)$-data.

4.1 PHENOMENOLOGICAL DESCRIPTION

The five curves in Fig. 10 represent the basic microwave field limitations applying to any superconductor [4]: the thermodynamic limit given by the superheating field B_{sh} (bold solid curve), the bulk (thin solid) and the Josephson (dashed) lower critical fields B_{c1} and B_{c1J}, and the critical ("quench") fields B_{Qg} or B_{Ql} determined by global heating

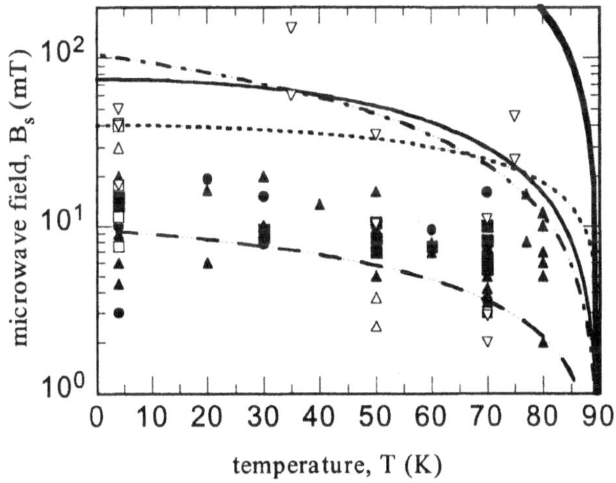

Figure 10. Theoretical and experimental critical microwave fields: Superheating field B_{sh} (bold solid curve), bulk lower critical field B_{c1} (thin solid), Josephson critical field B_{c1J} (dashed), global quench field B_{Qg} (dash-dotted), and local quench field B_{Ql} (dotted). The symbols denote field amplitudes B_s* measured at various laboratories with unpatterned (filled) and patterned (open) YBCO films on LAO (triangles), CbS (circles) and MgO (squares) (updated from [4]). Further explanations are given in the text.

on the entire film (dash-dotted) or by local heating at defects (dotted). The various critical fields can in general be distinguished by their absolute values and temperature dependences (drawn in Fig. 10 for YBCO, see below). However, as defects are involved, this distinction becomes ambiguous, and the associated time constants should be known to clarify the nature of the nonlinearity [4]. The time constants involved in the global magnetic field limitations ($B_c=B_{sh}$ or B_{c1}) scale with the overcriticality factor $1/\varepsilon=[B_s/B_c(T)-1]^{-1}$ and are thus expected to cause a smooth onset of nonlinearities. In contrast, the response time of Josephson junctions depends mainly on their characteristic voltage and can therefore be, in the microwave range, considered spontaneous [4,85]. Finally, the time constants of microwave heating are determined by the thermal properties of film, substrate and sample holder and can be checked experimentally [86], e. g., by varying the microwave duty cycle [87].

The superheating field is of the order of the thermodynamic critical field and thus well beyond the accessible field range for Nb_3Sn (≈0.5 T) and YBCO (≈1 T). It will therefore not be considered further. The remaining intrinsic *magnetic* limitation of the linear microwave response in the absence of defects is the lower critical field B_{c1}. The low, constant Z_s-values of *high-quality* films are likely limited by dynamic and hysteretic flux penetration into the superconductor at this field level. The temperature dependence of B_{c1} was found to follow $[1-(T/T_c)^2]$ as indicated in Fig. 10.

Josephson coupling is another *magnetic* mechanism, which causes nonlinear dissipation in HTS films at *weak links* due to locally reduced critical currents. The lower critical field of a long Josephson junction is [88]

$$\mu_0^{-1}B_{c1J}(T) = 2J_{cJ}(T)\times\lambda_J(T)\,, \qquad (8a)$$

with J_{cJ} and λ_J the critical current density and penetration depth of the junction. Figure 10 represents Eq. (8) assuming $J_{cJ}(T)\propto[1-(T/T_c)]$, $\lambda_L(T)\propto[1-(T/T_c)^2]^{-1/2}$, and an optimistic value $J_{cJ}=10^6$ A/cm^2. Nonlinearities in small junctions are expected to occur at

$$\mu_0^{-1}B_J(T) = J_{cJ}(T)\times\lambda_{eff}(T) \qquad (8b)$$

with $\lambda_{eff}=[\lambda_J^2+\lambda_L^2]^{1/2}$ an effective penetration depth. B_J corresponds to a microwave current exceeding J_{cJ}. However, Z_s-studies on engineered bicrystal Josephson junctions integrated into stripline resonators revealed that boundaries with misalignment angles between adjacent grains below about 10° did not induce enhanced nonlinearity [23,89].

Global heating, which is always present in microwave experiments due to the finite surface resistance, is the ultimate *thermal* field limitation of linear microwave response. The critical field B_{Qg} can be estimated by equating the dissipated microwave power to that heat flow which rises the temperature T of the sample to T_c:

$$\mu_0^{-2} B_{Qg}^2(T) = \frac{2(T_c - T)}{R_s(T) \times R_{th}(T)}. \tag{9}$$

R_{th} is the total thermal resistance of the film-interface-substrate sandwich, $R_{th}=d_F/\kappa_F+R_{bd}+d_S/\kappa_S$, with $d_{F,S}$ and $\kappa_{F,S}$ the thickness and the thermal conductivity of film ("F") and substrate ("S"), and R_{bd} is the boundary resistance between the two [4]. The temperature dependence of $B_{q,glob}$ in Fig. 10 resulted from numerical simulations [47].

Heating at *local defects* is likely to dominate the nonlinearities of films of *moderate quality* or at intermediate temperatures. Like granularity, local heating limits the linear microwave response at low fields. The local quench fields can be analytically approximated in analogy to global heating. For *isotropic* heat conduction from the defect to the substrate and a cold sink underneath at temperature T, the quench field $B_{Ql,iso}$ is

$$\mu_0^{-2} B_{Ql,iso}^2(T) \approx \frac{4\kappa_{ave} \times (T_c - T)}{R_{def}(T) \times r_{def}}, \tag{10a}$$

where κ_{ave} is the average thermal conductivity between T and T_c. R_{def} and r_{def} denote the surface resistance and the radius of the defect. The isotropic case applies to YBCO on $LaAlO_3$, whose thermal conductivities are comparable. Fig. 10 (dotted line) represents Eq. (10a) using $R_{def}\times r_{def}=5\mu\Omega m$ and $\kappa_{ave}=10$ W/Km [4,47]. For YBCO on substrates with high thermal conductivity like sapphire or MgO, the heat flow is *one-dimensional* since it is directed straight to the cold sink. The quench field for local heating is then

$$\mu_0^{-2} B_{Ql,one}^2(T) \approx \frac{2(T_c - T)}{R_{def}(T) \times R_{th}(T)} \tag{10b}$$

Equation (10b) reduces formally to Eq. (10a) for $R_{th}\approx r_{def}/2\kappa_{ave}$, but a quantitative comparison between the two local quench fields is complicated by the variable and temperature dependent thermal properties of film, substrate, and film-substrate boundary.

The field dependences of R_s for the two thermal (local or global heating) and two magnetic loss mechanisms (flux penetration into grain boundaries or grains) were subject of many publications. A quadratic increase of $R_s(B_s)\propto B_s^2$ described most data on classical as well as cuprate superconductors near the onset of nonlinearities [90,91], although with widely differing quadratic coefficients. At higher field levels, the quadratic behavior turned either into a steeper increase or tended to saturate (Figs. 11 and 14). This crossover reflects at least partially the feedback between magnetic and thermal effects. This can be illustrated for the defect-free case and a simplified field dependence

$$R_s(T, B_s) \approx R_s(T, 0) \times \{1 + [B_s / B_c(T)]^2\}. \tag{11a}$$

Heating induces an effective microwave field dependence of the sample temperature

$$T(B_s) \approx T(0) + R_s(T, B_s) \times R_{th}(T) \times B_s^2 / 2\mu_0^2. \tag{11b}$$

Combining both equations constitutes an implicit problem since R_s, R_{th} and B_c depend on temperature. The thermal-magnetic feedback becomes relevant in dissipative superconductors, i. e., at elevated temperatures and field amplitudes, and in films with low critical fields B_c. The resulting slope of $R_s(B_s)$ can be enhanced or suppressed depending on whether $R_{th}(T)$ increases or decreases with temperature. The measurable shapes of $R_s(B_s)$ depend therefore on the absolute temperature, the type of substrate, and on the cryogenic setup.

In addition to the theoretically expected critical microwave fields $B_c(T)$, Fig. 10 summarizes $B_s^*(T)$-results for YBCO obtained in various research labs with unpatterned films on $LaAlO_3$, MgO and sapphire (filled symbols), with coplanar resonators on $LaAlO_3$ and MgO (open triangles and squares), and with stripline resonators on $LaAlO_3$ (open inverted triangles) [4,47,92]. The scatter of the B_s^*-data reflects the variety of test methods and the variable quality of the investigated films. The highest field levels were achieved with patterned devices due to the field enhancement at the contours of the superconductor which, however, does not seem to alter the $R_s(B_s)$-performance [93]. An upper limit of the majority of B_s^*-data is given by $B_s^*(T) \leq B_0 \times [1-(T/T_c)^2]$ with $B_0 \approx 30$ mT. While consistent explanations of the various $R_s(B_s)$-dependences are lacking (Sec. 4.3), Fig. 10 points to the dominant field limitations. At temperatures above about 60 K, global magnetic and thermal effects bound the data fairly well [92]. Below 60 K, the critical field levels are more likely limited by heating at defects. This is especially the case in unpatterned films where the total area exposed to high microwave fields is usually larger than in patterned structures. However, the field levels achieved in stripline resonators [23,93] indicate that flux penetration and heating can be apparently weak enough so that $B_{c1}(T)$ can be exceeded.

No firm correlations have yet been established between B_s^* and other microwave properties (e. g., low-field R_s or frequency) or low-frequency transport properties (e. g., J_c or ΔT_c), but B_s^* seems to be related to the microstructure of the films (e. g., density of cracks or a-axis grains, Sec. 4.3 and [94]). Whatever the nature of the nonlinearities in YBCO, the apparent gap in Fig. 10 between the highest B_s^*-values achieved with small patterned devices and with large unpatterned films indicates a remaining potential for improving their homogeneity and defect density.

4.2 POLYCRYSTALLINE Nb_3Sn FILMS

The previous analysis implied that the nonlinear microwave response of superconductors is mostly determined by defects. This can be illustrated with $R_s(B_s)$-data on Nb_3Sn films prepared by tin vapor diffusion, whose grain structure could be controlled by varying the thickness of the Nb precursors (Sec. 3.1, [39,41]). Figure 11 displays typical results for a 1.2μm-thick 1"diamater film at 19 GHz. The onset of nonlinear $R_s(B_s)$ occurred at 4.2 K around 25 mT and thus well below $B_{c1}(4.2K) \approx 130$ mT, dropping rapidly at higher temperatures. Above $T=0.7T_c$, a linear increase of $R_s(B_s)$ developed (which was sometimes observed in YBCO films), followed by a plateau and another increase at higher field levels. The arrows indicate the fields $B_{on}(T)$ above which the time-domain signals, recorded during 400μs pulses (see Fig. 12), became distorted. Therefore, the shape of $R_s(B_s)$ at $B_s>B_{on}$ could not be accurately measured, but was assumed to correspond to the maxima of the transmission pulses. The inset to Fig. 11 displays $B_{on}(T)$ which followed in many cases the power law $[1-(T/T_c)^4]$, which is reminescent of the T-dependence of $\lambda^{-2}(T)$. B_{on} was found to be only slightly lower than the critical field B_s* defined in Sec. 4.1.

Figure 12 displays a typical set of pulse shapes monitored for another Nb_3Sn sample at various field strengths. Pulse distortion occurred around 17 mT as a shallow decrease of the transmitted power during the pulse. An oscillatory response with time decaying amplitudes developed at higher fields. The period of the oscillations decreased with increasing overcriticality as expected, but increased slightly during the pulses. Such a behavior agrees with the response expected for the fast recovery of a dissipative source (like heating at defects), superimposed by heating of the entire film. The fast oscillation between states of low and enhanced losses reflects the low temperature rise sufficient to exceed the T_c of the Nb_3Sn films. Accordingly, oscillatory pulses were rarely observed in YBCO films at low temperatures [c. f., Fig. 15 (a)]. Interpreting the decay time of about 100 μs as a thermal relaxation time, the heat diffusion seems to be limited by the

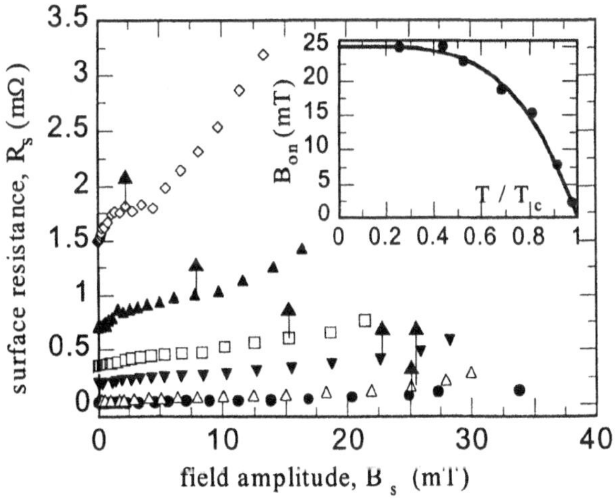

Figure 11. Typical $R_s(B_s)$-curves of a 1"diameter Nb_3Sn-film on sapphire at 19 GHz and various reduced temperatures (T_c=18K): circles-0.23, open triangles-0.39, inverted triangles-0.61, squares-0.72, filled triangles-0.83, diamonds-0.89. The arrows indicate the onset of pulse deformation at B_{on} (c.f., Fig. 12). The inset displays $B_{on}(T)$ (symbols) in comparison with a $(T/T_c)^4$-dependence (curve). After [39].

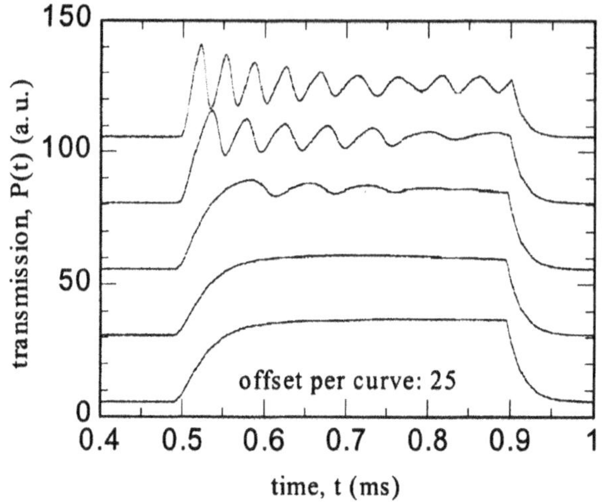

Figure 12. Sequence of pulse shapes (400 μs on, about 1 Hz repition rate) of the transmitted microwave power versus time for a 1"diameter Nb_3Sn film at 19 GHz and 4.2 K. The maximum power levels correspond to field amplitudes (in mT) of: 14-17-20-25-27 (from bottom to top). After [39].

low thermal conductivity of the Nb_3Sn film [95] or the interface between film and substrate (heat diffuses across a 0.5mm-thick sapphire substrate within microseconds, [4]). Consequently, local quench fields of about 20 mT are estimated from Eq. (10b) for Nb_3Sn, which are comparable to or even lower than for YBCO (Fig. 10).

Figure 13 summarizes B_{on}(4.2 K)-values for Nb_3Sn films of different thickness (filled circles). Pulse distortion was absent in thin films, d_F<1 μm. To avoid ambiguities, the onset field was therefore generally taken as the level of enhancing the low-field surface resistance by a factor of 3.5 [39]. The surprising result in Fig. 13 is the non-monotonic behavior of $B_{on}(d_F)$, displaying a maximum of 25 mT at 1.2 μm. This behavior results from the competition of two mechanisms with opposing trends. Curve 1 represents analytical B_{Ql}-estimations assuming a total surface resistance and a total radius of the defect(s) of R_{def}=13.7 mΩ and 180 μm, respectively, and neglecting the unknown boundary resistance R_{bd}. The local quench field decreases with increasing film thickness as expected from Eq. (10b), and provides a quantitative description of the B_{on}-data of "thick" films. In contrast, the increase of $B_{on}(d_F)$ of "thin" films was attributed to granularity [39,41]. This is illustrated by curve 2 which represents the scaling $B_{on} \propto J_{cJ}$ according to Eq. (8b), with J_{cJ} deduced from inductive AC measurements. J_{cJ} was found to increase in proportion to the average grain size, and hence to scale with the film thickness like $d_F^{1/2}$ ([41], Sec. 3.1). This square-root behavior describes $B_{on}(d_F)$ at d_F<1.5 μm very well. Further evidence for granular effects was obtained from analyzing the $R_s(B_s)$-curves in terms of the weakly coupled grain model [91].

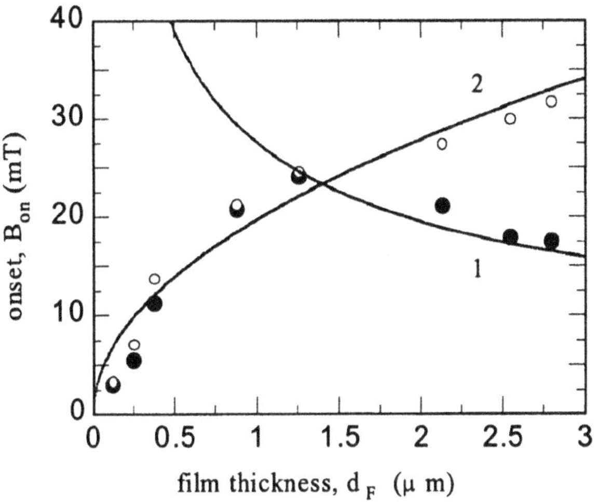

Figure 13. Measured onset fields B_{on} for nonlinearity, defined by $R_s(B_{on})=3.5R_s(B_s<<B_{on})$, of Nb_3Sn films of different thicknesses at 4.2 K (filled circles) [38,39]. The open circles denote B_{on} reconstructed from $R_s(B_s)$ assuming Josephson-coupled grains [91]. Curves 1 and 2 represent the results expected from local heating at a defect and from weak grain boundary coupling, respectively. Further details are given in the text.

The field dependence of R_s around B_{on} was described by a polynomial of degree two, where the quadratic coefficient b_R should be inversely proportional to J_{cJ}. The open circles in Fig. 13 represent the scaling of $1/b_R$ with d_F. It agrees well with the measured J_{cJ}-data. The involved parameters provided a consistent and quantitative representation of the $B_{on}(d_F)$-data and other low-frequency transport properties, extending over the entire range of film thickness. This agreement indicates that weakly coupled grains are subject to microwave heating due to their enhanced losses. Consequently, improved cooling or the use of epitaxial Nb_3Sn films should allow for higher B_{on}-values.

4.3 EPITAXIAL $YBa_2Cu_3O_{7-x}$ FILMS

Figure 14 shows $R_s(B_s)$ for epitaxial YBCO films in analogy to Fig. 11 for polycrystalline Nb_3Sn films. The scales for the microwave field level and the surface resistance as well as the qualitative $R_s(B_s)$ behavior are comparable, indicating that the origin of the nonlinearities, i. e., most likely weakly or non-superconducting defects, are similar for the two superconductors. The $R_s(B_s)$-curves below 60K are "S"-shaped with a range of constant R_s at low field levels. The trend of R_s to saturate at high field levels is associated with the thermal conductivity of sapphire, which increases with temperature, at least below 50 K. The time domain response remained undistorted up to the highest achieved field levels. The temperature dependent field range of linear response was eventually

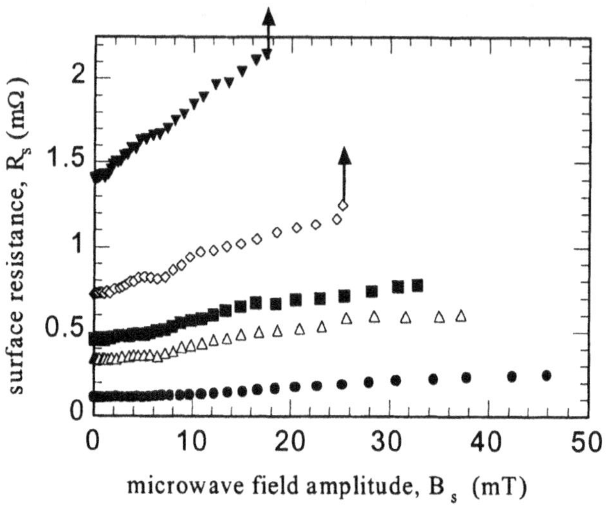

Figure 14. Typical $R_s(B_s)$-curves of a 550nm-thick 2"diameter YBCO film on sapphire at 19 GHz and various reduced temperatures (T_c=88K): circles-0.05, open triangles-0.57, squares-0.68, diamonds-0.80, filled triangles-0.88 [96]. The arrows indicate quenches.

limited as discussed in Sec. 4.1. Above 60 K, $R_s(B_s)$ increased step-like, as if switching defects into a lossy state. The curves were usually terminated by quenches (arrows) [47,92]. The critical field values B_s* which, above 60K, belong to the highest achieved with unpatterned films, are indicated in Fig. 10.

Figure 15 (a) displays a sequence of pulse shapes monitored in transmission for a YBCO film at 19 GHz with the resonator positioned above a bad spot [47]. The response of the film to the microwave pulse remained linear during the full period only up to B_s=2.6 mT (curve 1). At higher field levels, the transmitted signal started to break down on a time scale which decreased with increasing overcriticality [B_s(mT)-2.6] (curves 2 to 5). The relaxation became too slow at low power levels to be completed at the end of the pulses. Similar results have been reported frequently in the literature (e. g., [86]), and are considered representative of locally induced nonlinear losses which involve heating. The shapes of the transmitted pulses depend on the thermal properties of the film (Sec. 4.1, [86]), and are therefore expected for HTS to be distinctly different from those observed with Nb_3Sn (Fig. 12). However, the onset field levels for pulse deformation were found to decrease with increasing temperature like $[1-(T/T_c)^m]$, with m ranging between 2 and 4, similar to the results for Nb_3Sn.

Figure 15 (b) illustrates the effect of nonlinear microwave response on the transmitted power monitored in the frequency domain during a sweep across the resonance for two different sweep times (curve 1 and curves 2 to 4) at various power levels. The measurement conditions were otherwise the same as in Fig. 15 (a). At an input power

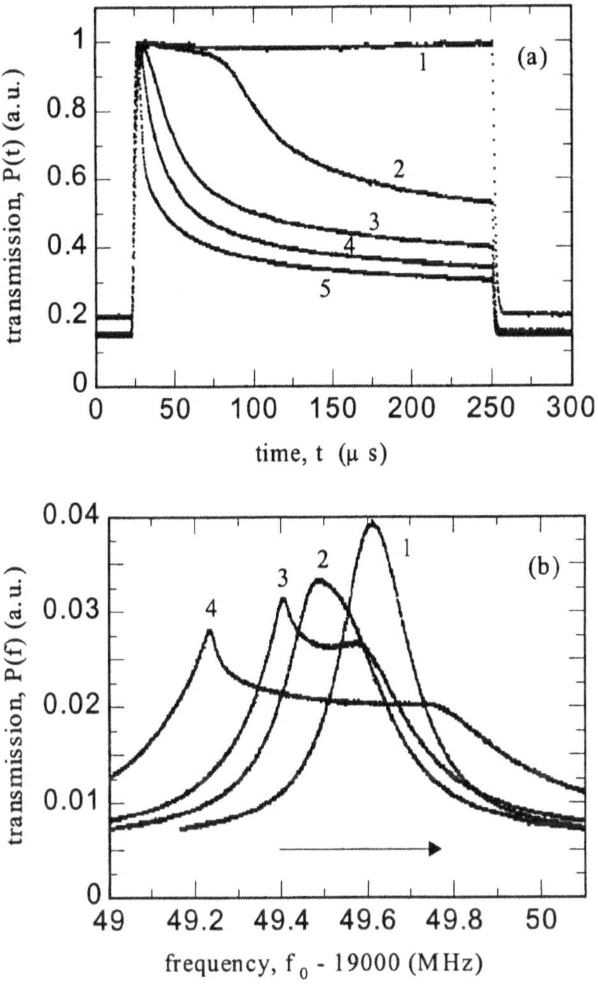

Figure 15. (a) Sequence of pulse shapes (225 μs on, about 1 Hz repition rate) of the transmitted microwave power, normalized to its maximum value, versus time for a 2"diameter YBCO film at 19 GHz and 77 K. The maximum power levels of curves 1 to 5 correspond to field amplitudes (in mT) of:2.6-3.1-3.6-4.1-4.6 [4,47]. (b) Sequence of resonances of the same film at 10 ms sweep time and 500 mW input power (1), and at 200 ms sweep time at 316 (2), 500 (3), and 1600 mW input power (4). The arrow indicates the sweep direction.

above 310 mW (curve 2), the resonance began to deviate from the expected Lorentzian shape. It became broader and split into separate peaks at higher power levels (curves 3 and 4). In some cases, the films suffered permanent damage after application of high microwave power, resulting in irreproducible P(f) shapes (e. g., [87,92]). In the case of distorted resonances, the determination of Q_L from the bandwidth (Sec. 2.2) becomes obviously ill-defined, especially since the distortion depends on the sweep rate (curves 1 and 3). While pulsed measurements with properly adjusted duty cycle are desirable to characterize the nonlinear response, microwave applications usually require CW operation and therefore demand the prevention of nonlinear effects [23,87].

There have been many efforts to search for correlations between the nonlinear microwave response and the microstructure of the investigated films. Except for granular effects in polycrystalline films [9], most efforts have not yet conclusive. There are no clear correlations between the onset field for nonlinearity and the film thickness, the low-field level of R_s, or structural parameters like the lattice mismatch between film and substrate, the width of the rocking curve, or the c-axis length [4,47]. Neither is there in general a degrading effect of patterning YBCO films into devices [93,97]. But systematic studies did reveal a trend of $B_s^*(77K)$ to increase on average from about 3 to 8 mT with the critical temperatures increasing from 87 to 91 K [47]. This behavior agrees quantitatively with the empirical temperature dependence $B_s^*(T)=B_s^*(0)\times[1-(T/T_c)^2]$ which yields, using $T_c\approx 90$ K and $B_s^*(0)\approx 80$ mT, an increase of $B_s^*(77K)$ by 5 mT for $\Delta T_c=+4$ K. Less obvious is the weak correlation found between $B_s^*(77K)$ [not visible for $B_s^*(4.2\ K)$] and the critical current density $J_c(77K)$, shown in Fig. 16 for epitaxial YBCO films on different substrates. Since there was no obvious correlation between $J_c(77K)$ and T_c (inset to Fig.16), the trend of B_s^* to increase with increasing critical current density indicates the relevance of magnetic rather than thermal nonlinear effects, as could be expected at such high T/T_c-values.

A rather pronounced and non-trivial correlation of B_s^* with film structure was recently reported for YBCO films laser ablated onto sapphire, during the optimization

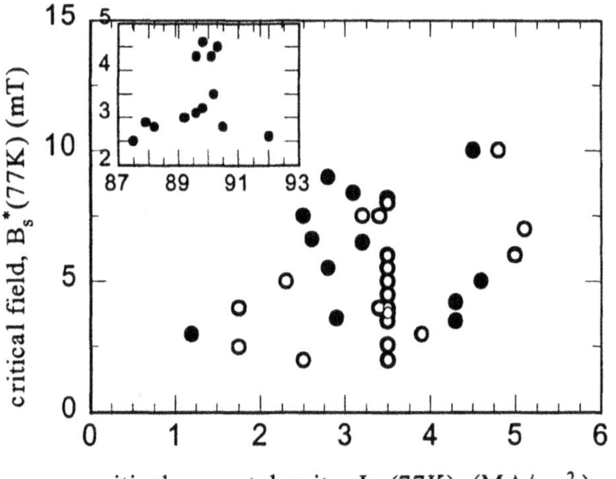

Figure 16. Dependence of the empirical critical microwave field B_s^* at 19 GHz and 77 K on the critical current density at 77 K for differently prepared YBCO films on various substrates (symbols) [47]. The inset displays the relation between $J_c(77K)$ and T_c of the samples represented by the filled symbols.

stage of the deposition process ([94,98], open symbols in Fig. 16). As shown in Fig. 17, B_s* depends strongly on the relative density of microcracks, as investigated by optical microscopy [99] and expressed by the ratio l/d of the average length l of a crack to the average separation d. According to this definition, the ratio l/d=1 represents a dense network of cracks. The origin of the cracks was attributed to the different thermal expansion coefficients of the HTS films and the sapphire substrates. Field enhancement along the contours of the cracks as well as microwave heating of their environment could account for the strong degradation of B_s* with increasing crack density. Fortunately, the formation of cracks and their deleterious effects could be avoided in optimized depositon techniques ([96,100], Figs. 14 and 16).

5. Summary

5.1 SURFACE IMPEDANCE AS A SPECIFIC MATERIAL PROPERTY

Section 3 illustrated methods to deduce microscopic parameters of superconductors from microwave surface impedance measurements. The comparative treatment of YBCO and Nb_3Sn highlighted the possibly quite subtle distinction between intrinsic properties and effects related to the preparation of the materials. Aiming at a theoretical description of the surface impedance of superconductors, strong-coupling effects, anisotropy, the high T_c-values, and possible two-band behavior involved approaches beyond the BCS theory.

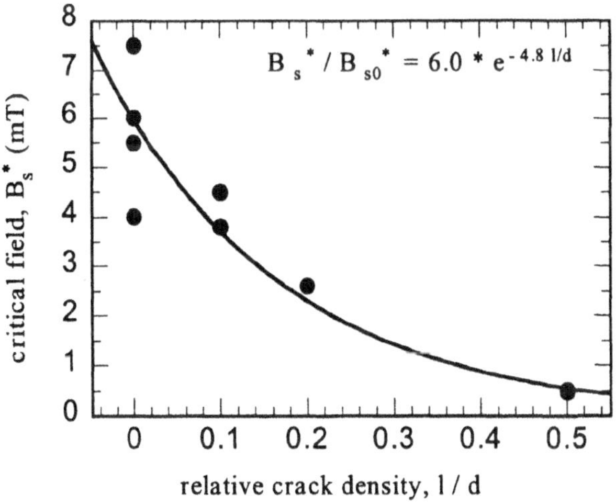

Figure 17. Dependence of the empirical critical microwave field B_s* on the relative crack density for 3"diameter YBCO films on CeO_2 buffered sapphire at 8.5 GHz and 77 K [47,94].

Simultaneous investigation of the real and the imaginary parts of Z_s or, equivalently, the complex conductivity σ (Sec. 2), provides information on the quasiparticle scattering, which can be inelastic (and possibly strongly temperature dependent) or elastic (depending on the type and density of defects) and coherent (as in Nb_3Sn below T_c) or incoherent (as in clean and strongly coupled superconductors like YBCO). The temperature dependent condensation of unpaired charge carriers into pairs could be analyzed in terms of the reduced inverse square of the penetration depth, providing valuable information on the quasiparticle density of states in the superconducting state. It was found that carrier doping and strain can strongly affect both scattering and pairing, especially if the band structures of the materials are taken into account (peaks in the density of states near E_F for Nb_3Sn and electronic interactions between CuO_2-planes and intermittent charge reservoirs for HTS).

The contribution of Z_s-measurements to the ongoing debate about the nature and the symmetry of the pair state in the cuprate superconductors is relevant in the following respects. 1. Power-law behavior of $Z_s(T)$ and large R_{res}-values are, by themselves, not sufficient to prove unconventional pairing or a d-wave symmetry of the order parameter. 2. Empirical analysis of $Z_s(T)$, which is independent of any theoretical background, provided evidence for two superconducting bands with different order parameters. The larger value governs the steep drop of surface resistance and penetration depth below T_c. The smaller value appears in the low-temperature behavior of the surface impedance. Indications for two-band behavior were concluded previously from other types of measurements like tunneling spectroscopy or positron annihilation [10]. 3. The observation of an exponential temperature dependence of $R_s(T)$ in strained YBCO films presents constraints for pairing mechanism in this superconductor, if not in all cuprates. The occurrence of a finite energy gap demonstrates the necessity for any theoretical approach in terms of d-, s-, or mixed d/s-wave symmetries, to evaluate the resulting density of states. 4. On the experimental side, the substrate-dependent $R_s(T)$-data connected the behavior of YBCO films to that of single-crystals, which were often considered to be the only type of material to study intrinsic properties. Finally, the obvious similarities in the nonlinear microwave response (Sec. 4) between conventional superconductors and cuprates emphasized the relevance of defects, which are partially still unidentified. The nonlinear response of defects masks unconventional behavior (like a nonlinear Meissner effect) until better samples and higher field levels become accessible.

Investigation of the surface impedance of any type of superconductors (A15, hole or electron doped cuprates, borocarbides, ruthenates, heavy fermions or organic compounds) will continue to be extremely important for the described reasons. Especially challenging is the study of Z_s at extremely low temperatures with high-resolution.

5.2 SURFACE IMPEDANCE AS A KEY TO MICROWAVE APPLICATIONS

The contributions to this summer school highlight the relevance of microwave superconductivity for applications, e. g., in communication, sensing, and medical technologies. Although superconductivity, like refrigeration, is an enabling technology and therefore probably not sufficient by itself to produce a major commercial impact, the optimization of the temperature, frequency, and especially microwave field dependent surface impedance is key to improving the performance of microwave devices, or to enable novel functions which are not available without superconductors, and hence to improving the conditions for a successful market penetration [4,12-15,24,27,30].

The linear and nonlinear microwave properties of YBCO and TlBaCaCuO (2212) have become good enough to develop laboratory demonstrators and prototype systems for real-world field trials or even operational system applications, and to judge the potential of novel devices accessible with these materials. However, the quality of HTS is still far from being sufficient to exploit this potential, or even from enabling the production of devices with reasonable yields. For instance, the statistical average of R_s(19GHz) of YBCO films at 60K is 0.74mΩ (on $LaAlO_3$), 0.55mΩ (on MgO) and 0.81mΩ (on sapphire). The corresponding values at 77K are 1.55 mΩ ($LaAlO_3$), 1.32mΩ (MgO) and 2.69mΩ (sapphire) [47]. These figures have to be compared with typical device specifications which require R_s<2mΩ at the desired temperature of operation. About 72% ($LaAlO_3$), 83% (MgO) and 22% (sapphire) of the samples meet this criterion presently at 77 K [47]. In terms of the nonlinear properties, the average critical fields B_s^*(77K) are 5.0 mT ($LaAlO_3$), 7.0 mT (MgO) and 5.1 mT (sapphire). Only about 44% ($LaAlO_3$), 80% (MgO) and 41 % (sapphire) of the samples meet the criterion B_s^*(77K)>5 mT, which is considered realistic for high-performance filter applications of HTS [4,47].

Possible improvements involve the development of microwave measurements with high signal and spatial resolution [31] to identify the sources of nonlinearities and to prevent them by improved fabrication, handling, packaging, and refrigeration. Enhanced T_c-values at otherwise sustained film quality can extend the range of operating temperatures suitable for high Q-factors and high circulating powers, and consequently lower the cryogenic burden. Finally, in order to compete with successfully developing alternative technologies, the development of high-permittivity and low-loss substrates that are compatible with the growth of epitaxial HTS films, and of materials suitable to tune the properties of superconductor devices, is essential.

New materials and technologies usually require sufficient time for investigation, acceptance and exploitation (and complicated materials and technologies even longer). Anticipating continuing progress of the HTS technology, this field remains equally fascinating, challenging, surprising, and promising at the beginning of the 21st century.

Acknowledgments

Valuable support from my colleagues at the Departments of Physics and Electrical Engineering at Wuppertal, and at the Schools of Physics and Astronomy, and Electronic and Electrical Engineering at Birmingham is highly appreciated. I am also grateful to S. Anlage, T. Dahm, J. Halbritter, R. Heidinger, H. Hilgenkamp, R. Humphreys, H. Kinder, M. Lorenz, M. Nisenoff, D. Oates, W. Prusseit, B. Stritzker, B. Willemsen, and R. Wördenweber.

References

1. Parks R. D. (ed.), (1969) *Superconductivity*, Marcel Dekker, New York.
2. Waldram, J. (1996) *Superconductivity of Metals and Cuprates*, Inst. of Physics Publishing, Bristol.
3. Trunin, M. (1998) Temperature dependence of microwave surface impedance in high-Tc single crystals: Experimental and theoretical aspects, *J. Supercond.* **11**, 381.
4. Hein, M.A. (1999) *High-Temperature Superconductor Thin Films at Microwave Frequencies*, Springer Tracts in Modern Physics **155**, Springer, Heidelberg.
5. Orlando, T.P., McNiff Jr., E.J., Foner, S., and Beasley, M.R. (1979) Critical fields, Pauli paramagnetic limiting, and material parameters of Nb_3Sn and V_3Si, *Phys. Rev.* **B19**, 4545-4561.
6. Cassinese, A., Hein, M.A., Hensen, S., and Müller, G. (2000) Effect of a temperature dependent effective quasiparticle mass on the surface impedance of $YBa_2Cu_3O_{7-x}$, *European Physical Journal* **B14**, 605-610.
7. Tinkham, M. (1996) *Introduction to Superconductivity*, McGraw-Hill, New York.
8. Turneaure, J.P., Halbritter, J., and Schwettman, H.A. (1991) The surface impedance of superconductors and normal conductors: The Mattis-Bardeen theory, *J. Supercond.* **4**, 341-355.
9. Portis, A.M. (1992) *Electrodynamics of High-Temperature Superconductors*, Lecture Notes in Physics, Vol. 48, World Scientific, Singapore.
10. Kresin, V.Z., Morawitz, H., and Wolf, S.A. (1993) *Mechanisms of conventional and high-T_c superconductivity*, Oxford University Press, New York.
11. Lancaster, M.J. (2000) Theory of RF superconductivity, contribution to this Summer School.
12. Talisa, S. (2000) Design of distributed 2d RF devices, ibid. Ref. [11].
13. Chaloupka, H.J. (2000) Design of lumped-element 2d RF devices, ibid. Ref. [11].
14. Mansour, R.R. (2000) HTS satellite applications, ibid. Ref. [11].
15. Willemsen, B. (2000) HTS wireless applications, ibid. Ref. [11].
16. Kolesov, S., Chaloupka, H., Baumfalk, A., and Kaiser, T. (1997) Planar HTS structures for high-power applications in communication systems, *J. Supercond.* **10**, 179-187.
17. Hong, J.S. Lancaster, M.J., Jedamzik, D., and Greed, R. B. (1998) 8-pole superconducting quasi-elliptic function filter for mobile communications application, *IEEE MTT-S International Microwave Symposium*, Baltimore, USA, June 1998.
18. Reppel, M., Chaloupka, H., Hong, J.S., Jedamzik, D., Lancaster, M., Mage, J.-C., Marcilhac, B. (1998) Superconducting preselect filters for base transceiver stations, in *Proceedings of the ACTS Mobile Commun. Summit*, Vol. 1, pp. 1-6.

19. Krupka, J., Geyer, R.G., Kuhn, M., and Hinken, J.H. (1994) Dielectric properties of single-crystals of Al_2O_3, $LaAlO_3$, $NdGaO_3$, $SrTiO_3$, and MgO at cryogenic temperatures, *IEEE Trans. Microwave Theory Techn.* **42**, 1886-1890.
20. Zuccaro, C., Ghosh, I., Urban, K., Klein, N., Penn, S., and Alford, N. (1997) Materials for HTS-shielded dielectric resonators, *IEEE Trans. Appl. Supercond.* **7**, 3715-3718; Zuccaro, C., Winter, M., Klein, N., and Urban, K. (1997) Microwave absorption in single crystals of lanthanum aluminate, *J. Appl. Phys.* **82**, 5695-5704.
21. Hein, M.A., Aminov, B.A., Baumfalk, A., Chaloupka, H.J., Hill, F., Kaiser, T., Kolesov, S., Müller, G., and Piel H. (1997) High-power high-Q YBaCuO disk resonator filter, in H. Rogalla and D.H.A. Blank (eds), *Applied Supercond. 1997*, IOP Publ., Bristol, No 158, p. 319.
22. Baumfalk, A., Chaloupka, H.J., Kolesov, S., Klauda, M., and Neumann, C. (1999) HTS power filters for output multiplexers in satellite communications, *IEEE Trans. Appl. Supercond.* **9**, 2857-2861.
23. Oates, D. E. (2000) Nonlinear behavior of HTS RF devices, ibid. Ref. [11].
24. Newman, N. and Lyons, W.G. (1993) High temperature superconducting microwave devices – fundamental issues in materials, physics, and engineering, *J. Supercond.* **6**, 119-160.
25. Samoilova, T.B. (1995) Nonlinear microwave effects in thin superconducting films, *Supercond. Sci. Technol.* **8**, 259-278.
26. VanDuzer, T. and Turner, C.W. (1981) *Principles of Superconductive Devices*, Elsevier, Amsterdam, North-Holland.
27. Shen, Z. Y. (1994) *HTS Microwave Circuits*, Artech House, Boston.
28. Hinken, J. (1988) *Supraleiterelektronik*, Springer, Berlin.
29. Ruggiero, S.T. and Rudman, D.A. (eds), (1990) *Superconducting Devices*, Academic Press, New York.
30. Lancaster, M.J. (1997) *Passive Microwave Device Applications of HTS*, Cambridge University Press, Cambridge.
31. Anlage, S.M. (2000) Microwave HTS spatial probes, ibid. Ref. [11].
32. Jenkins, A.P., Kale, K.S., and Dew-Hughes, D. (1996) Surface resistance measurement techniques, in A. V. Narlikar (ed.), *Studies of HTS*, Nova Sciences, New York, vol. 17, pp. 179-219.
33. Mazierska, J. (1997) Dielectric resonator as a possible standard for characterization of high temperature superconducting films for microwave applications, *J. Supercond.* **10**, 73-84.
34. Pond, J. (2000) Microwave measurements at cryogenic temperatures, ibid. Ref. [11].
35. Hensen, S., Müller, G., Rieck, C.T., and Scharnberg, K. (1997) In-plane surface impedance of epitaxial $YBa_2Cu_3O_{7-x}$ films: Comparison of experimental data taken at 87 GHz with d- and s-wave models of superconductivity, *Phys. Rev.* **B56**, 6237-6264.
36. Kaiser, T., Diete, W., Getta, M., Hein, M.A., Müller, G., Perpeet, M., and Piel, H. (1998) Niobium shielded sapphire resonator for field-dependent surface resistance measurements of superconducting films, *Particle Accel.* **60**, 171-182.
37. Labbé, J. and Friedel, J. (1966) Instabilité électronique et changement de phase cristalline des composés du type V_3Si a basse température, *J. de Physique* **27**, 153-165; (1966) Effet de la température sur l'instabilité électronique et le changement de phase cristalline des composés du type V_3Si a basse température, *J. de Physique* **27**, 303-308.
38. Perpeet, M., Hein, M.A., Müller, G., Piel, H., Pouryamout, J., and Diete, W. (1997) High-quality Nb_3Sn films on sapphire prepared by tin vapor diffusion, *J. Appl. Phys.* **82**, 5021.

39. Perpeet, M. (1999) Analysen zur Wechselwirkung phasenreiner und Ti-dotierter Nb_3Sn Filme auf Saphir mit elektromagnetischen Feldern im Mikrowellenbereich, Dissertation, Universität Wuppertal, WUB-DIS 99-18.
40. Marsiglio, F., Carbotte, J.P., Akis, R., Achkir, D. and Poirier, M. (1994) Eliashberg treatment of the microwave conductivity of niobium, *Phys. Rev.* **B50**, 7203-7206.
41. Perpeet, M., Cassinese, A., Hein, M.A., Kaiser, T., Müller, G., Piel, H., Pouryamout, J. (1999) Nb_3Sn films on sapphire – a promising alternative for superconductive microwave technology, *IEEE Trans. Appl. Supercond.* **9**, 2496-2499.
42. Flükiger, R., Küpfer, H., Jorda, J.L., and Muller, J. (1987) Effect of atomic ordering and composition changes on the electrical resistivity of Nb_3Al, Nb_3Sn, Nb_3Ge, Nb_3Ir, V_3Si and V_3Ga, *IEEE Trans. Magn.* **23**, 980.
43. Hardy, W.N., Kamal, S., and Bonn, D.A. (1998) Magnetic penetration depth in cuprates: a short review of measurement techniques and results, in J. Bok, G. Deutscher, D. Pavuna, and S.A. Wolf (eds), *The gap symmetry and fluctuations in high-T_c superconductors* NATO-ASI Series B: Physics, Plenum Press, New York, vol. 371.
44. Bonn, D.A., Kamal, S., Liang, R., Hardy, W.N., Homes, C.C., Basov, D.N., and Timusk, T. (1996) Surface impedance studies of YBCO, *Czech. J. Phys.* **46**, 3195.
45. Hein, M.A. (1998) in A. K. M. A. Islam (ed.), Physics and applications of high-T_c superconductors at microwave frequencies, International workshop on high-temperature superconductors, Rajshahi-University, Bangladesh, ISBN 984-31-0393-9, pp. 180-213.
46. Hein, M.A., Kaiser, T., and Müller, G. (2000) Surface resistance of epitaxial YBa2Cu3O7-x films on various substrates: Effects of pair condensation and quasiparticle scattering, *Phys.Rev.* **B61**, 640-647.
47. Kaiser, T. (1998) Nichtlineare Hochfrequenzeigenschaften von Hochtemperatursupraleiter-Filmen, Dissertation, Universität Wuppertal, Report WUB-DIS 98-13.
48. Hein, M.A., Cassinese, A., Hensen, S., Kaiser, T., Müller, G., and Perpeet, M. (1999) Pair and quasiparticle states of $YBa_2Cu_3O_{7-x}$ deduced from the surface impedance and a comparison with Nb_3Sn, *J. Supercond.* **12**, 129-138.
49. Bonn, D.A., Liang, R., Risemann, T.M., Baar, D.J., Morgan, D.C., Zhang, K., Dosanjh, P., Duty, T.L., MacFarlane A., Morris, G.D., Brewer, J.H., and Hardy, W.N. (1993) Microwave determination of the quasiparticle scattering time in $YBa_2Cu_3O_{6+x}$, *Phys. Rev.* **B47**, 11314-11328.
50. Dähne, U., Goncharov, Y., Klein, N., Tellmann, N., Kozlov, G., and Urban, K. (1995) Frequency and temperature dependence of the mm-wave conductivity of epitaxial $YBa_2Cu_3O_7$ films, *J. Supercond.* **8**, 129-134.
51. Zaitsev, A.G., Ockenfuss, G., Guggi, D., Wördenweber, R., and Krüger, U. (1997) Structural perfection of (001) CeO_2 thin films on $(1\underline{1}02)$ sapphire, *J. Appl. Phys.* **81**, 3069-3072.
52. Kinder, H., Berberich, P., Utz, B., and Prusseit, W. (1995) Double-sided YBCO films on 4" substrates by thermal evaporation, *IEEE Trans. Appl. Supercond.* **5**, 1575-1580.
53. Humphreys, R.G., Satchell, J.S., Chew, N.G., Edwards, J.A., Goodyear, S.W., Blekinsop, S.E., Dosser, O.D., and Cullis, A.G. (1990) Physical vapour deposition techniques for the growth of $YBa_2Cu_3O_7$ thin films, *Supercond. Sci. Technol.* **3**, 38-52.
54. Cassinese, A., Getta, M., Hein, M., Kaiser, T., Kürschner, H.G., Lehndorff, B., Müller, G., and Piel, H. (1999) Scanning Hall probe measurements on single- and double-sided sputtered YBCO films for microwave applications, *IEEE Trans. Appl. Supercond.* **9**, 1960-1963.

55. Mourachkine, A.P. and Barel, A.R.F. (1996) Microwave penetration depth measurement in high-T_c superconductos, in A. Narlikar (ed.), *Studies of high-temperature superconductors*, Nova Sciences, New York, Vol. 17, p. 221-246.

56. Klein, N., Tellmann, N., Schulz, H., Urban, K., Wolf, S.A., and Kresin, V.Z. (1993) Evidence of two-gap s-wave superconductivity in $YBa_2Cu_3O_{7-x}$ from microwave surface impedance measurements, *Phys. Rev. Lett.* **71**, 3355-3358.

57. Anlage, S.M. and Wu, D. H. (1992) Magnetic penetration depth measurements in cuprate superconductors, *J. Supercond.* **5**, 395-402.

58. Gasparov, V.A., Mkrtchyan, M.R., Obolensky, M.A., and Bondarenko, A.V. (1994) Anomalous temperature dependence of the electromagnetic penetration depth of $YBa_2Cu_3O_{7-\delta}$ crystals, *Physica C* **231**, 197-206.

59. Klein, N., Müller, G., Piel, H., and Schurr, J. (1989) Superconducting microwave resonators for physics experiments, *IEEE Trans. Magn.* **25**, 1362-1365.

60. Müller, G., Klein, N., Brust, A., Chaloupka, H., Hein, M., Orbach, S., Piel, H., and Reschke, D. (1990) Survey of microwave surface impedance data of high-T_c superconductors - evidence for nonpairing charge carriers, *J. Supercond.* **3**, 235.

61. Vendik, O.G., Kozyrev, A.B., and Popov, A.Yu. (1990) Properties of high-T_c superconductors at RF and microwaves – experimental data and some model notions, *Revue Phys. Appl.* **25**, 255-263.

62. Halbritter, J. (1990) RF residual losses, surface impedance, and granularity in superconducting cuprates, *J. Appl. Phys.* **68**, 6315-6326; and (1992) On extrinsic effects in the surface impedance of cuprate superconductors by weak links, *J. Appl. Phys.* **71**, 339-343.

63. Hirschfeld, P.J., Puttika, W.O., and Scalapino, D.J. (1993) Microwave conductivity of d-wave superconductors, *Phys. Rev. Lett.* **71**, 3705-3708.

64. Lee, P.A. (1993) Localized states in a d-wave superconductor, *Phys. Rev. Lett.* **71**, 1887-1890.

65. Keskin, E., Numssen, and Halbritter, J. (1999) Defects in YBCO relevant for RF superconductivity, *IEEE Trans. Appl. Supercond.* **9**, 2452-2455.

66. Scalapino, D.J. (1995) The case for d_{x2-y2} pairing in the cuprate superconductors, *Phys. Rep.* **250**, 329-365.

67. Beasley, M.R. (1995) Recent progress in high-T_c superconductivity – what would make a difference, *IEEE Trans. Appl. Supercond.* **5**, 141-151.

68. Adrian, S.D., Reeves, M.E., Wolf, S.A., and Kresin V.Z. (1995) Penetration depth in layered superconductors: Application to the cuprates and conventional multilayers, *Phys. Rev.* **B51**, 6800-6803.

69. Golubov, A.A., Trunin, M.R., Zhukov, A.A., Dolgov, O.V., and Shulga, S.V. (1996) Comparative description of the microwave surface impedance of Nb, BaKBiO and YBaCuO, *J. Phys. I France*, **6**, 2275-2290.

70. Eliashberg, G.M., Klimovitch, G.V., and Rylyakov, A.V. (1991) On the temperature dependence of the London penetration depth, *J. Supercond.* **4**, 393-396.

71. Carbotte, J.P. (1990) Properties of boson-exchange superconductors, *Rev. Mod. Phys.* **62**, 1027-1157.

72. Zhang, K., Bonn, D.A., Kamal, S., Liang, R., Baar, D.J., Hardy, W.N., Basov, D., and Timusk, T. (1994) Measurement of the ab plane anisotropy of microwave surface impedance of untwinned $YBa_2Cu_3O_{6.95}$ single crystals, *Phys. Rev. Lett.* **73**, 2484; Hosseini, A., Kamal, S., Bonn, D.A., Liang, R., and Hardy, W.N. (1998) c-axis electrodynamics of $YBa_2Cu_3O_{7-x}$, *Phys. Rev. Lett.* **81**, 1298.

73. Bonn, D.A., Kamal, S., Zhang, K., Liang, R., Baar, D.J., Klein, E., and Hardy, W.N. (1994) Comparison of the influence of Ni and Zn impurities on the electromagnetic properties of $YBa_2Cu_3O_{6.95}$, *Phys. Rev.* **B50**, 4051-4063.
74. Locquet, J.P., Perret, J., Fompeyrine, J., Mächler, E., Seo, J.W., and VanTendeloo, G. (1998) Doubling the critical temperature of $La_{1.9}Sr_{0.1}CuO_4$ using epitaxial strain, *Nature* **394**, 453-456.
75. Schuller, I.K. (1998) Superconductivity – strain yourself, *Nature* **394**, 419-420.
76. Kresin, V.Z., Bill, A., Wolf, S.A., and Ovchinnikov, Yu.N. (1998) High T_c oxides: two order parameters, magnetic scattering and upper limit of T_c, novel isotope effects, and the phonon-plasmon mechanism, ibid. Ref. [43].
77. Kresin, V.Z., Wolf, S.A., and Ovchinnikov, Yu.N. (1996) Effect of pressure on T_c and the "intrinsic" T_c of cuprates, *Phys. Rev.* **B53**, 11831.
78. Hirschfeld P.J. and Goldenfeld, N. (1993) Effect of strong scattering on the low-temperature penetration depth of a d-wave superconductor, *Phys. Rev.* **B48**, 4219-4222.
79. Hu, C.R. (1994) Midgap surface states as a novel signature for d_{x2}-d_{y2}-wave superconductivity, *Phys. Rev. Lett.* **72**, 1526-1529.
80. Matsumoto, M. and Shiba, H. (1995) On boundary effects in d-wave superconductors, *J. Phys. Soc. Jpn.* **64**, 1703; (1995) Coexistence of different symmetry order parameters near a surface in d-wave superconductors I, **64**, 3384; (1995) Coexistence of different symmetry order parameters near a surface in d-wave superconductors II, **64**, 4867.
81. A controversial discussion of this topic can be found in the Proceedings of the "Euroconf. on Polarons: Condensation, Pairing, Magnetism", Erice, Italy, in (1999) *J. Supercond.* **12**, No. 1, special issue.
82. Dahm, T. and Scalapino, D.J. (1997) Theory of intermodulation in a superconducting microstrip resonator, *J. Appl. Phys.* **81**, 2002-2009.
83. Dahm, T., and Scalapino, D.J. (1998), Analysis and optimization of intermodulation in high-T_c superconducting microwave filter design, *IEEE Trans. Appl. Supercond.* **8**, 149-157.
84. Dahm, T., (1999) MPI Dresden, Germany, private communication.
85. Halbritter, J. (1998) Nonlinear surface impedance in "low" and "high" T_c superconductors, *J. Supercond.* **10**, 91-96.
86. Wosik, J., Xie, L.M., Grabovickic, R., Hogan, T., and Long, S.A. (1999) Microwave power handling capability of HTS superconducting thin films: weak links and thermal effects induced limitations, *IEEE Trans Appl. Supercond.* **9**, 2456-2459.
87. Aminov, B.A., Piel, H., Hein, M.A., Kaiser, T., Müller, G., Baumfalk, A., Chaloupka, H.J., Kolesov, S., Medelius, H., and Wikborg, E. (1999) YBaCuO disk resonator filters operating at high power, *IEEE Trans. Appl. Supercond.* **9**, 4185-4188.
88. Barone, A. and Paterno, G. (1982) *Physics and Applications of the Josephson effect*, Wiley, New York.
89. Habib, Y.M., Oates, D.E., Dresselhaus, G., Dresselhaus, M.S., Vale, L.R., and Ono, R.H. (1998) Microwave power handling in engineered $YBa_2Cu_3O_{7-x}$ grain boundaries, *Appl. Phys. Lett.* **73**, 2200-2202.
90. Padamsee, H. (1983) Calculations for breakdown induced by large defects in superconducting Nb cavities, *IEEE Trans. Magn.* **19**, 1322.
91. Nguyen, P.P., Oates, D.E., Dresselhaus, G., and Dresselhaus, M.S. (1993) Nonlinear surface impedance of $YBa_2Cu_3O_{7-x}$ thin films: measurements and a coupled-grain model, *Phys. Rev.* **B48**, 6400.

92. Kaiser, T., Aminov, B.A., Baumfalk, A., Cassinese, A., Chaloupka, H.J., Hein, M.A., Kolesov, S., Medelius, H., Müller, G., Perpeet, M., Piel, H., and Wikborg, E. (1999) Nonlinear power handling of $YBa_2Cu_3O_{7-x}$ films and microwave devices, *J. Supercond.* **12**, 343-351.

93. Xin, H., Oates, D.E., Anderson, A.C., Slattery, R.L., Dresselhaus, G., and Dresselhaus, M.S. (1999) Comparison of power dependence of microwave surface resistance of unpatterned and patterned YBCO thin films, preprint.

94. Kästner, G., Schäfer, C., Senz, S., Kaiser, T., Hein, M., Lorenz, M., Hochmuth, H., and Hesse, D. (1999) Microstructure and microwave surface resistance of typical YBaCuO thin films on sapphire and $LaAlO_3$, *Supercond. Sci. Technol.* **12**, 366-375.

95. Cody, G.D., and Cohen, R.W. (1964) Thermal conductivity of Nb_3Sn, *Rev. Mod. Phys.* **36**, 121.

96. Zaitsev, A.B., Kutzner, R., Wördenweber, R., Kaiser, T., Hein, M.A., and Müller, G. (1998) Large-area $YBa_2Cu_3O_{7-x}$ films on sapphire with excellent microwave power handling capability, *J. Supercond.* **11**, 361-365.

97. Porch, A., Avenhaus, B. Wellhöfer, F., and Woodall, P. (1995) Microwave surface resistance of unpatterned and patterned $YBa_2Cu_3O_{7-\delta}$ thin films produced by pulsed laser deposition, in D. Dew-Hughes (ed.), *Applied Supercond. 1995*, IOP Publ., Bristol, No 148, p. 1039.

98. Lorenz, M., Hochmuth, H., Natusch, D., Börner, H., Lippold, G., and Kreher, K. (1996) Large-area double-sided pulsed laser deposition of $YBa_2Cu_3O_{7-x}$ thin films on 3-in. sapphire wafers, *Appl. Phys. Lett.* **68**, 3332-3334.

99. Kästner, G., Hesse, D., Lorenz, M., Scholz, R., Zakharov, N.D., and Kopperschmidt, P. (1995) Microcracks observed in epitaxial thin films of $YBa_2Cu_3O_{7-\delta}$ and $GdBa_2Cu_3O_{7-\delta}$, *Phys. Status Solidi a* **150**, 381-394.

100. Olsson, E., Gupta, A., Thouless, M.D., Segmüller, A., and Clarke, D.R. (1991) Crack formation in epitaxial [110] thin films of $YBa_2Cu_3O_{7-x}$ and $PrBa_2Cu_3O_{7-x}$ on [110] SrTiO3 substrates, *Appl. Phys. Lett.* **58**, 1682-1684.

GROWTH AND PROCESSING OF HIGH TEMPERATURE SUPERCONDUCTOR THIN FILMS

R.G.HUMPHREYS
DERA
Malvern
Worcestershire WR14 3PS
UK

1. Introduction

It is now twelve years since the discovery of superconductors with transition temperatures above the boiling point of liquid nitrogen. In that time a great deal of progress has been made in transforming these materials into a viable device technology.

I first reviewed the growth and characterisation of HTS films ten years ago [1]. Re-reading that paper, it is remarkable to see how rapid progress was in the first two years of the subject. At that time it was possible to know all the literature. Since then, it has expanded to the point where it is difficult to grasp in its entirety. As the field has evolved, many groups have left it and those of us that have remained have focussed on developing technology for a purpose, often at the expense of developing a detailed understanding of the underlying principles. In this article, I will try to focus on these. I will also aim to cite recent references through which it is possible to trace the earlier literature, rather than trace an idea from its originator, which is normal (and better) scientific practice, but the reference list would get out of control were I to attempt it. As a starting point, a single recent issue of Superconductor Science and Technology contains three rather fully referenced papers [2,3,4]. One of my aims in this article is to identify areas where further work is needed and to make some connections with literature not normally cited in HTS papers, but related to these areas.

Broadly speaking, the growth and processing technology necessary for making HTS passive microwave devices is in place. The big issues are cost, reproducibility and uniformity in production. Little has been published about these issues so any comments made will be largely conjecture. Improvements are sought in power handling. Often, if power handling is not a problem, the circuit should have been designed to be more compact and hence cheaper to make and to cool. As I see the state of the art, the ability to design compact circuits is lagging behind the ability to make them, but progress is being made and ultimately the materials technology should become the limiting factor. I will not say much about multilayer circuit technology as this is not really the main topic of this summer school (although the circuits also work at microwave or mm wave frequencies) but the situation there is much less advanced. For these, a deeper understanding of the growth and processing mechanisms is certainly needed.

H. Weinstock and M. Nisenoff (eds.), Microwave Superconductivity, 55–89.

2. Growth Techniques

In [1], the prerequisites for growth were identified as the right substrate, control of the cation composition, supplying sufficient oxygen and controlling the temperature. Since then the only changes in this list are the added constraints of cost and growth area. At that time, growth was often carried out *ex situ*, i.e. a film was first deposited under conditions resulting in amorphous material and it was subsequently annealed to form an epitaxial layer. Such processes are still used for Tl containing superconductors, but have been abandoned for the $RBa_2Cu_3O_7$ compounds (where R is a rare earth or Y) which are still the most widely used materials. The main reasons why *ex situ* techniques have been abandoned are the difficulty of getting a film to recrystallise fully into an epitaxial layer, the problems of obtaining thick, smooth films and the need for a benign substrate. It is worth remarking, however, that it is possible to envisage an *ex situ* process as being very cheap in production, and this has led to an increase in interest in *ex situ* techniques for high power applications [5].

I will focus on the growth of the $RBa_2Cu_3O_7$ compounds by *in situ* techniques, where the film is grown as an epitaxial layer in a single process. Most of the available results are for $YBa_2Cu_3O_{7-\delta}$ (YBCO). Five techniques are considered here. These are evaporation, laser ablation, metal-organic chemical vapour deposition (MOCVD), sputtering, and liquid phase epitaxy (LPE), Of these, evaporation and sputtering are currently used in production scale processes and MOCVD is being developed. Laser ablation is being scaled up. LPE is relatively immature, but is nevertheless of interest.

2.1. EVAPORATION

In evaporation, one evaporates the metals onto a heated substrate in the presence of oxygen. In this technique the films are uniform in thickness and composition. This has the advantage that the process designer has full control of the cation composition, but the disadvantage that a relatively complex control system is required. The main problem is that it is difficult to supply enough oxygen to the growing film while maintaining stable composition control. When too much oxygen is used in our system, the mass spectrometer sensitivity falls, and the metal evaporation rates and the chamber pressure interact leading to instability. We solve the problem by using atomic oxygen as efficiently as we can, but this is limited to about 2" diameter substrates with our current setup.

The most radical advance in growth technology for thin film HTS in the last 10 years has been the invention by Kinder and co-workers at Munich [6,7] of a scheme that allows evaporation to be used to cover large areas at a time. This uses the principle that although high oxygen activity is required during growth, it is not required all the time. Provided that the growing film is exposed sufficiently often to a high oxygen pressure, growth and oxygenation can be performed sequentially. They mount the substrates on a rotating holder that passes alternately into a region where the metals are evaporated and then into a region of high oxygen pressure. Achieving this requires some clever engineering, as a large pressure differential between the two regions can only be maintained if the gap between the wall of the "oxygen pocket" and the growth region is narrow, while rotating the hot substrate holder at ~5Hz. Nevertheless, this has been done and films are now produced by this technique from several sources in a very cost-effective way. According to Matijasevic et al. [8], who use a process derived from Kinder's idea, the film production costs are well below the cost of the substrates, and there is little advantage in reducing the costs further.

Because of the relatively low oxygen activity during growth, evaporated films often have more cation disorder than films grown by other techniques. Single crystals have much higher surface resistance than thin films and it is desirable to increase the quasiparticle scattering: *we are not looking for perfect material from which to make circuits*. It remains to be seen how many defects, and of what type, are best.

2.2. PULSED LASER DEPOSITION

Pulsed laser depostion seeks to achieve stoichiometric transfer of material from a target to a substrate. A UV laser is used, so that the absorption length of the light in the target is short, and it is imaged to illuminate uniformly an area of the target. When the laser is pulsed, a region of the target is ablated more or less adiabatically. A shock wave propagates through the oxygen gas in the chamber and visible luminescence can be seen as a plume normal to the target. The hot substrate is placed (often rotating) at a carefully chosen position with respect to the plume. In [1] I noted that the success rate with laser ablation was far higher then with other growth techniques, because of its simplicity, and concluded that it was the ideal process for prototyping new materials. I believe this remains true. Nevertheless, a great deal of complex chemistry and physics goes on in the plume [9]. One of the most surprising observations is the presence of large clusters (~100 atoms) which presumably condense in the vapour [10]. The main problems with the technique are that it depends on ablation from a point source, so measures must be taken to enforce film thickness uniformity (the substrate-target distance cannot be increased to the scales used in evaporation). The laser power must be scaled up if enough material is to be ablated to give large area coverage at useful rates, but modern lasers can achieve these power levels. Droplets are often produced due to target roughening, and a number of schemes to avoid a line of sight between the target and the substrate have been demonstrated to minimise this problem, but always, I believe, at a significant cost in growth rate. Achieving uniform large area coverage initially appeared very difficult, but a number of solutions have been published [11,12,13] using some combination of substrate motion, target motion, and laser beam splitting.

2.3. MOCVD

Metal-Organic Chemical Vapour Deposition (MOCVD) is a very attractive approach to growing films. The technique scales readily to large areas and it is possible to envisage simultaneous growth on both sides of a substrate. In MOCVD the metal cations are transported as volatile organic compounds. A factor slowing down progress has been learning to synthesise and purify these precursors [14]. On arriving at the hot substrate, the precursors decompose or "crack" to deposit the metals, and they are oxidised by molecular oxygen or some more reactive species (e.g. N_2O), possibly with the assistance of a plasma. The pressures in the reactor are quite high, and oxygen supply does not appear to be a problem. The biggest problem is composition control. The precursors (particularly for Ba) tend to form oligomers (an oligomer is a small size version of a polymer, containing a few molecules). Each oligomer has a different vapour pressure at a given temperature, so sources tend to age. Liquid injection into the gas stream can avoid difficulties with the vapour pressure differences, but one is still left with the different cracking behaviour of different oligomers. Fluorine containing precursors have been proposed as a solution to the oligomerisation problem. The dependence of the film morphology on composition is similar to that found in evaporation. Critical current densities can be as high as with any other technique. There are many variants of MOCVD, ranging from very simple single source approaches [15] to sophisticated reactors with complex control systems [16].

2.4. SPUTTERING

Sputtering has been widely used for the growth of HTS films. Sputtering uses a plasma generated by either a DC or RF discharge. Energetic ions from the plasma strike a stoichiometric ceramic target and atoms are sputtered off to be deposited on the hot substrate. The discharge is usually confined and stabilised by a magnetic field (magnetron sputtering). It was found at an early stage that the simplest configuration with a substrate opposite the source led to non-stoichiometric films due to negative ion resputtering. Two classes of solution have been found to work well. The first is to use a high pressure, and keep the substrate near the edge of the plasma. This leads to an experimental configuration that must be tightly controlled [17] and imposes a large heat load on the target, but has been developed to give large area films of excellent quality. The second is to orient the substrate perpendicular to the target [18] (off-axis sputtering). A variant of this is inverted cylindrical magnetron sputtering, which uses a hollow cylinder target [19]. Both of these solutions give a rather low growth rate. Scaling to large area has not been easy [17], but has been achieved on-axis [3] and off-axis [18].

2.5. LIQUID PHASE EPITAXY

Liquid phase epitaxy (LPE) is qualitatively different from the physical vapour deposition techniques described above, and has attracted much less attention. In LPE, a substrate is immersed in an off-stoichiometric melt, like those used to grow single crystals, and growth is induced by small temperature changes or gradients. The technique produces films of high crystallographic quality with smooth surfaces grown at a high rate and to arbitrary thickness. With these advantages, it seems to me to deserve more attention than it is getting. The films grow with low oxygen content and must be oxygenated by annealing. This often takes quite a long time because the films tend to be many microns thick. The films tend to crack on oxygenation, for the same reason [20]. On $SrTiO_3$ substrates, the critical thickness for crack formation is about 1.35μm [21], while films nearly 100μm thick can be grown on MgO substrates [22] where the thermal expansion mismatch is small. Although a nucleation layer of YBCO has often been used, it is apparently not essential [23], and on bare MgO, films grow with the YBCO (100) parallel to MgO (110), i.e. 45° rotated. The critical current densities often appear low. This is partly because LPE films have too few defects to pin flux strongly, but is also a consequence of the large film thickness [24]: critical current density tends to fall as film thickness increases for films produced by all techniques. Miura et al. [24] also showed that the microwave properties, both linear and non-linear, of LPE films are competitive. They may be too perfect in their current form (insufficient quasiparticle scattering) to give the lowest surface resistance, but it would be very interesting to see how the third order intercept scales with film thickness. LPE is only emerging now as a technology for HTS growth, but it is worth attention. At the very least, it gives a qualitatively different type of material against which to compare performance.

3. Substrates

A wide range of substrate materials has been used for the growth of HTS films. Lattice mismatch seems to be remarkably unimportant. Its main effect seems to be to control the critical thickness, i.e. the thickness that can be grown before the film relaxes to its unstrained lattice parameter through the introduction of dislocations. Chemical effects are important; for example some of the Al containing substrates poison the superconductor if it is grown directly on the substrate. This can be solved by interposing a chemically benign buffer layer. A substrate for microwave applications must have a low dielectric loss. I will leave it to the other speakers to discuss what dielectric constant is required. Some of the substrate materials are anisotropic: this makes the circuit design problem more difficult but I assume that this can be overcome.

Table 1 lists a few popular substrate materials. More extensive tables are given in [3, 25, 26]. The properties listed in Table 1 are intended to be indicative rather than definitive. I have not distinguished between dielectric constants measured at room temperature and low temperature. The tan δ values are all low temperature ones measured at microwave frequencies, but the precise frequency is not specified. It is worth noting that tan δ can easily be dominated by extrinsic defects in materials that have not been optimised: the low loss of sapphire is a result of considerable materials optimisation work, and is sample dependent. Nevertheless, there is an intrinsic limit in any material associated with the phonon lifetimes.

Some compounds like $LaAlO_3$ twin at a temperature below the growth temperature of the superconductor. I believe this excludes them as credible substrates in the long term. Such substrates have in effect an unpredictable and inhomogeneous dielectric constant, making precise design impossible. They also change shape when a superconducting layer is grown, so they can not be used for multilayer circuits. Some samples show an excess loss peak [34]. In spite of these difficulties, $LaAlO_3$ has been widely used because it is relatively cheap and available in large sizes. To retain these advantages and suppress the twinning, related materials like $LaSrAlO_4$ have been introduced.

In my group, we have used only MgO substrates since abandoning ex situ growth in 1989. It is cubic without phase transitions, has very low loss and is a good thermal expansion match to HTS. However, it is currently available only up to 2" –3"diameter, and because it cleaves easily, the larger substrate sizes are typically 500μm thick while we often prefer thicknesses closer to 300μm. MgO is often described in the HTS literature as hygroscopic (i.e. it absorbs water from the atmosphere). I have never seen

Table 1. Widely used Substrate Materials

The parameters listed are the lattice constant (a), coefficient of linear thermal expansion (α), relative permittivity (ε), loss tangent ($\tan\delta$) and the temperature of the orthorhombic to tetragonal; phase transition (T_t).

	a (Å)	α (10^{-6})	ε	$\tan\delta$	T_t °C	Ref
MgO	4.213	14	9.62	$<5\ 10^{-6}$		27, 28, 29
Al_2O_3	4.759	8.5	9.4	$<10^{-7}$		26, 30, 31
	12.991		11.6			
$LaAlO_3$	5.3646	10	24	$\sim5\ 10^{-6}$	527	28, 32, 33
	13.1096					
LaSr AlO_4	3.77	7.4	27	$<3 10^{-5}$		32
$(LaSr)(AlTa)O_3$	3.8685	10				33
Si		3.8				31

proper evidence to support this. It is easily established that the surfaces of MgO left lying around in the laboratory for extended periods degrade, but we have never found problems in water based lithography for MgO. It seems to me more likely that the degradation in surface quality in the atmosphere is due to reaction with CO_2 than water. This merely means that MgO is rather less sensitive to acids than the superconductor. MgO has the advantage that as an archetypal ionic compound it is very easy to grow as a thin film. Homoepitaxial growth works well at a temperatures above 140K (not C!) [35]. All our substrates are coated in situ with ~20nm of homoepitaxial MgO before we start to grow superconductor. This has the effect of burying any milling damage from ion cleaning and chemical contamination, e.g. from drying marks. It is not essential, but is cheap to do and results in a process that is tolerant of operator error.

Sapphire (Al_2O_3) has been intensively studied in spite of needing a buffer layer and being anisotropic because of its large area, low loss, high strength when polished thin and low cost. Its big disadvantage is that films thick enough for microwave devices usually crack during cooling after growth. This is usually attributed to the poor thermal expansion match to HTS (Fig. 1). The figure shows that it is only slightly worse than $LaAlO_3$. Is this difference enough to account for the dramatic difference between the two in cracking? Si is included in the figure not because it is a serious candidate as a substrate, but to illustrate why films on Si crack even more readily than those on sapphire [36]. On sapphire, this problem has been solved for one particular sputtering process, where, by reducing the growth temperature (and crystallographic quality) slightly, thick layers with low surface resistance can be grown [37]. I will come back to discussing why this might be possible in section 5.

Substrate polishing is important. The microwave designer would like to specify the thickness of the substrate to ~1μm, but substrate suppliers cannot polish to this tolerance. For high performance filters it may be necessary to measure the thickness of each substrate and use a slightly different mask for different thicknesses. Substrates polished in a batch are generally close to each other in thickness. Some polishing processes can lead to the substrates being curved which can make it awkward both to measure the thickness and to pattern precision circuits.

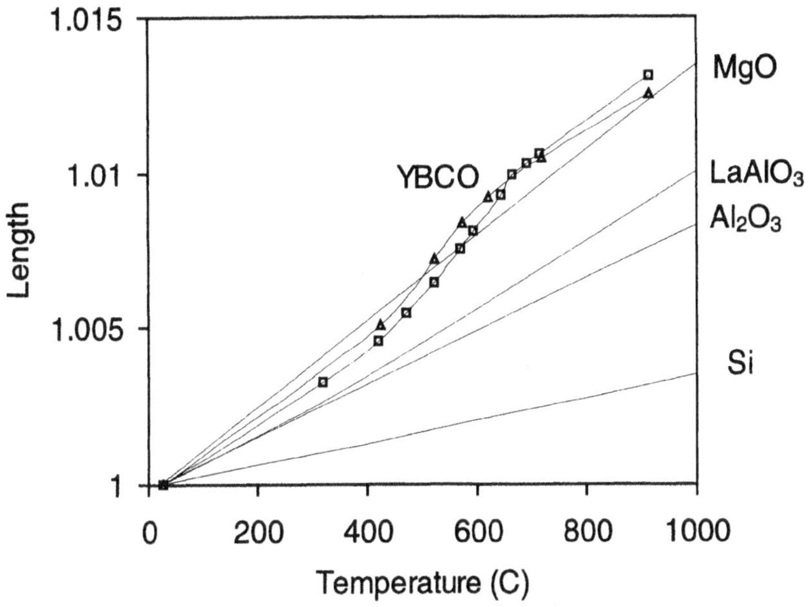

Fig. 1. Thermal expansion of YBCO compared to some common substrates. The YBCO data are from Jorgensen et al. [38] for oxygen pressures of 1bar and 0.02 bar. A more extensive set of data are given by Mizusaki et al. [39]. For YBCO, the average of the a and b lattice parameters is plotted, as is appropriate to densely twinned c-oriented films. The substrate thermal expansion curves are from the references of table 1.

4. Growth Conditions

Although many growth techniques have been used for HTS (the ones discussed in section 2 are the most popular, but not exhaustive), they have many aspects in common. In this section I address the important issues in the growth process, irrespective of which technique is used.

4.1. CATION STOICHIOMETRY

It is not surprising that the cation stoichiometry has important effects on film properties. We would expect a material like YBCO to behave rather differently to 1-2-3 compounds containing larger rare earths where substitution of the rare earth for the Ba is allowed. In the case of YBCO, it was established early on that excess Cu gives rise to precipitates and that smooth films can be grown by increasing the Y concentration [40] (Fig. 2). The Ba rich region is to be avoided if good electrical properties are sought.

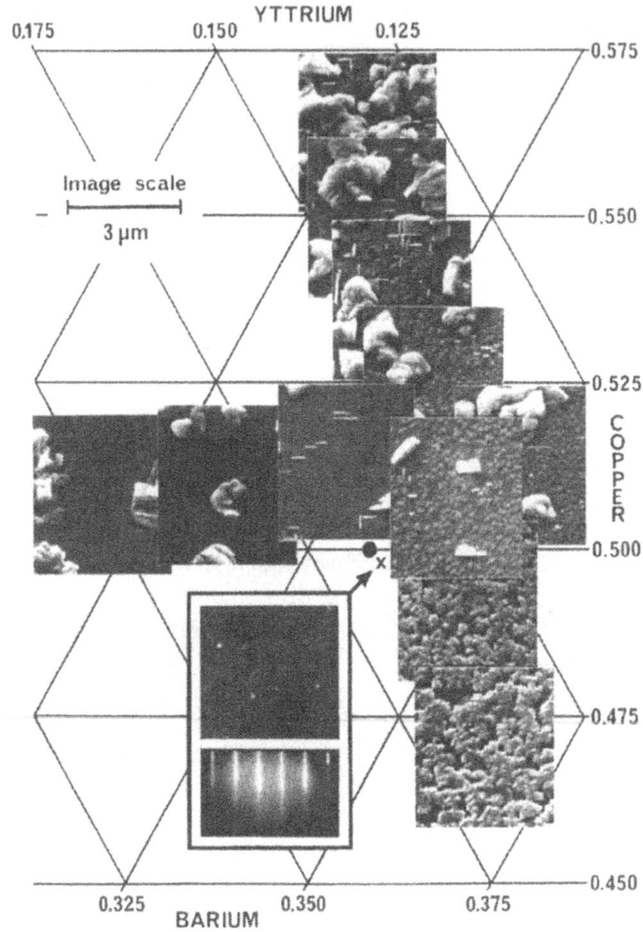

Fig. 2. Scanning electron microscope images of 0.35μm thick films superimposed on the composition diagram, from [40]. The inset is a film grown close to our optimum composition (•), together with its RHEED pattern after growth

Later AFM studies (Fig. 3) showed the individual grains of the material more clearly [41]. Ba rich growth produces a dendritic morphology while Y rich growth under the same conditions results in large grains with much less height variation. Ba rich and Cu deficient growth leads to very rough films. This is consistent with the limiting surface mobility being that of Y, so that an Y excess suppresses concentration gradients near a growing plane, while Y deficiency leads to an unstable growth front reminiscent of constitutional supercooling. We grow at different compositions for different purposes: near stoichiometric gives high T_c but we prefer Y rich for multilayer circuits so that the surfaces are smooth. Often we have chosen to grow slightly Cu rich for microwave devices, and to nucleate films Y rich for the first 20nm and then complete growth near stoichiometric composition. Y rich nucleation tends to suppress a-grain formation. When the composition is changed during growth in this way, the morphology observed is characteristic of the later layer and is not controlled by the nucleation conditions [41].

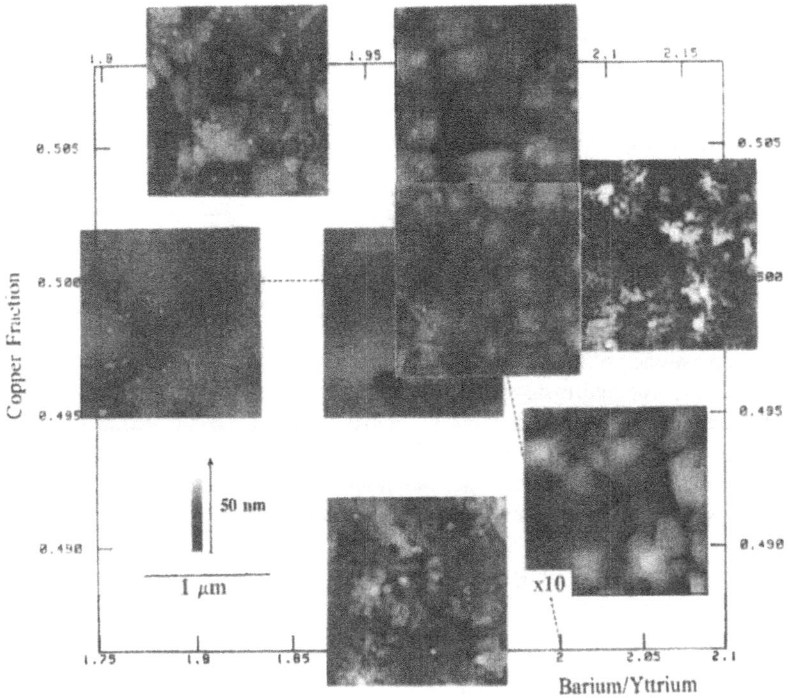

Fig. 3. Atomic force microscope images of morphologies of 0.35 μm thick films superimposed on the composition diagram. Note the 10x larger vertical scale on the bottom right image.

Morphologies like those of fig. 2 have been reported in films grown by all vapour deposition techniques, and seem to be quite general for YBCO. In our experience, the 1-2-3 compounds containing larger rare earths are easier to grow smooth, but control of the critical temperature is more difficult. Many groups have developed successful processes, however. In terms of the electrical properties of films, the biggest effect of composition variation in our process is through the way the morphology effects the film oxygenation. I will return to this point later.

4.2. TEMPERATURE CALIBRATION AND CONTROL DURING GROWTH

If the back side of the substrate is to be kept clean, as is essential if a layer is subsequently to be grown on it, the substrate is usually heated radiatively. It is notoriously difficult to control or even measure the temperature of radiatively heated substrates. For HTS growth, the problem is compounded by the fact that the clean substrate has a very low emissivity, which rises quickly as growth proceeds.

There are two approaches to measuring substrate temperature: contact thermometry and pyrometry. We use contact thermometry, embedding a fine thermocouple in a small hole in a substrate, with contact ensured by boric oxide. This method obviously has the drawback that it cannot be used in every growth run. We use it to calibrate the substrate temperature in terms of the temperature of our control thermocouple, which measures a radiation temperature behind the heater.

Pyrometry measures the radiation emitted by an object and is not easy to implement for HTS. Most oxide substrates are transparent from the ultraviolet to a wavelength of perhaps 8μm [42]. Beyond about 14μm they are highly reflecting [43]. Normally pyrometry is carried out at two wavelengths to compensate for system transmission (including coating of windows) but this is not possible within such a narrow band. The other problems and their solutions have been described by Choi et al. for $LaAlO_3$ substrates [44]. They use a quite complex model of the optical properties of the substrate coated by a growing film to calibrate the pyrometer.

The transparency of the substrate over much of the spectrum means that the substrate is only weakly coupled to its thermal environment. The situation changes radically as the superconducting layer starts to grow. In many cases, growth is initiated by opening a shutter, resulting in a drop in the substrate temperature. As growth proceeds, the absorption of the film increases the coupling to the heater and the temperature rises again. These large transients must be understood and controlled if we are to optimise the growth process.

Fig. 4 shows the temperature of a 25mm square substrate measured with a thermocouple in our evaporator as the shutter opens at a fixed heater temperature. The substrate temperature drops by over 50°C during the first few minutes, if growth is not taking place and the substrate is uncoated. If growth is taking place, the substrate temperature recovers as its emissivity increases. The temperature of an initially coated substrate falls much less, only about 20°C. For growth on 50mm square substrates, the temperature excursions are even larger: the temperature of an initially uncoated substrate falls by ~100°C in similar experiments.

With careful programming of the heater temperature to compensate for these effects, the substrate temperature can be held constant to within about ±10°C. The situation is greatly eased if the substrate is blackened on the back side, as is the case when growing on the second side of a substrate.

4.3. OXYGEN AND TEMPERATURE DURING GROWTH

Oxygen is important during nucleation, growth, during cool down after growth and in patterning. During growth, insufficient oxygen is responsible for cation disorder and reduced crystal quality [45]. Some degree of disorder seems desirable in microwave devices to increase quasiparticle scattering, but how much and of what type is not known. Single crystals have fewer structural defects than films, but have higher microwave surface resistance and lower critical current density.

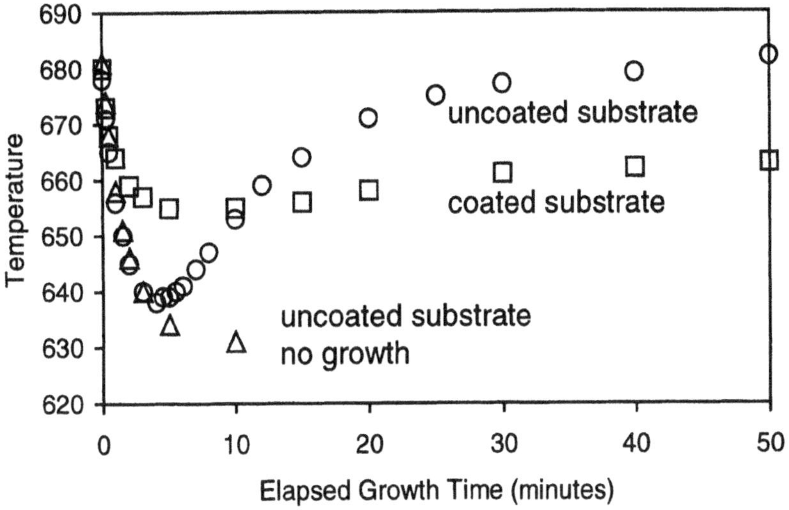

Fig. 4. Evolution of the substrate temperature after opening the shutter with the heater temperature constant. In each case the substrate temperature was constant at 680°C before the shutter opened. A film was growing during these experiments for the upper two curves.

In many techniques, films are believed to grow at a composition near $\delta = 1$, and they must be oxygenated to $\delta \sim 0$ before they are useful. In most techniques this is done during cooling after growth. Film edges may deoxygenate during patterning and if they do they may need reoxygenation afterwards. In my opinion the behaviour of oxygen and its effects are one of the least well understood aspect of HTS thin films.

The basic ideas behind the oxygen requirements during growth are shown in fig. 5. This shows the thermodynamic stability limit of YBCO as determined by three different groups [46-48], compared to experimental data [49]. Evidently the thermodynamic limit is quite difficult to determine accurately. Our experimental data are not perfect either; but they should be better than the discrepancies between the different estimates of the thermodynamic limit. The first thing to note is that we were able to grow with molecular oxygen in the region of the stability line determined by Lindemer et al. [48]. This does not necessarily mean that the Lindemer curve is right however; it is always possible that we were growing in a region where the YBCO was thermodynamically unstable but forms nevertheless for other reasons. Y_2O_3 has often been observed as small precipitates in the growth of YBCO, but it is not quoted by Byers and Ahn [47] as a contact phase. The reason for this has recently been resolved by Samolyenkov et al. [50,51]. They found that the relative stability of phases in thin films is not the same as in bulk material, as phases that form low energy interfaces with the superconductor are favoured. Thus phases that can grow epitaxially tend to form preferentially. This result means that stability lines determined using polycrystalline material are more useful as an indication of qualitative trends than a hard limit. Most of the points in fig. 5 used atomic oxygen, where successful growth extends beyond Lindemer et al.'s curve.

Prenucleating films at a lower temperature allowed successful growth at a temperature where growth would otherwise fail. We have also found that nucleation without atomic oxygen, under conditions that reproducibly produced good films if atomic oxygen is used, resulted in inhomogeneous films, in which substantial areas failed to crystallise in the superconducting phase [52]. The nucleation step is the most critical. This is not surprising. Different conditions for nucleation and growth have been widely used in the growth of a-oriented films, where a low temperature grown layer of PBCO can be used as a template. We believe it is useful much more generally.

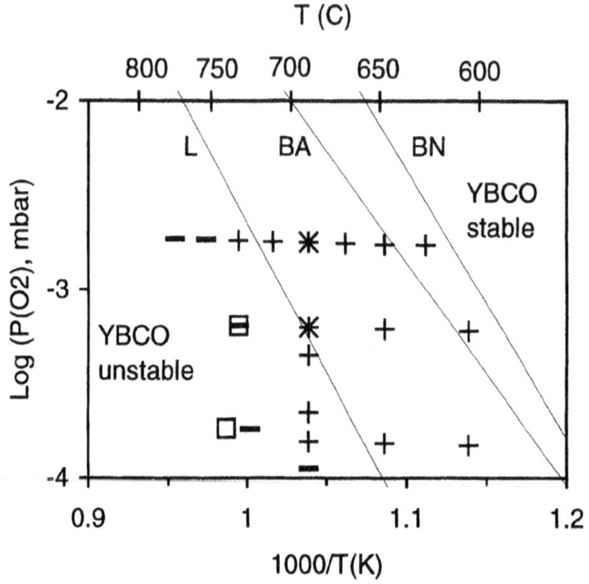

Fig. 5 Stability diagram for $YBa_2Cu_3O_6$. The solid lines are the stability limits of Bormann and Nölting (BN)[46], Beyers and Ahn (BA)[47] and Lindemer et al. (L)[48]. The experimental points are slightly corrected from Humphreys et al. [49]. The points indicate results of growing films by evaporation:

+ with an estimated 10% atomic oxygen

× with only molecular oxygen

□ with 10% atomic oxygen, after nucleating at 690°C

– failure

The criterion for "successful"growth here was not stringent. It was simply that the film should show a persistent current at 77K. At the lowest growth temperatures, the films were of poor quality.

Active oxygen (atoms, ions, excited molecules) is present not only when it is deliberately used but also in laser ablation and in sputtering. One of the most striking pieces of evidence of the high activity is the observation by Teshima et al. [53] that it is possible to cool in vacuum after growth by sputtering and still get a much higher oxygen content than the minimum – perhaps full oxygenation. Since the orthorhombic-tetragonal transition takes place close to δ = 0.4 at 735°C [39], the films were presumably orthorhombic as grown. This would imply that the oxygen activity under the growth conditions was equivalent to an atmosphere of molecular oxygen. It would be informative to see this work repeated and extended; we would like to know more about the properties of the films under the conditions of growth. Schlom et al. [54] have sought to estimate how active atomic oxygen is compared to molecular, while admitting that the calculations are flawed because the system is not at thermodynamic equilibrium. The equivalent pressures deduced range up to meaningless numbers, but for visualisation purposes it might not be extravagant to suggest that activities equivalent to Mbar pressures could be realisable. The only growth techniques where the absence of active oxygen can be assured are evaporation and MOCVD: both laser ablation and sputtering can create active oxygen. In evaporation, active oxygen is often introduced deliberately.

Table 2 summarises the conditions used by a number of laboratories for growing films with a diameter of 2" or greater, reflecting relatively mature processes. These have been chosen as representative; they are certainly not an exhaustive list. The pressure – temperature conditions from the table are plotted in fig. 6, compared the stability lines shown in fig. 5. It is interesting that many of the techniques fall on the Bormann and Nölting stability line. It seems more likely that this is because this is an "optimum" for the particular process rather than the conditions having been chosen to fit the line. Once again, however, this does not necessarily imply that the Bormann and Nölting line is right. The techniques that are far from this line are the high pressure sputtering processes, where excess oxygen is used to suppress negative ion resputtering, and our atomic oxygen assisted process where mass spectrometer rate control does not permit

Table 2. Large Area Growth Processes

Group	Technique	Diameter	Rate (nm/min)	Pressure (mbar)	Temp (°C)	Ref
Munich	Evaporation	9"	24	0.01	650	55
DERA	Evaporation	2"	6.6		680	
DuPont	Off-axis Sputtering	3"			700-750	56
Duke	Off-axis Sputtering	8"	2.9	0.15	735	57
Wuppertal	High pressure sputtering	2"	2.1	2	730	58
Jülich	High pressure sputtering	4"	1			59
Augsburg	Laser ablation	2.8 x 8"	5	0.08-0.15	~700	60
Leipzig	Laser ablation	3"	15			61
STI	MOCVD	2"	2.5	3	830-850	62

the use of more oxygen. The other way of looking at this graph is that several of the techniques favour a temperature close to 700°C, irrespective of the oxygenation conditions.

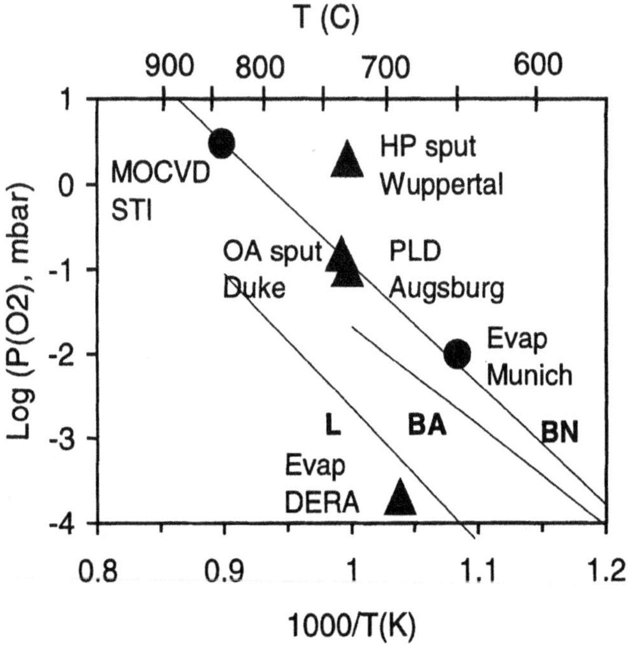

Fig. 6. Comparison of the temperature and pressure regimes of a number of large area growth processes for YBCO, taken from table 2. The lines are the stability lines labelled as in fig. 5. Triangles mark processes in which atomic oxygen is created and circles those in which the oxygen is molecular.

Our preferred process for microwave devices uses nucleation with excess Y at low temperature, followed by growth at higher temperature and with stoichiometric composition. Nucleation with excess Y suppresses nucleation of a-grains. Growth at higher temperature than we can nucleate YBCO on MgO tends to produce material with better crystallographic quality and larger grains [52]. In terms of microwave performance, the best looking films are not necessarily the best. There was general agreement at the ASI that rough films (containing precipitates or holes) tended to have better properties than smooth ones.

In most techniques, films need oxygenation during cool down after growth. In our early work, we found that atomic oxygen was important during nucleation and during cool-down, but not necessarily during growth. The most commonly used procedure is to increase the pressure of oxygen in the chamber after growth and cool relatively slowly. When the film is sufficiently oxygenated and cool enough, the oxygen atoms in the chains order and the films become orthorhombic. Because of the mechanical constraint imposed by the substrate, the twin size is small, some tens of nm, and varies as the square root of the film thickness [63]. The twins are organised in domains, within which the twins are parallel. The twin domain size is some microns, and they can be observed in the polarising microscope, usually most clearly seen in transmission between nearly crossed polarisers.

There is a considerable body of evidence that "fully oxygenated" YBCO is slightly overdoped [64]. This may be a good regime for devices because it means that a small amount of oxygen loss in processing has only minor consequences.

5. Nucleation and Growth

In discussing growth conditions, we have seen that nucleation is a more critical process than the subsequent growth. What is happening during the nucleation phase? I will describe here the picture described by Bauer et al. [65] of nucleation on MgO by evaporation. Earlier evaporation [66] and PLD results [67] are broadly consistent with these ideas. The first stage is island growth with a lattice constant that matches the substrate, i.e. the growing layer is strained by about 10% to match the substrate. After growth of a nominal 2nm film, islands occupy half the substrate area and are about 4nm high. There are about 100 islands/μm^2 after a film of this thickness has been cooled down, but this does not necessarily reflect the state of the film at the growth temperature. The maximum island size is governed by the balance between the strain energy and the energy required to form cracks; an island that is too large will split into two. Soon after this, the islands begin to coalesce and a layer relaxes very rapidly to its unstrained lattice constant by the introduction of a high density of edge dislocations (about one every 10 unit cells). Such misfit dislocations can be seen with unusual clarity in the TEM results of Kitamura et al. [23] for LPE material. The nucleation phase of growth is complete after a nominal 4nm thickness has been grown. Other substrates with better lattice matches will have larger critical thicknesses, but unless the match is very good, the lattice will relax. I believe that the main difference between growth on MgO and a well matched substrate is that in the latter case the film is continuous at the critical thickness, while for MgO the lattice relaxes as the islands coalesce.

After the nucleation phase we are dealing with homoepitaxy, but some defects formed in the nucleation phase propagate through the films. The most widely studied are screw dislocations, readily identified from characteristic spiral growth patterns. Some growth conditions [68] lead to 2D nucleation of new layers rather than spiral growth. For YBCO, both growth modes lead to a film surface consisting of a large number of hillocks, separated by quite deep valleys, often a significant fraction (~10%) of the film thickness deep. The crystal orientation is essentially the same on both sides of such a valley, so why do the valleys exist? When growth is 2D nucleated, there is some evidence they represent antiphase boundaries (Fig. 7), where adjacent grains meet with a height difference of 1/3 of a unit cell [69].

It is not very clear how common such boundaries are, or what their electrical properties might be. The best data so far come from studying films grown on misoriented substrates like $SrTiO_3$, where a growth mode that has been interpreted as step-flow growth has been observed. Actually, this is not step flow growth [70, 71], as adjacent steps are separated by antiphase boundaries. At low temperature, clear evidence of a higher critical current density parallel to the steps has been found, indicating increased pinning for vortex motion perpendicular to the steps [72]. These films also have other anisotropies, however. Twinning is partially suppressed, and the b-axis is preferentially aligned parallel to the steps, so the identification of the anisotropy of J_c with pinning at antiphase boundaries was made by calculating the expected pinning strength.

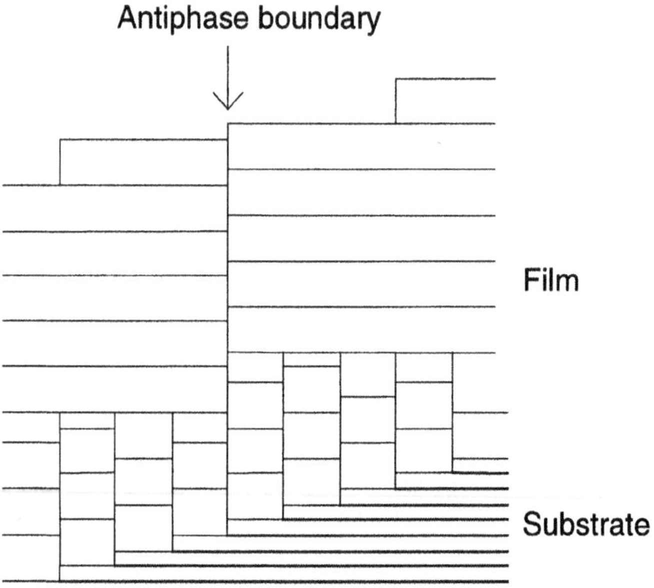

Fig. 7. Schematic of the join between two nuclei, with expanded vertical scale. Horizontal lines indicate unit cell spacings of the substrate and film. Vertical lines are antiphase boundaries. Within a nucleus, antiphase boundaries heal quickly [69]. Where two nuclei join, it is kinetically unfavourable to intercalate an extra layer and an antiphase boundary propagates through the film.

It is interesting to note that another sort of antiphase boundary is possible on MgO, which has the NaCl structure rather than a perovskite structure. The minimum step height on MgO is half a unit cell, so ,in principle, it is possible for APB's to form in YBCO with an offset of (1/2,1/2,1/6) in terms of the YBCO unit cell.

When growth is spiral, then the valleys are formed between adjacent spirals, but the origin of the spirals has never been satisfactorily explained. Here I would like to propose a new mechanism. For typical substrate misorientations, there are enough steps in the substrate that adjacent nuclei will not necessarily join up, but are quite likely to be misaligned by 1/3 of a unit cell perpendicular to the substrate, forming antiphase boundaries. Assuming that it is not possible for all the misalignments to correct themselves by (say) intercalation of an extra Cu layer, then growth could proceed from here in one of two ways. Either the antiphase boundaries propagate through the film as it grows or a screw dislocation forms at each point where three nuclei of different heights intersect, as depicted in fig. 8. I suggest that the former could be what happens in 2D growth and the latter represents spiral growth.

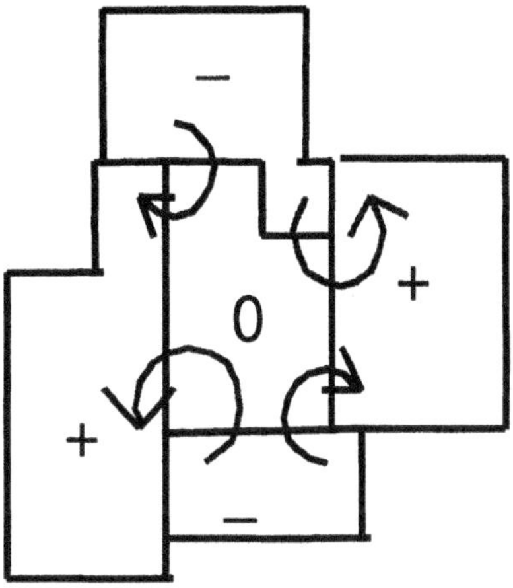

Fig. 8. A pattern of antiphase boundaries like that shown in fig. 2b of [70] for a film grown by 2D nucleation, i.e. without spirals. The central grain is arbitrarily designated to have height 0. Then the surrounding grains must have alternating positive and negative offsets with respect to it. At each intersection of three grains, a spiral direction is indicated by the arrows. It is proposed that under different growth conditions, the highest layers (labelled +) flow up and over the lowest layers (labelled -) thus annihilating the antiphase boundaries. Screw dislocations are then formed at the intersection points. Subsequent growth proceeds fastest at these points.

If this model is correct, there should be few, if any, antiphase boundaries in films grown entirely by a spiral mechanism. There should also be equal numbers of left and right handed spirals. There would be a tendency for neighbouring spirals to join as the film gets thicker, forming a dislocation half loop buried in the film and fewer spirals evident on the surface. It has recently been observed that in 2D growth, there are features that lead to etch pits around the edges of islands [73]. In this model, the etch pits would mark the line of intersection of three antiphase boundaries.

The study of Yeadon et al. [74] seems to have been done under growth conditions that are at the boundary between 2D and spiral growth, and may be more complicated as a result. I am not sure whether their detailed results are consistent with the model just proposed; if they are not then the model is wrong. However, their observation of a Burgher's vector of mixed (screw/edge) character could be consistent with a meeting of islands with (1/2,1/2,1/6) offset. In some samples they observed equal numbers of left and right hand spirals, as the model suggests, but thicker samples show a local imbalance. Whether this is consistent with the model is not clear. A proper symmetry analysis of how the local slope and half unit cell steps interact is required. It would be interesting to see their work continued in a simpler growth regime.

Once the film is grown, the next step is to cool it down. In most techniques, cooldown after growth is carried out in a relatively high pressure of oxygen and at a controlled rate to allow the film to oxygenate. This is done more for convenience than necessity. Oxygenation might proceed faster with a clean surface that has not been exposed to the atmosphere, and there is some reason to suppose that reasonably well oxygenated films are less reactive than oxygen deficient ones. However, in a production context it might well be attractive to cool films fast after growth and oxygenate them in a cheap furnace rather than an expensive growth machine.

We have already seen that films will crack if they are grown too thick for the thermal expansion mismatch with the substrate. There is plenty of evidence that when films crack, it takes place during cooling or, if films are cooled deoxygenated, during oxygenation [75]. For all the substrates of fig. 1, with the possible exception of MgO, the film is in tension after growth. The critical thickness for cracking is usually quoted as

$$h_c = \frac{K_{1c}^{\ 2}}{2\varepsilon^2 E^2}$$

where ε is the strain, E Young's modulus for the film and K_{1c} is the mode 1 fracture toughness. Kawashima et al [75] found that this explained the observed critical thickness for films on $SrTiO_3$ substrates, using K_{1c}=0.53 MPa $m^{1/2}$.

In a cracked film, the crack spacing l is given by

$$l = 8(h_c h)^{1/2}$$

This has been found to be in rather good agreement with the crack spacings observed in LPE a-oriented films on (110) $NdGaO_3$ using K_{1c} =1.48MPa $m^{1/2}$ [76]. It seems surprising that the fracture toughness should be larger for a-oriented films than c-oriented ones: one might suppose the reverse to be more likely. The correct value of the fracture toughness, and the application of the known theory of how films crack seems to me (as a non-expert) to need more attention.

In section 3, we saw that the thermal expansion of sapphire is only somewhat less than that of $LaAlO_3$, and that this difference seemed too slight to account for the different critical thicknesses in the two cases. The explanation may lie in the different elastic constants of the substrates. The theory of this situation has been studied by Hu and Evans [77]. At first sight, it appears that the effect is rather weak, but the publication deadline means that I cannot follow the problem further here. I have not found Young's modulus for $LaAlO_3$, but the value for sapphire [78] is over a factor of 2 larger than that of YBCO [79].

We are particularly interested to understand why films grown on sapphire crack, and why the sputtering process of Zaitsev [37] leads to high quality 0.7μm thick films on sapphire that do not crack. The authors propose an increased fracture strength due to cation disorder, but cracking remains a problem in evaporated films which are known to have significant cation disorder. I suggest that the explanation could lie in the oxygenation state of the film under the conditions of growth. The non-linear thermal expansion of the a-b plane in YBCO in fig. 1 is due to oxygen loading rather than ordering (fig. 10 of [39]). In section 4.3, we saw there was evidence [53] that sputtering may be able to grow films fully oxygenated, due to the active oxygen in the plasma. In this regime, the thermal expansion match to sapphire improves very significantly. It would be interesting to see the work of [53] extended. It is difficult to make an experiment that relies on back-filling the growth machine with inert gas completely convincing. A better approach might be to grow a capping layer that is impervious to oxygen on the surface to lock in the oxygen under the growth conditions. We have found that MgO fulfils this role, but many other insulators would do just as well.

In this discussion, we have encountered a number of types of defect. It may be helpful to summarise them in one place:

Point defects:
- Cation disorder (e.g. Y and Ba interchanged)
- Oxygen vacancies/interstitials

Line defects:
- Edge dislocations at the substrate-film interface
- Screw dislocations threading the film
- Intersections of antiphase boundaries

Planar defects:
- Stacking faults
- Anti-phase boundaries
- Twin boundaries
- Grain boundaries
- Twin domain boundaries

Volume defects
- Precipitates
- Oxygen non-uniformity
- Macroscopic non-uniformity, cosmetic defects

In general, point defects affect the order parameter and quasiparticle scattering. Line defects and planar defects are all candidates for pinning centres. Since pinning is due to a local reduction in order parameter, planar defects are also candidates as weak links.

6. Key Variables

Some years ago, in collaboration with Wuppertal and Birmingham, we undertook a systematic study of the factors affecting film quality [41]. This section outlines the main conclusions. The primary factors investigated were those of section 4, starting with composition. We have already seen the important role composition plays in controlling morphology. In examining the electrical and structural properties of the films shown in fig. 2, both before and after oxygen annealing, it was difficult to see direct effects of composition changes because the oxygen content varied with the morphology. About 20 different parameters were selected to characterise the film properties and a simple statistical correlation analysis was undertaken, looking for parameters with a linear correlation coefficient from a range of structural and electrical properties. This showed that a major random variable (at least for growth on MgO) was the substrate misorientation. Strong correlations were found between properties dependent on doping (c, T_c, J_c, penetration depth (λ) and resistivity (ρ_{100K})). T_c had a positive correlation with c and ρ_{100K} , indicating that the films tended to be overdoped. This came as a surprise to us at the time, although it should not have; there was already evidence in the literature for it, as cited in [41]. Cu rich samples (containing precipitates) lost oxygen during annealing, while smooth samples gained oxygen. We believe this is because the Cu rich samples oxygenate more fully in atomic oxygen during cool-down after growth, because they have many channels to bypass the slow c-axis diffusion. Microwave surface resistance (R_s) was remarkably insensitive to all the changes, except for the obvious correlation with T_c at high temperatures. The next part of the study investigated the effect of varying oxygen content. While the c-axis lattice parameter is a good relative measure of the oxygen content of a given film, the value of c corresponding to the peak T_c varied from film to film, and was much larger than the bulk value. The large c-axis lattice parameters in some films are usually attributed to cation disorder. This means that the lattice parameter cannot be used to deduce the oxygen content of a film using data from bulk material. The effect of growth temperature was the final parameter studied. It was striking that even when films were grown at relatively low substrate temperatures, so that their T_c's were seriously depressed and penetration depths increased, the microwave surface resistance at low temperature was little changed. We suppose this is because of compensating effect of increased quasiparticle scattering in disordered films.

More recently we have been interested in probes of film quality that offer good spatial resolution, as film non-uniformity is a likely source of performance limitations.

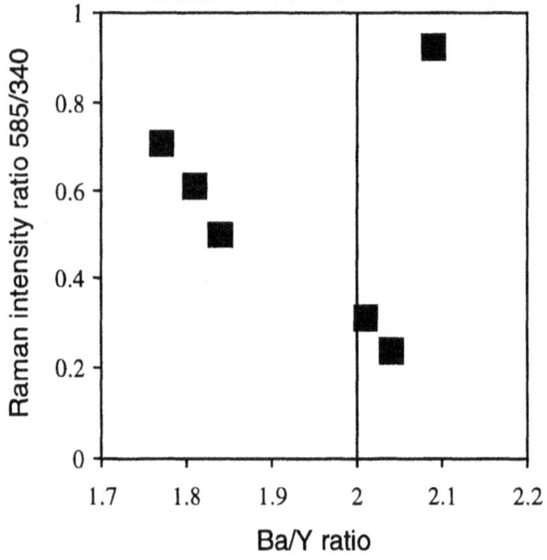

Fig. 9. Cation disorder measured by Raman scattering plotted as a function of film composition [45]. The vertical axis is the intensity of the 585cm^{-1} Raman line normalised to the intensity of the 340cm^{-1} line. The composition calibration in [45] was incorrect and has been revised based on energy dispersive X-ray analysis

An example of such a technique is Raman scattering [45], which has a spatial resolution down to ~1μm. The growth stoichiometry correlates strongly (fig. 10) with disorder which makes the Raman forbidden line at $585 cm^{-1}$ appear in the spectra. In principle, any disorder that breaks the inversion symmetry of the lattice can do this.

7. Oxygen Diffusion

The most important variable that we found in our study of film quality was oxygen content. Obtaining a uniform and controlled oxygen content in our films is important. When we come to discuss processing, we will see that there is evidence for sample degradation due to oxygen depletion near the edges of milled tracks. Whether or not there is a problem in practice is not clear, but it seems to me that the behaviour of oxygen in HTS is poorly understood. Oxygen gets into our films by diffusion, and it may be helpful to summarise the theory of diffusion as it applies to oxygen in YBCO, as the literature is not all that clear on the subject.

There are two main types of diffusion. The simplest is tracer diffusion in which one monitors the rate at which an isotope concentration gradient relaxes, with a uniform chemical concentration. These parameters have been determined for YBCO by Rothman et al. [80]. To simplify their conclusions, diffusion is slowest in the c-direction, and fastest in the b direction (i.e. parallel to the chains): $D_c \sim 10^{-4} D_a \sim 10^{-6} D_b$. The activation energy for diffusion is the same for all three directions, and is about 1eV over a wide temperature range. There is no strong dependence of tracer diffusion coefficient on oxygen concentration. For a twinned sample, D_{ab} is dominated by D_b: $D_{ab} \sim D_b/4$. This all sounds very straightforward.

However, in a practical process, we are interested in how fast chemical concentration gradients equilibrate. This is given by the chemical diffusion coefficient and is more complicated. What is the difference between this and the tracer diffusion coefficient? The following is simplified from Steele [81]. In a one–dimensional gradient of chemical potential, a diffusion current J can be written as

$$J = Cv = -CB\, d\mu / dx$$

where C is the oxygen concentration, v the mean drift velocity of the oxygen ions, B their mobility and μ the chemical potential, where

$$\mu = \mu_0 + kT \log(\gamma C) + \Delta\mu$$

Here γ is the factor relating oxygen activity to concentration, and Δμ is the chemical potential resulting from fields not directly related to concentration e.g. strain fields, electric fields. The diffusion coefficient D is defined by J = -D dC/dx, so

$$D = CB\frac{d}{dC}\left(kT\log(\gamma C) + \Delta\mu\right) = kTB\left(1 + \frac{d(\log\gamma)}{d(\log C)}\right) + B\frac{d(\Delta\mu)}{d(\log C)}$$

For tracer diffusion, γ and Δμ are independent of C, and the diffusion coefficient is just kTB. The term in d(log γ)/d(log C) is the correction for chemical diffusion. Its magnitude has been much discussed. Conder and Krüger [82] give a useful summary graph, showing a very large spread in the measured data. The term in Δμ is often assumed to vanish.

There is one more term that is only occasionally mentioned in the HTS literature, but I include it here for completeness. In a practical diffusion problem we are interested in the rate of change of the oxygen concentration at a point. Then conservation of matter gives

$$\frac{dC}{dt} = -\frac{dJ}{dx} = \frac{d}{dx}\left(D\frac{dC}{dx}\right) = D\frac{d^2C}{dx^2} + \frac{dD}{dx}\frac{dC}{dx} = D\frac{d^2C}{dx^2} + \frac{dD}{dC}\left(\frac{dC}{dx}\right)^2$$

If the diffusion coefficient is concentration dependent, then we expect deviations from Fick's law due to the term in $(dC/dx)^2$. Usually, diffusion data have been analysed without including this term. If it is ever significant, it will be in large concentration gradients.

So why is there such a spread in results of chemical diffusion measurements in HTS? First of all, the anisotropy is very large, and there is clearly a strong possibility that diffusion, particularly in the c-axis direction, is dominated by defects. This is believed to be the mechanism by which thin films oxygenate [83]. The second point is that if diffusion is the limiting mechanism, detailed balance requires that the rate of oxygenation should equal the rate of deoxygenation at sufficiently small deviations from equilibrium. This is seen in some sets of results (e.g, [84]) but by no means all. Where the two rates differ, either the driving force is too large for a simple analysis or the rate limiting step is not diffusion but at the surface of the sample [85]. Finally, Yamamoto et al. [86] have shown that oxygenation rates are much larger for atomic oxygen than molecular (we have observed similar effects). These experiments were in the large signal limit, but nevertheless the authors argue fairly convincingly that the most likely explanation is a chemical potential gradient (the term in $\Delta\mu$ above) due to lattice strain. Other results on deoxygenation during patterning appear to indicate that oxygen transport near room temperature may be faster than it is expected to be.

8. Processing High Temperature Superconductors

It has turned out that standard photolithography, as developed for semiconductor processing, and ion milling, using photoresist as a mask, perform remarkably well in patterning high temperature superconductors. In fact, it has seemed so easy that rather little systematic work has been published on the subject.

Early in the development of the technology, there were fears that wet processing would not be compatible with HTS materials, but the organic or alkaline solutions used in photolithography cause little harm to good quality material. Detailed studies are rare, however. Sheats et al. [87] found that processing could increase both R_s and J_c, and these changes were compounded rather than reversed on annealing. It would be good to see more systematic work in this area.

Ion milling has been entirely replaced for semiconductor processing by reactive ion etching, but no such processes have yet been found for superconductors, due to the difficulty of finding volatile species, particularly for Ba. The basic work on the ion milling process that was done 20 -30 years ago has been largely forgotten in the HTS literature. A good starting point is [88], which gives numerous references, although the papers themselves are rather theoretical in nature. Ion milling was abandoned for semiconductors because it introduces damage, differential etching rates between materials are limited, and high aspect ratios were difficult to achieve. Damage is much less critical in superconductors, and much of it can be annealed out. High aspect ratios are not usually required in HTS circuits, but the absence of good etch stops in milling is a problem for the HTS process designer. Thus these objections to milling are less strong for superconductors than for semiconductors.

In ion milling, an ion gun points at the sample and the ions (usually argon) strike the sample at some angle. In most cases the sample is rotated during milling for improved uniformity. In research a small 3 cm ion gun is often used. Our experience is that such a gun will often point 10 degrees or so away from its apparent direction, so a mounting that allows the gun to be tilted is desirable. The ion beam direction can change with time, and needs to be checked regularly using a probe or Faraday cup. For large area processing larger guns are used, and these issues become less critical.

Fig.10 shows the profile of ions produced by a 3cm gun for different sets of conditions. This figure shows that the gun produces a narrower beam at higher current and higher voltage. Thus doubling the beam current from 6 to 12 mA will produce a much larger increase in the milling rate. Reducing the voltage from 500 to 250 V will reduce the milling rate much more than just the reduction in milling yield. To some extent these effects will be less marked when a larger diameter gun is used. Finally the angular spread of the ion beam is much greater at low voltages and currents. This is important in milling sharp edges, and is not ameliorated by using a large gun.

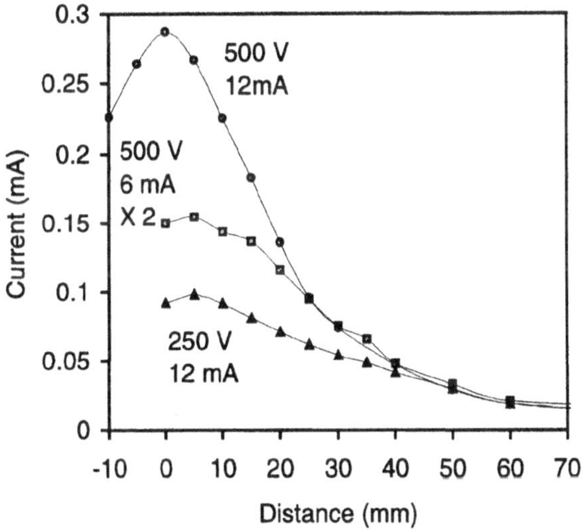

Fig. 10. Ion currrent collected by a probe biased at –3V ~15cm from a 3cm ion gun fed with 2 sccm argon. The probe was moved perpendicular to the beam. The accelerating voltage and total beam current are indicated. The current measured at 500V and 6 mA has been multiplied by 2. The results have been shifted slightly so that the peaks approximately coincide.

8.1. TEMPERATURE CONTROL IN ION MILLING

In the ion milling process, the ions are usually accelerated by several hundred volts and carry considerable kinetic energy. It is easy to show that any radiated heat input from the hot cathodes is negligible by comparison (we have verified this experimentally). If the sample is not heat-sunk, its temperature will rise during milling, leading to deformation or charring of the resist and possible deoxygenation of the superconductor. If the sample is to be kept cool during milling, it must be heat sunk. The substrate holder must either be massive enough that the temperature rise is small or it must be thermally linked to a cooling medium. A number of rotating sample holders are commercially available, although the design is quite tricky. So what can go wrong? The simplest mistake is to assume that lying the sample on top of a cold metal block implies that the sample is cold. In fact such a sample will be supported on only three points and thermal conduction between the two will be extremely poor - hardly better than vacuum. Let's look at the problem more quantitatively.

Consider a substrate of thermal conductivity σ_1 and thickness d_1 connected to a heat sink through a layer of thermal conductivity σ_2 and thickness d_2. The power density input at the front surface is neV, where n is the ion flux/unit area, e the electronic charge and V the accelerating voltage. If we assume that all the interfaces have negligible thermal impedance, the temperature differences (ΔT_i) across the layers are given by

$$\Delta T_i = neV \frac{d_i}{\sigma_i}$$

Typical milling conditions used for superconductors are n ~ 10^{15} ions/cm^2/s and V = 500V, so the input power is ~0.1 W/cm^2. The substrate thickness is typically ~ 0.5mm, and its thermal conductivity 0.1 W/cm/K. It is immediately apparent that ΔT_1, the temperature difference across the substrate, is negligible, <0.1 K for the above parameters. There are still occasional references in the literature to "substrate surface temperature" in milling processes. Clearly, the substrate is nearly isothermal and the temperature of the surface is essentially that of the substrate. If the substrate temperature rises it is because of poor thermal contact. For example, consider a heat sinking layer with σ_2 = 2.5 10^{-3} W/cm/K (this is about the lowest thermal conductivity available in room temperature solids, characteristic of polymers) and a thickness of 0.25mm. Its conductance is G = 0.1 W/cm^2/K and ΔT_2 = 1K, which is again negligible. If, on the other hand, no heat sinking layer is used, then the conductance between the wafer and its environment is the radiation conductance, which for small temperature differences is given by

$$G = 8\eta_s \sigma_{SB} T^3$$

where σ_{SB} is the Stephan-Boltzmann constant (5.67 10^{-12} W/cm^2/K^4) and the η_s is the sample emissivity. The factor is 8 because the sample has two surfaces. Even if the emissivity is high, G ~ 10^{-3} W/cm^2/K, about two orders of magnitude lower than the worst solid in the geometry considered above. If the sample is not heat sunk at all, its temperature rises by over 100°C (we have confirmed this experimentally). If we can arrange a thin heat sinking layer that is in good thermal contact everywhere, then the substrate temperature rise can be kept small. Good thermal contact is much more important than thermal conductivity. We use Teflon tape, as used for water pipe fittings, for this purpose. It has very poor thermal conductivity, but is thin and deforms well to give a good thermal contact.

Some groups use pulsed milling to limit the temperature rise. For a given average milling rate, the peak temperature of the substrate will be slightly larger if pulsed milling is used than if it is not. The only reason to use pulsed milling might be that the beam profile from the ion gun is better at a higher current. This is indeed the case, as we have seen, but is not the reason that is usually given for using pulsed milling.

8.2. THE ION MILLING PROCESS

Although little detailed work on ion milling has been reported in the HTS literature, there is a large literature on milling other materials. In this section we look at the fundamentals. The milling process is quite complicated [86, 87]. The milling rate depends on

(1) the mass and energy of the projectile ion
(2) the angle of incidence of the projectile ion on the surface being milled
(3) the sample temperature
(4) the rate of arrival of other species on the surface being milled
(5) the mass, chemical environment and physical environment of the layer being milled

Of these, the process designer has full control of the first three, and some control of the fourth.

The milling rate for a given species vanishes [90] below some energy, of the order of a few eV. Milling yields [91] at such energies are very low ($\sim 10^{-3}$ atoms/ion). At milling energies that are used in practice, the milling rate rises approximately linearly with energy, on a curve which extrapolates to zero milling rate between 50 and 100eV [90], with milling yields ~ 1atom/incident ion at around 500 - 1000eV.

In thinking about milling high temperature superconductors, comprising a number of different elements, we need to qualify the list of factors determining the milling rate. Evidently a sample has only one milling rate at a point, but each element in the surface will mill at a different rate. The obvious answer is that initially some elements mill faster than others, but quite quickly an equilibrium is established in which the surface being milled has a different composition from the bulk material. The thickness of this surface layer is of the order of the stopping distance of the projectile ion, say about 1nm for 400 eV Ar^+ ions [89]. It is well established that oxides tend to lose oxygen preferentially [92], and once the oxygen is lost, milling proceeds much more rapidly. Kelly [93] has sought to correlate those oxides that dissociate under milling with their dissociation vapour pressure at high temperatures, although it seems that the correlation was not perfect. An extra idea that needs to be considered is whether the residual gas in the milling chamber is important. In [92], it was shown that the milling rate of SiO_2 (and the differential milling rate compared to photoresist) is greatest for low ($<10^{-6}$ mbar) residual pressures of oxygen. It seems likely that water vapour (usually the dominant residual gas in vacuum systems) will play a similar role.

It is not enough only to consider milling a plane surface; the purpose of milling is to transfer a pattern through a mask (usually photoresist), and the shape of the surface evolves with time, often in a non-intuitive way. The suprising shapes that can result are due to the angle dependence of milling: the milling rate for superconductors and photoresist peaks at an angle of incidence of about 45°. Facets develop and there is a direct analogy with Frank's theory of crystal dissolution [94, 95]. There is a tendency for a curved surface to turn into two plane surfaces separated by a sharp cusp [96], in much the same way as facets develop on dissolving crystals. On a convex surface, the fastest milling facets develop, and on a concave surface the slowest milling ones. Under some conditions, a smooth surface can spontaneously roughen under grazing incidence milling [97], conditions that might intuitively be expected to polish it.

While the theories can explain some of the features seen in experiments, many ignore second order effects. Wilson et al. (98) identified three second order effects. These are ion reflection at grazing incidence, forward peaked sputtering at grazing incidence, and redeposition of sputtered atoms. They cite numerous further complications, but their three second order effects suffice to explain what is observed in patterning HTS. These effects are hard to model because they are non-local. Ion reflection and forward peaked sputtering are both potential contributors to the way morphology develops near a steep slope, and redeposition is observed almost anywhere that one looks for it.

The redeposition of milled material onto adjacent surfaces is of considerable practical importance. For example, if sputtered atoms are emitted at an angle θ from the surface normal with a cos θ distribution, the redeposition rate on a vertical wall being formed by milling is half the milling rate – hardly a negligible effect. Thus the net milling rate of a slope will often be significantly different from that measured on a planar sample at the same milling angle. In addition, both the redeposition and milling process have a random element; they comprise atoms or clusters rather than a continuum. Surfaces subjected to milling in the presence of a flux of redeposited atoms tend to roughen. The use of rotation to improve uniformity and to give a process that is isotropic makes it more difficult to visualise what is happening; some parts of the evolving surface may be temporarily shadowed during rotation. During this period they may accumulate redeposited material. When the surface emerges from shadow, the redeposited material must be milled off before any net material removal takes place on that part of the surface.

The second non-local effect is due to reflection of high energy ions from surfaces on which they are incident at near-grazing incidence. Little energy is lost in such a collision, and the reflected ion is still able to sputter. The result is that a shallow trench is formed at the base of a vertical wall as the reflected ions enhance the milling rate near the wall. This effect is known as trenching [99] and is usually suppressed by milling away from normal incidence. A competing effect is the tendency near grazing incidence for material to be milled off in the forward direction, leading to a reduced net milling rate [100], and a foot forms near the wall.

8.3. MILLING HIGH TEMPERATURE SUPERCONDUCTORS

Little systematic information has been published that might allow a comparison between the disparate (but invariably simple) materials used in the studies cited above, and high temperature superconductors. There are many papers that quote HTS milling rates for a specific set of conditions. These are broadly consistent with the above: milling yields are about 1 atom/ion at around 500eV, typical of slow milling material [90]. Photoresist mills somewhat faster than HTS. The secondary ions and energies produced in milling with 4KeV Ar^+ ions have been studied by Chenakin [101]. He found an initial transient in the milling rate consistent with the formation of a surface layer of changed stoichiometry and that secondary ion energies are ~10eV. Detailed SIMS spectra were also reported by Gauzzi et al. [102] using 5KeV Ar^+ ions and by Ilyinsky et al. [103] for lower energy ions. Ilyinsky et al. found that the addition of oxygen to the milling gas resulted in significant changes in the secondary ions produced.

Fig.11 shows the dependence of the milling rate on the angle of incidence of the Ar^+ ions for YBCO and other materials used in HTS circuits. Other data in the literature are broadly consistent with this picture, although there is considerable variation in detail.

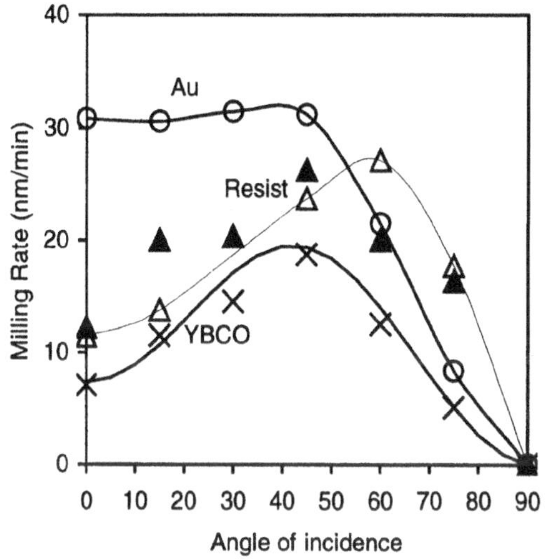

Fig. 11. Angle dependence of the milling rate of YBCO, and AZ1518 resist (solid triangles) from Goodyear [104] for 500eV Ar^+ ions at a beam current density ~ $0.3mA/cm^2$. These are compared to data for AZ1350 resist (open triangles) and Au from Johnson [100]. The relative vertical scales between data sets taken in different laboratories are my best guesses.

Because ion milling has only small differentials between materials, etch stop layers are not available and it is very useful to be able to monitor the composition of the material being milled. We introduced the use of secondary ion mass spectrometry (SIMS) for routine end point detection in ion milling of HTS [105] in 1991, and showed that a depth resolution of 4nm was easily obtainable in routine use. We found that the surface of the sample is fully hydrated for water vapour pressures in the milling chamber > 10^{-6} mbar. Below this pressure, the secondary ions detected contain fewer species like $(YOH)^+$. For the milling conditions used, the threshold for dehydration of the surface corresponds to a water vapour incidence rate comparable to the rate at which material is removed from the milled surface. Any effect of the water vapour pressure on the milling rate of the sample was slight, however.

We might expect only the yttrium (or rare earth) oxygen bond to be strong enough to hold onto oxygen, with the result that the surface becomes significantly yttrium rich. This is observed in practice. Fig. 12 shows the SIMS trace from milling through a layer of YBCO grown on a second YBCO film, where the surface of the first film has been prepared by ion milling *in situ* before growing the second layer. It is clear that the interface layer contains excess Y and is deficient in Ba and Cu. If the first film is not milled before growth of the second, no interface peak is observed. It is difficult to make such data quantitative, but as a rough estimate the Y excess at the interface corresponds to ~0.5nm of Y_2O_3. A similar change in surface stoichiometry has recently also been observed. by Wen et al. [106]. Reducing the milling energy is expected to reduce the depth of damage, but so far we have not identified such a trend in the SIMS data. Low energy milling is quite widely used for preparing the surfaces of edge barrier junctions, and there are many reports that this is necessary to prepare high Jc connections between layers, although the experience is not universal.

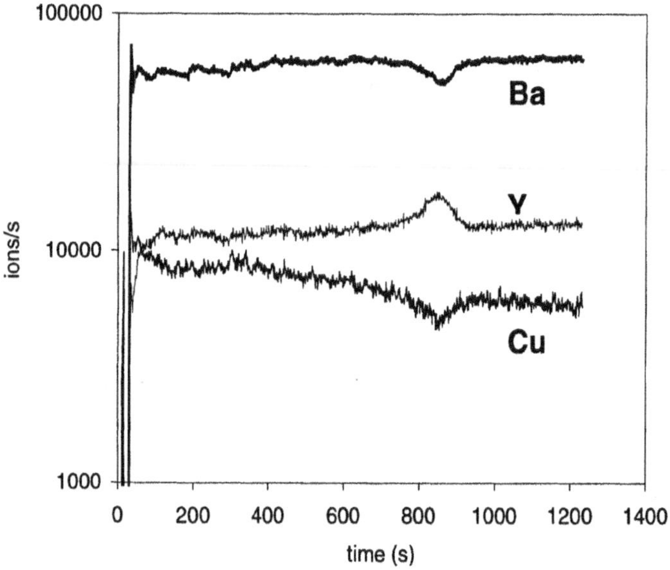

Fig. 12. Secondary ion mass spectrometry (SIMS) trace from milling through a layer of YBCO 0.1 μm thick grown on another YBCO layer. The surface of the lower YBCO layer was ion cleaned before growth at 500V for 2 minutes followed by 300V for 2 minutes.

It is well known in the preparation of TEM specimens that reducing the sample temperature during milling is advantageous for minimising damage, and some groups have cooled their samples below room temperature in HTS patterning. Unlike semiconductors, the thermal expansion coefficients of resist and oxides are sufficiently similar that the resist does not crack. For example Schneidewind et al. [107] have reported improvements in the critical currents of microbridges milled at low temperature. A significant improvement in J_c and T_c was found even for 20μm wide bridges. Bearing in mind that only the edges of the bridge are actually milled, this is hard to understand in terms of the expected diffusion coefficient of oxygen. We have not studied deoxygenation of milled edges in detail, but we have concluded that re-oxygenation of all circuits after patterning is the safest route. Ban et al. [108] have reported that introducing oxygen into the milling gas improves the quality of microwave resonators. Our attempts to look for similar effects [109] failed to identify any difference in properties between milled and wet etched resonators. Raman scattering [110] has been used to show deoxygenation effects extending about 2μm from an ion milled edge. This seems much larger than could be explained by measured diffusion coefficients assuming that the sample was effectively cooled during milling. To be quantitative, future work in this area would need to use well oriented films, a well characterised milling process with minimal redeposited material on the edge and an experimentally measured point spread function for the laser spot. Degradation on this scale is apparently implied by a number of transport measurements on narrow lines, e.g. [111]. However, in nearly all cases, the lines were long compared to their width and it is difficult to distinguish effects of film non-uniformity from deoxygenation. An exception is the work of de Nivelle et al. [112] who made very short bridges. Their T_c degraded from 87 K in the original film to 83K for a bridge width of 0.175μm. This suggests that at least in some material, deoxygenation is only significant on a length scale ~0.1μm.

Photoresist has about twice the milling rate of HTS, so it is not ideal as a milling mask. However, steep edges can be made routinely. This is often assisted by redeposition, which tends to suppress the rounding of the soft resist in milling. For very demanding applications, trilayer mask processes have been used in which the pattern is transferred using reactive ion etching from an organic resist to a slow-milling material. Carbon is a particularly convenient hard mask material that can be ashed away using an oxygen plasma without damaging the underlying superconductor [113]. At present, it does not seem that sophisticated processes like this are needed for passive microwave circuits. The resolution that can be achieved using a resist mask is at the 1μm level. This corresponds to a relative accuracy in resonator length $\leq 10^{-4}$ for example, and other factors like uncertainty in the substrate thickness and errors in the design software contribute larger errors.

The similarity in milling rates between YBCO and photoresist can be used to planarise film surfaces. Resist forms a smooth surface over a rough film or fine patterning features, and if the resist is then milled away, the resulting film surface is much smoother. This can be used to open contact holes [114] or produce planar multilayer circuits [115]. Another approach to planarisation is grazing incidence milling [116]. Milling using 1kV Xe^+ ions at an angle of incidence of 80° was found to reduce the surface roughness of post-annealed films. Planarisation can also be achieved by mechanical polishing [117].

Although nearly all the patterning of HTS films has used milling through a lithographically defined mask, some attempts at lift off have been reported [117]. In this approach, amorphous CaO was deposited at low room temperature over a photoresist pattern and the pattern lifted off with acetone. Then films could be grown through the CaO pattern and the unwanted superconductor lifted off by dissolving the CaO in water. I believe that processes like this have been abandoned because their resolution is poor.

9. The State of the Art

The objective of production is to fabricate circuits with the required performance at the lowest cost. The steps involved are not just growing the films, but rather a list like this:

Buy substrates
Screen for quality/thickness
Grow buffer layer
Grow superconductor
Pattern
Metallise
Anneal
Passivate
Dice
Package
Tune
Quality control
Vacuum encapsulate

The precise order of these steps may vary, and some may be omitted. For example, the need for passivation is far from clear, although some papers discuss it. In production, circuits are packaged in a vacuum encapsulation that is sealed for life, so any question of long term exposure to atmospheric CO_2 is irrelevant. In discussion at the ASI, B. Willemsen suggested that passivation was desirable to minimise damage to samples through handling in production, but it was not needed otherwise.

Some quality control will certainly be included, but it is undesirable to have to characterise wafers more often than absolutely necessary, unless the characterisation is very cheap. I have never seen a breakdown of the costs incurred by each step. Each step has an associated capital cost and manpower cost that contribute to the cost of the final product.

It is important to be able to make large area double sided films with high reproducibility. Growing a large area at a time (using multiple substrates) is advantageous because it reduces the cost of growth. Matijasevic [8] stated that the cost of double sided film growth using a Munich style evaporation technique was small compared to the cost of the substrate, and clearly there is little advantage in further reducing the costs of growth beyond this point.

Using large substrates is advantageous because it reduces the cost of the subsequent processing. Thus if it were practical to grow films on 8" sapphire wafers, the recurring cost of handling would be greatly reduced. On the other hand, the capital cost of processing equipment for larger areas is in general higher. The current industry standard for HTS is 2", but if the market continues to grow substrate size will increase.

I believe that some of the companies involved in production are now approaching throughputs of ~1000 wafers/year each. This means a throughput of several wafers/day. If equipment is not duplicated, each set up must have a throughput of this order. Duplication of low cost equipment may be attractive because it allows calibration and maintenance without stopping production.

Fig. 13 is an attempt to represent the economics of the different approaches to growth, based on the idea that the most efficient technique is the one that grows the largest volume (area times thickness) of superconductor per unit time. A typical film for microwave applications is ~0.7μm thick. At ~2nm/min, which seems to be typical of sputtering, it takes 6 hours to coat one side of the substrate. At the highest rate reported for the Munich process, growth takes only half an hour per side, and a large number of substrates are coated in a single growth run. In fact for evaporation the overall throughput is probably limited at present by the time taken to load and unload the growth machine, rather than the actual growth time. The evaporator may be more expensive to buy and maintain than the sputterer. It may also require more operator attention: if the sputterer can operate unattended overnight but the evaporator can not, for example, then some of the advantage of evaporation is lost.

For me the most surprising observation from this figure is the way laser ablation is

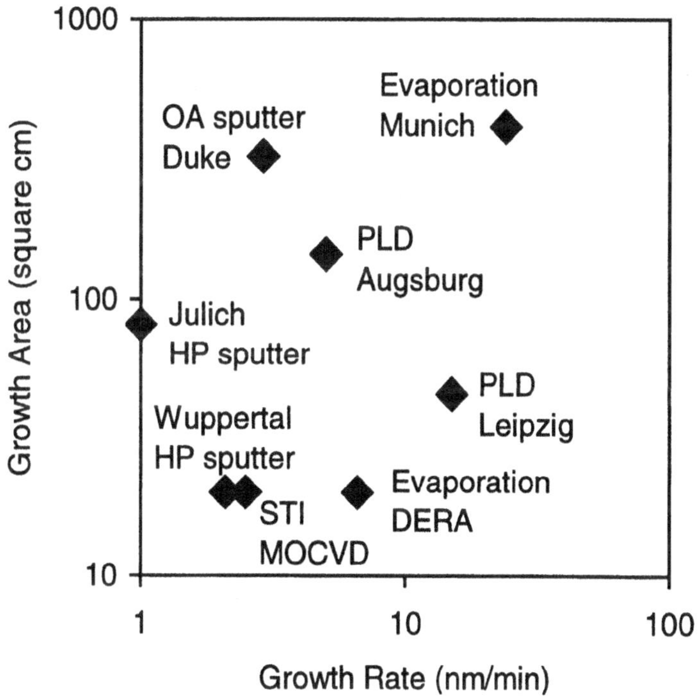

Fig.13. Growth area plotted against growth rate for the large area processes of Table 2.

keeping pace. In [1] and since, I have thought of laser ablation as an excellent technique for fast prototyping of new materials or processes, but unlikely to compete for real production. I am not sure that I have changed my mind, but at least it has made me think. Lastly, MOCVD is not fairly represented by fig. 13. There are few published data, but I believe MOCVD has good scaling properties to large area and it remains a competitor.

Throughout this paper I have focussed largely on YBCO grown by physical vapour deposition for the good reason that we have enough information to begin to understand the growth mechanisms. This does not necessarily mean that YBCO is the material of choice or that physical vapour deposition techniques will be the technology of choice. There is a strong economic pressure to use higher T_c materials: every watt saved in cooling is a selling point. One can also envisage other growth techniques with lower cost, e.g. *ex situ* techniques, but here I feel that for passive microwave devices, the pressure is less strong as the total subsystem cost is not dominated by the cost of growth. With good circuit design, quality is more important.

10. References

1. R G Humphreys, J S Satchell, N G Chew, J A Edwards, S W Goodyear, S E Blenkinsop, O D Dosser and A G Cullis. (1990) *Superconductor Sci. Technol.* **3**, 38
2. D J Rogers, P Bove and F Hosseini Teherani. (1999) *Superconductor Sci. Technol.* **12**, R75
3. R Wördenweber. (1999) *Superconductor Sci. Technol.* **12**, R86
4. G.Kästner, C.Schäfer, St. Senz, T.Kaiser, M.A.Hein, M.Lorenz, H.Hochmuth and D.Hesse. (1999) *Superconductor Sci. Technol.* **12**, 366
5. F Slolovyov, H J Wiesmann, L-J Wu, M Suenaga and R Feenstra. (1999) *IEEE Trans. Appl. Superconductivity* **9**, 1467
6. W Prusseit, B Utz, P Berberich and H Kinder. (1994) *J Superconductivity* **7**, 231
7. P Berberich, B Utz, W Prusseit and H Kinder. (1994) *Physica C* 219
8. V Matijasevic. (1998) *Applied Superconductivity Conference* unpublished
9. K Fukushima. (1995) *Int. J. Mod. Phys.* **B9**, 3625
10. C H Becker and J B Pallix. (1988) *J. Appl. Phys.* **64**, 5152
11. B Schey, T Bollmeier, M Kuhn, W Biegel and B Stritzker. (1998) *Rev. Sci. Insts.* 69, 474
12. J A Geer. (1992) *J. Vac. Sci. Technol.* **A10**, 1821
13. M Lorenz, H Hochmuth, D Natusch, H Börner, G Lippold, K Kreher and W Schmitz. (1996) *Appl. Phys. Lett.* **68**, 3332
14. D L Schulz and T J Marks. (1994) *Adv. Mater.* **6**, 719
15. R Hiskes, S A DiCarolis, J L Young, S S Laderman, R D Jacowitz and R C Tabor. (1991) *Appl. Phys. Lett.* **59**, 606
16. T Ushida, H Higa, K Higashiyama, I Hirabashi and S Tanaka. (1991) *Appl. Phys. Lett.* **59**, 860
17. U Krüger, R Kutzner, R Wördenweber, G Mank and A Kraemer-Flecken. (1993) *Appl. Phys. Lett.* **62**, 1559
18. D W Face, C Wilker, Z-Y Shen, P Pang and R J Small. (1995) *IEEE Trans. Appl. Superconductivity* **5**, 1581
19. X X Xi, G Linker, O Meyer, E Nold, B Obst, F Ratzel, R Smithey, B Strehlau, F Wechsenfelder and J Geerk. (1989) *Z. Phys. B Condensed matter* **74**, 13
20. T Aichele, P Görnert, R Uecker and M Mühlberg. (1999) *IEEE Trans. Appl. Superconductivity* **9**, 1510
21. J Kawashima, Y Yamada and I Hirabayashi. (1998) *Physica C* **306**, 114
22. A Takagi, U Mizutani, T Kitamura, S Taniguchi, Y Shiohara, I Hirabayashi, S Tanaka and Y Yamada. (1998) *IEEE Trans. Appl. Superconductivity* **7**, 1388
23. T Kitamura, S Taniguchi, Y Sugawara, Y Ikuhara, Y Shiohara, I Hirabayashi and S Tanaka. (1996)

Physica C **256**, 64
24. S Miura, K Hashimoto, J-G Wen, K Suzuki and T Morishita. (1998) *IEICE Trans. Electron.* **E81-C**, 1549
25. M J Lancaster (1997) *Passive microwave device applications of high temperature superconductors* Cambridge University Press.
26. E K Hollmann, O G Vendik, A G Zaitsev and B T Melekh. (1994) *Superconductor Sci. Technol.* **7**, 609
27. D Fincham, W C Mackrodt and P J Mitchell. (1994) *J. Phys. Condens. Matter* **6**, 393
28. T Konaka, M Sato, H Asano and S Kubo. (1991) *J Superconductivity* **4**, 283
29. E A Giess, R L Sandstrom, W J Gallgher, A Gupta, S L Shinde, R F Cook, E I Cooper, E J M O'Sullivan, J M Roldan, A P Segmüller and J Angilello. (1990) *IBM J. Res. Develop* **34**, 916
30. V B Braginsky, V S Ilchenko, and Kh S Bagdassarov. (1987) *Phys. Lett.* **120A**, 300
31. Landolt Bornstein *Numerical Data and Functional Relationships in Science and Technology* Group 3 Vol. 7 Part b1.
32. R Brown, V Pendrick, D Kalokitis and B H T Chai. (1990) *Appl. Phys. Lett.* **57**, 1351
33. B C Chakoumakos, D G Schlom, M Urbanik and J Luine. (1998)*J Appl. Phys.* **83**, 1979
34. C Zuccaro, M Winter, S Penn, N McN Alford, P Filhol, G Forterre and N Klein. (1997) *Proc. European Conf. on Appl. Superconductivity*
35. S.Yardavalli, M.H.Yang and C,P,Flynn. (1990-I) *Phys. Rev.* **B 41**, 7961
36. Y J Tian, S Linzen, F Schmidl, A Matthes, H Schneidewid and P Seidel. (1999) *Thin Solid Films* **338**, 224.
37. A G Zaitsev, R Kutzner, R Wördenweber, T Kaiser, M A Hein and G Müller. (1998) *J Superconductivity*, **11**, 361
38. J D Jorgensen, M A Beno, D G Hinks, L Soderholm, K J Volin, F L Hitterman, J D Grace and I K Schuller. (1987) *Phys. Rev.* **B 36**, 3608
39. J Mizusaki, H Tagawa, K Hayakawa and K Hirano. (1995) *J. Am. Ceram. Soc.* **78**, 1871
40. R G Humphreys, N G Chew, J S Satchell, S W Goodyear, J A Edwards and S E Blenkinsop. (1991) *IEEE Trans. Magnetics* **27**, 1357
41. N G Chew, J A Edwards, R G Humphreys, J S Satchell, S W Goodyear, B Dew, N J Exon, S Hensen, M Lenkens, G Müller and S Orbach-Werbig. (1995) *IEEE Trans. Appl. Superconductivity* **5**, 1167
42. J T Gourley and W A Runciman. (1973) *J. Phys. C* **6**, 583
43. J R Jasperse, A Kahan, J N Plendl and S Mitra. (1966) *Phys. Rev.* **146**, 526
44. B I Choi, A C Anderson, A C Westerheim and M I Flik. (1993) *J. Vac. Sci. Technol.* **A 11**, 3020
45. G Gibson, R G Humphreys, L F Cohen and J L McManus-Driscoll. (1999) *IEEE Trans Appl. Superconductivity* **9**, 1848 The Ba/Y ratios in table 2 and fig. 2 in this reference are too large by about 10%. This is corrected in fig. 9 of the present article.
46. R Bormann and J Nölting. (1989) *Appl. Phys. Lett.* **54**, 2148
47. R Beyers and B T Ahn. (1991) *Annu. Rev. Mater. Sci.* **21**, 335
48. T B Lindemer, F A Washburn, C S MacDougall, R Feenstra and O B Cavin. (1991) *Physica C* **178**, 93
49. R G Humphreys, J S Satchell, N G Chew, J A Edwards, S W Goodyear, M N Keene and S J Hedges. (1991) *Materials Sci. and Eng.* **B10**, 293
50. S V Samoylenkov, O Yu Gorbenko, I E Graboy, A R Kaul, H W Zandbergen and E Connolly. (1999) *Chem. Mater.* **11**, 2417
51. S V Samoylenkov, O Yu Gorbenko, I E Graboy, A R Kaul, O Stadel, G Wahl and H W Zandbergen. (1999) *Journal de Physique* **9**, Pr8-621
52. N G Chew (1997) PhD Thesis, University of Birmingham.
53. H Teshima, H Shimada, M Imafuku and K Tanaka. (1993) *Physica C* **206**, 103
54. D G Schlom, A F Marshall, J T Sizemore , Z J Cen, J N Eckstein, I Bozovic, K E von Dessoneck, J S Harris and J C Bravman. (1990) *J. Crystal Growth* **102**, 361
55. B Utz, R Semerad, M Bauer, W Prusseit, P Berberich and H Kinder. (1997) *IEEE Trans. Appl. Superconductivity* **7**, 1272
56. D W Face, C Wilker, Z-Y Shen, P Pang and R J Small. (1995) *IEEE Trans. Appl. Superconductivity IEEE Trans. Appl. Superconductivity* **5**, 1581
57. R A Rao, C B Eom, M Santer and S M Anlage (1997) *IEEE Trans. Appl. Superconductivity* **7**, 1278 See also P G Quigley, R A Rao and C B Eom. *J Vac Sci Technol.* **A15**, 2854 (1997)
58. G Müller, B Aschermann, H Chaloupka, W Diete, M Getta, B Gurzinski, M Hein, M Jeck, T Kaiser, S Kolesov, H Piel, H Schlick and R Theisejans. (1997) *IEEE Trans. Appl. Superconductivity* **7**, 1287
59. A G Zaitsev, G Ockenfuss, D Guggi, R Wördenweber and U Krüger. (1997) *J. Appl. Phys.* **81**, 3069
60. B Schey, T Bollmeier, M Kuhn, W Biegel and B Stritzker. (1998) *Rev. Sci. Inst.* **69**, 474

61. M Lorenz, H Hochmuth, D Natusch, H Börner, G Lippold, K Kreher and W Schmitz. (1996) *Appl. Phys. Lett.* **68**, 3332
62. E J Smith and J F Musolf. Applied Superconductivity Conference (1996) (unpublished)
63. S K Streiffer, E M Zielinski, B M Lairson and J C Bravman. (1991) *Appl. Phys. Lett.* **58**, 2171
64. J L Tallon and N E Fowler (1993) *Physica C* **204**, 237
65. M Bauer, F Baudenbacher and H Kinder. (1995) *Physica C.* **246**, 113
66. T Terashima, K Iijima, K Yamamoto, K Hirata, Y Bando and T Takada. (1989) *Japan. J. Appl. Phys.* **28**, L987
67. H U Krebs, Ch Krauns, X Yiang and U Geyer. (1991) *Appl. Phys. Lett.* **59**, 2180
68. B Dam, J H Rector, J M Huijbregtse and R Griessen. (1998) *Physica C.* **305**, 1
69. B Stäuble-Pümpin, V C Matijasevic, B Ilge, J E Moij, W J A M Peterse, P M L O Scholte, F Tuinstra, H V Venvik, D S Wai, C Traeholt, J G Wen and H Zandbergen. (1995-II) *Phys. Rev.* **B52**, 7604
70. B Dam, C Traeholt, B Stäuble-Pümpin, J Rector and D G de Groot. (1997) *J. Alloys and Compounds* **251**, 27
71. T Haage, J Zegenhagen, J Q Li, H-U Habermeier, M Cardona, Ch Jooss, R Warthmann, A Forkl and H Kronmüller. (1997-I) *Phys. Rev.* **B56**, 8404
72. Ch Jooss, R Warthmann, H Kronmüller, T Haage, H-U Habermeier and J Zegenhagen. (1999) *Phys. Rev. Lett.* **82**, 632.
73. J M Huibrigtse, F C Klassen, R C F van der Geest, B Dam and R Griessen. presented at MOS '99.
74. M Yeadon. M Aindow, F Wellhöfer and J S Abell. (1997)*J. Crystal Growth*, **172**, 145
75. J Kawashima, Y Yamada and I Hirabayashi. (1998) *Physica C.* **306**, 114
76. T Aichele, P Görnert, R Uecker and M Mühlberg. (1999) *IEEE Trans. Appl. Superconductivity* **9**, 1510
77. M S Hu and A G Evans. (1989) *Acta Metall.* **37**, 917
78. E S Zouboulis and M Grimsditch. (1991) *J. Appl. Phys.* **70**, 772
79. A S Raynes, S W Freiman, F W Gayle and D L Kaiser. (1991) *J. Appl. Phys.* **70**, 5254
80. S J Rothman, J L Routbort, U Welp and J E Baker. (1991-I) *Phys. Rev.* **B44**, 2326
81. B C H Steele (1973) in *Fast Ion Transport in Solids* Ed. W Oan Gool pp 103-118
82. K Conder and Ch Krüger. (1996) *Physica C.* **269**, 92
83. W A M Aarnink, R P J Ijsselsteijn, J Gao, A van Silfhout and H Rogalla. (1992) *Phys. Rev.* **B45**, 13002
84. V Dediu and F C Macotta. *Phys. Rev.* (1996-II) **B54**, 16259
85. S Kittelbereger, U Bolz, R P Huebener, B Hozapfel and L Mex. (1998) *Physica C.* **302**, 93
86. K Yamamoto, B M Lairson, J C Bravman and T H Geballe. (1991) *J. Appl. Phys.* **69**, 7189
87. J R Sheats, N Newman, R C Taber and P Merchant. (1994) *J. Vac. Sci. Technol.* **A12**, 388
88. *Ion bombardment modification of surfaces; Fundamentals and applications* (1984), O Auciello and R Kelly (eds.), Elsevier
89. J-C Pivin. (1983) J. Mat. Sci. **18**, 1267
90. E G Spencer and P H Schmidt. (1971)*J. Vac. Sci. Technol.* **8**, S52
91. N D Morgulis and V D Tischenko. (1957) *Bull. Acad. Sci. USSR* **20**, 1082
92. L Mader and J Hoepfner. (1976) *J. Electrochem. Soc.* **123**, 1893
93. R Kelly. (1978) *Nucl. Inst. and Methods* 149, 553
94. D J Barber, F C Frank, M Moss, J W Steeds and I S T Tsong. (1973).*J. Mat. Sci.* **8**, 1030
95. G Carter, J S Colligon and M J Nobes. (1973) *J. Mat. Sci.* **8**, 1473
96. M J Nobes, I V Katardjiev, G Carter and R Smith. (1987) *J. Phys. D* **20**, 870
97. P Sigmund. J. Mat. Sci. 8, 1545 (1973)
98. I H Wilson, J Belson and O Auciello (1984) in *Ion bombardment modification of surfaces; Fundamentals and applications* O Auciello and R Kelly (eds.), Elsevier, p.225
99. J P Ducommun, M Cantagrel and M Moulin. (1975) *J. Mat. Sci.* **10**, 52
100. L F Johnson (1984) in *Ion Beam Modification of Surfaces*, Ed. O Auciello and R Kelly Elsevier p. 361
101. S P Chenakin. (1990) *Vacuum* **42**, 139
102. A Gauzzi, H J Mathieu, J H James and B Kellett, (1990) *Vacuum* **41**, 870
103. L S Ilyinsky, B Emmoth, E K Hollmann, A G Zaitsev and A A Lavrentyev. (1995) *J. Phys. D* **28**, 996
104. S W Goodyear. (1994) PhD Thesis, Birmingham University
105. R G Humphreys, N G Chew, S F Morgan, J S Satchell, A G Cullis and P W Smith. (1992) *Appl. Phys. Lett.* **61**, 228
106. J G Wen, N Koshizuka, S Tanaka, T Satoh, M Hidaka and S Tahara. (1999) *Appl. Phys. Lett.* **75**, 2470
107. H Schneidewind, F Schmidl, L Linzen and P Seidel. (1995) *Physica C.* **250**, 191
108. M Ban, T Takenaka, K Hayashi, K Suzuki and Y Enomoto. (1996) *Japan. J. Appl. Phys.* **35**, 4318
109. A Porch, M J Lancaster and R G Humphreys (1995) *IEEE Trans. Microwave Theory and Techniques*

43, 306
110. J Jimenez, E Martin, P Martin, A Torres, C Belouet, and D Chambonnet. (1995) *Mat. Res. Bull.* **30**, 771
111. H Sato, H Akoh, K Nishihara, M Aoyagi and S Takada. (1992) *Japan. J Appl. Phys.* **31**, L1044
112. M J M de Nivelle, G J Gerritsma and H Rogalla. (1993) *Phys. Rev. Lett.* **70**, 1525
113. H R Yi, D Winkler, Z G Ivanov and T Claeson. (1995) *IEEE Trans. Appl. Superconductivity* **5**, 2778
114. S W Goodyear, N G Chew, R G Humphreys, J S Satchell and K Lander. (1995) *IEEE Trans. Appl. Superconductivity* **5**, 3143
115. A P Marathe, F Ludwig, T van Duzer and L Lee. (1995) *IEEE Trans. Appl. Superconductivity* **5**, 3135
116. A F Hebard, R M Fleming, K T Short, A E White, C E Rice, A F J Levi and R H Eick. (1989) *Appl. Phys. Lett.* **55**, 1915
117. P A Nilsson, G Brorson, J M Orliaguet, E Olsson, H Olin, M Gustafsson and T Claeson. (1995) *IEEE Trans. Appl. Superconductivity* **5**, 1653
118. B Roas (1991) *Appl. Phys. Lett.* **59**, 2594

HIGH TEMPERATURE SUPERCONDUCTING THICK FILMS FOR MICROWAVE APPLICATIONS

M. J. Lancaster and T. W. Button
The University of Birmingham
Edgbaston
Birmingham
UK
B152TT

1 Introduction

What is a superconducting thick film? There is no straightforward answer to this question and the answer does not necessarily relate to any particular thickness. We can say it is not a thin film, and a thin film is more easily defined. The processing of thin films aim at producing high quality growth on single crystal substrates, in many cases including high temperature superconducting thin films this is eptiaxial, single crystalline growth. To do this high vacuum technology is used with some method of vaporising the material to be deposited. The vapour is condensed and crystallised onto the substrate. Depending upon the method, good control of the growth of the film can be obtained. However, film thicknesses are usually limited to only about one micron and the quality of the film can reduce as the thickness is increased beyond this value.

On the other hand the fabrication of a superconducting thick film does not require high vacuum technology and does not aim at producing epitaxial deposition. The film is usually deposited on a polycrystalline substrate and therefore tends to be very granular in nature. Many materials have been produced in thick film form and they are widely used in the electronics industry. Thick film deposition techniques usually start with a powder, for simple metal films, this is just the metal itself. For more complicated films, different mixtures of the basic constituents can be used which react during the sintering process. The powder is mixed with polymers and solvents known as the vehicle. The powder and the vehicle form the thick film ink. The ink is deposited onto the required substrate. (Alumina is widely used in the electronics industry because of its low cost and good mechanical and electrical properties.) The circuit is then fired at high temperatures to remove the vehicle and optimise the electrical properties of the film. Such films can range in thickness from a few tens of microns to many hundreds depending upon the materials and techniques used.

There are, of course, other deposition processes that do not necessary fit into the classifications outlined above, such as sol-gel, electro-deposition and solution spraying.

H. Weinstock and M. Nisenoff (eds.), Microwave Superconductivity, 91–115.

Many bulk polycrystalline superconducting materials can be made by a similar powder route described above. The objects produced are solid superconductor and require no substrate; and tubes, rod, wires, cylinders and many other shapes can be easily produced [1].

The advantages of using thick films for microwave applications can be clearly seen. These are listed below:

1. Thick films can be produced on very large, flat polycrystalline substrates.
2. Curved surfaces can be covered enabling the production of large three-dimensional objects coated with superconductor.
3. Thick films are low cost, since the substrates, the processing equipment, and the initial HTS materials are all much lower in cost than for thin films.

The main disadvantage of thick superconducting films is that they tend to have a higher surface resistance than the thin film materials. This is discussed in a later section.

2 Superconducting thick films

Although there are many HTS materials, only a relatively few have been seriously investigated for thick film production and even less for potential candidates for microwave applications. Materials of interest are $YBa_2Cu_3O_x$ (YBCO 123)[2,3,4,5] $Tl_2Ba_2Ca_2Cu_3O_x$ (TBCCO 2223), $Tl_2Ba_2Ca_1Cu_2O_x$ (TBCCO 2212) [6,7] and $Bi_2(Pb)Sr_2Ca_2Cu_3O_x$ (BSCCO 2223) [8]. TABLE 1 shows the transition temperature of some of these materials.

TABLE 1 Transition temperature of some superconducting materials.

Material	Tc K
YBCO	93
BSCCO 2212	85
BSCCO 2223	110
TBCCO 2212	98
TBCCO 2223	125

YBCO has been shown to have a low surface resistance and thus have been used most widely in applications. It has been the most studied and therefore this review will concentrate on this material only.

3 Thick film production

There are many methods and many variants in techniques to produce superconducting thick films. No attempt will be made to discuss all of them here. Instead we outline the technique developed at The University of Birmingham which is based on a conventional thick film approach. This technology has been shown to produced HTS films with a relatively low surface resistance which have been used in microwave applications such as filters for base station receivers. The fabrication procedure or 'recipe' is as follows

1. Powders are made via a solid state route starting with of yttrium oxide, barium carbonate and copper oxide which are weighed out in the appropriate stoichiometric proportions.
2. The powders are mixed in a ball mill or vibroenergy mill in alcohol for 16 hours or until a homogeneous mixture is obtained and then dried in a rotary evaporator. These mills and evaporators are described below.
3. The powder is then calcined in an air furnace at a temperature of around 900°C for approximately 12 hours. This calcination step reacts the constituent oxides to form the desired phase according to the equation

$$2BaCO_3+1/2\ Y_2O_3+3CuO \Rightarrow YBa_2Cu_3O_x+2CO_2$$

4. As the calcination process takes place below the melting point of any of the constituents, and because the grains in the powder are only loosely connected by point contact between them, the process is sometimes repeated after a further mixing and drying stage.

5. The resultant friable 'crumble' is ground to a fine powder (particle size 2-5μm) in alcohol in a ball mill or vibroenergy mill using zirconia (to avoid contamination) milling media and then dried in a rotary evaporator.

6. The powder is then made into an ink using three-roll milling (see below) using a vehicle containing a mixture of solvents and polymers. A number of proprietary vehicles are available based on cellulose polymers and using terpineol as a solvent.

7. The ink is now placed onto the substrate. There are a number of techniques for doing this. If patterning is required at this stage it can be printed using screen-printing. This gives a uniform thickness and a well-defined pattern. Alternatively, the ink can be deposited onto the substrate using spin coating or 'doctor blading'. The former is done by placing a predetermined amount of ink on the substrate and spinning the substrate, which results in a layer of uniform thickness which is determined by the spin speed. Doctor blading is the formation of a layer of uniform thickness by running a blade at a uniform height above a thick layer of ink.

8. The circuit is now sintered at temperatures near 1000°C and the particles of the superconductor coalesce to form the solid thick film. The details of the sintering are crucial to the development of the film properties and is discussed in more detail below.

9. If the film has not already been patterned, it can be done after sintering. This can be done by an air abrasion technique, where high velocity particles of silicon carbide mechanically etch the superconductor. A rubber mask can be placed over the

superconductor where no etching is required. This mask can be screen printed or taped by hand for simple patterns.

3.1 EQUIPMENT FOR THICK FILM PRODUCTION

A number of pieces of equipment that have been mentioned above, will now be briefly described.

A three-roll mill is used for making the ink. Powder and vehicle, in a proportion containing approximately 80% powder by weight, are first hand-mixed to a paste using a spatula. Altering the powder-vehicle ratio and/or the nature of the vehicle itself can change the final viscosity characteristics of the ink. The paste is then introduced into the large gap between the rear rollers of the three-roll mill (see Figure 1). The rotating rollers force the paste through the smaller gap between the front rollers and the material is then removed onto the closely fitting take-off blade. This process is repeated two or three times until the ink has the correct viscosity and is free from any lumps or agglomerates.

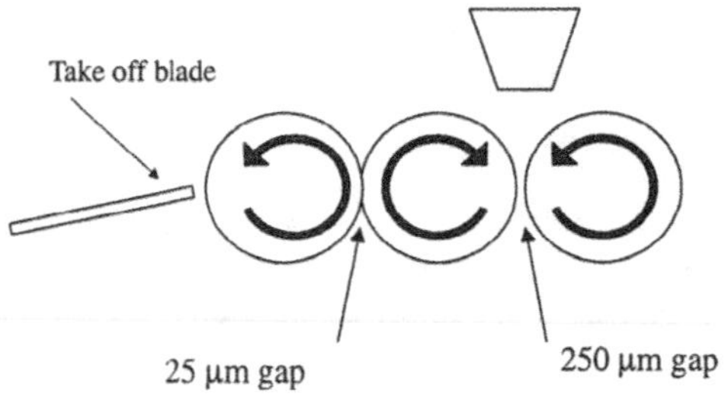

Figure 1 Schematic diagram of a three-roll mill.

Figure 2 shows a ball mill used for reducing particle size. It consists of two rotating cylinders that drive a cylindrical jar containing the powder, a liquid and some ceramic spheres. In this case, the spheres are zirconia to avoid any contamination to the YBCO. The process is timed to provide the appropriate degree of mixing and particle size reduction and can take many hours.

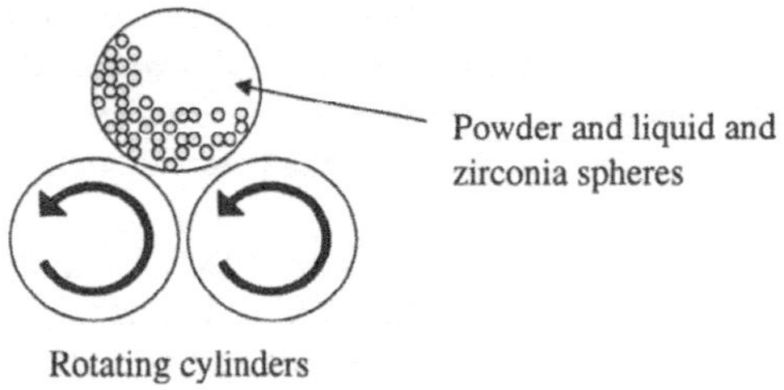

Figure 2 Schematic diagram of a ball mill.

Figure 3 shows a vibroenergy mill whose function is similar to the ball mill. However, instead of a rotation, a linear vibration is used to mix the powder, liquid and zirconia spheres. This is a much higher energy process resulting in faster particle comminution.

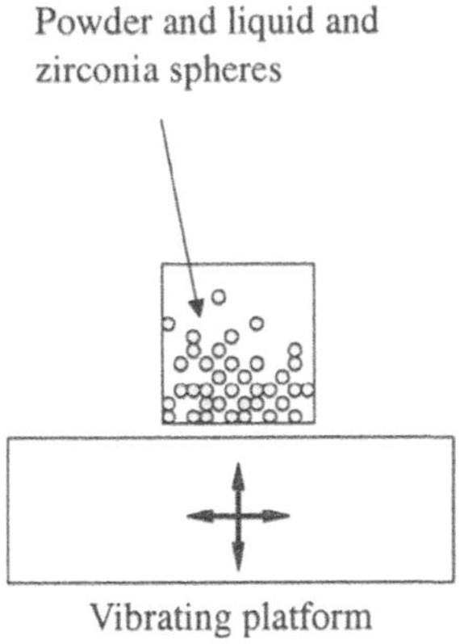

Figure 3 Schematic drawing of a vibroenergy mill.

Figure 4 shows a rotary evaporator. It consists of a glass bulb where the liquid with powder suspension is placed. A low heat is applied to the bulb via a water bath and the liquid is evaporated from the mixture, condensed in a water-cooled condenser in the neck of the tube, and collected in a secondary collecting bulb. The bulb containing the powder mixture is slowly rotated and operated at a low pressure thus reducing the boiling point of the liquid. Such apparatus allows the powder to be separated from the liquid in a very controlled manner without contaminants entering.

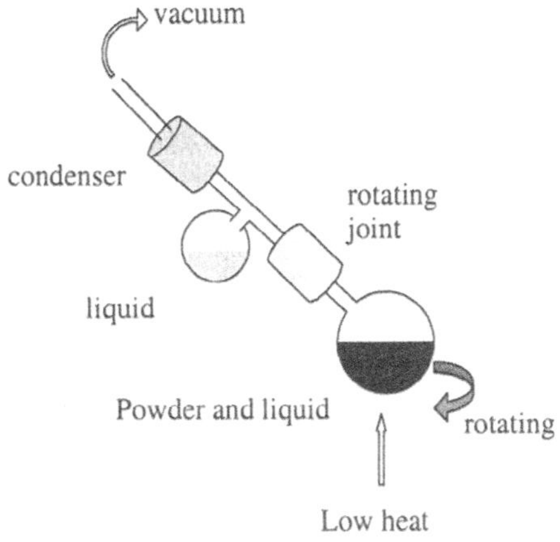

Figure 4 Schematic drawing of a rotary evaporator.

3.2 SINTERING

Sintering is the most crucial step in the production and optimisation of the properties of thick films. Once the particles of superconductor have been evenly deposited on the surface of the substrate, they must be encouraged to coalesce to form a uniform thick film. The first stage is simply the burn out of the organic components of the vehicle that is achieved by heating in air or oxygen at approximately 400°C. The second stage is more complex and is usually carried out in an oxygen atmosphere. Reference to the phase diagram of YBCO in Figure 5 [9,10,11] shows the existence of various phases as the temperature is increased. As the solid 123 is heated it reaches a peritectic temperature at about 1000°C. At this temperature the 123 separates into a solid phase and a liquid phase. The solid phase is 211 YBCO and the liquid is barium and copper rich. On cooling, the two phases react to produce the 123 phase. The presence of the liquid encourages grain growth and very large grains can be formed. In addition, the presence of 211 phase encourages the formation of dislocations which aid pinning and increase the critical current. Although YBCO can be sintered below the peritectic temperature, it has been found that both Rs and Jc are improved with sintering above the peritectic temperature. It is important to achieve the correct oxygen content in the YBCO phase for optimum superconducting properties. This is achieved by either slow cooling (1°Cmin^{-1}) from the sintering temperature in an oxygen atmosphere, or by a separate oxygen anneal at approximately 450°C for 12 hours.

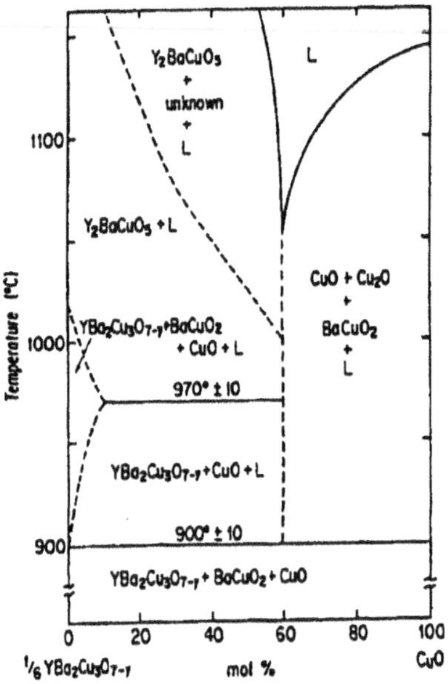

Figure 5 Phase diagram of YBCO

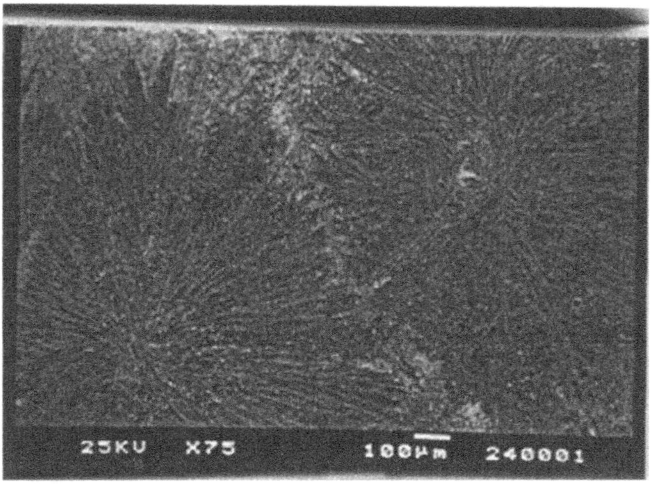

(a)

(b)

Figure 6 Microstructure of YBCO. The top picture (a) shows the large grain growth when the material is sintered above the peritectic temperature and the lower picture (b) shows the small grains when it is sintered below the peritectic temperature.

Figure 6 shows the difference in the microstructure of YBCO thick films resulting from sintering above and below the peritectic temperature. The large grain growth can be easily observed.

3.3 SUBSTRATES

The function of the substrate is to support the thick film. The following are important properties of the substrate;

1. Any reaction of the substrate with the film should not cause any degradation of the film. This usually means very little or no reaction with the substrate, although a small reaction can sometimes be helpful as it can cause good adhesion of the film to the substrate. Some of the substrates used for thin films, such as MgO or $LaAlO_3$, are unsuitable because of diffusion of species from the substrate to the film.
2. Very high temperatures are involved in the sintering of the films as described above. The substrate should be able to maintain its shape at these high temperatures.
3. From a microwave point of view, the loss tangent should be low if there are to be parts of the substrate that will be exposed to the electromagnetic field. The importance of the loss tangent varies greatly. For example, in a cylindrical cavity where none of the substrate is exposed, the loss tangent is immaterial.
4. The best dielectric constant depends greatly upon the type of device. It is immaterial for cavities, but becomes more important for antennas, particularly in their matching circuits. Although almost any dielectric constant can be tolerated, components may become too small if it is above about 50.
5. The substrate should be low cost.

Table 2 shows a number of substrate materials, commonly used for HTS thick film production, with their dielectric constants and loss tangents, together with the measurement frequency for these values. The measurements are made at room temperature. The dielectric constant does not change much as the temperature is reduced, however the loss tangent can be significantly reduced at cryogenic temperatures. The microwave measurements were made in a large cylindrical copper cavity with a dielectric puck of material placed inside. The puck is separated from the base of the cavity by a quartz spacer. Although there are quite a large number of suitable substrate materials, yttria stabilised zirconia (YSZ) is preferred for most applications.

Table 2 some dielectric substrates that are suitable for thick film YBCO fabrication.

Substrate Material	Dielectric Constant ε'	Dielectric Loss @ 300K - tan δ	Measurement frequency f
Y-stabilised zirconia [12]	25	10^{-3}	7.5GHz
YBa_2NbO_6 [12]	29	1.4×10^{-4}	4.9GHz
$NdBa_2NbO_6$ [12]	13.3	3×10^{-4}	8.6GHz
$BaZrO_3$ (90% of full density)[12]	12	2.5×10^{-4}	8GHz
$GdBa_2HfO_{5.5}$ [13]	20	5×10^{-3}	10MHz
$(Pr,Sm,Gd)Ba_2SbO_6$ [14]	12-16	$8\text{-}25 \times 10^{-4}$	10MHz
$YBa_2SnO_{5.5}$ [15]	10	5×10^{-2}	10MHz
$PrBa_2NbO_6$ [16]	15	2×10^{-4}	10MHz
$SmBa_2NbO_6$ [16]	9	1×10^{-4}	10MHz
$EuBa_2NbO_6$ [16]	11	2×10^{-4}	10MHz

Some work using metals as substrates has been successful, for example silver has been shown not to show any deleterious reaction with YBCO (or BSCCO and TBCCO), although the melting point of silver (~960°C) limits the maximum processing temperature which can be used. There have not been many measurements of the surface resistance of HTS on metals [17] but it has been found to be significantly greater than the values obtained for thick films on YSZ. For microwave devices the metal must be shielded by a complete layer of superconductor to avoid deterioration of the device performance.

Figure 7 shows a cross section of a YBCO film on an yttria stabilised zirconia substrate which has been processed above the peritectic temperature (i.e. film as shown in Figure 6(a)). The complex interface is clearly visible. Immediately above the YSZ substrate is a reaction layer containing barium zirconate with copper substitutions and copper oxide. Above this layer the YBCO tends towards single phase 123, but cracks can be observed near the surface. On the surface is the best YBCO occurs and contains some c axis texturing, which has the low surface resistance. It should be noted that microwave fields should not penetrate the lower surface, as the reaction layer is lossy. This rules out microstrip devices using YSZ as a substrate, although invented microstrip is a possibility.

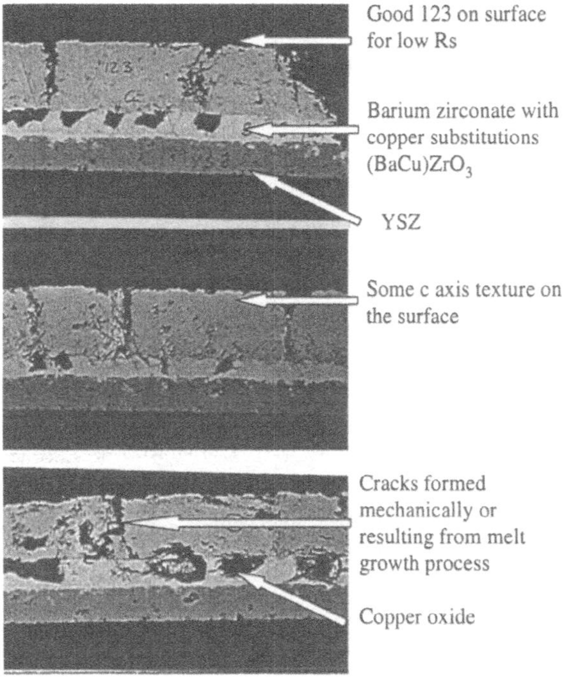

Figure 7 Three examples of a typical cross section of the substrate-YBCO interface for melt processed thick films. Each picture demonstrates different features.

3.4 CURRENT FLOW

The current flow in YBCO melt processed thick films has been studied in detail [18,19,20]. It has been found that the current carrying capabilities of YBCO thick films is strongly dependent upon the microstructure. The critical current is thus very dependent upon the processing conditions. The most important factor that limits the critical current to values well below the critical current of thin films is the grain boundaries where 'weak links' occur. These weak links have a much lower critical current than the grains themselves, limiting the overall critical current of the film. For the melt processed thick films, with the 'hub and spoke' microstructure of Figure 6(a), current is channelled through the spokes to the hub and out along the spokes again [21]. Experiments have been done by removing the hub and a change in critical current of the films occurs. Figure 8 shows the critical current as a function of applied magnetic field for these two situations. Note the large difference in the critical current between the thick film with and without the hubs of the grains.

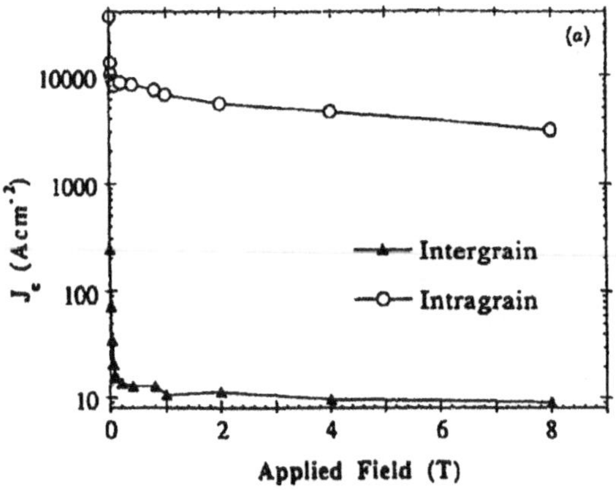

Figure 8 Transport critical current as a function of applied magnetic field at 4.2K. Intergrain and intragrain currents have been separated due to hub removal.

Since doping is known to alter the microstructure, it is therefore expected that doping will also alter the critical current. Platinium doping has been used successfully in bulk materials to increase the critical current and has also been shown to improve the properties of YBCO thick films [22]. Doping YBCO thick films with 0.1 wt% platinum powder caused an increase in the current carrying capacity of the material from J_c=1800Acm^{-2} to J_c=5000Acm^{-2}. However, the Rs values of these doped films are usually inferior to undoped materials.

4 Surface impedance of HTS thick films

The variation in the surface impedance of superconducting thick films is enormous. The lowest surface resistance has been obtained for the melt processed thick films described above. The best measurement of the surface resistance at 7.4GHz and 77K is 1.5mΩ (corresponding to 2.6mΩ at 10GHz), these films have a Jc of about 2000Acm^{-2}. This surface resistance is significantly higher than for thin films. Figure 9 shows the frequency dependence of the surface resistance for thin films, thick films, bulk polycrystalline materials and single crystals at 77K compared with copper at 77K and Niobium at the same reduced temperature (7.7K). The grey bands represent variation in reported values. The figure distinguishes between bulk polycrystalline and thick films, the latter generally having a lower surface resistance. Also, single crystals and thin films are distinguished the former being at the lowest edges of the grey band.

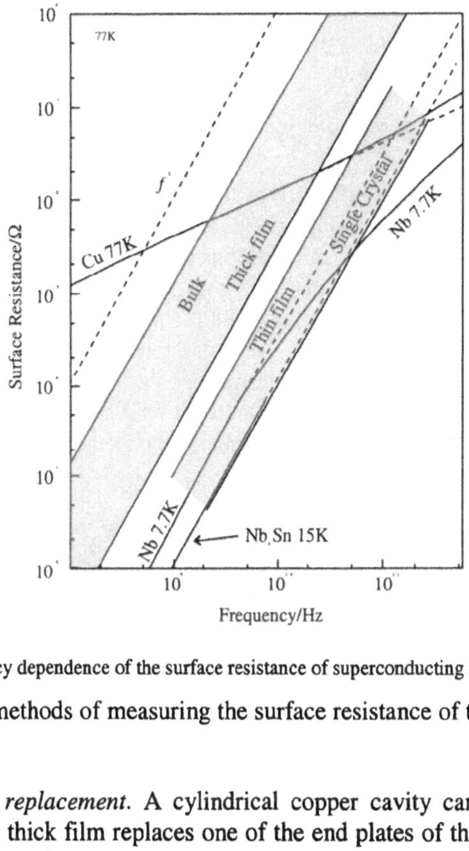

Figure 9 Frequency dependence of the surface resistance of superconducting materials at 77K.

There are many methods of measuring the surface resistance of thick films. Four are listed below:

1. *Cavity end wall replacement.* A cylindrical copper cavity can be used, where a superconducting thick film replaces one of the end plates of the cavity. The whole cavity is cooled to the measurement temperature. A measurement of the quality factor of the cavity with a copper end plate and then a superconducting end plate provides a measurement of the surface resistance of the HTS surface resistance. The problem with this method is that if the surface resistance of the superconductor is much lower than the copper, the sensitivity is greatly reduced.
2. *Complete HTS cavity.* Here a whole cavity can be produced with HTS, the measurement of the Q provides a direct measurement of the surface resistance. In this case the sensitivity is good and high Q values can be measured. However, care

must be taken to ensure good electrical connection between end plates and the cylindrical walls of the cavity. Also, if a number of films are being compared it is rather laborious to make whole cavities for each one.

3. *Microstrip resonator.* Placing two rectangular films facing each other with a small gap between them can make a microstrip resonator. Problems include the mechanical construction of such a device, two films are required and care must be taken to avoid radiation losses.
4. *Dielectric resonator.* A dielectric cylindrical puck can be placed on the HTS. The puck must have a low loss tangent giving it a high intrinsic Q. The Q of the puck is then reduced by losses in the HTS film. The surface resistance can be deduced from the Q of the structure by calculation and/or comparisons with materials of known surface resistance. Although this method is not easy to set up, it is probably the most accurate and convenient.

4.1 INFLUENCE OF MICROSTRUCTURE ON RS

It is possible to measure the influence of the microstructure on the surface resistance directly using the dielectric resonator technique described above [23]. If a high permitivity material is used, the puck becomes small. For example a rutile resonator operating in the TE_{011} mode at about 9GHz is only 3.4 mm across. If spacial variation of the granular structure is more than is 3.4mm, then the puck can be placed at an appropriate, interesting position of the HTS structure. For example, Figure 10 shows a sample of melt processed thick film, the rings show the position and size of the small rutile measurement puck. It can be seen that the lowest surface resistance is measured at the centre of the grain, whereas on the boundaries between grains, the surface resistance is significantly higher.

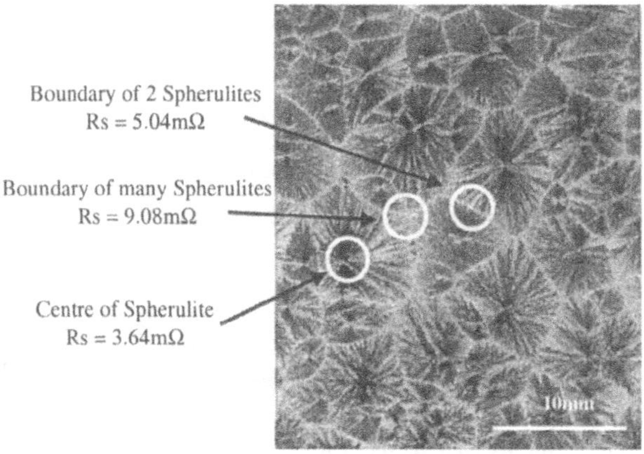

Figure 10 Sample of melt processed thick film showing the positions of a rutile puck and the corresponding values of surface resistance that were measured.

4.2 FREQUENCY DEPENDENCE OF R_S

Fundamental considerations in the two fluid model (see previous chapter) show that the surface resistance is proportional to the square of the frequency and the cube of the penetration depth, and is given by,

$$R_s = \frac{\omega^2 \mu^2 \sigma_1 \lambda^3}{2}$$

For the moment, assume that R_s is proportional to f^n (n=2 for the two fluid model) For single crystals and thin films n has been shown experimentally to be 2. However, for thick films, there is still some controversy. For very good films n approaches 2, while for poorer films n is less then 2, with values of 1.5 reported. However, these measurements are very difficult and care must be taken to retain accuracy over a wide frequency range. It may not be surprising that the poorer films have values of n less than 2 because of the non-superconducting inclusions present in the thick films.

It should also be noted that the penetration depth is much larger than for thin films or single crystals and can have values of several microns at 77K. This is due to the normal material inclusions and the granular nature of the films with large number of weak links.

4.3 POWER DEPENDENCE OF SURFACE RESISTANCE

The sample used in preparing Figure 10 has been used to determine the microwave power dependence of R_s for the thick film. As can be seen from Figure 11, the different regions of the microstructure have very different response to the increase in the microwave power. Where there are many grain boundaries, the low power surface resistance increases rapidly with the RF field. However, for the measurements of surface resistance made over the hub centre, not only are the low power surface resistance values lower, but the increase with increasing field is considerably less. The surface resistances of copper and a YBCO film are also shown, both have low power dependence.

4.4 MODELLING THE SURFACE RESISTANCE OF THICK FILMS

As discussed above, thick films are far from perfect, as they are composed of grains, inclusions of material other then superconductor, and voids. As the thick film materials improve, there are less voids and non-superconducting inclusions, but the presence of grains is inherent in the film production methods. This is a very difficult material to model accurately but some attempts have been made.

One model is to consider the superconducting material as an array of coupled superconducting grains [24]. The coupling between the grains is characterised by a reduced critical current and can be modelled by a resistively shunted Josephson junction, as depicted in Figure 12. The following discussion considers this problem with the intention of emphasizing the main mechanisms mathematically. However, because of the complex nature of the material a comparison with experimental measurements is strictly limited. However, the calculation does improve our understanding of the problem.

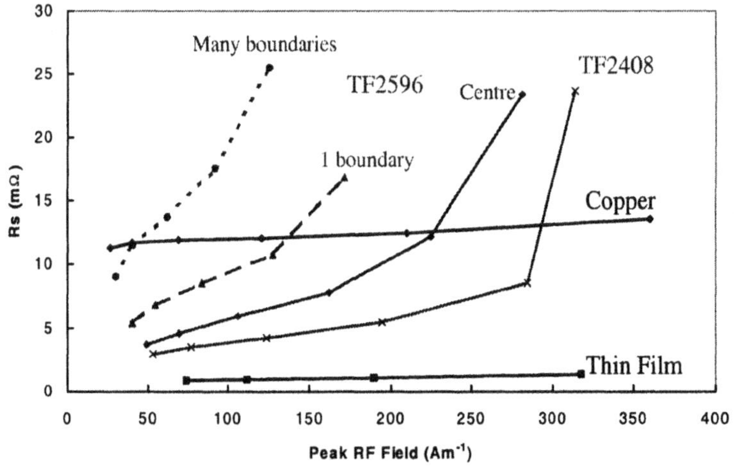

Figure 11 Power dependence measurements and the influence of microstructure for a YBCO thick film measured at 77K and 9.87GHz [23].

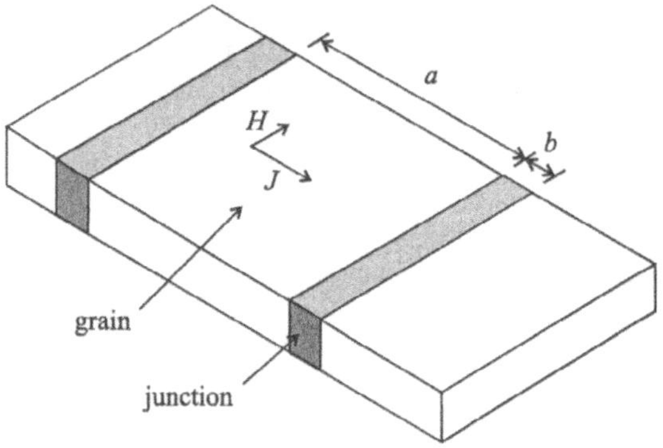

Figure 12 Schematic drawing of three superconducting grains connected by two weak links.

Consider the current through a Josephson junction shunted by a material with conductivity σ_j. The current density through the junction is given by [25,26,27]

$$J(t) = J_0 \sin(\delta(t)) + \sigma_j E(t), \qquad (1)$$

where the electric field is given by

$$E(t) = \frac{\Phi_0}{2\pi b_{eff}} \frac{\partial \delta}{\partial t} . \tag{2}$$

Here Φ_0 is the flux quantum

$$\Phi_0 = \frac{h}{2e} = 2.0678 \times 10^{-15} \text{ Wb} . \tag{3}$$

The phase difference, $\delta(t)$ of the macroscopic wave function across the junction

$$\delta(t) = \delta_0 + \frac{2\pi}{\Phi_0} Vt . \tag{4}$$

These are well know equations and the latter expression leads to an oscillating current at frequency $\omega = 2eV / \hbar$ for an applied constant voltage V across the junction. It should also be noted that in (2) the distance b_{eff} is given by $b_{eff}=b+2\lambda_g$, where b is the barrier thickness of the junction and λ_g is the penetration depth in the material on either surface of the junction. These equations assume that the microwave current density is small compared with the critical current density of the junction.

Now differentiate (1) with respect to time and then substitute into (2), resulting in

$$\frac{\partial J}{\partial t} = \frac{1}{l_j} E + \sigma_j E , \tag{5}$$

here l_j being the Josephson inductivity given by

$$l_j = \frac{\Phi_0}{2\pi b_{eff} J_0 \cos\delta} . \tag{6}$$

Given a sinusoidal time dependence in (5), the effective conductivity of a shunted junction can be written as

$$\sigma_{sj} = \frac{1}{j\omega l_j} + \sigma_j . \tag{7}$$

Simple effective medium theory can be used to consider the effect of grains of superconducting material in series with the junctions. Assume the current is forced to go through the junctions and the grains in series, then their effective resistivities can be added with appropriate weighting for the size of the grains and junctions.

An expression for the complex conductivity for such a medium then results in

$$\sigma = \frac{\sigma_g(1 + j\omega l_j \sigma_j)(a + b)}{a(1 + j\omega l_j \sigma_j) + j\omega l_j \sigma_g b}. \tag{8}$$

Here σ_g is the complex conductivity of the superconducting grain, and a and b represent the dimensions of the grain and junction as depicted in (2).

Now assume the conductance of the junction is dominated by the inductance ($\omega l_j >> 1/\sigma_j$) and that the grain is dominated by superconducting carriers ($\sigma_{1g} << \sigma_{2g}$). Then the complex conductivity of the material with small junctions ($b < a$) is approximately given by[24]

$$\sigma = \left[j\omega\mu_0\lambda_{eff}^2 + \omega^2\mu_0\lambda_j^2 \frac{\hbar}{2eI_c R} \right]^{-1}. \tag{9}$$

Here I_c is the critical current of the junction and R is its shunt resistance. The effective penetration depth is given by[24]

$$\lambda_{eff}^2 = \lambda_g^2 + \lambda_j^2, \tag{10}$$

where λ_g is the London penetration depth of the grain, and λ_j is the effective continuum penetration depth due to the network of grain boundary junctions alone. This is similar to the Josephson penetration depth but it is not the same. In this context it is given by

$$\lambda_j^2 = \frac{l_j}{\mu}\frac{b}{a} = \frac{\hbar}{a2eJ_0\mu_0}. \tag{11}$$

Assuming the above approximations, (9) and the two fluid model give the surface impedance of the granular material

$$Z_s = \omega\mu_0\lambda_{eff}\left[\frac{1}{2}\left(\frac{\lambda_j}{\lambda_{eff}}\right)^2 \frac{\hbar\omega}{2eI_c R} + j \right] \tag{12}$$

The surface impedance is dominated by the grain boundaries providing the size of the grains is small or if the junction critical current is small, making λ_{eff} approach λ_j. When the critical current is high then λ_{eff} approaches λ_g and the surface impedance is dominated by the grains. The magnetic field dependence of the conductivity and surface impedance of this composite structure can be studied by noting that there is a decrease in the critical current with increasing field. This section has emphasized a mechanism for the surface resistance of a granular film, and has shown mathematically some of the principle mechanisms. However, in practice the material is much more complex with defects and inclusion on many length scales. Comparison of the above result with practical results is therefore strictly limited.

5 Example thick film microwave circuits

This section is intended to give the reader some idea of the different shapes of thick film objects that have been produced, and some idea of what is achievable in terms of device performance. All the examples are of microwave components and a brief description of the performance is given. The following thick film microwave components have been demonstrated:

1. Thick film wires used for coaxial cable [28].
2. Wires in the shape of a helix for helical resonators [29].
3. Wires for antennas [30].
4. Thick film tracks printed on a zirconia cylinder in the shape of a helix [31].
5. Cylinders used for resonators [32,33].
6. Antennas of various shapes that have been patterned onto various planar substrates [34,35].
7. Planar microstrip and coplanar resonators have been patterned onto various substrates [36,37].
8. Thick film coaxial resonators [38]

5.1 SUPERCONDUCTING ANTENNAS

The first high temperature superconducting microwave component was an antenna built from bulk extruded wire in 1988 [39]. Since then most antennas have been made of thick films. Generally low frequency antennas or antenna arrays occupy a large area. For example HTS antennas operating near 1GHz have been patterned on thick film substrates a few centimetres square. An example with a meandered matching circuit is shown in Figure 13. This antenna is much smaller than a conventional antenna at the same frequency. If it were made out of conventional conductors the efficiency and therefore the gain of the antenna would be significantly worse. Although the gain increases over normal metal antennas of a similar size the bandwidth is reduced and this has been seen as a problem. However there are certain techniques that can be applied in order to increase the bandwidth [40]. Several small antennas similar to the one in Figure 13 have shown gain improvements of about 6db over copper counterparts and they are reported in reference [31].

Itoh [41] has examined a HTS electrically small ($\lambda/45$) helical antenna. The helical radiator works as an electric dipole and is joined directly to a ¼ wave matching circuit. This is made from parallel transmission lines in thick film HTS. The design was intended to improve radiation efficiency and to increase the bandwidth. This was achieved with about a 6dB increase in gain at a temperature of operation of 81K, only 11K below Tc.

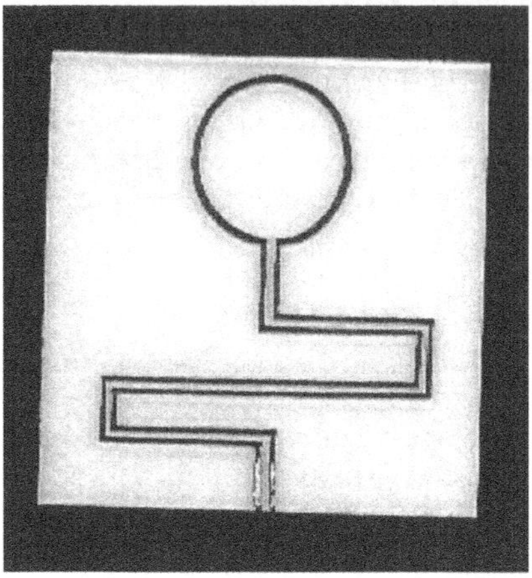

Figure 13 Thick film small antenna and matching circuit

Another application of thick films to antenna technology is in the area of superdirectional antenna arrays. These antennas have a high directivity while occupying a volume much less than a wavelength. This goes against intuition where it is well know from optics that as aperture gets smaller the beam diverges. Here, the opposite is occurring, the aperture is shrinking and a narrow beam is maintained. The reason is that the phase of the signal in the superdirectional antenna array can be controlled. From a theoretical point of view, any beam width can be obtained from any aperture even if it shrinks to a smaller and smaller size. The difficult part in the design of superdirectional antenna arrays is the layout of elements to control the phase in an appropriate manner. The reason why superconductors are used is that as antennas shrink in size their efficiency is reduced. Using superconductors improve this efficiency but at the expense of bandwidth. Figure 14 shows a four-element superdirectional antenna array, the one on the left is the superconducting one while the one on the right is a silver one for comparison [42,43]. Sixteen element arrays have also been demonstrated [44]. Superconductor is printed on both sides of the substrate and the bottom pattern can just be seen in the picture. The antennas are 100mm square with an operating frequency of about 1GHz so that the whole array is about one quarter of a free space wavelength. This is in contrast to conventional arrays where the elements are spaced by about a half wavelength. The superconducting antenna on the left shows an increase of 4.2dB in gain over the silver array on the right.

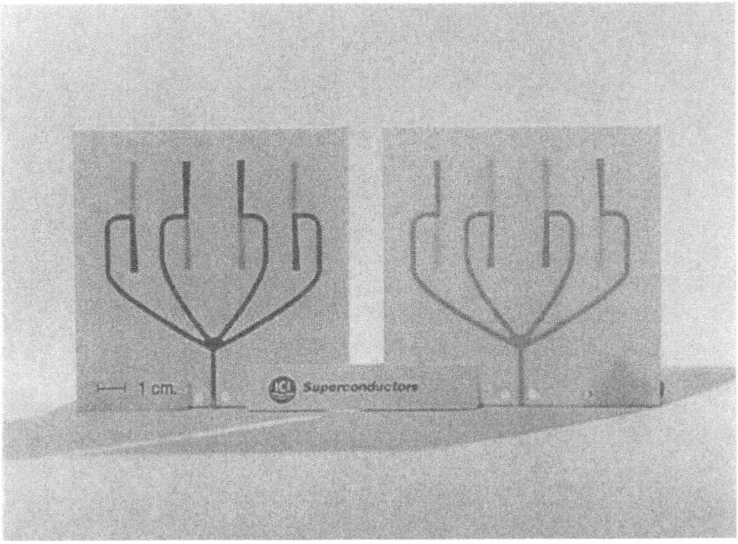

Figure 14 Thick film four element superdirectional antenna array, with the HTS version on the left and normal conducting on the right.

5.2 CYLINDRICAL CAVITY RESONATORS

Thick films can be used to coat the interior surfaces of cylindrical cavities, and at 10, 7.5 and 5.6GHz (TE_{011} mode), unloaded Q values of $2x10^5$, $5x10^5$ and $7.6x10^5$ have been measured at 77K. At 64K the unloaded Q is 10^6 at 7.5GHz [33]. An example of the parts that go into such a cavity is shown in Figure 15. The zirconia cylinder is internally coated with HTS and there are two end plates also coated on their inner surfaces. In one end plate there are two holes through which coaxial cables are inserted. These coaxial cables have small loop antennas on their end, which couple energy into and out of the cavity. The completed cavity is shown in Figure 16. This particular cavity operates at about 5.6GHz in the TE_{011} mode.

However, these cavities are far too large for applications at low frequency, for example at 900MHz such a cavity would require both diameter and length to be greater than 400mm and cooling a series of such large resonators would be difficult. One alternative geometry is to use HTS coaxial elements with either a normal metal or HTS outer shield. Some of these structures and other, smaller, quasi-lumped element devices have been manufactured using YBCO thick films on polycrystalline zirconia substrates. These coaxial resonators allow structures to be made electrically small and when implemented using HTS thick films, with dimensions of width 90mm and length 40mm, give Q values of over 86,000 at 77K and 900MHz [38]. Such Q values are sufficient for the construction of filters with low insertion loss and good rejection characteristics.

Figure 15 Parts of a thick film cavity

Figure 16 Completed thick film cavity (5.6GHz).

6 Examples of thick film systems

Two sub-systems that include superconducting thick film components will now be described.

6.1 FRONT END FILTER FOR MOBILE COMMUNICATIONS

One key area for applications of HTS thick films is in cellular base station front-end filters. There are number of companies looking into both thin film and thick film implementation of such filters. Products are now available in both technologies. For the thick film case, resonators based on the lumped element insert described in the previous section have been used. Superconducting filters have three advantages over conventional filters. They are smaller in size, have increased selectivity at the band edges, and have an increased sensitivity due to their lower insertion loss. This is all due to an increase in the quality factor of the resonators, which is up to 100,000 for the thick film cavities used at 900MHz.

An example of a frequency response for mobile communication applications is shown in Figure 17.

6.2 CRYOGENIC RECEIVER MODULE

An 11GHz cryogenic receiver has been demonstrated and is shown diagrammatically in Figure 18. The receiver consists of a thin film front-end filter followed by a silicon diode mixer. A thick film cylindrical cavity is configured with a HEMT transistor amplifier to function as an oscillator circuit to supply the local signal for the mixer for the down conversion to 1GHz. The oscillator power is 14dBm. The performance of the oscillator is shown in comparison with oscillators made using copper cavities in Table 3

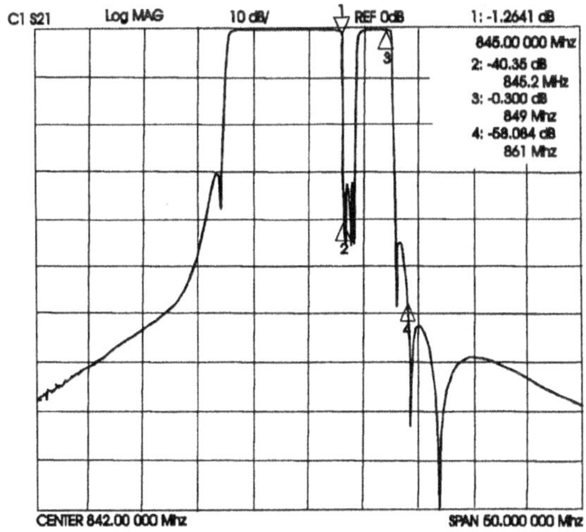

Figure 17 Performance of the thick-film YBCO filter manufactured by Illinois Superconductor Corporation. (ISC model Q45K-B3A). It is a 10 pole elliptic response bandpass filter with a 6-pole quasi-elliptic notch.

Table 3 Phase noise for HTS thick film oscillator [33].

Oscillator system	Measurement temperature (K)	Phase noise ($dBcHz^{-1}$)	
		1 kHz offset	10 kHz offset
Cu Cavity	300	-90	-120
Cu Cavity	77	-95	-124
HTS Cavity	77	-108	-135

The measurements of phase noise have been carried out on a 7.5GHz TE_{011} cavity in which both the barrel and endplates were coated with YBCO thick film. It should be noted that the measured phase noise of the HTS cavity is limited by the noise floor of the measurement system. It is likely that the phase noise of HTS thick films is lower but more precise measurements involving an alternative experimental procedure will be required.

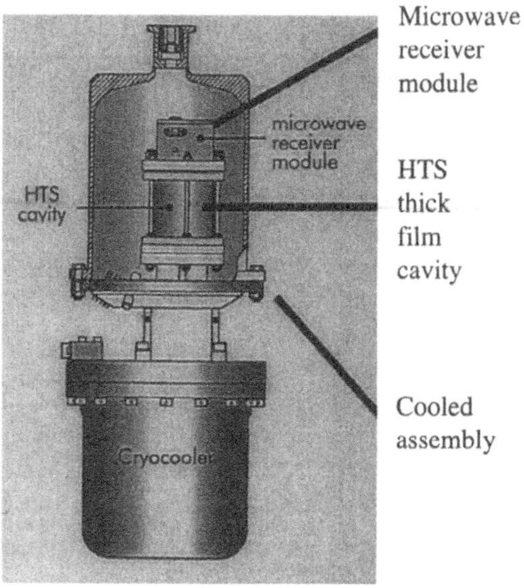

Figure 18 Cryogenic down converter.

A Stirling cryocooler is used to cool the sub-system to 60K.

7 Conclusion

High temperature superconducting thick film technology is now well developed. Good quality thick films are available with low surface resistance and good power handling capabilities. Their use has been demonstrated in many microwave components and they are being exploited commercially in the communications industry. The choice between thin films and thick films for a given application depends upon many factors. Although thin films have a significantly lower surface resistance and better power handling, this can be offset by the availability of larger area thick film components and exploiting the inherent three dimensional nature of the materials. The films have good prospects and it is expected that new, novel microwave components will emerge using their unique properties.

8 References

1 Alford, N. McN., Button, T. W., and Birchall J. D. (1990) Processing, Properties and Devices in High-Tc Superconductors, *Supercond Sci Technol* **1**, 1-7.
2 Alford, N McN., Button, T. W., Adams, M. J., Hedges, S., Nicholson, B., and Phillips, W. A. (1991) Low Surface Resistance $YBa_2Cu_3O_x$ Melt-Processed Thick Films, *Nature* **349,** 680-683.
3 Alford, N. McN., and Button, T. W. (1991) Surface Resistance of Thick Film $YBa_2Cu_3O_x$, *Adv. Mater.* **30,** 318-320.
4 Button, T. W., Alford, N. McN., Wellhofer, F., Shields, T. C., Abell, J. S, and Day M. (1991) Processing and Properties of High Tc Films, *IEEE Trans. Magn.* **27,** 1434-1437.
5 Bailey. A., Alvarez. G., Russell. G. J., and Taylor K. N. R. (1991) High critical current density thick films of (Y-Ba-Cu-O):Ag obtained by melt processing, *Bull. Mater. Sci.* **14,**111-116
6 He, Q., Christen, D. K., Klabunde, C. E., Traczyk, J. E., Lay, K.W., Paranthaman, M., Thompson, J. R., Goyal, A., Padraza, A. J., and Kroeger D. M. (1995) High-current superconducting $Tl_1Ba_2Ca_2Cu_3O_y$ thick films on polycrystalline Ag by spin coating, *Appl Phys Lett* **67,** 294-296.
7 Su, L.Y., Grovenor, C. R. M., Goringe, M. J., Dewhurst, C. D., Cardwell, D. A., Jenkins, R., Jones, H. (1994) Fabrication and properties of TBCCO films via air-atomising spray deposition of thallium-free precursors, *Physica C* **229,** 1270-78.
8 Holesinger, T. G., Phillips, D. S., Coulter, J. Y., Willis, J. O., Peterson, D. E. (1995) Isothermal melt processing of Bi-2212 thick films, *Physica C* **243,** no 1-2 93-102.
9 Oka, K., Nakane, K., Ito, M., Saito, M., and Unoki, H. (1988) Phase-equilibrium diagram in the ternary system Y_2O_3-BaO-CuO, *Jpn. J. Appl. Phys. lett* **27,** L1065-L1067
10 Maeda, M., Kadoi, M., and Ikeda, T. (1989) The phase diagram of the $YO_{1.5}$-BaO-CuO ternary system and growth of $YBa_2Cu_3O_7$ single crystals, *Jpn. J. Appl. Phys* **28,** no 8 1417-1420
11 Bourdillon, A., and Tan Bourdillon, N. X. (1994) *High Temperature Superconductors: Processing and Science,* Academic Press Harcourt Brace Jovanovich., San Diego
12 Alford, N. McN., Penn, S. J., and Button T. W. (1997) High temperature superconducting thick films, *Supercond. Sci. Technol.* **10,** 169-185.
13 Thomas, J. K., Koshy, J., Kurian, J. and Damodaran, A. D. (1995) Superconducting YBCO and YBCO-Ag thick films ($T_{c(0)}$=92 K) by dip coating on $GdBa_2HfO_{5.5}$, a new perovskite ceramic substrate, *Supercond. Sci. Technol.* **8,** 525-528.
14 Kurian, J., Koshy, J. K., Warir, P. R. S., Yadava, Y. P., and Damodaran A. D. J. (1995) Synthesis and characterization of rare-earth barium antimonates, a new group of complex perovskites suitable as substrates for $YBa2Cu_3O_{7-\delta}$ films, *Solid State Chem.* **116,** 193-198.
15 Koshy, J., Kumar, K. S., Kurian, J., Yadava, Y., and Damodaran, A. D. (1994) *Bull. Mater. Sci.* **17,** 577-584.
16 Koshy, J., Thomas, J. K., Kurian, J., Yadava, Y. P., and Damodaran A. D. (1993) Superconducting $YBa_2Cu_3O_{7-\delta}$ thick film ($T_{c(0)}$=92 K) on $SmBa_2NbO$: a newly developed perovskite ceramic substrate, *Physica C* **215,** 209-212.
17Ezura, E., Asano, K., Hayano, H., Hosoyama, K., Inagaki, S., Isagawa, S., Kabasawa, M., Kojima, Y., Kosuge, S., Mitsunobu, S., Momose, T., Nakada, K., Nakanishi, H., Shimbo, Y., Shishido, T., Tachikawa, K., Takahashi, T., Yoshihara, K. (1993) Microwave surface resistance of plasma-sprayed YBaCuO thick films on large-area metallic substrates, *Japan J. Appl. Phys. Part 1* **32,** 3435-3441.

18Cardwell, D. A., Jones, A. R., Ingle, N. J. C., Campbell, A. M., Button, T. W., Alford, N. McN., Wellhofer, F. and Abell, J.S. (1994) Investigation of critical state in melt processed $YBa_2Cu_3O_{7-d}$ thick films, *Cryogenics* **34,** 671-677.

19 Ingle, N. J. C., Cardwell, D. A., Jones, A. R., Wellhofer, F., Button, T. W. (1995) The Effect of Thickness on the Magnetic Properties of Melt Processed YBCO Thick Films,*Supercond Sci Technol* **8,** 282-290.

20 Dewhurst, C. D., Cardwell, D. A., and Alford N. McN. (1995) Investigation of the temperature dependence of the critical state in melt processed $YBa_2Cu_3O_7$ thick films, *Appl Phys Lett* **77,** 2067-2072.

21 Jones, A. R., Cardwell, D. A., Ingle, N. J. C., Ashworth, S. P., Campbell, A. M., Alford, N. McN., Button, T. W., Wellhofer, F. and Abell J. S. (1994) Correlation of transport and magnetic critical currents in melt-processed $YBa_2Cu_3O_{7-\delta}$ thick films *J. Appl Phys* **76** 1720-1725

22 Langhorn J., Bi Y. J. and Abell J. S. (1996) Platinum group metals as flux pinning additions in screen printed superconducting $YBa_2Cu_3O_7$ thick films *Physica C* **271** 164-170

23 Smith P. A., Bakar M. A., Porch A., and Button T. W. (1999) Microstructure and RF property correlatrion in HTS films. *IEEE Trans.on Applied Superconductivity* **9** 2175-2178

24 Hylton T. L., Kapitulnik A., Beasley M. R., Carini J. P., Drabeck L., and Grüner G. (1988) Weakly coupled grain model of high-frequency losses in high T_c superconducting thin films *Appl. Phys. Lett.* **53** 1343-1345

25 Franson J. D. and Mercereau J. E. (1976) Microwave impedance of superconducting weak links *J. Appl. Phys.* **47** 3261

26 Portis A. M. and Cooke D. W. (1992) Superconducting stripline model of intergranular flux penetration *Superconductor Science and Technology* **5** S395-S398

27 Portis A. M. (1992) *Electrodynamics of high temperature superconductors - Lecture Notes in Physics 48* World Scientific, Singapore

28 Woodall P., Lancaster M. J., Maclean T. S. M., Gough C. E., and Alford N. McN. (1991) Measurements of the surface resistance of $YBa_2Cu_3O_{7-\delta}$ by the use of a coaxial resonator. *IEEE Trans. on Magnetics.* **27** 1264-1267

29 Peterson G. E., Stawicki R. P. and Alford N. McN. (1989) Helical resonators containing high T_c ceramic superconductors *Appl. Phys. Lett.* **55** 1798-1800

30 Khamas S. K., Mehler M. J., Maclean T. S. M., Gough C. E., Alford N. McN. and Harmer M. A. (1988) High-T_c superconducting short dipole antenna *Electronics letters* **24** 460-461.

31 Lancaster M. J., Maclean T. S. M., Wu Z., Porch A., Woodall P., and Alford N. McN. (1992) Superconducting microwave resonators. *IEE Proc. H* **139** 149-156

32 Button T. W. and Alford, N. McN. (1992) High Q $YBa_2Cu_3O_x$ cavities, *Appl Phys Lett* **60,** 1378-1380.

33 Button, T. W. Smith, P. A., Alford, N. McN., Greed, R. B., Adams, M. J., and Nicholson, B. F. (1997) HTS Cavities for Low Phase Noise Oscillator Applications, *IEEE Transactions on Applied Superconductivity* **7,** 3064-3066.

34 Lancaster, M. J., Wu, Z., Maclean, T. S. M.,and Huang Y. (1992) Supercooled and superconducting small-loop and dipole antennas, *IEE Proc H* **139,** 264-270.

35 Ivrissimtzis, L. P., Lancaster, M. J., Maclean, T. S. T., and Alford, N. McN. (1994) On the design and performance of electrically small printed thick film $YBa_2Cu_3O_{7-x}$ antennas, *IEEE Trans. on Applied Superconductivity* **4,** 33-40.

36 Hosking, M. W., Proykova, Y. G., Hewitt, A., Tonkin, B. A., Alford, N. McN., Button, T. W., and Brouard K. C. (1993) Microwave Attenuation Factors and Surface Resistance of Thick Film Superconducting YBCO Transmission Line on Zirconia, in H C Freyhardt(eds.) *Applied Superconductivity 1993,* DGM Informationsgesellschaft mbH, Adenanerallee, Germany 1093-1096.

37 Hosking, M., Proykova, Y. G., Hewitt, A., Alford, N. McN., and Button, T. W. (1995) A Superconducting and Integrated Microwave Transmitter and Receiver in Microstrip Using Thick Film YBCO, *IEEE Trans Applied Superconductivity* **5,** 2287-2290.

38 Smith, P. A., Button, T. W., Holmes, J., Dolman, G., Meggs, C., Remillard, S. K., and Hodge, J. D.(1997) YBCO Thick Films for High Q Resonators, *IEEE Transactions on Applied Superconductivity* **7,** 1763-1765.

39 Khamas, S. K., Mehler, M., Maclean, T. S. M., Gough, C., Alford, N. McN., Harmer, M. (1988) *Electron Lett.* **24,** 460-461.

40 Lancaster, M. J., Wang, H. Y., Hong, J. S. (1999) Thin film HTS planar antennas, *IEEE Trans. on Applied Superconductivity* **8,** 168-177.

41 Itoh, K., Ishii, O., Koshimoto, Y., and Cho, K. (1992) High-Temperature Superconducting Small Helical Antenna, *IEICE Trans. Electron* **E75-C,** 246-250.

42 Ivrissimtzis, L. P., Lancaster, M. J., and Alford, N. McN. (1995) A supergain YBCO Antenna Array with integrated Feed and Balun, *IEEE Trans. on Applied Superconductivity* **5,** 3199-3202.
43 Ivrissimtzis, L. P., Lancaster, M. J., Maclean, T. S. M., and Alford, N. McN. (1994) High Gain Series-Fed Printed Arrays Made of High Temperature Superconductors, *IEEE trans. Antennas and Propagation* **42,** 1419-1429.
44 Ivrissimtzis, L. P., Lancaster, M. J., Maclean, T. S. M., and Alford, N. McN. (1994) High gain printed dipole array made of thick film high Tc superconducting material, *Electronics Letters* **30,** 92-93.

NONLINEAR BEHAVIOR OF SUPERCONDUCTING DEVICES

DANIEL E. OATES
MIT Lincoln Laboratory
244 Wood St.
Lexington MA 02420-9108
and
MIT Department of Physics
Cambridge MA 02139
U.S.A.

Abstract

Nonlinear effects in superconducting devices are an important limitation to applications of these devices. The phenomenology of the nonlinear behavior is presented. Methods of measurements and models of power dependence, intermodulation distortion, and harmonic generation are discussed. Currently, microscopic models do not explain all of the measurements. Measuring and modeling of the microwave properties of grain boundaries and grain-boundary Josephson junctions also are presented.

1. Introduction

Nonlinear surface impedance was noticed quite early in the development of the high-T_c materials. It remains still one of the most important limitations of the applications of these materials at microwave frequencies. It also is incompletely understood. For these reasons it is an interesting and fruitful area of research. The nonlinear surface impedance $Z_S(H_{\rm rf})$ is the dependence of Z_S on either the rf magnetic field $H_{\rm rf}$ or current $J_{\rm rf}$. This is also called power dependence because it leads to changes in the device response as the power through it changes. The surface impedance is

$$Z_S(H_{\rm rf}) = R_S(H_{\rm rf}) + iX_S(H_{\rm rf})$$

where both the surface resistance R_S and the surface reactance X_S are dependent on $H_{\rm rf}$. Using the usual two-fluid-model expressions for R_S and X_S, one can write

$$R_S(H_{\rm rf}) = \frac{\mu_0^2\omega^2\lambda^3(H_{\rm rf})\sigma_n(H_{\rm rf})}{2} \tag{1}$$

and

H. Weinstock and M. Nisenoff (eds.), Microwave Superconductivity, 117–148.

$$X_S(H_{rf}) = \omega\mu_0\lambda(H_{rf}),$$

where ω is the angular frequency, λ is the London penetration depth, and σ_n is the normal conductivity. Both λ and σ_n can be functions of the rf magnetic field.

While the low-power surface resistance is low enough to realize very low loss microwave devices such as filters and delay lines in applications where the power is low, the nonlinear surface impedance is still a serious limitation and excludes the use of superconducting devices in many applications that require either high power handling or highly linear operation. While the increase in R_S, and with it the loss, are the limiting factors for some systems, in other systems intermodulation distortion, which will be explained below, can be the limiting factor even without an observed increase in R_S. Efforts to explain the physics of the nonlinearities have been under way for several years and will be reviewed. Since, as discussed below, the nonlinearities are believed to result from extrinsic effects, and intrinsic nonlinearities are believed to be orders of magnitude lower, the aim of many of the studies is to identify the sources of the nonlinearities so that the materials can be improved. The understanding of nonlinear effects, however, is not yet sufficient to be able to do this. Moreover, the physics of some of the nonlinear effects is worth studying in its own right.

This review of the nonlinear effects in superconducting devices will focus on the nonlinearities in zero applied magnetic fields. We shall also focus on the high-T_c materials, although nonlinear behavior is observed in the low-T_c materials as well.

This paper is organized as follows. Section 2 describes the measurement methods for nonlinear surface impedance and presents some examples of the measurements. Section 3 presents a small selection of the models that have been used to interpret the results of the measurements. Section 4 presents measurements and models of harmonic and intermodulation distortion. Section 5 is concerned with measurements and modeling of grain boundaries and grain-boundary Josephson junctions. Section 6 presents the conclusions.

2. Measurement Methods of $Z_S(H_{rf})$

2.1. GENERAL

Nonlinear effects usually appear either as changes in the Z_S, or as generation of harmonics and intermodulation distortion. While these kinds of nonlinearity presumably have similar origins, they will be treated separately here. This section is concerned with measurements of $Z_S(H_{rf})$. Section 4 deals with harmonics and intermodulation distortion.

Most measurements of $Z_S(H_{rf})$ utilize the same resonators that are used to characterize the linear Z_S. The equivalent circuit model of a resonator is shown in Fig. 1. This is the usual basis for calculating the responses of linear resonators when all of the components are independent of current [1]. This model produces the usual Lorentzian resonance curve of transmitted power vs. frequency from which one can obtain the Q from the 3-dB points as discussed in other contributions [2,3] to this volume. When any of the components are not constant and are functions of the current, the resonator becomes nonlinear, and the response is no longer Lorentzian. Figure 2 presents the results of measurements of transmitted power vs. frequency with a YBCO

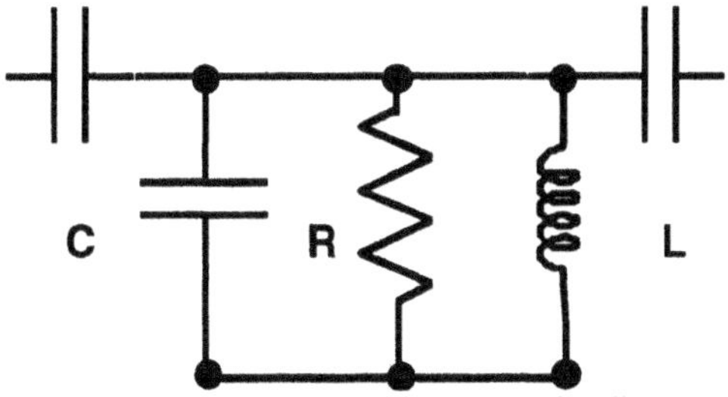

Figure 1. Equivalent circuit model used to describe resonators near resonance for both linear and nonlinear operation. This shows capacitive coupling. Inductive coupling also is possible. The resonant frequency is $f_0 = 1/2\pi\sqrt{LC}$ and $Q_0 = R/\omega L$.

resonator for a series of increasing power levels, showing the transition from linear to strongly nonlinear as a function of input power. Such effects have been successfully modeled using a nonlinear transmission line [4] and a lumped-element model [5], where the values of the components in Fig. 1 are taken as functions of current. Also shown as dashed lines in Fig. 2 are fits calculated with the transmission-line model [4], showing good agreement with the data.

When the nonlinear effects are strong enough that the resonance curve is no longer Lorentzian, the Q can not be determined from the 3-dB points of the resonance curve, and another method must be used. The parameters can be extracted from the models [4,5], or time-domain measurement methods can be used as described below in Sec. 2.2, where the initial decay slope defines the high-power Q. The simplest method, however, of obtaining the high-power Q rests on the observation that in Fig 1 at resonance, the point of maximum power transmission, the reactances in the circuit cancel, leaving only the resistance. Thus, from the value of the insertion loss at resonance one can obtain the resistance in the equivalent circuit from which the Q can be extracted [1]. The resonance frequency, the point where the insertion loss is smallest, can be related to the changes in λ [6].

For understanding and quantifying the nonlinear $Z_S(H_{rf})$, detailed knowledge of the current and magnetic field distribution is needed for the resonator being used for the measurement. Also needed is the I_{rf}, the total rf current flowing at resonance [1] that in the case of a transmission-line resonator is given by

$$I_{rf} = \sqrt{\frac{8r_v(1-r_v)Q_L P}{\pi n Z_0}},$$

where r_v is related to the insertion loss by IL = 20log r_v, Q_L is the loaded Q, P is the input power, n is the overtone mode number and Z_0 is the characteristic impedance. To be rigorously accurate in obtaining $Z_S(H_{rf})$, the following integral should be deconvolved.

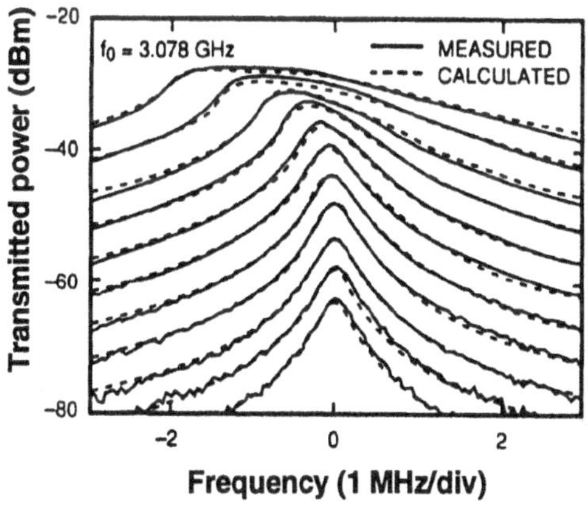

Figure 2. Transmitted power vs. frequency for a YBCO resonator as a function of input power. The maximum power is +30 dBm, and the curves are in 5-dBm steps. The frequency is 3 GHz, and the temperature is 77 K. Solid lines are measured data; dashed lines are calculated [4].

$$\frac{1}{Q}=\frac{2\int \lambda(J)R_S(J)J^2(x,y)dS}{\omega\int \mu_0 H^2 dV}, \tag{2}$$

but at the present time it is not known how to do this in a unique fashion. The usual presentation of the nonlinear Z_S is to plot the values of R_S and λ, obtained from the resonance curves using the low-power geometry factors vs. peak H_{rf}, the largest value in the spatial distribution of H_{rf} that is calculated from the current distribution. That is

$$R_S=\frac{\Gamma_R}{Q}$$

and

$$\frac{\Delta X_S}{X_S}=\Gamma_f\frac{\Delta f}{f},$$

where Γ_R and Γ_f are the low-power geometry factors discussed previously [2]. Thus, one normally plots R_S obtained from the Q with the assumption that the current distribution has not appreciably changed, an assumption supported by experiment [7]. This method also assumes that most of the nonlinear contribution is coming from the regions of the device where the H_{rf} is largest. This is a reasonable assumption for a stripline device because the current is peaked strongly in a small region, but it is not as accurate for a dielectric resonator. The role of the current distribution in extracting Z_S has been discussed previously, and for nonlinear measurements, knowledge of H_{rf} is needed as well.

2.2. MEASUREMENT METHODS FOR UNPATTERNED FILMS

Measurement methods of Z_S for unpatterned films have been discussed by Hein in another contribution to this volume [2]. Here we provide a brief review.

Measurements of $Z_S(H_{rf})$ in unpatterned films have generally been carried out using either dielectric resonators shielded with superconducting ground planes [8] (TE mode), parallel plate resonators [9], or disk resonators [10] (TM mode). Dielectric resonators are the preferred geometry because the current is zero at the edges of the film under test, and edge effects are eliminated. Parallel plate resonators are relatively simple to fabricate and use. Substrates can be small (1 cm x 1 cm), but heating effects are difficult to control, and the modes are not as well defined as in the dielectric case, so calculation of the peak fields is not as straightforward. In the following we discuss the use of shielded dielectric resonators to measure $Z_S(H_{rf})$. TM-mode resonators require 2-sided films.

There are several versions of dielectric resonators discussed in the literature, but most are derived from the design described by the Dupont group [8]. In general a dielectric cylinder of sapphire or other low-loss dielectric is employed whose permittivity $\varepsilon \geq 10$ in order to concentrate the fields in the dielectric and reduce fringing fields in the normal-metal housing of the resonator. Two superconducting films are used for the ground planes. The resultant Z_S is an average of the two. Frequencies have been reported from about 4 GHz up to 30 GHz, depending on the size and ε of the dielectric. Required wafer size ranges from 2-inch (50-mm) diameter for 8-10 GHz, to less than 1 inch for the highest frequencies.

For measurements of the nonlinear Z_S in the dielectric resonator there are two important considerations. First, because the volume is large and the current distribution is relatively uniform, obtaining sufficient circulating power to produce large enough H_{rf} requires high-power amplifiers. It is often necessary to apply greater than 10 W of microwave power to observe nonlinear effects. Amplifiers having this power output capability can become expensive. To maximize the circulating power, the input coupling must be adjustable *in situ*, so that critical coupling can be maintained at all temperatures and powers, even though the Q is changing. The output coupling is usually maintained at a weak level to maximize the circulating power because any loading of the output circuit will decrease the circulating power.

The second consideration related to nonlinear measurements is heating of the films due to the power dissipated in the resonator. With critical input coupling and weak output coupling, all of the input power is dissipated in the resonator, and since it is designed to be dominated by losses in the films, this power is dissipated in the film itself. Thus, means must be provided to efficiently conduct away from the film the heat produced, so that the film stays at the desired temperature. The heat conductivity of the film and the substrate are usually sufficient to prevent heating if proper thermal contact and a low thermal resistance path are provided to the back side of the wafer. Problems can arise when measurements are done in the vacuum space of a cryocooler, as is the case most often for these measurements. One side of the resonator can be thermally connected to the cold finger of the cryocooler, but extra provisions must be made to cool the other wafer. A version of dielectric resonator especially designed for measurements of the nonlinear Z_S was used by Xin *et al.* [11]. In this design a gold-plated copper foil is used as the thermal path for cooling the upper wafer. The lower wafer is in contact with a copper plate that is firmly bolted to the cold finger of the cryocooler.

Even with special thermal design, heating of the films can be a problem. Pulsed-power, time-domain measurements of the Q have been employed to alleviate heating effects with good results. As discussed previously, measurements of the resonator Q_L are made by measuring the 3-dB bandwidth Δf of the resonance curve. The $Q = f_0/\Delta f$, where f_0 is the center frequency of the resonance curve. In time-domain measurements a pulse of microwaves at the resonant frequency is applied to the input circuit of the resonator, and after the microwave power is turned off, the decay of the energy in the cavity is observed at the weakly coupled output port. The decay is exponential and related to the Q_L by

$$V(t) = V_0 \exp\left(-\frac{2\pi f_0 t}{Q_L}\right). \tag{3}$$

The pulse of microwaves should be as short as possible to minimize the average power and avoid heating but long enough so that the circulating power in the resonator can build to the steady-state value. From Eq. (3) it is clear that the time constant for the buildup is $2\pi f_0/Q_L$. A time-domain measurement generally allows higher input powers than a swept-frequency measurement. Even with all of the above provisions, it is sometimes difficult to reach high enough values of H_{rf} to observe nonlinear effects in Z_S.

2.3. MEASUREMENT METHODS FOR PATTERNED FILMS

By far the majority of measurements of $Z_S(H_{rf})$ have been carried out using patterned films, usually in one of the common planar transmission-line geometries such as microstrip, coplanar waveguide, or stripline. Examples of each of these transmission lines are given in Fig 3. Since the current distributions are spatially nonuniform and more strongly peaked in these patterned geometries [12,13] than in the dielectric resonators [11], for a given input power level, the H_{rf} is larger. The peak $J_{rf}(x)$ can be 10-times greater than the average $J_{rf} = I_{rf}/A$, where A is the cross-sectional area. Thus, it is easier and cheaper to reach rf fields high enough to cause nonlinear behavior in patterned films than for the unpatterned case. As an example of the stronger fields in patterned films, Fig. 4 shows the difference in current distribution between a stripline transmission line and a TE_{011} shielded dielectric resonator. The problem of heating in the measurement of the $Z_S(H_{rf})$ in the patterned film is usually less severe than in the dielectric resonator because the total dissipated power is lower. In patterned films the nonlinear effects take place at the edge within a relatively small area of the film.

The peaking of the current distribution has led to the supposition that the measurements of $Z_S(H_{rf})$ in patterned films are not typical of the entire film, and that damage to the edge in the patterning process can have a strong influence on the results. However, experience has shown that as film quality has improved, the nonlinearities as measured in patterned films have been also reduced [14], indicating that the patterned film measurements are a good measure of overall film quality.

An important reason to measure the nonlinear effects in patterned films is that many practical devices such as filters and delay lines are fabricated from patterned films. Thus this method characterizes the film as it is employed in applications. As will be discussed below in Sec. 4, almost all measurements of harmonic and intermodulation generation have been carried out on patterned films.

2.4. COMPARISON BETWEEN UNPATTERNED AND PATTERNED FILMS

Because of the question of edge damage, direct measurement of the $Z_S(H_{rf})$ in the same films, unpatterned and patterned, has recently been carried out by Xin *et al.* [11]. The $Z_S(H_{rf})$ of YBCO films were first measured in the dielectric resonator described above in Sec. 2.2. The nonlinear microwave-frequency surface resistance at 10.7 GHz of unpatterned, 2-inch-diameter, 400-nm-thickness YBCO films was measured. The unpatterned films were then patterned using standard photolithography and wet chemical etching with 0.1% phosphoric acid to make stripline resonators with resonant frequency of 1.5 GHz and with overtones at multiples of 1.5 GHz. The microwave surface resistance was then measured as a function of rf power for the patterned striplines. Comparing the

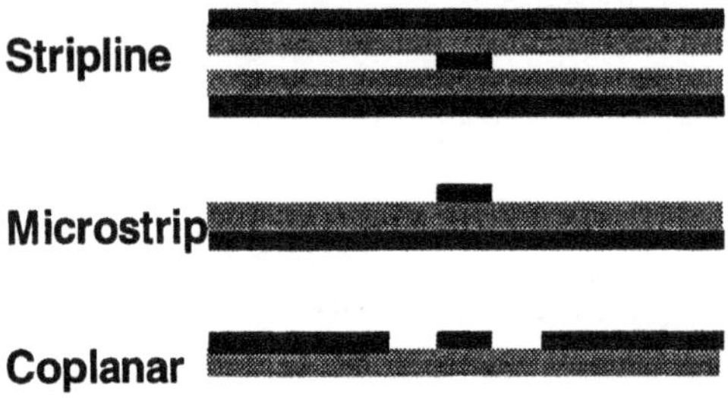

Figure 3. Transmission line geometries used for measurements of $Z_S(H_{rf})$ in patterned films shown in cross section. The microwave current flow is perpendicular to the plane of the figure. Black regions are the superconductor, and gray are the dielectric.

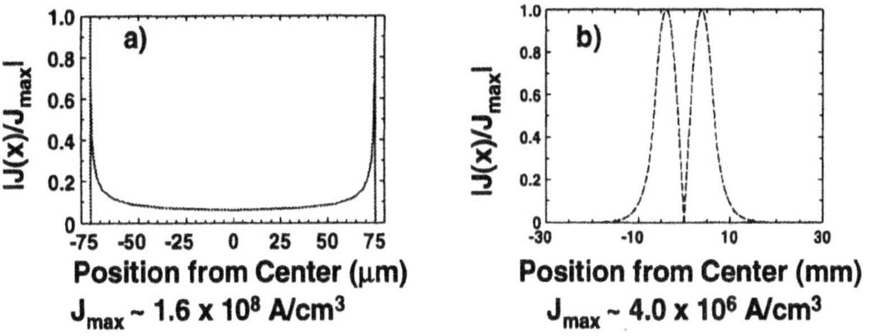

Figure 4. Comparison of current distributions (a) in a stripline resonator and (b) in a TE_{010} dielectric resonator.

results before and after the patterning process directly clarifies whether the patterning contributes to the power dependence of the microwave surface resistance observed for YBCO thin films.

The results for one set of films are shown in Fig 5. The points are the measured data and the solid lines are the data that were modeled for the two resonators assuming the same $Z_S(J_{rf})$ for both the patterned and unpatterned films. The films are modeled as

$$\rho(\vec{r}) = \rho_0 + \rho_2 J_{rf}^2(\vec{r}) \tag{4}$$

and

$$\lambda(\vec{r}) = \lambda_0 + \lambda_2 J_{rf}^2(\vec{r}), \tag{5}$$

where ρ_2 are λ_2 fitting parameters. The current distribution for a dielectric resonator is

$$J(r) = \frac{H(r)}{\lambda_0} = \begin{cases} A\dfrac{\beta}{\xi_1\lambda_0} J_1(\xi_1 r) & \text{for } 0 < r < a \\ A\dfrac{\beta}{\xi_2\lambda_0}\dfrac{J_0(\xi_1 a)}{K_0(\xi_2 a)} K_1(\xi_2 r) & \text{for } a < r < R \end{cases} \tag{6}$$

where β, ξ_1 and ξ_2 are constants determined by the geometry of the resonator, a is the radius of the sapphire puck and R is the radius of the superconducting films. The current density thus peaks near the center of the film. For the stripline resonator, $J(x)$ is taken as the analytic function [15],

$$J(x) = \begin{cases} J_s(0)\left[1-\left(\dfrac{2x}{w}\right)^2\right]^{\frac{1}{2}} & x << w \\ J_s(0)\left(\dfrac{1.165}{\lambda_0}\right)(wb)^{\frac{1}{2}} \exp\left(-\dfrac{(w/2-|x|b)}{\lambda_0^2}\right) & x \approx w \end{cases} \tag{7}$$

where w and b are, respectively, the width and thickness of the film. This analytic approximation agrees with the numerically calculated current distribution to within 5% in our case. The Q is obtained from Eq. (2) assuming ρ and λ are given by Eqs. (4) and (5) with the same parameters for both resonators. The current distributions are given in Eqs. (6) and (7), respectively. The solid lines in Fig 5 are calculated with the same

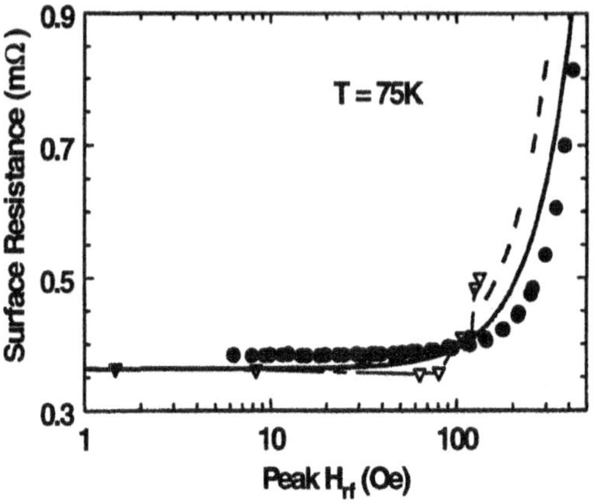

Figure 5. Comparison of patterned and unpatterned films. Points are measured data. Solid and dashed lines are calculated from a model described in the text for the stripline and dielectric resonators respectively.

fitting parameters, thus showing that the patterning process did not degrade the film properties.

2.5. SUMMARY OF RESULTS AND PHENOMENOLOGY OF $Z_S(H_{rf})$

In this section we attempt to summarize the phenomenology of the measurements of $Z_S(H_{rf})$. The predominant observation concerning the reports of measurements of $Z_S(H_{rf})$ is a large sample-to-sample variation. While samples fabricated by the same system tend to show some degree of consistency, even this is not always the case [16], and without question, samples from different deposition systems show large variations. Some measurements exhibit a quadratic dependence on H_{rf}; others show a linear dependence, while others follow a square-root type of curve, including a tendency to saturate at intermediate powers.

It is, however, fairly common that the results for $Z_S(H_{rf})$ show varying regions of nonlinearity. In many films a low-power region is present in which the Z_S remains linear. It then increases with some small power of H_{rf}. This is followed by a region of faster increase followed by breakdown that is probably thermal in origin. Not all films follow this general pattern, however, and in some of the very best films reported to date, little change in $Z_S(H_{rf})$ is observed before breakdown [10].

A striking finding concerning nonlinearity is that, to date, little correlation between the usual measures of film properties and nonlinear effects has been found [17]. The usual measures such as x-ray diffraction, mosaic spread, T_c, or J_c have little correlation with $Z_S(H_{rf})$ so long as a minimal level of film quality is maintained. Surprisingly, even the low-power linear Z_S does not correlate with nonlinear properties. Films with the lowest low-power surface resistance and penetration depth do not yield the best power dependence. It is, however, to be noted that improvements in nonlinear behavior have been demonstrated with the passage of time and with a general overall improvement of film quality, although no single measure has been found to be responsible for the improvement.

As discussed earlier, the frequency dependence of the low-power linear Z_S is well established, with for instance R_S varying as f^2, which can be derived from the two-fluid model as in Eq. (1). In the high-power nonlinear region, however, no simple frequency dependence has been established. This is in part because of the experimental difficulties encountered in measuring the frequency dependence. High-power amplifiers are often confined to a narrow band of operation. Additionally only transmission line resonators permit measurements at multiple frequencies on the same film in the same structure, so the amount of experimental evidence is small. As will be discussed below, measurements of the frequency dependence at high power that do exist clearly indicate that the low-power f^2 dependence does not carry over to the high-power region. The dependence is closer to f as measured by Nguyen *et al.* [6], or without simple power law as measured by Herd *et al.* [18,19].

3. Modeling the Nonlinear Surface Impedance

3.1. INTRINSIC vs. EXTRINSIC NONLINEARITY

It is generally agreed that the observed nonlinearities are due to defects in the material and not due to intrinsic processes [20,2122,23]. This assertion is supported by the large sample variation that is observed and by the declining strength of nonlinearity as the overall quality of the materials has improved with time. More convincingly, it is based on consideration of the magnitude of nonlinear effects that could result from the various intrinsic nonlinear processes that are discussed in the next paragraph.

The time-independent Ginzburg-Landau (G-L) equations [15] can be applied to the microwave electrodynamics of superconductors because the microwave period is long compared with any characteristic time in the superconductor. The nonlinearity of the G-L equations results from pair breaking associated with the transport currents. That is, the superfluid density is dependent on the current. Lam *et al.* [24] used the G-L equations to model the frequency shift in stripline resonators and found that unrealistically small values of the thermodynamic critical field had to be used to fit the data, indicating that the G-L nonlinearities were too small to explain the measurements. Also, Nguyen [25] used G-L theory to derive the following expression for the superfluid density,

$$n_s(T,J) = \lambda^{-2}(T,J) = n_s(T,0)\left[1 - b(T)J_s^2(\vec{r},t)\right],$$

where $b(T)$ is a temperature-dependent parameter, and he used it to calculate the R_s to lowest order in J^2. The results are shown in Fig 6 and compared with measured data and fits to the data using the coupled-grain model [6,25]. Clearly the G-L prediction is at least two orders of magnitude too small to predict the nonlinear R_s of even the very best films.

Dahm and Scalapino [26] have used the nonlinear Meissner effect to derive an expression for the nonlinear superfluid density. They find

$$\frac{n_s(T,j_s)}{n} = \left(\frac{\lambda(0)}{\lambda(T)}\right)^2\left[1 - b_\theta(T)\left(\frac{J_s}{J_c}\right)^2\right],$$

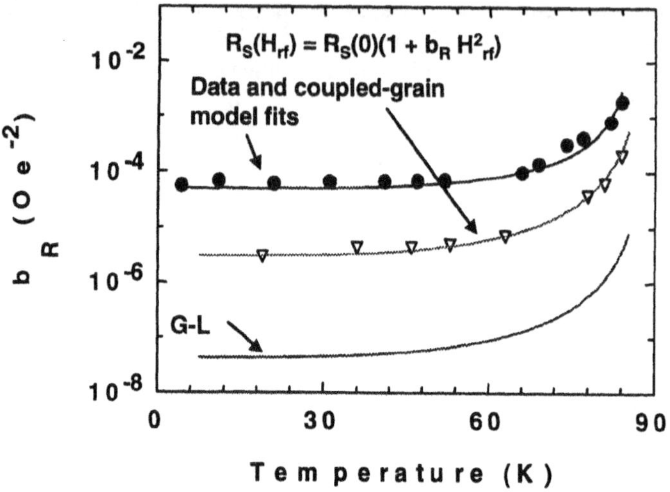

Figure 6. Comparison of G-L prediction with simple quadratic fit using coupled-grain model.

where j_c is the depairing critical current

$$j_c = ne\Delta_0 / p_f \sim 3\times10^8 \text{ A/cm}^2$$

and $b_\theta(T)$ is a temperature-dependent coefficient that reflects the order parameter symmetry. This expression is derived within the BCS framework and takes into account the presumed d-wave symmetry of the order parameter. The parameter b_θ is of order one and shown in Fig. 7 for the different symmetries of the order parameter. It was shown that because of the nodes in the energy gap, d-wave symmetry produces larger nonlinearity than s-wave, especially at low temperatures. But this analysis yields nonlinearities too small to explain the experiments, since experimentally nonlinearities occur at current densities two orders of magnitude smaller than the depairing current j_c.

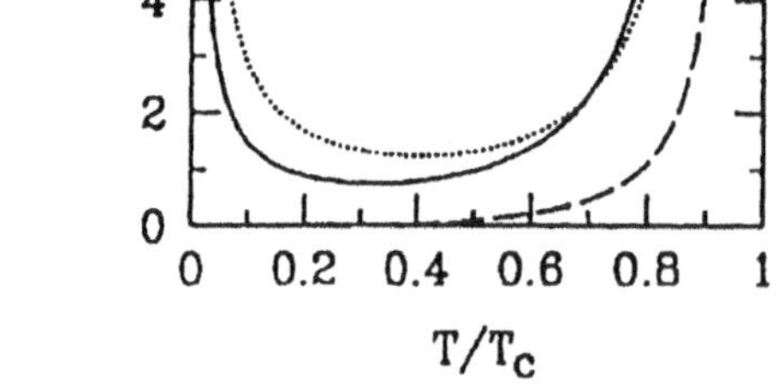

Figure 7. Plot of the calculated [26] values of b_θ for different symmetries of the order parameter.

The quantitative results of the Dahm and Scalapino analysis will be discussed in detail in Sec. 4.3 on intermodulation. Thus, both G-L and nonlinear Meissner theories give predictions that are too small to explain the data.

Since the intrinsic nonlinearities are not sufficient to explain the experimental results, most researchers attribute the causes to be various defects that have been observed in the materials. Examples are grain boundaries that can behave as Josephson junctions, twin boundaries, dislocations, inclusions of normal conducting material, or magnetic impurities. In the following we discuss two of the models that have been used to model the physics of the nonlinearities of Z_S.

3.2. COUPLED-GRAIN MODELS

Many hypotheses have been put forward to explain $Z_S(I_{rf})$. We do not attempt to survey all proposed models here, but review two representative examples. One such proposed explanation is the coupled-grain model that treats a film as an effective Josephson medium, a network of superconducting grains separated by grain boundaries which are Josephson-junction (JJ) weak links [6,21,27] that are responsible for the power dependence. Microstructural studies of YBCO thin films support the coupled-grain idea. They have revealed a granular composition independent of both deposition technique and substrate material [28,29,30]. Films with a *c*-axis orientation often contain some grains of *a*-axis-oriented material, twin boundaries, and in-plane grain boundaries resulting from basal-plane *a*-axis misorientation between *c*-axis-oriented grains. The in-plane grain boundaries can be characterized by a misorientation angle θ, which is the degree of rotation of the *a*-axis across the boundary separating two *c*-axis-oriented grains. In addition, experimental evidence [31] has been presented that even in high-quality films, whose power dependence is limited at $H_{rf} \approx H_{c1}$, one sometimes observes small increases in R_S at moderate powers. The observation [32] that intermodulation distortion has been measured at power levels well below the point where R_S increases, also suggests that low-angle grain boundaries might play an important role in the microwave properties of the materials by introducing harmonic generation.

A quantitative coupled-grain model has been introduced to calculate the linear surface impedance of HTS at low power levels [33,27] and later extended [34,35] to intermediate power levels where Z_S is nonlinear. The material was modeled as a series network of superconducting grains connected by resistively shunted Josephson junctions (RSJ). To include power dependence, analytical expressions were introduced for the complex resistivity of the RSJ, assuming a single type of junction behavior below the critical current I_c for the entire film. These assumptions limit the use of such a model to rf currents lower than I_c, which means for only low and intermediate power levels.

A perhaps more realistic extension of the coupled-grain model has been introduced [18,19] to include independent distributions of I_c and R_n values rather than the single value of the original coupled-grain model, and the complex impedance of the junctions is computed by numerically solving the RSJ equations. More detail of the solutions to the RSJ equations is given below in Sec. 5.1. The independent distributions of I_c and R_n allow a more realistic representation of the actual material properties, and the numerical solution allows the model to be applied at all values of rf current up to and beyond I_c.

The expected value of the RSJ impedance is computed as

$$\tilde{Z}_J(I_{rf}) = \mathrm{Ex}\left[Z_J(I_{rf})\right] = \int_0^\infty\int_0^\infty Z_J(I_{rf}, I_c, R_n) P(I_c) P(R_n) dI_c dR_n \,, \tag{8}$$

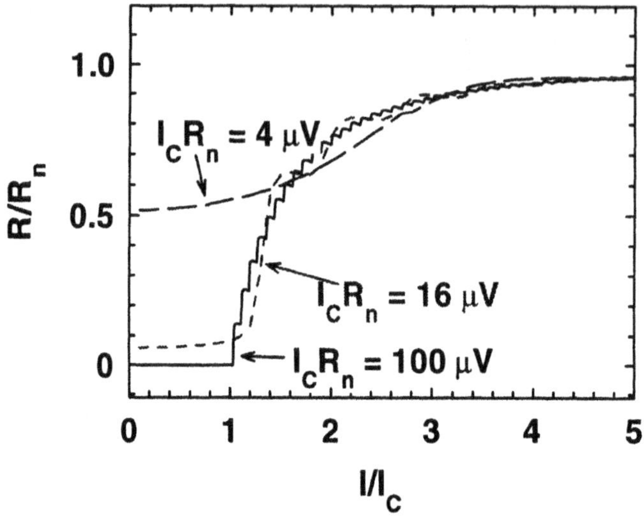

Figure 8. Results of the calculation of the resistance of a resistively shunted Josephson junction.

where $Z_J(I_{rf}, I_c, R_n)$ is the complex impedance of an individual Josephson junction given by the solution of the RSJ equations [15] at each value of I_c , R_n and I_{rf}. $P(I_c)$ and $P(R_n)$ are the probability density functions of the I_c's and R_n's in the material. The best agreement with experiment [19] was obtained by fixing R_n as a constant and taking $P(I_c)$ to be a broad distribution, tending to zero as I_c goes to zero, as shown in the inset of Fig. 9. The surface impedance of the coupled-grain/RSJ system is then computed as

$$Z_S(I_{rf}) = \sqrt{j\omega\mu_0(\rho_G + \tilde{\rho}_J(I_{rf})} ,$$

where ρ_G is the complex resistivity of the grains given as

$$\rho_G = (\omega\mu_0)^2 \lambda_G^4 \sigma_{G1} + j\omega\mu_0\lambda_G^2 ,$$

and $\tilde{\rho}_J(I_{rf})$ is the effective resistivity of the RSJ distribution given as

$$\tilde{\rho}_J(I_{rf}) = \tilde{Z}_J(I_{rf}) \cdot a ,$$

where a is the average grain diameter, and $\tilde{Z}_J(I_{rf})$ is given by (8). In the grains, σ_{G1} $<<\sigma_{G2}$. It is assumed that the grain resistivity is independent of rf power, and varies according to the two-fluid model with temperature. The frequency variation of $\tilde{\rho}_J(I_{rf})$ is implicitly calculated by the numerical solution of the RSJ equation.

Figure 8 shows the results of calculations of the resistance at 2 GHz of a resistively shunted Josephson junction for several values of I_cR_n product. When the rf current approaches I_c, the rf resistance rises from a low nonzero value at low current. When the rf current is many times I_c, the resistance asymptotically approaches the R_n of the junction. It is easily seen than that a distribution of I_c and R_n can lead to a steadily increasing R_S as is observed experimentally. In fact, a bimodal distribution of values could account for cases where the R_S shows a plateau in the curve vs. I_{rf}. More detail of the RSJ calculations is discussed in Sec. 5.1.

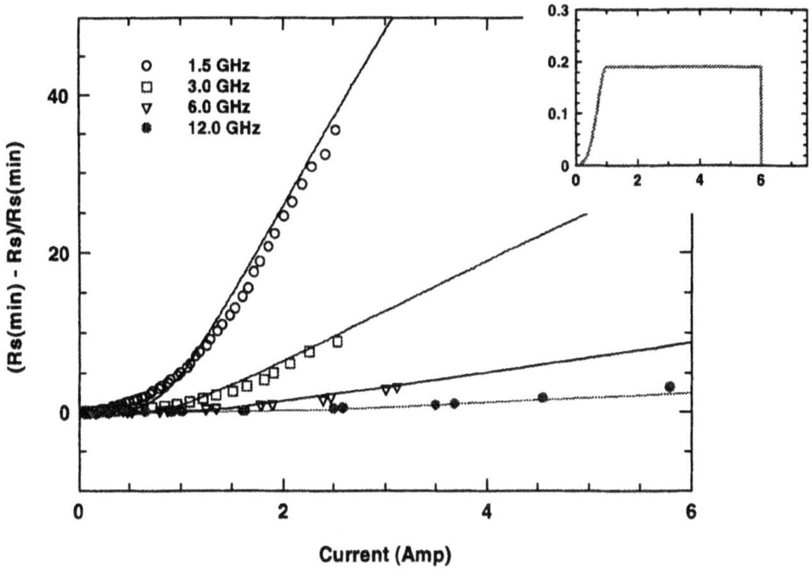

Figure 9. Comparison with experiment of the surface resistance calculated by the extended coupled-grain model. The inset shows $P(I_c)$.

Figure 9 shows a comparison of the extended coupled-grain model with measurements of R_S in a stripline resonator for various frequencies at 75 K. The model shows very good agreement including the variation in frequency. For more detail, see Ref. 19. The modeled reactance, compared with experiment in Fig. 10, did not show such good agreement, which the authors suggested may be due to a mechanism different from the coupled grains dominating the reactance. This lack of correlation of the resistance and reactance has been observed by others [36] and will be discussed in much greater detail in Sec. 3.4.

3.3. FLUX PENETRATION, BEAN CRITICAL-STATE MODELS

The well-known Bean critical-state model [37,38] has been used with success in modeling magnetic flux penetration in the low-frequency (≤ 100 kHz) case to explain magnetization behavior including hysteresis for a large variety of superconducting materials. Without transport current, the model assumes the current is either equal to J_c or zero everywhere and

$$|\nabla \times B(r)| = \mu_0 J_c . \tag{9}$$

It also is assumed that the J_c is independent of magnetic field. These assumptions are equivalent to strong pinning. Although developed to model the static case, it also has been applied by several authors [39,40,41,42] to the microwave case by including a transport current to calculate the nonlinear surface resistance resulting from penetration of the rf magnetic field into the material. The calculations take into account the exact geometry and detailed current distributions of the cases considered. The justification for the use of a quasi-static model at microwave frequencies is that the characteristic times for vortex creation and flux penetration are short compared with the period of the

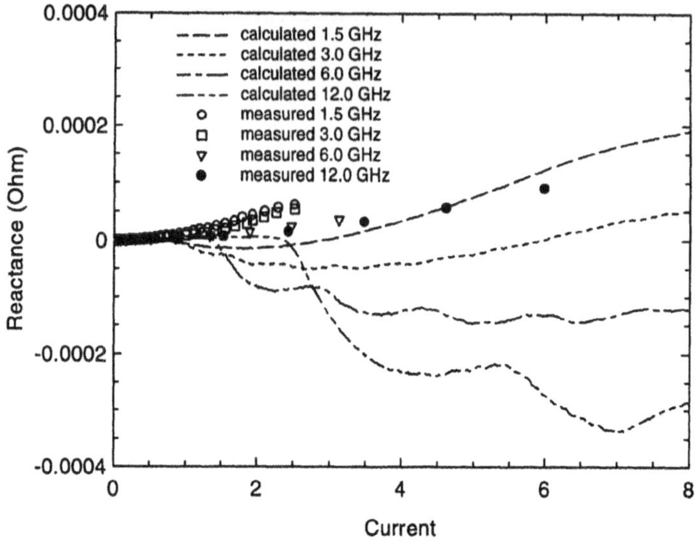

Figure 10. Comparison of the surface reactance calculated by the extended coupled-grain model with experiment. The following parameters were used: $a = 10^{-5}$ m, I_c min = 0.1 A, I_c max = 6.0 A, I_c mean = 1.0 A, variance = 0.3 A, and $R_n = 2.5\times10^{-5}\ \Omega$.

microwaves. The Bean model assumes that flux freely penetrates the superconductor, but doesn't differentiate whether it is in the form of Arbikosov vortices or Josephson vortices. Thus, the penetration could be in weak links between the grains of the coupled-grain model. It also assumes that the pinning is very strong so that the flux stays in the material until fields of the opposite sense cause flux of the opposite polarity to enter and cancel the initial flux.

Calculations of the surface resistance using the Bean model have been compared with experiment by Nguyen *et al.* [40] and found to give reasonable agreement in the high-power region where the R_S is increasing rapidly. Other authors have calculated similar results. The calculations of McDonald *et al.* [42] include results of harmonic and intermodulation distortion as discussed below in Sec. 4.3.

In the Bean model, the losses are hysteretic in nature, and as such, yield a simple frequency dependence with $R_S \sim f$. As discussed above frequency-dependence measurements are difficult and only a few have been carried out. Most measurements of the nonlinearity are at a single frequency. The R_S proportional to f is observed in some, but not all measurements. So the question of whether the Bean model gives the correct frequency dependence is still unresolved.

3.4. IMPEDANCE PLANE ANALYSIS, R-VALUE

It has been proposed [43,21] that a simple and valuable way to differentiate among the mechanisms of nonlinear surface impedance is examination of the quantity

$$r(H_{rf}) = \frac{\Delta X_S(H_{rf})}{\Delta R_S(H_{rf})},$$

where ΔX_S and ΔR_S are the changes in surface reactance and surface resistance, respectively, as a function of the microwave magnetic field H_{rf}. Each of the various physical mechanisms that have been proposed to describe the power dependence in high-T_c materials has a characteristic r value. Not only are the magnitudes of the r values different, but also the frequency and temperature dependence have characteristic signatures. Thus, the measured r values can in principle be used to differentiate among these mechanisms, especially by measuring as a function of frequency and temperature. For instance, pair breaking gives $r >> 1$, and $r \propto 1/\omega$, because reactance changes dominate. Weak links behave as $r \propto 1/\omega$, with r decreasing with temperature via the temperature dependence of the critical current of the weak link. However, when a distribution of weak-link parameters is considered, the r value can be quite different than for a single weak link. We do not discuss these questions any further here except to note that several authors have observed that R_S and X_S are not correlated [16,19], rendering the use of the r value questionable.

3.5. SUMMARY OF MODELING OF NONLINEAR SURFACE IMPEDANCE

The observed $Z_S(H_{rf})$ remains largely unexplained. It is, however, clear that the nonlinearities are extrinsic in nature as will be further verified in the next section on harmonic and intermodulation distortion. In spite of considerable effort, no model has been developed that explains all of the features observed in the measurements. While some of the modeling attempts have been able to explain individual experiments, the universal model does not yet exist. There are good indications that granularity and possibly Josephson-junction weak links play a role, but the detailed quantitative agreement with experiment has not yet been provided.

As discussed, the extended coupled-grain model is able to explain the nonlinear R_S and its frequency dependence, but not the X_S under the same circumstances. It also has recently been observed [16] that in a series of similar films the R_S and X_S are not correlated. Since all of the proposed models yield a definite relationship between R_S and X_S, they should be correlated. This seems to be significant. These difficulties to model the $Z_S(H_{rf})$ lead to a possible conclusion that more than one process is responsible for the experimental observations Perhaps one process is responsible for the $R_S(H_{rf})$, and another entirely different one is responsible for the $X_S(H_{rf})$. Such multiple cases could account for the lack of quantitative success in modeling the nonlinear effects.

4. Harmonic and Intermodulation Generation

4.1. GENERAL CONSIDERATIONS

Harmonic and intermodulation generation are sensitive probes of nonlinear behavior. As will be discussed in this section, they also have important implications for device performance. The usual way to introduce harmonic generation is to assume that the nonlinear behavior of the surface impedance can be expressed as a power series expansion in the current. If only the resistive part of the impedance is considered, then

$$V(t) = R \times I ,$$

where V is the voltage across the material, and R is the current-dependent resistance given by

$$R(I) = R_0 + aI + bI^2 + \ \ldots, \tag{10}$$

and the voltage is

$$V(t) = R_0 I(t) + aI^2(t) + bI^3(t)\,. \tag{11}$$

A corresponding expression for the reactive part of the impedance can be developed in a like manner.

In a superconductor it is expected that the resistance and inductance do not depend on the direction of current flow, so the odd-order terms of the expansion in Eq. (10) are zero. The exception is the case of an applied magnetic field that establishes a preferred spatial direction allowing dependence of the resistance and inductance on the odd powers of the current. If a single-frequency input is considered,

$$I(t) = A\cos\omega t\,.$$

Then, the power series expansion gives

$$I^2(t) = A^2\cos^2\omega t = \frac{A^2}{2}(1+\cos 2\omega t)\,,$$

and

$$I^3(t) = A^3\cos^3\omega t = A^3(3\cos\omega t + \cos 3\omega t)\,.$$

Thus, it is seen that harmonics, signals with frequencies that are multiples of the input signal frequency, are produced. Although harmonics are important to study, they generally fall outside of the frequency band of operation of real devices such as filters and resonators, so they are often not of practical importance. If, however, two tones of frequencies f_1 and f_2 are simultaneously present,

$$I(t) = A\cos\omega_1 t + B\cos\omega_2 t\,.$$

Then, the terms in Eq. (10) become

$$\begin{aligned} I^2(t) &= A^2\cos^2\omega_1 t + B^2\cos^2\omega_2 t + \\ &AB\cos(\omega_1 - \omega_2)t + AB\cos(\omega_1 + \omega_2)t \end{aligned}$$

and

$$\begin{aligned} I^3(t) &= A^3\cos^3\omega_1 t + A^2 B\cos(2\omega_1 - \omega_2)t + \\ &B^2 A\cos(2\omega_2 - \omega_1)t + \text{etc.} \end{aligned} \tag{12}$$

In this two-frequency case, in addition to the harmonics, sum and difference frequencies, known as intermodulation (IM), of various combinations of the input frequencies are produced. The third-order IM frequencies such as $2f_1$-f_2 are very important for device applications because they are close to the original frequencies and usually, in the case of a filter for instance, fall within the passband of the filter, producing spurious signals that degrade the performance of the device. Intermodulation has been the subject of much study in the high-T_c materials. Note that if A=B, then the third-order IM power increases with the cube of the power of the input signals. Third-order IM is important in active devices such as amplifiers, but is usually negligible in passive devices such as filters made from normal-metal conductors. In superconductors the IM is significant. The IM products are dependent on device geometry and on the design of the device, which in the case of a filter is given by the bandwidth and number

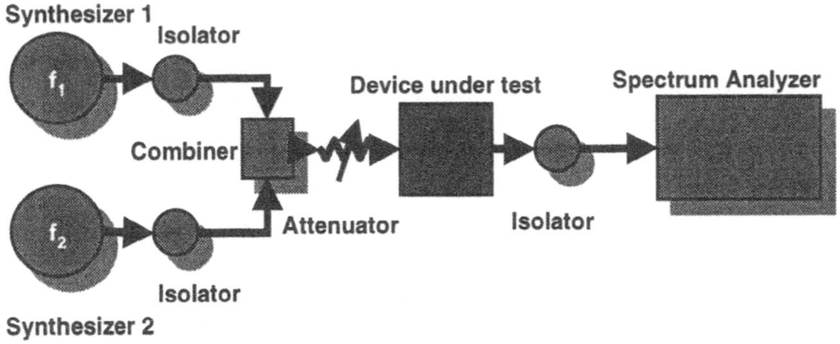

Figure 11. Method of measurement of intermodulation distortion in a filter.

of poles, with narrower band devices producing higher levels of IM because in general the *Q*s are higher and the circulating currents are larger.

4.2. MEASUREMENTS

4.2.1. Intermodulation measurements

Figure 11 shows the method of measurement of IM products. Two signals, usually of equal amplitude, both of whose frequencies are in the passband of the device to be tested, are applied through a summing network. Care must be taken to isolate the signal sources, usually synthesizers, from the filter because reflected signals can easily generate IM in the output stages of the synthesizers. The output power of the filter at the input (fundamental) frequencies and at the IM frequencies is measured in a spectrum analyzer as a function of the input power of the fundamental tones. Because

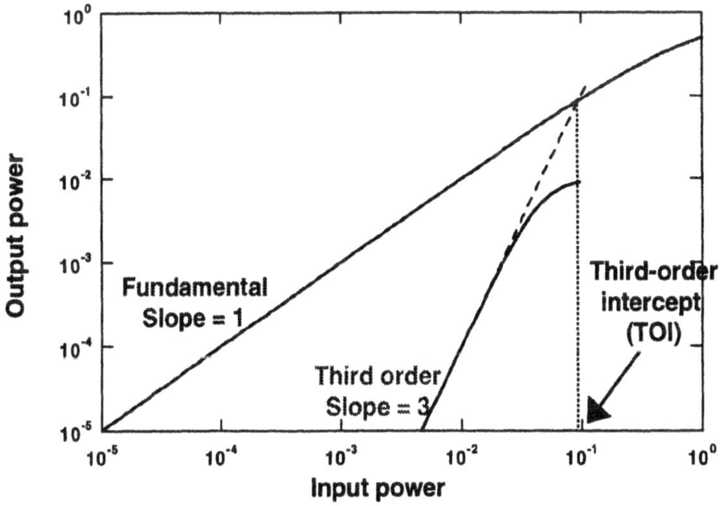

Figure 12. Schematic depiction of the results of the measurement of intermodulation products in a bandpass filter. Shown is the definition of the third-order intercept.

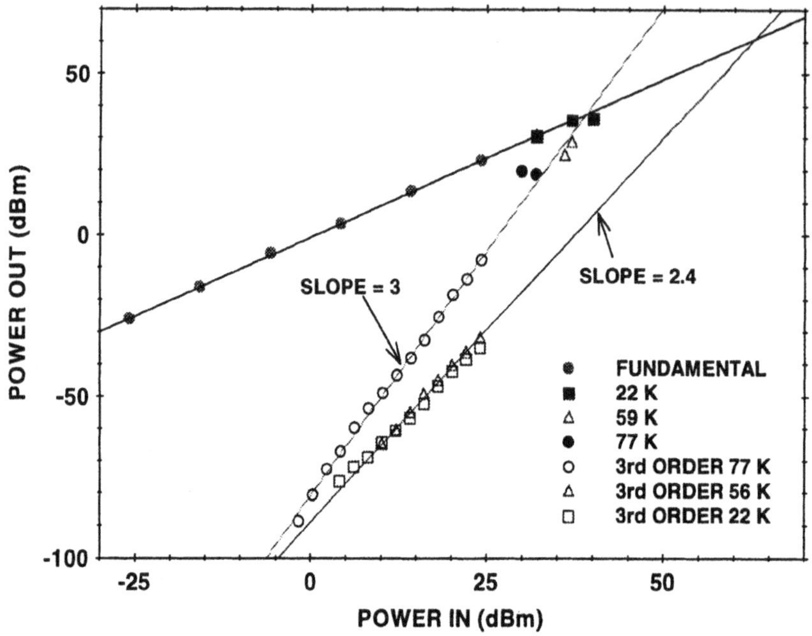

Figure 13. Actual measurements of IM in a high-T_c filter. This is a 5-pole filter designed with 10-Ω resonators for high-power-handling capability.

the spectrum analyzer also can internally produce IM of its own, power levels at the spectrum analyzer input must be kept within the linear operation range of the instrument. Attenuators at the input can be used for this purpose.

Figure 12 schematically shows typical results for the measurement of IM in a high-T_c filter. The output power is plotted against the input power, and it is assumed that the two input power levels are equal. Both the fundamental signal and the IM signals are plotted. On this log-log plot, the fundamental signal shows a slope of unity up to the maximum power that the filter can pass. The curve shows saturation effects when the maximum power is reached. The third-order IM signals are expected to increase with slope three on the log-log plot. The slope of three follows from Eq. (12). This is, however, not always the case with superconductors. The deviation from slope of three is discussed in more detail below. The IM signals also show saturation at sufficiently high power levels, in part due to energy conservation that is not included in Eq. (12).

Figure 12 also shows the definition of the third-order intercept (TOI, also called IP3) as the input power where the extrapolations of the low-power fundamental and IM signals intersect. The TOI is a convenient quantitative measure of the nonlinearity of a device such as a filter or amplifier. It also is used to characterize the third-harmonic measurements that we discuss below.

Figure 13 shows actual measurements of IM in a high-T_c filter [44]. This is for a 5-pole filter fabricated from YBCO deposited on $LaAlO_3$, designed with 10-Ω 1-mm-wide resonators for high-power-handling capability. This is a typical result for a filter

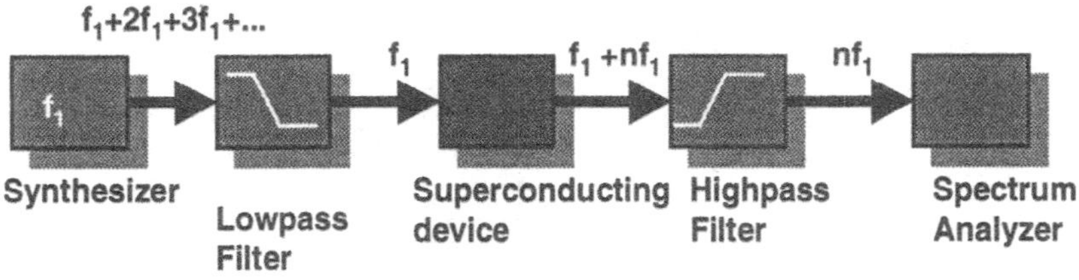

Figure 14. Block diagram of third-harmonic measurement.

and shows many of the features that are seen in general in high-T_c filters and resonators. The IM signal shows a slope of three at high temperature, as expected, but at low temperatures the slope is closer to two. Slopes different from three are found often in IM measurements in high-T_c devices. Usually the slope is less than three, but slopes greater than three also have been found. It should be noted that the measurable IM signals appear in the region of power where the filter behaves in a linear fashion at the fundamental frequency. That is, IM is visible and of sufficient magnitude that filter performance could be degraded in the region of power when the fundamental shows a slope of one. Thus, the IM is a more sensitive measure of nonlinearity than the measurements of loss, because in this region filter loss has not changed.

4.2.2. *Harmonic Measurements*

Third-harmonic measurements are a useful measure of nonlinearity, and while not as important practically as third-order IM, third-harmonic signals presumably arise from the same mechanism, and thus are a useful characterization tool. Their measurement, moreover, is simpler than IM measurements, thus providing an attractive alternative. Figure 14 shows the block diagram for measurement of third harmonics. Only one signal source is needed, but synthesizers are usually rich in harmonics, so a low-pass filter is needed to remove them. However, with the high-pass filter at the output of the device, signal levels at the spectrum analyzer are low, so a larger dynamic range is possible for this measurement method compared to IM measurements. Figure 15 shows the results of measurements of the third harmonic in coplanar delay lines [45]. The fundamental frequency in this case is 4 GHz, and the measurements are for YBCO on LAO. The figure shows results of measurements for several different delay lines of differing center conductor width and length in order to understand the effects of geometry. The figure caption provides the details of the geometry.

The second harmonic should not be generated in a superconductor in zero applied magnetic field because of symmetry; current flow in one direction is expected to have the same effect as that in the opposite direction, leading to only odd-order harmonic generation. A second harmonic can arise only if the forward and backward directions of current in the rf cycle are affected differently. Thus, second-harmonic generation is a reflection of spontaneous time-reversal symmetry breaking. A second harmonic was observed in YBCO in zero applied field and reported in Ref. [46]. No detailed explanation has been reported.

4.3. MODELING

Because of the practical importance, considerable effort has been made to model the IM results. McDonald *et al.* [42] have used the Bean model to calculate harmonics and IM.

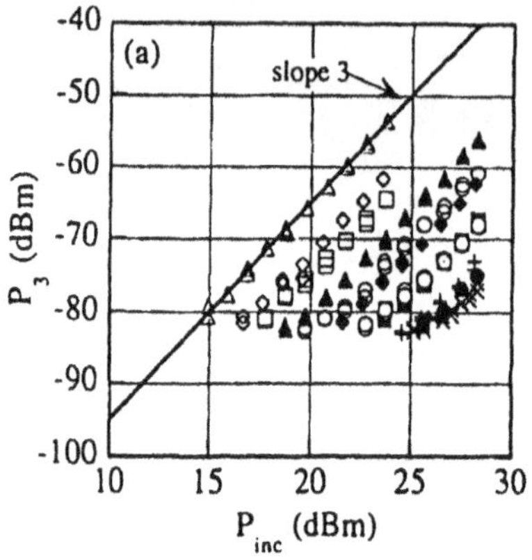

Figure 15. Measurements of third harmonic in a coplanar delay line.

The Bean model was discussed in Sec. 3.3. While the results of the calculations of McDonald *et al.* reproduce many of the features of the experimental data and fully incorporate the geometrical effects of the current distribution on the IM, extensive tests of the model have not yet been made.

Dahm and Scalapino [47,48] have developed a model of IM derived from the intrinsic nonlinearity arising from the d-wave symmetry of the superconducting order parameter. The derivation of the model was discussed above in Sec 3.1. Their model takes into account the geometry, and thereby, the current distribution in the device. Dahm and Scalapino have found that the intrinsic nonlinearity of the d-wave symmetry is insufficient to account for the measured IM. If, however, they assume a nonlinear penetration depth of the form [48]

$$\lambda(j,T) = \lambda(T)\left[1+\frac{1}{2}\left(\frac{j}{j_{\mathrm{IMD}}(T)}\right)^2\right], \tag{13}$$

where j_{IMD} is an empirically determined parameter, they are able to account for IM products in a variety of geometries and filter designs. Their modeling shows that the nonlinear surface resistance is less unimportant for IM generation than the nonlinear inductance resulting from the nonlinear penetration depth. This is because in most planar transmission-line circuits, the inductive reactance is much larger than the resistance, which is the same as saying that Q is large or the attenuation losses are low – almost always the case for superconducting devices. Thus, they model the inductance and resistance as

$$L = L_0\left[1+\Delta\mathcal{L}\left(\frac{I}{I_{\mathrm{IMD}}}\right)^2\right]$$

$$R = R_0\left[1+\Delta\mathcal{R}\left(\frac{I}{I_{\mathrm{IMD}}}\right)^2\right],$$

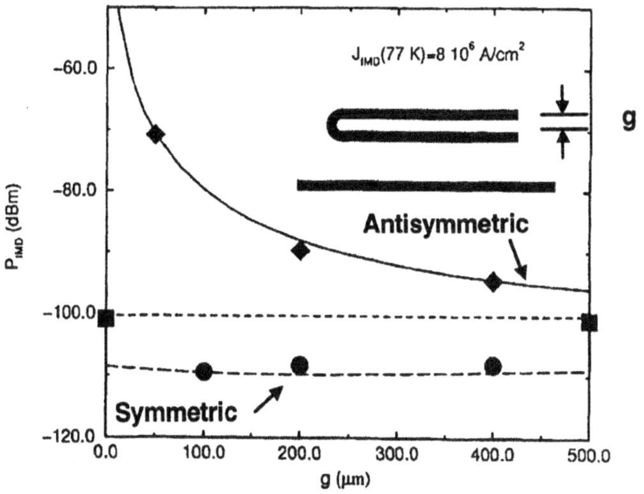

Figure 16. IM power as a function of gap in the geometry shown in the inset.

where $I_{IMD}=j_{IMD}A$, and A is the cross-sectional area. Using these nonlinear quantities, the intermodulation power P_{IMD} was found to be

$$P_{IMD} = Q^4 \frac{P_1^2 P_2}{P_c^2}\left[1+\left(\frac{\Delta\mathcal{R}}{Q\Delta\mathcal{L}}\right)^2\right], \tag{14}$$

with

$$P_c = \omega L_0 I_{IMD}^2 / \Delta\mathcal{L},$$

where Q is the resonator Q and P_1 and P_2 are the input powers of the two signals. Thus, if Q is sufficiently large, then the second term in the brackets of Eq. (14) is negligible.

Dahm *et al.* [48] have tested this phenomenological theory using different geometries with substantially different current distributions and obtained good agreement using the same value of j_{IMD}. Figure 16 shows a schematic of the experiment. Measurements were done as a function of the gap spacing between the microstrip line and the u-shaped resonator for frequencies at the fundamental of the resonator and the first overtone. In the former case the current distribution in the two legs of the resonator is antisymmetric, and in the latter, symmetric with significantly smaller current peaks. Agreement was good for the two extremes of current distribution using the same value of j_{IMD}.

Booth *et al.* [45] also have used the phenomenological model of Dahm *et al.* to explain the dependence on geometry of the third-harmonic measurements shown in Fig. 15. The result is shown in Fig. 17 where all of the geometries are scaled accordingly. From this it is seen that the model of Dahm *et al.* also is able to explain the third-harmonic measurements of Booth *et al.* This both verifies the model and verifies that the third-order IM and third-harmonic generation have a common origin as expected *a priori* and as assumed in the introduction to this section.

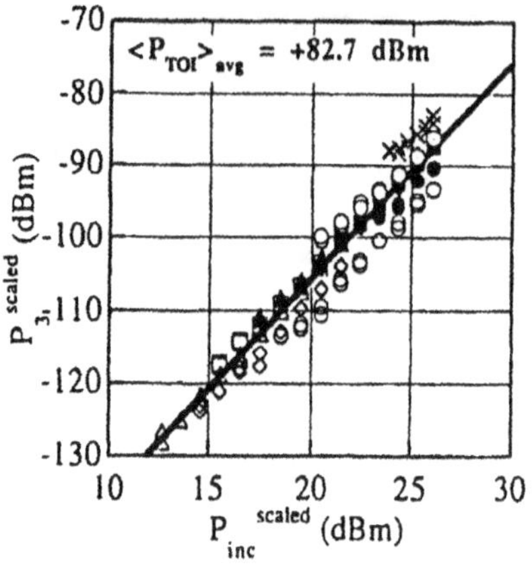

Figure 17. Third-harmonic measurements of Fig. 15 where the geometry effects have been scaled according to Eq. (13) using just one value of J_c.

It should be emphasized that the model of IM and third-harmonic generation proposed by Dahm *et al.* does not identify the origins of the nonlinearities. It only postulates that the nonlinearities cause a nonlinear inductance. The microscopic origins of J_{IMD} used in this model remain unidentified.

5. Measurements of the Microwave Properties of Grain Boundaries and Josephson Junctions

5.1. BACKGROUND

In this section we discuss the properties of Josephson junctions (JJs) at microwave frequencies. As discussed above in Sec. 3.2 under coupled-grain models and in Sec. 4.3 under IM modeling, weak links, a kind of JJ possibly in grain boundaries, may play an important role in the nonlinear effects observed in superconductors. It is known that large-angle grain boundaries behave as JJs at low frequencies near dc. Also, in the discussion of the Bean model, it was shown that flux penetration including Josephson fluxons also could play an important role. Thus, it is of interest to consider the properties of JJs at microwave frequencies. It is beyond the scope of this article to discuss the dc properties, and they can be found in most textbooks on superconductivity. See for instance, Ref. 15. Another reason to study the properties of JJs at microwave frequencies is the potential application as mixing devices in which the nonlinearity of a fabricated JJ is used to mix signals and translate frequencies.

A JJ is usually characterized by its critical current I_c and normal resistance R_n. The natural Josephson frequency f_j is given by [15]

$$f_j = \frac{I_c R_n}{\Phi_0},$$

where

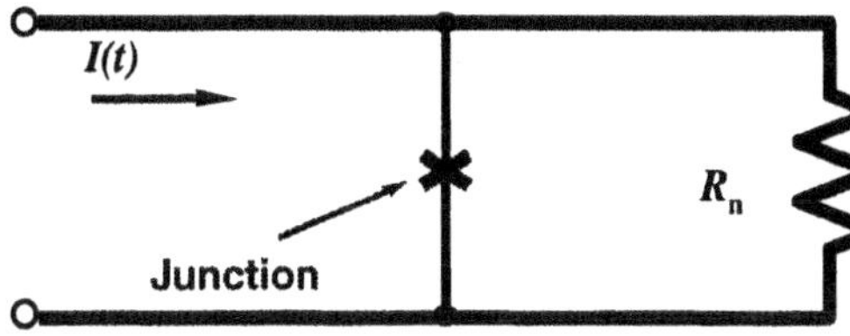

Figure 18. Equivalent circuit of a Josephson junction.

$$\Phi_0 = \frac{h}{2e}$$

is the flux quantum. For typical I_cR_n products, the f_j is usually >> 10 GHz, so at microwave frequencies the kinds of junctions that we are interested in can be treated using the same formalism as for dc. The equivalent circuit model of a JJ appropriate for microwave frequencies is given in Fig. 18, assuming that all dimensions are small compared to a Josephson penetration depth λ_j given by

$$\lambda_j = \sqrt{\frac{\hbar}{2eJ_c\mu_0(2\lambda + d)}},$$

where J_c is the critical current density, λ is the London penetration depth and d is the junction thickness. For high-T_c materials the capacitance of the junction can usually be neglected and has been omitted from Fig 18, yielding the resistively shunted junction model (RSJ), valid for small junctions with uniform current distribution.

The equations governing the RSJ model are

$$\frac{L_0}{R_n}\frac{d\phi}{dt} + \sin\phi = \frac{I(t)}{I_c}$$

$$I_j = I_c \sin\phi$$

where $L_0 = \Phi_0 / 2\pi I_c$, ϕ is the phase of the junction, and I_j is the current through the junction. If one assumes that the input is $I(t)=\sin\,\omega t$, where ω is the microwave frequency, the RSJ equations can be integrated numerically to obtain the current and voltage V across the junction. The effective impedance of the junction is then $Z_{eff} = V(t)/I(t) = R_{eff} + iX_{eff}$, where V is complex, taking into account the components of V that are in and out of phase with the current, and R_{eff} and X_{eff} are the microwave resistance and reactance of the junction. Figure 8 shows the results of this calculation for three values of the I_cR_n product for a frequency of 2 GHz. The general behavior is a low, but finite, rf resistance at low currents followed by a rise as the microwave current approaches the junction critical current. The sharpness of the rise is determined by the I_cR_n product. For large current, the resistance asymptotically approaches R_n. The reactance decreases at I_c, and approaches zero at large currents [18]. Thus, at rf currents several times the critical current, the junction behaves as a normal resistor of value R_n. The steps in the curves occur each time the phase of the junction changes by 2π.

A junction whose extent is greater than λ_j is known as a long junction, and spatial current distributions must be taken into account. For high-T_c materials, λ_j is typically a few μm, so long-junction effects are expected in μm-sized devices. The solutions are governed by the sine-Gordon equation given by

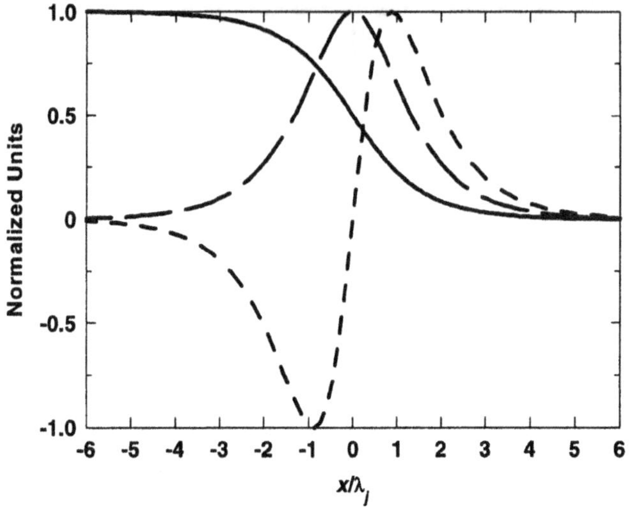

Figure 19. Josephson vortex quantities in normalized units vs. position in the junction. Short-dashed line: current density normalized to J_C; long-dashed line: magnetic field; solid line: Josephson phase ϕ.

$$\lambda_j^2 \nabla^2 \phi = \sin\phi + \tau_j \frac{\partial \phi}{\partial t},$$

where $\tau_j = 1/f_j$, and we have again neglected the junction capacitance. For films thin compared to the Josephson penetration depth, we get the following,

$$\frac{J(x,t)}{J_c} = \lambda_j^2 \frac{\partial^2 \phi}{\partial x^2} = \sin\phi + \tau_j \frac{\partial \phi}{\partial t}. \qquad (15)$$

This must be integrated numerically subject to the proper boundary conditions, which usually include the microwave current distribution. For an isolated junction, the time-independent solutions to Eq. (15) can be carried out analytically [38] and include Josephson vortices whose current distribution is shown in Fig. 19. With this introduction to the microwave properties of JJs, we now review the experimental situation.

5.2. MEASUREMENTS OF GRAIN BOUNDARIES

Because of the proposed role of JJ weak links and specifically weak links in grain boundaries, direct measurements of the microwave properties of isolated grain boundaries as a function of the angular misalignment of the two grains are needed to corroborate the models of nonlinearity and give insight into the behavior of Josephson vortices and flux penetration at microwave frequencies. These topics have not been studied extensively due to the experimental difficulties of isolating single grain boundaries at microwave frequencies. Understanding of the grain boundaries may contribute to understanding the complex nonlinear behavior of these materials. It is worth noting that the dc properties of grain boundaries also are a subject of intense

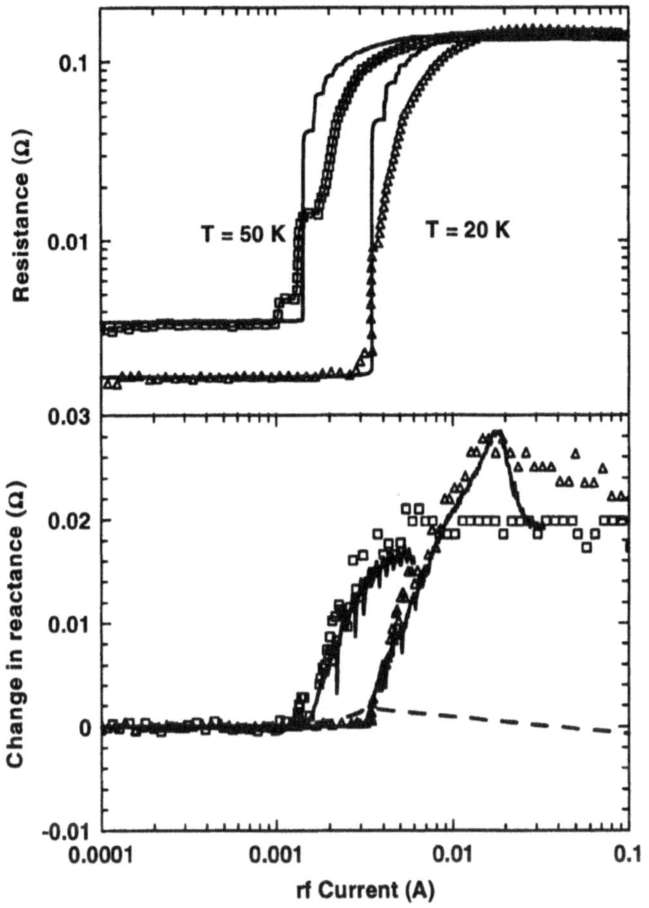

Figure 20. Plot of R and Δf/f for a 24° grain boundary each for two temperatures. The solid lines are for the model discussed in the text.

study because they are not well understood, and they play a vital role in the dc transport properties of bulk and film materials [49].

Several groups have studied the properties of Josephson edge junctions and bicrystal grain-boundary junctions at microwave frequencies using cavity resonators, transmission-line resonators, and transmission lines [50,51,52,53,54]. A systematic study of the microwave impedance of YBCO grain boundaries (GB) as a function of GB misorientation angle from 2 to 24° also has been reported [55,56,57] in which a patterned transmission-line-resonator technique was used (described in the following paragraph) that allows the separation of the effects of the background film from those of the GB.

The stripline-resonator measurement technique exploits the properties of standing waves to separate the effects of the GBs from those of the rest of the material. The resonator line is patterned with a GB located at the midpoint of the line, spanning the entire width [55,56]. The measurements use YBCO deposited on a sapphire bicrystal substrate. A controlled GB is formed by the rotation of the a-b plane of the YBCO film at the bicrystal junction. The standing wave at resonance has a current maximum at the GB for the fundamental resonant mode (f_1 = 3 GHz), while the second resonant mode (f_2 = 6 GHz) has a current node at the GB. Any effect the GB has on the rf power handling will be detectable at the fundamental mode, and absent in the second mode. The Q and fundamental resonant frequency f_1 of each stripline resonator was measured as a function of rf input power at temperatures from 10 to 70 K. The inverse unloaded quality factor $1/Q_u$ and normalized resonant frequency shift $\Delta f_1/f_1(0)$ as a function of peak rf current are shown in Fig. 20 at two temperatures for the 24° GB. This shows distinct Josephson-junction behavior as seen in the steep increase in R_J and the asymptotic approach to a constant R_n. The solid line is the result of the model calculation discussed below. The steps result from Josephson vortex nucleation, as has been discussed in Ref. 57, and will be discussed further in a later section. The R_n values are independent of temperature. The value of I_{rf} at which the resistance increases sharply is related to the I_c of the junction. The rf current distribution is strongly peaked at the edge of the stripline [55], and when the rf current at the edge exceeds the local critical-current density of the junction, the resistance increases sharply. These measurements show the same general features as the RSJ model and support strongly the use of the RSJ results in the coupled-grain model discussed in Sec. 3.2

Figure 21 shows measurements of the $1/Q_u$ vs. I_{rf} for 2, 5, 10, 24° GBs, and for a film on a single-crystal substrate. This particular plot shows results at 50 K. Similar results have been obtained at other temperatures. The figure shows that the 10 and 24° GBs exhibit behavior very different from that of the film fabricated on a single-crystal substrate, as the I_c is significantly suppressed. But the 2 and 5° GBs do not show $1/Q$ values significantly different from that of the film on the single-crystal substrate. The

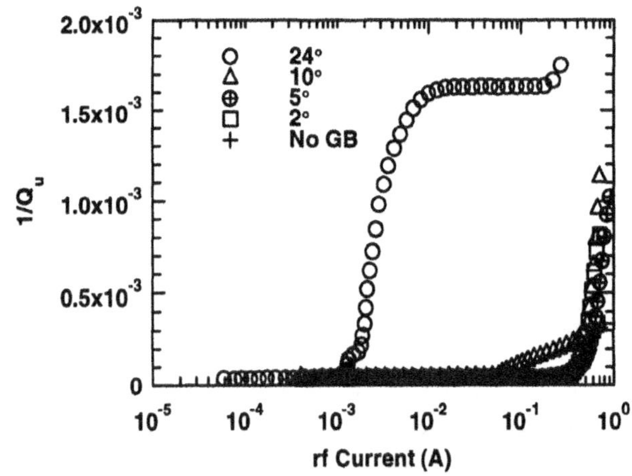

Figure 21. Plot of $1/Q_u$ as a function of rf current at T = 50 K. The curves are largely indistinguishable until the rf current reaches a level where the GB causes the $1/Q_u$ to increase compared to the film on the single-crystal substrate. This increase is seen only in the 24° and 10° GBs. O is the θ = 24° GB, Δ is the θ = 10° GB, ⊕ is the θ = 5° GB, □ is the θ = 2° GB, and + is the film on the single-crystal substrate without an engineered GB.

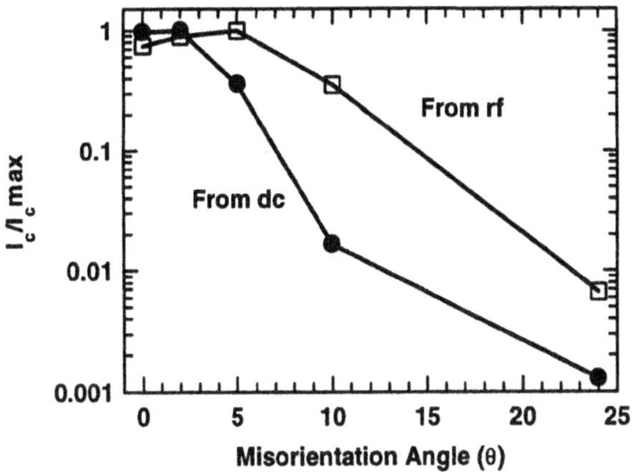

Figure 22. Plot of I_c vs. misorientation angle for the dc case and the rf case.

curves for these angles lie on top of one another, indicating that the I_c as determined by the rf measurements is the same in the engineered grain boundary as in the rest of the film. Figure 22 shows the dependence of I_c on the misorientation angle of the grain boundary θ, showing explicitly that the low-angle grain boundaries do not suppress the rf critical current and that the dc I_c is a much stronger function of the misalignment angle than I_c derived from the rf measurements. At this time the difference between the dc and rf behaviors is not understood.

5.3. JOSEPHSON VORTICES

The 24° GB has been successfully modeled in detail by using an extended resistively shunted junction (ERSJ) model [55] that solves the sine-Gordon equation using an equivalent circuit and incorporates the current distribution of the stripline. The GB is a long Josephson junction, since the length of the JJ (150 μm) is large compared with the Josephson penetration depth (≈ 5 μm). To account for long-junction effects, the GB is modeled as a parallel equivalent circuit of small junctions coupled by inductors that simulate the magnetic coupling of the finite Josephson penetration depth.

The results of the model calculation are compared with experiment for the 24° GB in Fig. 20. The agreement, while not exact, is generally quite good. It is clear that the model is able to duplicate the observed steps in the R vs. I_{rf} data. These steps are caused by the nucleation of Josephson vortices. Each new step is the creation of an extra vortex on each rf cycle as the rf current is increased. The model, as well as the experiment, approach a constant resistance equal to the R_n of the junction at high values of the rf current.

The ERSJ model reveals the dynamics of the Josephson vortices under the influence of the rf currents. As described in Refs. 55 and 57, the microwave current creates vortices at the edge of the junction when the rf current density exceeds the local critical-current density. For modeling, the nonuniform rf current density, which is strongly peaked at the edge of the junction, must be taken into account. The vortices thus created at the edge then propagate into the junction over the course of the rf cycle.

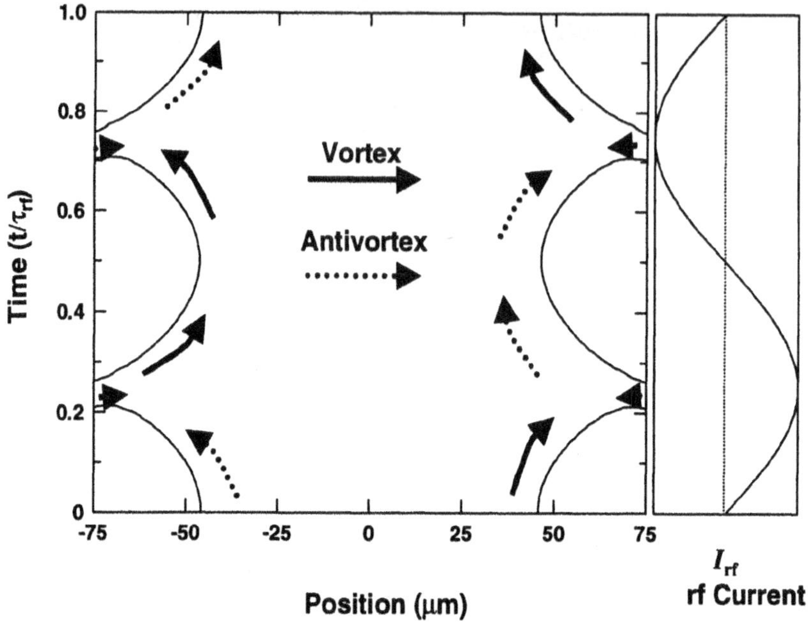

Figure 23. Plot of vortex trajectories for the case of rf current where the current density is slightly greater than the critical current density.

As the rf current reverses direction, the vortices then reverse their direction of motion and leave the junction through annihilation by antivortices as the rf current reverses itself on the negative half of the cycle. When the amplitude of the rf current increases, the number of vortices created during each rf cycle increases, and the effective resistance and reactance of the junction also increase. It is the generation and propagation of the Josephson vortices that produce the losses, and thus, the effective resistance of the junction. Examination of the dynamics shows that the losses are maximum during the creation and annihilation events. For more detail on the vortex dynamics, Ref. 57 should be consulted.

Figure 23 shows the trajectories of Josephson vortices for the case of J_{rf} slightly greater than J_c at the edge of the junction. Detailed examination shows that the losses are associated with creation and annihilation events. More detail on the fascinating physics of Josephson vortices is given in Ref. 57. As expected, the nonlinear impedance of the JJ shown in Fig. 20 also produces harmonics that are shown in Fig. 24.

The exact role of Josephson vortices in nonlinearities is not yet understood in detail. It is likely that it is important to incorporate vortex physics into the models of nonlinear impedance and harmonic generation. This work has not yet been carried out, but it is anticipated to be a fruitful area of investigation.

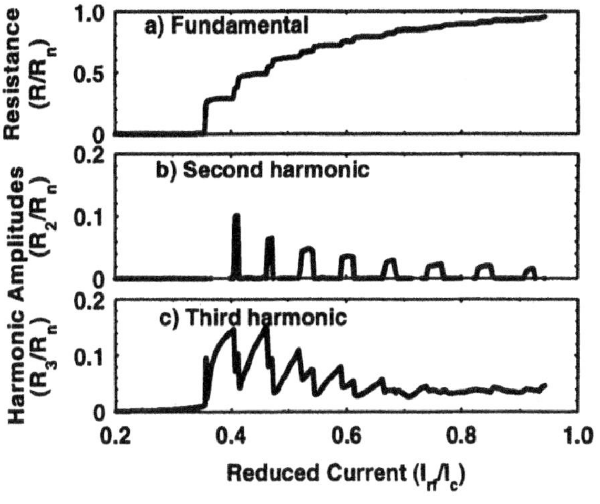

Figure 24. Harmonic generation as calculated from the model described in the text and in Ref. 57.

6. Discussion and Conclusions

We have surveyed the rich field of nonlinear effects in superconductors, including experimental methods and modeling. The concentration has been on high-T_c thin-film materials. We have indicated the important practical implications of nonlinear effects in devices. Because of practical implications, this field remains one of great interest. There are a number of applications that would become possible if the nonlinear effects could be reduced. For instance, very low loss bandpass filters in the transmit section of both radar and communications systems are needed to suppress spurious out-of-band signals generated by the power amplifiers in the output stage of these systems. Such filters need the low loss that superconductors can provide to save expense in the power amplifier and avoid heating in the filter, but the limitations in power handling of superconductors preclude widespread use. A factor of ten improvement in power handling would, however, be enough to open many applications. There also are systems requiring highly linear operation in the receive filters where superconductors could be employed. IM distortion in a receive filter can lead to false targets in radar systems, especially when large signals are present while small targets are tracked. A reduction of a factor of ten in the nonlinearity of the material results in four orders of magnitude reduction of spurious signals as is seen from Eq. (14).

Since, as we have discussed in Sec. 3.1, the intrinsic nonlinearities are believed to be more than one order of magnitude lower than measured nonlinearities, it is expected that the observed nonlinearities are extrinsic and can be eliminated though improved material science. Thus, the aim of much of the research in nonlinear effects is to understand the mechanisms, identify the sources, and provide the information that material growers need to fabricate better material with fewer nonlinearities. This goal has had only limited success to date, and the problem remains ongoing.

The discussion of Sec. 3 showed that while the proposed models describe some aspects of the observed nonlinearities, no single model is able to explain the majority of the results. Heating effects as a possible source were discussed by Hein [2]. The discussion and results presented in Sec. 5 indicate that large-angle grain boundaries and Josephson vortices can account for some of the nonlinear effects. It seems most likely that some combination of weak links and flux penetration is responsible for the nonthermal part of the nonlinearities. The flux penetration could be in the form of Josephson vortices or Josephson-Abrikosov vortices, as discussed in more detail by Golosovsky [23]. A quantitative model incorporating the results of the investigations of JJs has not yet been formulated.

To conclude then, challenging problems remain for this field. Understanding of the nonlinear effects needs to be improved, and thorough understanding has been elusive, probably because several effects are contributing. But the rewards are great. Better materials and better devices are the goal. Along the way, important physics and important material science need to be discovered and investigated.

7. Acknowledgements

This work was supported by the Air Force Office of Scientific Research. The author wishes to thank many colleagues for fruitful discussions, including: A. Anderson, J. Halbritter, J. Wosik, M Hein, and H. Xin.

8. References

[1] Oates D. E., Anderson A. C., and Mankiewich P. M. (1990) Measurement of the surface resistance of $YBa_2Cu_3O_{7-x}$ thin films using stripline resonators, *J. Supercond.* **3**, 251-259.

[2] M. Hein this NATO ASI proceedings.

[3] M. Lancaster this NATO ASI proceedings

[4] Oates, J. H., Shin, R. T., Oates, D. E., Tsuk, M. J., and Nguyen, P. P. (1993) A nonlinear transmission line model for superconducting stripline resonators, *IEEE Trans. Appl. Supercond.* **3**, 17-22.

[5] Wosik, J., Xie, L.-M., Grabovickic, R., Hogan, T., and Long, S. A., (1999) Microwave power handling capability of HTS superconducting thin films: weak links and thermal effects induced limitation, *IEEE Trans. Appl. Supercond.* **9**, 2456-2459.

[6] Nguyen, P. P., Oates, D. E., Dresselhaus, G., and Dresselhaus, M. S., (1993) Nonlinear surface impedance of $YBa_2Cu_3O_{7-x}$ thin films: measurement and a coupled-grain model, *Phys. Rev. B* **48**, 6400-6412.

[7] Cullbertson, J. V., Newman, H. S., and Wilker, C., (1998) Optical probe of microwave current distributions in high temperature superconducting transmission lines, *J. Appl. Phys.*, **84**, 2768-2787.

[8] Shen, Z. Y., Wilker, C., Pang, P., Holstein, W. L., Face, D., and Kountz, D. J., (1991) High-T_c superconductor-sapphire microwave resonator with extremely high Q-values up to 90 K, *IEEE Trans. MTT*, **40**, 2424-2431..

[9] Cohen, L. F., Cowie, A., Gallop, J. C., Ghosh, I. S., and Goncharov, I. M., (1997) Microwave power dependence in Gd 123 and Tl 2212 thin films: Examining the evidence for limiting behavior, *J Supercond.* **10**, 85-90.

[10] Kaiser, Th., Aminov, B. A., Baumfalk, A., Cassinese, A., Chalouoka, H. J., Hein, M. A., Kilesov, S., Medelius, H., Müller, G., Perpeet, M., Piel, H., and Wikborg, E. (1999) Nonlinear power handling of $YBa_2Cu_3O_{7-\delta}$ films and microwave devices, *J. Supercond.*, **12**, 343-351.

[11] Xin, H., Oates, D. E., Anderson, A. C., Slattery, R. L., Dresselhaus, G., and Dresselhaus, M. S. (1999) Comparison of power dependence of microwave surface resistance of unpatterned and patterned YBCO thin films, Submitted to *IEEE Trans. Microwave Theory Tech.*

[12] Sheen, D. M., Ali, S. M., Oates, D. E., Withers R. S., and Kong, J. A. (1991) Current Distribution, Resistance, and Inductance for Superconducting Strip Transmission Lines, *IEEE Trans. Appl. Supercond.* **1,** 108-115.

[13] Oates, D. E., Anderson, Alfredo C.,. Sheen, D. M., and Ali, S. M. (1991) Stripline resonator measurements of Z_S versus H_{rf} in thin films, *IEEE Trans. Microwave Theory Tech.* **39,** 1522-1529.

[14] Hammond, R. (1998) presentation at Fifth Symposium on High-Temperature Superconductors in High Frequency Fields, Stockholm Sweden

[15] T. Van Duzer and C. W. Turner (1981) *Principles of Superconductive Devices and Circuits,* Elsevier North Holland, New York.

[16] Karel, A. P., Velichko, A. B., Powell, J. R., Porch, A., Lancaster, M. J., and Humphreys, R. G. (1998) Unusual features in the nonlinear microwave surface impedance of Y-Ba-Cu-O thin films, *Phys. Rev. B,* **58,** 11198-11192.

[17] Westerheim, A. C., Anderson, A. C,. Oates, D. E., Basu, S. N., Bhatt, D., and Cima, M. J. (1994) Relation Between Electrical Properties and Microstructure of $YBa_2Cu_3O_{7-x}$ thin Films Deposited by Single-Target Off-Axis Sputtering, *J. Appl. Phys.* **75** 393-403.

[18] Herd, J. S., Oates, D. E., and Halbritter, J. (1997) Identification and modeling of microwave loss mechanisms in $YBa_2Cu_3O_{7-x}$, *IEEE Trans. Appl. Supercond.* **7,** 1299-1302.

[19] Herd, J. S., Oates, D. E., Xin, H., and Berkowitz, S. J. (1999) Coupled-grain/RSJ series array for modeling of nonlinear microwave surface impedance of YBCO thin films, *IEEE Trans. Appl. Supercond.* **9,** 2117-2120.

[20] Halbritter, J. (1990) RF residual losses, surface impedance, and granularity in superconducting cuprates, *J. Appl. Phys.,* **68,** 6315-6320.

[21] Halbritter J., (1995) Granular superconductors and their intrinsic and extrinsic surface impedance, J. Supercond. 8, 691-695.

[22] Anlage, S., Hu, W., Vlahacos, C. P., Steinhauer, D., Feenstra, B. J., Dutta, S. K., Thanawalla, A., and Wellstood, F. C. (1999) Microwave nonlinearities in high-T_c superconductors: the truth is out there, *J. Supercond.* **12,** 353-361.

[23] Golosovsky, M. (1998) Physical mechanisms causing nonlinear microwave losses in high-T_c superconductors, *Particle Accelerators* **61,** [351-370]/87-106.

[24] Lam, C.-W., Sheen, D. M., Ali, S. M., and Oates, D. E. (1992) Modeling the nonlinearity of superconducting strip transmission lines, *IEEE Trans. Appl. Supercond.* **2,** 58-66.

[25] Nguyen, P. (1994) Ph.D. Thesis M.I.T. Physics Dept, Cambridge, MA.

[26] Dahm, T. and Scalapino, D. J. (1997) Theory of intermodulation is a superconducting microstrip resonator, *J. Appl. Phys* **81,** 2002-2009.

[27] Hylton, T. L. and M. R. Beasley, (1989) Effect of grain boundaries on magnetic field penetration on polycrystalline superconductors, *Phys. Rev. B* **39,** 9042-9048.

[28] Martinez, O., Jimenez, J,. Martin, P., Prieto, A. C., Degoy, S., Chambonnet, D., Belout, C., Nicoletti, S., and Correra, L., (1996) A microraman study of the structural properties of PLD high T_c superconducting thin films, *Physica C* **270,** 144-154.

[29] Kästner, G., Hesse, D., Scholz, R., Koch, H., Ludwig, F., Lorenz, M., and Kittel, H. (1995) Microstructure defects in YBCO films, *Physica C* **243,** 281-293.

[30] Marshall, A. F. and Ramesh, R. (1993) Microstructure of interfaces in $YBa_2Cu_3O_{7-x}$ thin films, in *Interfaces in High-T_c Superconducting Systems*, edited by S. L. Shinde and D. A. Rudman (Springer-Verlag, N.Y.), p. 83.

[31] Kaiser, T., Hein, M. A., Müller, G., Perpeet, M., and Piel, H. (1999) Microwave field limitations of YBaCuO films by inhomogeneities and local defects, paper MPB-04 1998 Applied Superconductivity Conference, Palm Desert, CA.

[32] Willamsen, B., King, B. H., Dahm, T., and Scalipino, D. J. (1999) Microwave intermodulation in superconducting microstrip resonators, *IEEE Trans. Appl. Supercond.* **9,** 4181-4184.

[33] Hylton, T. L., Kapitulnik, A., Beasley, M. R., Carini, J. P., Drabeck, L., and Gruner, G. (1988) Weakly conducting, coupled grain model of high frequency losses in high T_c superconducting thin films, *Appl. Phys. Lett.* **53,** 1343-1345.

[34] Attanasio, C., Mariato, L., and Vaglio, R. (1991) Residual surface resistance of polycrystalline superconductors, *Phys. Rev. B* **43,** 6128-6131.

[35] Nguyen, P. P., Oates, D. E., Dresselhaus, G., and Dresselhaus, M. S. (1993) Nonlinear surface impedance for $YBa_2Cu_3O_{7-x}$ thin films; Measurements and a coupled-grain model, *Phys. Rev. B* **48,** 6400-6412.

[36] Karel, A. P., Velichko, A. B., Powell, J. R., Porch, A., Lancaster, M. J., and Humphreys, R. G. (1998) Unusual features in the nonlinear microwave surface impedance of Y-Ba-Cu-O thin films, *Phys. Rev. B*, **58**, 11198-11192.

[37] Bean, C. P. (1962) Magnetization of hard superconductors, *Phys. Rev. Lett.* **8**, 250-253.

[38] Orlando, T. P. and Delin, K. A. (1991) *Foundations of Applied Superconductivity*, Addison-Wesley Publishing Co. Reading, MA

[39] Norris, W. T. (1970) Calculation of hysteresis losses in hard superconductors carrying ac: isolated conductors and edges of thin sheets, *J. Phys D: Appl Phys.* **3**, 489-507.

[40] Nguyen, P. P., Oates, D. E., Dresselhaus, G., Dresselhaus, M. S., and Anderson, A. C. (1995) Microwave hysteretic losses in $YBa_2Cu_3O_{7-x}$ and NbN thin films, *Phys. Rev. B* **51**, 6686-6695.

[41] Sridhar, S. (1994) Non-linear microwave impedance of superconductors and critical state ac response, *Appl. Phys. Lett.* **65**, 1054-1056.

[42] McDonald, J., Clem, J. R., and Oates, D. E. (1998) Critical-state model for intermodulation distortion in a superconducting microwave resonator, *J. Appl. Phys.* **83**, 5307-5312,

[43] Golosovsky, M. A., Snortland H. J., and Beasley, M. R. (1995) Nonlinear microwave properties of superconducting Nb microstrip resonators, *Phys. Rev. B* **51**, 6462-6469.

[44] Liang, G.-C., Zhang, D., Shih, C.-F., Johansson, M. E., Withers, R. S., Ruby, W., Oates, D. E., Anderson, A. C., Polakos, P. A., Mankiewich, P. M., DeObaldia, E., and Miller, R. E. (1995) High-power HTS microstrip filters for wireless communication, *IEEE Trans. Microwave Theory Tech.* **43**, 3020-3029.

[45] Booth, James C., Beall, J. A., Rudman, D. A., Vale, L. R., and Ono, R. H. (1999) Geometry dependence of nonlinear effects in high temperature superconducting transmission lines at microwave frequencies, To be published in *J. Appl. Phys.*

[46] Oates, D. E., Habib, Y. M., Lehner, C. J., Vale, L. R., Ono, R .H., Dresselhaus, G., and Dresselhaus, M. S. (1999) Measurements and modeling of microwave impedance of high-T_c grain boundaries, *IEEE Trans. Appl. Supercond.* **9**, 2446-2451.

[47] Dahm, T. and Scalapino, D. J. (1997) Theory of intermodulation in a superconducting microstrip resonator, *J. Appl. Phys*, **81**, 2002-2009.

[48] Dahm, T., Scalapino, D. J., and Willemsen, B. A. (1999) Phenomenological theory of intermodulation in HTS resonators and filters, *J. Supercon.* **12**, 339.342.

[49] Mannhart, J. and Hilgenkamp, H. (1998) Possible influence of band bending on the normal state properties of grain boundaries in high-T_c superconductors, *Mater. Sci. Eng.* **B56**, 77-85.

[50] Oates, D. E., Nguyen, P. P., Habib, Y. M., Dresselhaus, G., Dresselhaus, M. S., Koren, G., and Polturak, E. (1996) Microwave power dependence of $YBa_2Cu_3O_{7-x}$ thin-film josephson junction edge junctions, *Appl. Phys. Lett.*, **68**, 705-707.

[51] Hein, M. A., Beuven, S., Gottschlich, M., Perpeet, M., Piel, H., and Siegel, M. (1996) Experimental evaluation of the nonlinear surface resistance of $YBa_2\,Cu_3O_{7-\delta}$ step-edge grain boundary Josephson junctions, *J. Supercond.* **9**, 233-236.

[52] Xie, L. M., Wosik, J., and Wolfe, J. C. (1996) Nonlinear microwave absorption in weak-link Josephson junctions, *Phys. Rev. B* **54**, 15494-15499.

[53] Cowie, A., Cohen, L. F., and Denhoff, M. W. (1999) The microwave power handling of an FIB generated weak link an aYBCO film, *Supercond. Sci. Technol.* **12**, 431-435.

[54] Shimakage, H., Booth, J. C. Vale, L. R., and R. H. Ono, (1999) Third harmonics generation from Y-Ba-Cu-O bicrystal Josephson junctions in coplanar waveguides, *Extended Abstracts, 7th International Superconducitve Electronics Conference*, Berkeley CA, 191-193.

[55] Habib, Y., Lehner, C. J., Oates, D. E., Vale, L. R., Ono, R. H., Dresselhaus, G., and Dresselhaus, M. S. (1998) Measurements and modeling of the microwave impedance in high-T_c Grain-Boundary Josephson junctions: fluxon generation and rf josephson-vortex dynamics, *Phys. Rev. B* **57**, 13833-13844.

[56] Habib, Y., Lehner, C. J., Oates, D. E., Vale, L. R., Ono, R. H., Dresselhaus, G., and Dresselhaus, M. S. (1998) Microwave power handling in engineered $YBa_2Cu_3O_{7-\delta}$ grain boundaries, *Appl. Phys. Lett.* **73**, 2200-2202.

[57] Lehner, C. J., Oates, D. E., Habib, Y., Dresselhaus, G., and Dresselhaus, M. S. (1999) Modeling the microwave impedance of high-T_c long Josephson junctions, *J. Supercond.* **12**, 363-375.

HTS RESONATORS AND FILTERS

Salvador H. Talisa
Northrop Grumman Corporation
Electronic Sensors and Systems Division
Baltimore, MD 21203

1. Introduction

Among the most important applications of HTS microwave technology are high-performance passband filters because they can be made in planar configurations. Filters are often the dominant contributors to system volume, in particular when banks of low-loss filters are required. In order to realize passband filters with low loss, narrow bandwidth and high selectivity, it is necessary to use high-Q structures which, in conventional technology, can only be obtained at the expense of high volume. HTS planar configurations like microstrip or coplanar waveguide, on the other hand, have Q comparable to microwave cavities at a much smaller volume (1) and so HTS is an attractive approach to reducing the volume of high-performance filters.

2. From Resonators to Filters

A straightforward way of thinking of band-pass filters is as coupled resonators. The performance of a resonator is characterized by its quality factor Q. When the resonator forms part of an electrical circuit, the circuit delivers and takes back energy from the resonator, affecting its characteristics. The unloaded Q of a resonator, Q_u, is its intrinsic quality factor, without the effects of an external circuit. A first-order filter consists of a single resonator. Its bandwidth can be adjusted by the degree of coupling between the resonator and the external circuit. In a lightly coupled resonator, little disturbance is introduced by the input and output circuits and its resonant conditions and bandwidth are close to those of an ideal, unloaded resonator. When coupling into the resonator is strong, the disturbance is large and the Q is now dominated by both the resonator and the external circuit, making the total or loaded Q lower than the ideal, unloaded Q, and therefore the filter bandwidth is now wider. For a higher-order filter composed of several resonators coupled together, the idea is the same. For narrow band filters, the resonators are loosely coupled and the minimum bandwidth is limited by the unloaded Q of the resonators. Wider bandwidth filters will have tighter coupling among resonators. Clearly, then, narrow-band filters are a desirable application for HTS because their inherent high Q enables narrow bandpass filters with low loss and small volume.

H. Weinstock and M. Nisenoff (eds.), Microwave Superconductivity, 149–159.

There is a trade off between bandwidth, filter order (number of resonators) and insertion loss. The following is an approximate expression for the mid-band insertion loss of a filter (2), in dB, which reflects this trade off:

$$L_{dB,n} \cong \frac{4.34}{B} \cdot \sum_{k=1}^{n} \frac{g_k}{Q_{uk}} \quad (1)$$

Here, n is the filter order, g_k are the normalized series inductance and shunt capacitance values of the low-pass prototype filter (2), B is the filter bandwidth expressed as a fraction of the center frequency and Q_{uk} is the unloaded Q of the k-th resonator. For purposes of estimation, it is assumed that all the resonators in the filter will have the same Q_u. Figure 1 illustrates the trade-off between insertion loss, bandwidth and filter order as a function of resonator Q_u. It shows how expression (1) can be used to estimate the potential of a certain filter technology, in this case HTS, and understand its limitations. The insertion loss was estimated for Chebychev-type filters (2) of the 4th and 5th orders, respectively, and for 1% and 0.1% fractional bandwidths. The purpose of this chart is to point out the difference in loss caused by increasing the filter order by one and by increasing the fractional bandwidth by a factor of ten. Figure 1 complements Figure 5 of (1) by making a connection between the insertion loss of a filter of a given order and bandwidth and a specific structure and its volume. The information provided by these two figures can readily be extended to cover other structures and technologies.

3. Design Considerations

As discussed above, one of the most important applications of superconductors is in narrow passband filters because these can be realized in planar technology, which result in small structures that can be readily integrated with other filters and circuitry. It was also discussed that the coupling between the resonators that make up a narrow band filter needs to be weak. The mechanism for implementing weak coupling between resonators must allow for control and predictability so that robust filter designs that are relatively intolerant of external spurious coupling mechanisms can be carried out. An example illustrating this point can be found in the parallel-coupled line filter topology. Figure 2 shows this topology, which is well known to be suited for microstrip filters with relative bandwidths below 15% (2). An analysis based on Figure 1, however, shows that if HTS is used, bandwidths below 1% are possible from the loss standpoint. Indeed, several research groups used this structure to make initial HTS filter demonstrations with 1 to 2% fractional bandwidths (3, 4). Table 1 shows the couplings required to achieve a 1.25% bandwidth, 4th-order Chebychev filter with 0.1 dB ripple at 4 GHz (5), as well as the spacing between resonators (see Figure 2). This spacing was calculated using commercial software based on coupled microstrip line circuit models and a simple Look-Up Table technique (5), which was generated using a two-dimensional electromagnetic field solver as a tool to refine the results of the circuit-model-based software. The effectiveness of this technique was demonstrated experimentally (5). Notice in Table 1 that two of the three required couplings are less than -30 dB and that the error in estimating the coupled-line distances given by the circuit-model software

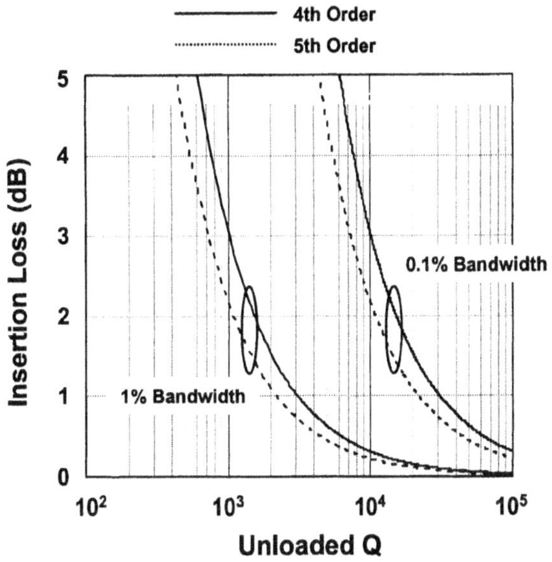

Figure 1. Estimated insertion loss of 4th- and 5th-order Chebychev passband filters of 1% and 0.1% fractional bandwidths as a function of the unloaded Q (Q_u) of the resonators that make up the filter. It was assumed that all the resonators have the same Q_u. The chart shows the increase in insertion loss caused by increasing the filter order by one and by reducing the fractional bandwidth by a factor of ten.

increases as the coupling gets weaker. For narrower bandwidths, weaker couplings are needed, which would result in larger separation between resonators and hence increased difficulty in accurately predicting and controlling the required coupling. Bandwidths of less than 1% with this filter topology could probably be achieved with great difficulty and very low yield because of the practical issues associated with controlling the required weak coupling (6).

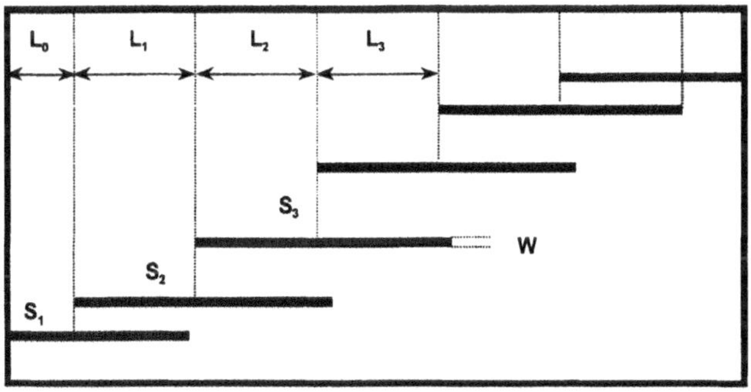

Figure 2. Parallel-coupled microstrip or stripline filter topology. The lines are λ/2 resonators coupled as λ/4 backward-coupled sections. This is a well-known configuration suitable for fractional bandwidths of less than 15% and early demonstrations of HTS filters made use of it. However, control of the weak coupling required for bandwidths less than 1%, which the low loss of HTS allows, is very difficult and results in designs that are very sensitive to material and geometrical tolerances.

TABLE 1 - Comparison of separations between resonators in a parallel-coupled microstrip filter calculated using conventional and look-up table approaches (see Figure 2)

$f_o = 4$ GHz
W = 0.176 mm
$L_1 \cong L_2 \cong L_3 \cong 4.788$ mm

Parameter	Required Coupling (dB)	Conventional (mm)	Look-Up Table (mm)
S_1	-17.6	0.572	0.530
S_2	-35.8	2.367	1.931
S_3	-37.8	2.772	2.161

Workers in this field have realized this fundamental problem and have identified alternative structures that allow significantly better control of weak inter-resonator coupling. Recognizing that in microstrip backward-coupled resonators, such as those in Figure 2, the problem is compounded by the presence of spurious forward coupling, researchers have demonstrated good control of coupling using forward coupling alone in microstrip (7) and backward coupling alone in stripline (8). Also, the use of planar lumped elements (9), and inductive coupling in coplanar line (10, 11) have been demonstrated successfully. Today, HTS filters are being made with fractional bandwidths of 1% or less by commercial companies for the base-station wireless market as pre-production prototypes (12, 13).

4. Complex Structures

The potential for filters with electrical performance similar to bulkier waveguide components but at significantly smaller sizes can be most readily fulfilled for the case of banks of filters, whether switched or multiplexed. Because HTS technology is planar, a relatively higher level of integration is possible compared to waveguide or dielectric resonator filter technology. HTS filter banks have been demonstrated by several groups in the form of band-pass multiplexers (5, 14 - 16), or banks of switched band-reject filters (17).

4.1 EXAMPLE: FOUR-CHANNEL PASS-BAND MULTIPLEXER

Figure 3 shows a diagram of a multiplexer architecture (5, 14). The number of channels that can be accommodated by this approach is limited in principle only by the bandwidth of the 90° hybrid coupler. Input microwave energy is equally split at the first coupler. If the frequency is within the passband of the two identical filters connected to this coupler, it passes through the filters and adds in phase at the output port of the output hybrid coupler for Channel 1. If the frequency is not within the passband of the Channel 1 filters, the signal is reflected back to the input coupler where it recombines such that it is out-of-phase at the input port of the first coupler and in-phase at the input port of the second channel hybrid coupler. The process then repeats itself until the signal exits the device through the appropriate channel port. Figure 4 presents details of the implementation of this device (5, 18), showing one input connector, four output connectors and a through port terminated in an external (coaxial) load. Additional channels could be connected to this port provided they are still within the bandwidth (about 10%) of the hybrid coupler used in this demonstration. Figure 4 also shows a detail of the assembly, which includes the internal HTS interconnections between filters and the integrated thin-film resistive terminations at the out-of-phase port of the output hybrid in each channel.

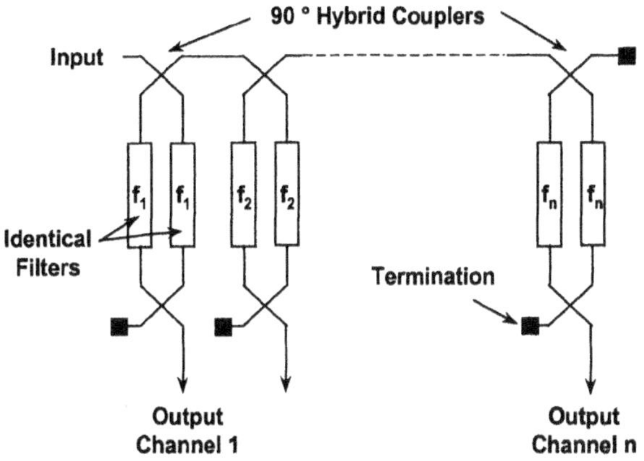

Figure 3. Multiplexer architecture used to demonstrate a 4-channel HTS microstrip device. This maximum number of channels that this configuration allows is limited by the bandwidth covered by the 90-degree hybrid. Each filter is terminated in 50 Ω and is essentially isolated from the others. Other schemes require that the impedance termination in each filter be adjusted to account for the presence of all the filters in the multiplexer, practically limiting the maximum number of channels to ten or twelve.

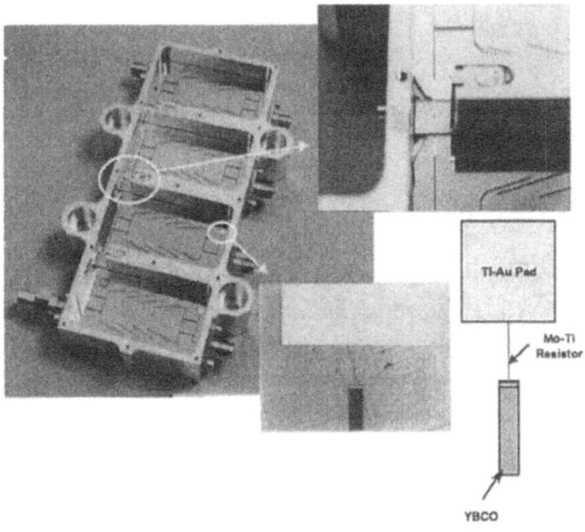

Figure 4. Photo montage of a four-channel YBCO-on-LAO microstrip multiplexer demonstrated under the U. S. Navy's High-Temperature Superconductivity Space Experiment II (HTSSE-II). Details of the design, fabrication and assembly of this device can be found in references (8) and (47).

The HTS material used for this work was YBCO thin film deposited on 500-μm-thick lanthanum aluminate (LAO) substrates. The package included an aluminum frame holding the external coaxial connectors and the niobium carriers onto which the LAO substrate pieces were mounted. Niobium has a good thermal expansion match to LAO, so the electrical and mechanical integrity of the device was preserved during thermal cyclings from ambient temperature to near 77 K. The substrate-carrier assemblies were mounted on the aluminum frame using a beryllium-copper spring arrangement. Figure 5 shows the measured performance. The low-frequency skirts of channels 2, 3 and 4 show some level of interaction between channels that could be eliminated using a wider guard-band between channels. References (5, 18) include a full discussion of the design, fabrication and measurements on this device.

This unit was one of a number of demonstration devices delivered to the U. S. Navy's High-Temperature Superconductivity Space Experiment II program by several contractors for inclusion into the space package (19).

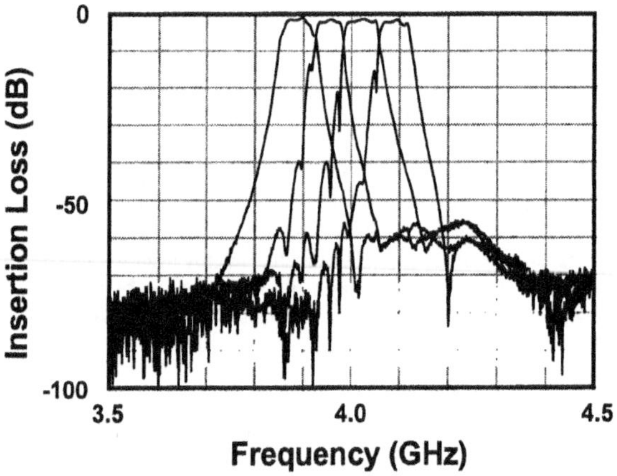

Figure 5. Measured response of the four-channel HTS multiplexer. Further details can be found in (8, 47).

5. High-Power-Handling Designs

Some of the most promising applications require that HTS filters handle sufficiently high signal power levels to preserve linearity over a significant dynamic range. As explained earlier, current crowding near the edges of typical planar transmission lines (e.g., microstrip, stripline and coplanar waveguide) ultimately limits the maximum power level that the filter can handle. Increasing the quality of the material and improving the design of filter structures has been a major endeavor at several institutions. Improved filter designs are based on planar structures which minimize current crowding at the edges, as is the case of low-impedance microstrip lines (20). Another promising approach is the use of planar resonator structures based on the circular TM_{010} mode (21-23). The salient features of this approach are shown in Figures 6(a) and (b), which show the electromagnetic fields and current profile in a microstrip and a disk resonator, respectively. In the latter, the RF magnetic fields do not close above the substrate but within it, under the disk. Thus the current density does not peak at the edges of the resonator and its distribution is fairly uniform. The only possible drawback of this approach is that the fields are so confined to the disk resonator that coupling between resonators to form a filter may be more difficult, perhaps requiring three dimensional structures for proper control of the coupling. This would reduce some of the advantages of planar integration. However, planar filters using disk resonators have been successfully demonstrated (21).

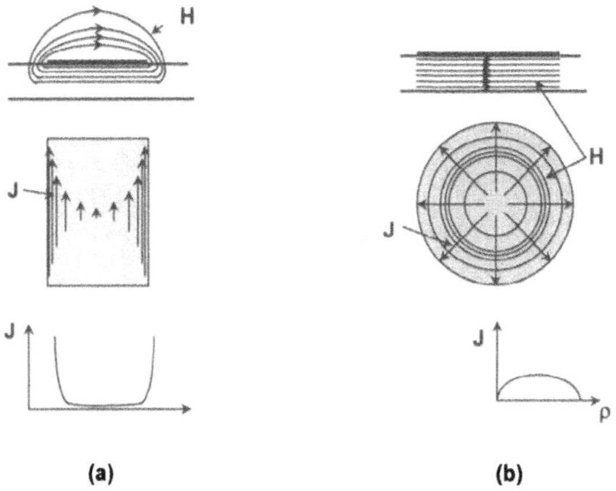

Figure 6. Diagram due to Dr. Z-Y Shen (E. I. Du Pont de Nemours and Co.) (29), showing the magnetic field and current distribution in a $\lambda/2$ microstrip resonator (a) and a TM_{010} disk resonator (b). In the disk, the magnetic field lines are circular and remain on the plane of the disk, so the current distribution is much more uniform than in the case of the regular microstrip resonator. The advantage of the disk is that it can handle much higher power levels (21-23).

6. Cryogenic Packaging

Key to the insertion of superconducting microwave circuits into electronic systems is the integration of the HTS components with a cryogenic refrigerator and its associated control electronics. Clearly, for HTS technology to be successful, the cryogenic aspects of the system must be made transparent to the user by using long-life-time cryocoolers and optimally small cryogenic packages with standard envelop characteristics and interfaces (e.g. 19-inch rack mounts and back-plane blind-mate connectors).

Many important considerations enter into the design of a cryogenic package suitable for a microwave HTS subsystem. Figure 7 is a schematic representation of this package, showing its main elements and the various heat inputs that must be considered for an appropriate thermal design. Reference (13) provides specific details on the cryogenic package for a communications filter subsystem.

The choice of a cryocooler will depend on the particular system under consideration and its cooling requirements. An airborne application may require the use of a small Stirling-cycle cooler because of volume restrictions. On the other hand, a communications ground station in a remote location that needs to operate unattended for a long period of time, may require a larger, more reliable refrigerator of the Gifford-McMahon type. Cooling requirements are imposed by the component or subsystem to be cooled, which will determine the amount of cooling power required at the operating temperature.

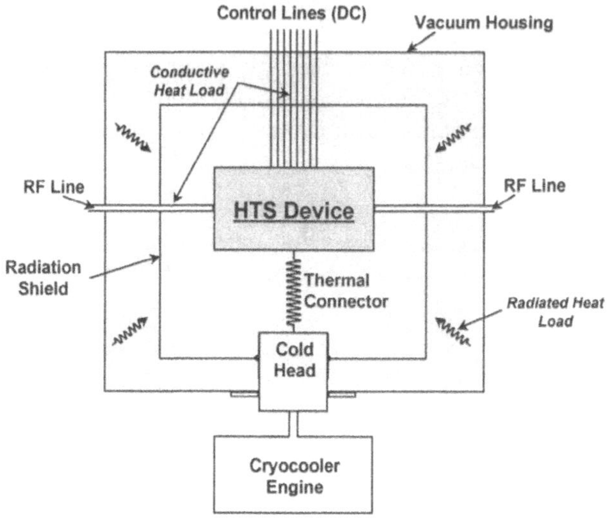

Figure 7. Schematic diagram of the cryogenic package for a hypothetical HTS device showing conducted heat inputs through input/output RF and DC control lines and the mechanical support of the cold head, as well as the radiated heat input from the (warm) wall of the vacuum housing. The purpose of this diagram is to show the main elements that affect the design of the cryogenic package.

Cryocoolers likely to be used in microwave HTS technology will typically have from 1 to 5 Watts of cooling capacity. A primary concern that systems designers have is the reliability of cryogenic refrigerators, which varies greatly depending on their type and size. Leveraging developments in other fields, such as infrared detectors, the reliability of small cryocoolers has steadily increased in recent years, with some manufacturers claiming up to 20,000 hours of Mean Time To Failure (MTTF). On the other hand, larger laboratory or industrial units and specialized coolers for aerospace applications operate for 5 to 10 years and require minimal or no servicing. Table 2 lists some cryocooler types of interest. The intent is not to be all-inclusive but to provide a basic reference to the type of coolers most likely to be employed in HTS microwave technology. Reference (24) is a good source of the latest developments in cryocooler technology. Figure 8 is a photograph of a commercial HTS filter subsystem, showing the cryocooler and associated electronics in their open package.

7. Conclusion

HTS microwave filters are a promising technology, especially at frequencies below 3 GHz where the loss in conventional microwave materials force high-performance filters to be very large in order to achieve the required low insertion losses. Leveraging developments in infrared imaging detector technology and perhaps new developments of cooled semiconductor components for fast computer workstations, cryocooler technology is progressing to the point where long lifetimes and cryocoolers of small size and weight are now available. Great interest in the U. S. and abroad exists in the wireless commercial communications market and several companies are building and testing base-station receiver front-ends consisting of cryogenically-cooled HTS filter-LNA subassemblies.

TABLE 2 - Some cryogenic refrigerator types likely to be used in HTS technology

Cooler Type	Nominal Heat-Lift Range Available at 80 K	Comments
Split Stirling	0.5 - 3 W	Available from many manufacturers; used primarily for infrared detectors. Has a cold head separated from a compressor by a metallic pressure line up to 15 cm long.
Integral Stirling	0.5 - 5 W	Also used in infrared detectors; at least one version is being used in an HTS development prototype. The compressor and cold finger are integrated into one unit.
Gifford-McMahon	2 - > 200 W	Widely used in the support of vacuum systems for semiconductor industry; highly reliable and versatile. The compressor and cold head are separate units connected by gas lines that can be several meters long.
Throttle-Cycle	~ 4 W	Reliable and low cost. The compressor and cold head are separate units connected by gas lines that can be several meters long.
Joule-Thompson	0.5 - 2 W	Generally used as an open-cycle cooling system for IR detector applications.
Pulse Tube	0.5 - 10W	Emerging technology, low cold-head vibration and long lifetime potential

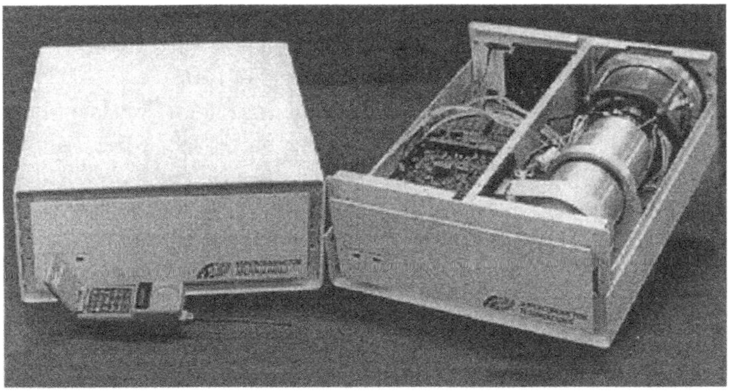

Figure 8. Photograph of a HTS filter assembly for commercial wireless applications, courtesy of Superconductor Technology, Inc

8. References

1. Talisa, S. H. Design of HTS Distributed Two-Dimensional Devices, this volume.
2. Matthaei,G. L., Young, L. Jones, E. M. T. (1980) *Microwave Filters, Impedance-Matching Networks, and Coupling Structures*, Artech House, Dedham, MA.
3. Talisa, S. H., Janocko, M. A., Moskowitz, C., Talvacchio, J., Billing, J. F., Brown, R., Buck, D. C., Jones,C. K., McAvoy,B. R., and Wagner, G. R., (1991) Low- and High-Temperature Superconducting Microwave Filters, *IEEE Trans. Microw. Theory Tech.* **39,** 1448 – 1454.
4. Lyons, W. G., Bonetti, R. R., Williams, A. E., Mankiewich, P. M., O'Malley, M. L., Hamm, J. M., Anderson, A. C., Withers, R. S., Meulenberg, A. and Howard R. E. (1991) High-T_c Superconductive Microwave Filters, *IEEE Trans. Magn.* **27**, 2537 - 2539.
5. Talisa, S. H., Janocko, M. A., Meier, D. L., Talvacchio, J., Moskowitz, C., Buck, D. C., Nye, R. S., Pieseski, S. J. and Wagner, G. R. (1996) High-Temperature Superconducting Space-Qualified Multiplexers and Delay Lines, *IEEE Trans. Microw. Theory Tech.* **44**, 1229-1239.
6. Lyons, W. G. and Lee, L. H. *Accuracy Issues and Design Techniques for Superconducting Microwave Filters*, presentation at the Computer-Aided Design of Superconducting Microwave Components Workshop, part of the 1994 IEEE International Microwave Symposium. The authors are with Lincoln Laboratory, Massachusetts Institute of Technology.
7. Zhang, D., Liang, G. C., Shih, C. F., Withers, R. S., Johansson, M. E., and DelaCruz, A (1995) Compact Forward-Coupled Superconducting Microstrip Filters for Cellular Communications, *IEEE Trans on Appl. Supercond.* **5,** 2656-2659.
8. Matthaei G. L. and Hey-Shipton, G. L. (1993) Novel Staggered Resonator Array Superconducting 2.3-GHz Bandpass Filter, *IEEE Trans. Microw. Theory Tech.* **41**, 2345 – 2352.
9. Swanson, D. G., Forse, R. and Nilsson, B. J. L. (1992) A 10 GHz Thin Film Lumped Element High Temperature Superconductor Filter, *IEEE MTT-S Int. Microw. Symp. Digest,* 1191 - 1193.
10. Swanson, D. G. and Forse, R. (1994) "An HTS End-Coupled CPW Filter at 35 GHz," IEEE MTT-S Int. Microw. Symp. Digest 199 - 202.
11. Vogt A. and Jutzi W., (1997) An HTS Narrow Bandwidth Coplanar Shunt Inductively Coupled Microwave Bandpass Filter on $LaAlO_3$, *IEEE Trans. Microw. Theory Tech.* **45**, 493 – 497.
12. Koepf, G. (1998) Superconductors Improve Coverage in Wireless Networks, *Microwaves and RF,* 63 – 74.
13. Scharen, M. J. Chase, D. R. Ho, A. M. O'Baid, A. Raihn, K. R. Forse, R. J. (1997) Filter Subsystems for Wireless Communications, *IEEE Trans. Appl. Supercond.* **7**, 3744 – 3749.
14. Mansour,R. R. Ye, S. Dokas, V. Jolley, B. Thomson, G. Tang, W.-C. Kudsia, C. M. (1996) Design Considerations of Superconductive Input Multiplexers for Satellite Applications, *IEEE Trans. Microw. Theory Tech.,* **44**, 1213 – 1228.
15. Rauscher, C. Pond, J. M. and Tait, G. B. (1996) Cryogenic Microwave Channelized Receiver, *IEEE Trans. Microw. Theory Tech.* **44,** 1240 – 1247.

16. Fiedziuszko, S. J., Curtis, J. A., Holme, S. C. and Kwok, R. S. (1996) Low Loss Multiplexers with Planar Dual Mode HTS Resonators, *IEEE Trans. Microw. Theory Tech.*, **44**, 1248 – 1257.
17. Fenzi, N. O., Rhain, K. F., Negrete, G. V., Soares, E. R. and Matthaei, G. L. (1994) An optically switched bank of HTS bandstop filters, *IEEE MTT-S Int. Microwave Symp. Digest*, 195 - 198.
18. Talisa, S. H., Janocko,M. A., Meier, D. L., Moskowitz, C., Grassel, R. L., Talvacchio, J., LePage, P., Hira, G., Buck, D. C., Pieseski, S. J., Brown J. C. and Wagner, G. R. (1995) High-Temperature Superconducting Four-Channel Filterbanks, *IEEE Trans. Appl. Supercond* **5**, 2079-2082.
19. Kawecki, T. G., Golba, G. A., Price, G. E., Rose, V. S. and Meyers, W. J. (1996) The High Temperature Superconductivity Space Experiment (HTSSE-II) Design, *IEEE Trans. Microw. Theory Tech.*, **44**, 1198-1212.
20. Liang, G. -C., Zhang, D., Shih, C. -F., Johansson, M. E., Withers, R., Anderson, A. C., Oates, D. E., Polakos, P., Mankiewich,P., de Obaldia, E. and Miller, R. E. (1995) High-temperature superconducting microstrip filters with high power handling capability, *IEEE MTT-S Int. Microwave Symp. Digest,* 191 - 194.
21. Shen, Z-Y, Wilker,C., Pang, P., Face, D. W., Carter, C. F., Harrington, C. M. (1997) Power Handling Capability Improvement of High-Temperature Superconducting Microwave Circuits, *IEEE Trans. Appl. Supercond* **7**, 2446 – 2453.
22. Chaloupka, H., Jeck, M., Gurzinski, B. and Kolesov, S. (1996) Superconducting planar disk resonators and filters with high power handling capability, *Electronics Letters* **32,** 1735 – 1736.
23. Kolesov, S., Chaloupka, H., Baumfalk, A. and Kaiser,T. (1997) "Planar HTS Structures for High-Power Applications in Communication Systems," J. Superconductivity, **10,** 179 - 187.
24. Ross, R. G. Jr., editor, (1997), *Cryocoolers 9, A publication of the International Cryocooler Conference*, Plenum Press, New York,

THREE-DIMENSIONAL CRYOGENIC FILTERS

RAAFAT R. MANSOUR
University of Waterloo
Waterloo, Ontario, Canada, N2L 3G1

1.0 Introduction

Microwave filter networks represent a critical and substantive portion of any communication systems. Such a system, be it wireless or satellite, requires filters to group the signals received into channels for amplification and processing. The phenomenal growth in telecommunication industry in recent years has brought significant advances in filter technology as new communication systems emerged demanding equipment miniaturization while requiring more stringent filter characteristics. In particular, the growth of satellite communication industry has spurred tremendous activity in the area of microwave filter design and has been responsible for many advances made in this field.

The commercial availability of low loss Dielectric Resonator (DR) materials with excellent temperature stability in early 80's and the emergence of the High Temperature Superconductor (HTS) technology in late 80's have also brought new exciting possibilities for filter miniaturization. The HTS technology, in particular, has offered the possibility, for the first time, to build filters that can compete with traditional waveguide filters not only in size but also in Q. The dielectric resonator technology is currently the technology of choice for passive microwave filters for wireless and satellite applications, while widespread use of HTS filter technology has not been yet realized.

Passive microwave filters can be divided into three main categories based on the resonator configuration whether it is a one-dimensional (1D), two-dimensional (2D) or three-dimensional (3D). Lumped elements filters may be classified as 1D filters while planar filters are classified as 2D filters. The category of 3D filters include waveguide filters, dielectric resonator filters and HTS filters, whose resonators have a 3D configuration.

In this chapter, we address the general category of 3D cryogenic filters, which includes 3D HTS filters, and other 3D filters that are designed to operate at cryogenic temperatures. We first discuss the concept of 3D resonators reviewing traditional 3D waveguide and dielectric resonator filters. Then, we present in details the performance and design considerations for three 3D cryogenic filters: cryogenic dielectric resonator filters, hybrid DR/HTS filters and HTS thick film coated filters.

H. Weinstock and M. Nisenoff (eds.), Microwave Superconductivity, 161–188.

2.0 Microwave Three Dimensional (3D) Resonators

At microwave frequencies, a resonator can be formed by enclosing a volume of space with conducting materials. Figure 1 illustrates the gradual transition from a conventional LC resonance circuit to a microwave cavity. The inductance is first replaced by just one turn, a further reduction in the inductance is accomplished by connecting a number of single turns in parallel. The reduction in the inductance is needed to construct a high frequency resonator. In the limit, these parallel-connected inductances form a cavity around the original lumped capacitor as shown in Figure 1d. A cavity resonator could have any arbitrary shape, the most common configurations are cylindrical and rectangular cavities.

The volume within the cavity can support an infinite number of electromagnetic field configurations or modes. These modes are typically classified as TE, TM or HE modes [1]-[2]. The stored electric and magnetic energies in the electromagnetic field of these modes determine the equivalent inductance and capacitance of the cavity. The resonance frequencies of a cavity are identified as the frequencies at which the stored electric energy is equal to the stored magnetic energy. The energy stored can be controlled either by varying the cavity dimensions or by inserting a tuning element into the cavity to perturb the field distributions. Thus, the resonance frequency of 3D cavity resonators is mainly determined by the cavity dimensions and can be easily adjusted by the use of a tuning element inserted inside the cavity.

By a proper choice of the cavity shape and its dimensions, it is possible to construct a cavity that will have a single mode (and thus one electrical resonator at the required frequency of operation) or have degenerate modes (i.e. modes having the same resonance frequency). The latter case allows the realization of two electrical resonators within the same physical cavity (dual-mode cavities) or three electrical resonators within the same cavity (triple-mode cavities). The main advantage of operating in dual-mode or triple-mode configurations is size reduction. These modes however, have an impact on the unloaded Q, spurious performance and power handling capability of the cavity resonator. Table 1 gives a summary of the features of each mode of operation.

Loading the cavity with dielectric materials can lead to reduction in the cavity size. Additionally, it allows better control of the temperature drift of the resonance frequency. On the other hand, dielectric loading reduces the unloaded Q and degrades the power handling capability of the resonator. The size reduction however is very attractive justifying the use of dielectric resonator filters in current commercial satellite and wireless base-station applications. The unloaded Q of a 3D dielectric–loaded cavity resonator can be written as:

$$Qu = (Rs/G + F \tan\delta)^{-1}$$

where G is a factor determined by the cavity's geometry, which typically increases as the cavity dimensions increase, while F is a factor determined by the fraction of the electrical energy of the cavity stored in the dielectric materials. Rs is the surface resistance of the cavity housing and $\tan\delta$ is the loss tangent of the dielectric material. It is clear that the unloaded Qu can be increased by reducing either Rs and $\tan\delta$ or both of them.

The HTS thick-film coated cavity filters, presented in Section 6, rely on achieving high unloaded Q values by reducing Rs. On the other hand, cryogenic dielectric filters, presented in Section 4, achieve high Q values by using particular dielectric materials whose loss tangent decreases as the temperature is lowered. The hybrid DR/HTS filters, presented in Section 5, combines the advantages of both approaches by reducing both Rs and tanδ.

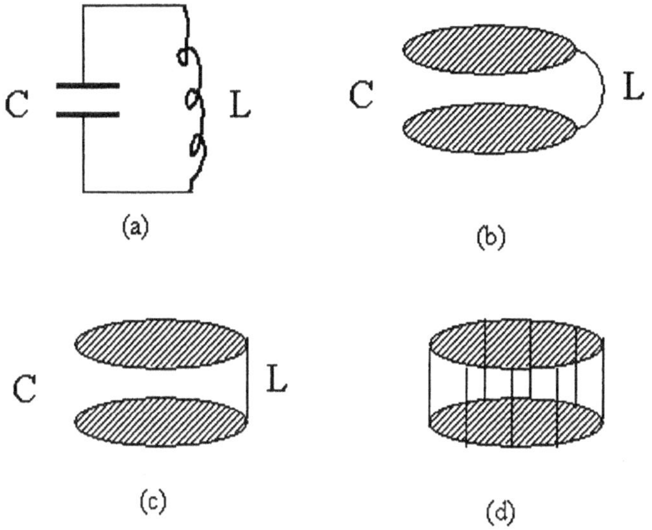

Figure 1 Schematic drawing showing the gradual transition from an open tuned circuit with lumped constants to a cylindrical resonant cavity

Table 1 Comparison between various modes of operation

Parameter	Single-mode	Dual-mode	Triple-mode
Unloaded Q	High	Medium	Medium
Size	Large	Medium	Small
Power Capability	High	Medium	Medium
Spurious Response	Good	Fair	Fair
Design Complexity	Low	Medium	High

3.0 Conventional 3D Filters

3.1 BASIC CONCEPTS OF MICROWAVE FILTERS

Any bandpass filter can be viewed as a network composed of an interconnection of resonators [3]. The interconnection (or coupling) represents a mean by which electromagnetic energy can be exchanged between the different resonators. A general network of an n-pole filter is shown in Figure 2. The resonators are numbered 1, 2,...n with the input applied to cavity 1 and the output taken from cavity n. the coupling coefficient between the ith and jth resonators is donated by M_{ij}. The M_{ij}'s are real (positive, negative or zero) numbers whose values are independent of frequency [3].

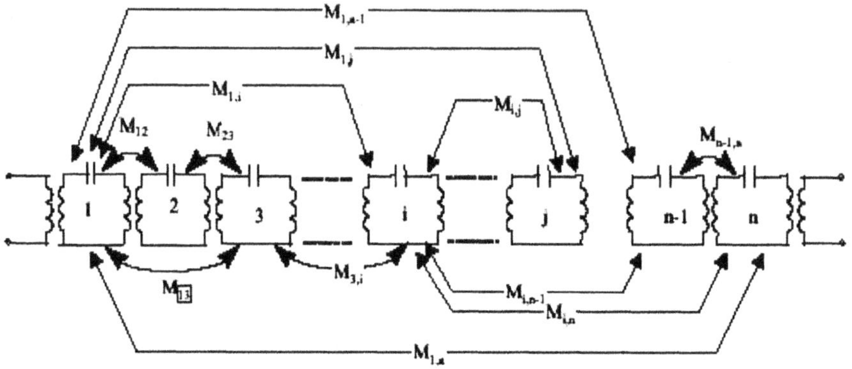

Figure 2 A general filter network of an n-pole filter.

The currents flowing in the n resonators are described by equation 1. where r = wL/Q represents the loss due to the finite Q of the resonator elements. R_1 and R_n are the impedances of the source and load of the circuit and $\lambda = (\omega - 1/\omega)$, ω is the normalized frequency [3]. The coupling elements M_{ij} (i = 1..... n, j =1......n), are calculated for Chebyshev or elliptic filters using the analysis given in [4]. The input voltage e_1 is determined by the input power level.

$$\begin{bmatrix} i_1 \\ i_2 \\ \cdot \\ \cdot \\ \cdot \\ \cdot \\ i_n \end{bmatrix} = \begin{bmatrix} j\lambda + R_1 + r & j\omega M_{12} & \cdots\cdots & j\omega M_{1n} \\ j\omega M_{12} & j\lambda + r & \cdots\cdots & j\omega M_{2n} \\ \cdot & & & \cdot \\ \cdot & & & \cdot \\ \cdot & & & \cdot \\ \cdot & & & \cdot \\ j\omega M_{1n} & j\omega M_{2n} & \cdots\cdots & j\lambda + R_n + r \end{bmatrix}^{-1} * \begin{bmatrix} e_1 \\ 0 \\ \cdot \\ \cdot \\ \cdot \\ \cdot \\ 0 \end{bmatrix} \qquad (1)$$

The coupling values are divided into two categories:

i) Coupling between adjacent resonators known as sequential coupling, e.g. M_{12}, M_{23} M_{34} ...etc.

ii) Coupling between non-adjacent resonators known as cross coupling e.g. M_{13}, M_{14}, M_{15}... M_{24}, M_{25} etc.

The coupling values (both sequential and cross coupling) play the key role in determining the filter bandwidth and its function response whether Chebyshev, elliptic or self equalized. Figure 3 shows the signal flow diagram of 4-pole and 8-pole

Chebyshev and elliptic function filters. It illustrates the impact of the coupling between non-adjacent resonators on the filter response. The coupling between non-adjacent resonators (cross coupling) can be realized with the use of folded filter structures [5] or with the use of the dual-mode concept, which is described in Section 3.

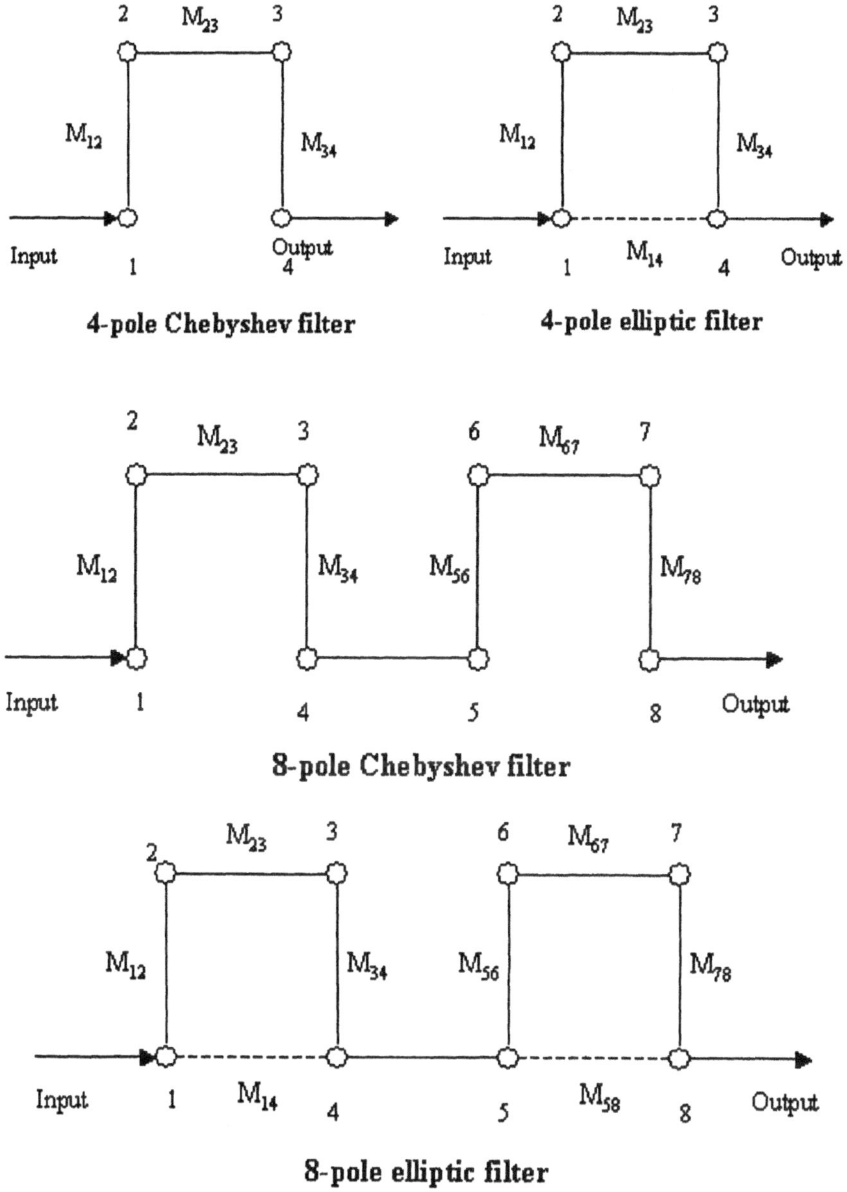

Figure 3. The signal flow diagram of 4-pole and 8-pole Chebyshev and elliptic function filters.

Figure 4 shows a comparison between microwave filters realized using single-mode waveguide technology, dual-mode waveguide technology, dielectric resonator technology and superconductor technology. All filters shown in Figure 4 operate at 4 GHz. Two types of superconductive filters are included in this Figure: one uses the hybrid DR/HTS technology and the other is based on HTS thin film technology. The following subsections briefly describe the conventional waveguide and dielectric resonator filter structures shown in Figure 4, while in Sections 4-6 we describe various configurations for 3D cryogenic filters.

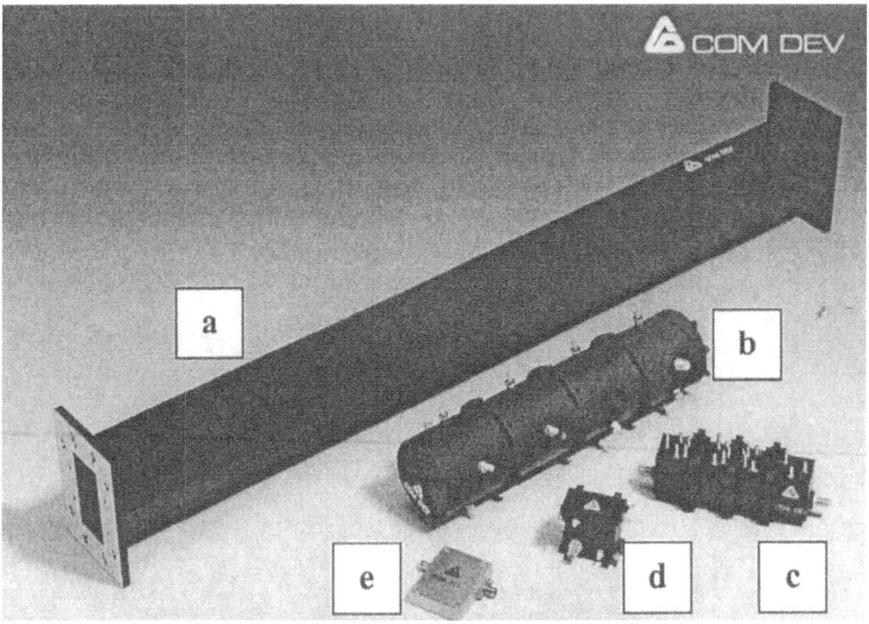

Figure 4 A comparison between C-band input channel filters. a) single-mode waveguide filter, b) dual- mode wave guide filter , c) Dual-mode DR filter, d) hybrid DR/HTS filter and e) HTS thin film filter.

3.2 WAVEGUIDE FILTERS

Figure 5 illustrates a schematic layout of a single–mode waveguide filter. The structure consists of waveguide resonators operating in the TE_{101} modes coupled by irises [6]. This type of filter has been used in many microwave applications because of its low-loss and high power handling capability and amenability to low-cost manufacturing. Satellites in the early days, e.g. Intelsat IV, used conventional Chebyshev rectangular waveguide filters for input and output multiplexers. Other versions of this filter have been also widely used, examples are: inductive post filters [7] and E-plane septum filters [8].

The idea of using degenerate modes in waveguide cavities was first proposed by Regan [9] in 1948, who described the use of dual modes in cavities of square and circular cross sections. Later in 1951 Lin [10] showed how this could be extended to 3 – 5 pole Chebyshev filters in a single cylindrical cavity. In 1953 Currie [11] described a fifth-order Chebyshev filter in a single spherical cavity.

Early in 1970's, researchers from COMSAT published a series of papers [12]-[13] establishing the theory and feasibility of using the concept of dual-modes to construct elliptic function waveguide filters. The concept of dual-mode filters had a tremendous impact on filter technology. It reduced the number of physical cavities required for a filter by a factor of 2 with commensurate reduction in mass and volume. Such promise, coupled with the burgeoning demands of satellite industry, resulted in considerable R&D activity in the area of multi-mode filter networks.

Figure 6 illustrates a typical dual-mode filter realized in circular waveguides. Each cavity supports two orthogonal modes in a single physical cavity. These modes are independently tuned with the use of two tuning screws spaced apart by 90 degree. Coupling between these two modes is achieved via a metallic screw located at 45 degree between the two tuning screws, while coupling between modes in different cavities is achieved through irises. The cross-shaped iris provides the necessary sequential coupling between modes 2 and 3 and necessary cross coupling between modes 1 and 4. In addition to size and mass reduction, the dual mode structure makes it easy to couple between non-adjacent modes allowing the realization of elliptic and self-equalization function filters.

The only drawback of such type of filters compared to standard rectangular single-mode waveguide filters is that square or circular cavities support a relatively higher number of modes limiting the achievable spurious free window of these filters. This problem however can be circumvented by controlling the dimensions of the cavities and irises.

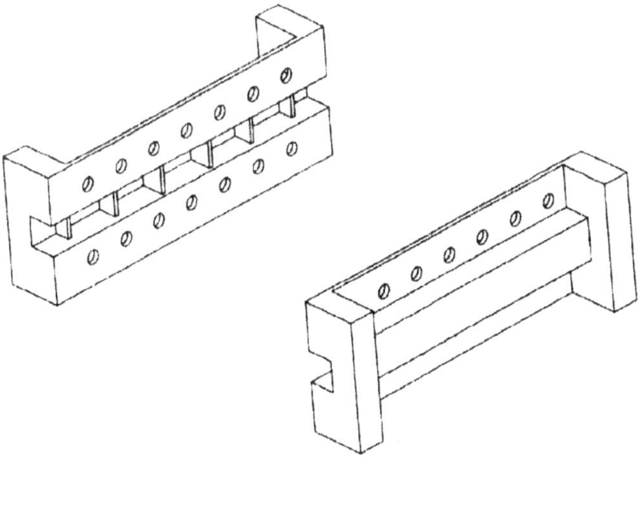

Figure 5 Layout of an inductive iris single-mode filter

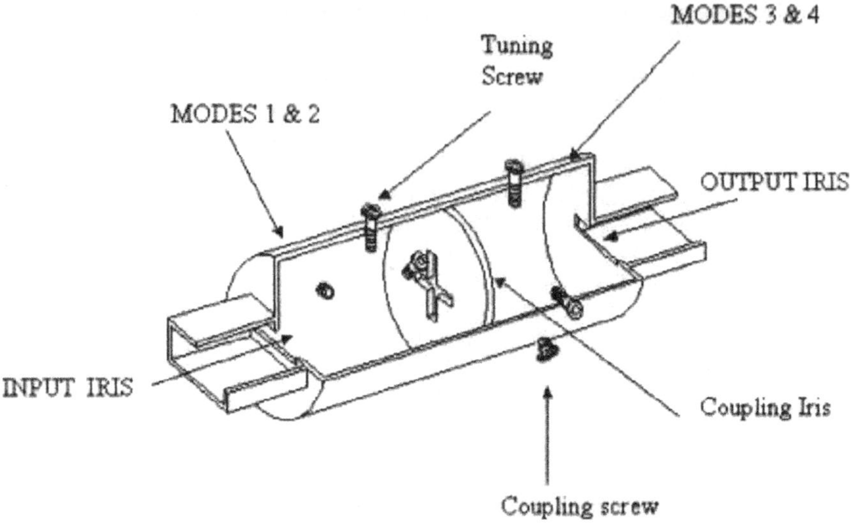

Figure 6 Layout of an inductive iris single-mode filter

3.3 DIELECTRIC RESONATOR FILTERS

In the 1960's several pioneering work were published on the behavior of dielectrics at microwave frequencies including the first reports on microwave dielectric resonator filter by Cohn in 1965 [14]-[15] and by Harrison in 1967 [16]. However, such filters were not implemented in practical applications because of the poor thermal stability of the dielectric resonator materials available at that time. The dielectric constant of these materials exhibited a large change in their values as the temperature is varied. This, in turn, caused the temperature drift of the filter center frequency to be as large as 500 ppm/C.

Advances in the 70's and 80's in dielectric material technology have made it possible to combine high unloaded Q, small temperature drift and high dielectric constant in materials suitable for use at microwave frequencies. Over the past two decades several different types of low loss dielectric composition for use in microwave filters have been developed with improvements in all areas; including loss tangent and range of dielectric constant. High Q dielectric materials with dielectric constant ranging from 20 to 90 and with a temperature drift of –6 to + 6 ppm/C are now commercially available from various manufacturers.

Dielectric resonator filters are widely used in cellular and satellite applications [17] - [19]. A typical dielectric resonator filter consists of a number of dielectric loaded cavities operating either in single TE or TM modes or HE dual-modes [19]-[20]. Each cavity consists of a solid cylindrical dielectric resonator mounted inside a metallic structure operating below cut-off. The dielectric resonators are supported inside the cavity either in an axial or in a planar configuration as shown in Figure 7 and Figure 8 respectively.

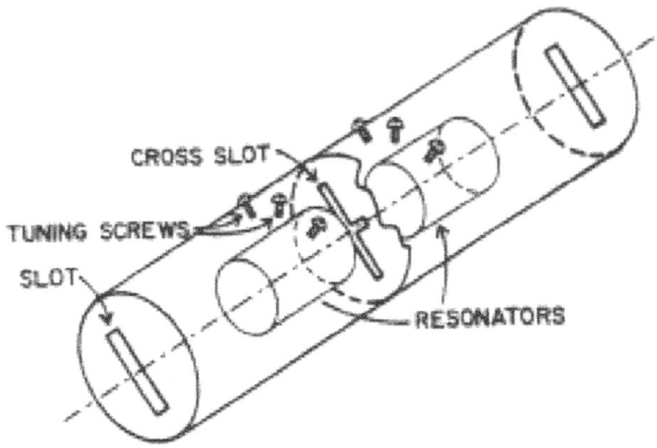

Figure 7 Schematic drawing of a dual-mode dielectric resonator filter in an axial configuration

Figure 8 A dielectric resonator filter in a planar configuration

4.0 Cryogenic Dielectric Resonator Filters

Various dielectric materials exhibit a considerable improvement in loss tangent *tanδ* upon cooling to cryogenic temperatures. For example high purity sapphire ($\varepsilon_r = 9.4$) at X-band frequencies exhibits *tanδ* of 10^{-5} at 300 K and *tanδ of* 10^{-7} at 77 K [21]. Other dielectric material such as magnesium oxide MgO ($\varepsilon_r = 9.7$), rutile ($\varepsilon_r = 105$) and lanthium aluminate $LaAlO_3$ ($\varepsilon_r = 23.4$) exhibit a similar loss tangent improvement at cryogenic temperatures [22] as compared to room temperature operation.

The observed temperature dependence of *tanδ* is consistent with the theoretical model given in [23] which describes the source of the intrinsic losses in dielectric materials. Unfortunately intrinsic losses are not the dominant contribution, extrinsic losses due to defects also have a strong impact on the behavior of *tanδ* at cryogenic temperatures. Figure 9 illustrates the temperature dependence of the loss tangent of two $LaAlO_3$ crystals grown by Czrichalski and Verneuil techniques [21]. Both samples show improvement in loss tangent as the temperature is decreased from 300 K to 77 K, however, the crystals grown by the Czrichalski techniques exhibit a peak in loss tangent around 77 K. Verneuil crystals on the other hand, exhibit a relatively small peak yielding a loss tangent at 77 K which is 5 times smaller than the loss tangent observed at 300 K.

Some of the dielectric resonator materials, which were developed for conventional room temperature filter applications, exhibit low loss at cryogenic temperature while maintaining very small temperature drift. Example is Ba-Mg-Ta-O compound manufactured by Murata [24] with a permittivity around 22. Figure 10 illustrates the measured loss tangent versus temperature [25] of BMT. Other dielectric materials manufactured by Murata and other suppliers exhibit a similar loss tangent improvement at cryogenic temperatures. The results indicate the potential of improving the loss tangent of dielectric materials filters when operating at cryogenic temperatures.

The above discussions also suggest that with a proper choice of dielectric material, conventional dielectric resonator filters can be cooled to improve their insertion loss performance. As with conventional DR filters, these filters will need to be tuned at cryogenic temperatures. A special tuning set-up will be required to allow the access of tuning screws at cryogenic temperatures. There have been reports [26] suggesting the possibility of designing cryogenic filters that are operational at both room temperatures and cryogenic temperatures. These filters however, will require having a temperature compensation mechanism to account for the frequency drift as the temperature changes from room temperature to cryogenic temperature.

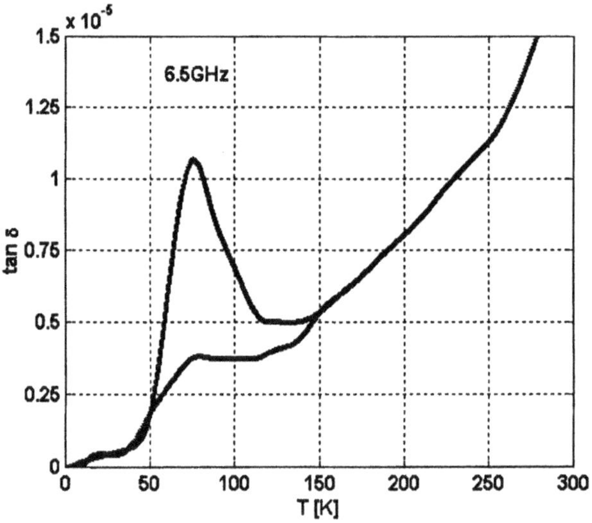

Figure 9. The temperature dependence of the loss tangent of two $LaALO_3$ crystals grown by Czrichalski and Verneuil techniques

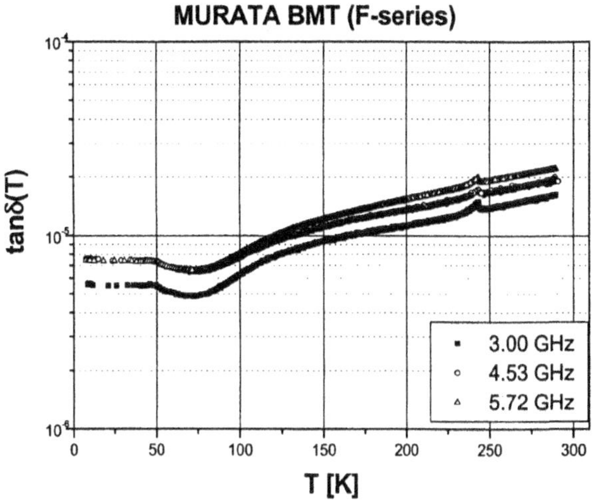

Figure 10. The measured loss tangent versus temperature of BMT material

A fully automated measuring system to measure Q and dielectric constant, over the temperature range 10 - 310 K, has been proposed in [27]. Figure 11 illustrates the DR cavity used in the measurement. It consists of a dielectric resonator supported on Quartz spacer in a copper cavity. Tuning is achieved by using a top plate that is adjustable by means of a screw. The TE_{01} fundamental mode is used for measurement. This mode is used because the Q of the mode is easily related to the dielectric loss tangent. Also, the electric field is symmetric with the geometry of the sample and cavity, which helps to reduce the sources of loss due to the cavity.

Measurements are made using an RF probe that is weakly coupled to the cavity so that errors in the insertion loss due to uncertainly of the cable loss at different temperatures do not contribute to an error in the unloaded Q. The assembly is placed on the cold head of a cryocooler, which allows the temperature to be varied between 20 and 300 K. Table 2 illustrates the measured Q of four different ceramic compounds at 20 K, 77 K and 300 K.

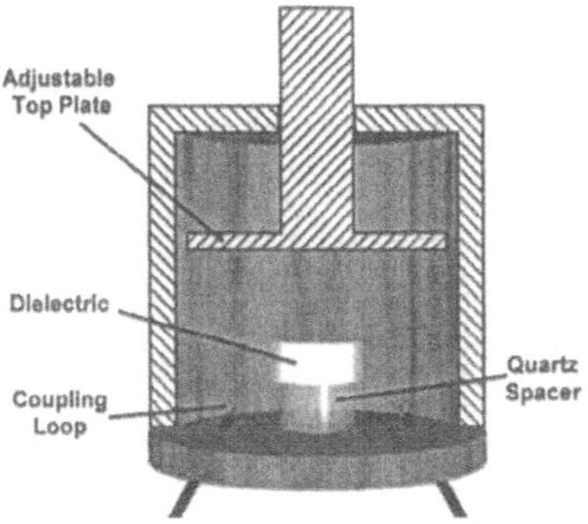

Figure 11 A dielectric resonator cavity for measuring unloaded Q

Table 2 The measured Q values for different ceramic compounds.

Material	Qu at 20 K	Qu at 77 K	Qu at 300 K
Alumina	704,000	269,000	40,000
BaMgTaO (BMT)	87,000	37,000	21,000
ZrSnTiO	56,000	17,000	5,300
TiO_2	77,000	30,000	4,400

Any conventional dielectric resonator filter can be potentially designed to operate at cryogenic temperatures. The key design considerations are the proper selection of the dielectric materials, and the proper selection of the resonator support structure. The designer needs to pay attention to thermal stability of the support structure.

Figure 12 shows a schematic view of a cryogenic dielectric resonator dual-mode 2-pole filter proposed in [21] for high power applications. The resonator has a hemispherical shape and is made of $LaAlO_3$. The resonator is placed in a cylindrically shaped metallic cavity and is isolated from the ground plane by a dielectric spacer made also of $LaAlO_3$. Unloaded Q of 46,000 at 3.7 GHz and at 77 K has been reported in [21] for such resonators.

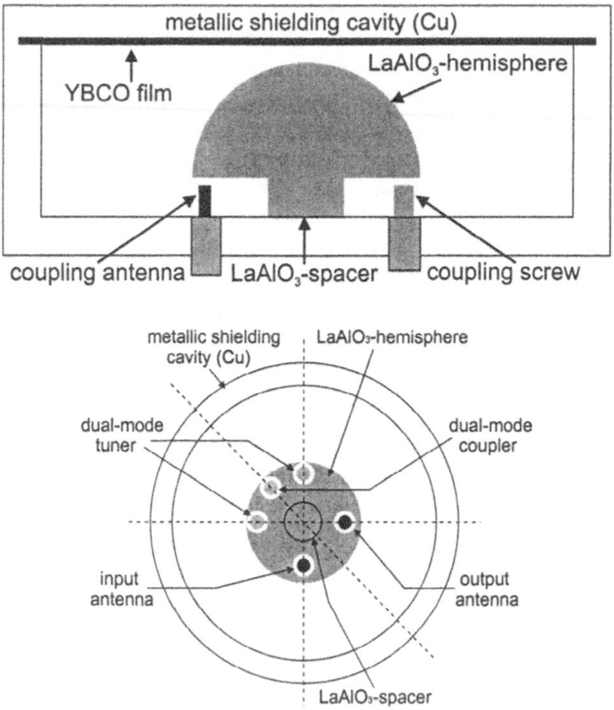

Figure 12 A cryogenic 2-pole filter built using a dielectric resonator having a semi spherical shape

5.0 Hybrid Dielectric/HTS Filters

For the conventional dielectric resonator shown in Figure 13a, the tangential electric field of the HEE_{11} mode vanishes at the plane z=0 {20]. As a consequence, introducing a conducting wall at this plane will slightly perturb the field distribution of this mode. Thus the size of the dielectric resonator filter operating in the HEE_{11} mode may be reduced by supporting the resonator directly on a conducting surface (image-plate) as shown in Figure 13c. With the use of dielectric support that fills the whole cavity, further size reduction can be achieved as shown in Figure 13d.

A normal-conducting image-plate however, significantly degrades the Q of the resonator, which in turn makes the approach unfeasible for high-Q applications. On the other hand, if the image-plate is replaced by a surface made of high temperature superconductor materials, the original unloaded resonator Q will be only slightly affected allowing the use of this resonator structure to realize compact size filters with a superior loss performance for cryogenic applications. The reduction in size is attributed to the use of the image-plate, while the improvement in loss performance is attributed to the use of the HTS image-plate as well as to the increase in the unloaded Q of the dielectric resonator as the temperature is lowered from 300 K to 77 K.

Several hybrid DR/HTS filters, which utilize the concept illustrated in Figure 13 were built and tested [28]- [29]. Figure 14 shows a detailed configuration of a 4-pole hybrid DR/HTS filter [29]. The filter consists of two dielectric resonators operating in image-type dual-modes. The resonators are supported inside the filter housing using two blocks made of low-loss low-dielectric constant ceramic materials. The HTS material is in the form of small wafers (0.5 "x 0.5" at C-band), which are diced from a standard 2" HTS thin film deposited on 0.02 inch thick wafer of Lanthium Aluminate. The HTS films are kept in contact with the dielectric resonators using springs and metallic plates bolted to the filter housing. Both the housing and the ceramic blocks have holes to allow the filter to be tuned using standard metallic screws. It is to be noted that the shorting plate is not limited to be square in shape; it could be rectangular, circular or any other shape of any size as long as it covers the circular faces of the dielectric resonators.

Details of the support configuration are shown in Figure 15. The support block is made of ceramic materials with low loss tangent such as Transteck D4 ($\varepsilon_r = 4$) or Sapphire ($\varepsilon_r = 9.4$). The use of dielectric support with a higher dielectric constant will help to further reduce the size of the filter. The support block is machined to allow a snug fit between the dielectric resonators and the HTS image plate. The support may also have other holes, which are designed to be in alignment with probes and tuning screws. Figure 16 illustrates an 8-pole version of the dual-mode hybrid DR/HTS filters. The measured performance of this filter is given in Figure 17. This type of filters can be also built to operate as single-mode filters. A detailed description of the single-mode version of this type of filters is given in [29].

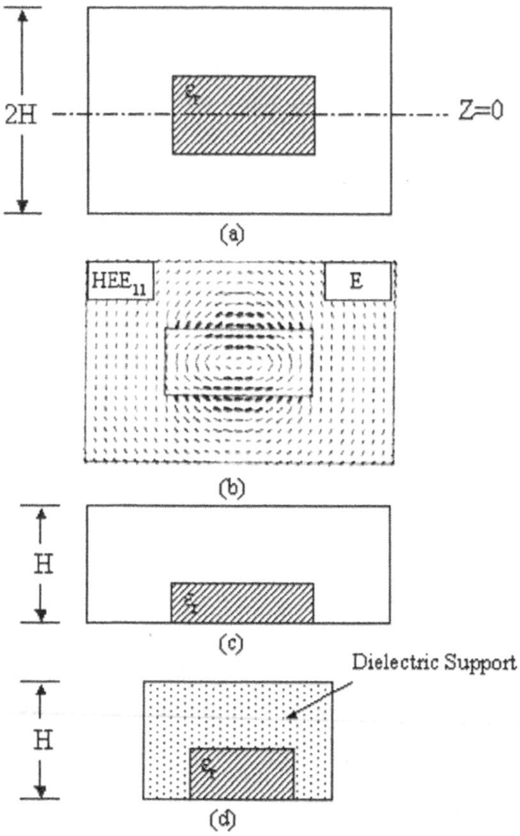

Figure 13 Transition from conventional DR cavity to an image DR cavity. A) conventional DR resonator , b) field distribution c) Image DR resonator, d) Image DR resonator with a DR support.

The hybrid DR/HTS filter can handle extremely high power levels in the range 50-100 watts [30]. The insertion loss of this type of filters is mainly determined by the loss tangent of the dielectric resonators and the ceramic blocks as well as the surface resistance of the HTS materials. The loss tangent of the dielectric material improves at lower temperatures, which helps to lower the insertion loss of the filter. The advantages of hybrid DR/HTS filters over conventional HTS thin film planar filters are:

i) No etching is required for the HTS material. Etching may reduce the power handling capability of the HTS material and degrade its surface resistance. The design also eliminates the need to deposit gold contacts, which are required for thin film filters. Also there is no need to generate circuit and contact masks.
ii) In patterned thin film filters, a good portion of the HTS materials is wasted due to etching process, while in hybrid DR/HTS filters the wafer is diced into small pieces "shorting plates". The wafer is used efficiently.
iii) The filters can be easily tuned using conventional tuning screws.

iv) With the availability of low-loss dielectric resonators with high dielectric constant in the future, such filters can be potentially made smaller in size than their counterparts HTS thin film filters.

v) The filter has a good spurious performance. The image plate not only helps to reduce the filter size but it also helps to eliminate spurious modes making it possible to design filter of this type with a spurious free window of more than 2 GHz width centred at a frequency of 4 GHz.

The major drawback of this class of filter is the mechanical design complexity: the filter must be thermally stable to ensure performance repeatability as the temperature changes from cryogenic to room temperature (for shipping or storage) and then back to cryogenic temperature (for testing and operation).

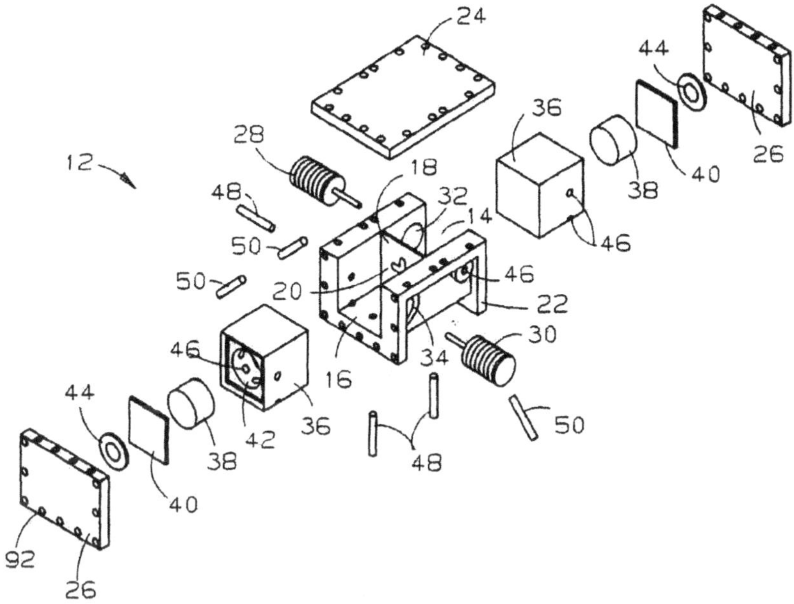

Figure 14 A detailed description of a 4-pole Hybrid DR/HTS filter. 38's are dielectric resonators. 40's are HTS wafers diced from a standard 2" wafer, .36's are dielectric supports . 44's are spring washers. 22, 24 and 26 are elements of the filter housing.

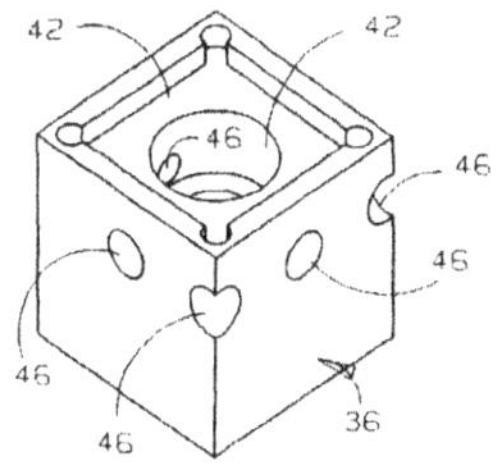

Figure 15 A detailed description of a the dielectric support structure 36 shown in Figure 14

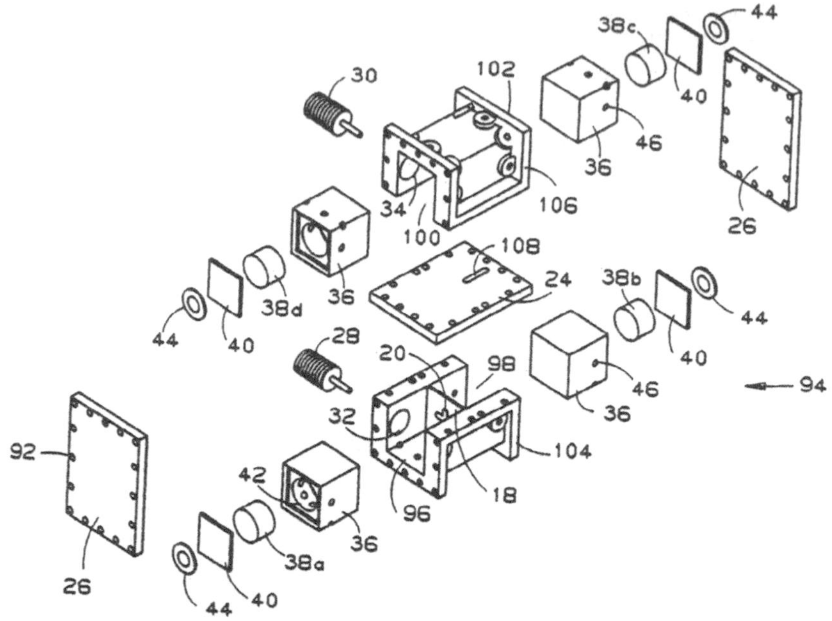

Figure 16 A layout of the 8-pole Hybrid DR/HTS filter:

- 38's are dielectric resonators
- 40's are HTS wafers diced from a standard 2" wafer
- 36's are dielectric supports
- 44's are spring washers
- 26's , 102 and 104 are element of the filter housing
- 28's are input/output probes

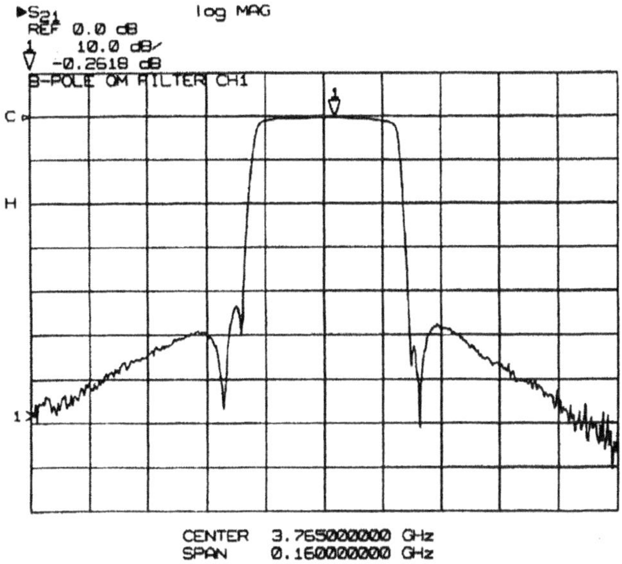

Figure 17 The measured RF isolation of the 8-pole hybrid DR/HTS filter shown in Figure 4.3.

The overall Q-factor of the hybrid DR/HTS resonator can be written as:

$$1/Q_{total}=1/Q_{DR}+1/Q_{sidewall}+1/Q_{topwall}+1/Q_{ground}$$

The factors Q_{DR} , $Q_{sidewall}$, $Q_{topwall}$ and Q_{ground} represent the losses in dielectric resonator and support structure, sidewalls, top wall and bottom wall respectively. Figure 18 illustrates the resonant frequency and unloaded Q as a function of the ratio of the diameter to length ratio for the image dielectric resonator configuration shown in Figure 13c. Two modes are considered, HE and TE. For each mode the loss contributions from sidewalls, top walls, bottom walls, and dielectric losses were evaluated in [28]. Although the TE mode offers a higher unloaded Q, the HE dual-mode was chosen since it allows the realization of miniature filter structures. Figure 19 shows the unloaded-Q results for three resonators:

i) hybrid dielectric/gold thin film at 300 K.

ii) hybrid dielectric/gold thin film at 77 K.

iii) hybrid dielectric/HTS thin film at 77 K.

In computing these results, loss tangent of 0.00005 and 0.00001 were assumed for the dielectric resonator materials at 300 K and 77 K respectively. It can be seen that the use of HTS thin films offers a significantly higher unloaded Q. The measured results of the 8-pole filters presented in Figure 17 indicate an unloaded Q in the range of 8000-10000 [28]. This degradation is attributed to additional looses associated with the support structure and the tuning screws.

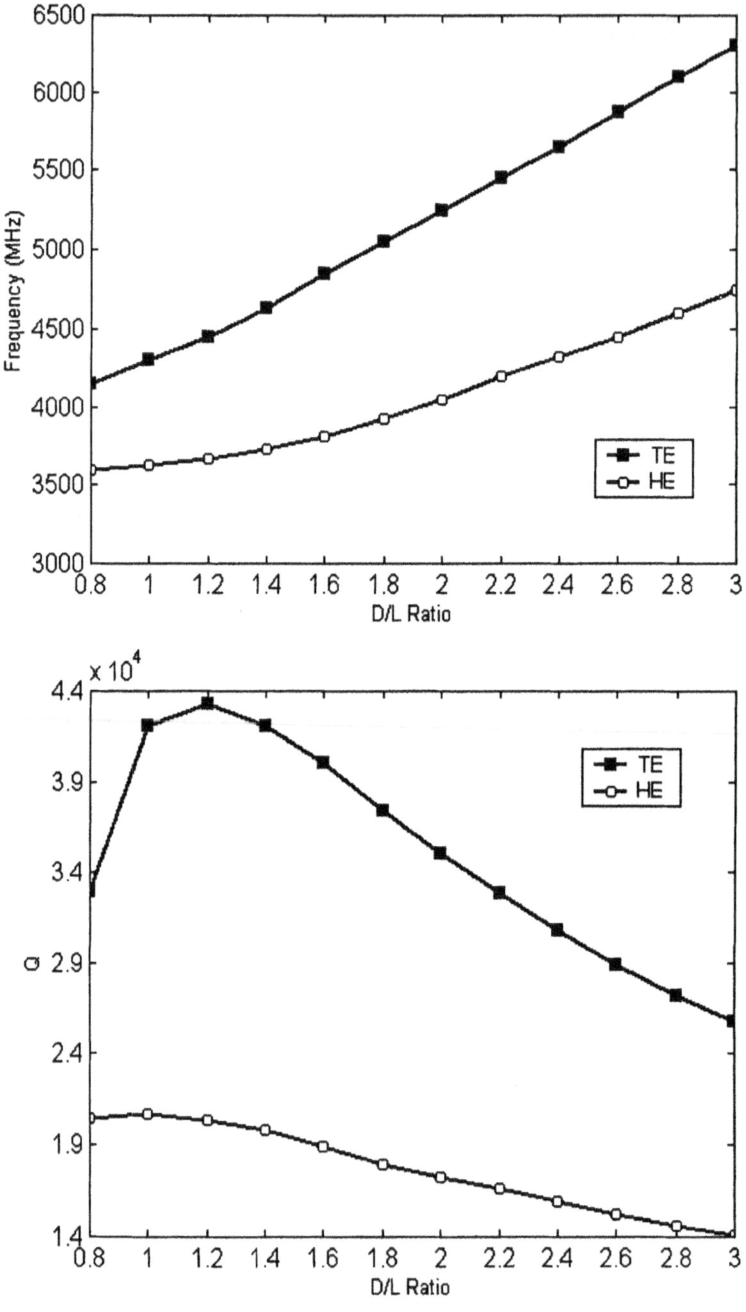

Figure 18 The resonant frequency and unloaded Q as function of the ratio of diameter to length of dielectric puck for the dielectric resonator configuration shown in Figure 13c.

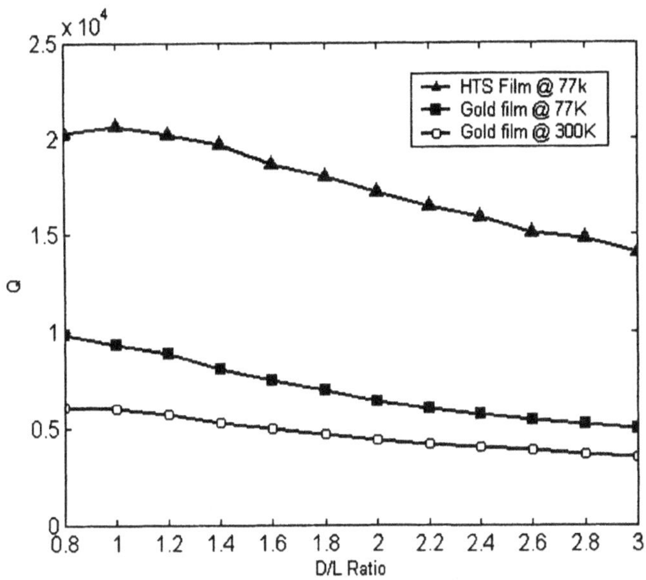

Figure 19 The unloaded-Q results as function of ratio diameter to length of dielectric puck for three resonators

6.0 Thick Film HTS-Coated Filters

Melt processed YBCO thick films display low surface resistance and can be used to coat three-dimensional structures for microwave applications [31]-[32]. YBCO thick films typically have a current density of 2000 A cm^{-2}, posses a far better capability of handling high levels of magnetic field than bulk sintered HTS and have a relatively small value of surface resistance of approximately 2 mΩ (10 GHz at 77 K). Figure 20 illustrates a comparison between the surface resistances of bulk YBCO, metal processed YBCO thick film and YBCO thin films.

The feasibility of constructing a thick film YBCO 5.66 GHz cavity resonator operating in TE_{011} modes with unloaded Q values of 715,000 at 77 K has been demonstrated in [33]. Measured results of cavities operating at 10 GHz and 7.5 GHz have been also reported in [34] . However, the size of TE_{011} cavities is too bulky to be used at frequencies below about 1-2 GHz. At lower frequencies, two resonator structures have been proposed for thick film coating: coaxial resonators and split-ring resonators. Thick film coaxial resonators have found limited use because of the need to coat all of the surfaces in the cavity to achieve reasonably high Q values. On the other hand, the split-ring resonator has been used by Illinois Superconductor Corporation to manufacture thick film HTS filters for wireless applications [35]- [36].

The split-ring resonator was first proposed in 1976 [37] as an accelerating structure. It was also proposed in [38] for use in magnetic resonance applications. Figure 21 illustrates the configuration for a split-ring resonator proposed in [39] for constructing filters for low frequency applications. The resonator consists of a rectangular plate bent to form a hollow cylinder with a gap running from the top to the bottom of the cylinder at the ends of the bent plate. The resonator is then inserted in a larger metallic cavity to provide shielding. The typical characteristic dimensions are of the order of 1/10 of the resonant wavelength, and typical Q's are of the order 2000 in the frequency range of 1-2 GHz. For example, a silver-coated split-ring resonator at 1413 MHz would have a Q of 2020 and a length of 17.8 mm [39]. The Q would be significantly higher by using YBCO coating instead of silver coating (more than 10 times at 5 GHz). Figure 22 illustrates two 3-pole filters employing the split-ring resonators concept. Coupling between resonators in the filter shown in Figure 22a is achieved through the use of a waveguide section below cutoff while the filter shown in Figure 22b employs irises to provide the right coupling between resonators. The resonators in Figure 22a have a relatively large length to increase the Q. Details on the YBCO thick film version of this filter can be found in two patents filed by Illinois Superconductors [35] and [36].

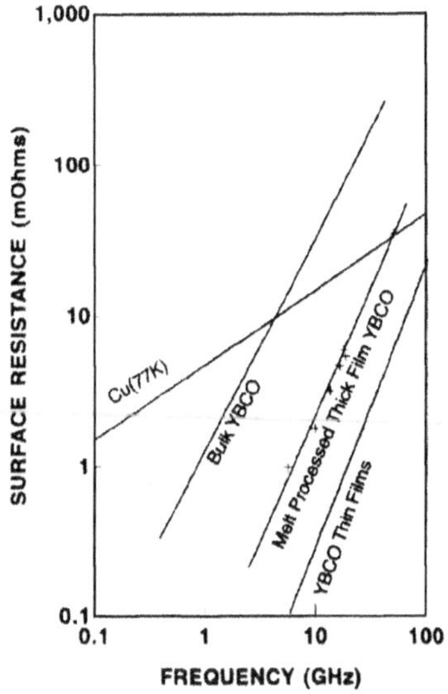

Figure 20 A comparison between the surface resistance of Bulk YBCO, Metal processed YBCO thick film and YBCO thin films.

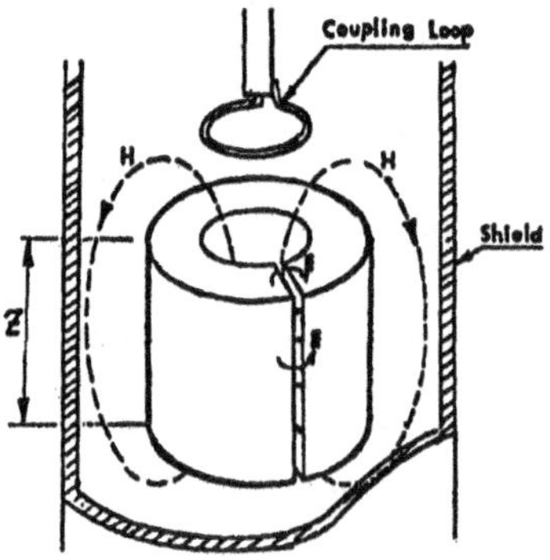

Figure 21 A schematic drawing of a split ring-resonator

(a)

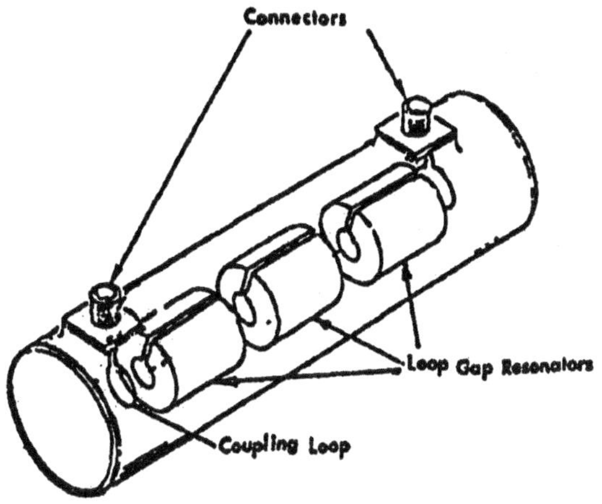

(b)

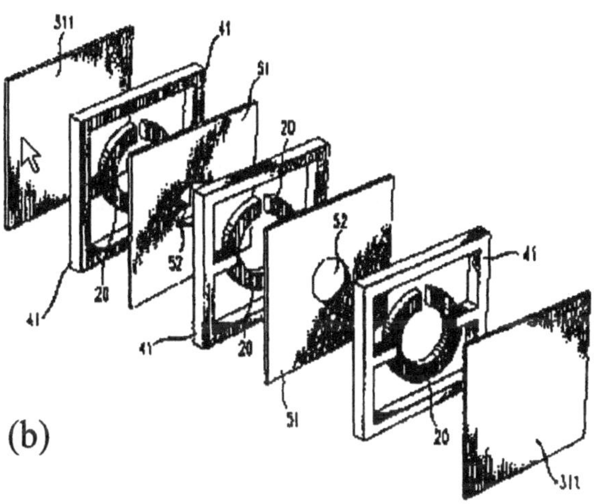

Figure 22 Two 3 -pole filters employing split-ring resonators. a) coupling between resonators is achieved using a below-cutoff waveguide section, b) coupling between resonators is achieved by using irises.

7.0 Design and Tuning of Cryogenic 3D Filters

With the use of an accurate ElectroMagnetic (EM) simulator and an optimization software package any of the 3D filters described in this Chapter can be accurately designed. However this approach may be very computation-intensive. For example, it will take months, using today's computer workstation, to completely optimize the performance of the 8-pole hybrid DR/HTS filter shown in Figure 16. Additionally as will be explained in Section 7.3, tuning is considered a must for any filter that is designed to meet the stringent wireless and satellite requirements. Therefore designers of these types of filters basically restrict the design process to the use of filter synthesis techniques to calculate the coupling matrix and use the EM-simulator to calculate resonance frequency (i.e. resonator physical dimensions) and coupling values (i.e. iris physical dimensions). This approach neglects higher order mode interaction and does not take into consideration the input/output coupling. The designers however typically rely on tuning to complete the design process. In the following sections we address the circuit synthesis design techniques, ElectroMagnetic (EM) simulation techniques tools and the tuning techniques, which are typically used in the design of these filters.

7.1 FILTER SYNTHESIS TOOLS

There are numerous techniques for filter synthesis. The majority of these techniques however, are limited to certain filter response functions and fail to describe the generalized filter structure described in Figure 2. These techniques deal only with resonators coupled in a single cascade configuration from which only Chebyshev function responses can be produced. Coupling between non-adjacent resonators need to be considered when more general filter characteristics are desired.

Two techniques are published in [40]-[4] to deal with the generalized filter structure shown in Figure 2. These two techniques are used to evaluate the coupling matrix M, allowing the realization of the desired filter response with minimum number of coupling elements. The techniques are capable of analyzing completely general type of response characteristics including Chebyshev, elliptic, quasi-elliptic and self-equalization.

7.2 EM SIMULATION TOOLS

Once the coupling matrix is determined, the next step is to find the physical dimensions of the resonator and the coupling elements in order to construct the filter. The resonators described in Sections 4, 5 and 6 of this Chapter have irregular shapes but can be accurately characterized using techniques such as Finite Element Methods (FEM) and Finite-Difference Time Domain (FDTD) techniques.

Over the past five years, several EM simulation software packages have become available commercially to characterize arbitrarily shaped microwave structures. The list includes HFSS [41]-[42] and Microwave Studio [43]. These packages are capable of calculating the resonance frequencies of all modes, the field distribution and coupling between resonators. The software package developed by University of Maryland [44] is also very useful in designing dielectric resonator filters. The package is capable of analyzing coupling between two dielectric resonators placed in circular and rectangular cavities.

As shown in Sections 4–7, coupling between resonators in 3D filters is usually achieved by means of irises. With the use of the electric/magnetic wall symmetry concept illustrated in Figure 23, the problem of calculating the coupling value between two resonators can be reduced to the problem of calculating the resonance frequencies.

With the knowledge of the two resonant frequencies f_m and f_e, the filter center frequency f_o and the bandwidth BW, the coupling value M_{ij} can be written as

$$M_{ij} = (f_0/BW) \cdot (f_m^2 - f_e^2) / (f_m^2 + f_e^2) \qquad (2)$$

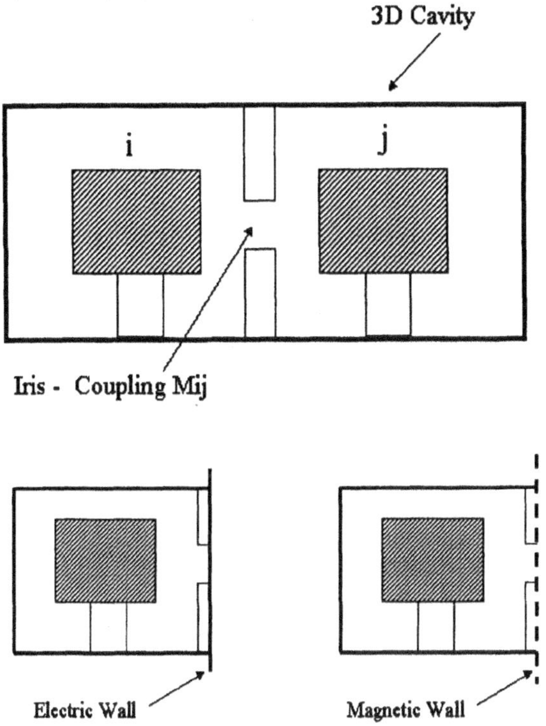

Figure 23 The electric/magnetic wall symmetry concept for calculating the coupling between two cavity resonators

Coupling between two modes in one resonator is realized using a coupling screw. Typically, no analysis is required for such type of coupling since the coupling screws are adjustable allowing the realization of the desired coupling. The location of the input/output probes is selected based on the field distribution of the mode of operation. Typically, the input/output probes are also adjustable, allowing the realization of the correct input and output coupling during the tuning process.

7.3 TUNING OF 3D FILTERS

Given the electrical performance required for wireless and satellite applications, the minimum order for the filter function is between 8th to 10th order. A large number of variables need to be adjusted during tuning, some are continuously changeable, such as the tuning screws and coupling screws. Some are fixed and can only be changed by being physically removed from the filter and machined to a different dimension, such as the irises.

For example, a hybrid DR/HTS 8-pole filter, of the type shown in Figure 16, has 15 tuning elements of which 12 are adjustable by screws and 3 are fixed by iris coupling.

Traditional methods of tuning are very difficult and time-consuming process. On the other hand, exact design of the channel filter to eliminate tuning adjustment is not yet possible for the following reasons:

i) The dimensional tolerance needs to be better than 0.0005"

ii) Modelling of cavity coupling screws at the present time does not have the resolution and accuracy necessary to achieve an exact design.

iii) The variation in dielectric constant of the dielectric materials from batch to batch is quite large

The use of tuning elements is therefore essential to fine tune the filter performance. Tuning can be efficiently implemented with the use of Computer-Aided Tools (CAT), which could considerably reduce the cost associated with the tuning process of these filters.

8.0 High Power Handling Capability of 3D Cryogenic Filters

Three-dimensional HTS filters are generally capable of handling more RF power than their counterpart 2-D planar HTS thin film filters. As the case of any HTS filter, the power handling capability of 3D filters is governed by the quality of the HTS materials (critical current density, grain boundaries, etc...) as well as the filter geometry. For the thick-film coated HTS filters the power handling capability may be limited by the achievable critical current density values for the melt-processed HTS thick film material. The cryogenic DR filters and the hybrid DR/HTS filters do not suffer from the same problem and therefore are capable of handling relatively high power levels [30]. Figure 24 illustrates the measured insertion loss of a hybrid DR/HTS filter at input power levels of 10, 20, 30 and 40 watts. Only very small negligible change in insertion loss is observed as the power is increased from 10 to 40 watts.

Passive intermodulation products (PIM) have been known to be a potential problem in high power filters. In conventional waveguide and dielectric filters, PIM arises primarily due to the formation of very thin oxide layers on metal surfaces, mechanical imperfections on the joints or both. Non-linearity in such components could also arise due to micro-cracks and voids in metal structures or due to metal particles.

In HTS components the main source of non-linearity is the HTS material itself: the surface resistance of HTS films is not only a function of temperature and frequency but also a function of local RF fields. This in turn can result in harmonic generation [30]-[45]. Figure 25 illustrates the passive intermodulation results for a 4-pole hybrid DR/HTS filter of the type discussed in Section 5. The filter has a third-order intercept point close to 100 dBm, which is good enough for many high power applications.

The heat generated inside 3D HTS filters is determined by the same parameters, which impact the Q factor of resonator, namely; field distribution, loss tangent of dielectric resonator and support structure and surface resistance of the HTS materials used inside the cavity. Generally, all the above parameters define the regions where there are high RF fields, and these are the same regions where heating is likely to be largest and thermal run-away may be initiated. If such high RF field regions exist close to the HTS films it can degrade the surface resistance of the HTS materials and can even drive the HTS materials out of the superconducting state. The power handling capability of the HTS resonator can thus be improved by optimizing the above parameters to minimize the temperature rise in areas close to the HTS films.

A finite element software package was used in [30] to simulate electromagnetic field distribution and power loss densities within hybrid dielectric/HTS resonators operating in different modes. The power loss densities deduced from the software package were then used as inputs to determine the temperature variation inside the resonator. With the optimum choice of the thermal conductivity of the different materials one can minimize the temperature rise where the HTS films are located inside the 3-D filter.

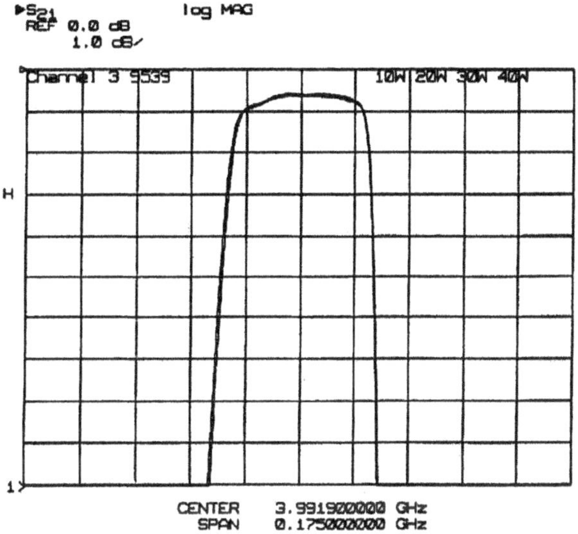

Figure 24 The measured insertion loss of a hybrid DR/HTS filter at 10, 20, 30 and 40 watts.

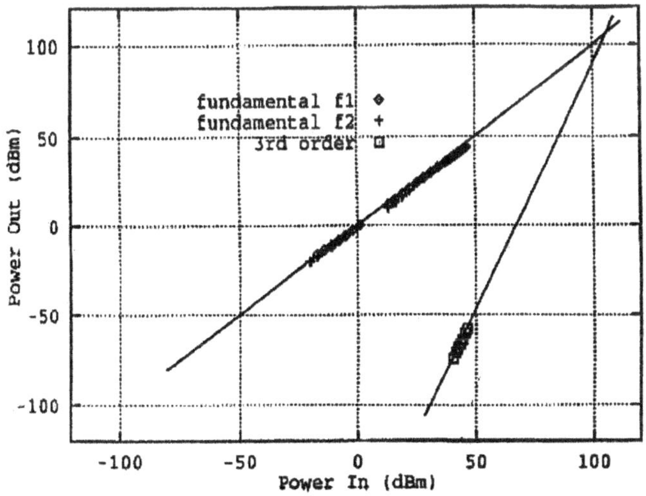

Figure 25 The passive intermodulation results for the 4-pole hybrid DR/HTS filter

9.0 Conclusions

This Chapter has addressed the design and performance of three 3-D cryogenic filters: cryogenic dielectric resonator (DR) filters, hybrid DR/HTS filters and thick-film coated filters. One unique advantage of 3-D cryogenic filters over 2-D HTS thin film planar filters is that they are amenable to the well established design tools and techniques which have been used for years by the industry in the design of traditional waveguide and dielectric resonator filters.

The improvement in Q of cryogenic DR filters is attributed to the reduction of the loss tangent of the dielectric materials with decreasing temperatures. In contrast to HTS filters which need to operate below the critical temperature of the HTS materials, cryogenic DR filters can operate at a relatively higher temperature ranges (for example 120 K – 160 K) while achieving a reasonable improvement in loss performance. Operation at higher cryogenic temperatures will reduce the load on the cryocooler. While the majority of dielectric materials exhibit a lower loss tangent at lower temperatures, for some dielectric materials the reduction in loss tangent with temperature is not monotonic. The designer needs to characterize the Q of the dielectric resonator to identify the temperature range where the dielectric material exhibits the lowest value of loss tangent.

The outlook of applications of cryogenic DR filters is in high power systems and in applications where graceful degradation after cryocooler failures is required. In this case, the filter will remain operational, but with a lower Q value. Such a feature requires the implementation of a temperature compensation mechanism that would maintain the center frequency of the filter and satisfy most of the other specifications both at cryogenic temperatures and at room temperatures.

The successful development of HTS thick film coating techniques has opened the possibility of three dimensional microwave resonators with superior loss performance. However, the achievable current density of HTS thick films is relatively low, at the present time, thus limiting the use of HTS thick film coated filters to low power applications. Additionally, thick film coating has been successful only for specific structures, which have smooth surfaces such as split ring resonators. Illinois Superconductor is currently commercializing HTS thick film coated filters for wireless applications.

In comparison with traditional room temperature 3-D filters, both cryogenic DR filters and HTS thick film coated filters only offer improvement in loss performance but not size reduction. The hybrid DR/HTS filter can offer both loss improvement and size reduction. Hybrid DR/HTS filters also exhibit a much higher power handling capability than HTS thin film and HTS thick film filters. No HTS film etching is required to build hybrid DR/HTS filters, This in turn reduces the fabrication cost and eliminates the waste in HTS materials which typically occurs during etching. This type of filter was used to build an input multiplexer that meets the typical requirements of satellite C-band input multiplexers. The results of the HTSSE-II program [46] have demonstrated the suitability of such type of filters to survive launch and space environment.

With the development of advanced dielectric resonators with high dielectric constant (60-100) and with a low loss tangent $(10^{-6} - 10^{-7})$ at 77 K, hybrid DR/HTS filters can potentially be built to have a smaller size and a superior RF performance compared to HTS thin film planar filters.

Acknowledgement

I would like to express my gratitude to Dr. Marty Nisenoff for reviewing the manuscript and for his constructive and helpful comments. I would like also to thank my colleagues at COM DEV Ltd., V. Dokas, S. Ye, B. Jolley, G. Thomson, S. Peik, W. C. Tang, and C. M. Kudsia for their support, help with analysis, simulation, design and testing. I would like to thank the management of COM DEV Ltd for their support and encouragement.

References

[1] Collin R. E. (1966) Foundations for microwave engineering, McGraw-Hill, New York.
[2] Kajfez and Guilon P. (1986) Dielectric resonators, Artech House Inc. MA.
[3] Atia A. E and Williams A. E. (1972) Narrow bandpass waveguide filters, IEEE Trans. Microwave Theory Tech., vol. MTT-20, pp.238-265.
[4] Cameron, R. J. (1980) General prototype network synthesis methods for microwave filters", ESA Journal, vol, 6 pp. 1018-1028.
[5] Atia A. E. and Williams A. E. (1971) New types of waveguide bandpass filters for satellite transponders, Comsat Technical Review, Vol 1, No 1, pp.21-43.
[6] Cohn S. B. (1957) Direct coupled resonators filters, Proc. IRE, vol, 45 pp.187-197.
[7} Matthaei G. L., Young L. and Jones E. M. T. (1980) Microwave filters impedance matching networks and coupling structures, Artech House, MA.
[8] Vahldieck R., Bornomenn and Arndt F. (1983), Optimized wave guide E-plane metal insert filters for millimeter wave applications, IEEE Trans. Microwave Theory Tech., vol. MTT-31, pp.65-69.
[9] Regan G. L. (1948) Microwave Transmission Circuits (MIT Radiation Laboratory Series, Vol 9, McGraw-Hill, New York, pp.673-677.
[10] Lin W. (1951) Microwave filters employing a single cavity excited in more than one mode , Journal of Applied Physics, Vol. 22, No. 8, pp.989-1001.
[11] Currie M. R. (1948) The utilization of degenerate modes in a spherical cavity, Journal of Applied Physics, Vol 24, No.8, pp. 998-1003.
[12] Williams A. E. (1970) A four-cavity Elliptic waveguide filter, IEEE Trans. Microwave Theory Tech., vol. MTT-20, pp.1109-1114.
[13] Atia A. E. and Williams A. E. (1971) New types of waveguide bandpass filters for satellite transponders, COMSAT technical Review, Vol 1, No 1, pp.21-43.
[14] Cohn S. B. and Targow E. N (1965) Investigation of microwave dielectric resonator filters, Rantec Div., Emerson Electric Co.., Internal report.
[15] Cohn S. B (1967) Microwave bandpass filters containing high Q dielectric resonators, IEEE Trans. Microwave Theory and Tech. Vol , MTT-16, pp.218-227.
[16] Harrison W. H. (1967) Miniature high-Q bandpass filter employing dielectric resonators, IEEE Trans. Microwave Theory and Tech. Vol. MTT-16, pp. 210-218.
[17] Wakino K. et al. (1975), Microwave bandpass filters containing dielectric resonators with improved temperature stability and spurious response, IEEE MTT Symposium Dig. pp.63-65.
[18] Fiedziuszko S. J. (1982), Dual mode dielectric resonator loaded cavity filter, IEEE Trans. Microwave Theory and Tech. Vol. MTT-30 pp.1311-1316.
[19] Kudsia C., Cameron R. and Tang W. C. (1992) Innovations in microwave filters and multiplexing networks for communications satellite systems, IEEE Trans. Microwave Theory and Tech. Vol. MTT-40 pp.1133-1149.
[20] Zaki K. A. and C. Chen (1986) New results in dielectric loaded resonators, IEEE Trans. Microwave Theory Tech. vol. MTT-34 pp. 815-824.

[21] Klein N. et al. (1999) High-Q dielectric resonator devices at cryogenic temperatures, IEEE Transactions on Applied Superconductivity, Vol. 9, No. 2 pp. 3573-3576.
[22] Klein, N., Scholen, A., Tellmann N., Zuccaro, C. and Urban, K.W (1996) Properties and applications of HTS-shielded dielectric resonators: a state-of-the-art report, IEEE Transactions on Microwave Theory and Techniques, Vol. 44, pp. 1369-1373.
[23] Krupka J. et al (1994) Dielectric properties of single crystals of Al2O3,$LaAliO_3$,$NdCuO_3$,$SrTiO_3$ and MgO at cryogenic temperatures, IEEE Transactions on Microwave Theory and Techniques, Vol. 42 pp. 1890-1890.
[24] Murata Manufacturing Co., Ltd., Kyoto, Japan.
[25] Private communication with Norbert Klein.
[26] K&L Microwave Inc. Salisbury, MD, USA.
[27} Penn S. and Alford N. High dielectric constant, low loss dielectric resonator materials, EPSRC Final report GR/K70649 , EEIE, South Bank University, London, UK.
[28] Mansour R. R., Dokas V., Thomson G., Tang W. C. and Kudsia C. (1994) A C-band superconductive input multiplexer for communication satellites", IEEE Trans. Microwave Theory Tech. vol. MTT-42 pp. 2472-2478.
[29] Mansour R. R. and Dokas V. (1996) US Patent # 5,498,771, Miniaturized dielectric resonator filters and method of operation thereof at cryogenic temperatures.
[30] Mansour R. R, Jolley B., Ye S. Thomson, F. S. and Dokas V (1996) On the power handling capability of high temperature superconductive filters", IEEE Trans., Microwave Theory and Tech., Vol. MTT-44, pp. 1322-1338.
[31] Alford N. M et al. (1991) Low surface resistance in YBCO melt-processed thick films, Nature, Vol. 349, p.680.
[32] Button T. W. et al (1991), The processing and properties of high Tc thick films, IEEE Trans. Magn.etics, vol.27, pp. 1434-1437.
[33] Button T. W. and Alford N. M. (1992) High Q YBCO cavities, Applied Phys. Lett vol. 60., pp. 1378-1380.
[34] Button T. W. et al. (1996) Properties and applications of thick film high temperature superconductors, IEEE Trans. Microwave Theory and Tech Vol, MTT-44, pp. 1356-1360.
[35] Lithgow R. D. and Peters J. M. (1997), Electromagnetic resonant filter comprising cylindrically curved split ring resonators, US Patent 6,616,540A.
[36] Lithgow R. D., Koh E., Lines M. E. and Peterson E (1997) Electromagnetic resonator comprised of annular resonant bodies disposed between confinement plates, US patent 5,629,266A.
[37] Dick G. J. at al. (1976) The superconducting split ring resonator as an accelerating structure, Nuclear Instruments and Methods, vol. 138, No. 2, pp. 24-26.
[38] Hardy W. N. et al. (1981) Split-ring resonator for use in magnetic resonance from 200-2000 MHz. Rev Sci, Instrument , Vol .52 (2), p.213.
[39] Mehdizadeh M. et al, (1983) Loop gap resonator - a lumped microwave resonant structure, IEEE Trans. Microwave Theory and Tech. Vol. MTT-31, pp. 1059-1063.
[40] Atia A. E., Williams A. E. and R. W. Newcomb (1974) Narrow-band multiple-coupled cavity synthesis, IEEE Trans. on Circuits and Systems, pp. 649-655.
[41] Ansoft HFSS software package, Pittsburgh, PA, USA.
[42] Agilent HFSS Software package, Agilent Technologies, Palo-Alto, CA, USA.
[43] CST, MAFIA software, release 4.0, www.cst.de.
[44] Zaki K. A. and Chen C. (1990) A novel coupling method for dual mode waveguide or dielectric resonator filter IEEE Trans. Microwave Theory Tech. vol. MTT-38 pp. 1885-1893.
[45] Shen Z. Y. (1994) High temperature superconducting microwave circuits, Artech House Inc. Norwood, MA 02062.
[46] Nisenoff M. and Meyers W. J (2001) On-orbit status of the high temperature superconductivity experiment, IEEE Applied Superconductivity Transactions. Vol 11. pp. 799-805.

DESIGN OF HTS DISTRIBUTED TWO-DIMENSIONAL DEVICES

Salvador H. Talisa
Northrop Grumman Corporation
Electronic Sensors and Systems Division
Baltimore, MD 21203

1. Introduction

The advent of high-temperature superconducting (HTS) materials has enabled a number of important applications in passive microwave electronics. Superconductors exhibit very low electrical losses at microwave frequencies and, although finite, at the practical operating temperature of 77 K (boiling point of liquid nitrogen at atmospheric pressure), these losses are more than two orders of magnitude lower than that of normal conductors at frequencies of 10 GHz and below. This has opened the possibility for high-performance planar microwave components since high-quality epitaxial films can be deposited on both sides of low microwave loss, single-crystal substrates, allowing the fabrication of components in planar configurations such as microstrip, stripline and coplanar waveguide. These configurations are widely used throughout the microwave community in a variety of technologies ranging from GaAs microwave monolithic integrated circuits (MMIC) to integrated circuits using ceramic and laminated substrates. Figure 1 shows a schematic cross section of these three common planar microwave structures. The use of HTS in these circuit configurations results in passive devices such

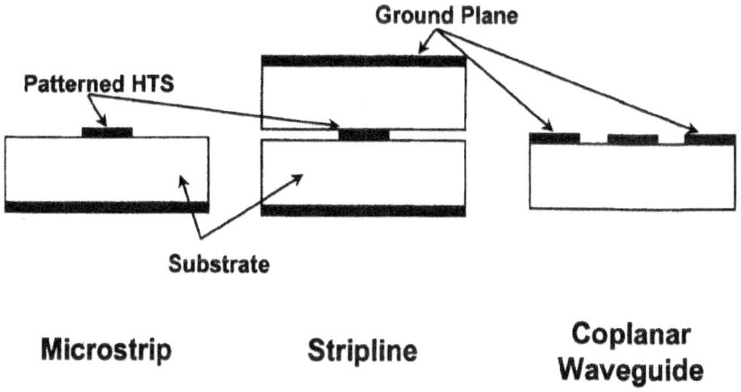

Figure 1. HTS technology can be used to the fabrication of microwave planar devices such as those schematically depicted here in cross section. One advantage of coplanar waveguide (and its variant slot-line, not shown) is that only one HTS film is needed to make a device. One disadvantage of stripline is that it is difficult to package. HTS devices have been demonstrated in all three structures

H. Weinstock and M. Nisenoff (eds.), Microwave Superconductivity, 189–204.

as resonators, filters and delay lines with performance far superior to conventional planar technology at greatly reduced mass and volume and with the attractive feature that many well established design techniques can be used for HTS circuits.

The discussion in this article will focus on HTS microwave technology with the understanding that conventional low-temperature superconductors (LTS), e.g. Nb or NbN, can also be used in the same fashion. Practical LTS materials operate typically at 4.2 K, the boiling temperature of liquid He. Furthermore, it should be kept in mind that a larger variety of substrates can be used in LTS technology because it does not require single-crystal epitaxial films. For example, Nb microwave and digital circuits have been demonstrated on Si and single-crystal sapphire substrates (1).

2. High-Temperature Superconducting Film Processing Technology

The basic elements of HTS thin film technology from the point of view of microwave applications will be discussed here briefly. Only epitaxial thin films will be considered because, to date, significantly better properties can be obtained from epitaxial material than from non-epitaxial materials. Most, if not all, of the developments covered in this article, however, are also valid for the case of non-epitaxial materials. The main advantage of these latter materials is cost and the ability to coat non-planar surfaces such as the inside of a cylinder to form a high-Q cavity resonator (2)

Among the various high-temperature superconductors there are two that have achieved a level of maturity and acceptance in the industry: (1) $YBa_2Cu_3O_7$, usually referred to as YBCO, and (2) $Tl_2Ba_2CaCu_2O_8$ or TBCCO. High-quality epitaxial thin films of these materials can be deposited on both sides of a variety of low microwave loss, single-crystal substrates by physical vapor deposition techniques, mainly sputtering (3) and laser ablation (4), and also by metal-organic chemical vapor deposition (MOCVD) (5) and co-evaporation (6). The two most common substrates to date are $LaAlO_3$ (LAO) and MgO. Circuits on sapphire have been demonstrated but the crystal lattice mismatch between HTS and sapphire is large enough to restrict the film thickness to below desirable values of at least 500 nm. Both LAO and MgO substrates are available commercially from a number of suppliers around the world and can be readily obtained in circular wafers up to 7.6 cm in diameter and 250 to 500 μm in thickness. At 77 K and for frequencies between 1 and 10 GHz the loss factor (tan δ) of LAO and MgO is less than 10^{-5}. This is 10 to 100 times smaller than that of most practical microwave substrates and is compatible with the low conductor loss of HTS.

YBCO is typically grown in-situ at elevated temperatures near 700°C as an epitaxial single-crystal film, with thickness ranging from 400 to 600 nm. This means that the desired crystalline phase of the material is formed as the film grows by suitably adjusting the growth conditions. In contrast, TBCCO films are prepared by first depositing an amorphous film at near ambient temperature and then annealing at near 750°C to form the desired crystalline phase. One advantage of the in-situ deposition of YBCO films is the ability to deposit other crystalline layers such as insulators or normal conductors, needed for the fabrication of Josephson junctions. HTS microwave technology can leverage the research on Josephson junctions by making use of insulating layers in other ways such as lumped element capacitors, for example. TBCCO films have the advantage of a higher critical temperature (Table 1), offering a greater operating temperature margin than YBCO films. In addition, TBCCO films are typically grown with thickness close to 1 μm, allowing practical devices to be made with thicknesses two or three times greater than the London penetration depth (200 to 300 nm) at 77 K. This results in increased operational stability with respect to small changes in temperature, lower insertion loss performance, and greater device repeatability.

The most important characteristics or these two HTS materials and their commonly used substrates are given in Table 1. The properties listed are those that a microwave designer would want to know if designing an HTS device.

2.1 DEVICE PROCESSING

Fabrication of microwave devices using either YBCO and TBCCO follows straightforward photolithographic techniques. After deposition of an HTS film, patterning of the superconducting layer is typically accomplished by using Ar-ion milling. After HTS film patterning, low-resistivity normal metal is deposited and patterned to provide contacts for interfacing with other devices or instrumentation. The contact layer can be made by depositing a thin (200 - 300 nm) film of gold or silver followed by an annealing step at 500 to 600°C (7). Interfacing with other devices via coaxial connectors or directly to other substrates, superconducting or otherwise, can be accomplished using gold wire or ribbon attached to the low-resistivity contacts by ultrasonic thermal compression bonding or gap-welding (ribbons). Fabrication details of filters and delay lines made at Northrop Grumman can be found in (8). Other institutions follow similar procedures.

TABLE 1 - Basic characteristics of the most commonly used HTS thin films and dielectric substrates for microwave applications (both types of HTS films can be grown in either substrate)

HTS Films:	YBCO	TBCCO
Critical Temperature (T_c)	90K	105K
Surface Resistance (R_s) at 77 K and 10 GHz (f^2 dependence)	0.2 - 0.5 mΩ	0.2 - 0.5 mΩ
Film Thickness	400 - 600 nm	800 - 1000 nm
Critical DC current density (J_c) at 77 K	10^6 A/cm^2	10^6 A/cm^2
Penetration Depth at T = 0 K (nm)	200 nm	200 nm
Substrates	$LaAlO_3$ (LAO)	MgO
Relative Dielectric Constant (ε_r) at 77 K	23.4	9.7
Dissipation Factor (tan δ) at 77 K at 10 GHz	$< 10^{-5}$	$< 10^{-5}$
Typical dimensions	5, 7.5 and 10-cm diameter; 250 and 500-μm thick	5 and 7.5-cm diameter; 250 and 500-μm thick

2.2 MICROWAVE PACKAGING OF HTS DEVICES

Proper packaging of cryogenic microwave devices is critical to the success of the technology. In general, planar microwave devices consist of a dielectric or semi-insulating substrate mounted on a metallic package or carrier by means of conducting epoxy or soldering. The mounted device then interfaces with other devices or measurement instrumentation usually via coaxial connections. In this case the metallic carrier forms part of the ground terminal by connecting to both the outer jacket of the coaxial interconnection and the ground plane defined on the substrate. This means that microwave currents flow into the planar device through the center conductor of the coaxial interface and back out through the ground plane on the substrate, the metallic carrier and its connection to the coaxial connector outer jacket. This current flow into the packaging assembly must occur without appreciable impedance mismatch or discontinuity. Figure 2 illustrates these points schematically with an example for a microstrip device.

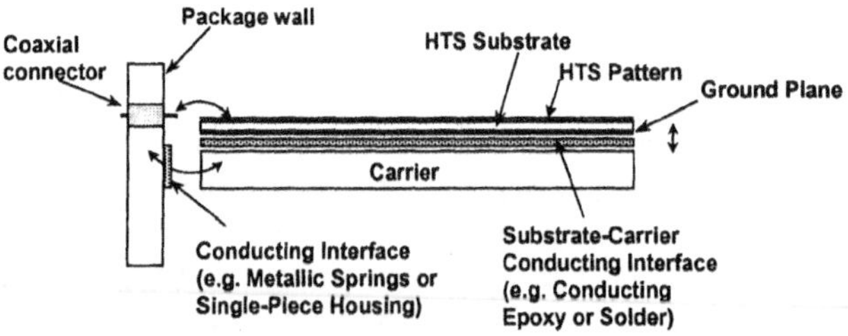

Figure 2. Microwave packaging of HTS devices is challenging because mechanical and electrical integrity must be maintained when the device is cycled from ambient to cryogenic temperatures. Special attention must be paid to the ground-current return path so as not to introduce parasitic reactances that could severely affect the performance.

The criticality of these connections between substrate, carrier and connectors is exacerbated when cryogenic cooling of the microwave package is required. A large thermal mismatch between the various components may cause cracking of the substrate or degradation of the interface between substrate and carrier or connector. Furthermore, the quality of the interface between the substrate and the outside world must be preserved through many temperature cycles between ambient and cryogenic temperature. This is required to allow for repeated testing of the device and, once integrated into a system, to allow for an operational environment that may require that the device warm up to room temperature while not in use. Suitable microwave package designs must therefore include one or more of the following elements: (1) the use of thermally matched materials, (2) adhesives that remain sufficiently pliable at cryogenic temperatures and (3) configurations that allow the various parts of the package to contract and expand freely while maintaining good electrical contact.

When considering materials that are thermally matched to HTS substrates the important parameter is the total contraction between room temperature and the operating temperature, e.g. 77 K, and not the thermal coefficient of expansion. Figure 3 shows the relative thermal contraction for several materials, including LAO and MgO, in μm/cm (that is, at a given temperature, a 1-cm-long piece of material shrinks so many μm from its length at 300K). As can be seen, the slope of this curve, i.e. the thermal coefficient of expansion, varies greatly with temperature over the range of interest, making it a practically useless parameter for the selection of appropriate materials. Although there are some differences in the rate of contraction as the materials shrink from room temperature, the key parameter is the total contraction at 77 K, relative to room temperature. For example, bulk Nb and LAO are fairly well matched at 77 K, even though their rates of contraction as a function of temperature are slightly different. This is borne out by extensive experimentation (8). Other substrate/carrier material pairs have been successfully used as well.

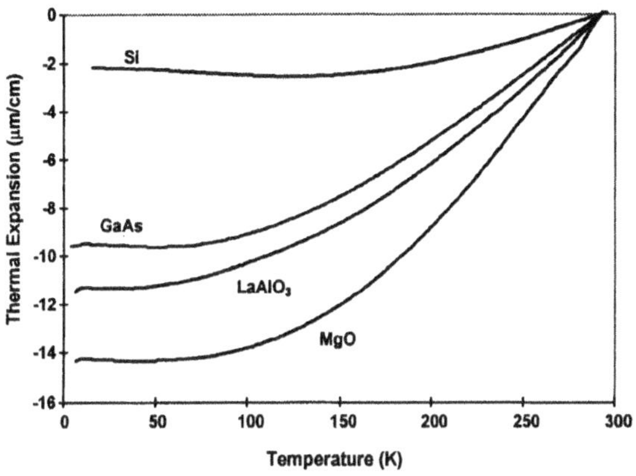

Figure 3 Measured relative thermal contraction of several materials of interest in HTS technology, including Si and GaAs. This or similar data should be used in the selection of carrier materials and the packaging of hybrid HTS-semiconductor components. If a good thermal match between parts that must remain in intimate mechanical and electrical contact cannot be obtained, sufficiently pliable conductive adhesives may be adequate for some applications, particularly if the mismatched parts are small.

3. Passive Superconducting Microwave Device Fundamentals

The key reason for developing a microwave HTS technology is the exploitation of the lower loss afforded by HTS compared with conventional metals like gold and copper. High performance, low loss devices using conventional materials can generally be made at the expense of high volume, by using hollow or partially dielectric-filled waveguide components. In contrast, HTS components could be made with the same or better performance in a much reduced volume, which must include the cryocooler.

Figure 4 shows a comparison between calculated losses for several common types of transmission line, including HTS microstrip on LAO, for the parameters listed in Table 1. For calibration, included in this comparison is a Au microstrip at 77 K, also on LAO. For the microstrip cases only the conductor and dielectric losses were calculated; radiation losses or coupling to spurious surface modes were ignored. Also included are the losses in X-Band waveguide, which are lower than HTS for the parameters chosen, although HTS microstrip offers the advantage of much wider bandwidth and ready integration of several components into a small volume. Notice that the useful frequency range for X-band waveguide is typically only 8 - 12 GHz. Integration capability combined with low loss are key advantages of HTS technology because more systems insertion opportunities will arise for applications offering significant size advantages with respect to conventional approaches.

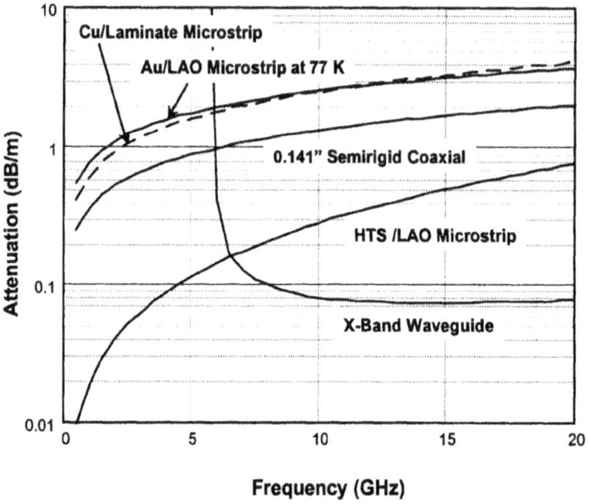

Figure 4. Comparison of calculated attenuation for various transmission line types including HTS and gold microstrip on 500-μm-thick LAO substrates. The impedance for each line was 50 Ω. All microstrip substrates were assumed to be 500 μm thick. Notice that X-band waveguide has lower loss than HTS microstrip (for the parameters chosen). However, HTS microstrip has broader bandwidth and the potential for smaller volume because it facilitates the integration with other planar microwave components.

A different but often useful way of obtaining insight into the advantages of HTS planar circuits is from the point of view of resonant structures, which form the basic building block of passband filters. This can best be discussed in terms of quality factor, Q, which is the ratio between stored and dissipated energy.

$$Q = \frac{StoredEnergy}{DissipatedEnergy} \tag{1}$$

For empty electromagnetic cavity-type resonators this is in essence a figure of merit measuring the degree of compromise between volume (stored energy) and surface area (microwave losses on the conducting surfaces). In general, both stored and dissipated energy depend on the dielectric constant and the geometric configuration used, with the dissipated energy depending also on the surface resistance of the (super)conducting surfaces and the losses in the dielectric. In practice, there could be other types of losses such as radiation which, for simplicity, will be neglected in this discussion. In principle, the performance of any passive device can be projected from the Q of the type of structure used to make the device. For example, a resonator made up of a section of microstrip line can be calculated from well known expressions (9). The insertion loss of a filter can, in turn, be estimated from the Q of the resonators that make up the filter (10).

Figure 5 is a plot comparing the Q and volume, as a function of frequency, of resonators made from a microstrip line section and an empty metallic cube, respectively.

Since the dimensions of the resonator are specified at each frequency, the volume calculated is that of the smallest cube capable of resonating at a given frequency. The microstrip HTS resonator volume was calculated assuming it is mounted in an enclosure with cross section as shown in the figure, where the walls and the lid are sufficiently far away from the superconducting strip that their contribution to the loss of the resonator is negligible. The Q of HTS microstrip, although higher, is within the same order of magnitude as the Q of the cube resonator. However the estimated volume can be two or more orders of magnitude smaller, especially at the lower microwave frequencies.

Although, to first order, the main feature distinguishing HTS from conventional planar microwave passive devices is low loss, other important differences exist and are discussed in the following sections.

3.1 SURFACE IMPEDANCE AND PENETRATION DEPTH

The surface or internal impedance of a conductor is the characteristic impedance seen by a plane wave incident perpendicularly upon a planar (super)conducting surface. For both normal metals (e.g. copper, gold) and superconductors, the surface impedance per unit length and width is given by (11,12; also, refer to article by M. Lancaster, this volume):

$$Z_S = R_S + jX_S = \sqrt{\frac{j\omega\mu}{\sigma}} \tag{2}$$

where ω is the angular frequency, μ is the permeability and σ is the conductivity, which is real for normal conductors but is complex for superconductors. In both cases, the RF fields decay exponentially inside the material, defining a field penetration depth. In the case of superconductors, however, this parameter is independent of frequency at microwave frequencies and is orders of magnitude smaller than the normal conductor penetration depth (usually referred to as the skin depth). The superconducting penetration depth can be derived from the perfect diamagnetism of superconductors, the so-called Meissner effect, and explained by the Gorter-Casimir and London two-fluid model of superconductivity (11, 13).

Table 2 summarizes the differences between normal and superconductors from the point of view of their microwave surface impedance. Notice that the surface resistance of superconductors, R_s, has a frequency-squared (f^2) dependence. In contrast, normal conductors depend on the square root of frequency ($\sqrt{f}$). Figure 6 shows the difference in surface resistance between copper at 300 and 77 K, and HTS at 77 K. This must be taken into consideration especially when designing wide band components. The figure also highlights the large differences between copper and HTS at frequencies below 1 GHz.

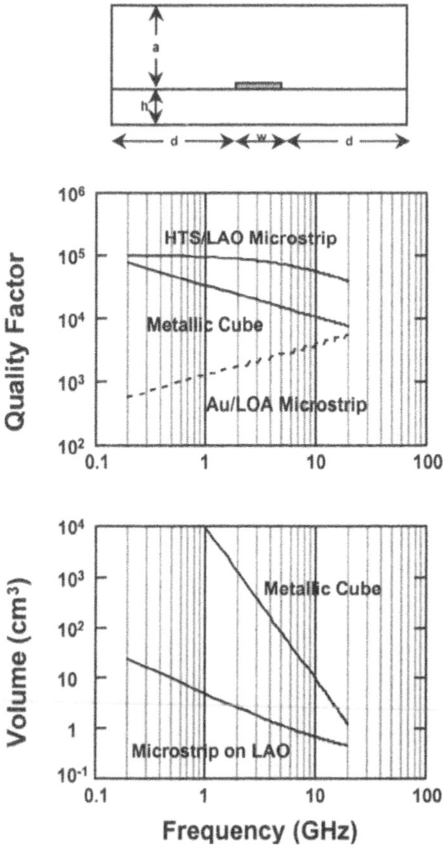

Figure 5. Calculated comparison of quality factor (Q) and volume for two types of resonators: 50-Ω HTS microstrip on 500-μm-thick LAO and a metallic cube. The Q of a gold-on-LAO microstrip at 77 K was also calculated as a reference. The metallic cube is representative of a simple cavity resonator and was chosen because its Q is easy to calculate (12). The volume for the microstrip was chosen as a packaged device with the cross section shown in the figure with a/h = 10, d/h = 20 and a length of ($\lambda/2 + 4h$).

As with planar microwave devices using normal conductors, best performance control is obtained when the geometric distributed inductance of the circuit dominates the kinetic (or internal) inductance of the superconductor. That is, from a practical point of view, the thickness of the superconductor must be at least two to three times larger than the penetration depth at the temperature of operation. The penetration depth is a strong function of temperature and is given approximately by (11, 13)

$$\lambda_L(T) = \frac{\lambda_L(0)}{\sqrt{1-\left(\frac{T}{T_C}\right)^4}} \tag{3}$$

where $\lambda_L(0)$ is the penetration depth at 0 K and is a fundamental parameter of the material, T is the operating temperature and T_C is the critical temperature. For YBCO,

$\lambda_L(0) \cong 150$ nm, which results in $\lambda_L(77\ K) \cong 214$ nm. Thus, for operation at 77 K, the HTS film must be at least 500 to 600 nm thick in order for the kinetic inductance effects to be negligible with respect to the total inductance of the circuit.

For practical microwave design purposes, this allows treating the superconductor as a normal conductor with a surface resistance that can be obtained from measured values and a frequency-squared scaling factor. It has become customary for workers in the field to normalize the surface resistance to 10 GHz and 77 K, even though measured data may have been taken at a different frequency. Devices where the kinetic inductance is allowed to dominate have been demonstrated (14). However, they are lossy, difficult to fabricate and quite dependent on temperature because of the strong temperature dependence of the penetration depth.

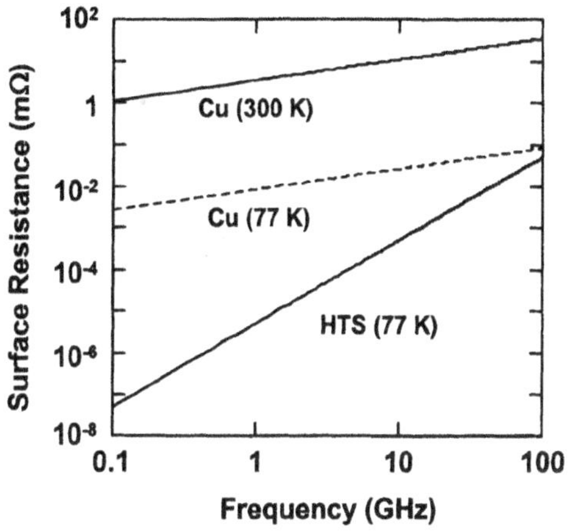

Figure 6. Nominal surface resistance of HTS at 77 K and copper at 77 K and 300 K as a function of frequency. The surface resistance of copper scales with frequency as $f^{1/2}$; for HTS it scales as f^2.

3.2 $LaAlO_3$ (LAO) SUBSTRATE PROPERTIES

The relative dielectric constant of LAO, one of the preferred HTS substrates, is ε_r = 23.4 at 77 K. This is a higher value than most common microwave substrates whose dielectric constants usually do not exceed 10 (ε_r = 9.7 for MgO, another preferred HTS microwave substrate). The significance of this is that common planar component design techniques are based on empirical circuit models whose validity may not extend to the relatively high dielectric constant of LAO. For example, the design of microstrip parallel-coupled-line filters involves the use of quarter-wave coupled-section models that generally are not valid for $\varepsilon_r > 18$. This requires that more sophisticated design techniques be developed, making use, for example, of electromagnetic field solvers, or empirically extending the range of existing models.

LAO has a cubic crystal structure above about 450°C. Below that temperature it transforms to a rhombohedral structure which is only a slight distortion of the cubic structure consisting of a very small stretching of the cubic unit cell along its diagonal. In order to release stress, the crystal will form twin structures, symmetrically related regions oriented in different directions. Non-cubic crystals are anisotropic and as a result of this twinning the LAO substrate is composed of slightly anisotropic regions randomly distributed throughout the substrate (11). The net result is an average dielectric constant with a uniformity of approximately 1%. This results in a 2% variation in the resonant frequency of a planar resonator and there are many filter applications, for example, where this is not acceptable. In contrast, MgO, which is cubic, yields a dielectric constant uniformity of 0.1% (15).

TABLE 2 - Microwave surface impedance comparison between normal conductors and superconductors (Note: Parameter f is frequency in GHz)

$$Z_S = R_S + jX_S = \sqrt{\frac{j\omega\mu}{\sigma}}$$

Normal Conductors	Superconductors
σ = Real	σ = Complex
Penetration (Skin) Depth: $\delta = \sqrt{\frac{1}{\pi f\mu\sigma}}$	Penetration Depth: $\lambda_L(T) = \frac{\lambda_L(0)}{\sqrt{1-\left(\frac{T}{T_C}\right)^4}}$ $\lambda_L(0) \cong 150$ nm for YBCO
$R_S = X_S = \frac{1}{\sigma\delta} \propto \sqrt{f}$	$R_S \propto f^2$ $X_S \cong 2\pi f\mu\lambda_L$
Copper at 300 K: $\delta = \frac{2.1}{\sqrt{f}}$ µm $R_S = 8.24\sqrt{f}$ mΩ	YBCO at 77 K: $\lambda_L(77) \cong 0.2$ µm $R_S = 5f^2$ µΩ
Copper at 77 K: $\delta = \frac{0.9}{\sqrt{f}}$ µm $R_S = 3.4\sqrt{f}$ mΩ	

3.3 DYNAMIC RANGE CONSIDERATIONS: NOISE FIGURE

An important characteristic of an electronic device is its dynamic range, or range of signal power levels over which the device will operate properly. In the case of passive HTS devices, they are expected to be linear over a certain dynamic range, limited at low signal power levels by noise and at high signal power levels by the onset of nonlinear behavior.

Starting at the lower end, the noise generated in a passive device will generally be of a thermal nature. A measure of how much noise any device generates is given by the noise figure (16), which is, by definition, related to the excess noise generated in the device when a matched resistor at 290 K (ambient temperature) is connected to the input. The noise figure would be equal to 1 (or, equivalently, 0 dB) if the device were perfectly noiseless or if it were an ideally lossless passive device. The accepted noise figure definition for a passive device as a function of device temperature is (16)

$$F_{dB} = 10 \cdot \log\left[1 + (L-1) \cdot \frac{T}{290}\right]. \qquad (4)$$

Here, L is the insertion loss of the device expressed as a number greater than or equal to 1 (i.e. 10log(L) ≥ 0 dB) and T is the operating temperature in Kelvin. For a passive, lossy device at 290 K, the noise figure is equal to its insertion loss, a rule that system designers commonly use when dealing with passive components such as filters or transmission lines. HTS devices, however, because they operate at cryogenic temperatures (77 K, typically), will have a lower noise figure, according to the accepted definition (4). Figure 7 shows the noise figure as a function of the insertion loss of the passive device when operated at 77 K and 290 K (ambient temperature). Thus, in considering the dynamic range of HTS devices, the lower end of the range will tend to be lower than for conventional devices, not only because of their inherent lower loss but because they operate at cryogenic temperatures. Measurements reported in the literature (17) confirm, to first order at least, that the noise in HTS passive devices is indeed thermal in nature.

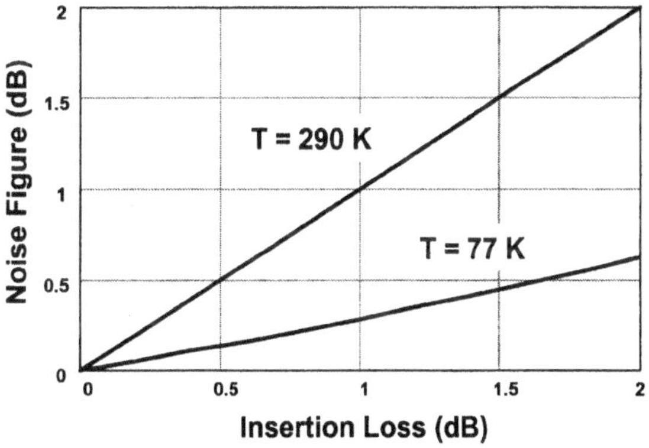

Figure 7. Noise figure of a cryogenically cooled passive device as a function of insertion loss. Notice that the noise figure equals the loss at 290 K (by definition), but it is lower for devices operating at a lower temperature.

3.4 DYNAMIC RANGE CONSIDERATIONS: NONLINEARITY AND POWER HANDLING

The dynamic range of HTS passive devices is limited at large signal levels by nonlinearities in the superconductor. This is in contrast to conventional technology, for which this upper limit could be orders of magnitude higher, generally limited by phenomena such as the voltage breakdown of air or the dielectric used, or melting of the metallic pattern due to high currents. On the other hand, HTS materials are fundamentally limited by the critical magnetic field, above which the material loses its superconducting properties. Nonlinear behavior occurs as the RF magnetic field approaches its critical value. Furthermore, in practical devices the microwave fields tend to be nonuniformly distributed so that critical values are exceeded first in selected areas of the device where the currents or fields are the highest. For example, a microstrip line has a much higher current density near the edges of the line than along the center. Thus, as the signal power level is increased, the current density at the line edges will approach critical state first, generating nonlinearities and increased losses, degrading the performance of the device.

Since HTS epitaxial films are not always single crystals, another source of nonlinear behavior stems from current flow across grain boundaries. Crystalline grain boundaries are in effect Josephson junctions with a critical current that is lower than the bulk crystal and which, when exceeded, becomes nonlinear.

When a device is nonlinear it gives rise to what is called intermodulation distortion; that is, two signals of different frequencies f_1 and f_2 applied to the device will generate mixing products. In general, the largest mixing products are those of the third order, namely, those that occur at frequencies $2f_1 - f_2$ and $2f_2 - f_1$. The upper end of the dynamic range is then reached when the power level of the applied signals is such that the third-order products rise above the noise floor and can be mistaken and processed by the system as real signals. The nonlinear behavior in a device is characterized by a parameter called the third-order intercept point (TOI) (18), usually expressed in dBm (decibels referred to 1 mW of power).

A system dynamic range can, in turn, be determined from that of its components. An important example is that of a microwave receiver front-end, usually consisting of a low-noise amplifier (LNA) placed after the antenna, which is then followed by one or more downconversion stages. The dynamic range of the receiver is primarily determined by the noise figure, gain and TOI of the LNA, with the components that follow having much less influence. Many applications demand a preselector filter between the antenna and the LNA to reject strong out-of-band interfering signals that could generate unwanted spurs through mixing due to the nonlinearity of the LNA (17). The preselector filter must not significantly degrade the receiver dynamic range and so it must have low insertion loss (i.e. low noise figure) and a TOI sufficiently higher than that of the LNA with respect to both in-band signals and the rejected out-of-band interfering signals. This is an important example because HTS filter technology is a strong candidate for this type of preselector in some applications like wireless communications base-station receivers (19).

The linearity condition, which may be a relatively minor issue with conventional filter technology, is, in contrast, very important in the case of HTS filter technology. Specific developments in this area will be discussed in the chapter devoted to HTS Resonators and Filters in this volume.

A fundamental characterization of the nonlinear behavior and power handling in HTS materials is through the surface impedance and its dependence on the RF magnetic field H_{RF} (20 - 25):

$$Z_s(H_{RF}) = R_s(H_{RF}) + jX_S(H_{RF}). \qquad (5)$$

The essence of the nonlinear dependence of the surface impedance on signal power level or, equivalently, H_{RF}, is readily understood by observing the response of a microstrip resonator, shown in Figure 8. As the input power is increased the resonator Q degrades (R_s dependence on H_{RF}) and the resonance shifts to lower frequencies (X_s dependence on H_{RF}). Several regimes have been identified in the study of nonlinear phenomena in HTS (22): A linear region at sufficiently low power levels, a region where nonlinear behavior is dominated by grain-boundary weak links (Josephson-junction-like defects in the crystalline make-up of the HTS film) and a strongly nonlinear region dominated by hysteretic vortex penetration. Above this regime, breakdown of the superconducting state occurs, with the surface resistance increasing abruptly due to heating and the formation of normal-state domains (22). (See D. Oates' article in this volume for a general discussion of nonlinear behavior in HTS.)

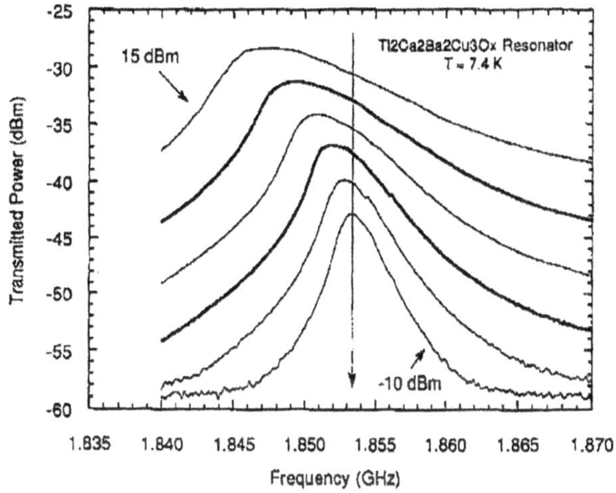

Figure 8. Effect of increasing the input power level on a superconducting microstrip resonator. This measurement, courtesy of Dr. M. Golosovsky, Hebrew University of Jerusalem, captures the essence of the nonlinear RF power dependence of the surface impedance $Z_s = R_s + jX_s$. As power level increases so does R_s and the resonance Q decreases. On the other hand, the effect of increased power level on X_s manifests itself on a shift of the resonance towards lower frequencies (22).

When the magnetic field exceeds its critical value the material becomes a normal conductor, causing the dissipated heat in the material to increase dramatically. This may damage the device if the cryocooling system cannot remove the excess heat. If no damage has occurred, the device will revert to the superconducting state after the incident power is reduced below the critical value. The related topic of intentionally inducing a superconducting-to-normal transition as a switching mechanism has been studied extensively (26-28).

Systematic studies of intermodulation distortion in HTS samples and devices have been reported in the literature (21, 29-32). TOI values in excess of +70 dBm at 1.3 GHz and 80 K have been obtained (29) for HTS planar transmission lines a few millimeters long. This value is well above that of most semiconductor low-noise amplifiers, for example. However, it must be kept in mind that intermodulation distortion is a function of the stored energy in the device, i.e. the group delay. This is important when considering the performance of HTS passband filters because filters have a delay characteristic that is lower near the center of the passband and higher toward the edges, the ratio depending on the filter order and type of response. Therefore the intermodulation distortion caused by two signals near the middle of the passband could be much lower than that produced by two signals near the edges of the passband. It is important that this be taken into consideration when characterizing the nonlinear behavior of a filter.

4. Conclusion

High-Temperature Superconductor microwave technology offers unique advantages based on the low microwave loss of HTS materials and the inherent low thermal noise in cryogenically cooled components. The main applications to date, likely to have a significant impact on electronic systems, are related to increased microwave receiver sensitivity and selectivity. Great interest in the U. S. and abroad exists in the wireless commercial communications market and several companies are testing base-station receiver front-ends consisting of cryogenically cooled filter-LNA subassemblies.

5. References

1. See for example, Withers R. S. and Ralston, R. W. (1989) Superconductive analog signal processing devices, Proc. IEEE, **77**, 1247-1263. This paper contains many references to earlier work by the authors.
2. Smith, P. A., Button,T. W., Holmes J., Dolman, G., Meggs, C., Remillard, S. K. and Hodge, J. D. (1997) YBCO Thick Films for High Q Resonators, *IEEE Trans. Appl. Supercond.* **7**, 1763 – 1765.
3. Face, D. W., Wilker, C., Shen, Z-Y, Pang, P. and Small, R. J. (1995) Large Area $YBa_2Cu_3O_7$ Films for High Power Microwave Applications, *IEEE Trans. Appl. Supercond.* **5**, 1581 - 1586.
4. The following reference is representative of good microwave-quality films grown by laser ablation, even though the substrate used was ferrimagnetic: Pique, A., Harshavardhan, K. S., Moses, J., Mathur, M., Belahoubek, E., Venkatesan, T., Denlinger, E. J., Kalokitis, D., Fathy, A., Pendrick,V., Rajeswari, M. and Jiang, W. (1995) Microwave compatible $YBa_2Cu_3O_{7-x}$ films on ferrimagnetic garnet substrates, *Appl. Phys. Lett.* **67**, 1778 - 1780.
5. Smith E. J., Musolf; J., Soares, E. (1997) *Composition controlled MOCVD as a route to high Q HTS thin film devices,* Proc. Third Joint ISTEC/MRS International Workshop on Superconductivity: Suitable Materials and Processing for HTS Applications Towards the Next Decade, 145.
6. Utz, B., Semerad, R., Bauer, M., Prusseit, W., Berberich P. and Kinder, H. (1997) Deposition of YBCO and NBCO Films on Areas of 9 Inches In Diameter, *IEEE Trans. Appl. Supercond.* **7**, 1272 - 1277.
7. Ekin, J. W., Panson, A. J. and Blankenship, B. A. (1988) Method for making low-resistivity contacts to high T_c superconductors, *Appl. Phys. Lett.,* **52**, 331-333.
8. Talisa, S. H., Janocko, M. A., Meier, D. L., Talvacchio, J., Moskowitz, C., Buck, D. C., Nye, R. S., Piesski, S. J. and Wagner, G. R. (1996) High-Temperature

Superconducting Space-Qualified Multiplexers and Delay Lines *IEEE Trans. Microw. Theory Tech.*, **44**, 1229-1239.
9. Hoffmann, R. K. (1987) *Handbook of Microwave Integrated Circuits*, Artech House, Norwood, MA
10. Matthaei, G. L., Young, L., Jones, E. M. T. (1980) *Microwave Filters, Impedance-Matching Networks, and Coupling Structures*, Artech House, Dedham, MA.
11. Shen, Zhi-Yuan (1994) *High-Temperature Superconducting Microwave Circuits*, Artech House, Norwood, MA.
12. Ramo, S. Whinnery, J. R. and Van Duzer, T. (1965), *Fields and Waves in Communication Electronics*, John Wiley & Sons International Edition.
13. Vendik, I. Vendik O. (ed. Kollberg E.), (1997), *High Temperature Superconducting Devices for Microwave Signal Processing, in 3 parts*, Skladen, St.-Petersburg, Russia.
14. Carroll, K. R., Pond, M J. and Cukauskas, E. J. (1993) Superconducting Kinetic-Inductance Microwave Filters, *IEEE Trans. Appl. Supercond.*, **3**. 8 – 16.
15. Unpublished data obtained jointly by Northrop Grumman, NASA Lewis Research Center and Superconductor Technology, Inc.
16. Mumford W.W. and Scheibe, E. H. (1968) *Noise Performance Factors in Communication Systems*, Horizon House - Microwave, Inc, Dedham, MA, UMI Out-of-Print Books on Demand.
17. Talisa, S. H., Robertson, M. A., Meier, B. J. and Sluz, J. E. (1994) Dynamic range considerations for high-temperature superconducting filter applications to receiver front-ends, *IEEE MTT-S Int. Microwave Symp. Digest,* 997-1000.
18. Bao-Yen Tsui, James (1992) *Microwave Receivers with Electronic Warfare Applications*, Krieger Publishing Co. Malabar, FL.
19. Koepf, G. (1998) *Superconductors Improve Coverage in Wireless Networks,* Microwaves and RF, 63 – 74.
20. Oates, D. E. Nguyen, P. P. Dresselhaus, G. Dresselhaus, M. S. and Chin, C. C. (1993) *IEEE Trans. Appl. Supercond.* **3**, 1114 – 1118.
21. Wilker, C., Shen, Z-Y, Pang, P., Holstein, W. L., and Face, D. W. (1995) Nonlinear Effects in High-Temperature Superconductors: 3rd Order Intercept from Harmonic Generation, *IEEE Trans. Appl. Supercond.* **5**, 1665 – 1670.
22. Golosovsky, M. (1997) *Physical Mechanisms Causing Nonlinear Microwave Losses in High-T_c Superconductors*, (Invited Paper), 8th Workshop on RF Superconductivity, Abano Terme, Italy,
23. Sridhar, S. (1994) Non-linear Microwave Response of Superconductors and ac Response of the Critical State, *Appl. Phys. Lett.* **65**, 1054 – 1056.
24. Oates, J. H., Shin, R. T., Oates, D. E., Tsuk, M. J., and Nguyen, P. P. (1993) A Nonlinear transmission Line Model for Superconducting Stripline Resonators, *IEEE Trans. Appl. Supercond.* **7**, 17 – 22.
25. Oates, D. E. Nguyen, P. P. Habib,Y. Dresselhaus, G. Dresselhaus, M. S. Koren, G. and Polturak, E. (1996) Microwave power dependence of $YBa_2Cu_3O_7$ thin-film Josephson edge junctions, Appl. Phys. Lett. **68**, 705 – 707.
26. Reference (13) Part II, Chapter 7.
27. Karasik, B. S. Zorin, M. A. Milostnaya, I. I. Elantev, A. I. G. Gol'tsman, N. (1995) Subnanosecond switching of YBaCuO films between superconducting and normal states induced by current pulse, J. Appl. Phys. **77**, 4064 – 4070.

28. Vendik, I Kozyrev,A. Gaidukov, M. Sherman, V. Svishchev, A. Razumov, S. Dovgan, D. Boholi, S. Dirassen, B. Zaitsev, A. Woerdenweber, R. (1997) The superconducting microwave devices based on S-N transition in HTS films, 27th European Microwave Conference Proceedings, 909 – 914.
29. Shen, Z-Y Wilker,C. Pang, P. Face, D. W. Carter, C. F. Harrington, C. M. (1997) Power Handling Capability Improvement of High-Temperature Superconducting Microwave Circuits, IEEE Trans. Appl. Supercond. **7**, 2446 – 2453.
30. Ma, Z. de Obaldia, E. Hampel, G. Polakos, P. Mankiewich, P. Batlogg, B. Prusseit, W. Kinder, H. Anderson, A. Oates D. E., Ono,R. Beall, J. (1997) RF Power Dependence Study of Large Area YBCO Thin Films, IEEE Trans. Appl. Supercond. **7**, 1911 – 1916.
31. Dahm T. and Scalapino, D. J. (1997) Theory of intermodulation in a superconducting microstrip resonator, J. Appl. Phys. **81**, 2002 - 2009.
32. Vendik, O. G. Vendik, I. B. and Samoilova, T. B. (1997) Nonlinearity of Superconducting Transmission Line and Microstrip Resonator, IEEE Trans. Microw. Theory Tech. **45**, 173 – 178.

DESIGN OF LUMPED-ELEMENT 2D RF DEVICES

H.J. CHALOUPKA and S. KOLESOV
University of Wuppertal
Fuhlrottstrasse 10 /D 42097 Wuppertal, Germany

Abstract. This chapter deals with the design of passive circuits for the frequency range between about 0.1 to 5 GHz, comprising lumped elements (inductors and capacitors) in high-temperature superconductor (HTS) thin-film technology. Major emphasis is given to realization and to limits in the achievable range of values of the lumped elements. The design of bandpass filters from realizable HTS lumped elements is considered in some detail. Furthermore, applications to allpass structures and hybrid couplers are discussed.

Table of contents:

H. Weinstock and M. Nisenoff (eds.), Microwave Superconductivity, 205–238.

1. Introduction

1.1 SIGNIFICANCE OF LUMPED-ELEMENT HTS CIRCUITS

Analog microwave circuits comprising normal-conducting passive components and semiconductor active components are extensively used in many subsystems for consumer electronics, communications, radar and electronic warfare applications. In conventional technology, the need for cost reduction and smaller size and weight are the main drivers for the use of hybrid microwave integrated circuits (MIC) and especially of monolithic microwave integrated circuits (MMIC) with planar (2-D) passive components instead of 3-D components such as coaxial and waveguide components [1-3]. Planar lumped-element (LE) circuits present a subset of planar circuits and offer an even higher degree of miniaturization than planar circuits with distributed elements [4-9]. However, severe restrictions must be taken into account, if large coaxial and waveguide components are substituted with normal-conducting planar lumped-element components. Limiting restrictions associated with this substitution are primarily a significant reduction of the quality factor as well as reduced power-handling capability.

With the availability of HTS thin-films on single-crystalline substrates, the trade-off between size reduction and a high quality factor can be overcome, since planar HTS resonators possess (at temperatures below about 70 K) values of Q on the order of 30,000 to 100,000, which are comparable or even exceed the Q values of large, bulky cavity resonators or room-temperature dielectric resonators - see, e.g., [4]. The second drawback of LE circuits, namely the low power-handling capability, cannot be overcome by the use of HTS structures. Although the power limitation due to heating typical for normal-conducting LE circuits is relaxed due to much smaller losses in HTS structures, one must cope with the specific nonlinearities of the HTS conductivity, which may cause even lower limits to the maximum input power. Furthermore, at present no concepts exist for a monolithic integration of semiconductor active components with HTS passive components.

All passive components which are candidates for HTS applications are characterized by a relatively large group delay and perform their function by means of controlled storage of electric and magnetic field energy [4]. Examples of such components are bandpass and bandstop filters, matching networks, dispersive and non-dispersive delay lines and networks for antenna arrays. These components are composed of "building blocks" such as resonators, transmission lines, inductors and capacitors, where field-energy storage takes place. In general, the storage of electric field energy is associated with the storage of magnetic field energy and vice-versa. A half-wave transmission-line resonator may serve as an example of a building block, where both electric and magnetic field energy are stored and are not locally separated. In (reactive) lumped elements only one type of field energy is stored, namely magnetic field energy in inductors and electric field energy in capacitors. Lumped-element circuits, e.g., lumped-element bandpass filters, are composed of inductors and capacitors, whereas for distributed-element components, resonators and transmission lines are used as building blocks.

As in the case of normal-conducting planar circuits, lumped-element circuits (LE circuits) offer the following advantages in comparison to distributed-element circuits for the same frequency and function:

- Considerably smaller size
- Spurious response shifted to higher frequencies, leading to a wider bandwidth
- More degrees of freedom and more sophisticated synthesis procedures for L-C circuits than for coupled-resonator-circuits
- Lumped-element circuits better suited to encompass switching, e.g., micro-electromechanical switches (MEMS) and tuning elements, e.g., ferroelectric varactors

In comparison to lumped-element normal-conducting circuits, HTS LE circuits possess a dramatically higher Q, but suffer from the fact that monolithic integration with semiconductor components is not achievable currently. Furthermore, limitations in power-handling , based on the specific nonlinear response of superconductors, represent a severe limitation.

Consequently, HTS LE circuits are candidates mainly for high-performance passive circuits where both high Q and size much smaller than a wavelength are required. This applies to the frequency range below about 5 GHz and in particular to frequencies in the UHF (0.3 to 3 GHz) and upper VHF (0.1 to 0.3 GHz) regimes. The realization of planar circuits with many HTS lumped elements for frequencies below about 50 MHz becomes impractical because of the large wafer size required (e.g., larger than 7 to 10 cm). However, a combination of HTS lumped elements with low-loss commercial chip capacitors allows the extension of HTS circuitry down to frequencies of some tens of MHz. With respect to the miniaturization of the entire subsystem, the additional volume of a closed-cycle refrigerator must be taken into account. This constraint can be minimized if a "large" number of circuits share a common cooler.

1.2 GENERAL CONSIDERATION OF LUMPED ELEMENTS

In order to illustrate the difference between a distributed and a lumped-element resonator, the transmission-line configuration shown in Fig. 1.1 is considered. It consists of two lossless transmission-line sections of length $l/2$ with (in general) different characteristic impedances Z_{01} and Z_{02}, but equal phase velocity v_{ph}. The section on the left-hand side is shorted, whereas the section on the right-hand side is open-ended. It is easily shown that the lowest resonant frequency f_r of this configuration is (with $\lambda_r = v_{ph}/f_r$) given by

$$\tan^2\left(\frac{\pi}{\lambda_r}l\right) = \frac{Z_{02}}{Z_{01}} \tag{1.1}$$

For a homogeneous transmission line with $Z_{01} = Z_{02}$, resonance appears when the length l equals a quarter of the corresponding guided wavelength (quarter-wave resonator). From Eq. 1.1 it is deduced that the same resonant frequency can be obtained with transmission-line resonators of much smaller length l, if the ratio Z_{02}/Z_{01} can be made considerably smaller than 1. This miniaturized type of resonator is known as a stepped-impedance resonator. If the ratio is made larger than 1, the total resonator length becomes larger than that of a quarter-wave resonator. This dependence of the required length on the ratio is depicted in Fig. 1.2.

The ratio of the mean stored electric field energy $W_e^{(1)}$ to the mean magnetic field energy $W_m^{(1)}$ is, for the shorted transmission-line section of length $l/2$, equal to the the ratio of the mean stored magnetic field energy $W_m^{(2)}$ to the mean electric field energy $W_e^{(2)}$ for an open-ended transmission-line section of length $l/2$. These ratios are given by

$$\frac{W_e^{(1)}}{W_m^{(1)}} = \frac{W_m^{(2)}}{W_e^{(2)}} = \frac{1 - si\left(\frac{2\pi}{\lambda} l\right)}{1 + si\left(\frac{2\pi}{\lambda} l\right)} = \Lambda_W , \tag{1.2}$$

with $si(x) = \sin(x)/x$.

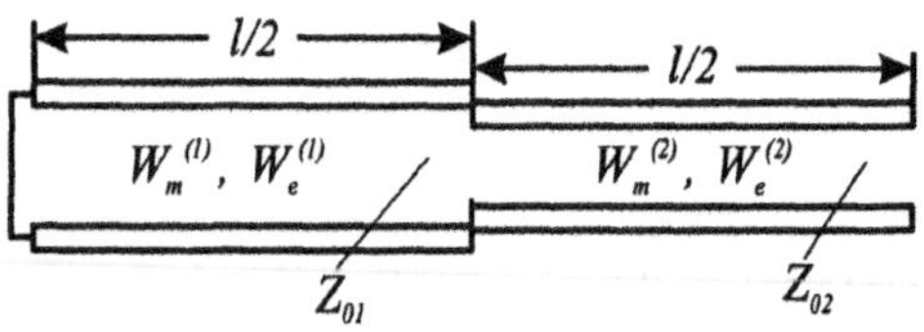

Figure 1.1. Resonator built from two uniform transmission-line sections of length $l/2$. The transmission-line sections shown schematically are characterized by different characteristic impedances. The left section is shorted, and the right section is open-ended.

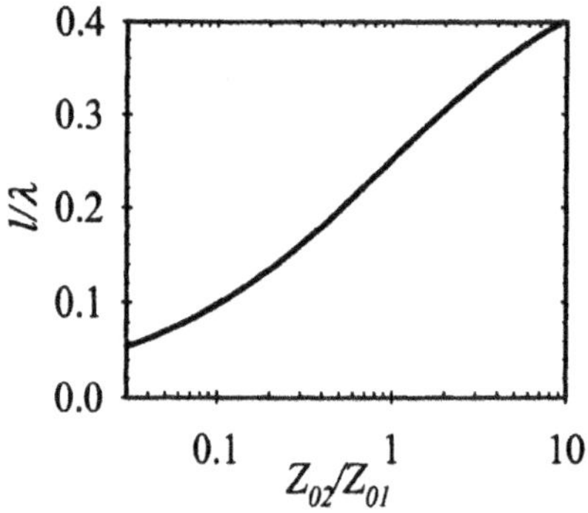

Figure 1.2. Required length of a stepped-impedance resonator as a function of Z_{02}/Z_{01} for a quarter-wave resonator with $Z_{02}/Z_{01} = 1$ and $l = \lambda_r/4$.

As in all types of resonators, the mean value of the total stored electric field energy W_e is equal to the mean value of the total stored magnetic field energy W_m, so that for the total field energy

$$W=W_m+W_e=2W_e=2W_m \tag{1.3}$$

follows. The distribution of stored field energy to the two parts of the resonator is illustrated in Fig. 1.3 as a function of the normalized length, assuming a fixed resonant frequency. The corresponding equations are

$$\frac{W_m^{(1)}}{W}=\frac{W_e^{(2)}}{W}=\frac{1}{1+\Lambda_W}, \quad \text{and} \quad \frac{W_m^{(2)}}{W}=\frac{W_e^{(1)}}{W}=\frac{\Lambda_W}{1+\Lambda_W} \tag{1.4}$$

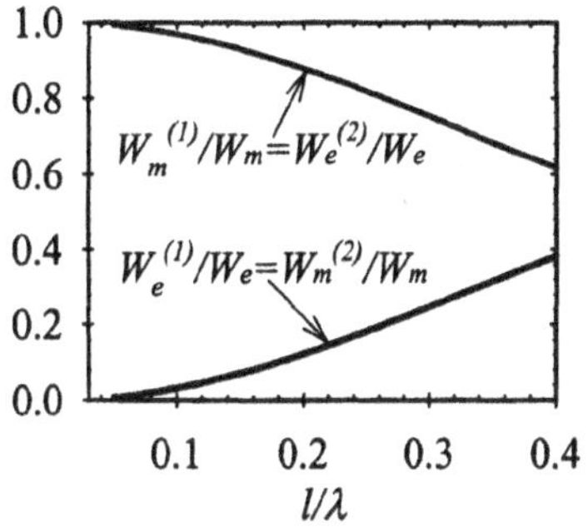

Figure 1.3. Distribution of the stored field energy to the two parts of the stepped-impedance resonator - see Fig. 1.1 – as a function of the normalized length for a fixed resonant frequency.

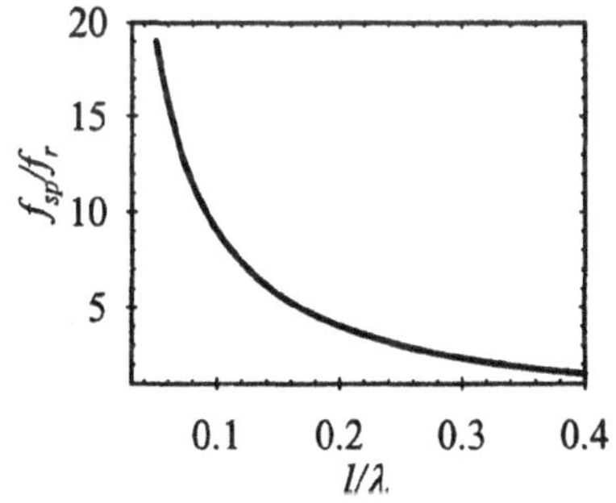

Figure 1.4. Normalized resonant frequency of the spurious mode as function of the length of the resonator from Fig. 1.1. Resonant frequency f_r is assumed to be constant, so that a reduction of the length requires a decreased ratio Z_{02}/Z_{01}.

It is clearly seen from Eq. 1.4 and Fig. 1.3 that in the case of a quarter-wave resonator, considerable portions of electric and magnetic field energy are stored in both line sections ($\Lambda_W = 0.22$); for the case $l \to 0$, the magnetic field energy is exclusively stored in the shorted section and electric field energy is stored in the open-ended section. From the latter it can be deduced that for $l \to 0$ the line sections represent lumped elements, namely the shorted section represents an inductor, and the open-ended section, a capacitor.

From Eq. (1.1) the spurious resonant frequency f_{sp} closest to the desired resonant frequency f_r is found to be

$$\frac{f_{sp}}{f_r} = \frac{v_{ph}}{f_r \cdot l} - 1 \quad . \tag{1.5}$$

From the plot of Eq. 1.5 in Fig. 1.4 it becomes clear that with a reduction of the length associated with a higher value of Z_{02}/Z_{01}, the ratio of spurious to desired resonant frequency can be increased.

The example presented has confirmed two statements which were made in section 1.1, namely

(a) that lumped-element circuits possess a smaller size than distributed circuits of the same resonant frequency,

(b) and that lumped-element circuits offer a wider bandwidth (i.e., spurious resonant frequencies higher) than distributed circuits.

By means of the example considered above, it was furthermore shown that for $l > 0$ there are always both types of (electric and magnetic) field energy present in each of the two parts of the resonant circuit. Hence, a lumped element which by definition would ideally require the storage of only one type of field energy, exists in reality only as an approximation. If one requires that the ratio Λ_W - see Eq. 1.2 - is smaller than about 0.14, one finds the following necessary condition for a structure to be considered as a lumped-element:

$$a < \lambda / 10. \tag{1.6}$$

Here λ denotes the guided wavelength, and $a = l/2$ is the length of the structure under consideration.

Figure 1.5 illustrates the approximation by a capacitance of a short open-ended line (with characteristic impedance Z_0 and phase velocity v_{ph}) of length a. The input impedance $Z(\omega)$ can be evaluated as

$$Z(\omega) = \frac{Z_0}{j \tan\left(\frac{\omega}{v_{ph}} a\right)} = \frac{1}{j\omega C_1} + \frac{1}{\frac{1}{j\omega L_2} + \frac{1}{\frac{1}{j\omega C_3} + \cdots}} \tag{1.7}$$

with capacitance and inductance values given by

$$C_1 = \frac{a}{v_{ph} Z_0}, C_3 = \frac{C_1}{3}, \text{ and } L_2 = \frac{a Z_0}{3 v_{ph}}. \tag{1.8}$$

Eq. (1.7) corresponds to a lumped-element equivalent circuit for an open-ended transmission line with an infinite number of lumped elements. This circuit is depicted on the left side of Fig. 1.5. For $f < v_{ph}/10\, l$ (equivalent to $a < \lambda/10$) the infinite network can be approximated by a single capacitance.

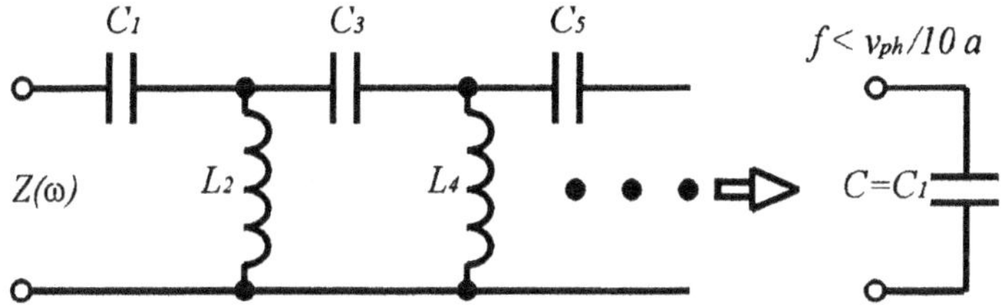

Figure 1.5. Left: Equivalent lumped-element circuit for an open-ended transmission line of length a with an infinite number of elements. Right: Approximation by a single capacitance (lumped element) for $f < v_{ph}/10\, a$.

Figure 1.6 shows one of a large variety of possible realizations of a resonator according to the schematic concept of Fig. 1.1. On the left side a (distributed) quarter-wave resonator realized as a "hairpin resonator" with strip conductors on a substrate without a ground plane is shown. Its geometrical length l is given by about a quarter of the guided wavelength $\lambda = \lambda_0/((\varepsilon_r+1)/2)^{0.5}$. On the right side this hairpin resonator has been modified by employing a relatively high characteristic impedance for that part of the line which is shorted, and a relatively small value of the characteristic impedance for the open-ended half of the transmission line. Although this resonator possesses the same resonant frequency as the hairpin resonator, its length $\tilde{l}$ is considerably shorter. For $\tilde{l}/2 < \lambda/10$ the transmission-line sections can be considered as an inductor (left) and a capacitor (right).

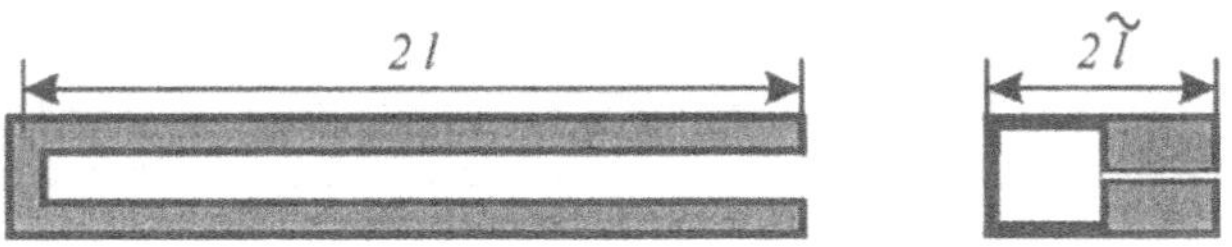

Figure 1.6. Hairpin resonator (left) as an example of a distributed-element circuit and a shorter lumped-element circuit (right) with the same resonant frequency. The lumped-element circuit is derived from a stepped-impedance approach with $\tilde{l} < \lambda/10$.

Planar thin-film structures are always based on a planar substrate with conductive structures. The conductors are placed either on one or on both sides of the substrate.

Inductances and capacitances are two-terminal devices (also referred to as one-ports) with two "connecting points." There are therefore two types of lumped elements (LE) to be distinguished:

(i) lumped elements with both connecting points situated on the same side of the substrate;

(ii) lumped elements with connecting points on different sides of the substrate.

Type (i) requires planar conductors on one side of the substrate only. However, a conductive ground plane is often employed for shielding. In the case of type (ii) a ground plane is often used as a second connecting point.

As shown in Fig. 1.7, with a ground plane the structures must be considered as three-terminal (two-port) devices instead of two-terminal devices. Lumped-element two-ports are to be described by 3 elements in a Π -see Fig. 1.7- or T –circuit.

In Fig. 1.7, no ground plane is employed in structures (a1) and (b1) , whereas structures (a2) and (b2) are build with a ground plane. If the "conducting loop" of structure (a2) is connected between 1 and 2, it represents an inductance L in parallel to a capacitance $C/2$. For sufficiently small size, the parallel capacitance can be neglected, so that one ends up with an inductance L. On the other hand, if structure (a2) is connected between 1 and 1′, one obtains a capacitance C in parallel to a series L-C circuit. If L is negligible, (a2) becomes a capacitance $2C$. In case of structure (b2) the capacitance becomes $C_a+C_b/2$ between 1-2, but $C_b+C_aC_b/(C_a+C_b)$ between 1-1′.

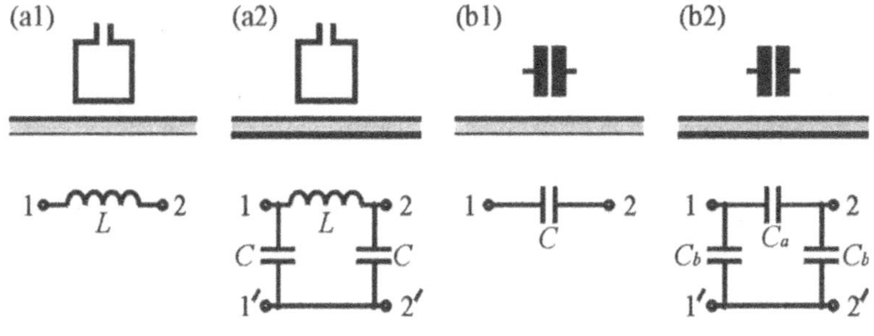

Fig. 1.7. Equivalent-circuit characterization of lumped-element structures without (structures (a1) and (b1)) and with (structures (a2) and (b2)) ground plane.

2. Planar lumped elements in HTS technology

2.1 REALIZATION OF PLANAR CAPACITORS AND INDUCTORS

This section deals with the realization of lumped elements for HTS thin-film circuits with operational frequencies in the range between about 0.1 to 5 GHz.

Figure 2.1 depicts three types of capacitors available in HTS thin-film technology, namely (a) gap capacitors, (b) interdigital capacitors and (c) patch capacitors.

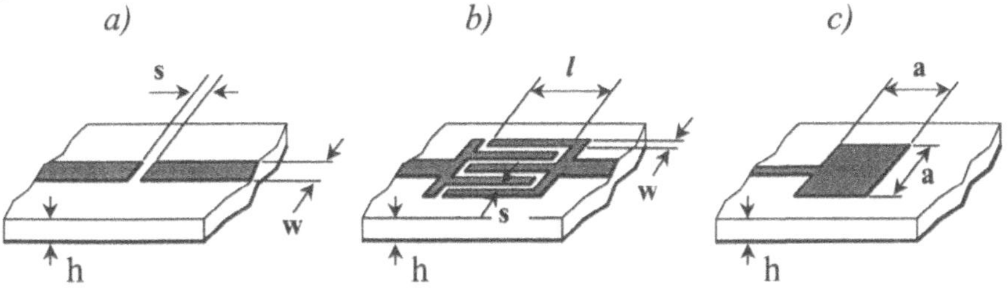

Figure 2.1. Different types of planar capacitors

For low capacitance values (0.02 to 0.5 pF) gap capacitors can be used - see Fig.2.1a. Their capacitance can be estimated via [10]

$$C/\mathrm{pF} \approx 3\cdot 10^{-3}(\varepsilon_r+1)\cdot\frac{w}{\mathrm{mm}}\cdot\ln\left(\frac{2h}{\pi s}\right). \tag{2.1}$$

Here the gap width and transverse dimension are denoted by s and w, respectively. Other parameters are thickness h and relative permittivity ε_r of the substrate with a conductive groundplane. With a narrow gap (s = 20 μm) and a 0.5-mm-thick sapphire or lanthanum aluminate (LAO) substrate, one obtains a capacitance value of about 0.1 to 0.2 pF per millimeter of transverse gap size w. Taking the restriction $w < 0.1\ \lambda_0/\sqrt{\varepsilon_r}$ into account, one finds at 2 GHz (with w < 5 mm for sapphire and w < 3 mm for LAO) capacitance values on the order of up to 0.5 pF.

For values greater than about 0.5 pF, interdigital capacitors must be used – see Fig. 2.1b. In the special case with finger width w equal to the gap width s between the fingers, the capacitance is given by

$$C/\mathrm{pF} \approx 3.5\cdot 10^{-3}(\varepsilon_r+1)\cdot(N-1)\cdot(l/\mathrm{mm}+w/\mathrm{mm}) \quad \text{for } l > w\ . \tag{2.2}$$

Here l denotes the length of the N fingers comprising the interdigital capacitor. It is interesting to note that the capacitance value does not explicitly depend on the gap width s, but on the ratio s/w (here taken to be 1). However, for a given transverse dimension of the interdigital capacitor, an increase in the number of fingers N requires the gap width to be reduced to relatively low values. For an interdigital capacitor with

N = 10 to 20 fingers which occupies an area with diameter of about 1 to 2 mm, capacitance values on the order of 1 to 5 pF can be achieved. In this case, a typical value on the order of 20 to 200 μm is used for the width w of the fingers.

Capacitances to the ground are realized by means of patch capacitors – see Fig. 2.1c. With a linear size $a >> h$, the capacitance can be approximated by

$$C/\mathrm{pF} \approx 8.8 \cdot 10^{-3} \varepsilon_r \frac{(a/\mathrm{mm})^2}{h/\mathrm{mm}} K(a/h, \varepsilon_r) \quad . \tag{2.3a}$$

Here the function $K(a/h, \varepsilon_r)$ accounts for the fringing field and can be approximated via

$$K(a/h, \varepsilon_r) \approx 1.04 + \frac{3.04\varepsilon_r + 1.21}{\varepsilon_r} \cdot \frac{h}{a} \quad . \tag{2.3b}$$

The highest achievable value for the capacitance of a patch capacitor is limited by two restrictions:

(i) Dimension a is limited by the condition $a < 0.1\lambda_0/\sqrt{\varepsilon_r}$ for lumped elements. Due to this condition C_{max} decreases with the square of the operational frequency f_0.

(ii) If only condition (i) is taken into account, the absolute size a of the patch becomes relatively large for frequencies below about 1 GHz. Hence, depending on the available size of the wafer and the number of elements (capacitances and inductances) which must be placed on the wafer, there is a limitation a_{max} for the absolute size of the patch.

Fig. 2.2 shows the maximum capacity of a patch capacitor as a function of the operational frequency f_0, the absolute size limit a_{max} and the substrate material (LAO with $\varepsilon_r \approx 24$ and MgO or sapphire with $\varepsilon_r \approx 10$).

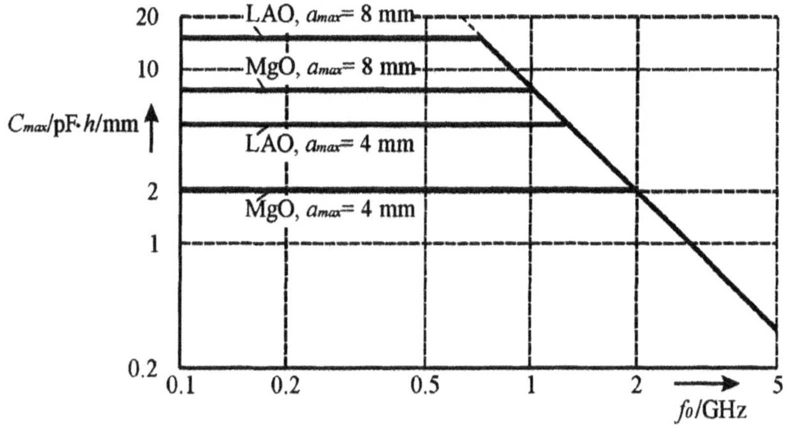

Figure 2.2. Maximum capacitance of a patch capacitor as a function of the operational frequency f_0, the absolute size limit a_{max} and the substrate material.

The main conclusions from Fig. 2.2 are that
(a) for a given h the capacitance C_{max} is independent of the substrate material in the frequency regime where condition (i) governs the size limitation, and
(b) with higher substrate permittivity ε_r, higher capacitance values can be achieved in this frequency regime, where the absolute limit a_{max} must be taken into account.

Single-layer inductors are shown in Fig. 2.3. For values of the inductance below about 0.5 nH a narrow strip – see Fig. 2.3(a) - is sufficient. Its inductance is estimated via

$$L/nH \approx \frac{0.8}{\pi}\frac{l}{\text{mm}}\ln\left(\frac{2l}{\pi w}\right) . \quad (2.4)$$

With respect to the strip width w, a trade-off between the maximum achievable inductance and a high onset level for non-linear effects such as intermodulation - see also [4] - must be taken into account. Although considerably higher values than 0.5 nH are predicted from Eq. (2.4) if "very narrow" strips are employed, the width should be restricted to about $w > 0.1$ mm in order to avoid strong non-linear effects at a relatively low value of the stored magnetic field energy.

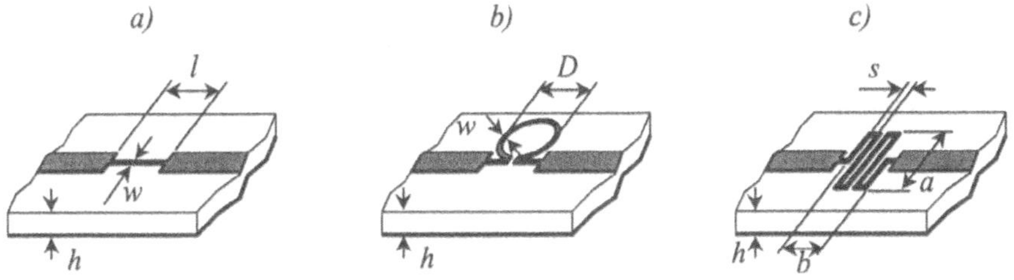

Figure 2.3. Different types of single-layer inductors.

Loop inductors, as shown in Fig. 2.3(b), with diameter D of the loop and strip width w, possess an inductance of about

$$L/nH \approx 0.6\frac{D}{\text{mm}}\ln\left(\frac{2D}{w}-1\right) . \quad (2.5)$$

In order to provide the highest inductance values for a given occupied area A, a spiral inductor - see below - should be used, but this requires a multi-layer technology. The single-layer meander inductor - see Fig. 2.3(c) - represents the second-best solution for a high L at a given surface. It is characterized by outer linear dimensions a and b, strip width w and spacing s (strip center to strip center) between adjacent meander arms. The inductance can roughly be estimated via

$$L/\mathrm{nH} \approx 0.8\frac{a/\mathrm{mm}\cdot b/\mathrm{mm}}{w/\mathrm{mm}}F\left(\frac{s}{w}\right) \quad \text{with } F\left(\frac{s}{w}\right)=\frac{w}{s}\ln\left(\frac{2s}{w}-1\right). \tag{2.6}$$

In order to discuss Eqs. (2.6), a given outer size b and given strip width w (e.g., $w = 0.1$ mm) is assumed. For a low number of meander arms the ratio $\beta = s/w$ is much larger than 1. By increasing the number of meander arms associated with a decreased value of β, the inductance increases, but only up to a value of $\beta \approx 2.3$. This is because the increase in magnetic flux due to the increased length of the meander is overcompensated by the reduction of the area between the meander arms - note $L = 0$ for $s = w$. The flat maximum between $b = 2$ and 2.6 is on the order of $F_{max} = 0.55$.

Until now the influence of the ground plane has been neglected. Due to image currents, the presence of a ground plane always leads to a reduction of the inductance by a factor $q < 1$ which in a crude approximation is given by

$$q = \frac{1}{1 + w/h} \tag{2.7}$$

For $a \leq \lambda_g/10$ and $w > 0.1$ mm (for sufficient power handling capability) and an absolute size restriction of $a = 4$ mm, one obtains for LAO at operational frequencies higher than about 1.5 GHz, a maximum inductance of $L_{\max}/nH \approx (8.5/f/GHz)^2$, and $L_{max} \approx 36$ nH for lower frequencies. The corresponding values for sapphire and MgO are $L_{\max}/nH \approx (13.4/f/GHz)^2$ above 2.1 GHz, and 41 nH for lower frequencies.

The fabrication of all HTS structures for capacitors and inductors considered up to this point is based on well-known, single-layer deposition technology. If multi-layer technology with thin (d about 20 to 80 µm) insulating layers between HTS layers can be employed, conductor-insulator-conductor (CIC) capacitors as shown in the left-hand part of Fig. 2.4 will become available with capacitances up to about 100 pF. As already mentioned, spiral inductors offer the highest inductance value for a given occupied area, but they require a crossover for the interconnect to the center of the spiral -see right-hand side of Fig. 2.4. Multi-layer technology permits the realization of an all-HTS circuit.

a) b)

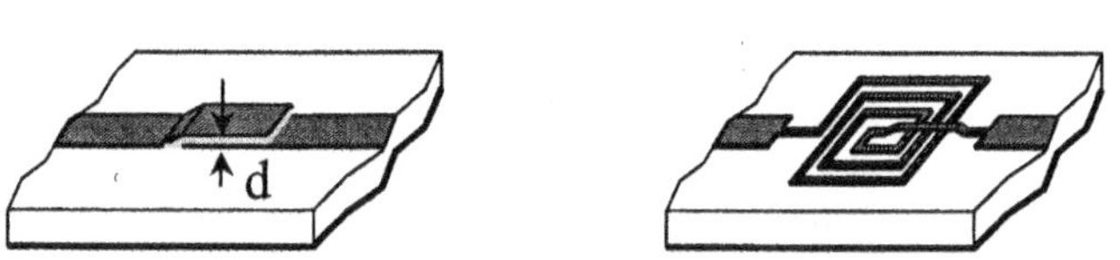

Figure 2.4. Lumped elements which require multi-layer technology: Left: Conductor-insulator-conductor capacitor. Right: Spiral inductor.

2.2 INTRODUCTION OF A CHARACTERISTIC FIGURE OF MERIT

As will be explained in section 3.1.3, the required element values for the realization of narrowband passive circuits can to some extent be varied by means of impedance transformers. However, impedance transformers change the internal impedance level by simultaneously changing inductances and capacitances. If the inductance is decreased, the capacitance is increased by the same factor. In other words, the product of capacitances and inductances remains constant. Consequently, the product of the largest inductance and the largest capacitance can serve as a figure of merit for the realizability of a circuit. Note, that in this definition, the capacitor with the highest capacitance and the inductor with the highest inductance can be situated at remote positions within the circuit. If the $L_{max}C_{max}$ product is normalized to the center frequency of the circuit, one obtains the figure of merit

$$\Lambda = \omega_0^2 L_{max} C_{max} \ . \tag{2.8}$$

In the following, a crude estimation for the achievable figure of merit Λ is given. It uses the limits for patch capacitors shown in Fig. 2.2 and the limits for meander capacitors. As in the preceding discussion two limits for the element values are taken into account, namely that the linear size must be (i) smaller than 1/10 of a guided wavelength and (ii) smaller than an absolute value taken to be 30 mm/$\sqrt{\varepsilon_r}$. The result for the figure of merit Λ, normalized to the substrate thickness, is depicted in Fig. 2.5. The shaded area shows the order of an achievable value as a function of the frequency. For example, for a 0.5-mm-thick substrate, a value of about 40 can be achieved for operational frequencies of about 1 GHz. For lower and higher frequencies, the achievable value drops due to absolute size limitations and due to the condition imposed by Eq. 1.6, respectively.

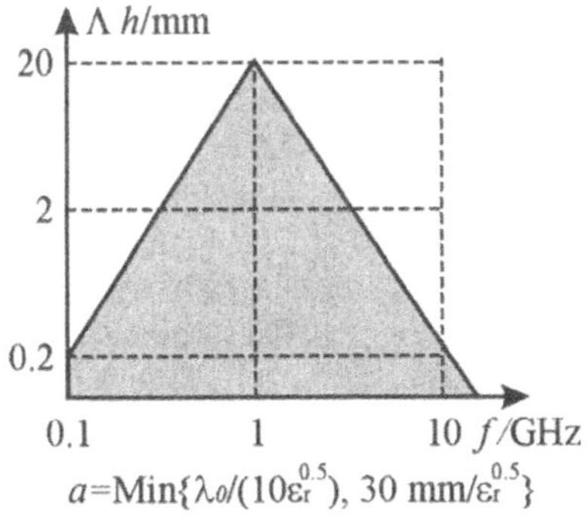

Figure 2.5. Upper limit for the characteristic number as function of frequency. Shaded area indicates approximate region of achievable values.

Some authors - see, e.g., [13] - use the approach of "frequency-dependent inductances" in order to realize an "effective inductance" higher than the available "physical" inductance. This approach is illustrated in Fig. 2.6. An inductance L, which may be part of a bandpass filter, is loaded by a parallel capacitor C, so that the resonant frequency $\omega_r=1/\sqrt{LC}$ is higher than the highest operational frequency ω_{max}. The effective inductance then becomes

$$L_{eff}(\omega)=\frac{dX}{d\omega}=L\left[1+(\omega/\omega_r)^2\right]\cdot\left[1-(\omega/\omega_r)^2\right]^{-2} \tag{2.9}$$

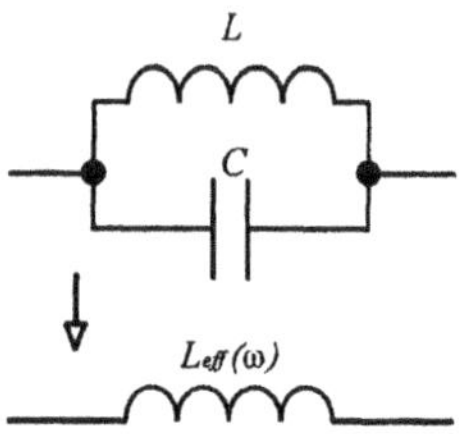

Figure 2.6. Frequency-dependent inductance

Because this inductance is frequency dependent, large deviations between design response and real response can occur if the constant inductance in the design is replaced with a frequency-dependent inductance. This requires modified design rules to be applied for frequency-variable inductances.

2.3 UNLOADED QUALITY FACTOR AND POWER HANDLING CAPABILITY OF HTS LUMPED ELEMENTS

The resistance R_{diss} of an inductor L which accounts for the dissipative losses in the conductor is given as

$$R_{diss}=\frac{\omega_0 L}{Q} \tag{2.10}$$

A precise estimate of Q for typical HTS inductors as shown in Fig. 2.3 is not possible. However, from experimental results obtained for YBCO structures on LAO or sapphire at temperatures below 70 K and frequencies around 1 to 2 GHz, the unloaded quality factor is estimated to be in the range of 20,000 and 50,000 if strip conductors narrower than about 100 μm are avoided [10].

The unloaded quality factor of capacitors is in a first approximation (via $Q=1/\tan\delta$) determined by the loss tangent of the substrate material.

The power-handling capability of a lumped element can be characterized by the maximum stored field energy W_{max} causing no significant degradation of Q. If the maximum stored field energy is multiplied by the (circular) frequency, the maximum "oscillating power" $P_{osc}=\omega_0 W_{max}$ is obtained - see e.g. [10].

Experiments with YBCO structures on LAO or sapphire at temperatures of about 70 K and frequencies of about 1 to 2 GHz, indicate a power handling of lumped HTS inductors on the order of 5 to 50 W oscillating power [10]. For lower temperatures significantly higher power handling occurs. From the maximum oscillating power of its inductors, the power handling of a filter can be deduced [4]. With $\Delta f/f_0$ as the fractional bandwidth of a bandpass filter and P_{inc} as the power incident to the input port, the oscillating power per inductor is estimated as

$$\omega W = \frac{f_0}{\Delta f} \varsigma \cdot P_{inc}\,, \tag{2.11}$$

with ς depending on the filter type (e.g., Chebyshev, Cauer, etc.), the position of the inductor in the circuit and on the deviation of the frequency from the center frequency. For a very crude approximation, the maximum value of ς is for a Chebyshev filter taken to be on the order of 2.5, so that for a filter with 1% fractional bandwidth, an oscillating power of 5 to 50 W transforms into a maximum input power (e.g., coincident with the 1-dB compression point) of about 20 to 200 mW . If the simple model of a memory-less nonlinear model applies, the 3rd-order intercept point is about 10 dB above the 1-dB compression point [4]. Under this assumption a 3-dB compression point on the order of 23 to 33 dBm can be expected.

3. Design of planar lumped-element HTS circuits

3.1 BANDPASS FILTERS

3.1.1 *General considerations*

As for all linear two-ports, filters are characterized by the frequency dependence of the scattering parameters, including the transmission and input reflection coefficient,

$$S_{21}(f) = |S_{21}(f)| \cdot \exp(-j\Phi_{21}(f)) \tag{3.1a}$$

and

$$S_{11}(f) = |S_{11}(f)| \cdot \exp(-j\Phi_{21}(f))\,, \tag{3.1b}$$

respectively. From these scattering parameters, insertion loss $L(f)$, return loss $L_r(f)$ in dB and group delay $\tau(f)$ are derived via

$$L(f) = 20\log(1/|S_{21}(f)|),\, L_r(f) = 20\log(1/|S_{11}(f)|)\,. \tag{3.2a}$$

and

$$\tau(f) = d\Phi_{21}(f)/df \quad . \tag{3.2b}$$

In order to execute a well-defined function in a subsystem (e.g., preselection in a receiver front end [4]), amplitude and group delay response must meet requirements, specified in a manner as shown in Fig. 3.1, for the amplitude response.

As seen from Fig. 3.1, the most important specification parameters for a bandpass are the passband width Δf and the widths $(\delta f)_1$ and $(\delta f)_2$ of the transition regions between the passband and the stopbands. Further parameters are the maximum insertion loss $L_{0,max}$ and the minimum return loss $L_{r,min}$ in the passband and minimum attenuation $L_{s1,min}$ and $L_{s2,min}$ in the stopbands. In many system applications, additional specifications for the group delay response are given. Equalization of the group delay response by means of an allpass is briefly discussed in section 3.2.

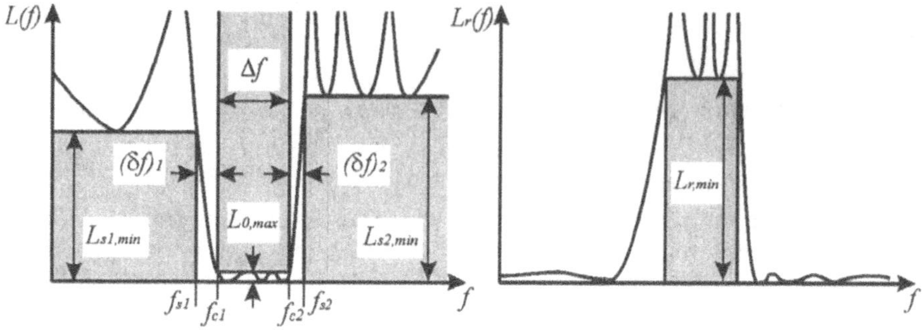

Figure 3.1. Specification parameters for a bandpass filter (shaded rectangular areas) and one specific approximation of this specification. Specification parameters are passband width Δf and widths of the transitions $(\delta f)_1$ and $(\delta f)_2$, as well as maximum insertion loss $L_{0,max}$ and minimum return loss $L_{r,min}$ in the passband, and minimum attenuation $L_{s,min}$ in the stopbands.

With respect to the specific subject of this chapter, synthesis procedures for circuits consisting of inductors and capacitors in HTS technology must be provided. This synthesis is carried out in several steps. In the first step, a response function is selected which both meets the specifications and belongs to the class of response functions of finite lumped-element networks. The choice of this response function ("approximation") is not unique. Thus, in principle, numerical approximations of the response by rational functions can be employed directly in the bandpass (BP) region. As well, one can use standard analytic Chebyshev or Cauer lowpass (LP) approximations with a subsequent LP-BP transformation.

The second step consists of the synthesis of an LC circuit which realizes the selected response function. A straightforward circuit synthesis usually fails to deliver values of the capacitances and inductances which are available in HTS technology. Therefore, in a third step, circuit modifications become necessary. This includes insertion of impedance transformers, as well as the application of equivalence transformations to circuit sections. Typically the first part of the synthesis procedure is based on non-dissipative lumped elements, leading to a nominal insertion-loss response. In a subsequent step, the filter is analyzed, taking losses and tolerances into account.

For an approximation of the degradation of the amplitude response (difference between "real" and nominal response) due to dissipation, one considers a uniform dissipation with all inductors and capacitors having the same unloaded quality factor Q. For this case, dissipation results in a frequency-dependent increase of insertion loss according to

$$\Delta L_{diss} = 8.7 \frac{2\pi f \cdot \tau(f)}{Q} \quad . \tag{3.3}$$

The strongest degradation occurs close to the band edges where the group delay τ is maximum. The maximum value of the group delay can be related approximately to the steepness of the filter skirts via

$$\tau_{\max} = \beta_0 \frac{L_{s,\min} + L_{r,\max}}{\delta f} \quad . \tag{3.4}$$

The parameter β_0 depends on the approximation chosen and is estimated to about 0.015/dB for a Chebyshev and 0.005/dB for an elliptic (Cauer) response. Thus, the unloaded quality factor limits the maximum achievable steepness of the filter skirts. In the theoretical case of a lossless filter structure with given passband width Δf, any arbitrarily high steepness can be obtained by increasing the filter order N to a sufficiently high value. In a real filter with dissipation, the increase of steepness with filter order is limited according to Eqs. 3.3 and 3.4. Given the approximation type and the passband width, increasing the filter order N results in steeper skirts only up to a value N_{opt}. The choice of a higher order $N > N_{opt}$ results in a degradation instead of an improvement of the steepness.

Since HTS lumped-element circuits offer much higher Q, steeper filter skirts have become available. For a given passband width, a higher filter order N_{opt} can be realized, which in turn provides significantly steeper skirts. However, the range of possible element values is lower than in the case of normal-conducting structures.

3.1.2 Impact of filter architecture on transmission pole and null location

General conclusions on the frequency response of a filter can be drawn from the knowledge of its "architecture." For simplicity, the discussion of these results is based on the assumption of a lossless filter structure.

Filter networks are composed of reactance (LC) one-ports, characterized by reactance functions $X(s)$ or susceptance functions $B(s)= -1/X(s)$ of the complex frequency $s=\sigma+j\omega$. The complex frequency formulation means that a current $i(t) = \mathrm{Re}\{I(s)\exp[st]\}$ through a reactance two-port produces a voltage

$$v(t) = \mathrm{Re}\{jX(s)I(s)\exp[st]\} = \mathrm{Re}\{(I(s)/jB(s))\exp[st]\}.$$

Poles of $X(s)$ at $s_p=\sigma_p+j\omega_p$ represent eigen oscillations

$$v(t) = v_0 \exp(\sigma_p t)\cos(\omega_p t + \varphi)$$

of the open one-port, and poles of $B(s)$ (coincident with nulls of $X(s)$) correspond to eigen oscillations of the shorted one-port. Since the (ideal) reactance one-port contains no losses at all, the eigen oscillations are undamped Consequently, all poles and nulls of reactance functions possess $\sigma_p = 0$ and are therefore located on the imaginary axis of the s-plane. Furthermore, it can be shown that poles and nulls are single and alternate along the imaginary axis.

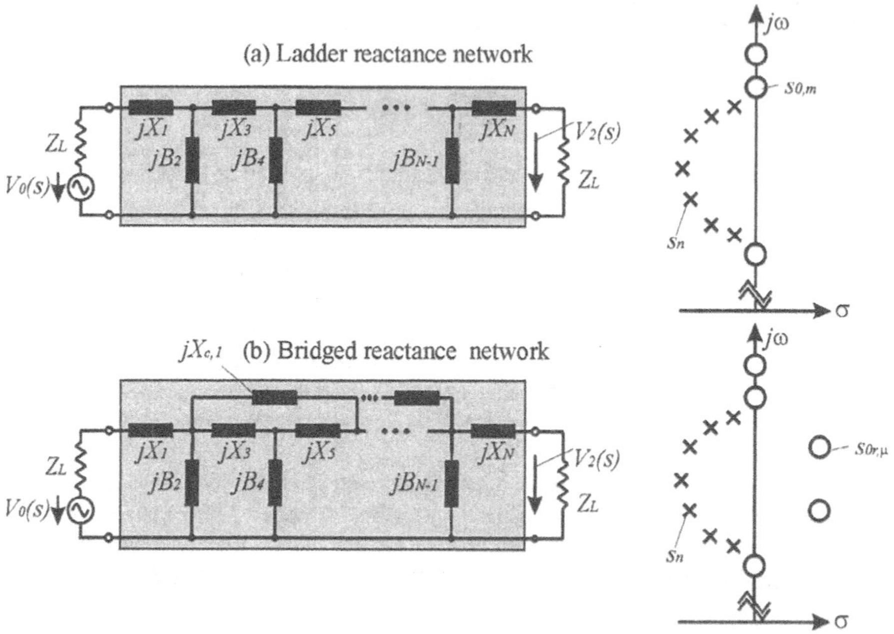

Figure 3.2. (a) Ladder reactance network with pole-null distribution in the upper half of the complex s-plane. (b) Bridged reactance network with additional nulls in the right half-plane.

A reactance (LC) filter is a two-port which consists of suitably interconnected reactance one-ports. The manner in which these interconnections are performed is referred to as "architecture." The terminations at the two ports of the filter are assumed to be equal and purely resistive. With this assumption the transmission coefficient relates the output voltage V_2 to the source voltage V_0 according to

$$S_{21}(s) = 2V_2(s)/V_0(s). \tag{3.5}$$

The transmission coefficient is a rational function of s and exhibits poles and nulls in the complex s-plane. Poles of the transmission coefficient function correspond to eigen oscillations of the terminated two-port (with driving source voltage of zero). Due to the presence of resistive terminations, eigen oscillations are damped, and this, in turn, requires all poles to be situated in the left half-plane ($\sigma < 0$). Poles which are not located on the real axis, occur as conjugated pole pairs.

The networks can be classified with respect their "architecture." The upper part of Fig. 3.2 depicts a ladder network with alternating series and shunt branches. Each series reactance $X_\mu(s)$ and each shunt susceptance $B_\mu(s)$ (both shown as reactangular black-filled boxes) may be composed of an arbitrary finite number of inductances and capacitances. In a ladder network, nulls of the transfer functions can occur only at poles of the series reactances and at poles of the shunt susceptances. Since these poles occur on the imaginary axis only, nulls of the transfer function (denoted by $s_{0,m}$ in the upper right part of Fig. 3.2) can be located on the imaginary axis only.

The transmission factor of a lossless ladder reactance two-port has minimum phase. This means that the group delay response $\tau(\omega)$ is given by

$$\tau(\omega) = \frac{0.115}{\pi} \int_{-\infty}^{\infty} L(\omega') \frac{\omega'^2 + \omega^2}{\left(\omega'^2 - \omega^2\right)^2} d\omega' \quad , \tag{3.6}$$

and is completely determined by the frequency response of the insertion loss $L(\omega)$. As already seen from Eq. 3.4, a steep filter skirt is always associated with a large value of the group delay near the band edge. In the case of a high-order ladder-type filter, a sharp increase of the group delay from the band center to the band edge occurs. If an application requires (nearly) constant group delay, two distinct possibilities are known to meet this demand: (a) A cascaded allpass ("equalizer") filter can be added to the ladder-type filter with the specified insertion loss response, but with strongly frequency-dependent group delay. The group delay of the allpass is chosen to produce, together with the group delay of the ladder network, an approximately constant response. (b) A "self-compensated" filter with flat group delay response can be designed.

For both allpass filters, as well as for self-compensated bandpass filters, transmission nulls must be placed in the right half-plane in order to produce a group delay different from that given in Eq. 3.6. This, in turn, requires the employment of architectures different from that of the ladder type. Since lattice networks are difficult to be realized in planar technology, bridged networks, as depicted in part (b) of Fig. 3.2, are employed. In section 3.2 of this chapter an example will be presented of an HTS allpass filter utilizing a bridged LC circuit.

At this point, one difference between (distributed) coupled-resonator filters and lumped-element LC filters should be mentioned. Since coupled-resonator filters, with coupling only between adjacent resonators, possess no transmission nulls at finite frequencies, the realization of attenuation poles at finite frequencies, as well as group delay equalization, requires cross-coupling (equivalent to "bridges") between non-adjacent resonators. In contrast, LC ladder networks allow the realization of attenuation poles at finite frequencies via the use of series reactances and/or shunt susceptances with poles at finite frequencies. Therefore, bridged structures (coupling between non-adjacent branches) are required for self-compensated filters or allpass filters only.

3.1.3 Design of Chebyshev bandpass filters

LP prototype. In the traditional design approach, the circuit architecture, as well as the values of inductances and capacitances of the bandpass (BP) filter, are derived from the low-pass (LP) prototype via an LP-BP transformation and impedance scaling. The frequency response of the power transfer function $|S_{21}|^2$ and the corresponding circuit of a Chebyshev LP prototype are depicted in Fig. 3.3. The appropriate values of the filter coefficients g_μ are given in tables or can be calculated from closed-form formulas. The prototype filter is normalized with respect to the passband edge-frequency f_c, so that for the normalized dimension-free frequency, $\Omega = f/f_c$ the passband extends up to $\Omega = 1$. Furthermore, all impedances are normalized to the (real) resistance Z_L of the terminating load at the input port. Hence, all inductances and capacitances of the prototype LP are dimension-free and equal the filter coefficients - see Fig. 3.3. From the normalized stopband edge frequency, Ω_s, the minimum stopband attenuation, $L_{s,min}$, and the minimum passband return loss, $L_{r,min}$, the required filter order N is estimated to be

$$N \geq 0.05 \cdot \left(L_{s,\min} + L_{r,\min} + 6\right) \Big/ \log\left(\Omega_s + \sqrt{\Omega_s^2 - 1}\right) . \tag{3.7}$$

Due to the "equiripple" response the nominal insertion loss vanishes at N distinct frequencies in the range $-1 < \Omega < 1$ and reaches (within this passband) its maximum (N-1) times.

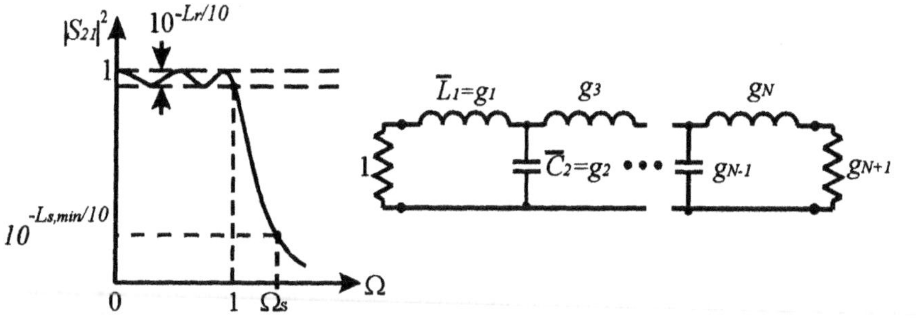

Figure 3.3. Chebyshev low-pass prototype filter.

BP resulting from direct LP-BP transformation. With given center frequency f_0 and passband width Δf, the prototype LP is transformed into the BP via

$$\Omega = \frac{f_0}{\Delta f}\left(\frac{f}{f_0} - \frac{f_0}{f}\right) = \frac{\omega_0}{\Delta\omega}\left(\frac{\omega}{\omega_0} - \frac{\omega_0}{\omega}\right) . \tag{3.8}$$

Furthermore, impedance scaling requires that all impedances are multiplied by Z_L, the real load resistance. With a direct application of the LP-BP transformation and impedance scaling of the circuit in Fig. 3.3, the inductances g_μ with $\mu = 1, 3, \ldots$ are transformed into series resonant circuits with inductance and capacitance given by

$$L_{s,\mu} = g_\mu \frac{Z_L}{\Delta\omega} \quad \text{and} \quad C_{s,\mu} = \frac{1}{g_\mu} \cdot \frac{\Delta\omega}{\omega_0^2 Z_L} , \tag{3.9a}$$

and the capacitances g_ν with $\nu = 2,4,\ldots$ are transformed into parallel resonant circuits with

$$C_{p,v} = g_v \frac{1}{\Delta\omega \cdot Z_L} \quad \text{and} \quad L_{p,v} = \frac{1}{g_v} \cdot \frac{\Delta\omega \cdot Z_L}{\omega_0^2}. \tag{3.9b}$$

There are several reasons why the so-obtained filter circuit cannot be realized with HTS lumped elements as described in section 2.1. Two of these reasons are that (i) shunt inductances cannot be fabricated and that (ii) the figure of merit Λ as introduced in section 2.2 becomes

$$\Lambda = \Lambda^{(A)} = \omega_0^2 \cdot L_{\max} \cdot C_{\max} = \omega_0^2 \cdot L_{s,\max} \cdot C_{p,\max} = g_{\mu,\max} \cdot g_{v,\max} \left(\frac{\omega_0}{\Delta\omega}\right)^2 \approx \left(\frac{\omega_0}{\Delta\omega}\right)^2 >> 1.$$

In order to overcome these problems a modified synthesis procedure must be applied which employs "impedance inverters" and "impedance transformers." These "unphysical" two-ports are used to simplify the procedure, but later will be eliminated by means of LC circuits.

Impedance inverters and transformers. For the conversion of the circuit into a modified circuit without shunt inductances, impedance inverters are introduced. By definition, the admittance matrix of an impedance inverter is given by

$$\begin{bmatrix} I_1 \\ I_2 \end{bmatrix} = \frac{j}{K} \begin{bmatrix} 0 & 1 \\ 1 & 0 \end{bmatrix} \cdot \begin{bmatrix} V_1 \\ V_2 \end{bmatrix} \tag{3.10}$$

with a real-valued frequency-independent inverter constant K, so that an output impedance Z_2 is transformed via

$$Z_1 = \frac{K^2}{Z_2}$$

into an input impedance Z_1.

The impedance inverter can be composed of 3 (non-physical) frequency-independent reactances as shown in the center segment of Fig. 3.4. If, for the operational range of the designed filter, $|f-f_0| << f_0$ holds, the frequency-independent reactances can be replaced with capacitances - see right segment of Fig. 3.4.

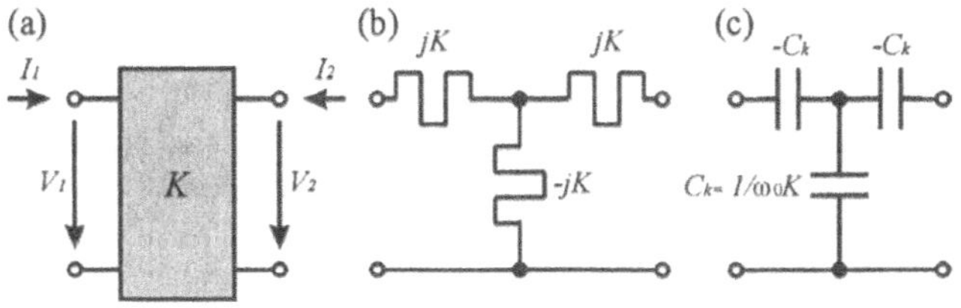

Figure 3.4. (a) Impedance inverter as a two-port. (b) Representation by means of frequency-independent reactances. (c) Approximation.

The formal introduction of another two-port, namely of an impedance transformer, provides additional degrees of freedom in the choice of values for the inductances and capacitances. As shown in Fig. 3.5, a resistance R is transformed into a resistance R/n^2, with $n>1$ as the transformation ratio. An approximate realization by means of positive and negative capacitances is shown in the right part of Fig.3.5. This approximation requires $\omega C_t << 1/R$.

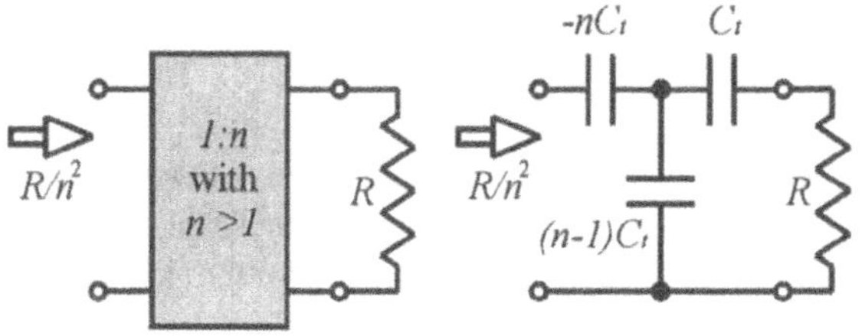

Figure 3.5. Impedance transformer

Bandpass filter resulting from the introduction of impedance inverters and transformers. If N-1 impedance inverters with arbitrarily real values of the inverter constants are introduced , the circuit according to Fig. 3.3 is converted into the circuit shown in Fig. 3.6 with the new parameters

$$\bar{g}_{2n+1}=g_{2n+1}\frac{\prod_{v=1}^{n}K^2_{2v,2v+1}}{\prod_{v=1}^{n}K^2_{2v-1,2v}}\;,\quad \bar{g}_{2n}=g_{2n}\frac{\prod_{v=1}^{n}K^2_{2v-1,2v}}{\prod_{v=2}^{n}K^2_{2v-2,2v-1}}\;. \tag{3.11}$$

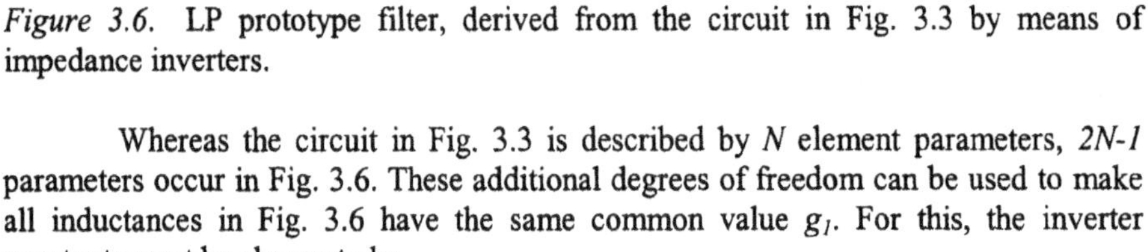

Figure 3.6. LP prototype filter, derived from the circuit in Fig. 3.3 by means of impedance inverters.

Whereas the circuit in Fig. 3.3 is described by N element parameters, $2N$-1 parameters occur in Fig. 3.6. These additional degrees of freedom can be used to make all inductances in Fig. 3.6 have the same common value g_1. For this, the inverter constants must be chosen to be

$$K_{mn}=\frac{g_1}{\sqrt{g_m\cdot g_n}}\,, \tag{3.12}$$

which leads to a symmetric structure with $K_{N-m,N-m+1}=K_{mn}$ and $Z_{L2}=1$.

$$K_{N-m,N-m+1}=K_{m,m+1} \tag{3.13}$$

In the following, a step-by-step synthesis procedure for bandpass filters is derived. It starts with the prototype network in Fig. 3.6 and employs an LP-BP transformation and the concept of impedance transformers.

(a) First, the inverter constants in Fig. 3.6 are chosen according to Eq. 3.12 , resulting in a structure for which the values of all inductances are identical and equal to g_1. Furthermore, the normalized output load resistance $\bar{g}_{N+1}$ becomes unity.

(b) By an LP-BP transformation (Eq. 3.8) and scaling of all impedances to Z_L , the inductances are transformed into series resonant circuits with values $L = g_1 Z_L/\Delta\omega$ and $C = \Delta\omega/(g_1\omega_0^2 Z_L)$. Impedance inverters and load resistors remain unchanged under the LP-BP transformation, but with impedance scaling, their values are multiplied by Z_L.

(c) The N-1 impedance inverters are replaced with capacitances as shown in Fig. 3.4c. As a result of this step, there are N-1 positive shunt capacitors. Furthermore, negative capacitances occur in series with the series resonant circuits obtained from step (b). In the case of the two outer circuits (adjacent to the ports), one negative capacitance is series connected, but two negative capacitances are series connected to each inner resonant circuit ($\nu = 2,\ldots, N$-1). All series negative capacitances are merged with the adjacent positive series capacitors to yield a net positive capacitance. Whereas after step (b) all series resonant circuits are identical, the "absorption" of the negative capacitances in step (c) results in different positive net capacitances.

(d) In order to gain an additional degree of freedom, at both ports impedance transformers - see Fig. 3.5 - are inserted between the ports and the outer series resonant circuits, such that the positive capacitance $C_t > 0$ in the realization of the transformer is placed at the port. The transformation ratio n is (at this stage) still considered as variable. The scattering parameters of the filter two-port are kept unchanged in comparison to the two-port without impedance transformers if the internal impedance level is decreased by a factor of n^2. The negative series capacitance $-nC_t$ (in Fig. 3.5) can be absorbed into the positive series capacitance of the adjacent resonant circuit.

As a result of steps (a) to (d) the circuit shown in Fig. 3.7 is obtained. Note, that the capacitances adjacent to the ports C_t and $(n\text{-}1)C_t$ originate from the impedance transformers. All N inductance values are identical and given by L_s. The N-1 shunt capacitances depend on the filter coefficients g_m - see Fig. 3.3. One has

$$L_s = \frac{g_1 \cdot Z_L}{n^2 \Delta\omega} \qquad \text{and} \qquad C^{(p)}_{m,m+1} = \frac{\sqrt{g_m \cdot g_{m+1}}}{g_1} \frac{n^2}{\omega_0 Z_L} \quad . \tag{3.14}$$

The series capacitances $C_{s,\nu}$ of the N-2 "inner" resonant circuits ($\nu = 2\ldots N$-1) become

$$C_{s,\nu} = \frac{\Delta\omega}{\omega_0} \frac{n^2}{g_1 \cdot \omega_0 Z_L} \cdot \frac{1}{1 - \dfrac{\Delta\omega}{\omega_0}\left(\dfrac{1}{\sqrt{g_{\nu-1} \cdot g_\nu}} + \dfrac{1}{\sqrt{g_\nu \cdot g_{\nu+1}}} \right)} \quad . \tag{3.15}$$

The series capacitances of the outer series resonant circuits ($\nu = 1$ and N) are chosen to be

$$C_{s,1} = C_{s,N} = \frac{C_A}{1 - \dfrac{C_A}{nC_t}} \quad \text{with} \quad C_A = \frac{\Delta\omega}{g_1 \omega_0^2 Z_L} \cdot \frac{n^2}{1 - \dfrac{\Delta\omega}{\omega_0 \sqrt{g_1 \cdot g_2}}}. \tag{3.16}$$

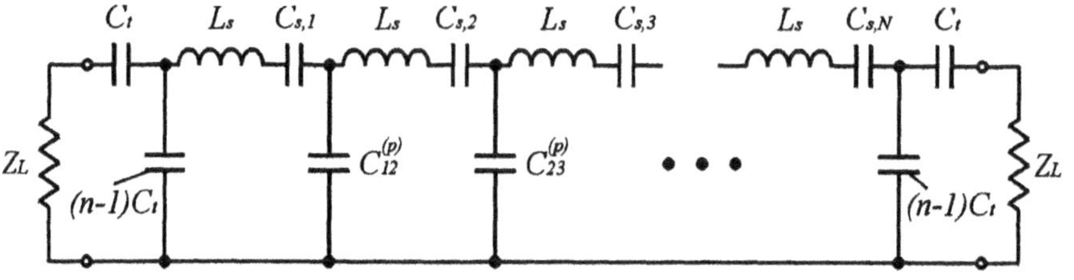

Figure 3.7. Bandpass filter based on impedance inverters and impedance transformers.

It is important to note that for a given filter response specification, all parameters in Fig. 3.7 are prescribed, except for the value of the transformation ratio n. This permits (within the range of realizable n values) a variation of the absolute value of the required inductances or capacitances. However, the products of inductances and capacitances remain unchanged with n. In order to check the realizability of a filter structure the figure of merit Λ – see Eq. 2.8 - must be derived from the maximum required inductance and the maximum required capacitance. From Eqs. 3.14 and 3.15, it follows that the ratio of the shunt capacitances to the series capacitances $C_{m,m+1}^{(p)}/C_{s,\nu}$ is on the order of $\omega_0/\Delta\omega$. Thus, the maximum shunt capacitance must be multiplied by the maximum series inductance in order to derive the figure of merit.

$$\Lambda = \Lambda^{(B)} = \omega_0^2 L_{\max} \cdot C_{\max} \approx \sqrt{(g_\mu \cdot g_\nu)_{\max}} \, \frac{\omega_0}{\Delta\omega} > \frac{\omega_0}{\Delta\omega} >> 1. \tag{3.17}$$

The figure of merit is, for the case of the circuit in Fig. 3.7, on the order of $\omega_0/\Delta\omega$. Note, that in the case of the bandpass which is derived from a direct application of the LP-BP transformation to the prototype in Fig. 3.3, the characteristic number is on the order of $(\omega_0/\Delta\omega)^2$. Hence, the circuit in Fig. 3.7 is "much closer" to a realization than is the circuit due to a direct LP-BP transformation. However, it was shown in section 2.2 that a figure of merit larger than about 20-40 is difficult to achieve. Consequently, for filters with a fractional bandwidth of several percent, a modification of the circuit is needed. This modification is explained in the following section.

Circuit modification by means of T-Π transformation. In order to achieve a realizable filter, the required figure of merit Λ must be reduced. A possible way for achieving this goal is a reduction of the shunt capacitances by means of a T-Π transformation. This approach is illustrated in Fig. 3.8. In a first step the capacitances $C_{s,\nu}$ of the series resonant circuits in Fig. 3.7 are split into two capacitances $2C_{s,\nu}$, as shown in the upper part of Fig. 3.8. The obtained T-sections consisting of three capacitances are transformed into equivalent Π-sections. The capacitances are transformed according to

$$C^{(s)}_{m,m+1} = \frac{4C_{s,m} \cdot C_{s,m+1}}{2(C_{s,m} + C_{s,m+1}) + C^{(p)}_{m,m+1}} \tag{3.18a}$$

$$C^{(a)}_{m,m+1} = \frac{2C_{s,m} \cdot C^{(p)}_{m,m+1}}{2(C_{s,m} + C_{s,m+1}) + C^{(p)}_{m,m+1}} . \tag{3.18b}$$

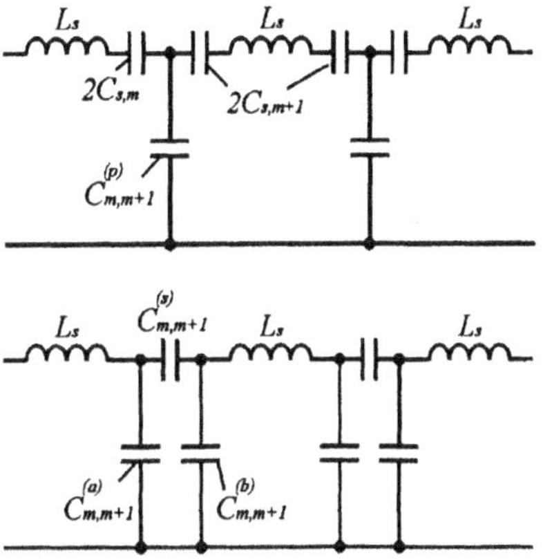

Figure 3.8. Modification of circuit Fig. 3.7 by means of T-Π transformation.

Taking into account that the shunt capacitances in Fig. 3.7 are much larger than the series capacitances, one finds from Eq. 3.18b that the shunt capacitances in the lower part of Fig.3.8 are about twice as large as the series capacitances in Fig. 3.7. Furthermore, in the lower part of Fig. 3.8 the series capacitances are by a factor $\Delta\omega/\omega_0$ lower than the shunt capacitances. By multiplying the largest value of the shunt capacitance in the Π-structure with the largest series inductance, one obtains a value of about $\Lambda = 2$ for the figure of merit. Because of this, the circuit in Fig. 3.9 is suitable to be realized as an HTS lumped-element circuit.

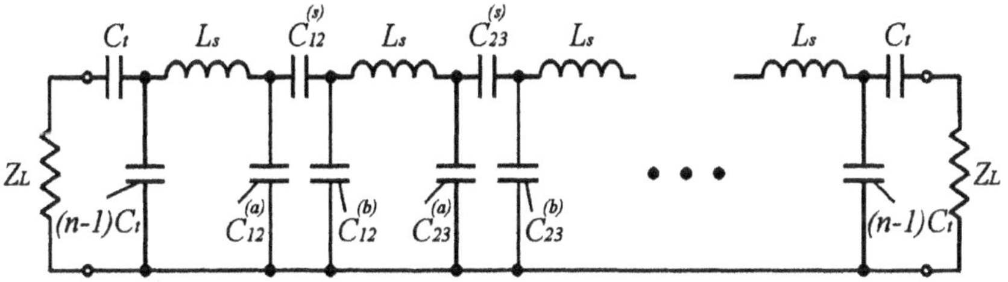

Figure 3.9. Realizable bandpass filter circuit

3.1.4 Examples of actual lumped-element HTS filters

Almost all lumped-element HTS filters so far reported in the literature [11-19] are different topological realizations of the common architecture according to Fig. 3.9, which was derived in section 3.1.3. The choice of this architecture is justified by a number of attractive features which are advantageous for use in HTS technology. The structure consists of an alternating sequence of Π capacitor networks and series inductors which can be realized easily in planar technology. Moreover, the shunt capacitors can take up "parasitic" capacitance to the ground of the inductors and series capacitors. The series capacitors (responsible for the coupling between the resonant sections) should be reasonably small for narrow-band filters.

Filters based on this architecture have been realized over a very wide frequency range from 264 MHz [19] to 10 GHz [11], of different order N from $N = 2$ [14] to $N = 6$ [12, 15, 16, 18], and of different fractional bandwidth from 0.27% [13, 14] to 3% [11]. YBCO [12-19] as well as TlCBCO [11] thin one-sided [11, 12, 15, 16] and double-sided [13, 14, 17-19] HTS films on LAO [11-17], sapphire [18], and MgO [19] substrates were used for the fabrication of the filters.

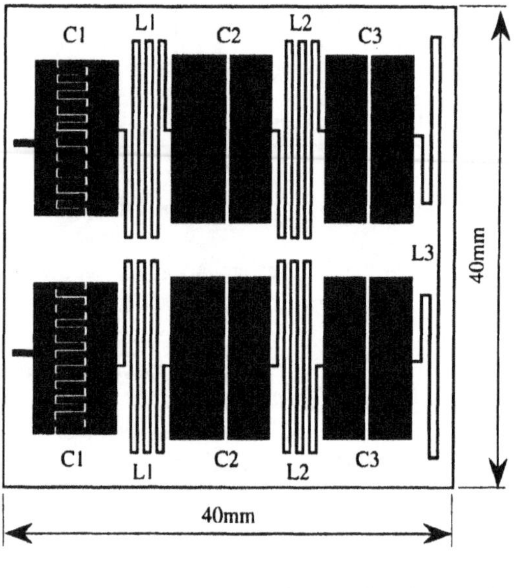

Figure 3.10. Layout of a 5-pole lumped-element band-pass filter [19] for a center frequency of 246 MHz and fractional bandwidth of 0.76 %.

In all filters referenced, the Π capacitor networks were realized as two patches with a gap between them. In the case of very narrow (less than 1%) band filters and/or relatively high center frequency (more than about 6 GHz), the required weak coupling between the resonant sections results in an unacceptably large gap. This problem has been overcome by adding a redundant patch to each Π-capacitor network [11], or by introducing a frequency-dependent inductor (a half-loop inductor in parallel with an interdigital capacitor) in place of a pure inductor [13, 14]. The inductors were realized as square or rectangular loops [12-18], as meander lines [19], and as spiral inductors with normal-conducting crossover [11]. To provide the required external Q factor, interdigital capacitors were used in the Π-capacitor networks next to the ports [13, 14, 19].

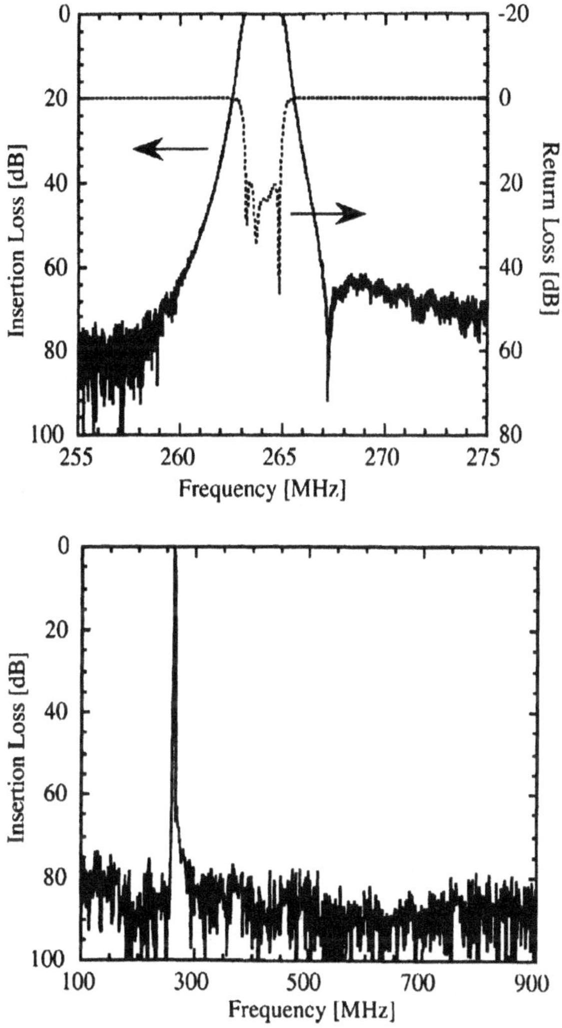

Figure 3.11 Measured frequency response of the filter in Fig. 3.10 [19].

A five-pole filter with a center frequency of 264 MHz and with fractional bandwidth of 0.76% has been fabricated on a 40x40 mm^2 MgO substrate [19]. The layout of the filter is shown in Fig. 3.10. At 70 K this filter showed a maximum insertion loss of 0.1 dB and minimum return loss of 20 dB. The frequency response was free from spurious modes up to 900 MHz. Fig. 3.11 shows the measured response of this filter for two different frequency scales. Even for such a low frequency, the necessity of taking into account the frequency dependence of the circuit elements while simulating the circuit, should be emphasized.

Another five-pole filter with center frequency at 900 MHz and a fractional bandwidth of 0.27% has been fabricated on an LAO substrate about 1-inch long [13, 14]. At 77 K a maximum insertion loss of 1.2 dB and a minimum return loss of 20 dB were observed.

A six-pole 1857-MHz filter with 0.8% fractional bandwidth was realized on a 1.7x0.4 inch^2 sapphire substrate [18]. From a maximum insertion loss of 0.5 dB and minimum return loss of 25 dB at 77 K, an unloaded Q-factor of about 7,000 can be deduced. For a lower temperature of 60 K the passband insertion loss of this filter was reduced to 0.2 dB, and the minimum return loss became 18 dB (without further trimming), corresponding to an unloaded Q factor of the resonators of about 35,000.

A six-pole filter with a center frequency of 4050 MHz and with fractional bandwidth of 2% has been fabricated on a 15x10 mm^2 LAO substrate [12, 15, 16]. At 77 K this filter showed a maximum insertion loss of 0.5 dB.

Finally, a relatively wideband (3% fractional bandwidth) 10-GHz four-pole filter should be mentioned. It was implemented on a 7.62x2.54 mm^2 LAO substrate [11].

Although the above-described architecture - see Fig. 3.9 - is suitable for realization of Chebyshev filter response only, most of the referenced filters had at least one uncontrollable transmission zero either in the low [18] or in the high [11, 19] frequency stop bands due to the coupling between non-adjacent resonators separated by an odd number of resonators. The filter designed in [12, 15, 16] exhibited two transmission zeroes due to the coupling between non-adjacent resonators separated by an even number of resonators.

An attempt to build an elliptic filter using lumped-element HTS technology was made in [22]. Here, series LC resonators in shunt branches lead to transmission zeroes in the stop band of the filter. The three-pole filter with a center frequency of about 2 GHz possessed a fractional bandwidth of 2%. The LAO substrates had dimensions of 12x10 mm^2. At 77 K this filter showed a maximum insertion loss of 0.5 dB and minimum return loss of 20 dB, corresponding to an unloaded Q factor of the resonators of about 10,000. The filter was derived from an equivalent circuit in order to obtain two transmission zeroes of third order each, one at the low and one at the high frequency stop band. Due to parasitic couplings between elements of the filter, the positions of the zeroes were shifted. This fact calls for additional work and new ideas in realizing elliptic lumped-element filters.

In [21] a lumped-element coplanar-line HTS one-pole stop-band filter was presented for a center frequency of 5.6 GHz. It consists of a 20-finger interdigital capacitor with the central finger that connects both sides of the capacitor acting as an inductor. The filter fits to an MgO substrate of 10x10 mm^2. The unloaded Q factor of this structure ranged from 9,400 at 15 K to 1,300 at 77 K. This filter also can operate as a switch. With the application of a bias current the inductor is switched from the superconducting to the normal state.

3.2 ALLPASS FILTER SECTIONS

Whereas the bandpass filters of section 3.1 were designed to provide a specified amplitude response, another class of filters are needed to possess a prescribed phase response. Allpass filters exhibit a nominal transmission coefficient with (theoretically) unity magnitude (zero insertion loss) over the entire frequency range, but they possess a pronounced group delay response. Allpass filters with frequency-dependent group delay find applications as "equalizers." The function of an equalizer is to compensate the group-delay distortions in highly selective bandpass filters. Further applications are in pulse-compression filters for analog RF signal processing [4].

The allpass response requires that each transmission pole s_n in the left half-plane is accompanied by a transmission zero $s_{0,n} = - s_n$ in the right half-plane. In Fig. 3.12 a bridged-T network with 4 degrees of freedom (C_1, C_2, L_1, L_2) is depicted.

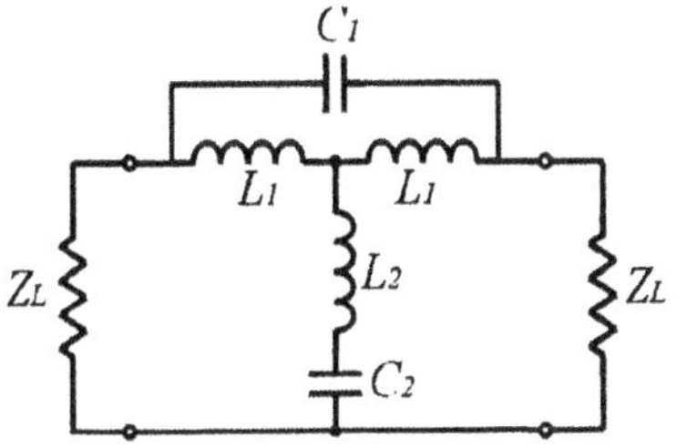

Figure 3.12. Bridged-T network as an allpass filter.

This LC network exhibits a resonance with even current distribution (with respect to the symmetry plane) as well as a resonance with odd current distribution. The complex-valued frequencies for the even and odd modes are

$$|s_e|^2 = \omega_e^2 + \sigma_e^2 = \frac{2}{C_2(L_1 + 2L_2)}, \text{ with } \sigma_e = -\frac{Z_L}{2(L_1 + 2L_2)} \tag{3.19a}$$

and

$$|s_d|^2 = \omega_d^2 + \sigma_d^2 = \frac{1}{2C_1 L_1}, \quad \text{with } \sigma_d = -\frac{1}{4Z_L C_1}, \tag{3.19b}$$

respectively.

The bridged-T network becomes an allpass if the poles of the even and odd modes are coincident, with $\omega_e = \omega_d = \omega_0$ and $\sigma_e = \sigma_d = \sigma_0$. Two degrees of freedom in the element parameters can be used to meet this condition. The remaining two parameters are used to adjust the values of ω_0 and σ_0. For $\sigma_0 << \omega_0$, the group delay has a maximum near $\omega = \omega_0$ with $\tau_{max} \approx 2/\sigma_0$. Since the area under the $\tau(\omega)$ curve is a constant, a high-valued maximum is always a "narrow" maximum. Due to dissipation in the lumped elements, the insertion loss is not negligible if there is a violation of the condition $\omega_0 \cdot \tau_{max} << Q$, with Q being the unloaded quality factor of the circuit. Thus, the use of HTS structures allows higher group delays to be realized.

In [23] a bridged-T according to Fig. 3.12 was realized from thin-film YBCO on a 10x10 mm^2 area and 0.5-mm-thick LAO substrate. The layout and dimensions are shown in the left part of Fig. 3.13. The measured and computer-simulated response of the group delay are depicted in the right part of Fig. 3.13.

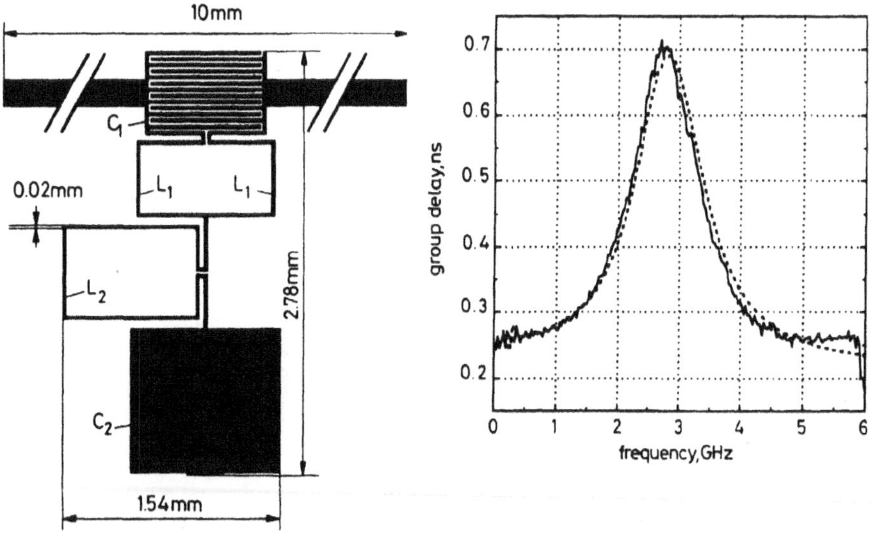

Figure 3.13. Bridged-T network as an LC allpass [23] Left: Layout showing realization using lumped elements and microstrip techniques. Right: Frequency response of group delay, derived from measured data at 77 K (full line) and from computer simulations (dashed line).

3.3 LUMPED-ELEMENT HYBRID COUPLERS

Hybrid couplers may serve as another example of HTS lumped-element circuits. A 90-degree coupler represents a reciprocal and (ideally) lossless four-port. It possesses in the ideal case the following properties: If a wave is incident at one of the 4 ports (denoted as port 1) and the remaining three ports are terminated by matched loads, the power is split into two equal parts which appear at two other ports (ports 3 and 4) with a 90-degree phase shift between ports 3 and 4. The remaining port (port 2) is decoupled.

The simplest and most popular realization of a 90-degree coupler is the branch-line coupler - see e.g. [3]. It can be built in planar technology as a ring consisting of four quarter-wave transmission-line sections. Single couplers are (in contrast to highly-selective bandpass filters and allpass filters) not characterized by a high-frequency, group-delay product $\omega_0\tau$. Because of this, these structures are not "very sensitive" to dissipative losses. Utilization of HTS technology is therefore justified only for those cases in which a circuit with a relatively large number of cascaded couplers must be implemented. Butler matrices for analog RF beamforming networks of multi-beam antennas serve as an example of networks with a large number of couplers and therefore a high frequency-group-delay product. Simultaneous beamforming - see, e.g. [24] - of M=2^m beams for an array with M radiation elements, requires m·2^{m-1} hybrid couplers. For example, for M=16 simultaneous beams, 32 couplers are needed. If , furthermore, cross-overs are avoided by employing pairs of (0-dB) couplers, the number of these additional couplers also strongly increases with M. For the case of these large numbers of couplers and for frequencies in the range between about 0.1 to 3 GHz, where quarter-wave line sections are relatively large, miniaturized couplers are highly desired.

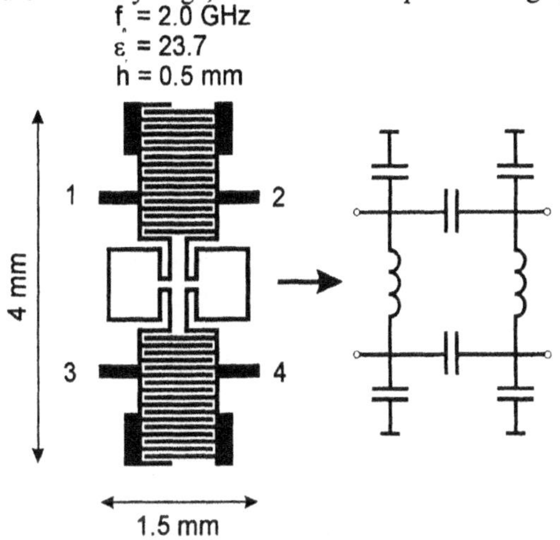

Figure 3.14. 2-GHz lumped-element 90-degree hybrid coupler, realized from YBCO thin films on a 0.5-mm-thick LAO substrate.

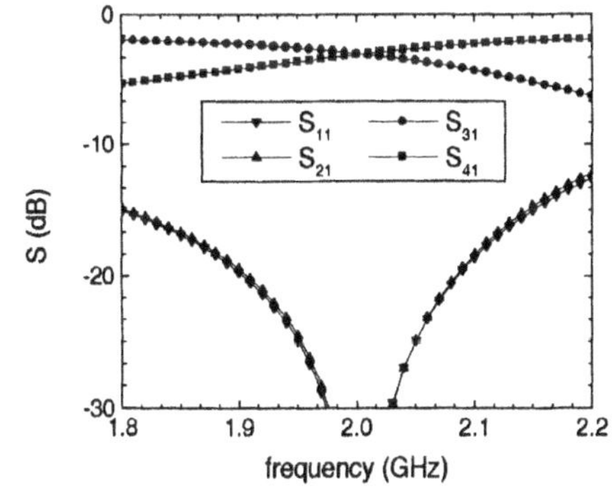

Figure 3.15. Frequency-dependence of the scattering parameters for the lumped-element hybrid coupler shown in Fig. 3.14.

Fig. 3.14 shows an example of a 2-GHz hybrid realized as a lumped-element LC circuit in HTS technology. The scattering parameters of this hybrid coupler are given in Fig. 3.15, indicating that at center frequency (2 GHz) the ports are perfectly matched and that port 2 is decoupled from port 1. Furthermore, it is seen that at center frequency the power incident at port 1 is split into equal parts, available at ports 3 and 4. For frequencies apart from the center frequency, deviations from ideal behavior can be seen.

4. Conclusions

The trade-off between miniaturization and a high quality factor for passive RF components can be overcome by means of planar HTS circuits. In the frequency range between about 0.1 to 2 GHz an even higher degree of miniaturization is obtained if lumped HTS elements (LE) instead of distributed HTS elements are employed.

The design of LE circuits is governed by upper limits for the achievable capacitance and inductance values. These upper limits can be characterized by means of a characteristic figure of merit Λ. A significant extension to higher capacitances and higher inductances, and therefore to lower Λ values, could be obtained if an HTS multi-layer technology for the fabrication of low-loss conductor-insulator-conductor capacitors and spiral inductors were available.

A straightforward design procedure for Chebyshev bandpass filters composed of lumped capacitors and inductors exists. It starts with the conversion of a low-pass prototype filter into an LC bandpass by means of an LP-BP transformation. This circuit must be transformed into a modified circuit with achievable component values. This transformation employs impedance inverters and transformers, as well as a T-Π conversion.

Higher LC filter architectures include filters with transmission nulls at finite frequencies, as well as filters with "self-compensated" group-delay response. However, until now, there have been no attempts to design and fabricate these filters of higher complexity in HTS technology.

HTS lumped-element LC circuits also can be built to serve as allpass filters or as hybrid couplers in RF beamforming networks, such as Butler matrices.

5. References

1. Howe, H. (1984) Microwave integrated circuits - an historical perspective, *IEEE Trans. Microwave Theory Tech.* **32**, 991-996.
2. Sobohl, H. (1971) Application of integrated circuit technology to microwave frequencies, *Proc. IEEE*, **59**, 1200-1211.
3. Bahl, I. and Bhartia, P. (1988) *Microwave solid state circuit design*, John Wiley and Sons, New York.
4. Chaloupka, H. (2000) Microwave applications of high-temperature superconductors, *Applications of Superconductivity*, in H. Weinstock (ed.) Kluwer Academic Publishers.
5. Alley, G. D.(1970) Interdigital capacitors and their applications to lumped-element microwave integrated circuits, *IEEE Trans. Microwave Theory Tech.* **18**, 1028-1033.
6. Daly, D. A., et al. (1967) Lumped elements in microwave integrated circuits *IEEE Trans. Microwave Theory Tech.* **15**, 713-721.
7. Aitchison, C. S. (1971) Lumped components for microwave frequencies, *Philips Tech. Rev.* **32**, 305-314.
8. Chaddock, R. E. (1974) The application of lumped-element techniques to high frequency hybrid integrated circuits, *Radio Electron. Eng.* **44**, 414-420.
9. Wolff, I. and Knoppik N. (1974) Rectangular and circular microstrip disc capacitors and resonators, *IEEE Trans. Microwave Theory Tech.* **22**, 857-864.
10. Chaloupka H., Kolesov S. (1999), Internal Report, University of Wuppertal.
11. Swanson, Jr., D. G. , Forse, R. and Nilsson J. L. (1992) A 10-GHz thin film lumped element high temperature superconductor filter, *IEEE MTT-S Digest 1992*, 1191-1193.
12. Ye S. and Mansour R. R. (1994) Design of manifold-coupled multiplexers using superconductive lumped element filters, *IEEE MTT-S Digest 1994*, 191-194.
13. Zhang, D. , Liang G. -C., Shih C. F. and Withers R. S. (1995) Narrow-band lumped-element microstrip filters using capacitively-loaded inductors, *IEEE MTT –S Digest 1995*, 379-382.
14. Zhang, D. , Liang G.-C., Shih C. F., Johansson M. E. and Withers R. S. (1995) Narrow-band lumped-element microstrip filters using capacitively-loaded inductors, *IEEE Trans. Microwave Theory Tech.* **43**, 3030-3036.
15. Mansour R. R., Ye S., Dokas V., Jolley B., Thomson G., Tang W.-C. and Kudsia C. M. (1996) Design considerations of superconductive input multiplexers for satellite applications, *IEEE Trans. Microwave Theory Tech.* **44**, 1213-1227.
16. Mansour R. R., Jolley B., Ye S., Thomson F. S. and Dokas V. (1996) On the power handling capability of high temperature superconductive filters, *IEEE Trans. Microwave Theory Tech.* **44**, 1322-1338.

17. Matthaei G. L., Rohlfing S. M. and Forse R. J. (1996) Design of HTS, lumped-element, manifold-type microwave multiplexers, *IEEE Trans. Microwave Theory Tech.* **44**, 1313-1321.
18. Kwok R. S., Zhang D., Huang Q., Kaplan T. S., Lu J. and Liang G.-C. (1999) Superconducting quasi-lumped element filter on R-plane sapphire, *IEEE Trans. Microwave Theory Tech.* **47**, 586-591.
19. Saito K., Sakakibara N., Ueno Y., Kobayashi Y., Yamaguchi D., Satou K., Mimura T. (2000) 246 MHz HTS lumped element bandpass filter, *IEICE Trans. Electron.* E83-C, SCE99-11, 15-19.
20. Lancaster M.J., Li J., Porch A. and Chew N.G. (1993) HTS lumped element resonator, *Electronics Lett.* **29**, 1728-1729.
21. Lancaster M. J., Huang F., Porch A., Avenhaus B., Hong J.-S. and Hung D. (1996) Miniature superconducting filters, *IEEE Trans. Microwave Theory Tech.* **44**,1339-1346.
22. Reppel, M., Chaloupka H. and Kolesov S. (1998) Highly miniaturized superconducting lumped-element filter, *Electronics Lett.* **34**, 929-930.
23. Patzelt, T., Aschermann, B. , Chaloupka H., Jagodzinski U. and Roas, B. (1993) High-temperature superconductive lumped-element microwave allpass sections, *Electronics Lett.* **29**, 1578-1579.
24. Hansen, R. C. (1998), Phased-array antennas, John Wiley & Sons, Inc.

NEAR-FIELD MICROWAVE MICROSCOPY OF MATERIALS PROPERTIES

STEVEN M. ANLAGE, D. E. STEINHAUER, B. J. FEENSTRA, C. P. VLAHACOS AND F. C. WELLSTOOD
Center for Superconductivity Research, and Materials Research Science and Engineering Center, Department of Physics, University of Maryland, College Park, Maryland 20742-4111 USA

Abstract. Near-field microwave microscopy has created the opportunity for a new class of electrodynamics experiments of materials. Freed from the constraints of traditional microwave optics, experiments can be carried out at high spatial resolution over a broad frequency range. In addition, the measurements can be done quantitatively so that images of microwave materials properties can be created. We review the five major types of near-field microwave microscopes and discuss our own form of microscopy in detail. Quantitative images of microwave sheet resistance, dielectric constant, and dielectric tunability are presented and discussed. Future prospects for near-field measurements of microwave electrodynamic properties are also presented.

1. Introduction

Measurement of the electromagnetic response of materials at microwave frequencies is important for both fundamental and practical reasons. The complex conductivity of metals and superconductors gives insights into the physics of the quasiparticle excitations and collective charge properties of the material. Interesting collective behavior which can be investigated through the complex conductivity includes superconductivity, spin and charge density waves, Josephson plasmons, etc. The dielectric properties of materials give insights into the polarization dynamics of insulators and ferroelectrics. A class of interesting insulating materials, the parent compounds of cuprate superconductors, combine both dielectric and antiferromagnetic behavior because of strong correlations between the charge car-

H. Weinstock and M. Nisenoff (eds.), Microwave Superconductivity, 239–269.

riers. Ferromagnetic materials interact with microwaves in several unusual ways, including ferromagnetic resonance and anti-resonance, and spin-wave resonance. In addition, a series of discoveries over the past several decades have found many materials which display coexistence of superconductivity with either antiferromagnetism, or modified forms of ferromagnetism. This wide range of interesting electromagnetic behavior of contemporary materials requires that experimentalists working in this field master many diverse measurement techniques and have a broad understanding of condensed matter physics.

On the practical side, electromagnetic measurements are essential for creating new device technologies and optimizing existing devices and processes. For example, nonlinearities in superconducting materials limit their applications at microwave frequencies to relatively low power uses. Although intrinsic nonlinearities of superconductors will eventually limit their utility, most practical nonlinearities are caused by extrinsic defects in the material or device. Careful characterization of these materials on the length scale of the extrinsic inhomogeneities is required to tackle this problem. Another practical issue is the optimization of materials for frequency-agile applications. Here the issue is measurement of dielectric or magnetic properties which can be tuned with an electric or magnetic field, often in thin-film materials. Yet another important application of electromagnetic measurements is in diagnostic measurements. For example, in a semiconductor integrated circuit process, one needs to measure and control quantities such as sheet resistance, doping profiles, and electromigration in wires. As the speed of microprocessors continues to climb, the microwave properties of materials become increasingly important. Finally, if high-speed integrated circuits fail, it is important to locate the fault (often an open or short) as accurately as possible, and to correct the problem quickly to maintain a high yield. Electromagnetic measurements are thus an important foundation for many emerging technologies.

In this paper we present a new paradigm for electromagnetic measurements of materials. We first briefly review the traditional methods of microwave measurements and point out some of their important limitations. We then present an alternative approach to these measurements through quantitative near-field microwave microscopy. The remainder of the paper is devoted to reviewing the progress we have made in this exciting new field of research.

2. Traditional Microwave Measurements of Electromagnetic Properties

The fundamental electrodynamic quantities of greatest interest are the surface impedance Z_s, the conductivity σ, the dielectric permittivity ε, and the magnetic permeability μ. A review of the definitions of conductivity and surface impedance of normal metals and superconductors is given elsewhere in this book. All of these quantities are complex and in general are a function of many variables, including frequency, temperature, magnetic field, electric field, etc.

Traditional microwave measurements of these quantities are typically done on length scales of the free-space wavelength of the signal, which is about 3 cm at 10 GHz. The earliest microwave measurements on superconductors by Pippard were done with the sample acting as a quarter-wavelength resonator [1]. In this case the electromagnetic properties measured are actually an average of the properties along the sample, weighted by the form of the standing-wave resonance pattern. In the case of elemental Pb or Sn this is not much of an issue, but in the case of complicated multi-element oxides it can give rise to misleading results.

A refinement of this technique came through the use of cavity perturbation methods. In this case the sample is immersed in a larger electromagnetic cavity in a region of uniform electric or magnetic field [2, 3]. The properties of the sample can be deduced by comparing the resonant frequency shift and quality factor change from a well-characterized initial (unperturbed) state of the cavity. In this case one measures an average of the electromagnetic properties of the sample, again weighted by the distribution of fields and currents created in the sample [3]. These field and current distributions can be quite complicated for most practical samples [4, 5], becoming simple only for carefully shaped ellipsoids of revolution.

Other resonant techniques have been developed in recent years, particularly for examination of thin-film superconducting materials. These include the parallel-plate resonator technique, in which two congruent films form a transmission line resonator with a length scale on the order of the microwave wavelength [6, 7]. Dielectric resonators are also sensitive to thin-film surface impedance but are averaged over an area on the order of the wavelength [8, 9]. Far-field surface-impedance microscopy techniques use a diffraction-limited beam in a confocal geometry to locally measure the surface impedance at millimeter-wave frequencies, but have a spatial resolution limited to a few millimeters [10, 11].

Non-resonant methods can also be employed to measure electrodynamic properties. For instance, waveguide transmission through thin films has been successful at determining the absolute value of the penetration depth, but provides an average of the properties over the film, again on the length scale of the microwave wavelength.

2.1. LIMITATIONS OF TRADITIONAL TECHNIQUES

Although traditional electrodynamics measurement techniques have been very successful, they do suffer from some fundamental limitations. First, they tend to measure a weighted average over large areas of the sample, on the scale of millimeters to centimeters. In the case of oxide thin films, it is often difficult to prepare a material which is homogeneous on such long length scales. For oxide single crystals, doping inhomogeneities have been shown to exist on many length scales from the nanometer to the millimeter range [13]. In addition, surface morphology and second phases (such as flux particles) complicate the response of the crystal to electromagnetic radiation. Hence, traditional electrodynamics measurements yield only an average or overall picture of the sample properties. This means that the intrinsic behavior of the material can be masked by the response of a relatively small fraction of extrinsic or second-phase material. When new materials are measured, there is often a lingering doubt whether the response is due to new intrinsic physics or some unforeseen extrinsic effect.

A second limitation of traditional techniques is the generation of large screening currents, especially near edges, and in the case of superconductors, the subsequent admission of rf magnetic flux into the sample. The rf flux will significantly increase the surface impedance and can easily dominate the response of the material. Such issues have arisen many times in the exploration of electromagnetic properties of new superconducting materials. For instance, early measurements of the penetration depth in thin films of YBCO were corrupted by vortex entry into the films [14]; harmonic generation in superconducting crystals can be dominated by edge effects [15]; and the nonlinear Meissner effect in superconductors can be swamped easily by vortex entry problems [16, 17].

2.2. A NEW PARADIGM FOR ELECTRODYNAMICS MEASUREMENTS

To overcome the limitations of traditional electrodynamics measurements, we have developed a new family of near-field microwave microscopes which permit local quantitative measurements of surface impedance and conductivity, as well as dielectric and magnetic properties. Near-field techniques also relieve us from the "light-cone constraint", in which the length scales that we can probe with electromagnetic radiation are dictated by the frequency. For instance, this permits broadband electrodynamics measurements to be performed on very fine length scales. Because the spatial resolution is determined by geometrical features which we can control, a class of altogether new electromagnetic experiments can be performed. For example, we now can measure the local tunability of high-frequency electrical properties and make new connections between microstructure and its associated physical properties.

3. An Overview of Microwave Microscopy Techniques

Near-field microwave microscopy is an art which has formed gradually over the years from many different sources. In principle the intellectual founder of the near-field microscopy effort was Synge in his prescient paper of 1928 [18]. However, most practitioners of the art are not familiar with this seminal work despite its importance. Many near-field techniques were developed empirically, some, in fact, developed to solve practical problems in manufacturing. The earliest effort to perform high resolution quantitative microwave measurements seems to be from the ferromagnetic resonance community, led by the work of Frait [19], and later Soohoo [20]. However, not all subsequent developments can be traced back to these roots.

To codify the great variety of work on near-field microwave microscopy of materials, we present a simple picture of the five basic types of microwave microscopy in Fig. 1. Although these classifications are somewhat gross, they allow us to cleanly distinguish the main efforts in the field.

Fig. 1(a) illustrates a traditional microwave cavity resonator with a small hole in one wall. The sample is placed in close proximity to the wall; a small region of the sample, defined by the hole diameter, perturbs the resonant frequency (causing a frequency shift Δf) and quality factor (Q) of the resonator. Because the hole is so small (0.5 mm diameter for Frait [19]) and the sample makes such a small perturbation to the cavity, one must examine a highly lossy property of the sample, such as ferromagnetic resonance. Ferromagnetic resonance (FMR) gives information about the local internal fields and magnetization of the sample. Frait [19], Soohoo [20] and Bhagat [21] have successfully used this microscope for FMR imaging, while Ikeya [22] has used it for electron spin resonance imaging.

An important variation on this theme was developed by Ash and Nichols [23]. They used an open hemispherical resonator with a flat-plate reflector. The plate had a small hole in it and the sample was positioned close to this hole, but outside the resonator. They did not study FMR, so in order to recover a perturbation signal from the resonator, the sample separation from the hole was modulated at a fixed frequency. They were able to phase-sensitively recover the reflected signal from the resonator to construct a qualitative image of the sample as it was scanned under the hole. These cavity methods all make use of an evanescent mode to couple the resonant cavity to a local part of the sample. In that sense these methods resemble

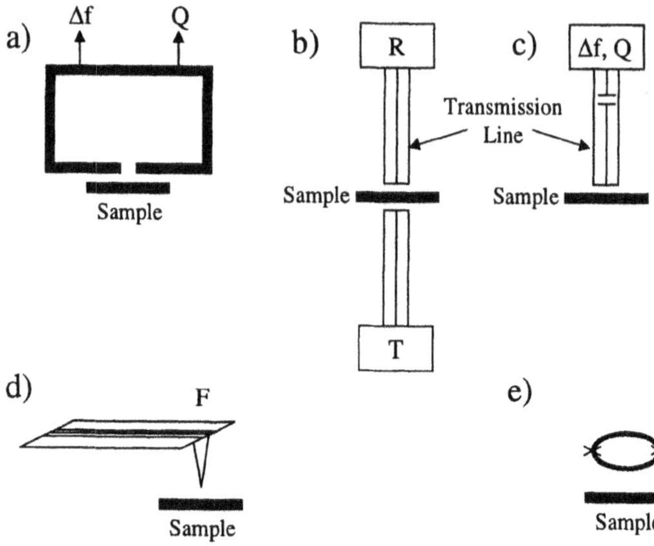

Figure 1. Schematic of the five main classes of near-field microwave microscopes; a) shows a microwave resonator with a small hole in one wall. The frequency shift Δf and quality factor Q are monitored; b) shows the transmission line reflection (R) and transmission (T) method; c) shows the scanned transmission-line resonator technique in which Δf and Q are monitored; d) shows the cantilevered sharp-tip method where a force (F) or other quantities are monitored; e) shows the scanning SQUID method.

the tapered optical fiber with aperture method of near-field scanning optical microscopy (NSOM) [24, 25].

Fig. 1(b) illustrates a class of non-resonant microscopes in which the sample is placed on or near the end of a microwave transmission line. The complex reflectivity R or transmission coefficient T is measured, and properties of the sample are deduced. The most common technique is to measure reflectivity from a coaxial transmission line terminated (possibly with an air gap) by the sample [26, 27, 28, 29, 30, 31, 32, 33]. Variations include the use of waveguides [34, 35], waveguides covered by a resonant slit [36, 37] and microstrip lines [38] in reflection. Transmission measurements also have been done in the coaxial [31, 39] and waveguide [37] geometries. These techniques have mainly been used to map metallic conductivity or sheet resistance, and dielectric constant. In some cases the measurements are done quantitatively.

Fig. 1(c) shows schematically one of the most sensitive forms of near-field microwave microscopy. The idea is that the sample is put near the open end of a transmission-line resonator, and changes in the resonant frequency and quality factor are monitored as the sample is scanned. This class of techniques differs from that shown in Fig. 1(a) through the use of a "field enhancing" feature at the end of the transmission line, rather than an evanescent aperture in the resonator. The field-enhancing feature sets the scale for the spatial resolution of this class of microscopes. The first embodiment of this concept was conceived for measuring the moisture content of paper [40]. Other embodiments use coaxial transmission lines with the sample in contact with the open end [41, 42, 43, 44, 45, 46, 47, 48, 49, 50, 51, 52, 53, 54, 55, 56], or with an air gap between the probe and the sample [43, 44, 45, 46, 47, 48, 49, 50, 51, 52, 53, 54, 55]. A related far-field technique has been used to image surface resistance and nonlinearity with a scanned dielectric resonator in contact with the sample [8].

The resonant transmission-line technique has proven to be the most quantitative of all the near-field microwave microscopy methods. Quantitative imaging of topography, sheet resistance, dielectric constant and loss, and other quantities has been achieved. In addition, the use of field enhancing features has pushed the spatial resolution of resonant and non-resonant microscopes to the sub-μm domain while maintaining the quantitative nature of the measurements.

The last two classes of near-field microscopy techniques are less developed, but show great promise for the future. Fig. 1(d) shows a cantilever with a sharp tip positioned over the sample. This geometry can be used to perform atomic force microscopy to determine the topography of the sample, in addition to microwave microscopy. There are three sub-classes of scanning probe measurements of materials properties at microwave frequencies. The first is to perform localized electron-spin resonance measurements [57]. The tip is used for STM, and a magnetic field is applied to the sample. It is found that a local signal at the Larmor precession frequency can be extracted from the tip and converted into an image of ESR response of the sample. A second embodiment is to create a magnetic field gradient on the sample (e.g., with a small magnetic particle on the tip) while immersing it in an rf magnetic field. The sample will locally satisfy the magnetic resonance condition and exert a force on the cantilever [58, 59]. A third scanning probe method is to simply use a sharp metallized tip to perform "apertureless" near-field microscopy. The sharp tip in close proximity to a metallic sample will locally enhance radiation introduced by a focused far-field beam. If an additional phenomenon can take place in this localized region due to the enhanced field strength, and it can be detected, one has a local microscopic probe of the physics associated with that phenomenon. One important example of this kind of microscopy is apertureless infrared microscopy, where the reflected signal from the region of the probe tip is measured while the tip is periodically dithered up and down [60].

Fig. 1(e) shows the scanning superconducting quantum interference device (SQUID) method of microwave microscopy. The SQUID will generate circulating rf currents when a dc bias is placed across the loop. The frequency of these currents is directly proportional to the applied bias voltage. The currents generate rf magnetic fields which then impinge on the sample. The sample will generate its own response currents which in turn modify the inductance of the SQUID loop. By monitoring the magnetic-field feedback signal required to keep the SQUID in a constant flux state, one can map the electromagnetic response of the sample [61]. This method has the advantage of being very broadband (in principle from rf up to the gap frequency of the superconductor used to make the SQUID, ~100 GHz or more) and quantitative.

3.1. DETAILED DESCRIPTION OF THE MARYLAND MICROSCOPE

Our group has developed a novel form of near-field scanning microwave microscopy based on the scanned resonator technique (Fig. 1(c)). As shown in Fig. 2, our microscope consists of a resonant coaxial cable which is weakly coupled to a microwave generator on one end through a decoupling capacitor C_d, and coupled to a sample through an open-ended coaxial probe on the other end. In the absence of a sample, the microscope is a half-wave resonator (Fig. 4). As a metallic sample approaches the open end of the coaxial probe, the boundary condition there goes progressively from open circuit to short circuit (e.g., for the blunt probe shown in Fig. 3(a)). In the limit of the sample touching the coaxial probe, the microscope becomes a quarter-wave resonator (Fig. 4). In the process, the resonant frequency shifts by one half the spacing between neighboring modes, which is on the order of 100 MHz. Hence the microscope frequency shift is very sensitive to the probe/sample separation, as well as the electrical properties of the sample (i.e., the conductivity, dielectric constant and magnetic permeability).

Consider again the limit of the sample very far from the probe. In this case the quality factor of the resonator, Q, is high since only internal dissipation processes (and radiation losses) are active. As the sample approaches the open end of the coaxial probe it must be considered part of the resonant circuit. As such, it will add losses to the system and in general reduce the Q of the microscope. The Q thus gives insight into the additional loss mechanisms introduced by the sample.

As the sample is scanned beneath the probe, the probe-sample separation will vary (depending upon the topography of the sample), causing the capacitive coupling to the sample, C_x, to vary. This has the result of changing the resonant frequency of the coaxial-cable resonator. Also, as the local electrical properties of the sample vary, so will the resonant frequency and

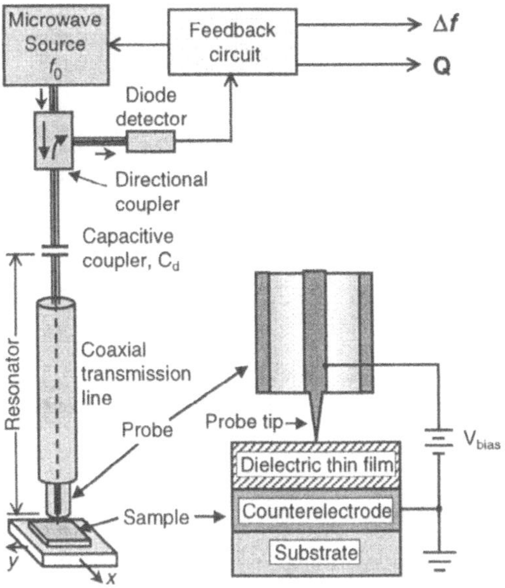

Figure 2. A schematic diagram of the scanned resonator near-field microwave microscope developed at the University of Maryland. The microwave source is kept on a microscope resonance by means of a feedback loop which supplies the frequency-shift error signal Δf. The quality factor, Q, of the microscope resonator is also measured. Shown in the inset is an expanded view of the tip/sample interaction and a dc bias circuit used to measure local dielectric tunability.

quality factor, Q, of the resonant cable. The goal then is to turn the measured quantities Δf and Q versus position over the sample into quantitative images of sample electrical properties.

A circuit is used to force the microwave generator to follow a single resonant mode of the cable, and a second circuit is used to measure the Q of the circuit, both in real time (Fig. 2) [46, 47]. The microwave source is frequency modulated by an external oscillator at a rate $f_{FM} \sim 3$ kHz. The electric field at the probe tip is perturbed by the region of the sample beneath the probe's center conductor. We monitor these perturbations using a diode detector which produces a voltage signal proportional to the power reflected from the resonator. A feedback circuit [47] (Fig. 2) keeps the microwave source locked to a resonant frequency of the transmission line, and has a voltage output which is proportional to shifts in the resonant frequency due to the sample. Hence, as the sample is scanned below the open-ended coaxial probe, the frequency shift and Q signals are collected by a computer. The circuit runs quickly enough to accurately record at

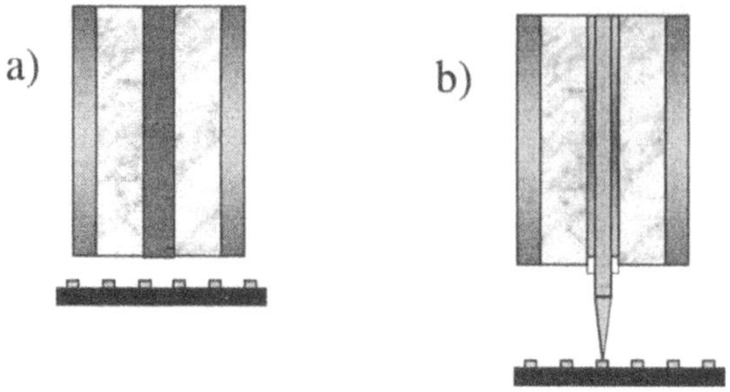

Figure 3. Probe tips used in the scanning near-field microwave microscope. a) shows a blunt electric-field probe tip in non-contact mode, and b) shows the STM-tip electric-field probe in contact mode.

scan speeds of up to 25 mm/sec.

To determine the quality factor Q of the resonant circuit, a lock-in amplifier, referenced at $2f_{FM}$, gives an output voltage $V_{2f_{FM}}$ which is related to the curvature of the reflected power vs. frequency curve on resonance, and hence to Q. To relate $V_{2f_{FM}}$ and Q, we perform a separate experiment, in which we vary Q using a microwave absorber at various heights below the probe tip, and measure the absolute reflection coefficient $|\rho|^2$ of the resonator. If $|\rho_0|^2$ is the reflection coefficient at a resonant frequency f_0, then the coupling coefficient between the source and the resonator is $\beta = (1 - |\rho_0|) / (1 + |\rho_0|)$. The loaded quality factor of the resonator [62, 63] is $Q_L = f_0/\Delta f$, where Δf is the difference in frequency between the two points where $|\rho|^2 = (1+\beta^2) / (1+\beta)^2$. The unloaded quality factor, which is the Q of the resonator without coupling to the microwave source and detector, is then $Q_0 = Q_L (1+\beta)$. We also measure $V_{2f_{FM}}$, and find that there is a unique functional relationship between Q_0 and $V_{2f_{FM}}$; thus, we need to calibrate this relationship only once for a given microscope resonance. In a typical scan, we record $V_{2f_{FM}}$, and afterward convert $V_{2f_{FM}}$ to Q_0.

4. Properties of the Near-Field Microwave Microscope

4.1. SPATIAL RESOLUTION

The spatial resolution of the microscope has been demonstrated to be the larger of the probe-sample separation and the diameter of the inner conductor wire in the open-ended coaxial cable [45]. As with other forms of near-field microscopy, the probe is placed well within one wavelength of the sample under study. This is particularly easy to accomplish at rf, microwave and millimeter-wave frequencies because the wavelength ranges from meters to millimeters. However, the high spatial resolution comes from a "field-enhancement" feature of the probe. It has been found that sharp tips produce higher spatial resolution than blunt probes. Figure 3 contrasts these two types of probes in a transmission-line microscope, like those shown in Fig. 1(b) and (c). Figure 3(a) shows a blunt electric field probe, while Fig. 3(b) shows an STM-tipped probe. The blunt probe has a spatial resolution given by the area of the center conductor when operated in non-contact mode at a height smaller than the inner conductor diameter. The STM tip has a spatial resolution in contact mode on the order of 1 μm, depending somewhat on the bluntness of the tip.

The "lightning-rod effect" is responsible for the increased spatial resolution with a sharp-tip probe. This same principle is used in apertureless near-field scanning optical microscopy. The basic idea is that the electromagnetic fields in the near-field environment can be treated in the static approximation. In electrostatic equilibrium there will be a large electric field at sharp corners, with the local field being inversely proportional to the radius of curvature. This enhancement means that the response of the sample immediately below the sharp tip will dominate the signal. Quantitative calculations of dielectric response from an STM-tipped probe bear out this qualitative picture [64].

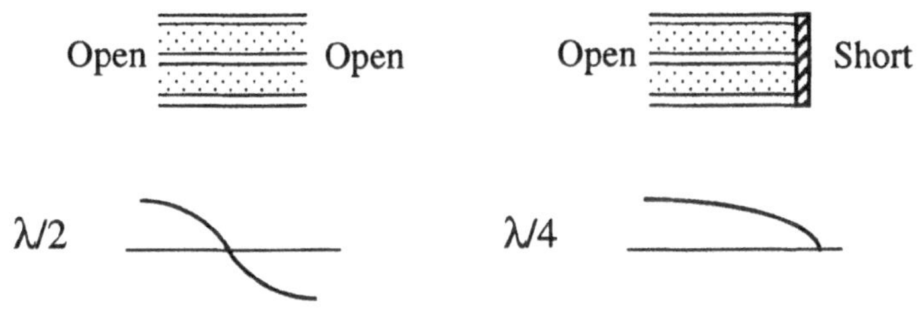

Figure 4. Microwave microscope in the two extremes of no sample (a) and with a metallic sample shorted to the probe (b). In (a) the microscope is a half-wave resonator, while in (b) it is a quarter-wave resonator.

In addition to achieving high spatial resolution it also is important to vary the spatial resolution while maintaining quantitative measurement capability. There are times when a gross characterization or measurement of a sample property is required, for instance on the scale of a wafer or thin-film. There are physical property features on many length scales in materials. Hence, a microscope must adapt itself to imaging all of these scales. Our microscope is well adapted for this purpose and has been used with spatial resolutions of 500 μm, 200 μm, 100 μm, 12 μm and 1 μm.

4.1.1. *Evanescence and Microwave Microscopy*

An evanescent wave is one which does not propagate, but is exponentially attenuated with distance. Such behavior can be created with microwaves in a variety of situations. The simplest is to use a single-conductor cylindrical wave guide and to reduce its lateral dimensions such that the cutoff frequency is made greater than the frequency of the propagating mode. The wave is then attenuated on the length scale of the diameter of the waveguide [65]. This is the method employed by the microscopes in Fig. 1(a). The attractive feature of evanescence is the fact that the wave equation for the electromagnetic fields reduces to Laplace's equation in regions small compared to the wavelength of the radiation. This allows one to use low-frequency analysis to understand the distribution of the fields in the evanescent near-field region.

Evanescence through an aperture brings with it several distinct disadvantages. First, evanescent waveguides are characterized by a purely reactive impedance [65], hence they reflect signals back to the source very efficiently. This leads to the problem of a very poor signal-to-noise ratio, as is encountered in traditional NSOM through tapered optical fibers. The near-field optical community is now moving towards the "apertureless NSOM" method in which no evanescent waves are used. Instead they rely on a field-concentrating feature to enhance the signal from a very localized part of the sample. This is similar to the methods which we employ in our microscopes.

Another disadvantage of evanescence through an apeture at microwave frequencies is the long length scale of the evanescent decay. Evanescence is of great utility in scanning tunneling microscopy because the decay length is so short that it permits atomic resolution imaging. However, for radiation coming out of a hole beyond cutoff, the decay length scale is on the order of the hole diameter. This can be made no smaller than a few hundred microns in practical microwave microscopes without reducing the signal-to-noise ratio to unity.

However, techniques of field enhancement can improve the spatial resolution to much smaller values without compromising the signal-to-noise. Here it may be that short-length-scale evanescent waves associated with the field enhancement feature may dramatically improve the spatial resolution [66]. These evanescent waves are commonly encountered at the boundary between two different waveguides [67, 68]. The details of how these evanescent waves contribute to the imaging of electromagnetic properties on short length scales remain to be determined.

4.2. CONTRAST MECHANISMS

The microwave microscope generates contrast for many different electromagnetic properties of materials. These include topography, conductivity, dielectric constant and magnetic permeability. In addition, other properties such as ferromagnetic and antiferromagnetic resonance produce contrast for the microscope.

In order to quantitatively extract this information from the microscope images, it is necessary to model the microscope and its interaction with the sample. Here we begin with a model of our microscope, and defer the discussion of sample-specific modeling to the next section.

The heart of the microscope is the microwave resonator (Fig. 5). It is coupled to the generator by the capacitor C_d and to the sample through the capacitor C_x (in the case of an electric-field probe). The sample can be modeled as an impedance Z_x which includes resistive, capacitive and inductive loads which it presents to the microscope. Also included in the model are the directional coupler and detector which are used to measure the reflected signal from the resonator.

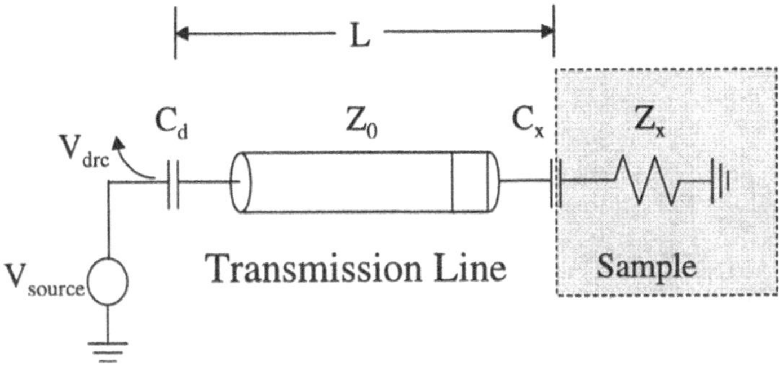

Figure 5. Electrical schematic of a simple model used to describe the near-field microwave microscope. The source has voltage V_{source}, the coupling capacitor is C_d, the resonator has characteristic impedance Z_0 and interacts with the sample through capacitance C_x, and the sample has complex impedance Z_x. The measured directional coupler detector output voltage is V_{drc}.

One can calculate the signal measured in the detector V_{drc} in terms of the source driving voltage, V_{source}. In the simplest case, one finds the following expression for the reflectivity of the resonator:

$$V_{drc} = \frac{Z_3 - Z_0}{Z_3 - Z_0} P(1 - P) V_{source}, \quad (1)$$

where $Z_3 = Z_d + Z_0 \frac{(Z_x+Z_0)e^{\gamma L}+(Z_x-Z_0)e^{-\gamma L}}{(Z_x+Z_0)e^{\gamma L}-(Z_x-Z_0)e^{-\gamma L}}$, Z_0 is the characteristic impedance of the transmission line, P is the directional coupler voltage fraction, $Z_d = 1/i\omega C_d$, γ is the complex propagation constant of the transmission line of length L, and Z_x is the impedance of the capacitor C_x in series with the sample.

The presence of the sample modifies the resonance of the microscope in several ways. First, as the probe-sample distance varies, so does the capacitance C_x. Consider a metallic sample below a blunt probe (e.g., Fig. 3(a)). If the sample is much closer than the probe diameter, there is a parallel-plate capacitance between the probe and the sample. As the sample approaches the probe, this capacitance will increase, and the resonant frequency will decrease. This is because the increased capacitance at the end of the transmission line will effectively lengthen the resonator. Ultimately as the sample comes into contact with the probe, a short-circuit condition is achieved and the microscope becomes a quarter-wave resonator. Hence, the greatest frequency shift expected is that from converting a half-wave to a quarter-wave resonator, which is half the distance between neighboring modes in either the open circuit or short-circuit microscope.

4.3. FREQUENCY COVERAGE

Because the microscope is a transmission-line resonator with many closely spaced modes, and because each mode can be used for imaging, the microscope is a tremendously broadband instrument. In a typical design, the transmission-line resonator has a length of approximately 1 m, giving a fundamental mode at about 125 MHz, and harmonic modes at integer multiples of this frequency. Hence, imaging can be done up to frequencies where higher-order modes begin to propagate in the coaxial cable, which can be over 100 GHz for small diameter cable. The only other limitation is the microwave electronics required to operate the microscope, such as the source, directional coupler, detector and connectors. However, using modern sources, electronics and the 1-mm standard connector, a 100+ GHz imaging bandwidth can be achieved.

4.4. TEMPERATURE RANGE

Our microscopes work well at room temperature. However, we also have constructed a cryogenic version of the microwave microscope. In this case only the probe and part of the resonator is actually held at cryogenic temperatures, along with the sample. Many images have been acquired at liquid-nitrogen temperature (77 K) in both materials and device-imaging modes [51, 74, 76]. We see no reason why this cannot be extended to pumped helium temperatures (1.2 K). At the other extreme, high-temperature coaxial cables exist which can be used for imaging up to 1000°C [69].

4.5. OTHER INDEPENDENT PARAMETERS

One can imagine introducing other quantities into the sample during microwave microscopy. For instance, light illumination is an obvious degree of freedom. We have found that measurements of tunability are very interesting with the scanning near-field microwave microscope. In this case the use of external electric and magnetic fields is most important.

4.5.1. *Electric field*

We have developed a technique to introduce a local and tunable electric field in our microwave microscope when imaging in contact mode (Fig. 2). A dc electric bias voltage can be applied to the center conductor of the microwave probe through a bias tee. This will create an electric field between the probe tip and the surrounding grounded surfaces. By introducing a suitable ground plane, controlled electric fields can be applied to the sample being imaged at microwave frequencies [48].

4.5.2. *Magnetic field*

We also can introduce a static magnetic field on samples being imaged by the microwave microscope using an electromagnet. The field can be modulated for imaging magnetic resonant phenomena in the sample.

5. Quantitative Imaging with the Near-Field Microwave Microscope

Here we present some of the results on quantitative imaging with our microscope. Very simply, one expects lossy properties of the sample shown in the model of Fig. 5 to affect the Q of the microscope, while reactive properties and topography will primarily affect the resonant frequency.

5.1. TOPOGRAPHY

Images obtained from resonant near-field scanning microwave microscopes are the result of two distinct contrast mechanisms: shifts in the resonant frequency due to electrical coupling between the probe and the sample, and changes in the quality factor Q of the resonator due to losses in the sample [46, 47]. Such images will inevitably contain intrinsic information about a sample (such as dielectric constant or surface resistance) as well as extrinsic information (such as surface topography). To facilitate quantitative imaging of intrinsic material properties, it is essential to be able to accurately account for the effects of finite probe-sample separation and topography.

The interaction between the probe and a metallic sample can be represented by a capacitance C_x between the sample and the inner conductor of the probe, analogous to the mechanism in scanning capacitance microscopy [44, 55]. For a highly conducting sample and a small probe-sample capacitance C_x, i.e., $C_x \ll L/cZ_0$, the reflected voltage V_{drc} at the output of the directional coupler can be written as above in Eq. (1) with $Z_3 = Z_d + \frac{Z_0[\cosh(\gamma L)+i\omega C_x Z_0 \sinh(\gamma L)]}{\sinh(\gamma L)+i\omega C_x Z_0 \cosh(\gamma L)}$, where c is the speed of light, and ω is the angular frequency of the source. A plot of Eq. (1) versus frequency would show a series of dips corresponding to enhanced absorption at the resonant frequencies [44]. For C_x=0 and C_d=0, the resonant frequencies are $f_n = nc/(2\sqrt{\varepsilon_r}L)$, for n=1,2, ..., where ε_r is the dielectric constant of the transmission line; for our system a resonance occurs every 125 MHz. For small C_x, the n-th resonant frequency changes by:

$$\Delta f_n \approx -f_n \frac{cC_x Z_0}{L}.$$

Since C_x depends on the distance between the inner conductor and the sample, this equation implies that Δf can be used to determine the topography of the sample. When the sample is very close to the probe, one expects that $C_x \approx A\varepsilon_0/h$, so that $\Delta f \sim -f_n c \varepsilon_0 A Z_0/hL$, where A is the area of the center conductor of the probe. On the other hand, when the sample is far from the probe, one must resort to numerical simulation to find $C_x(h)$. To find $C_x(h)$ for the coaxial probe geometry, we solved Poisson's equation using a finite difference method on a 200 by 150 cell grid. The overall experimental behavior is qualitatively well described by the numerical simulation, including the weak frequency shifts observed at large separation. For convenience, we parameterize our measured calibration curves with an empirical function to easily transform measured frequency shifts to absolute heights.

The topographic imaging capabilities of the system were demonstrated by imaging an uneven, metallic sample: a U.S. quarter-dollar coin. First, the frequency shift was recorded as a function of position over the entire sample. We next measured the frequency shift versus height at a fixed position above a flat part of the sample - see Fig. 2 of reference [55] - and determined the transfer function $h(\Delta f)$. Using $h(\Delta f)$, we then transformed the frequency-shift image into a topographic surface plot - see Fig. 3(b) of reference [55].

As a result of this work, we can quantitatively account for the contribution of topography to frequency-shift images. Our technique allows for a height resolution of 55 nm for a 30-μm probe-sample separation and about 40 μm at a separation of 1.75 mm [55]. The technique is simple and should be readily extendible to non-metallic samples, smaller probes and closer or farther separations.

5.2. SHEET RESISTANCE

Based on simple reasoning, we expect that the losses of the sample will affect the quality factor Q of the microscope. A simple situation to study is that of a resistive thin-film on a dielectric substrate with a thickness much less than the skin depth. In this case one can show that the sample presents a sheet resistance $R_x = \rho/t$ to the microscope, where ρ is the thin-film resistivity and t is the local thickness of the film. To determine the relationship between Q_0 and sample sheet resistance (R_X), we used a variable-thickness aluminum thin-film on a glass substrate [46]. The cross-section of the thin-film is wedge-shaped, implying a spatially-varying sheet resistance. Using a probe with a 500-μm-diameter center conductor in non-contact mode, and selecting a resonance of the microscope with a frequency of 7.5 GHz, we acquired frequency-shift and Q_0 data. We then cut the sample into narrow strips to take two-point resistance measurements and determine the local sheet resistance. We find that Q_0 reaches a maximum as $R_X \to 0$; as R_X increases, Q_0 drops due to loss from currents induced in the sample, reaching a minimum around $R_X = 660\ \Omega/\square$ for a height of 50 μm. Similarly, as $R_X \to \infty$, Q_0 increases due to diminishing currents in the sample. This means that R_X is a double-valued function of Q_0. This presents a problem for converting the measured Q_0 to R_X. However, R_X is a single-valued function of the frequency shift [46], allowing one to use the frequency-shift data to determine which branch of the $R_X(Q)$ curve should be used.

To explore the capabilities of our system, we scanned a thin-film of $YBa_2Cu_3O_{7-\delta}$ (YBCO) on a 5-cm-diameter sapphire substrate at room temperature [47]. The film was deposited using pulsed laser deposition with the sample temperature controlled by radiant heating. The sample was rotated about its center during deposition, with the ~3-cm-diameter plume held at a position halfway between the center and the edge. The thickness of the YBCO thin-film varied from about 100 nm at the edge to 200 nm near the center.

Figure 6 shows two microwave images of the YBCO sample. The frequency shift [Fig. 6(a)] and Q_0 [Fig. 6(b)] were acquired simultaneously, using a probe with a 500-μm-diameter center conductor at a height of approximately 50 μm above the sample. The scan took approximately 10 minutes to complete, with raster lines 0.5 mm apart. The frequency shifts in Fig. 6(a) are relative to the resonant frequency of 7.5 GHz when the probe was far away (>1 mm) from the sample; the resonant frequency shifted downward by more than 2.2 MHz when the probe was above the center of the sample. Noting that the resonant frequency drops monotonically between the edge and the center of the film, and that the resonant frequency is a monotonically increasing function of sheet resistance [46], we conclude that the sheet resistance decreases monotonically between the edge and the center.

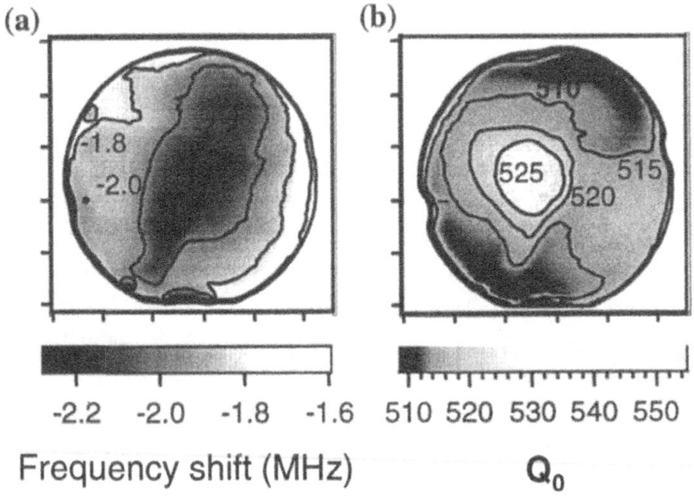

Figure 6. Images of (a) frequency shift and (b) quality factor Q of a 2-inch-diameter variable thickness $YBa_2Cu_3O_7$ thin film on sapphire. The frequency-shift image is relative to the probe far away (> 1 mm) from the sample. A probe with a 500-μm-diameter center conductor was used at a height of approximately 50 μm, at a frequency of 7.5 GHz.

The frequency shift and Q_0 images [Fig. 6(a) and (b)] differ slightly in the shape of the contour lines. This is due to the 300-μm-thick substrate being warped, causing a variation of a few microns in the probe-sample separation during the scan. In practice, the frequency shift (Δf) is dominated by topography, but also has contributions from R_x. The change in Q is dominated by sheet resistance, but also has contributions from the topography. To deal with this, we have devised a way to deconvolve these influences and produce images of sheet resistance which are not contaminated by topographic features, and topographic images which are not contaminated by sheet resistance variations. The results are shown in Fig. 7. Figure 7(b) confirms that the film does, indeed, have a lower sheet resistance near the center, as was intended when the film was deposited. We note that the sheet resistance does not have a simple radial dependence, due to either non-stoichiometry or defects in the film. Figure 7(a) shows that the wafer is, indeed, warped, with a higher ridge running from upper right to lower left.

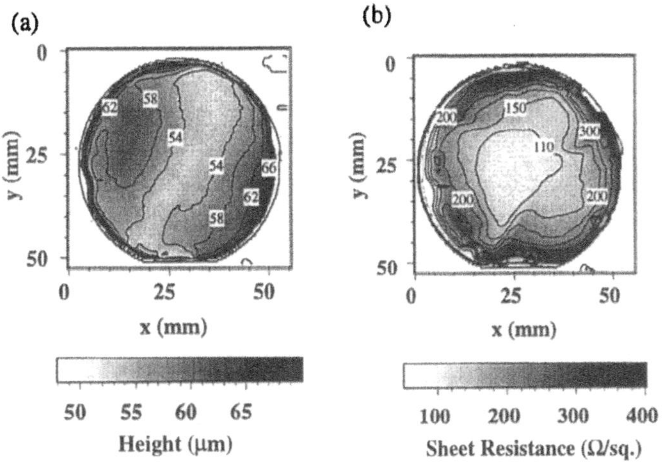

Figure 7. Images of (a) topography (in μm below the probe tip) and (b) sheet resistance of a 2-inch-diameter variable-thickness $YBa_2Cu_3O_7$ thin film on sapphire.

After scanning the YBCO film, we patterned it and made four-point dc resistance measurements throughout the wafer. The dc sheet resistance had a spatial dependence identical to the microwave data in Fig. 7(b) [51]. However, the absolute values were approximately twice as large as the microwave results, most likely due to degradation of the film during patterning. We also found from surface profilometer traces that the topography image of Fig. 7(a) is quantitatively correct.

We have estimated the sheet resistance sensitivity for the microscope as $\Delta R_X/R_X = 6.4 \times 10^{-3}$, for $R_X = 100\ \Omega/\square$ using a probe with a 500-μm-diameter center conductor at a height of 50 μm and a frequency of 7.5 GHz [47]. The sensitivity scales with the capacitance between the probe center conductor and the sample (C_X); increasing the diameter of the probe center conductor and/or decreasing the probe-sample separation would improve the sensitivity.

5.3. DIELECTRIC CONSTANT

The ability to image variations in relative permittivity or dielectric constant ε_r is useful for both fundamental and applied reasons. For example, the dielectric properties of thin-film ferroelectric materials are of interest in studying the finite-size effect on the ferroelectric phase transition. In thin-film microelectronics, testing for variations in dielectric constant can be used for quality control or to develop better growth techniques. Also, knowledge of the dielectric constant at microwave frequencies is of great importance for the design of broadband circuits and future generations of high-speed microprocessors.

5.3.1. *Linear dielectric response - Non-Contact Mode*

We began our studies of dielectric materials with measurements in non-contact mode with a blunt probe (Fig. 3(a)) [53]. To calibrate the system for dielectric measurements, we constructed a test sample by placing six pieces of different dielectric material into the bottom of a square plastic mold and pouring epoxy into the mold. In addition, silicone adhesive was used to hold down each piece. After the epoxy cured, the test sample was removed from the mold, polished and positioned on the XY table. The materials embedded in the epoxy were silicon, glass microscope slide, $SrTiO_3$, Teflon, sapphire and $LaAlO_3$. All six pieces were approximately 500-μm thick and about 6 mm x 8 mm in size. The overall thickness of the test sample was 6 mm.

We measured the frequency shift Δf versus height h above the six pieces, which have dielectric constants ranging from 2.1 to about 230. We also tested the epoxy which has an unknown dielectric constant. Each piece, as well as the probe, was flat and smooth on the scale of 5 μm, as judged by an optical microscope. For these measurements, we used a probe with a 480-μm center-conductor diameter and a source frequency of 9.08 GHz. For each scan, the probe was first brought in contact with a dielectric and the frequency shift Δf was recorded as the height was systematically increased. Samples with the largest dielectric constant produced the largest frequency shift, as expected [53]. The largest shift we observed was -26.2 MHz, when the probe was in contact with a $SrTiO_3$ sample with $\varepsilon_r \sim 230$ [53]. The smallest shift we found was -1.2 MHz when the probe was in contact with a Teflon sample with $\varepsilon_r \sim 2.1$ [53]. The frequency shift is essentially zero above 1 mm and saturates when the probe-sample distance is smaller than a few microns.

We used the above information to construct an empirical calibration curve that directly relates the frequency shift to the dielectric constant ε_r. In order to construct the calibration curve we took the difference between the frequency shift at two different heights, h_1 and h_2, i.e., $f_d=\Delta f(h_2)-\Delta f(h_1)$, where h_2 is far away ($h_2 > 1000$ μm). By taking the difference, we eliminated the effect of drift in the microwave source frequency. Proceeding this way for the test samples, we constructed two calibration curves of f_d versus ε_r - see Fig. 2 of reference [53] - one curve for h_1=10 μm and h_2=1.1 mm, and the other for h_1=100 μm and h_2=1.1 mm. We then parameterized each calibration curve with an empirical function, allowing us to easily transform any measured frequency shift to a dielectric constant. From these curves we found that we can enhance the sensitivity to the dielectric constant considerably by using a small probe height. On the other hand, at closer probe-sample separations the influence of topographic features will be enhanced. Hence, it is very important to either control the height of the probe or to de-convolve the topography from the resulting frequency shift and Q images taken in non-contact mode.

5.3.2. *Linear dielectric response - Contact Mode*

To avoid the issue of topographic features corrupting the dielectric imaging, we have developed a contact-mode version of dielectric imaging [48]. We use the near-field scanning microwave microscope with a sharp protruding center conductor. The probe tip, which has a radius of curvature $\sim 1\ \mu$m - see inset to Fig. 2 and Fig. 3(b) - is held fixed, while the sample is supported by a spring-loaded cantilever applying a controlled normal force of about 50 μN between the probe tip and the sample [48]. Due to the concentration of the microwave fields at the tip, the boundary condition of the resonator, and hence, the resonant frequency f_0 and quality factor Q are perturbed depending on the dielectric properties of the region of the sample immediately beneath the probe tip. We have shown that the spatial resolution of the microscope in this mode of operation is about 1 μm [45].

To observe the microscope's response to sample dielectric permittivity ϵ_r, we monitored the frequency-shift signal while scanning samples with known ϵ_r. With well-characterized 500-μm-thick bulk dielectrics, we observed that the microscope frequency shift monotonically increases in the negative direction with increasing sample permittivity, as noted above [53]. We observed similar behavior with thin-film dielectric samples.

For comparison, we did a finite-element calculation of the rf electric field near the probe tip. Because the probe-tip radius is much less than the wavelength ($\lambda \sim 4$ cm at 7 GHz), a static calculation of the electric field is sufficient [65]. Cylindrical symmetry further simplifies the problem to two dimensions. We represent the probe tip as a cone with a blunt end, held at a potential of $\phi = 1$ V. Using relaxation methods we solved Poisson's equation for the potential, $\nabla^2\phi = 0$, on a rectangular grid representing the region around the probe tip. The two variables we used to represent the properties of the probe were the aspect ratio of the probe tip ($\alpha \equiv dz/dr$) and the radius r_0 of the blunt end.

Since the sample represents a small perturbation to the resonator, we can use perturbation theory [48] to find the change in the resonant frequency:

$$\frac{\Delta f}{f} \approx \frac{\epsilon_0}{4W} \int_{V_S} (\epsilon_{r2} - \epsilon_{r1}) \mathbf{E}_1 \cdot \mathbf{E}_2 dV, \tag{2}$$

where $\mathbf{E}_1$ and $\mathbf{E}_2$, and ϵ_{r1} and ϵ_{r2} are the unperturbed and perturbed electric fields, and relative permittivities of two samples, respectively; W is the energy stored in the resonator, and the integral is over the volume of the sample. We compute an approximate W using the fact that the loaded quality factor [47] of the resonator is $Q_L = \omega_0 W/P_{loss}$, where ω_0 is the resonant frequency, and P_{loss} is the power loss in the resonator. Using four bulk samples with known relative dielectric permittivities between 2.1 and 305, and fixing $r_0 = (0.6\ \mu\text{m})/\alpha$, we used α as a fitting parameter to obtain agreement between the model results from Eq. (2) and our data at 7.2 GHz; we found agreement to within 10 % for several different probe tips, with $1.0 < \alpha < 1.7$.

To extend this calibration model to thin films, we extend the finite-element calculation to include a thin-film on top of the dielectric sample substrate. Once a probe's α parameter is determined using the bulk calibration described above, we use the thin-film model combined with Eq. (2), integrating over the volume of the thin-film, to obtain a functional relationship between Δf and the dielectric permittivity of the thin-film. Using the thin-film model, we found that for high-permittivity ($\epsilon_r > 50$) thin films, the microwave microscope is primarily sensitive to the in-plane component of the permittivity tensor.

Figure 8(a) shows a quantitative permittivity ϵ_r image of a sample consisting of a 440-nm $SrTiO_3$ (STO) thin-film on a 500-μm $LaAlO_3$ (LAO) substrate. The film was made by pulsed laser deposition at 700° C, in 200 mTorr of O_2. The film is paraelectric at room temperature. In the microwave (7.2 GHz) permittivity image [Fig. 8(a)], the film shows a dielectric constant on the order of 180 over most of its area (20 μm by 20 μm here). Several low-permittivity defects on the film also are visible. From atomic force microscopy measurements we find that these features are second-phase laser particles deposited on top of the STO film. Their greater thickness is part of the reason the relative permittivity is lower in the image.

Figure 8(b) shows a quantitative permittivity ϵ_r image of a sample consisting of a 275-nm $Ba_{0.60}Sr_{0.40}TiO_3$ (BST) thin-film on a 500-μm LAO substrate. The film also was made by pulsed laser deposition at 700° C, in 200 mTorr of O_2. The film is paraelectric at room temperature. It shows a larger average dielectric constant than STO, but also contains laser particles. The fine structure in Figs. 8 may be indicative of varying dielectric properties from grain to grain in the films.

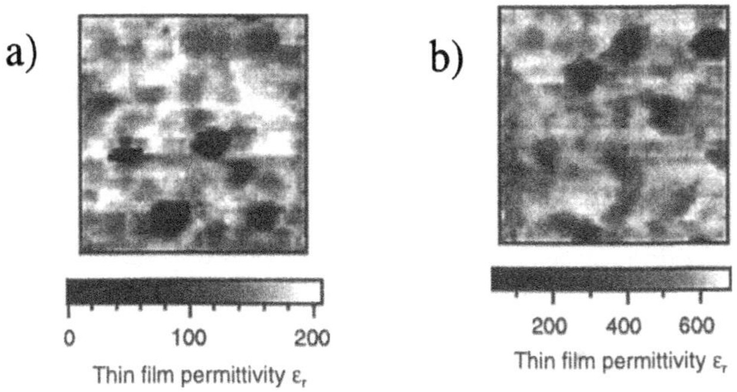

Figure 8. a) Dielectric constant of thin-film $SrTiO_3$ over a 20-μm by 20-μm area taken at 7.2 GHz. The substrate is $LaAlO_3$, and the 440-nm-thick film was made by pulsed laser deposition. b) Dielectric constant of thin-film $Ba_{0.60}Sr_{0.40}TiO_3$ over a 20-μm by 20-μm area taken at 7.2 GHz. The substrate is $LaAlO_3$ and the 275-nm-thick film was made by pulsed laser deposition

5.3.3. *Nonlinear dielectric response in contact mode*

In order to measure the local dielectric tunability of thin films, we apply a dc electric field to the sample by voltage biasing (V_{bias}) the probe tip - see Fig. 2. A grounded metallic counterelectrode layer immediately beneath the dielectric thin-film acts as a ground plane. To prevent the counterelectrode from dominating the microwave measurement, thus minimizing its effect on the microwave fields - we ignored the counterelectrode in our static field model - the sheet resistance of the counterelectrode should be as high as possible. In our case, we use low carrier density $La_{0.95}Sr_{0.05}CoO_3$ with a thickness of 100 nm, giving a sheet resistance $\sim$ 400 $\Omega/\Box$. We have confirmed by experiment and model calculation [46, 48] that the contribution of the counterelectrode to the frequency shift is small ($\Delta f < 30$ kHz) relative to the contribution from a dielectric thin-film with thickness > 100 nm ($\Delta f > 200$ kHz).

We can examine the tunability of the dielectric properties at a fixed point on the sample and generate a hysteresis loop [48, 64]. As expected, the permittivity goes down and the Q goes up when a voltage is applied. To image the tunability of the sample, we change the bias voltage applied to the probe tip and image the dielectric constant again. A tunability figure of merit can be defined with this data. In general, one wants a highly tunable material which is not very lossy. Hence, the figure of merit to maximize is K = $(\Delta\varepsilon_r/\varepsilon_r)/\tan\delta$, where $\Delta\varepsilon_r$ is the change in dielectric constant upon tuning with some fixed electric field, and $\tan\delta$ is the dielectric loss at zero bias. In our case we can define a similar figure of merit with the raw data as K = $(\Delta f/f)Q$, where Δf is the frequency shift due to tuning a 400-

nm-thick $Ba_{0.60}Sr_{0.40}TiO_3$ (BST) thin-film with a bias of 2.5 V, and Q is the unbiased quality factor of the microscope. Figure 9 shows the resulting figure of merit image over a 20-μm by 13-μm area. The main region of very low tunability (white area) is most likely a laser-deposited particle of non-paraelectric material. However, there are smaller regions of low tunability which are separated by just a few microns from regions of very good tunability. This demonstrates the power of the scanning microwave microscope to delve into the microstructure-property relationship on the microscopic scale.

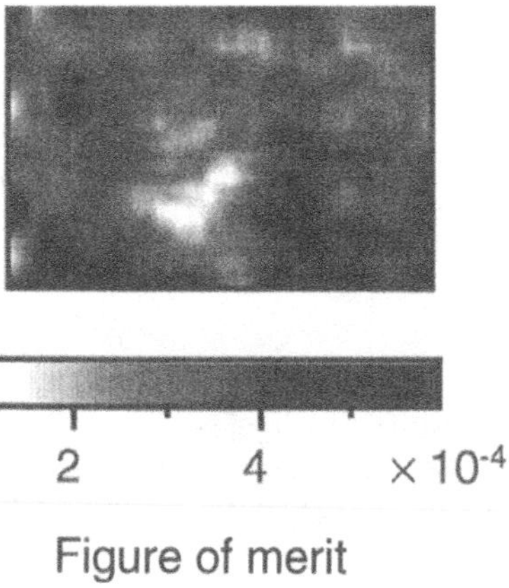

Figure 9. Figure of merit of a 400-nm-thick $Ba_{0.60}Sr_{0.40}TiO_3$ film on $LaAlO_3$ made by pulsed laser deposition. The measurement is made at 7.2 GHz and shows a 20-μm by 13-μm area. The figure of merit is defined as the relative frequency shift of the microscope due to dc electric field bias multiplied by the Q signal.

The microwave technique we use is sensitive to both film thickness and permittivity. As a result, the permittivity of the large defects in Fig. 8(a) is underestimated due to the change in film thickness at these locations, which we have confirmed with our model. However, by examining an AFM image, topographic features can be readily distinguished from permittivity features. We have further shown that in most of the images the topography is too small to account for a significant contribution to the observed permittivity contrast. We can calculate the sensitivity of the microwave microscope by observing the noise in the dielectric permittivity and tunability data. For a 370-nm-thick film on a 500-μm-thick LAO substrate, with an averaging time of 40 ms, we find that the relative dielectric permittivity sensitivity is $\Delta\epsilon_r = 2$ at $\epsilon_r = 500$, and the tunability sensitivity is $\Delta(d\epsilon_r/dV) = 0.03\ V^{-1}$ [48, 64]

6. Future Prospects

Having demonstrated quantitative imaging of losses, topography and dielectric constant on the micron length scale and below, what is next for this impressive instrument? Ongoing research is focused on a number of issues. The first is an effort to improve the spatial resolution of the microscope while maintaining its quantitative capabilities. The frontier seems to be improved field-enhancement techniques, similar to those now being pursued in near-field scanning optical microscopy. The exploitation of resonant electromagnetic phenomena combined with cantilever techniques is another tool for improving spatial resolution. Another important issue is the utilization of the microwave microscope's broad frequency bandwidth. This can be used to examine frequency-dependent phenomena, such as electronic and magnetic dynamics in correlated electron systems.

Superconducting materials present an interesting problem for near-field microwave microscopes. Already progress has been made in imaging the transition temperature in superconducting thin films [43, 70, 71, 72]. Although a superconducting microscope is probably required to image surface impedance on the micron scale, it may be possible to learn more from conventional microscopes which study nonlinearities [52, 54, 73]. Our group also has imaged electric fields in the near-field of operating superconducting microwave devices at 77 K and above [73, 74, 75, 76]. The effort is focused on isolating the local sources of nonlinearity under operating conditions.

Many other possibilities of electrodynamics experiments remain to be exploited. What is clear is that a new era of local electromagnetic experiments has begun.

7. Conclusions

We have given a short introduction to the new field of near-field microwave electrodynamics measurements. Five major developments of near-field microwave microscopy have been identified. We have presented a thorough introduction to the form of microscopy we have developed at the University of Maryland. Quantitative images of metallic sheet resistance, topography, dielectric constant and dielectric tunability have been presented. Great potential exists for new quantitative electrodynamics measurements on ever finer length scales.

Acknowledgments

This work could not have been possible without the assistance of S. K. Dutta, B. J. Feenstra, W. Hu, A. Schwartz, J. Lee and A. Thanawalla. This work has been supported by the Maryland/NSF Materials Research Science and Engineering Center on Oxide Thin Films DMR 9632521, as well as NSF grant # ECS-9632811, an NSF SBIR (DMI-9710717) subcontract from Neocera, Inc. and by the Maryland Center for Superconductivity Research.

References

1. A. B. Pippard, "The surface impedance of superconductors and normal metals at high frequencies: I. Resistance of superconducting tin and mercury at 1200 Mcyc./sec.," Proc. Roy. Soc. A **191**, 370-384 (1947).
2. J. C. Slater, "Microwave Electronics," Rev. Mod. Phys. 18, 441-521 (1946); L. C. Maier, Jr. and J. C. Slater, "Field Strength Measurements in Resonant Cavities," J. Appl. Phys. **23**, 68-83 (1952).
3. O. Klein, S. Donovan, M. Dressel, and G. Grüner, "Microwave Cavity Perturbation Technique: Part I: Principles," Int. J. Infrared and Millimeter Waves **14**, 2423- 2457 (1993); S. Donovan, O. Klein, M. Dressel, K. Holczer, and G. Grüner, "Microwave Cavity Perturbation Technique: Part II: Experimental Scheme," Int. J. Infrared and Millimeter Waves **14**, 2459-2487 (1993); M. Dressel, O. Klein, S. Donovan, and G. Grüner, "Microwave Cavity Perturbation Technique: Part III: Applications," Int. J. Infrared and Millimeter Waves **14**, 2489-2517 (1993).
4. H. Ning, H. Duan, P. D. Kirven, A. M. Hermann, and T. Datta, "Magnetic Penetration Depth in High-Tc Superconducting $Tl_2Ca_1Ba_2Cu_2O_{8-\delta}$ Single Crystals," J. Super. **5**, 503-509 (1992).
5. C. E. Gough, and N. J. Exon, "Microwave response of anisotropic high-temperature-superconductor crystals," Phys. Rev. B **50**, 488-495 (1994).
6. R. C. Taber, "A parallel plate resonator technique for microwave loss measurements on superconductors," Rev. Sci. Instrum. **61**, 2200-2206 (1990).
7. V. V. Talanov, L. V. Mercaldo, S. M. Anlage and J. H. Claassen, "Measurement of the absolute penetration depth and surface resistance of superconductors and normal metals with the variable spacing parallel plate resonator," to be published in Rev. Sci. Instrum. (2000).
8. John Gallop, L. Hao, F. Abbas, "Spatially Resolved Measurements of HTS Microwave Surface Impedance," Physica C **282-287**, 1579-1580 (1997); L. Hao, J. C. Gallop, "Spatially Resolved Measurements of HTS Microwave Surface Impedance," IEEE Trans. Appl. Supercond. **9**, 1944-1947 (1999).
9. C. Wilker, Z-Y. Shen, V. X. Nguyen, and M. S. Brenner, "A sapphire resonator for microwave characterization of superconducting thin films," IEEE Trans. Appl. Supercond. **3**, 1457-1460 (1993).
10. Steve Hogan, Sigurd Wagner, and Frank S. Barnes, "Resistivity measurement of thin semiconductor films on metallic substrates," Appl. Phys. Lett. **35**, 77-79 (1979).
11. J. S. Martens, V. M. Hietala, D. S. Ginley, T. E. Zipperian, and G. K. G. Hohenwarter, "Confocal resonators for measuring the surface resistance of high- temperature superconducting films," Appl. Phys. Lett. **58**, 2543-2545 (1991).
12. E. Keskin, K. Numssen, and J. Halbritter, "Defects in YBCO relevant for rf superconductivity: T-, f- and H-dependencies," IEEE Trans. Appl. Supercon. **9**, 2452 (1999).
13. E. F. Skelton, A. R. Drews, M. S. Osofsky, S. B. Qadri, J. Z. Hu, T. A. Vanderah, J. L. Peng, and R. L. Greene, "Direct observation of Microscopic inhomogeneities with energy-dispersive diffraction of synchrotron-produced x-rays," Science **263**, 1416-1418 (1994).
14. A. F. Hebard, A. T. Fiory, M. P. Siegal, J. M. Phillips, and R. C. Haddon, "Vortex-pair nucleation at defects: A mechanism for anoalous temperature dependence in the superconducting screening length," Phys. Rev. B **44**, 9753-9756 (1991).
15. G. Hampel, B. Batlogg, K. Krishana, N. P. Ong, W. Prusseit, H. Kinder, A. C.

Anderson, "Third-order nonlinear microwave response of $YBa_2Cu_3O_{7-\delta}$ thin films and single crystals," Appl. Phys. Lett. **71**, 3904-3906 (1997).

16. C. P. Bidinosti, W. N. Hardy, D. A. Bonn, and R. Liang, "Measurements of the Magnetic Field Dependence of l in $YBa_2Cu_3O_{6.95}$: Results as a Function of Temperature and Field Orientation," Phys. Rev. Lett. **83**, 3277-3280 (1999).
17. A. Carrington, R. W. Giannetta, J. T. Kim, and J. Giapintzakis, "Absence of nonlinear Meissner effect in $YBa_2Cu_3O_{6.95}$," Phys. Rev. B **59**, R14173-14176 (1999).
18. E. A. Synge, "A suggested method for extending microscopic resolution into the ultra-microscopic region," Phil. Mag. C **6**, 356-362 (1928).
19. Zdenek Frait, "The use of high frequency modulation in studying ferromagnetic resonance," Czeck. J. Phys. **9**, 403-404 (1959); Z. Frait, V. Kambersky, Z. Malek, and M. Ondris, "Local variations of uniaxial anisotropy in thin films," Czeck. J. Phys. B**10**, 616-617 (1960).
20. R. F. Soohoo, "A Microwave Magnetic Microscope," J. Appl. Phys. **33**, 1276-1277 (1962).
21. S. E. Lofland, S. M. Bhagat, H. L. Ju, G. C. Xiong, T. Venkatesan, and R. L. Greene, "Ferromagnetic resonance and magnetic homogeneity in a giant-magnetoresistance material $La_{2/3}Ba_{1/3}MnO_3$," Phys. Rev. B **52**, 15058-15061 (1995).
22. M. Ikeya and T. Miki, "ESR Microscopic Imaging with Microfabricated Field Gradient Coils," Jap. J. Appl. Phys. **26**, L929-L931 (1987); M. Ikeya, M. Furusawa, and M. Kasuyai, "Near-field scanning electron spin resonance microscopy," Scanning Microscopy **4**, 245-248 (1990).
23. E. A. Ash and G. Nicholls, "Super-resolution Aperture Scanning Microscope," Nature **237**, 510-512 (1972).
24. D. W. Pohl, W. Denk, and M. Lanz, "Optical stethoscopy: Image recording with resolution $\lambda/20$," Appl. Phys. Lett. **44**, 651-653 (1984).
25. E. Betzig, M. Isaacson and A. Lewis, "Collection mode near-field scanning optical microscopy," Appl. Phys. Lett. **51**, 2088-2090 (1987).
26. C. A. Bryant and J. B. Gunn, "Noncontact Technique of the Local Measurement of Semiconductor Resistivity," Rev. Sci. Instrum. **36**, 1614-1617 (1965).
27. Y. S. Xu and R. G. Bosisio, "Nondestructive Measurements of the Resistivity of Thin Conductive Films and the Dielectric Constant of Thin Substrates Using an Open-Ended Coaxila Line," IEE Proc. H **139**, 500-506 (1992).
28. M. A. Stuchly and S. S. Stuchly, "Coaxial Line Reflection Methods for Measuring Dielectric Properties of Biological Substances at Radio and Microwave Frequencies - A Review," IEEE Trans. Instrum. and Meas. **IM-29**, 176-183 (1980); M. A. Stuchly, M. M. Brady, S. S. Stuchly and G. Gajda, "Equivalent Circuit of an Open-Ended Coaxial Line in a Lossy Dielectric," IEEE Trans. Instrum. and Meas. **IM-31**, 116-119 (1982); T. W. Athey, M. A. Stuchly and S. S. Stuchly, "Measurement of Radio Frequency Permittivity of Biological Tissues with an Open-Ended Coaxial Line: Part I," IEEE Trans. Microwave Theory and Tech. **MTT-30**, 82-86 (1982); M. A. Stuchly, T. W. Athey, G. M. Samaras and G. E. Taylor, "Measurement of Radio Frequency Permittivity of Biological Tissues with an Open-Ended Coaxial Line: Part II - Experimental Results," IEEE Trans. Microwave Theory and Tech. **MTT-30**, 87-92 (1982); G. B. Gajda and S. S. Stuchly, "Numerical Analysis of Open-Ended Coaxial Lines," IEEE Trans. Microwave Theory and Tech. **MTT-31**, 380-384 (1983).
29. E. C. Burdette, F. L. Cain, and J. Seals, "*In Vivo* Probe Measurement Technique for Determining Dielectric Properties at VHF Through Microwave Frequencies," IEEE Trans. Microwave Theory Tech. **MTT-28**, 414-427 (1980).
30. M. Fee, S. Chu and T. W. Hänsch, "Scanning electromagnetic transmission line microscope with sub-wavelength resolution," Optics Communications **69**, 219-224 (1989).
31. S. J. Stranick and P. S. Weiss, "A versatile microwave-frequency-compatible scanning tunneling microscope," Rev. Sci. Instrum. **64**, 1232-1234 (1993); S. J. Stranick

and P. S. Weiss, "A tunable microwave frequency alternating current scanning tunneling microscope," Rev. Sci. Instrum. **65**, 918-921 (1994); L. A. Bumm and P. S. Weiss, "Small cavity nonresonant tunable microwave-frequency alternating current scanning tunneling microscope," Rev. Sci. Instrum. **66**, 4140-4145 (1995).

32. G. Q. Jiang, W. H. Wong, E. Y. Raskovich, W. G. Clark, W. A. Hines, J. Sanny, "Open- ended coaxial-line technique for the measurement of the microwave dielectric constant for low-loss solids and liquids," Rev. Sci. Instrum. **64**, 1614-1621 (1993).
33. K. Asami, "The scanning dielectric microscope," Meas. Sci. Technol. **5**, 589-592 (1994).
34. R. J. Gutman and J. M. Borrego, "Microwave scanning microscopy for planar structure diagnostics," IEEE MTT Digest, 281-284 (1987); Bhimnathwala and J. M. Borrego, "Measurement of the sheet resistance of doped layers in semiconductors by microwave reflection," J. Vac. Sci. Technol. B **12** , 395-398 (1994).
35. N. Qaddoumi and R. Zoughi, "Preliminary study of the influences of effective dielectric constant and nonuniform probe apeture field distribution on near field microwave images," Materials Evaluation, Oct., 1169-1173 (1997).
36. M. Golosovsky and D. Davidov, "Novel millimeter-wave near-field resistivity microscope," Appl. Phys. Lett. **68**, 1579-1581 (1996); M. Golosovsky, A. Galkin, and D. Davidov, "High-Spatial Resolution Resistivity Mapping of Large-Area YBCO Films by a Near-Field Millimeter-Wave Microscope," IEEE MTT **44**, 1390-1392 (1996); M. Golosovsky, A. Lann, and D. Davidov, "A millimeter-wave near-field scanning probe with an optical distance control," Ultramicroscopy **71**, 133-141 (1998); A. F. Lann, M. Golosovsky, D. Davidov, and A. Frenkel, "Combined millimeter-wave near-field microscope and capacitance distance control for the quantitative mapping of sheet resistance of conducting layers," Appl. Phys. Lett. **73**, 2832-2834 (1998); A. F. Lann, M. Golosovsky, D. Davidov, and A. Frenkel, "Microwave near-field polarimetry," Appl. Phys. Lett. **75**, 603-605 (1999).
37. J. Bae, T. Okamoto, T. Fujii, K. Mizuno, T. Nozokido, "Experimental demonstration for scanning near-field optical microscopy using a metal micro-slit probe at millimeter wavelengths," Appl. Phys. Lett. **71**, 3581-3583 (1997).
38. M. Tabib-Azar, N. Shoemaker and S. Harris, "Non-destructive characterization of materials by evanescent microwaves," Meas. Sci. Tech., **4**, 583-590 (1993); M. Tabib-Azar, D. -P. Su, A. Pohar, S. R. LeClair, and G. Ponchak, "0.4 μm spatial resolution with 1 GHz ($\lambda = 30$ cm) evanescent microwave probe," Rev. Sci. Instrum., **70**, 1725-1729 (1999); M. Tabib-Azar, P. S. Pathak, G. Ponchak, and S. LeClair, "Nondestructive superresolution imaging of defects and nonuniformities in metals, semiconductors, dielectrics, composites, and plants using evanescent microwaves," Rev. Sci. Instrum., **70**, 2783-2792 (1999); M. Tabib-Azar, R. Ciocan, G. Ponchak, and S. R. LeClair, "Transient thermography using evanescent microwave microscope," Rev. Sci. Instrum., **70**, 3387-3390 (1999); G. Ponchak, D. Akinwande, R. Ciocan, S. R. LeClair and M. Tabib-Azar, "Evanescent Microwave Probes Using Coplanar Waveguide and Stripline for Super-Resolution Imaging of Materials," IEEE MTT-S Digest, (1999).
39. F. Keilmann, US Patent 4,994,818, filed Oct. 27, 1989; R. Merz, F. Keilmann, R. J. Haug, and K. Ploog, "Nonequilibrium Edge-State Transport Resolved by Far-Infrared Microscopy," Phys. Rev. Lett. **70**, 651-653 (1993); F. Keilmann, "FIR Microscopy," Infrared Phys. Technol. **36**, 217-224 (1995); F. Keilmann, D. W. van der Weide, T. Eickelkamp, R. Merz, and D. Stöckle, "Extreme sub-wavlength resolution with a scanning radio-frequency transmission microscope," Optics Commun. **129**, 15-18 (1996); B. Knoll, F. Keilmann, A. Kramer, and R. Guckenberger, "Contrast of microwave near-field microscopy," Appl. Phys. Lett. **70**, 2667-2669 (1997).
40. R. G. Bosisio, M. Giroux, and D. Couderc, "Paper Sheet Moisture Measurements by Microwave Phase Perturbation Techniques," J. Microwave Power **5**, 25-34 (1970).
41. E. Tanabe and W. T. Joines, "A Nondestructive Method for Measuring the Complex Permittivity of Dielectric Materials at Microwave Frequencies Using an Open

Transmission Line Resonator," IEEE Trans. Instrum. and Meas. **IM-25**, 222-226 (1976).

42. Y. Cho, A. Kirihara and T. Saeki, "Scanning nonlinear dielectric microscope," Rev. Sci. Instrum. **67**, 2297-2303 (1996); Y. Cho, S. Kazuta, and K. Matsuura, "Scanning nonlinear dielectric microscopy with nanometer resolution," Appl. Phys. Lett. **75**, 2833-2835 (1999).
43. T. Wei, X.-D. Xiang, W. G. Wallace-Freedman and P. G. Schultz, "Scanning tip microwave near-field microscope," Appl. Phys. Lett. **68**, 3506-3508 (1996); Y. Lu, T. Wei, F. Duewer, Y. Lu, N. Ming, P. G. Schultz and X.-D. Xiang, "Nondestructive Imaging of Dielectric-Constant Profiles and Ferroelectric Domains with a Scanning-Tip Microwave Near-Field Microscope," Science **276**, 2004-2006 (1997); C. Gao, T. Wei, F. Duewer, Y. Lu and X.-D. Xiang, "High spatial resolution quantitative microwave impedance microscopy by a scanning tip microwave near-field microscope," Appl. Phys. Lett. **71**, 1872-1874 (1997); I. Takeuchi, T. Wei, F. Duewer, Y. K. Yoo, X.-D. Xiang, V. Talyansky, S. P. Pai, G. J. Chen, and T. Venkatesan, "Low temperature scanning-tip microwave near-field microscopy of YBCO films," Appl. Phys. Lett. **71**, 2026-2028 (1997); H. Chang, C. Gao, I. Takeuchi, Y. Yoo, J. Wang, P. G. Schultz, X.-D. Xiang, R. P. Sharma, M. Downes, and T. Venkatesan, "Combinatorial synthesis and high throughput evaluation of ferroelectric/dielectric thin-film libraries for microwave applications," Appl. Phys. Lett. **72**, 2185-2187 (1998); C. Gao, and X.-D. Xiang, "Quantitative microwave near-field microscopy of dielectric properties," Rev. Sci. Instrum. **69**, 3846-3851 (1998).
44. C. P. Vlahacos, R. C. Black, S. M. Anlage and F. C. Wellstood, "Near-field Scanning Microwave Microscope with 100 μm Resolution," Appl. Phys. Lett. **69**, 3272-3274 (1996).
45. Steven M. Anlage, C. P. Vlahacos, Sudeep Dutta, and F. C. Wellstood, "Scanning Microwave Microscopy of Active Superconducting Microwave Devices," IEEE Trans. Appl. Supercond. **7**, 3686-3689 (1997).
46. D. E. Steinhauer, C. P. Vlahacos, Sudeep Dutta, F. C. Wellstood, and Steven M. Anlage, "Surface Resistance Imaging with a Scanning Near-Field Microwave Microscope," Appl. Phys. Lett. **71**, 1736-1738 (1997). cond-mat/9712142.
47. D. E. Steinhauer, C. P. Vlahacos, S. K. Dutta, B. J. Feenstra, F. C. Wellstood, and Steven M. Anlage, "Quantitative Imaging of Sheet Resistance with a Scanning Near-Field Microwave Microscope," Appl. Phys. Lett. **72**, 861-863 (1998). cond-mat/9712171.
48. D. E. Steinhauer, C. P. Vlahacos, C. Canedy, A. Stanishevski, J. Melngailis, R. Ramesh, F. C. Wellstood, and S. M. Anlage, "Imaging of Microwave Permittivity, Tunability, and Damage Recovery in (Ba,Sr)TiO_3 Thin Films," Appl. Phys. Lett. **75**, 3180-3182 (1999).
49. B. J. Feenstra, C. P. Vlahacos, Ashfaq S. Thanawalla, D. E. Steinhauer, S. K. Dutta, F. C. Wellstood and Steven M. Anlage, "Near-Field Scanning Microwave Microscopy: Measuring Local Microwave Properties and Electric Field Distributions," IEEE MTT-S Int. Microwave Symp. Digest, p. 965-966 (1998). cond-mat/9802293.
50. Steven M. Anlage, C. P. Vlahacos, D. E. Steinhauer, S. K. Dutta, B. J. Feenstra, A. Thanawalla, and F. C. Wellstood, "Low Power Superconducting Microwave Applications and Microwave Microscopy," Particle Accelerators **61**, [321-336]/57-72 (1998). cond-mat/9808195
51. Steven M. Anlage, D. E. Steinhauer, C. P. Vlahacos, B. J. Feenstra, A. S. Thanawalla, Wensheng Hu, Sudeep K. Dutta, and F. C. Wellstood, "Superconducting Material Diagnostics using a Scanning Near-Field Microwave Microscope," IEEE Trans. Appl. Supercond. **9**, 4127-4132 (1999). cond-mat/9811158.
52. Steven M. Anlage, Wensheng Hu, C. P. Vlahacos, David Steinhauer, B. J. Feenstra, Sudeep K. Dutta, Ashfaq Thanawalla, and F. C. Wellstood, "Microwave Nonlinearities in High-T_c Superconductors: The Truth Is Out There," J. Supercond. **12**, 353-362 (1999). cond-mat/9808194.

53. C. P. Vlahacos, D. E. Steinhauer, S. K. Dutta, B. J. Feenstra, Steven M. Anlage, and F. C. Wellstood, "Non-Contact Imaging of Dielectric Constant with a Near-Field Scanning Microwave Microscope," The Americas Microscopy and Analysis, January, 5-7, (2000).
54. Steven M. Anlage, A. S. Thanawalla, A. P. Zhuravel', W. Hu, C. P. Vlahacos, D. E. Steinhauer, S. K. Dutta, and F. C. Wellstood, "Near-Field Scanning Microwave Microscopy of Superconducting Materials and Devices," in *Advances in Superconductivity XI*, ed. by N. Koshizuka and S. Tajima, (Springer-Verlag, Tokyo, 1999), pp. 1079 - 1084.
55. C. P. Vlahacos, D. E. Steinhauer, S. K. Dutta, B. J. Feenstra, Steven M. Anlage, and F. C. Wellstood, "Quantitative Topographic Imaging Using a Near-Field Scanning Microwave Microscope," Appl. Phys. Lett., **72**, 1778-1780 (1998). cond-mat/9802139.
56. M. J. Werner and R. J. King, "Mapping the ε'' of conducting solid films *in situ*," MRS Proc. (1996); U.S. Patent #5,334,941, "Microwave reflection resonator sensors," issued August 2, 1994 to R. J. King.
57. Y. Manassen, "Scanning Probe Microscopy and Magnetic Resonance," Adv. Mater. **6**, 401- 404 (1994).
58. Z. Zhang, P. C. Hammel and P. Wigen, "Observation of ferromagnetic resonance in a microscopic sample using magnetic resonance force microscopy," Appl. Phys. Lett. **68**, 2005-2007 (1996); Z. Zhang, P. C. Hammel, M. Midzor, M. L. Roukes, and J. R. Childress, "Ferromagnetic resonance force microscopy on microscopic cobalt single layer films," Appl. Phys. Lett. **73**, 2036-2038 (1998).
59. K. Wago, D. Botkin, C. S. Yannoni, and D. Rugar, "Paramagnetic and ferromagnetic resonance imaging with a tip-on-cantilever magnetic resonance force microscope," Appl. Phys. Lett. **72**, 2757-2759 (1998).
60. B. Knoll, and F. Keilmann, "Near-field probing of vibrational absorption for chemical microscopy," Nature **399**, 134-137 (1999).
61. R. C. Black, F. C. Wellstood, E. Dantsker, A. H. Miklich, D. T. Nemeth, D. Koelle, F. Ludwig, and J. Clarke, "Microwave microscopy using a superconducting quantum interference device," Appl. Phys. Lett. **66**, 99-101 (1995); R. C. Black, F. C. Wellstood, E. Dantsker, A. H. Miklich, D. Koelle, F. Ludwig, and J. Clarke, "High-frequency magnetic microscopy using a high-T_c SQUID," IEEE Trans. Appl. Supercon. **5**, 2137-2141 (1995).
62. J. E. Aitken, "Swept Frequency Microwave Q-Factor Measurement," Proc. IEE **123**, 855-862 (1976).
63. K. Zaki and G. J. Chen, private communication.
64. D. E. Steinhauer, C. P. Vlahacos, C. Canedy, A. Stanishevski, J. Melngailis, R. Ramesh, F. C. Wellstood, and S. M. Anlage, "Quantitative Imaging of Permittivity and Tunability with a Near-Field Scanning Microwave Microscope," submitted to Rev. Sci. Instrum. (1999).
65. S. Ramo, J. R. Whinnery and T. van Duzer, *Fields and Waves in Communication Electronics*, second edition, (Wiley, New York, 1984), p. 445.
66. Ichiro Takeuchi, personal communication.
67. J. C. Booth, Ph.D. Thesis, "Novel Measurements of the Frequency Dependent Microwave Surface Impedance of Cuprate Thin Film Superconductors," University of Maryland, 1996.
68. G. L. James, "Analysis and Design of TE_{11}-to-HE_{11} Corrugated Cylindrical Waveguide Mode Converters," IEEE Trans. Microwave Theory Tech. **29**, 1059-1066 (1981).
69. D. Gershon, J. P. Calame, Y. Carmel, T. M. Antonsen Jr., and R. M. Hutchen, "Open-ended coaxial probe for high-temperature and broad-band dielectric measurements," IEEE Trans. Microwave Theory Tech. **47**, 1640-1648 (1999).
70. B. J. Feenstra, Ashfaq S. Thanawalla, Wensheng Hu, D. E. Steinhauer, Steven M. Anlage, F. C. Wellstood, "Local measurements of normal and superconducting state properties of high T_c superconductors at microwave frequencies," Bull. Am. Phys.

Soc. **44**, 1479 (1999).
71. A. F. Lann, M. Abu-Teir, M. Golosovsky, D. Davidov, A. Goldgirsch, and V. Berlin, "Magnetic-field-modulated microwave reflectivity of high-T_c superconductors studied by near-field mm-wave microscopy," Appl. Phys. Lett. **75**, 1766-1768 (1999).
72. A. F. Lann, M. Abu-Teir, M. Golosovsky, D. Davidov, S. Djordjevic, N. Bontemps, and L. F. Cohen, "A cryogenic microwave scanning near-field probe: Application to study of high-T_c superconductors," Rev. Sci. Instrum. **70**, 4348-4355 (1999).
73. Wensheng Hu, B. J. Feenstra, A. S. Thanawalla, F. C. Wellstood, and Steven M. Anlage, "Imaging of Microwave Intermodulation Fields in a Superconducting Microstrip Resonator," Appl. Phys. Lett. **75**, 2824-2826 (1999).
74. Ashfaq S. Thanawalla, S. K. Dutta, C. P. Vlahacos, D. E. Steinhauer, B. J. Feenstra, Steven M. Anlage, and F. C. Wellstood, "Microwave Near-Field Imaging of Electric Fields in a Superconducting Microstrip Resonator," Appl. Phys. Lett. **73**, 2491-2493 (1998). cond-mat/9805239.
75. S. K. Dutta, C. P. Vlahacos, D. E. Steinhauer, Ashfaq S. Thanawalla, B. J. Feenstra, F. C. Wellstood, Steven M. Anlage, and Harvey S. Newman, "Imaging Microwave Electric Fields Using a Near-Field Scanning Microwave Microscope," Appl. Phys. Lett. **74**, 156-158 (1999). cond-mat/9811140.
76. Ashfaq S. Thanawalla, W. Hu, D. E. Steinhauer, S. K. Dutta, B. J. Feenstra, Steven M. Anlage, F. C. Wellstood, and Robert B. Hammond, "Frequency Following Imaging of the Electric Field around Resonant Superconducting Devices using a Near-Field Scanning Microwave Microscope," IEEE Trans. Appl. Supercond. **9**, 3042-3045 (1999). cond-mat/9811141.

COOLING AND PACKAGING OF RF COMPONENTS

H.J.M. TER BRAKE
University of Twente, Faculty of Applied Physics
P.O.Box 217, 7500 AE, Enschede, The Netherlands

1. Introduction

The ideal cryogenic refrigerator ("cryocooler" or "cooler") for cooling superconducting RF components should have a cooling power of a few Watts at the desired operating temperature, with an input power of also only a few Watts. Furthermore, it would be nice if it had infinite operating lifetime and costs only a few dollars. More wishes can be made, but it is obviously clear that such an ideal cooler, unfortunately, does not exist. Therefore, in the design of the complete system, a trade-off has to be made to compromise these wishes or requirements. In this trade-off process, device design and cryopackaging design have to interact. For example, a lower operating temperature may be desirable from the point of view of device performance while a lower temperature will necessitate a cooler that is bigger, heavier and more expensive. In this chapter, the cryopackaging aspects are considered. The main questions to be answered are:

- How do available coolers work?
- What are trends in cooling?
- What are the important aspects in packaging?

The discussion is focused primarily on 80 K as the operating temperature, and front-end High-Temperature Superconducting (HTS) filters as the application. A more general discussion on cryogenic systems for superconducting devices was presented in a previous NATO-ASI contribution [1]. In that publication, the available cooler technology was described in much more detail than in this chapter.

An important parameter of a cooler is the cooling capacity with respect to its input power. This performance parameter is considered in the following section. In section 3, the requirements for the cryogenic system are reviewed.

H. Weinstock and M. Nisenoff (eds.), Microwave Superconductivity, 271–304.

The available coolers and trends in cooling are discussed in section 4. Finally, the packaging (including cooler interference) is treated in section 5.

2. Coefficient of Performance, Efficiency

In a cryocooler, work W is performed on a thermodynamic system (e.g., compression of gas in a cylinder) in which heat Q_L is lifted from a cold reservoir at temperature T_L (e.g., by expanding the gas) and heat Q_H is rejected to a hot reservoir at temperature T_H (e.g., the heat of compression). The *coefficient of performance* (*COP*) of the cooler is defined as [2,3]:

$$COP = Q_L/W \tag{1}$$

In the literature, the *efficiency* of a cooler is often given as the ratio of the actual cooler *COP* to the *COP* of an ideal thermodynamic cycle [4,5]:

$$\eta = COP_{cooler} / COP_{ideal} \tag{2}$$

In an ideal thermodynamic system all processes are assumed to be reversible (i.e., without losses) and, therefore, the entropy of the system does not change. As a result, the *COP* of an ideal cooler can be derived as [1]:

$$COP_{ideal} = T_L / (T_H - T_L) \tag{3}$$

A typical example of an ideal cycle is the Carnot cycle that consists of four processes, two isothermal and two adiabatic (or isentropic). Another example of an ideal thermodynamic cycle is the Stirling cycle, which will be considered in section 4.2.

Unfortunately, the ideal case cannot be realized in practise. In coolers, irreversible processes and associated losses occur e.g., heat leak from the warm end to the cold end via conduction and radiation, friction losses in the cooling medium and many others. As a measure of the irreversibility, the efficiency of the cooler relative to Carnot, as defined in Eq. (2), can be used. The percent-Carnot efficiency of a large range of coolers was investigated by Strobridge [6] and more recently by Bruning *et al.* [7]. Because parasitic losses become more relevant as the cooler capacity, and hence dimensions, are scaled down, it is obvious that the percent-Carnot efficiency is poorer the smaller the cooler. The efficiencies of low-power coolers that, for instance, are used for cooling infrared sensors are only a few percent of Carnot, whereas for extremely large facilities in the kW-range, that are used for nuclear particle accelerators, efficiencies may well approach 30%.

3. Requirements

A cooler can be characterized by various parameters, such as operating temperature, weight, energy consumption, etc. Many of these parameters may be common for different applications, but each application may have its own list of priorities and required features. In Table 1, the parameters are given for two illustrative applications: HTS filters for telecommunication (base stations), and HTS SQUIDs for fetal magnetocardiography. Some of these are considered in more detail.

Operating temperature: Obviously, a superconductor must be cooled to below its transition or critical temperature T_c; preferably even to lower temperatures since the superconducting properties become less temperature dependant and, as a result, the required temperature stability is less severe. The London penetration depth and the surface resistance of HTS materials are rapidly decreasing as the operating temperature is lowered [8,9]. Below about 0.6 T_c, they are relatively temperature insensitive. However, a reduction of the operating temperature is not without penalty since a lower temperature requires a larger, heavier, and more expensive cooler [1].

Cooling power: The required cooling power depends on the size of the system and the number of connections to the outside world (see section 5.2). Here, again, trade-off's have to be made in the design since larger cooling power directly translates into larger and more expensive coolers.

TABLE 1. Typical cryogenic system requirements for HTS front-end filters [10,11] and HTS SQUIDs for fetal magnetocardiography [12]

	HTS filters	HTS SQUIDs for fMCG
Operating temperature	60 – 77 K	55 – 80 K
Cooling power	3 – 10 W	0.5 – 2 W
Cool-down time	< 2 h with 0.4 kg Cu	< 8 h with 0.3 kg Cu
Lifetime	> 5 years	10 years (10 h per day)
Maintenance	> 5 years	< 500 US$ per year; down time < 1 week
Cost	< 3,000 US$	< 5,000 US$
Temperature stability	< 0.1 K	< 0.1 K @ 77 K, < 0.5 K @ 55 K
Size	19" rack mount	not specified
Weight	< 30 kg	< ca. 10 kg
EMI: susceptibility	not specified	not specified
EMI: emissivity	not specified	< 0.6 pT_{RMS} (at 15 cm)
Vibrations: susceptibility	DOT standard*	not specified
Vibrations: emissivity	not specified	< 2.10^{-5} rad/√Hz
Shock	45 cm drop	not specified

*Standard of the US Department of Trade

Cost: The acceptable price of a cooler is related to that of the total system. In general, cryocooler costs of less than 10% of the total is considered to be reasonable. This, usually, implies cooler prices of a few thousand dollars, whereas the current generation of "1 W at 80 K" coolers typically cost 15,000 US dollars per unit. Therefore, cost reduction is a clear issue (see also section 4.4). Furthermore, the cost per cooler is related to the quantity sold. At a production of 10,000 coolers per year, the cost is expected to come down to about 1,000 US dollars (e.g., [13]).

Temperature stability: The properties of a superconducting device are temperature dependent, especially when the device is operated close to the critical temperature of the superconducting material. For example, the temperature dependency of the London penetration depth can change the effective sensing area of a high-T_c SQUID magnetometer that could be interpreted, mistakenly, as an apparent magnetic-field fluctuation [18].

Size and weight: For most applications this is not an important parameter, although exceptions exist, for example, hand-held systems (such as a SQUID magnetometer for airplane structural inspection). In the case of space applications, size and weight are at the top of the priority list since the costs for launching a satellite into earth orbit is very high, roughly 25.000 US dollars per kilogram of mass [14].

EMI, electromagnetic interference: In EMI two different aspects are distinguished: EM-susceptibility and EM-emissivity. Susceptibility relates to the sensitivity of the system to EM fields generated in the environment (e.g., is a cooler affected by stray fields of a superconducting magnet?). Emissivity relates to the EM fields that the cooling system generates itself, which may affect equipment that is to be cooled (e.g., a cooler compressor will generate magnetic fields that can disturb a SQUID magnetometer). It is clear that the requirements with respect to EMI are determined by the specific application.

Mechanical vibrations: A number of applications have fairly stringent requirements with regard to the vibrations that are generated by the cooler (e.g., because of an oscillating compressor piston). A good example is a SQUID magnetometer operating in the earth's magnetic field [12]. High-T_c parallel plate resonators are also known to be sensitive to microphonics [15]. The susceptibility of a cryogenic system to vibrations may also be problematic, such as in the case of space applications, where during launch, accelerations of 40 g to 50 g occur (mainly in the frequency band 200-700 Hz) [14].

4. Coolers

4.1. INTRODUCTION

The most attractive feature of a cryocooler compared to a cryogen, such as liquid helium or nitrogen, is its turn-key operation. No operator expertise or access to a cryogenic facility is required. This dramatically lowers the reluctance on the part of the user towards the use of superconducting devices outside the cryogenic laboratories. A liquid cryogen has a temperature that is fixed by its boiling point, and its temperature can only be varied by changing the vapour pressure above the liquid, e.g., by pumping on the bath. Cryocoolers are much more flexible in their operating temperature and can provide all cryogenic temperatures down to below 1 K. The operating temperature of a given cooler can simply be varied by changing the input power. Temperature stability and low interference are clear advantages of cryogens although methods are available to reduce the temperature fluctuations and interference associated with the use of cryocoolers (see section 5.3).

As mentioned above, cryocoolers are very "friendly" to the user because of their turn-key operation. On the other hand, compared to liquid cryogens, they are rather "unfriendly" to the manufacturer because of their complex construction. The latter aspect directly translates into higher costs. As to reliability, liquid-cryogen cooling is quite reliable, although the availability of the liquid may be a problem. A big advantage of cryocoolers compared to liquid cryogens is the weight required for long-term operation. This is especially relevant in space applications (see also section 3).

This section starts with a description of the four most widely used types of gas-expansion coolers: Linde-Hampson with Joule-Thomson expansion in 4.2.1, Stirling in 4.2.2, Gifford-McMahon in 4.2.3, and pulse tube coolers in 4.2.4. An extensive review of gas-expansion coolers was prepared by Walker [4,16]. In section 4.3, some other cooling methods are briefly introduced. Finally, the trends in cryocooling are considered in section 4.4.

4.2. GAS-EXPANSION COOLERS

4.2.1. *Linde-Hampson cycle with Joule-Thomson expansion*

In the early 1850s two British scientists J.P. Joule and William Thomson (later Lord Kelvin) performed experiments on the expansion of gases. They expanded gases at a high pressure through a porous plug under adiabatic conditions. Under these conditions, the expansion is also isenthalpic and the temperature of the gas increases or decreases depending on the sign and magnitude of the differential $(\partial T/\partial P)_H$ [2]. This differential is referred to as the Joule-Thomson coefficient μ. For an ideal gas $\mu = 0$, i.e., no temperature change under isenthalpic expansion. For real gases μ is positive below the so-called inversion temperature and cooling can be obtained as the result of an expansion. This inversion temperature depends on the pressure, and the maximum value (i.e., at $P = 0$) for helium is 40 K, for hydrogen 205 K, for neon 250 K, and for nitrogen 621 K.

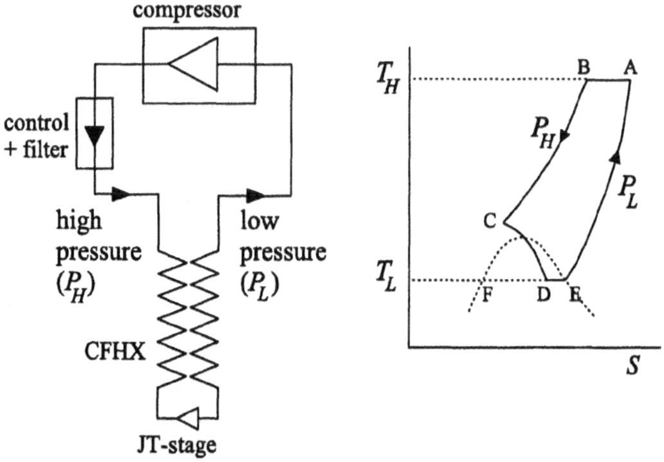

Figure 1. Linde-Hampson cooler (symbols: see text)

A Linde-Hampson cooler [17], which is schematically depicted in Figure 1, consists of a compressor, a control and filter unit, a counterflow heat exchanger (CFHX), and a Joule-Thomson expansion stage (JT). The compressor compresses the gas from the low pressure P_L in state A to the high pressure P_H in state B. The heat of compression is rejected to the environment so the compression can be considered isothermal at temperature T_H. Usually, the pressure ratios (P_H/P_L) are so high that multistage compressors are used. After compression, the gas is cleaned by various filters where solid particles (such as dust) and droplets of compressor lubricant or water are removed. These particulates would, otherwise, be trapped in the CFHX or the JT-stage and thus block the cooler. The flow of gas into the CFHX is manually or automatically controlled. In the CFHX the compressed fluid at pressure P_H is cooled from state B to state C by heat transfer to the low-pressure return fluid, at pressure P_L, heating the return flow from state E to state A. The high-pressure fluid in state C expands isenthalpically to state D inside the liquid-vapour envelope. Part of the fluid becomes liquid in state F and part is vapour in state E. The liquid in state F can take up heat from its environment and vaporize into state E. The cooling power of the cryocooler can be expressed as:

$$\dot{Q}_L = \dot{m}(h_E - h_D) \tag{4}$$

Here, h is the specific enthalpy of the gas (per unit of mass) and $\dot{m}$ is the mass flow rate of the gas. For the ideal case of a perfect heat exchanger, the enthalpy change on the high-pressure side exactly compensates that on the low-pressure side: $h_B - h_C = h_A - h_E$. Because the expansion itself is isenthalpic ($h_C = h_D$), the cooling power can directly be related to the enthalpy change on the warm side of the CFHX:

$$\text{perfect CFHX:} \quad \dot{Q}_L = \dot{m}(h_A - h_B) \tag{5}$$

It is clear that for cooling $h_A > h_B$ is required, corresponding to the warm-end temperature of the CFHX being below the inversion temperature of the working fluid.

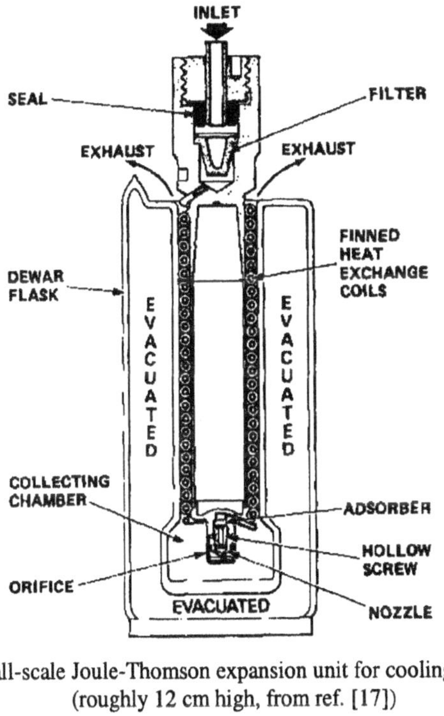

Figure 2. Small-scale Joule-Thomson expansion unit for cooling infrared sensors (roughly 12 cm high, from ref. [17])

A typical small-scale Joule-Thomson expansion unit for cooling infrared sensors is shown in Figure 2. The type of CFHX that is used in this unit is called a Giaque-Hampson coil tube. A length of fine-bore circular tube with copper fins soldered to the outside is wound in the form of a closely coiled helix around a cylindrical mandrel [17]. The compressed gas flows through the coiled tube and expands through the Joule-Thomson nozzle at the cold end of the heat exchanger. The expanded low-pressure, cold gas flows back transversely over the coiled tube thus establishing a good heat transfer to the warm, high-pressure gas. The JT-nozzle can be realized in a number of ways such as by flattening the end of the high-pressure tube, or by inserting a small rod or wire into the end of the tube [17].

Mixed refrigerants: Thus far, the description of the Linde-Hampson cycle with Joule-Thomson expansion has assumed a single-component working fluid (e.g. nitrogen). It is, however, advantageous to mix refrigerants with higher boiling points into the working fluid. The beneficial effect is twofold: firstly, larger enthalpy differences between the high pressure and the low pressure states are obtained and, secondly, because a portion of the fluid liquefies, the heat exchange in the counterflow heat exchanger is more efficient. At the high-pressure side, all the fluid can be liquefied before the expansion. In that case, the expansion valve is usually referred to as a throttling valve rather than a Joule-Thomson valve [18]. Missimer [19] and Radebaugh [20] have reviewed the history of mixed refrigerants. The first to realize a temperature below 100 K were Alfeev *et al.* who achieved a temperature of 78 K with a mixture of 30 mol% nitrogen, 30 mol% methane, 20 mol% ethane and 20 mol% propane with an efficiency of 10 times that of pure nitrogen [21].

The use of a refrigerant mixture in a single flow stream expansion has been developed by APD Cryogenics into the commercial cooler known as the "Cryotiger" [22]. The mixture is compressed with an air-cooled oil-lubricated compressor to 2.0 - 2.5 MPa and expanded to 0.3 MPa through a throttling valve. The compressor input power is between 300 and 500 W. The compositions of the mixtures used are proprietary and depend on the required range of operating temperatures: for example, mixture PT 13 has a cooling power of 0.75 W at 75 K and 1.50 W at 77 K, whereas mixture PT 12 can provide 2.0 W at 83 K and 5.0 W at 88 K. The Cryotiger is designed to operate continuously for 5 years with no maintenance.

Another approach is to use phase separation. In the liquefaction of natural gas, a cascade of separate refrigerant streams with a series of heat exchangers and phase separators has always been used to separate out the impurities in the natural gas. In 1959 Kleemenko proposed a single flow stream concept that consisted of a gas mixture for the liquefaction of natural gas [23]. MMR has commercialized this cycle into a 120 K cooler designed for the storage of biological samples (MMR Bio 120 [24]). This unit, with dimensions of 40 cm × 27 cm × 26 cm, including the compressor, can lift 5 W at 120 K with 180 W input power, which corresponds to an efficiency of 4.2% of Carnot. The advantages of this system are that it requires zero maintenance and it is low cost. Little and Sapozhnikov claim at least one order of magnitude lower cost than that of any other cooler for the same temperature range and that material costs are roughly $300 [25]. In September 1999, 3 units were running under DARPA tests, each for more than 15,000 operating hours without maintenance [26].

4.2.2. *Stirling cooler*

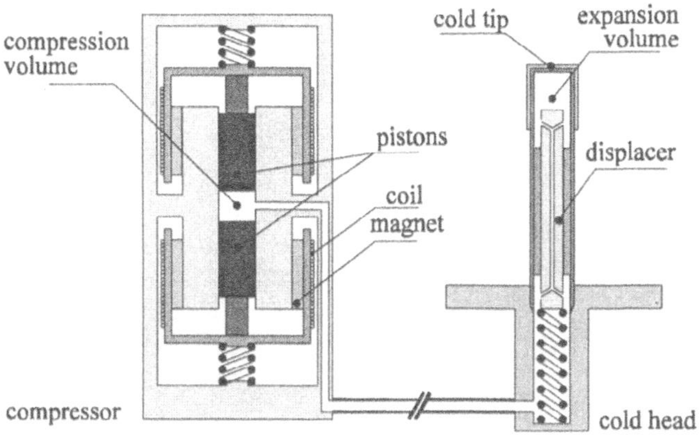

Figure 3. Cross-section of compressor and cold head of split-Stirling cryocooler, compressor height typically 12 cm (Courtesy of Signaal Usfa)

A schematic of a so-called split-type Stirling cooler [17] is presented in Figure 3. In the compressor, two pistons are driven by linear motors. The motion of each piston is produced by a time-varying magnetic field of a coil that interacts with a permanent magnetic field around the coil. Because the coil is connected to the piston, a pressure wave is established in the gas that acts on a spring-loaded displacer in the cold head. The operation is explained in Figure 4; the gas is compressed isothermally in step a $\rightarrow$ b and the heat of compression is rejected to the environment. After the compression, the displacer moves down and displaces gas from the compression space to the expansion space. In this process, the gas is forced to flow through the interior of the displacer that holds the regenerator. This is a material with a large heat capacity (e.g., copper screens or lead spheres) along which a temperature gradient is established from the compression space at T_H to the expansion space at T_L. The gas exchanges heat with the regenerator material and thus is cooled to a temperature close to that of the cold tip (isochoric step b $\rightarrow$ c). Then, the gas is allowed to expand isothermally (step c $\rightarrow$ d). Because the gas is in good thermal contact with the environment the heat of expansion Q_L is absorbed from the environment and the process can be considered as isothermal. This is the actual cooling step in the cycle. After the expansion, the gas is returned to the compression space by motion of the displacer. Passing the regenerator the gas warms up to the compression space temperature T_H (isochoric step d → a). Although an ideal Stirling cycle has the same *COP* as the Carnot cycle [1], practical Stirling coolers unfortunately exhibit various loss mechanisms that decrease the efficiency [17]. The efficiency for small coolers is only a few percent with respect to Carnot.

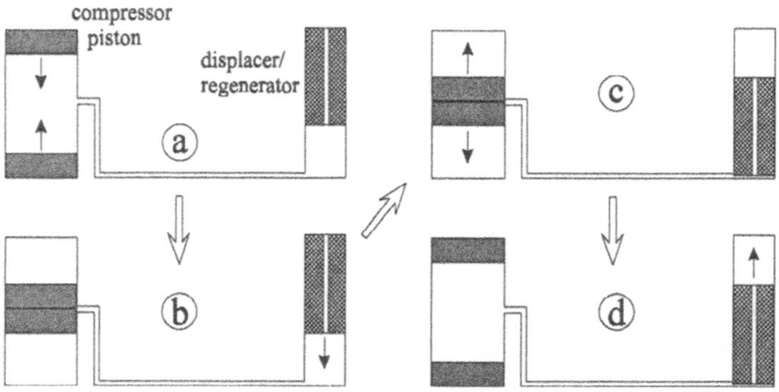

Figure 4. Split-Stirling cycle

Since the late 1950's, small Stirling coolers with helium as the working gas were especially developed for cooling infrared detectors in military applications. At present, thousands of units a month are produced worldwide for cooling these detectors in night vision equipment and missile guidance systems [27]. There are a number of companies that manufacture coolers for this market with cooling powers ranging from 0.2 W to 2 W at 80 K and an environmental temperature of 20 ^{o}C. The required input power is typically 50 W (resulting in Carnot efficiencies ranging from 1% to 10%). The dimensions of the coolers for infrared detectors are standardized: cold tip diameter 5 to 10 mm, length of the cold finger typically 6 cm, compressor diameter 55 mm and length 12 cm. Smaller coolers were developed aiming at integration with the infrared detector. The smallest cooler that is available on the market at the time of this writing is the Inframetrics cooler that weighs only 300 g and is able to lift 0.15 W at 77 K with an input power of less than 3.5 W [28]. Stirling coolers have also been developed for cooling HTS filters, in conformance with the requirements given in Table 1. Two examples of such coolers are depicted in Figure 5.

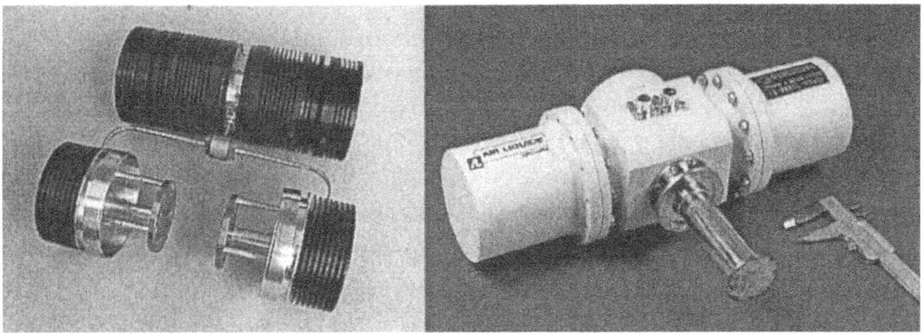

Figure 5. Two examples of Stirling coolers for HTS filter applications. Left: Leybold SUCOMS cooler; right: Air Liquide/Ricor K535. Both coolers have a cooling power of about 6 W at 80 K with an input power of roughly 200 W (Courtesy of Leybold and Air Liquide/Ricor).

4.2.3. *Gifford - McMahon cooler*

One disadvantage of the Stirling cooler is the inability to separate, by a large distance, the compression and expansion sections, thus limiting the flexibility of the cooler [29]. At the end of the 1950's, Gifford and McMahon developed a concept for separating the compressor and expander [30]. In their approach, the compressor is a high-pressure generator that is connected to the expansion unit by flexible gas lines and a valve unit that generates the pressure wave. The separation can be several meters. The basic operation of the Gifford - McMahon cooler (referred to as the GM-cooler) is explained in Figure 6. The cold head is similar to that of the Stirling cooler in Figure 4 except that the displacer/regenerator is driven by a separate electric motor.

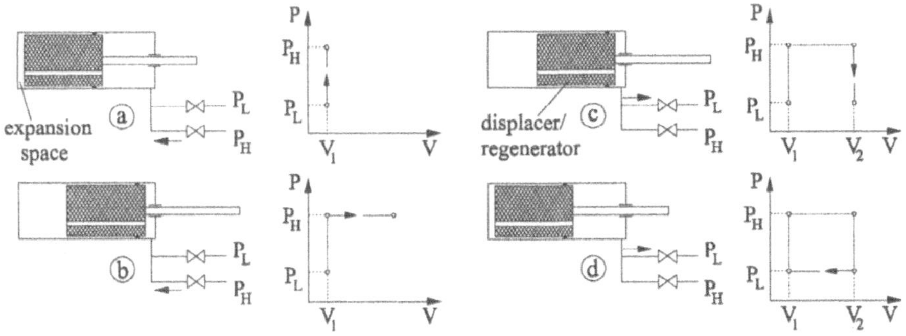

Figure 6. Idealized Gifford- McMahon cycle

First, the high-pressure valve is opened and the cold head is filled with high-pressure gas. The volume *V* of the expansion space is minimum V_1 (step a). Then, the displacer moves (to the right in Figure 6(a)) and forces the high-pressure gas to flow to the expansion space. In passing through the regenerator the gas cools and the pressure reduces. As a result, more gas enters the cold head from the compressor via the inlet valve so that the maximum cycle pressure is maintained. The volume of the expansion space now is maximum V_2 (step b). Cooling is obtained by closing the high-pressure valve and opening the low-pressure exhaust valve. The gas expands causing a temperature decrease. Heat can then be absorbed from the surroundings of the cold tip. The expansion space volume is still maximum but the pressure drops to the minimum value (step c). The next step is again a regenerative step in which the displacer moves the gas from the cold end to the warm end of the cylinder. The gas exchanges heat with the regenerator and warms up. The expansion space again is at minimum volume (step d).

GM-coolers with helium gas as the working fluid were primarily developed for cryopumping and are now available from a large number of suppliers among which are the biggest suppliers of vacuum equipment. Cryopumps are the largest commercial application of cryocoolers (roughly 20,000 units per year [31]). In addition, GM-coolers are installed in MRI systems for reducing the boil-off rate of the liquid helium bath. These coolers are commonly manufactured in single-stage and double-stage versions. The latter type has cooling powers of several watts at the coldest stage (typically at 20 K) and several tens of Watts at the other stage (at 80 K) with an input of a few kilowatts [1].

4.2.4. *Pulse tube coolers*

An important drawback of both Stirling and GM coolers is the presence of a moving displacer in the cold part of the cooler. This displacer limits the lifetime (due to wear) and, because of its movement, it can generate noise (vibrations and EMI [1]). In 1963, Gifford and Longsworth discovered a regenerative refrigeration technique that did not need a displacer [32]. They called their discovery a pulse tube refrigerator.

An improved concept was introduced by Mikulin *et al.* in 1984 [33] and later adapted by Radebaugh and co-workers [34]. In this concept, an orifice was inserted at the warm end of the pulse tube to allow some gas to flow into a relatively large reservoir volume. The ideal operation of the orifice pulse tube refrigerator is explained in Figure 7. Similar to the Stirling cycle, gas is compressed in step a $\rightarrow$ b with rejection of heat Q_{HI} to the environment followed by a regenerative step b $\rightarrow$ c. The expansion of the working gas (c $\rightarrow$ d) is not against a piston as is the case in a Stirling cooler but against a compressible gas volume in the pulse tube. This gas volume can be considered as a freely moving piston. The heat of expansion Q_L is absorbed at the cold heat exchanger. By means of the gas piston the mechanical work of expansion is transferred to the warm end of the tube. The associated heat Q_{H2} (which equals Q_L) is rejected to the warm environment. The cycle is closed with the regenerative step d → a. The orifice and the reservoir are needed to establish the proper phasing between the gas movement in the tube and the pressure wave imposed by the compressor.

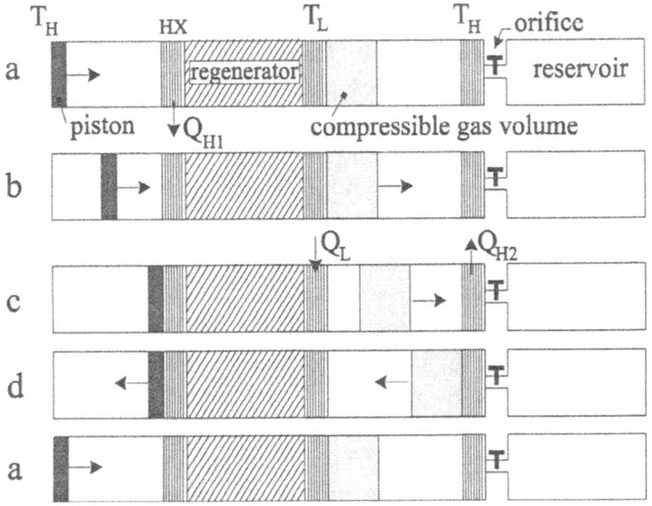

Figure 7. Schematic representation of an ideal orifice pulse tube refrigerator

Because the gas does not expand against a piston, as in the case of a Stirling cooler, the heat of expansion is not 'used' as work input in the cooler but is rejected to the environment and thus lost. Therefore, contrary to an ideal Stirling cooler, a pulse tube cooler can never achieve Carnot efficiency [1]. However, as already mentioned, because of various loss mechanisms, the efficiency of practical Stirling coolers is far below Carnot. In addition, in the last few years, pulse tube coolers have been improved to such a level that their efficiencies at cryogenic temperatures are comparable to those of Stirling coolers [31]. The most important improvement was a so-called double inlet that was introduced by Zhu *et al.* in 1990 [35]. They added a by-pass from the warm end of the pulse tube to the inlet of the regenerator. In that by-pass, a second orifice was placed for which the benefit is two-fold. First of all, it gives an extra parameter to optimize the phasing between mass flow in the tube and pressure wave at the inlet. Secondly, part of the gas flow that is needed to expand and compress at the warm end of the pulse tube is taken directly from the inlet instead of passing through the regenerator and the pulse tube. A reduction in mass flow through the regenerator results in lower regenerator losses, and thus a higher efficiency especially at higher operating frequencies. The configuration with the second orifice is called a double-inlet orifice pulse tube refrigerator. Very efficient coolers can be realized in this way as, for instance, reported by Chan *et al.* [36] in 1993. They reported on a cooler with a heat lift of 0.53 W at 80 K with an efficiency of 7.7% relative to Carnot.

However, there is an important problem related to the double inlet approach. It appears that when the second orifice is opened the system can stabilize in multiple equilibrium states with very different cold end temperatures. As shown in various papers (e.g. [37-39]), these temperature instabilities are caused by the fact that a closed loop fluid path is established. Via the second orifice, a DC fluid current can flow unregenerated from ambient temperatures to the cold end. This can give rise to a significant heat load on the cold stage. A further "secondary-streaming" problem at low operating frequencies (a few Hz) is the effect of gravity on the gas flow in the tube. Depending on the orientation of the tube in the gravitational field, Thummes *et al.* measured temperatures between 50 K and 170 K [40]. At higher frequencies (e.g. 50 Hz) this problem does not occur because the cycle is too short for the flow pattern to be affected by gravity.

Apart from these problems, it is clear that the pulse tube cooler has great advantages over the Stirling or the GM cooler: the absence of moving parts in the cold results in longer life, lower interference, and lower cost. Moreover, a pulse tube is less sensitive to side loads and less sensitive to contamination coming from the compressor. For that reason, pulse tube coolers are being used with high-T_c superconducting devices like SQUIDs [41,42] and filters in cellular telephone applications [43,44].

4.3. OTHER COOLERS

Besides the gas-expansion coolers considered in section 4.2, a variety of coolers are possible starting with the basic equation $dQ=TdS$; Any system that can be transformed from an 'ordered' state with relatively low entropy S_o to a 'disordered' state with relatively high entropy S_d is able to take up heat at a certain temperature T equal to $T\ (S_d - S_o)$. In a liquid-cryogen cryostat, for instance, heat is absorbed in the transformation of the cryogen from the ordered liquid-state to the disordered vapour-state. Analogously, in the gas-expansion coolers of section 4.2, heat is taken from the environment because the entropy of the gas increases as it expands (lower pressure: less ordered). This entropy concept can be applied in a very general sense. Magnetocaloric cooling is a good example. Here, a material is magnetized in good thermal contact with the warm end of the cooler. The magnetic domains in the material align and the system transfers to a relatively high order. Heat is rejected to the warm reservoir. Then, the material is brought into thermal contact with the cold end of the cooler, where the superconducting device is located. As the material is allowed to demagnetize, thermal energy is absorbed to transform the system back into the disordered state. Heat is absorbed at the cold end and the device is cooled. Another example is a dilution refrigerator in which He^3 is pumped from the pure He^3 phase into the diluted phase of 6.4% He^3 in He^4 and thus experiences an increase in entropy. Thermoelectric cooling can also be explained using the entropy concept. Electrons in a conductor or a semiconductor behave as a Fermi-fluid in which ordering occurs below the Fermi temperature. This temperature can be lowered, and thus the entropy increased, by reducing the electron density. Therefore, a transport of electrons from a metal to a semiconductor gives an increase in electron entropy and heat is absorbed. These cooling principles are considered in more detail elsewhere [1,45].

Two other cooling concepts are also worth mentioning in this context. Firstly, thermionic refrigeration in which a voltage is applied across a barrier that separates two electrodes. The electrons that are most thermally excited will escape over the barrier. Since hot electrons are leaving, the cathode cools. This phenomenon is not new as Thomas Edisson observed it in 1883 in experiments with his newly invented light bulb. However, the effect was not seriously considered for cooling purposes until a paper by Mahan was published in 1994 [46]. For a single junction, a temperature difference of about 1 K can be established. Therefore, a stack of devices is required if several tens of degrees Kelvin differential is desired [47,48].

The second interesting alternative cooling concept is cooling by means of anti-Stokes fluorescence. The basis of this cooling principle can best be explained by considering a three-level atom imbedded in an otherwise transparent solid. Laser light of photon energy $E_L = E_3-E_2$ pumps an atom from energy level 2 to level 3. A subsequent radiative deexcitation moves the atom either back to level 2 or to the ground state, level 1. In the former case, the emitted photon has the same energy E_L as the absorbed laser photon and the system is unchanged. In the latter case, the fluorescing photon carries away energy $E_F = E_3-E_1 > E_L$ and there is a net shift of excitation from level 2 to level 1. The relative populations of levels 1 and 2 are thus pushed out of thermal equilibrium. To restore thermal equilibrium, the atoms must make a transition from level 1 to 2 by absorbing a phonon of energy $E_P= E_2-E_1$, thereby decreasing the solid's thermal energy and thus cooling the transparent solid.

This idea of optical cooling is also not new and dates back to 1929 [49]. The most important requirement for establishing net cooling is that the fluorescent efficiency is close to unity. That is, an atom of level 3 may not decay nonradiatively to a lower level by producing phonons. High fluorescent efficiencies are possible if the energy gap E_3-E_2 is relatively large, and if the transparent solid is of extreme purity. It was only quite recent that such materials became available. This work was pioneered at Los Alamos National Laboratory by Epstein and co-workers, who used a heavy-metal fluoride glass doped with trivalent ytterbium ions. Their proof of principle was published in Nature in 1995, with a reported temperature drop of 0.3 K [50]. Gradually, the experimental results have been improved and the best result at the moment of this writing is a temperature difference of 56 K [51]. Based on this work, a system was designed to achieve a temperature of 80 K with a *COP* of 1% and a specific mass of 4 kg per Watt of cooling power [52].

4.4. TRENDS IN COOLING

In the developments toward improved coolers, during the last few years, the following requirements have been addressed one by one: high reliability, low cost, and small size.

High reliability. Three to five years ago, the most important issue was reliability. Available coolers had specified lifetimes of the order of one year, whereas the requirement for civil applications was more than 5 years [53]. Developmental work was done on the compressor and on the cold head (the gas-expansion unit). Concerning the latter, the main attention has been on eliminating the moving parts in the cold. Typical examples are the pulse tube refrigerator and the developments in mixed-gas Joule-Thomson cooling. In the case of the compressor, the key issue was the rubbing contact between the compressor piston and the cylinder. Wearing of the rubbing seals was the limiting factor in compressor lifetime. Now, these rubbing contacts can be eliminated by using flexure bearings that support the piston and the displacer inside their respective cylinders without any contact. By very accurate machining, the gap between piston and cylinder can be reduced to a few μm. The flow impedance of this clearance gap is so high that it acts as a dynamic seal for the helium gas. Because of the extremely narrow gaps, the bearings that support the pistons have to be stiff in the radial direction and weak in the axial direction. This is realized with so-called flexure bearings. A typical compressor with these bearings is shown in Figure 8. The MTTF of these compressors can exceed 50000 hours, whereas standard Stirling compressors have MTTF levels of less than 10000 hours [43,54,55]. The work on these long-life compressors was pioneered by Davey of the University of Oxford in the early and mid 80s [56]. These were developed primarily for space applications with extremely long life (10 year) and in very small quantities (one or two units at a time) and, therefore, at high cost. At the present time, compressors for terrestrial applications using flexure bearings are under development for highly reliable long-life pulse tube coolers (mostly for cooling HTS filters) [e.g. 43].

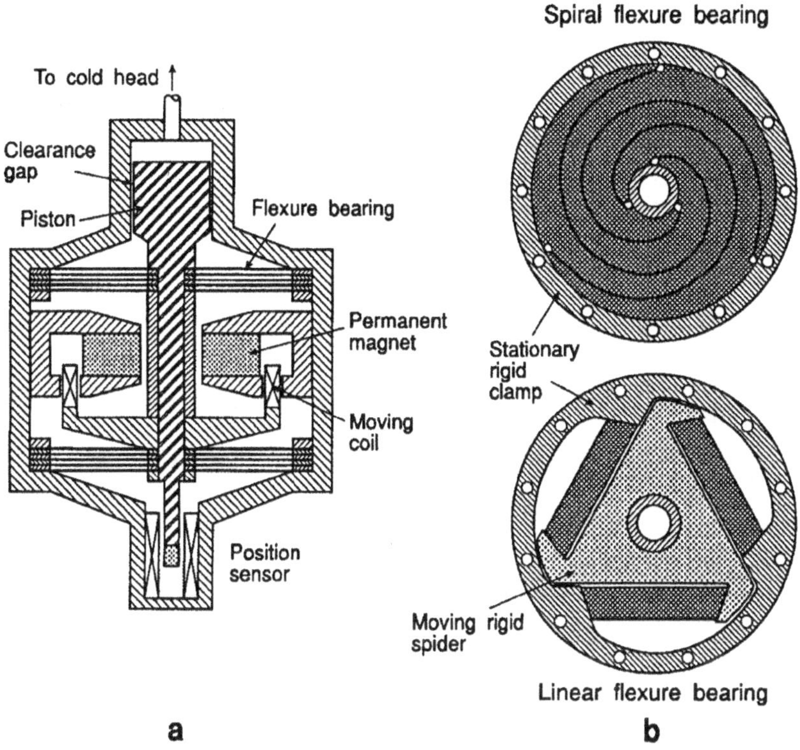

Figure 8. (a) Cross-section of a linear compressor with clearance seals and flexure bearings. (b) Two types of flexure bearings (from ref. [20]).

Low cost. The next challenge was to realize highly reliable coolers at a sufficiently low price [11]. As mentioned in section 3, a price of 10% of the total system cost is generally accepted, but of course, the lower the better. Cooler manufacturers are continuously improving the design of the coolers and manufacturing them with cost being the main driver. A nice description of such a process was presented by Superconductor Technologies Inc. [13]. The US DARPA program that sponsored research aiming at $ 1,000 coolers largely stimulated these movements towards lower cryocooler cost. The production volume is also an important cost determining parameter. As a rough indication, the price per cooler goes down by a factor of two when the number of coolers produced is increased by one order of magnitude [13].

Small size. At present, more and more attention is paid to the size of the cooler. This parameter is also related to the total system and something like 10% is considered acceptable. It will, for instance, be very hard to make a commercial product of a cryogenically cooled microprocessor if the required cooler has the size of a full computer. Since devices get smaller and smaller, cooler miniaturization becomes more and more important. Because the dissipation associated with electronic devices can be very small (e.g., in superconducting devices), there is also a need for extremely small coolers (with dimensions of centimetres) with small cold heads (order of mm) and low cooling powers (order of mWs). Extreme miniaturization is possible by using micromechanical techniques. Walker first suggested this in 1990 [57]. Patents already exist on a micro-pulse-tube cooler [58] and on a micro-Stirling-cooler [59]. A project called microcooling was started at the University of Twente [60,61].

5. Packaging

5.1. INTRODUCTION

Packaging of electronic devices, usually, refers to sealing the devices against negative influences from outside (e.g., moisture). This is done by coating the device with a protective (passivation) layer, or by placing it in a hermetically sealed ceramic package. With respect to superconducting devices, however, packaging also includes the 'cryopackage'. That means that the device, the cooler, and the thermal insulation have to be considered as a single package.

The protection of high-T_c superconducting devices against moisture is especially relevant if Josephson junctions are used, as for instance, in SQUID magnetometers or digital circuits. The performance of these devices is determined by the properties of the junctions, which can deteriorate dramatically after prolonged exposure to moisture. One of the deteriorating mechanisms in this respect is diffusion of oxygen out of the region near the Josephson barrier. Various protective layers are applied including amorphous SiO_2 [62,63], and occasionally ceramic packages are used [64]. However, when the device is cooled by a cryocooler, it is more or less sealed from the environment as the device is usually mounted in a vacuum space to provide thermal insulation (see 5.2). Thus, in operation it cannot come into contact with moisture and specific chip-packaging is not necessary.

The thermal design of the cryopackage is considered in section 5.2. and is focused on high-T_c filters. A special aspect of cryopackaging is the reduction of cooler interference, which is discussed in section 5.3.

5.2. CRYOPACKAGE THERMAL DESIGN

5.2.1. *Introduction*

An important design issue for the cryopackage is the reduction of the thermal load on the cooler. A higher load means that a bigger, heavier, and more expensive cooler, which requires more input power, is needed. In this section, the four contributions to this heat load are considered: conductive flow via gas in the vacuum space, conductive flow via the mechanical supports, conductive flow via the RF cables, and thermal radiation. In order to quantify and compare these contributions, a practical case is considered, that is based on a filter system built by Superconductor Technologies Inc. [13,65]. In this example, the outer casing is 15 cm in diameter and 15 cm high (A_{out} = 0.11 m^2). The cryocooler-cooled part of the system at 80 K is assumed to have half the surface area (i.e., A_{in} = 0.055 m^2), and the vacuum space to have a volume of 1700 cc.

5.2.2. *Residual gas in vacuum space*

Acceptable pressure. Consider, as a starting point, the vacuum space to be filled with air at a pressure of 10^5 Pa. At this high pressure, the thermal conduction is in the so-called viscous regime [66,67]. Assuming a value for the thermal conductivity of 0.024 Wm^{-1}K^{-1} [66], a heat load of roughly 20 W results! For a practical package this load should be two orders of magnitude lower. This implies that the gas conductivity must be in the molecular regime, where the heat-transfer coefficient h is given by [66,67]:

$$h = \alpha_{\text{eff}} \frac{\gamma + 1}{\gamma - 1} p \sqrt{\frac{\mathrm{R}}{8\pi MT}} \tag{6}$$

Here, R is the universal gas constant, T the temperature, p the pressure, M the molecular mass of the gas and γ the specific heat ratio (c_p/c_v). The most relevant gases are diatomic (O_2, H_2, N_2, CO, [68]), so $\gamma = 7/5$. Further, in Eq. (6), α_{eff} is an effective accommodation coefficient (to account for incomplete energy exchange between a wall and a molecule). Hydrogen is the biggest problem because of the small mass, but also because the other gases are cryotrapped. For hydrogen α_{eff} is roughly 0.5 [66] so that the conductive load in our example can be written as 40×p(Pa) W. Assume an overall heat load of 2 W to be acceptable with 0.4 W as the maximum value for the conductive gas flow. Then, the end-of-life vacuum pressure is 10^{-2} Pa, or 8×10^{-5} torr. This is close to the value of 10^{-4} torr specified by Conductus [69].

Gas sources. Firstly, gas can flow from the outside into the vacuum space via real leaks. One possibility is gas pathways through physical damage e.g., cracks in ceramic feedthroughs and poor weld joints. These can be prevented by proper manufacturing processes. A second possibility of real leaks is gas permeation. However, by selecting suitable good materials [e.g. 66] and proper design, permeation should also not be a problem.

Secondly, gas can be accumulated in the vacuum space from virtual leaks: trapped gas pockets, volatile materials and material outgassing. In assembling the system, gas can be trapped in screw threads, between large surfaces that are in contact with one another, and in voids in epoxy or solder joints. Although, in the design, attention should be paid to this point, it is not a severe problem because of cryotrapping. Under the assumption that gases other than helium, neon and hydrogen are condensed on the cold 80 K surfaces, it can be shown that, in our example, 50 cc of trapped air is needed in order to generate a pressure of 10^{-2} Pa. Volatile materials refer to oil that is left behind in the manufacturing process and to finger prints and skin oil from assemblers. It is obvious that this virtual gas leak can be prevented by careful processing and clean materials. Outgassing is the desorption of gas from a material that is placed in a vacuum. It is affected by pressure, temperature, material, and surface properties. Outgassing rates for different materials can be found in literature [e.g. 66]. In general, plastics have outgassing rates that are two to three orders of magnitude higher than those of metals. In the example, the outside wall is made of stainless steel and the cold parts of copper (because of the required high thermal conductivity). Here, the outgassing rate is determined by the copper surfaces [66]. Assume the vacuum space is closed after two hours of pumping. Then, copper outgases at a rate of 8×10^{-6} $Pa.m^3/(sec.m^2)$ [66]. With a copper area of 0.2 m^2 and a vacuum space volume of 1700 cc, this results in a pressure increase in the vacuum space of 9×10^{-4} Pa/sec. However, most of the desorbed gas is cryotrapped and the gases that are not cryotrapped (He, Ne, H_2) will hardly be adsorbed in the first place. A complication is that hydrogen is chemisorbed in metals and diffuses out to vacuum. Assuming a partial pressure of 0.3 Pa of He, Ne and H_2 in air, the end-of-life pressure of 10^{-2} Pa would be reached in about 40 days. This vacuum life can be extended by pumping longer, and at an elevated temperature. Since the outgassing rate is roughly proportional to the pumping time [66], pumping for 1 week instead of only two hours means an improvement by a factor of 80. Furthermore, the rate of desorption is proportional with temperature [66]. Therefore, pumping at e.g., 150 °C instead of 20 °C results in about 40% improvement. All together, the vacuum life is thus increased to 14 years! A further positive effect is that, in operation, the copper surfaces are at relatively low temperatures, and thus the desorption rate is proportionally lower.

Gettering. As shown above, the major source of vacuum degradation is outgassing. The most likely gases are hydrogen and carbon monoxide [68]. To maintain the low pressure in the vacuum space, chemically active getters, which are materials that chemisorb gases, can be used. For cryogenic dewars, non-evaporable getters have been developed in the shape of small pills. A very effective getter alloy is ZrV-Fe, which is available from SAES as St 707™ (Zr: 70%, V: 24.6%, Fe: 5.4%) [70,71]. This getter has to be activated at, typically, 450 °C for 10 minutes. After proper activation, the pumping speed for hydrogen is in the range of 1 m^3s^{-1} per kg of sorber material, with a sorption capacity of up to 1 $Pa.m^3kg^{-1}$ [70]. Consider as an example, a St 707 pill 6 mm in diameter and 4 mm high, which has a mass of 550 mg and is, therefore, able to absorb roughly 6×10^{-4} $Pa.m^3$ of hydrogen. This small pill can recover the vacuum space of 1700 cc, starting from the end-of-life pressure of 10^{-2} Pa, 35 times. In other words: the vacuum life is extended by a factor of 35!

Furthermore, residual gases can be cryotrapped in an effective way by mounting charcoal or a zeolite to the cold surfaces. A disadvantage of cooled getters, however, is that the sorbed gases, to a large extent, are released back into the vacuum when the system is warmed up to room temperature.

5.2.3. *Mechanical supports*

The conductive heat flow through a support of cross-sectional area A and length l can be derived from the basic Fourier-Biot equation as

$$\dot{Q}_{con} = \frac{A}{l}\int_{T_L}^{T_H} \lambda(T)dT \qquad (7)$$

Here, $\lambda(T)$ is the temperature-dependant thermal conductivity of the support material; T_H and T_L are the high and low end temperatures, respectively. From Eq. (7) it is obvious that the conductive load can be reduced firstly by reducing A/l (a geometric approach) and secondly, by reducing λ (material selection). However, these approaches should not result in a too low mechanical strength, since the structures have to mechanically support the inner parts of the cryopackage. A useful parameter for the material selection is λ/σ_y , in which σ_y is the yield strength. In Figure 9 the thermal conductivity of various matcrials is plotted versus the yield strength (at 300 K). It is clear that the engineering composites (epoxies reinforced with fibres of glass, carbon, Kevlar etc.) are the best candidates. Usually, small composite straps are applied with unidirectional fibres that are loaded under tension.

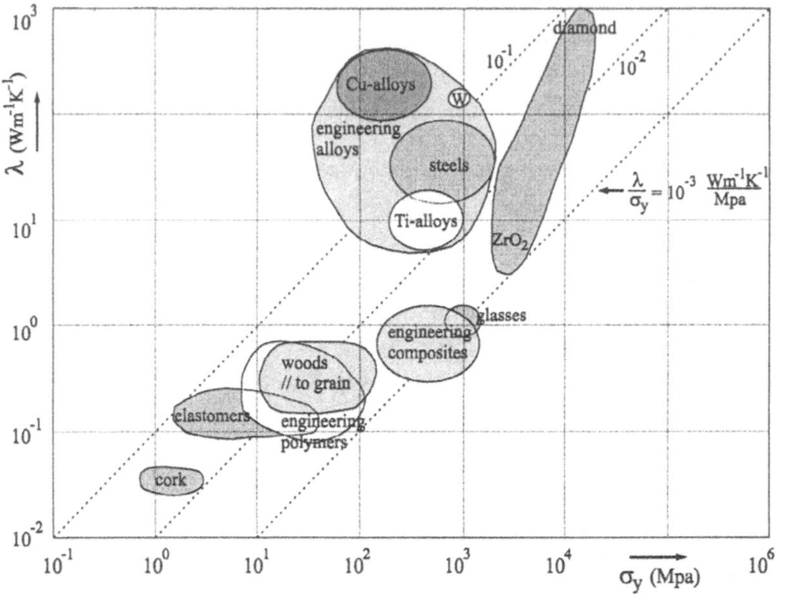

Figure 9. Thermal conductivity versus yield strength at 300 K for various materials [72-74].

Material properties at low temperatures can be found at numerous places in literature and are, to some extent, available in software packages. The reader is especially referred to the review presented by Reed and Clark [73] and the software that is commercialized by Cryodata [74]. An update on the cryogenic properties of unidirectional composites was published by Reed and Golda in 1994 [75], and more recently by Schutz [76]. A review focused on cryogenic composite supports was written by Reed and Golda in 1997 [77].

5.2.4. *RF-cables*

RF-cables are required for the communication between the superconducting device and the outside world. Similar to the mechanical supports, the conductive heat flow through these cables can be reduced by reducing the diameter-to-length ratio (i.e., smaller A/l in Eq. (7)) or by using materials with a low thermal conductivity λ. However, both measures result in higher electrical losses. According to the Wiedemann-Franz-Lorenz law (WFL), the thermal conductivity λ of a material, its electrical conductivity σ, and the absolute temperature T, are related by the ratio $\lambda/\sigma = LT$, in which L is a constant, the Lorenz ratio [73]. This relation implies that lower thermal conductivity automatically means lower electrical conductivity and, therefore, higher electrical losses. In choosing the optimum materials and cable configuration, a compromise has to be made between thermal load and electrical losses. In this respect a figure of merit (FOM) for RF cables is defined as the product of the signal attenuation and the heat conduction integral (in dB W) [78]. Note that this figure of merit should be as small as possible, which is somewhat different from most common figures of merit. Some commercial coaxial transmission lines are discussed in a paper by Wedenig *et al.* [78]. The results are summarized in Table 2.

TABLE 2. RF transmission lines: configuration, attenuation and figure of merit

	AN50085 (1)	AS50085 (1)	JS50085 (1)	UT-34-M (2)	UT-20-SS (2)	AMTEL (4)
inner cond. O.D. [mm]	0.51	0.51	0.51	0.2	0.11	0.51
inner cond. material	SCW	SCW	SBC	SPCW	SPCW	SPCW
dielectric	PTFE	PTFE	PTFE	PTFE	PTFE	PTFE
outer cond. O.D. [mm]	2.16	2.16	2.16	0.95	0.51	1.68
outer cond. material	SS	SS	SS	Cu	SS	5 μm Cu
att. [dB/m] @ 1 GHZ	0.66	1.5	1.88	1.5	5.96	1.3 (5)
FOM [dBW](0-100 K)	0.064	0.082	0.006	0.045	0.011	---
FOM[dBW](80-300 K)	≈ 0.2 (3)	≈ 0.3 (3)	≈ 0.02 (3)	≈ 0.15 (3)	≈ 0.05 (3)	0.007 (5)

1: Uniform Tubes Inc. Micro Coax line, data from [78]
2: Precision Tubes Inc. Coaxitube line, data from [78]
3: estimate based on Wiedemann-Franz-Lorenz law, see text
4: data from [79]
5: estimate based on 2 GHZ data presented in [79] and $\sqrt{f}$ behaviour of losses [78]
SCW: silvered copper clad steel wire; SBC: silvered beryllium copper wire; SPCW: silver plated copper clad steel wire; PTFE: polytetrafluorethylene; SS: stainless steel; Cu: copper

In Table 2 cable AN50085 is similar to AS50085 but it has a solid copper lining inside the stainless steel outer tube. As a result, it has a lower attenuation but a higher heat conduction, yielding only a slight difference in FOM. The data from [78] presented in Table 2, are based on the heat conductivity integrals from 0 to 100 K. Following the above-mentioned WFL-law, at higher temperatures the figures of merit will be higher, but the relative differences between the cables will be the same.

Included in Table 2 are data of a cable developed by AMTEL [44]. This cable uses the center conductor and the dielectric of a standard UT-85 size microcoax cable. The outer conductor is formed by evaporating 5 μm of copper on the dielectric. The argument is that, at GHz frequencies, the penetration depth of copper is of the order of only a few μm. More copper, therefore, does not result in lower electrical losses, whereas it does increase the thermal load.

Another interesting approach is a flexible ribbon cable that was recently described by Tighe *et al.* [80]. This type of cable is especially suited for a large number of relatively short lines. Tighe *et al.* manufactured a prototype cable of one inch width, carrying 127 signal lines. The dielectric, consisting of a 50 μm thick polyimide film, was coated on one side with a 0.6 μm thick sputtered copper ground plane. The signal lines, each 80 μm in width, were patterned from a 4 μm thick plated-copper film on the other side of the dielectric. The measured insertion loss per inch of cable length at 1 GHz is 0.50 dB at 300 K, and 0.13 dB at 77 K [80]. With an average thermal conductivity of copper of 400 $Wm^{-1}K^{-1}$, the heat flow through a cable of 1 inch long can be evaluated as 0.25 W. This is for the total of 127 lines. For each individual signal line this means a FOM at 1 GHz of about 7×10^{-4} dB W! As mentioned before, this approach is especially suited for short connections since the losses per meter length are in the range of 10 to 20 dB/m.

5.2.5. *Radiation*

The radiative load Φ_{rad} to the inner part of the package can be evaluated by means of the following expression:

$$\Phi_{rad} = \frac{A_{in} \times \sigma \left(T_{out}^4 - T_{in}^4\right)}{\frac{1}{\varepsilon_{in}} + \frac{A_{in}}{A_{out}}\left(\frac{1}{\varepsilon_{out}} - 1\right)} \tag{8}$$

Eq. (8) gives the radiative heat flow from the outer surface with area A_{out}, temperature T_{out}, and emissivity ε_{out} to the inner surface with analogue parameters A_{in}, T_{in}, and ε_{in}. Here, σ is the Stefan-Boltzmann constant. Assuming 'normal' oxidized surfaces (i.e., not dirty and not polished) a value of 0.2 can be used for ε. Then, in our example a radiative load of 3.6 W results. The first way to further reduce the radiative load is to lower the emissivities of the surfaces. By polishing the copper surfaces, the ε can be reduced to 0.03 and that of stainless steel to 0.05 [1]. By applying a silver coating, an emissivity of 0.01 can be established at 80 K and 0.02 at 300 K [1]. Assume $\varepsilon_{in}=\varepsilon_{out}= 0.05$, then $\Phi_{rad} = 0.85$ W. With $\varepsilon_{in}=\varepsilon_{out}= 0.02$, a further reduction to 0.34 W is obtainable.

A next step may be the incorporation of a thermal radiation shield. This is a thin metallic shield with a low emissivity (usually polished copper or aluminium). The thermal radiation that is emitted by the 300 K outside wall is for a large part reflected by this shield. The remainder is emitted as thermal radiation to the 80 K surface inside the package. The heat flow and the shield temperature can be determined from a simple heat balance based on Eq. (8). A shield temperature of about 260 K is obtained, which is only slightly dependent on specific values of emissivities, shield position or shield size. In contrast, the heat flow strongly depends on these parameters as a direct result of Eq. (8). This flow can be minimized by reducing the shield area and the emissivities. Consider a shield with ε = 0.02, positioned close to the inner part of the package such that A_{shield}/A_{out} = 0.55. In the original case with $\varepsilon_{in}=\varepsilon_{out}$= 0.2 the load of 3.6 W is reduced to 0.26 W by the shield! In the other extreme case with $\varepsilon_{in}=\varepsilon_{out}$= 0.02, the load of 0.34 W is reduced to 0.15 W. This reduction is not that dramatic because the radiative heat load is already quite small because of the low emissivity values.

A further reduction is possible by increasing the number of shields. It can be shown that for N shields, all surfaces having equal emissivities, the radiative load decreases by a factor of N+1 [81]. The temperatures of the shields can again be evaluated from simple heat balances. It is striking that these shield temperatures are all relatively high. E.g., for N=10 the innermost, and coldest, shield has a temperature close to 170 K. Usually, thin aluminized Mylar foil is used for the purpose of these shields. Superinsulation, or multilayer insulation (MLI), consists of alternating layers of this highly reflecting aluminized Mylar and a low-conductivity spacer material such as nylon net. The superinsulation known as NRC-2 is crinkled so that successive layers make occasional point contact and no spacer material is needed [82]. The optimum layer density of MLI is roughly 30 layers per cm [81,83]. The advantage of using MLI is the further reduction of the radiative load by roughly one order of magnitude. The disadvantages, on the other hand, are the higher costs (material and labour) and an increase of the outgassing rate.

5.3. COOLER INTERFERENCE

5.3.1. *Introduction*

A special aspect of cryopackaging is the interference from the cooler, and ways to minimize it. Depending on the specific device application, this interference can be electro-magnetic interference (EMI), mechanical vibrations, and/or temperature fluctuations. EMI can be caused by currents flowing through the compressor coils or the motors that drive the valves or the displacer, and by moving magnetic material. Mechanical vibrations are caused by the movement of pistons, displacers and valves. Also, gas flow or pressure oscillations may cause mechanical vibrations. The cold-tip temperature may fluctuate with a characteristic frequency (in regenerative coolers, the operating frequency). For a Cryotiger cooler, a 0.08 Hz temperature oscillation was measured [84]. Besides the oscillations, the temperature may also exhibit a low-frequency drift. In this section, methods are discussed to reduce the effect of these interference mechanisms. A more detailed review was presented elsewhere [1].

5.3.2. *Time separation*

The simplest approach for the suppression of cooler interference is to switch off the cooler when the superconducting device is to be used. This is the separation of cooler and device in time. In the next sub-section, separation in space is considered.

There are two problems in time separation: firstly, the cooler should not be switched off too frequently because that could seriously limit its lifetime, and secondly, the temperature of the device will slowly increase while the cooler is switched off (thermal drift). Both problems can be solved by incorporating a sufficiently large thermal buffer or thermal storage unit (TSU). To most efficiently stabilize the temperature, it is best to provide a buffer or thermal storage unit with a phase change material (PCM, e.g., a boiling liquid). Using this approach, Longsworth *et al.* equipped a Linde-Hampson cooler with a 1.4 litre helium reservoir [85]. After switching off the cooler, this buffer maintained a temperature of 4.2 K for a period of 10 hours. In Europe, time separation has been studied at the Friedrich-Schiller University of Jena by Kaiser *et al.* They worked with neon in the liquid-vapour transition at 27 K and also suggested a gas mixture of propane, ethane and methane in the solid-liquid transition at 66 K [86]. Bugby *et al.* have considered a TSU for space applications, operating at roughly 60 K [87]. They selected nitrogen trifluoride as the PCM, which undergoes a solid-solid transition at 56.7 K (realignment of the crystal structure) with a heat of transition of 21.3 J/g. As an alternative, nitrogen was suggested that has a melt transition at the triple point of 63.1 K with a heat of melting of 25.7 J/g.

5.3.3. *Space separation*

The problem to be solved in the case of space separation is how to transport the heat from the superconducting device to the cooler. Three possible interfaces can be conceived: a heat pipe, a forced gas circulation and a thermal strap.

Heat pipe. A standard heat pipe is a tube with grooves along the inside wall oriented in the axial direction. The tube is filled with a fluid like ammonia or, for lower temperatures, nitrogen or oxygen, and at one end it is connected to a cold station (i.e., the cryocooler) where the fluid condenses. By capillary forces, the fluid is pumped into the grooves and transported to the other end of the tube, where it is connected to the device to be cooled. Heat is absorbed from the device and the fluid evaporates. Because of the pressure built up, the vapour can return via the center part of the tube to the cold condenser region [88]. The attractive point of the heat pipe is the more or less isothermal behaviour, that is, high heat fluxes can be realized over relatively large distances at very small temperature drops. Furthermore, it has no moving parts, is highly reliable and has a relatively low mass. Therefore, heat pipes have been developed especially for applications in space. In ground applications gravity affects the fluid motion and, in general, special configurations have to be made. Either gravity is applied as an extra driving force (e.g. [89]), or larger capillary forces are established (e.g. [90]).

Forced gas circulation. Heat can also be transferred by means of a gas that is circulated in a closed-cycle system. This method has been investigated at the University of Twente by Van den Bosch *et al.* [91,92] who used helium gas that was circulated by a diaphragm pump and cooled by a two-stage GM-cooler. Using a 2.5 m long gas line, a heat exchanger was cooled on which a superconducting device could be mounted. A minimum temperature of 31 K was established with a helium gas flow of 6×10^{-6} kg/s and a cooler second-stage temperature of 14 K. This configuration was adapted by Tröll and Heiden who used a flexible PFTE gas line with a length of 2 m [93]. Because of the relatively high gas permeability of helium through PFTE materials, neon was used as the circulating gas. With a cooler cold-stage temperature of 45 K, a temperature of 60 to 70 K was achieved depending on the applied mass flow (10^{-4} and 10^{-5} kg/s, respectively). Another interesting configuration was developed at Conductus for an NMR spectrometer with HTS receiver coils [94]. Instead of using a separate gas circulation loop driven by a pump, they installed a secondary gas flow loop that uses a small quantity of gas tapped off of the GM compressor. A flexible transfer line is used with a length of 1.5 m. A temperature below 20 K was achieved in somewhat less than 1 hour.

Thermal strap. If the separation between cooler and superconducting device can be relatively small (roughly 0.5 m or less) a thermal strap interface is a reasonable option, e.g., a flexible copper lead [e.g. 95]. For example, the application of a copper-braid cable with a cross-sectional area of 40 mm^2, and a length of 15 cm would have a 10 K temperature difference over the cable per Watt of heat transported.

5.3.4. *Low-noise coolers*

By far the largest part of the noise that a cooler generates arises from moving parts. These parts are driven by motors with current-carrying coils that generate EMI and vibrations may also result. Therefore, the elimination of moving parts in the cold stage results in low-noise cold heads, as, for instance, is the case in JT-coolers and pulse tube refrigerators. Alternatively, non-magnetic or even non-metallic components can be used in order to reduce EMI [96-98].

It is obvious that, besides the cold heads, it is also attractive to develop compressors without moving parts. The main effort in this respect is the development of sorption compressors. In such a compressor a number of cells with adsorber material (for instance charcoal) are alternately cooled to adsorb a working gas (e.g. nitrogen) and heated to desorb it. A gas flow can be established in this way from a heated desorbing sorption cell to a cooled adsorbing cell. This flow can be used in a Linde-Hampson cycle to obtain cooling from a JT-valve. By far the largest amount of research in this field has been carried out at Jet Propulsion Laboratory, mainly on metal hydrides with a hydrogen cycle [99]. More recently, work has also been performed at the University of Twente [61].

5.3.5. *Noise suppression techniques*

Direct EMI: Coolers usually operate at a well-defined frequency and generate periodic EMI signals. Therefore, various filtering techniques can be applied. Zimmerman, for instance, suggested using commutative filters that were synchronized with the cooler frequency [100]. Also, more or less sophisticated digital signal processes can be applied in which templates of the cryocooler noise are used (e.g. [101-104]).

However, a disadvantage of these filtering techniques is that all information, including the desired data, at the cooler operating frequency and its higher harmonics is lost. An alternative procedure for the reduction of direct EMI in a SQUID magnetometer was suggested by Van den Bosch *et al.* [95]; Direct EMI can be related to the currents that drive the coolers. In general, the relative configuration of coolers and SQUIDs is fixed and the transfer functions relating currents to detected fields at all SQUIDs can be measured without the subject present. Then, during the actual measurements on the subject, the cooler currents can be recorded and using the known transfer functions, the current-induced field contributions can be calculated and subtracted from the data. The disadvantage of this approach is that it only works for a fixed configuration.

Mechanical vibrations: Mechanical vibrations can be reduced in various ways: proper mechanical design, passive damping and active damping. The design of the system should be such that the cooler is mechanically uncoupled from the device as well as possible. Large masses can be placed at critical places to reduce the vibration amplitudes. A cooler usually operates at a fixed frequency and the vibrations of a system element can be damped in a passive way by connecting a mass-spring system to it that vibrates in anti-phase at that specific frequency. This method initially was used to damp the vibrations caused by linear motor compressors. Later, a second piston was included in the compressor that was driven in anti-phase in a back-to-back configuration. This resulted in the dual-opposed piston compressor as depicted in Figure 3. The vibrations can also be minimized in an active way by applying compensating forces [e.g. 105-109].

Temperature fluctuations: The simplest way to reduce temperature fluctuations is to use a passive thermal filter which is a piece of material placed between the cold tip of the cooler and the superconducting device to be cooled. This material conducts the heat but it also damps the thermal oscillations as a first-order low-pass filter. For example, Hohmann *et al.* attenuated the 0.08 Hz oscillations of the Cryotiger by a factor of 50 using a 5 mm thick piece of PMMA [84].

The temperature can also be controlled by incorporating an active feedback loop. Clappier *et al.* measured the temperature of their instrument and this signal was used to automatically control the electrical input power to the cooler so that the temperature was stabilized [110]. They thus reduced the long-term fluctuations (over 24 h) from the K-range to less than 10 mK. More sophisticated master-slave type feedback configurations can also be conceived [1].

6. Conclusion

Several coolers were discussed of which the operation is based on the expansion of a compressed working fluid. These coolers are available with cooling capacities in the range of 0.1 – 10 W at 80 K. The *coefficient of performance* is of the order of 10^{-2}, typically 1 W of cooling power for 50 W input. Comparing the alternatives, a pulse tube refrigerator is most attractive in terms of lifetime and interference level.

The main trends in cooling are aiming at longer lifetimes of the coolers (> 5 years) and at lower costs (few k$). Also, the size of coolers is important and needs to be reduced and, in this respect, MEMS technologies are explored. Furthermore, alternative cooling technologies are investigated as for instance fluorescent cooling.

In order to thermally insulate the HTS device from the environment, a vacuum space is crucial. The vacuum life is determined by outgassing. Maximum life can be realized by careful design and manufacturing, extended pumping at an elevated temperature, and by the incorporation of getter material. Then, the thermal load in HTS front-end filters is determined by thermal radiation and conduction via the RF cables.

The HTS device may experience interference from the cooler in terms of electromagnetic noise, vibrations, or temperature fluctuations. Various methods were considered to reduce the effect of these interference mechanisms.

Acknowledgements

I would like to express my thanks to Marty Nisenoff and Matthias Hein for sending me material and discussing with me various parts of this chapter. Also, I am indebted to a number of industries for sending me useful material: Leybold GmbH, AMTEL Inc., Sunpower, Airliquide/Ricor, and Signaal USFA.

References

1. Ter Brake, H.J.M.(1999) Cryogenic systems for superconducting devices, in H. Weinstock (ed.), Kluwer Academic Publ. Dordrecht.
2. Callen, H.B. (1985) *Thermodynamics and an Introduction to Thermostatistics*, John Wiley & Sons, New York.
3. Finn, C.P.B. (1986) *Thermal Physics*, Routledge, London.
4. Walker , G. (1983) *Cryocoolers, part 1: Fundamentals*, Plenum Press, New York.
5. Sheahen, T.P. (1994) *Introduction to high-temperature superconductivity*, Plenum Press, New York.
6. Strobridge, T.R. (1974) Cryogenic Refrigerators - An updated Survey, National Bureau of Standards Technical Note 655 (Supt. Documents, U.S. Govt. Printing Off.).
7. Bruning, J., Torrison, R., Radebaugh, R. and Nisenoff, M. (2000) The 1999 cryogenic refrigeration survey, to be published in *Advances in Cryogenic Engineering* **45**.
8. Ter Brake, H.J.M., Aarnink, W.A.M., Van den Bosch, P.J., Hilgenkamp, J.W.M., Flokstra, J. and Rogalla, H. (1997) Temperature dependence of the effective sensing area of high-Tc dc SQUIDs, *Superconductor Science and Technology* **10**, 512-515.
9. Hein, M. RF properties of superconductors, this volume.
10. Workshop on "Military and Commercial Applications of Low Cost Cryocoolers"(M-CALC), San Diego, January 1996 and January 1998.
11. Workshop on "Cellular Base Station" at the 10th International Cryocooler Conference, Monterey, May 26th 1998.
12. Rijpma, A.P., Ter Brake, H.J.M., *et al.* (2000) Cryogenic aspects of a fetal heart monitor based on high-T_c SQUIDs, to be published in *Advances in Cryogenic Engineering* **45**.
13. Loung, V., O'Baid and Harper, S. (1997) Path to low cost and high reliability Stirling coolers, in R.G. Ross, Jr. (ed.), *Cryocoolers* **9**, Plenum Press, New York, pp. 97-108.
14. Private communication May 1997 Fokker Space Leiden, The Netherlands.
15. Klein, N. (1996) HTS-shielded dielectric resonators for oscillators and filters, Low Power Cryocoolers, Bad Honnef, Germany 3-5 June 1996.
16. Walker , G. (1983) *Cryocoolers, part 2: Applications*, Plenum Press, New York.
17. Walker, G. (1989) *Miniature Refrigerators for Cryogenic Sensors and Cold Electronics*, Clarendon Press, Oxford.
18. Longsworth, R.C., Boiarski, M.J. and Klusmier, L.A. (1995) 80 K Closed-cycle throttle refrigerator, in R.G. Ross, Jr. (ed.), *Cryocoolers* **8**, Plenum Press, New York, pp. 537-541.
19. Missimer, D.J. (1994) Auto-Refrigerating Cascade (ARC) Systems - an overview, Tenth Intersociety Cryogenic Symposium, AIChE Spring National Meeting, March 1994.
20. Radebaugh, R. (1995) Recent developments in cryocoolers, in *Proc. 19th International Congress of Refrigeration*, Den Haag, pp. 973-989.
21. Alfeev, V.N., Brodyanski, V.M., Yogadin, V.M., Nikolsky, V.A. and Ivantsov, A.V. (1973) Refrigerant for a cryogenic throttling unit, UK patent 1,336,892.

22. APD Cryogenics Inc., 450 Old Niskayuna Road, P.O.Box 461, Latham NY, USA.
23. Kleemenko, A.P. (1960) One flow cascade cycle (in scheme of natural gas liquefaction and separation, in *Proc. 10th Int. Congr. of Refr.*, Copenhagen 1959, Pergamon Pres, pp. 34-39.
24. MMR Technologies, Inc., 1400 North Shoreline Boulevard Suite A-5, Mountain View, CA 4043-1346, USA.
25. Little, W.A. and Sapozhnikov, I. (1997) Low cost cryocoolers for cryoelectronics, in R.G. Ross, Jr. (ed.), *Cryocoolers* **9**, Plenum Press, New York, pp. 509-513.
26. Private communication M. Nisenoff, September 1999.
27. Walker, G. and Bingham E.R. (1994) *Low-Capacity Cryogenic Refrigeration*, Clarendon Press, Oxford.
28. Inframetrics, Inc. 16 Esquire Road, No. Billerica, MA 01862-2598, USA.
29. McMahon, H.O (1960) Recent developments in gas cryogenics, *Cryogenics* **1**, 65-70.
30. Gifford, W.E. and Mc Mahon, H.O. (1959) A low temperature heat pump, in Proceedings. 10th International Congress of. Refrigeration. Vol. 1 (Copenhagen 1959).
31. Radebaugh, R. (1997) Advances in cryocoolers, in T. Haruyama, T. Mitsui and K. Yamafuji (eds.), *Proceedings 16th International Cryogenic Engineering Conference*, Kitakyushu, Japan, 20-24 May 1996, Elsevier Science, New York, pp. 33-44.
32. Gifford, W.E. and Longsworth, R.C. (1963) Pulse-tube refrigeration, ASME paper No. 63-WA-290 presented at Winter Annual Meeting of the American Society of Mechanical Engineers, Philadelphia, Pennsylvania (Nov. 17-22, 1963).
33. Mikulin, E.I., Tarasov, A.A. and Shkrebyonock (1984) Low-temperature expansion pulse tubes, *Advances in Cryogenic Engineering* **29**, Plenum Press, New York, pp. 629.
34. Radebaugh, R., Zimmerman, J., Smith, D.R. and Louie, B. (1986) A comparison of three types of pulse tube refrigerators: new methods for reaching 60 K, in R.W. Fast (ed.), *Advances in Cryogenic Engineering* **31**, Plenum Press, New York, pp. 779-789.
35. Zhu, S., Wu, P. and Chen, Z (1990) Double inlet pulse tube refrigerators: an important improvement, *Cryogenics* **30**, pp. 514.
36. Chan, C.K., Jaco, C.B., *et al.* (1993) Miniature pulse tube cooler, in *Proceedings 7th International Cryocooler Conference*, 17-19 November 1992, Santa Fe, pp. 113-124.
37. Duband, L., Charles, I., Ravex, A., Miquet, L. and Jewell, C. (1999) Experimental results on inertance and permanent flow in pulse tube coolers, in R.G. Ross, Jr. (ed.), *Cryocoolers* **10**, Plenum Press, New York, pp. 281-290.
38. Kotsubo, V., Huang, P. and Nast, T.C. (1999) Observation of DC flows in a double inlet pulse tube, in R.G. Ross, Jr. (ed.), *Cryocoolers* **10**, Plenum Press, New York, pp. 299-305.
39. Wang, C., Thummes, G. and Heiden, C. (1998) Control of DC gas flow in a single-stage double-inlet pulse tube cooler, *Cryogenics* **38**, 843-847.
40. Thummes, G., Schreiber, M., Landgraf, R. and Heiden, C. (1997), Convective heat losses in pulse tube coolers: effect of pulse tube inclination, in R.G. Ross, Jr. (ed.), *Cryocoolers* **9**, Plenum Press, New York, pp. 393-402.
41. Thummes, G., Landgraf, R., Mück, M., Klundt, K. and Heiden, C. (1997) Operation of a high-T_c SQUID gradiometer by use of a pulse tube refrigerator, in T. Haruyama, T. Mitsui and K. Yamafuji (eds.) *Proceedings 16th International Cryogenic Engineering Conference*, Kitakyushu, Japan, 20-24 May 1996, Elsevier Science, New York, pp. 283-286.
42. Saez, S.*et al.* (1997) Progress towards a hand portable pulse tube refrigerator for high T_c SQUID operation, in H. Koch and S. Knappe (eds.), Extended abstracts of 6th International Supercond. Electronics Conf., Berlin, 25-28 June 1997, PTB Braunschweig, pp.432-434.
43. Nast, T., Champagne, P. and Kotsubo, V. (1998) Development of a low-cost unlimited-life pulse-tube cryocooler for commercial applications, in P. Kittel (ed.), *Advances in Cryogenic Engineering* **43**, Plenum Press, New York, pp. 2047-2053.

44. Advanced Mobile Telecommunication Technology Inc. (AMTEL), 500-1 Minamiyama, Komenoki-cho, Nisshin-si Aichi-ken, 470-0111 Japan.
45. Radebaugh R. (1983) Fundamentals of alternate cooling, in G. Walker, *Cryocoolers part 2: Applications*, Plenum Press, New York, pp. 129-175.
46. Mahan, G.D. (1994) Thermionic refrigeration, *Journal of Applied Physics* **76**, 4362-4366.
47. Zhou, R., Dagel, D. and Lo, Y.H. (1999) Multilayer thermionic cooler with a varying current density, *Applied Physics letters* **74**, 1767-1769.
48. Mahan, G.D. and Woods, L.M. (1998) Multilayer thermionic refrigeration, *Physical Review Letters* **80**, 4016-4019.
49. Pringsheim, P. (1929) Zwei Bemerkungen über den Unterschied von Lumineszenz- und Temperaturstrahlung, *Zeitung der Physik* **57**, 739-746.
50. Epstein, R.I., Buchwald, M.I., Edwards, B.C., Gosnell, T.R. and Mungan, C.E. (1995) Observation of laser-induced fluorescent cooling of a solid, *Nature* **377**, 500-502.
51. Edwards, B.E., Anderson, J.E., Epstein, R.I., Mills, G.L. and Mord, A.J. (1999) Demonstration of a solid-state optical cooler; A new approach to cryogenic refrigeration, *Journal of Applied Physics* **86**, 6489-6493 (48 K step reported, 56 K was claimed in private communication with R.I. Epstein, November 3rd 1999).
52. Epstein, R.I., Edwards, B.C., Mungan, C.E. *et al.* (1997) The Los Alamos solid-state optical refrigerator, in R.G. Ross, Jr. (ed.), *Cryocoolers* **9**, Plenum Press, New York, pp. 681-686.
53. Workshop on Cryogenic packaging of High Temperature Superconducting (HTS) Electronic Devices, Boston USA, 16 October 1994, (1995) *Cryogenics* **35**, 405-406.
54. Jones, B.G. (1995) Development for space use of BAe's improved single-stage Stirling cycle cooler for applications in the range 50-80 K, in R.G. Ross, Jr. (ed.), *Cryocoolers* **8**, Plenum Press, New York, pp. 1-11.
55. Radebaugh, R. (1990) A review of pulse tube refrigeration, in R.W. Fast (ed.), *Advances in Cryogenic Engineering* **35**, Plenum Press, New York, pp. 1191-1205.
56. Davey, G. (1990) Review of the Oxford cryocooler, in R.W. Fast (ed.), *Advances in Cryogenic Engineering* **35**, Plenum Press, New York, pp. 1423-1430.
57. Walker, G. and Bingham, E.R. (1990) Micro and nanno cryocoolers: speculation on future development, in *Proceedings of the 6th International Cryocooler Conference*, Plymouth, Mass., October 1990, pp.363-375.
58. Crete, D., Cabanel, R. and Friederich, A. (1995) Refroidisseur à gaz pulsé, European patent publication 0 672 873 A1.
59. Bowman, L., Berchowitz, D.M. and Urieli, I. (1994) Microminiature Stirling cycle cryocoolers and engines, US patent 5,457,956.
60. Burger, J.F.*et al.* (1997) Microcooling: study on the application of micromechanical techniques, in R.G. Ross, Jr. (ed.), *Cryocoolers* **9**, Plenum Press, New York, pp. 687-696.
61. Burger, J.F. *et al.* (1999) Thermodynamic considerations on a microminiature sorption cooler, in R.G. Ross, Jr. (ed.), *Cryocoolers* **10**, Plenum Press, New York, pp. 553-563.
62. Jansman, A.B.M. (1999) *High-T_c dc SQUIDs for use in a background field*, Ph.D. thesis University of Twente, Enschede.
63. Prusseit, W. (2000) Protective coatings for $Yba_2Cu_3O_7$-thin film devices, in X. Obradors, *et al.* (eds.) *Applied Superconductivity 1999* **2**, IOP London, pp. 69-72.
64. Ter Brake, H.J.M., Karunanithi, R., Holland, H.J., Flokstra, J. *et al.* (1997) A seven-channel high-T_c SQUID-based heart scanner, *Measurement Science and technology* **8**, 927-931.
65. Scharen, M.J., Chase, D.R., Ho, A.M., O'Baid, A, *et al.* (1997) Filter subsystems for wireless communications, *IEEE Transactions on Applied Superconductivity* **7**, 3744-3749.
66. Roth, A. (1990) *Vacuum Technology*, Elsevier, Amsterdam.
67. Burger, J.F., Holland, H.J., Van Egmond, H..*et al.* (1999) Fast gas-gap heat switch for a microcooler, in R.G. Ross, Jr. (ed.), *Cryocoolers* **10**, Plenum Press, New York, pp. 565-574.

68. Della Porta, P. (1996) Gas problem and gettering in sealed-off vacuum devices, *Vacuum* **47**, 771-777.
69. Lehmann, G., Ramsden, J., Sochor, J. and Beeck, G. (1998) Cryopackaging for real world products, in P. Kittel (ed.), *Adv. in Cryog. Eng.* **43**, Plenum Press, New York, pp. 865-870.
70. Boffito, C., Ferrario, B., Rosai, L. and Doni, F. (1987) Gettering in cryogenic applications, *Journal of Vacuum Science Technology* **A5**, 3442-3445.
71. SAES Getters S.p.A., Via Gallarate 215, 20151 Milano, Italy (www.saesgettes.com).
72. Waterman, N.A. and Ashby M.F. (1997) *Materials selector*, Chapman&Hall, London.
73. Reed, R.P. and Clark, A.F. (1983) *Materials at Low Temperatures*, Am. Soc. of Metals.
74. Cryodata Inc., P.O.Box 558, Niwot, Colorado, USA.
75. Reed, R.P. and Golda, M. (1994) Cryogenic properties of unidirectional composites, *Cryogenics* **34**, 909-928.
76. Schutz, J.B. (1998) Properties of composite materials for cryogenic applications, *Cryogenics* **38**, 3-12.
77. Reed, R.P. and Golda, M. (1997) Cryogenic composite supports: a review of strap and strut properties, *Cryogenics* **37**, 233-250.
78. Wedenig, R., Knoblauch Ch. and Niinikoski, T.O. (1998) RF transmission line for cryogenic applications, *Cryogenics* **38**, 239-245.
79. Kubota, H. and Takeuchi,H. (1999) Low thermal leakage coaxial cable, *IEEE Transactions on Applied Superconductivity* **9**, 3117-3120.
80. Tighe, T.S., Akerling, G. and Smith, A.D. (1999) Cryogenic packaging for multi-GHz electronics, *IEEE Transactions on Applied Superconductivity* **9**, 3173-3176.
81. Ter Brake, H.J.M. and Flokstra, J. (1988) Computer aided cryostat design: recent developments R.G. Scurlock and C.A. Bailey (eds.), Proceedings 12th International Cryogenic Engineering Conference, Butterworth, UK, pp. 88-92.
82. Superinsulation NRC-2 Oxford Instruments Ltd, Eynsham, Oxford OX8 1TL, UK.
83. Barron, R. (1966) *Cryogenic Systems*, McGraw-Hill Book Company, New York.
84. Hohmann, R., Lucía, M.L., Soltner, H., *et al.* (1997) Integration of HTS SQUIDs with portable cooling devices for the detection of materials defects in non-destructive evaluation, in R.G. Ross, Jr. (ed.), *Cryocoolers* **9**, Plenum Press, New York, pp. 925-934.
85. Longsworth, R.C. (1980) Servicable refrigerator system for small superconducting devices, in J.E. Zimmerman, D.B. Sullivan and S.E. Mc Carthy (eds.), *Refrigeration for cryogenic sensors and electronic systems*, Proc. NBS Boulder, Oct. 6-7 1980, pp. 82-92.
86. Kaiser, G., Thürk, M. and Seidel, P. (1996) Closed cycle cryocoolers with a phase-change system for the cooling of high-T_c SQUIDs, in P. Kittel (ed.), *Advances in Cryogenic Engineering* **41**, Plenum Press, New York, pp. 1247-1253.
87. Bugby, D.C., Bettini, R.G., Stouffer, C.J..*et al.* (1997) Development of a 60 K thermal storage unit, in R.G. Ross, Jr. (ed.), *Cryocoolers* **9**, Plenum Press, New York, pp. 747-764.
88. Dunn, P.D. and Reay, D.A. (1982) *Heat Pipes*, Pergamon Press 3rd ed., Oxford.
89. Prenger, F.C., Hill, D.D..*et al.* (1997) Heat pipes for enhanced cooldown of cryogenic systems, in R.G. Ross, Jr. (ed.), *Cryocoolers* **9**, Plenum Press, New York, pp. 831-839.
90. Rosenfeld, J.H., Wolf, D.A. and Buchko, M.T. (1995) Emerging technologies for cryocooler interfaces, in R.G. Ross, Jr. (ed.), *Cryocoolers* **8**, Plenum Press, New York, pp. 743-753.
91. Van den Bosch, P.J. (1996) *Cryocooler-operated high-T_c SQUID system for magneto-cardiography in an unshielded environment*, Ph.D. thesis University of Twente, Enschede.
92. Van den Bosch, P.J., Holland, H.J., Ter Brake, H.J.M. and Rogalla, H. (1995) Closed-cycle gas flow sytem for cooling of high-Tc dc SQUID magnetometers, *Cryogenics* **35**, 109-116.
93. Troell, J. and Heiden, C. (1997) Low noise gas flow cryosystem for cooling high-T_c SQUID, in T. Haruyama, *et al.* (eds.), *Proceedings 16th International Cryogenic Engineering Conference*, Kitakyushu, Japan, 20-24 May 1996, Elsevier Science, New York, pp. 453-456.

94. Kotsubo, V. and Black, R.D. (1996) Apparatus for cooling NMR coils, U.S. patent 5,508,613.
95. Van den Bosch, P.J., Ter Brake, H.J.M., Holland, H.J., De Boer, H.A., Verberne, J.F.C. and Rogalla, H. (1997) Cryogenic design of a high-T_c SQUID-based heart scanner cooled by small Stirling cryocoolers, *Cryogenics* **37**, 139-151.
96. Sullivan, D.B., Zimmerman, J.E. and Ives, J.T. (1980) Operation of a practical SQUID gradiometer in a low-power Stirling cryocooler, in J.E. Zimmerman, D.B. Sullivan, S.E. Mc Carthy (eds.), *Proc. Conf. Refrigeration for Cryogenic Sensors and Electronic Systems*, NBS Boulder, 6-7 October 1980, pp. 186-194.
97. Klundt, K., Lienerth, C., Thummes, G., Steinmeyer, F., *et al.* (1998) Use of a pulse tube refrigerator for cooling a HTS-antenna for magnetic resonance imaging, in P. Kittel (ed.), *Advances in Cryogenic Engineering* **43**, Plenum Press, New York, pp. 2085-2092.
98. Gerster, J., Kaiser, G., Reißig, Thürk, M. and Seidel, P. (1998) Low noise cold head of a four-valve pulse tube refrigerator, in P. Kittel (ed.), *Advances in Cryogenic Engineering* **43**, Plenum Press, New York, pp. 2077-2084.
99. Wade, L.A. (1992) An overview of the development of sorption refrigeration, in R.W. Fast (ed.), *Advances in Cryogenic Engineering* **37**, Plenum Press, New York, pp. 1095-1106.
100. Zimmerman, J.E. (1980) Cryogenics for SQUIDs, in H.D. Hahlbohm and H. Lübbig (eds.) *Proceedings of the second International Conference on Superconducting Quantum Devices*, Springer-Verlag, Berlin, pp. 423-443.
101. Kazami, K., Takada, Y., Fujimoto, S., *et al.*. (1994) A Drung-type magnetometer mounted on a GM cryocooler, *Superconductor Science & Technology* **7**, 256-259
102. Kazami, K., Takada, Y., Yoshida, T., Ogata, H. and Kado, H. (1995) Cooling of SQUIDs using a Gifford-McMahon cryocooler containing magnetic regenerative material to measure biomagnetism, *Cryogenics* **35** 143-148.
103. Sata, K., Fujimoto, S., Fukui, N., Haraguchi, E., Kido, T., Nishiguchi, K. and Kang, Y.-M. (1997) A 61-channel SQUID system for MEG measurement cooled by a GM/JT cryocooler, *IEEE Transactions on Applied Superconductivity* **7**, 2526-2529.
104. Sata, K., Fujimoto, S., Fukui, N., Haraguchi, E., Kido, T., Nishiguchi, K. and Kang, Y-M. (1997) Development of SQUID based systems cooled by GM/JT cryocoolers, in T. Haruyama, *et al.* (eds.), *Proceedings 16th Int. Cryogenic Engineering Conference*, Kitakyushu, Japan, 20-24 May 1996, Elsevier Science, New York, pp. 1173-1176.
105. Wu, Y.A. (1994) Active vibration control algorithm for cryocooler, in P. Kittel (ed.), *Advances in Cryogenic Engineering* **39**, Plenum Press, New York, pp. 1271-1280.
106. Collins, S.A., Paduano, J.D. and Von Flotow, A.H. (1995) Active multi-axis vibration cancellation for split-Stirling cryocoolers, in R.G. Ross, Jr. (ed.), *Cryocoolers* **8**, Plenum Press, New York, pp. 437-448.
107. Mon, G.R., Smedley, G.T., Johnson, D.L. and Ross Jr., R.G. (1995) Vibration characteristics of Stirling cycle cryocoolers for space application, in R.G. Ross, Jr. (ed.), *Cryocoolers* **8**, Plenum Press, New York, pp. 197-208.
108. Kieffer, M., Wu, A. and Champion S. (1997) Summary and results of Hughes improved standard spacecraft cryocooler vibration suppression experiment, in R.G. Ross, Jr. (ed.), *Cryocoolers* **9**, Plenum Press, New York, pp. 705-710.
109. Rijpma, A.P., Verberne, J.F.C., Witbreuk, E.H.R., Bruins, P.C. and Ter Brake, H.J.M. (1998) Adaptive periodic disturbance cancellation in a set-up of two cryocoolers, *Journal of Sound and Vibration* **217**, 419-434.
110. Clappier, R.R. and Kline-Schoder (1994) Precision temperature control of Stirling cryocoolers, in P. Kittel (ed.), *Adv. in Cryog. Eng.* **39**, Plenum Press, New York, pp. 1177-1184.

ANALOG DEVICES IN MILITARY SYSTEMS

DANIEL E. OATES
MIT Lincoln Laboratory
244 Wood St.
Lexington MA 02420-9108
U.S.A.

Abstract

Superconducting analog signal processing devices have potential applications in many military systems. Several promising applications are examined here, including delay lines, tapped-delay-line filters, dispersive delay lines, phase shifters, and low-noise oscillators. For each application, the devices and how they fit into a system are discussed. The advantages of superconductors over conventional technology are examined.

1. Introduction

Why analog? What relevance does analog signal processing have in a world where signal processing increasingly means digital signal processing to most people? There are, of course several answers. First, the world is really analog, and most signals in communication and radar systems are received as analog signals. To be processed digitally, an analog signal must be converted to a digital form by some sort of analog-to-digital conversion (A/D) process that is limited by the speed of the hardware that does the conversion. The speed of the A/D limits the information bandwidth of the signal to be processed. To preserve the information content of an analog signal, A/D conversion must take place on a sampled version of the input signal, sampled at a rate that is twice the inverse bandwidth of the input. The A/D converter must then digitize the samples into however many bits n that are required for the dynamic range necessary in the system, where the dynamic range of a digital signal is given by 20 log 2^n. Thus, to digitize a signal of 2-GHz bandwidth, it must be sampled at a rate of 4 Gigasamples per second (GS/s). If eight bits are required for a dynamic range of 48 dB then the overall bit rate of the conversion process is 8x4=32 Gbits/s. Such processing

H. Weinstock and M. Nisenoff (eds.), Microwave Superconductivity, 305–335.

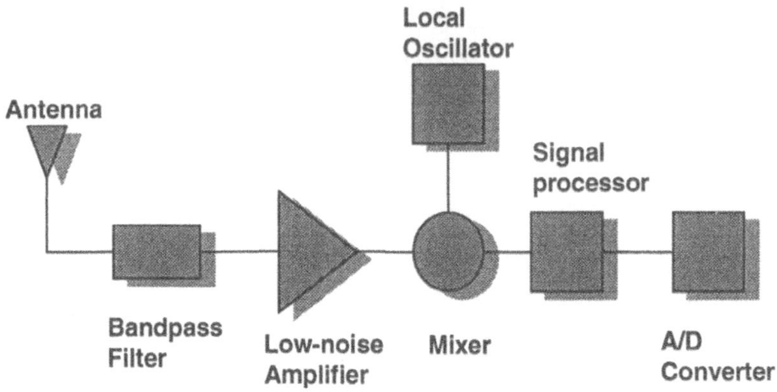

Figure 1. Typical receiver front end.

speeds are difficult to implement, expensive and consume significant power. For instance, the current state of the art in integrated circuit A/D devices is represented by the Maxim [1] MAX104 that operates at 1 GS/s (capable of digitizing a 500-MHz bandwidth analog signal), with 7.5-bit accuracy and 5-W power consumption. In many cases, it is more efficient in size, weight, and power for wideband signals to be first processed by analog signal processing subsystems. For further detail about the A/D conversion process and digital signal processing, any one of a large number of books may be consulted. See for instance Ref. 2. A/D converters also are an important potential application of superconducting electronics, but this topic is beyond the scope of this chapter.

To illustrate the importance of analog signals, Fig. 1 shows a receiver front end typical of a radar, communication, or other wireless system. The microwave, radio frequency (rf) signals are received by the antenna. A bandpass filter removes out-of-band interference before the signals are amplified in the low-noise amplifier. The filter serves to protect the low-noise amplifier from overloading due to large out-of-band signals. Many systems, however, do not employ a narrowband filter because losses in such a filter degrade the system noise figure and sensitivity, which leads to a shorter range of operation in radar or communication systems. This filter loss is the reason for the large interest in superconducting front-end filters that can perform the filter function with losses substantially less than 1 dB. After amplification, the signals are either processed by a wideband rf system, such as the chirp transform system discussed below, or are mixed with a local oscillator to produce a lower intermediate frequency (IF). At this point, further signal processing may occur. However, the information bandwidth of the signal is limited to the bandwidth of the IF system. This is why very wideband signals are processed at the rf. Once the signals have been down converted in frequency, either an analog processing subsystem acts on the signal to extract information, or the IF signal undergoes A/D conversion which is followed by a digital signal-processing system. The advantage of digital signal processing is that the operations are defined by software rather than hardware. The software algorithms are easier to implement and test than hardware signal processing, and they are more flexible and can be updated in the future when new problems or needs are encountered. Thus, for flexibility and accuracy, the A/D conversion takes place as close to the front end as possible. Digital processing can be more accurate than analog, but accuracy cannot be arbitrarily large since noise is always present in real systems. An A/D converter, in addition, has a finite signal-to-noise ratio and can introduce harmonic and intermodulation distortion. These are important limitations of A/D converters, and in applications requiring large dynamic range, the limitations can be sufficiently great that analog signal processing is the better alternative.

Another reason for using analog devices is processing power. It is often the case that analog processing can accomplish the equivalent of a digital processor using less power and/or in a more compact package. As will be discussed below, the spectrum analysis performed by an all-analog superconducting chirp transform system accomplishes the equivalent of $>200\text{x}10^9$ digital floating-point operations per second, (>200 Gflops), while consuming 20 W of power. A digital system with equivalent processing power would consume an estimated 4 kW.

The approach in this chapter is to give several examples of passive, analog, signal-processing devices and applications of these devices. We discuss the principles of operation, as well as the advantages of superconductors when they are used in these kinds of systems. One illustrative implementation is discussed for each of the applications. In Sec. 2 we discuss delay lines, in Sec. 3 phase shifters, and in Sec. 4 low-noise oscillators. The intent is to be pedagogical. We do not intend this chapter to be an exhaustive review of the literature of analog superconducting devices. However, the references given should be a good start to investigate the literature of any particular subject.

2. Delay Lines

2.1. GENERAL CONSIDERATIONS

One of the simplest analog signal processing devices is the delay line, which, as the name suggests, simply adds a time delay to a signal applied to the input. The output of the ideal delay line is exactly the same as the input, and it appears after some amount of time delay. Mathematically,

$$A(t) = B(t-\tau),$$

where A is the output signal, B is the input, and τ is the delay. In the frequency domain this can be written as

$$H(f) = A(f)G(f)e^{i\phi(f)}, \tag{1}$$

where H is the output signal and G the input, $A(f)$ is the transfer function (which in the ideal case is unity), and $\phi(f)$ is the phase shift, which for an ideal delay line is given by

$$\phi(f) = 2\pi f\tau, \tag{2}$$

where τ is the delay. Notice that the idealized simple delay line expresses its presence only in the phase and not the amplitude of the transmitted signal. The phase shift in the delay line is a linear function of the frequency. As will be seen later, real delay

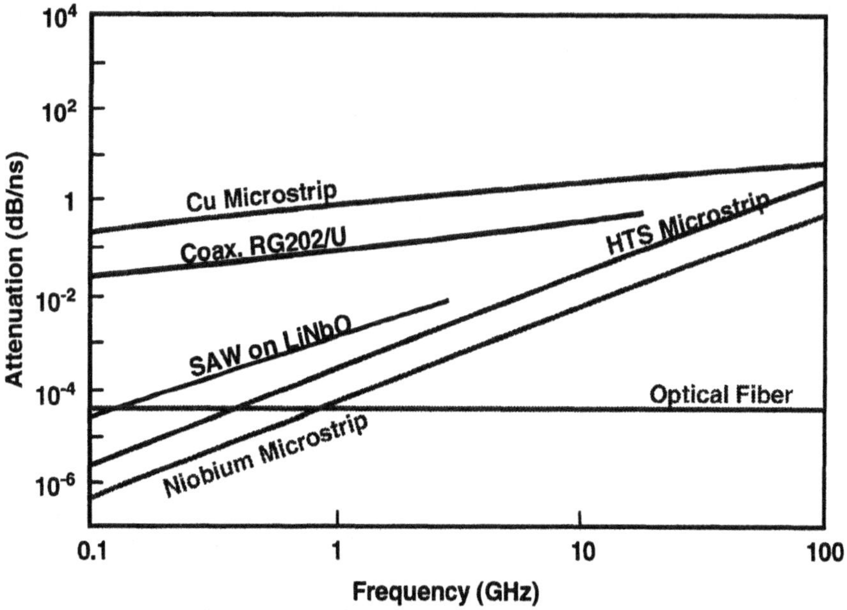

Figure 2. Typical values of attenuation vs. delay for the various technologies shown.

lines can affect the amplitude as well as the phase, so it is customary to speak of separate amplitude and phase responses of these devices. Real delay lines always introduce some attenuation and deviation from the linear phase shift.

While delay lines have several applications directly as simple delays to synchronize signal arrival times, for instance, they also are the basis for more sophisticated devices such as chirp filters and transversal filters. As discussed below, the advantage of superconductors in delay lines is that a long delay can be obtained in a compact geometry, yet with low propagation losses because of the low surface resistance of superconductors in comparison with ordinary metals. The figure of merit for signal processing devices is the time-bandwidth product that for superconducting delay lines can be quite large, as each of the quantities, the time delay and the bandwidth, can be much larger than in conventional technologies.

Other technologies used for delay lines are acoustic, using either bulk or surface ultrasonic waves, and optical fibers, where the signal of interest is modulated onto an optical carrier which is then transmitted over the fiber and demodulated at the output to regain the original signal. Of course, coaxial cable and other normal metal techniques also are used for providing delay. In most cases, the limitations are the propagation losses. Figure 2 shows typical values of insertion loss vs. delay for various technologies. Here the advantage of HTS delay lines over copper coaxial cable is clear. At low frequencies, surface acoustic wave (SAW) delay lines on $LiNbO_3$ have comparable performance, but manufacturing high-frequency, wide-bandwidth SAW devices is extremely difficult, and in such applications superconductors are very attractive. Optical fiber has low loss, but signals must be converted to optical and then

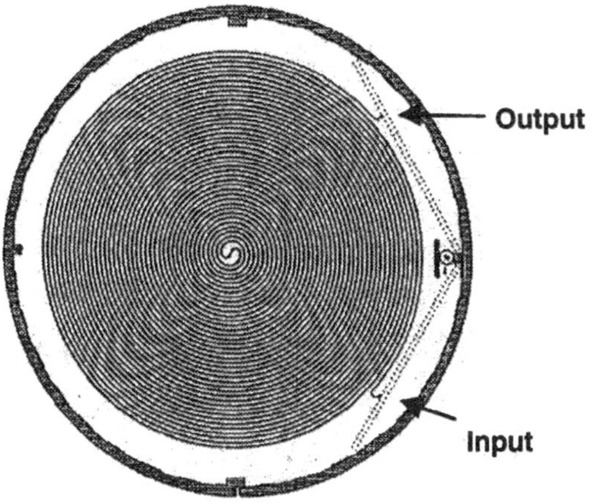

Figure 3. Picture of a spiral delay line.[4] This is a 44-ns delay line with YBCO on LAO. (The moiré patterns are not real, but a consequence of the digitization of the photograph.)

back to electrical after the delay. Additionally, optical fibers are difficult to use in tapped delay lines as discussed below.

For ease of fabrication and integration with other components, superconducting delay lines are usually fabricated using planar transmission-line circuit geometries such as stripline, microstrip, or coplanar waveguide. However, coaxial-cable implementations have found applications previously [3]. As in all microwave circuits, the transmission line must be designed with a reasonable value of impedance that is not too far from 50 Ω so that wideband impedance matching in and out of the line is relatively simple. In order to obtain the maximum delay in the minimum amount of space, the transmission line is most often made in a meander or spiral to optimize the available delay in a given substrate size. Figure 3 is a photograph [4] of a spiral delay line that provides 44 ns of delay in stripline. This shows a 2-inch-diameter $LaAlO_3$ wafer with YBCO deposited on it. The line is 100-μm wide.

The turns of the meander or spiral, however, must not couple adjacent parts of the line, or uneven frequency and phase response will result, degrading the performance of the delay line. To prevent adjacent-line coupling, the lines must be maintained at a minimum separation, which is related to substrate thickness. Usually, depending on geometry, the required spacing is some multiple of the substrate thickness. This requirement for separation is in competition with obtaining the maximum time delay. Line separation is unused real estate. More delay, however, can be obtained on a given size of substrate by using thinner substrates and narrower lines to satisfy the impedance level requirements [5]. The narrower lines in thinner substrates can be packed more tightly together, thus giving more delay. The achievable

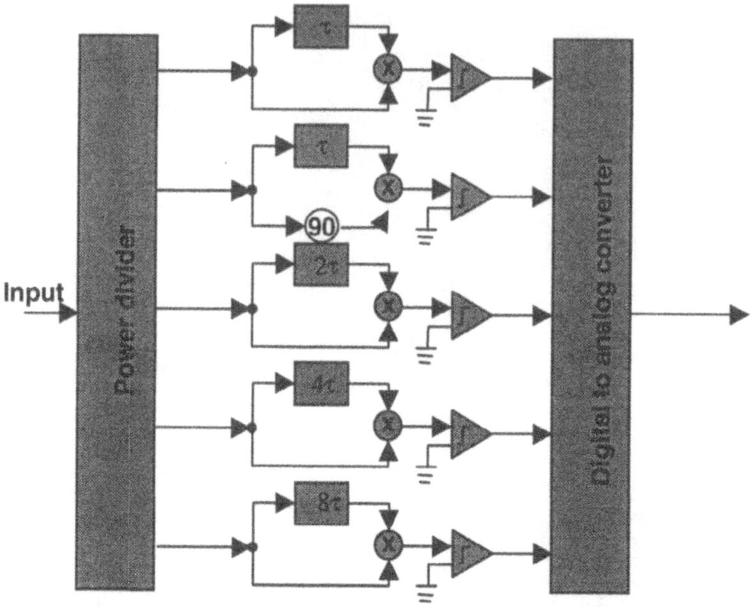

Figure 4. Schematic diagram of a digital instantaneous-frequency measurement system.

delay then scales approximately with $1/t$, where t is the substrate thickness. The required spacing can be expressed, as some multiple of substrate thicknesses with the exact value dependent on the amount of isolation between lines that is required by the system application. The required isolation is related to the system-specified frequency- and phase-response flatness. For instance, in stripline geometry to obtain 60 dB of line-to-line isolation, the spacing must be at least 10 substrate thicknesses. There is no analytical solution for this spacing; each geometry must be evaluated separately, usually numerically. However, the substrate thickness cannot become too small because of the insertion loss. It is instructive to consider a parallel-plate transmission line [6], while the delay scales with $1/t$, the attenuation constant also is proportional to $1/t$, so that the delay cannot be increased arbitrarily without encountering unacceptable losses.

A properly designed transmission-line geometry will not limit the bandwidth of the delay line, which is usually limited by other factors such as the connectors in and out of the package in which the delay line is mounted.

Dispersion, the dependence of propagation delay on frequency due to dependence of propagation velocity on frequency, can be one of the limitations of delay-line performance. Dispersion, for instance, will distort the shape of a pulse as it propagates through the delay line and cause a departure from the linear dependence of phase on frequency in Eq. (2). For delay-line applications, superconductors have the advantage of no material dispersion, while normal metals introduce dispersion through the frequency dependence of the skin depth. In superconductors the penetration depth is independent of frequency. Of course, the geometry chosen to implement the delay line can introduce mode dispersion, as is the case for microstrip and coplanar

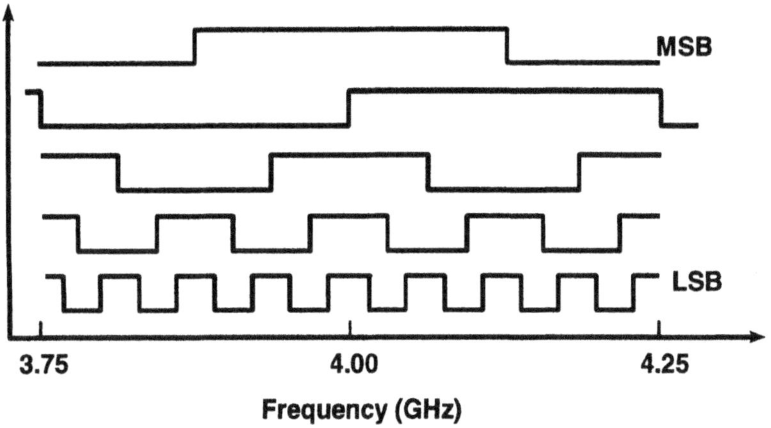

Figure 5. Output of an ideal DIFM system.

waveguide. Stripline, being a geometry that supports a pure TEM wave, exhibits no mode dispersion.

2.2. INSTANTANEOUS FREQUENCY METER

One use for simple delay lines is an instantaneous frequency-measuring (IFM) system [4,7]. Such systems are widely used in electronic warfare and electronic intelligence systems for determining the frequency of unknown signals over a broad band. The system employs multiple delay lines and functions by comparing the phase of an unknown signal with time-delayed versions of the same signal. The performance depends on the length of the delay and the accuracy of the phase shift of the signal transmitted through the lines, because deviations from the linear phase dependence on frequency, shown in Eq. (2), will degrade performance. Thus, superconducting delay lines are good candidates for an IFM system because long delays can be obtained with minimal loss and very low dispersion, as discussed above.

A schematic view of a 5-bit, digital IFM (DIFM) system is shown in Fig. 4, as reported in Ref. 4. An incoming signal is split five ways and each of the five signals is split again to be applied to a phase discriminator subsystem that works by mixing the undelayed signal with a delayed version. The output of the mixer produces sum and difference frequencies in the usual way, and is proportional to

$$\cos \omega t \times \cos(\omega t - \omega \tau_n) = \cos(2\omega - \omega \tau_n) + \cos(\omega \tau_n),$$

where τ_n is the delay of the nth delay line. This is low-pass filtered to eliminate the high-frequency sum signal, and then used as the input to a comparator whose output becomes one of the bits of the output. The five channels have delays that increase by powers of two to produce the binary-encoded digital output with the exception of the most significant bits which, for ease of decoding, are the same length, but 90° out of phase. This configuration produces a Gray code. The output of an ideal DIFM is shown for the design parameters of Ref. 4 in Fig. 5. The resolution of the frequency determination is related to the delay of the longest delay line, and is $\pm(8\tau_{max})^{-1}$.

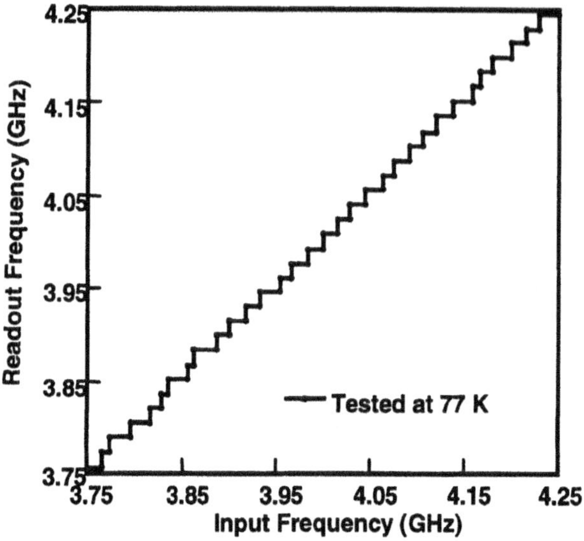

Figure 6. Results of DIFM system as reported in Ref. 4.

The bandwidth needed to be processed determines the shortest delay. The processing time required to produce a valid output is the time delay of the least significant bit, which is the delay of the longest delay line. This processing time also determines the shortest pulse that can be measured in the system. The performance of the system depends on the precision of the delays and on the linearity of the phase in each of the separate delay lines. A more complete discussion of the errors is given in Ref. 4. In the presence of multiple simultaneous signals, the ideal IFM receiver will provide accurate data on the strongest observed signal and suppress other signals that may be present simultaneously. This is both the advantage of the IFM and its limitation. It cannot identify simultaneous frequencies and cannot determine signal strength.

Figure 6 shows results for a DIFM system [4] that was built for the High Temperature Superconductivity Space Experiment II (HTSSE II), a satellite payload to test high-T_C devices [8]. The longest delay for this system is 16 ns. The delay lines were implemented with YBCO on LAO. Stripline geometry was used to minimize dispersion, since stripline supports a pure TEM mode. A 45-Ω-impedance line was used. Stripline also allows dense packaging with minimal radiation loss and cross talk between devices. The authors quote an average error of 3.1 MHz from the design values for the transitions between output states.

2.3. TAPPED DELAY LINES

A filter of arbitrary response can be fabricated from a delay line by periodically tapping a portion of the signal, weighting the amplitude and phase of the tapped signals, and summing them. This is normally called a transversal filter. Mathematically, this operation is expressed as

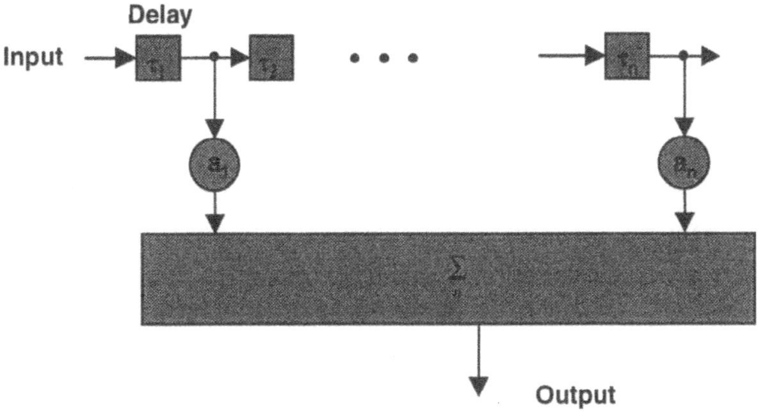

Figure 7. Schematic diagram of the operation of a tapped delay line transversal filter. Time-delayed samples of the input signal are weighted in phase and amplitude, and then coherently summed to provide the output of the filter.

$$A(t) = \sum_{n=1}^{N} a_n B(t - \tau_n) e^{j\omega(t-\tau_n)} ,$$

where n is the nth tap, a_n is the complex weight of the nth tap providing both amplitude and phase weighting, and τ_n is the delay of the nth tap. Figure 7 shows schematically the operation of a transversal filter. The filter is synthesized by the impulse response, which is given by the tap weights and tap delays. Remembering that the frequency response of a device is the Fourier transform of the impulse response, one advantage of a transversal filter is its relatively straightforward design procedure. One designs the impulse response and implements it in the a_n and τ_n accordingly. Another advantage of the transversal filter is that it allows separated amplitude and phase weighting and can provide very precise control of the phase response of the filter. Other implementations of filters do not normally allow such precise control.

While various implementations of transversal filters have been reported, the most useful example is the so-called chirp filter. In this implementation, the taps are placed such that the delays τ_n increase linearly with n. The amplitude of the taps can be flat or weighted as desired. The impulse response of such a filter is a "chirp, " a tone whose frequency changes linearly with time. This is shown by the downchirp impulse response in Fig. 8. Upchirps also are possible with the opposite arrangement of the taps. The amplitude weights of the taps can vary with position to provide a desired shaping of the frequency response. The phase response of the filter is a quadratic function of the frequency. That is, $\phi(f) \sim f^2$.

An application of chirp filters is in pulse-compression radars. Most radars operate in a peak-power-limited mode where the output power is limited by the transmitter. In order to illuminate the targets with more energy, and thus obtain a larger return signal, the radar must emit a signal of finite length in time. The long pulse, if it is a single frequency, reduces the range resolution of the radar. To solve the

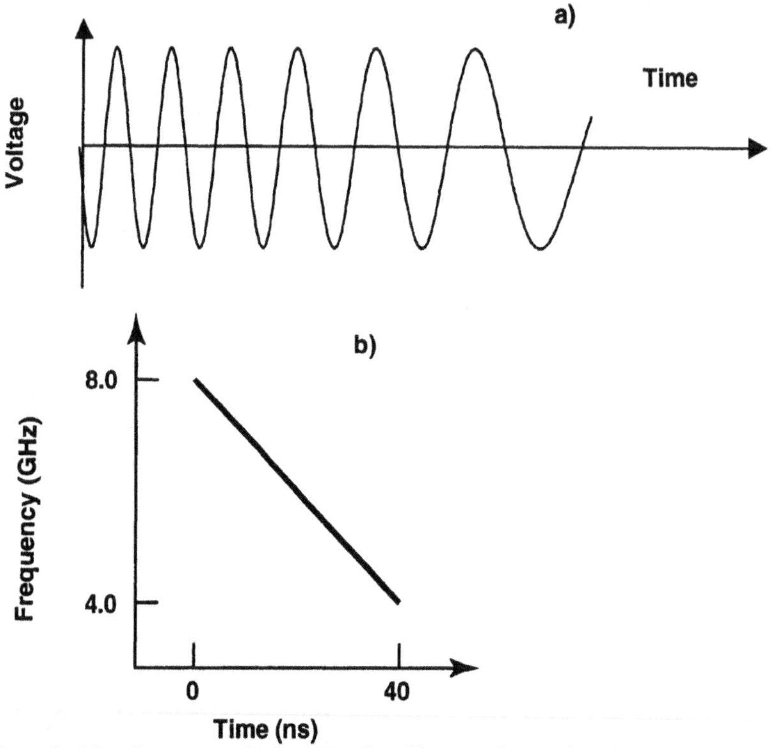

Figure. 8. a) Impulse response of a downchirp filter. The output is a quasi-continuous-wave tone whose frequency decreases linearly with time. Upchirp implementations also are possible. b) Plot of instantaneous frequency vs. time of impulse response of a).

problem of range resolution, the output pulse is chirped, that is, the system emits a pulse whose frequency changes in a linear fashion in time. This is called linear FM. If the return signal is then compressed in a chirp filter of the opposite sense as the output, that is, if the emitted signal is an upchirp , then the receive filter must be a downchirp. Then the output of the filter is a short time pulse, and the range resolution of the radar is enhanced.

While pulse-compression radars are in wide use, superconducting technology is not yet a good candidate for application in this area because the bandwidth requirements for such radars are generally considerably less than 1 GHz. Superconducting delay lines as discussed above show the largest advantages over conventional technology when extremely wide bandwidth is required. In the following, an implementation and application of superconducting chirp filters is discussed, showing a great advantage over conventional technology.

2.4. CHIRP TRANSFORM SYSTEM

Figure 9 shows schematically the implementation of a chirp filter that has been reported by workers at MIT Lincoln Laboratory [5]. The delay-line structure is

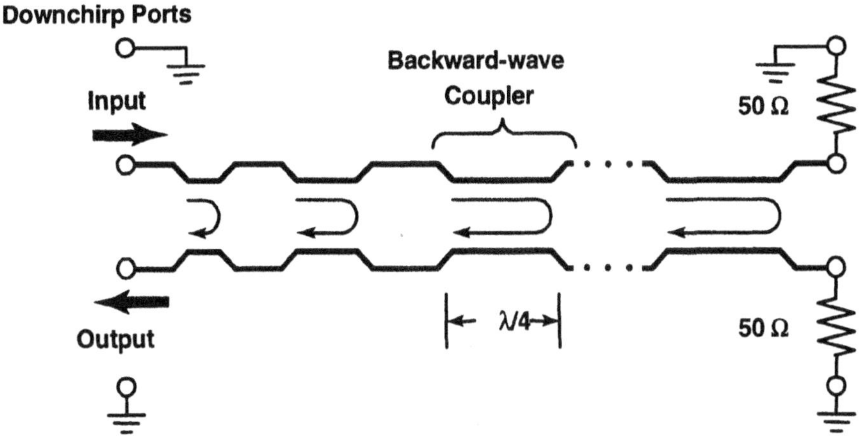

Figure 9. Structure and operation of a proximity-tapped chirp filter.

stripline geometry. The filter consists of two delay lines, one for the input signal and one that provides the summing network for the output. The tapping of the input line is implemented with λ/4 backward-wave couplers to the output line. (A backward-wave coupler is a standard microwave circuit component, and can be found in most microwave engineering textbooks – see for example Ref. 9.) Briefly, it is a four-port device, as shown in Fig. 10. Port 1 is the input, and energy is coupled to port 3 depending on the strength of the coupling between the λ/4 sections of line determined by the spacing. Port 2 is the through connection, where the input signal energy that remains continues down the line, and no energy is coupled to port 4, which is simply terminated in the characteristic impedance. Each coupler has a peak in its response at the frequency at which the coupler is λ/4 long, where λ is the electromagnetic wavelength of the signal on the transmission line. By making the reciprocal of the length of each coupler a linear function of the position along the line, the peak frequency response of the backward-wave couplers varies as a linear function of the delay. This then produces the desired linear chirp shown in Fig. 9. Figure 11 is a photograph of a chirp filter reported by Lincoln Laboratory [5]. In this implementation the two delay lines are laid out in a spiral geometry to optimize delay. This is a two-port device, and the unused ends of the input and output transmission lines are terminated as in Fig. 9 to prevent unwanted reflections. The performance parameters of this device are given in the figure caption.

The chirp filters of this design have been used to implement a spectrum-analyzer chirp-transform system that is based on the chirp-transform algorithm given by the following. The Fourier transform of a signal *h(t)* is given by

$$H(\omega) = \int_{-\infty}^{\infty} h(\tau) e^{-i\omega\tau} d\tau . \tag{3}$$

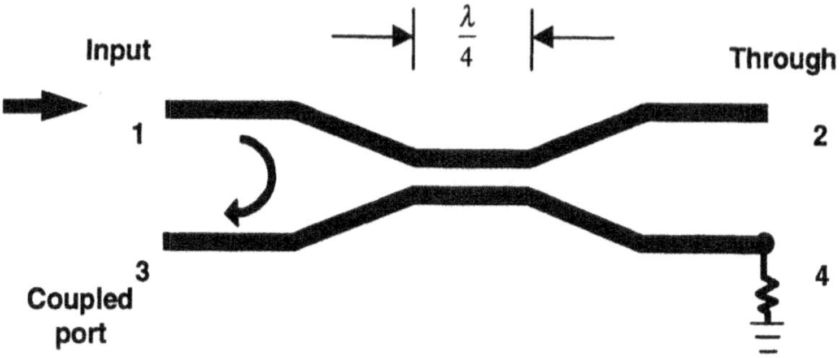

Figure 10. Backward-wave coupler.

This is the usual definition of a Fourier transform. Then let $\omega=\mu\tau$, as is the case in a chirp filter where μ is the chirp slope. The following substitution for the exponent in the integral is made:

$$-i\omega\tau = -i\mu\tau^2 = i\mu\frac{(t-\tau)^2}{2} - i\mu\frac{t^2}{2} - i\mu\frac{\tau^2}{2}, \tag{4}$$

and substituted in Eq. (3). Then the Fourier transform becomes

$$H(\mu t) = \exp(-i\mu\frac{t^2}{2})\int_{-\infty}^{\infty}\left[h(\tau)\exp(-i\mu\frac{\tau^2}{2})\right]\exp\left[i\mu\frac{(t-\tau)^2}{2}\right]d\tau\,. \tag{5}$$

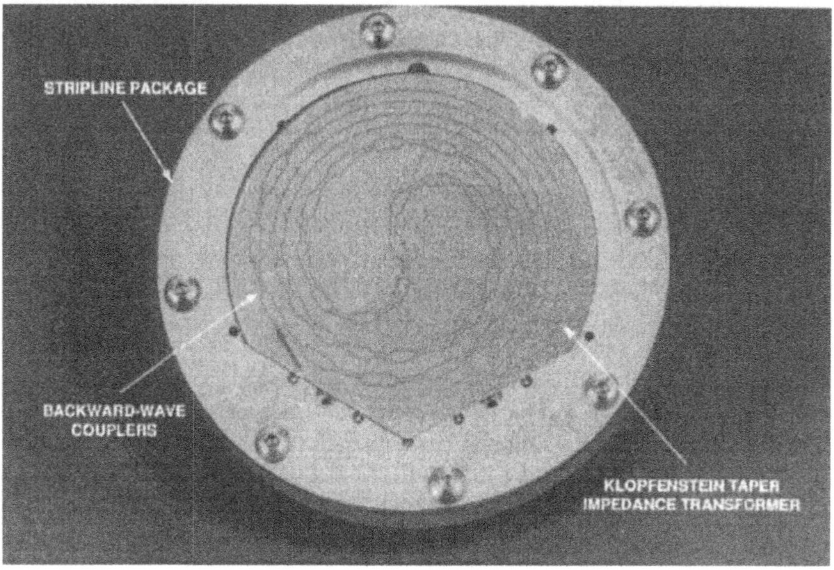

Figure 11. Photograph of a chirp filter from Lincoln Laboratory. This example has a 12-ns chirp length. The relatively short chirp allows the various backward-wave couplers to be visible in the photograph. This device uses YBCO on LAO.

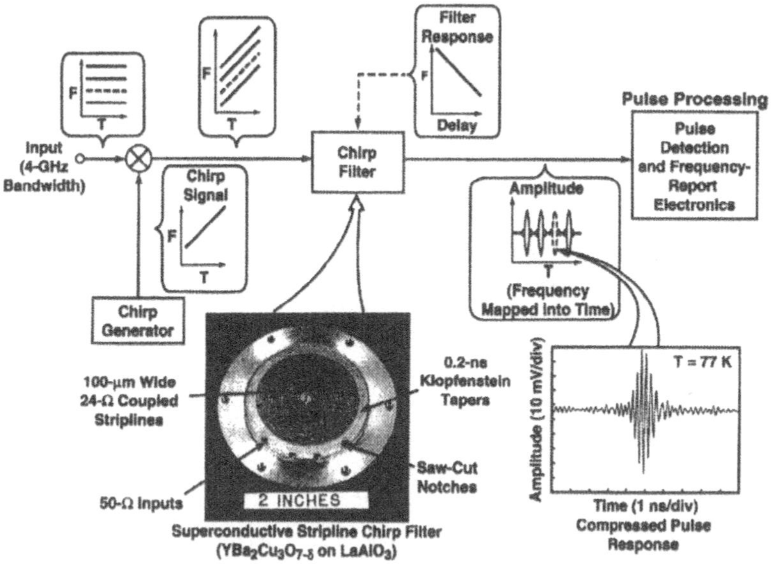

Figure 12. System diagram of a compressive receiver.

In Eq. (5) the first term in square brackets under the integral represents the multiplication of the signal $h(\tau)$ by a chirp, $\exp(i\mu\tau^2/2)$, which varies linearly in frequency over time and therefore has a quadratic phase vs. time. The integration over τ in Eq. (5) is the convolution of the signal multiplied by the chirp, with a chirp of the opposite sense from that which first multiplied it. Convolution in time is the same as passing the signal through the filter with this impulse response. Further, Eq. (5) completes the Fourier transform with a multiplication of the chirp-filtered signal by another chirp with the same chirp slope as that of the original multiplication. Equation (5) represents the multiplication-convolution-multiplication (M-C-M) chirp transform and yields the complex Fourier transform of $h(\tau)$, complete with amplitude and phase, all mapped into the time domain in the output of the chirp transform system. Other variations of the chirp transform are possible, but will not be discussed here [5]. If the phase information on the Fourier transform is not needed, as is the case when one is interested in spectrum analysis, then the last chirp multiplication step is omitted. This yields the frequency and amplitude of the input signal. This is called the multiply-convolve (M-C) transform, and is the basis of a compressive receiver system reported by Lincoln Laboratory. Figure 12 shows the reported system that has been included on the HTSSE II [8] satellite. The chirp filters used in that system are fabricated from YBCO deposited on lanthanum aluminate (LAO). The chirp bandwidth is 3 GHz, and the delay length is 12 ns. The performance parameters of such a system are given in Table 1. The analysis bandwidth is given by the bandwidth of the chirp filter, and the frequency resolution is determined by the total delay of the chirp filter. As can be seen in Fig. 12, the chirp that multiplies the input signal is generated by conventional

Table 1

	Reported in Ref. 5	Projected
Analysis bandwidth	3.0 GHz	4 GHz
Frequency resolution	110 MHz	4 MHz
Frequency bins	28	120
Analysis time	128 ns	40 ns

electronics. Although it is possible to generate the chirp by impulsing a chirp filter of the appropriate chirp slope, it turns out to be simpler to generate the chirp conventionally using a voltage-controlled oscillator.

This system is intended for use in the front end of an airborne electronic-warfare warning system to identify signals that might illuminate the aircraft and might constitute a threat. The advantage of the superconducting chirp filters in this application is the wide bandwidth that is possible and the very high duty cycle that the analog chirp transform can provide. The described system has a 50% duty cycle, so 100% time coverage is possible with two parallel systems. Such an arrangement could provide detection of radar signals that are operated at very low duty cycle in order to avoid detection.

As discussed above the frequency resolution of a chirp transform system is a function of the delay. Longer delays give better frequency resolution in direct proportion. Essentially $\Delta f \sim 1/\tau$, where Δf is the frequency resolution and τ is the delay. Therefore, to obtain higher resolution, longer delays are desirable. As discussed above, the delay is related to the substrate size and thickness, given that a fixed amount of line-to-line isolation is needed. It has been pointed out that 12 ns is the limit of available delay on standard 2-inch-diameter, 0.020-inch-thick LAO wafers. However, a 40-ns chirp filter has been demonstrated10 using LAO thinned to 0.005 inch and bonded to a carrier substrate that provides mechanical strength for the necessary fabrication steps. Such a system has the projected performance shown in Table 1. This performance would be better than competing technologies for the signal-intercept application.

3. Phase Shifters

Electrically adjustable phase shifters are a potentially important application of superconductors. Phase shifters have many applications, the most important of which is in phased-array antennas.Figure 13 shows how a phase shifter is used in a phased-array antenna to provide the capability to electronically steer the beam of the antenna. Under system control, a specific amount of phase shift is adjusted into each phase shifter. The collection of individual phase shifts determines the direction in which the beam of the antenna points. If, for instance, the phase shifts are all equal, the antenna beam points straight ahead. If there is a 360° phase difference along the length, then the angle of the beam is

$$\theta = \tan^{-1} \lambda/D,$$

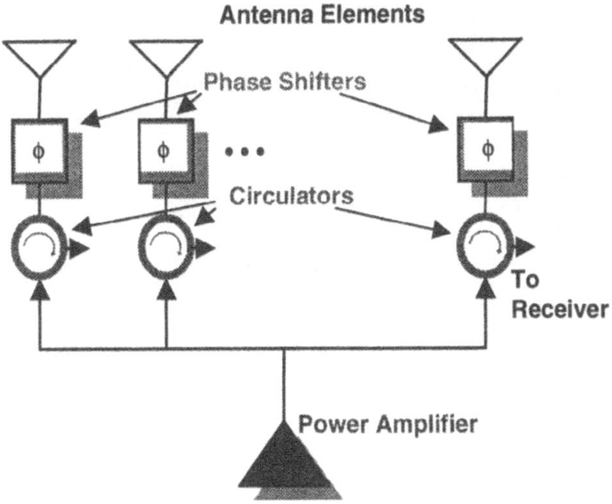

Figure 13. Schematic diagram of a phased-array antenna showing the application of a phase shifter.

where λ is the free-space wavelength of the radiation, and *D* is the antenna aperture. Obviously, greater or lesser amounts of phase shift can continuously sweep the beam over a large angle. Electronic control allows far more rapid switching of the beam in random directions than is possible by mechanical means. The phase-shift adjustments also can alter the radiation pattern of the antenna so that a null in the spatial response can be arranged to reduce the effects of an interfering signal.

A phase shifter is required at each element of the antenna shown. Large high-resolution phased-array antennas may have as many as 10^4 elements, so the phase shifters must be small and, for many applications, consume low power. It is important that the phase shifter have extremely low loss, because in receive mode any loss in the phase shifter will impact the system noise figure and degrade the system sensitivity. Any losses in transmit mode added by the phase shifter must be compensated with extra amplification which adds size, weight, expense, and power consumption. For system insertion, loss in the phase shifter should be less than 0.2 dB. It is the low-loss and compact-size requirements that make superconductors a good candidate for phase-shifter applications. The figure of merit that is usually applied to phase shifters is degrees of phase shift per dB of insertion loss, deg/dB.

It also is required that the phase shifter be rapidly tunable. An antenna may be required to switch the pointing direction in times of 1 ms or less. Thus, the phase shifter also must be capable of responding in this time, so only electronically adjustable phase shift can be considered.

The phase shift is most simply implemented by changing the propagation constant in a transmission line, and most of the examples of superconducting phase shifters use this principle. The propagation velocity of a transmission line is

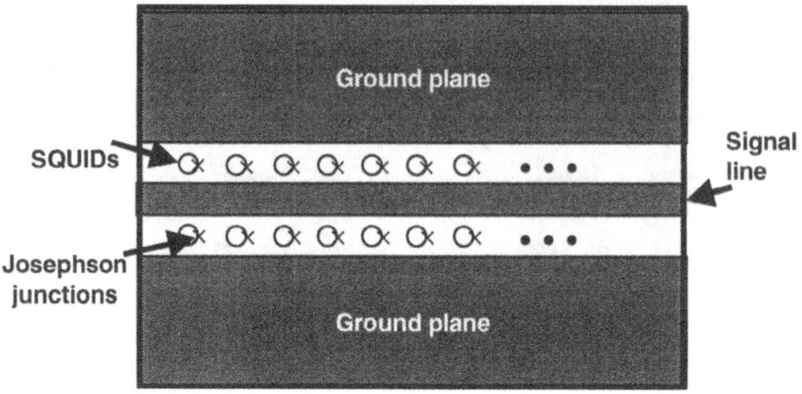

Figure 14. Schematic view of distributed Josephson-junction phase shifter. Coplanar waveguide was used for the implementation reported in Ref. 11.

$$v = \sqrt{\frac{1}{LC}}, \tag{6}$$

where L is the inductance per unit length, and C is the capacitance per unit length. The phase shift is

$$\Delta\phi = \omega\tau = \omega l / v,$$

where l is the length, and τ is the delay, equal to l/v. Thus the phase shift can be varied by changing v through either the inductance or capacitance of the transmission line.

Several designs of superconducting phase shifters have been reported; here three of the most promising designs will be discussed, those employing Josephson

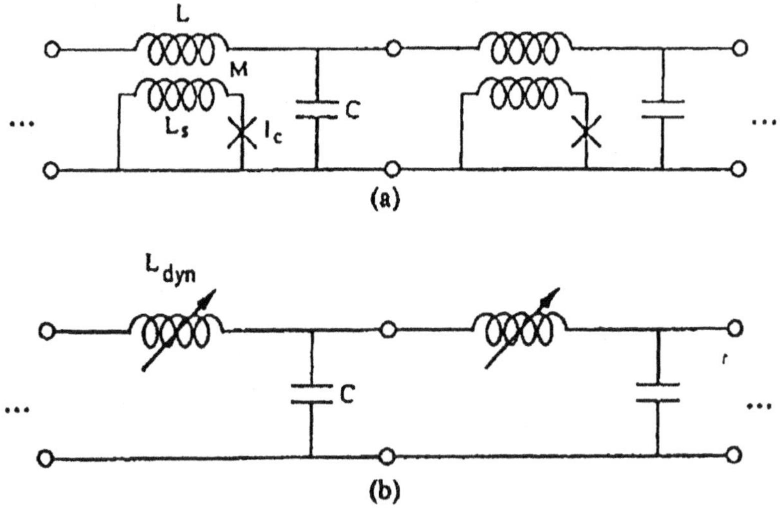

Figure 15. Equivalent circuit of the distributed Josephson-junction phase shifter.

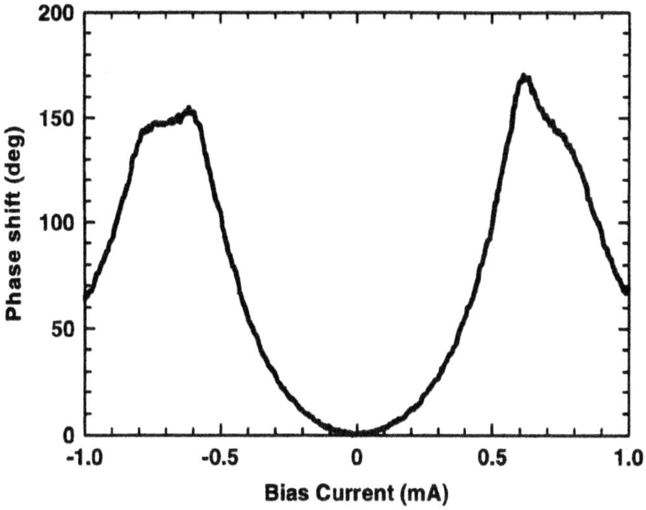

Figure 16. Results of phase shift vs. bias current for high-T_c phase shifter at 9 GHz and 68 K.

junctions [11], ferroelectric materials [12], and ferrite materials [13,14]. Other methods, such as alteration of the kinetic inductance by large currents [15] or by heating [16], also have been reported, but in general do not give the performance of the three methods discussed below.

3.1. JOSEPHSON-JUNCTION-BASED PHASE SHIFTERS

The distributed Josephson-junction phase shifter consists of a length of superconducting planar transmission line that is magnetically coupled to an array of single-junction SQUIDs along its length. Figure 14 shows a schematic view of the phase shifter (as reported by the TRW group [11]) that uses coplanar waveguide (CPW) geometry. The dynamic (rf) inductance of a SQUID depends on the amount of magnetic flux in the geometric loop. As discussed above, the propagation velocity is determined by the capacitance per unit length, which is fixed, and the inductance per unit length that is affected by the SQUIDs. The velocity can be electronically altered by changing the amount of magnetic flux in the SQUID loops, controlled by a dc current in the center conductor of the CPW that is superimposed on the rf current. By varying the dc current, the inductance is varied, altering the propagation velocity, and thus the phase shift through the device. Figure 15 shows the equivalent circuit for a transmission line and indicates how the inductance of the SQUID influences the transmission line. Durand *et al.* [11] have calculated the dynamic inductance L_{dyn} per cell of the transmission line, including the SQUID contribution, as a function of the dc current, and obtain

$$L_{dyn} = L\left[1 - K^2 \frac{\beta\cos\theta}{1+\beta\cos\theta}\right],$$

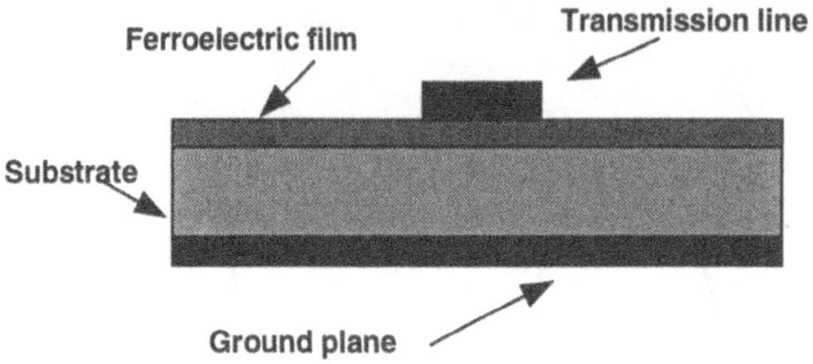

Figure 17. Cross section of a simple ferroelectrically tuned transmission line.

where L is the inductance per cell of the transmission line without the SQUID, β is the SQUID parameter $\beta = 2\pi L_s I_c/\Phi_0$, where L_s is the geometric inductance of the SQUID loop, I_c is the critical current, Φ_0 is the flux quantum, and $K^2 = M^2/LL_s$ where M is the mutual inductance between the transmission line and one SQUID.

In this geometry, the SQUIDs are loosely coupled to the transmission line, so losses can be kept low. It also is tolerant of defects in the junctions, since there are many SQUIDS (>10^4) [17]. A few nonfunctioning devices will not be noticed, and there are no stringent requirements on junction uniformity. The results for a YBCO phase shifter are shown in Fig. 16. For details of the design and performance, see Ref. 17. The phase shift shown is approximately 180 degrees. The authors do not state a loss figure, but in earlier niobium work [11], the TRW group reported losses less than 3 dB for a similar geometry. The losses were not intrinsic, but resulted from the test fixture and impedance mismatches. Presumably, a figure of merit of 200 is a lower limit of what can be done with this technology.

3.2. FERROELECTRIC-FILM-BASED PHASE SHIFTERS

The Josephson-junction-based phase shifter achieves adjustable phase shift from the variation of the propagation velocity from changes in the inductance per unit length of

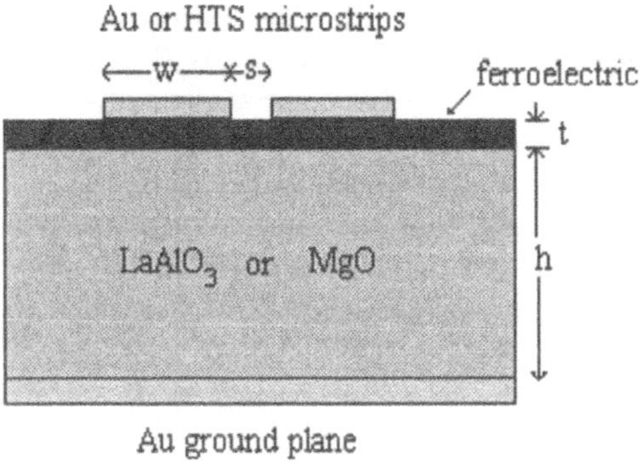

Figure 18. Schematic view of phase shifter reported in Ref. 12.

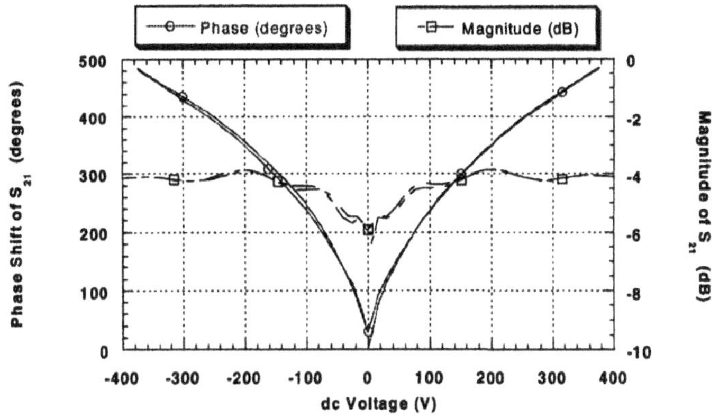

Figure 19. Results of phase shifter shown in Fig. 18 as reported in Ref. 12

a transmission line. Equation (6) shows that a variable capacitance also can be used to vary the propagation velocity. The voltage dependence of the permittivity of ferroelectric materials $\varepsilon(V)$ can be used to produce the variable capacitance. This use of ferroelectrics is similar to the applications in tunable filters that have already been discussed in another contribution to this volume [18]. A schematic view of a simple structure that illustrates phase-shifter geometry is shown in Fig. 17. By applying a dc voltage between the transmission line and the ground plane, the ε of the ferroelectric layer is changed, the capacitance of the microstrip line is changed, and the velocity of propagation will be altered. If the transmission line is superconducting, then the usual benefits of a superconductor are gained, namely, low loss in a compact geometry.

A strontium titanate thin film is an excellent candidate for the ferroelectric film. Properly stated, $SrTiO_3$ (STO) is a nonlinear dielectric, not a ferroelectric, but we will follow the usual terminology and refer to it as a ferroelectric, as is done in most of the literature. STO is compatible with YBCO thin films, as each can be grown

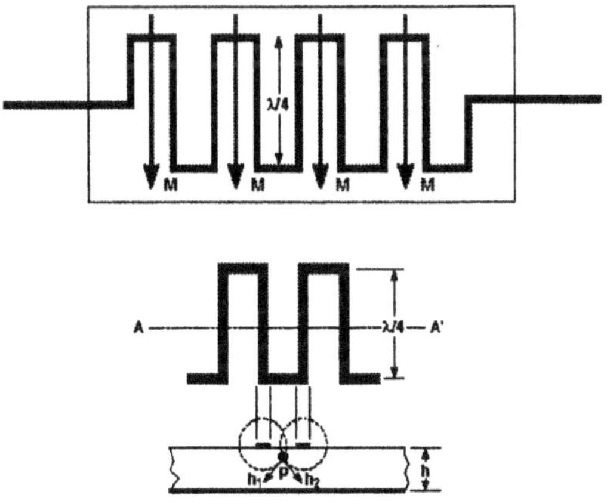

Figure 20. Meanderline structure showing the generation of circular polarization.

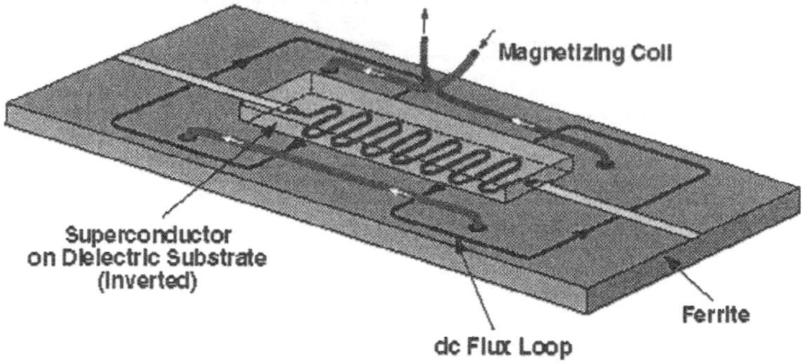

Figure 21. Schematic view of phase shifter. YBCO is deposited on LAO. The ferrite is held in close proximity to the YBCO meanderline. The magnetic circuit is contained in the plane of the ferrite substrate.

epitaxially on the other [19,20], making the structure in Fig. 17 possible to fabricate using standard deposition techniques.

One example of a ferroelectric phase shifter that has been reported [12] uses a structure that is a variation of that of Fig. 17, and is shown in Fig. 18. This uses coupled lines in order to maximize the phase shift. When the coupled lines are excited in the so-called odd mode, in which the rf voltages are out of phase with respect to each other in the two lines, the electric field in the ferroelectric layer is enhanced over the case of the even mode in which the voltages are in phase. The odd mode gives stronger interaction with the thin STO film than in the even mode. The stronger interaction leads to larger changes in phase shift.

Figure 19 shows the results of phase shift vs. voltage for the phase shifter shown in Fig. 18 at 16 GHz and three different temperatures [12]. The insertion loss is about 10 dB. This shows adequate phase shift. The figure of merit for this device is ~ 30 deg/dB.

3.3. FERRITE-BASED PHASE SHIFTERS

Ferrites are materials whose permeability can be changed by the application of a magnetic field. They have been applied to nonreciprocal devices such as circulators and to tunable devices such as phase shifters and filters, using conventional designs with waveguides and normal metals [21]. In planar circuits, however, normal-metal ohmic losses become significant. With low-loss superconductors replacing the normal metals, miniature devices based on microstrip circuitry can become attractive replacements for bulky waveguide phase shifters. Because superconductor surface resistances deteriorate in magnetic fields, however, the use of superconductors to solve the conduction loss problem has not been employed often. For designs where significant magnetic fields penetrate the circuit, superconductivity may have no advantage over the normal conductivity of copper at 77 K. However, devices designed to exploit the low ohmic losses of high-T_c superconductors and at the same time to eliminate the deleterious effects of magnetic field penetration of the superconductor have been reported [13]. These devices confine the magnetic flux entirely within the

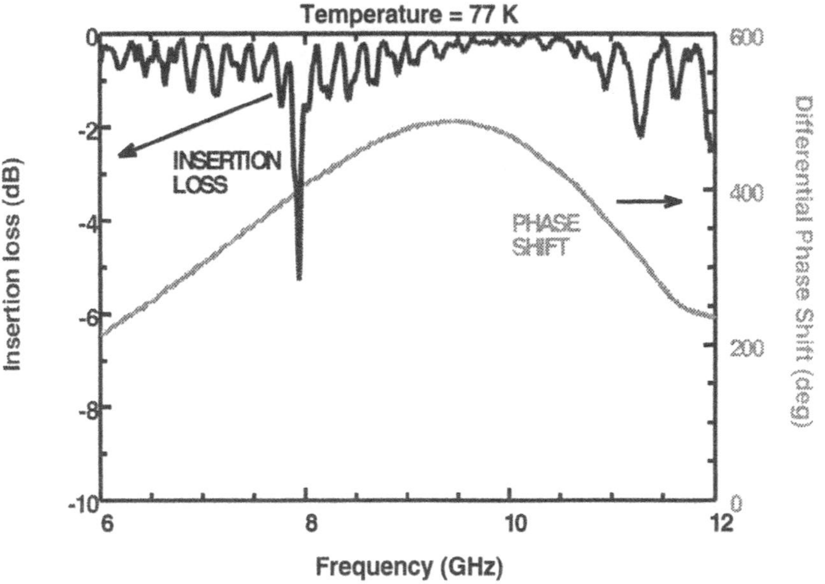

Figure 22. Insertion loss and differential phase shift versus frequency for a YBCO meanderline circuit on LAO substrate of Fig. 21.

ferrite in the shape of a toroid, thereby avoiding DC flux penetration of the superconductor while permitting rf signal penetration of the ferrite.

The most effective interaction with the ferrite is obtained with circularly polarized rf fields, which are produced by combining two linearly polarized waves separated in phase by 90 deg. A meanderline structure in which the meanders are one-quarter-wavelength long produces circular polarization midway between the meanders, and elliptical polarization elsewhere. This is the simplest way to produce circular polarization in a planar structure [13]. The relationships between ferrite magnetization, meander design, and resulting rf magnetic fields are shown in Fig. 20.

The strongest superconductor-to-ferrite coupling and easiest-to-assemble devices are obtained when the superconductor is deposited directly on the ferrite substrate. While the ability to deposit YBCO films on single-crystal ferrite has been demonstrated [22], deposition on polycrystalline material is more difficult. Thus, a hybrid structure is required in which the YBCO is deposited on LAO and pressed against the polycrystalline ferrite with the YBCO side in contact with the ferrite substrate. This structure is shown in Fig. 21.

The phase-shifter toroid structure is approximated by a planar ferrite substrate with four holes in it to allow for coils that produce the closed flux path completely in the plane of the substrate. The coils are arranged so that the magnetization is approximately parallel to the meanders of the YBCO stripline that is deposited on the LAO. The return path of the flux is outside of the meanderline. The ferrite used (Trans Tech 1210M) has a saturation magnetization $4\pi M_S$ = 1200 G (300-K value). The

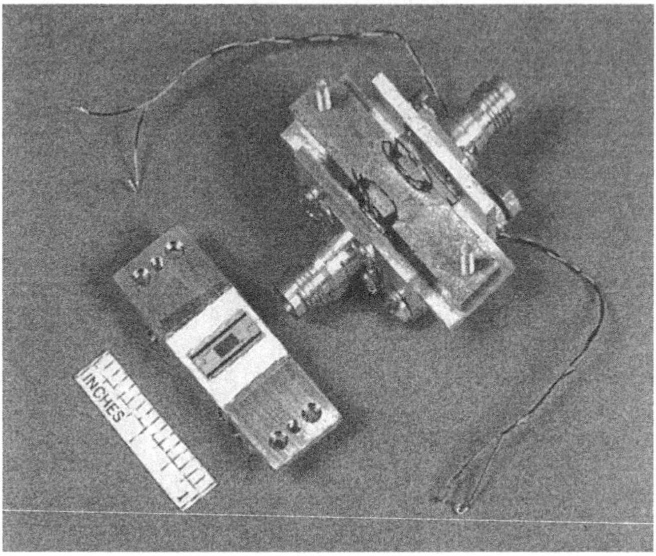

Figure 23. Photograph of ferrite/superconductor phase shifter.

amount of phase shift can be controlled by the magnitude and direction of the magnetization of the ferrite.

Figure 22 presents maximum differential phase shift and insertion loss data for the device [13]. Any value of phase shift between zero and the maximum is possible by varying the magnetization. The insertion loss is independent of the state of magnetization. The frequency dependence of the phase shift results from the frequency dependence of the circular polarization in the meanderline.

The insertion loss, corrected for probe losses, is less than 0.2 dB, and the phase-shift peak is approximately 500 deg. The variations in insertion loss are due to small impedance mismatches in the measurement probe and the device. and did not change at lower temperatures indicating that the observed losses are due to reflections rather than ohmic losses in the YBCO. From the known surface resistance R_S of the YBCO, the insertion loss is expected to be less than 0.1 dB.

The figure of merit for this device is 5000 deg/dB. The figure-of-merit values obtained in these experiments may be compared with the 130-deg/dB value at X band reported by Roome [23] and comparable values by Hansson *et al.* [24], both with copper meanderline circuits at room temperature. Even at cryogenic temperatures, where the insertion loss from the copper circuit would decrease by about a factor of two, the device employing superconductors would have at least a factor-of-ten improvement in figure of merit. If the theoretical limits of the ferrite properties can be realized in practice, figure-of-merit values of 10,000 deg/dB could be reached.

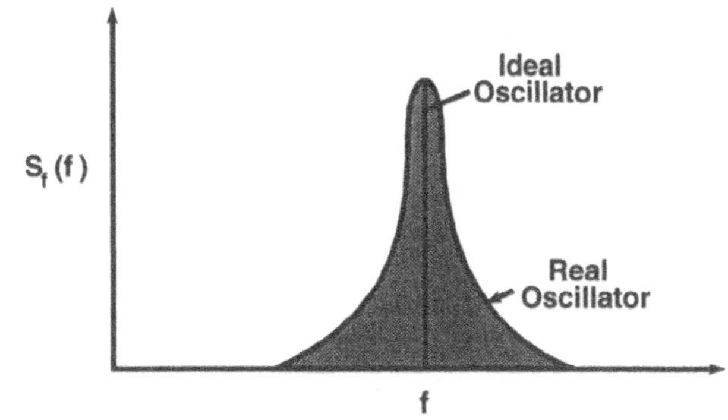

Figure 24. Schematic depiction of the broadening of the spectrum of an oscillator resulting from the presence of phase noise.

4. Low-Noise Oscillators

This section discusses an important potential application of superconductors to frequency-stable, low-noise oscillators. We restrict the discussion to applications where short-term (<1 s) stability is important. Oscillators with good long-term (>1 s up to the order of days) stability are usually called clocks, but clocks will not be covered here. For a discussion of clock applications of superconductors, see Ref. [25].

Almost every receiver system contains a local oscillator (LO) to mix with a received signal to produce a resulting signal at a fixed (usually lower) intermediate frequency (IF) for further amplification, filtering, and signal processing. Figure 1 shows a typical system as was discussed earlier, and shows where the LO is located in the system. The oscillator must be stable in frequency so as not to affect system performance. Any instabilities or noise produced by the LO will appear in the IF signal and can degrade system performance. The primary deviation from ideal behavior introduced by the LO is phase noise. At this point, a few words of introduction to phase noise are needed. An ideal oscillator produces a pure sine wave,

$$V(t) = A \sin \omega_0 t\,,$$

where A is a constant amplitude, and $\omega_0 = 2\pi f_0$ is the angular frequency of the oscillator. But, any real oscillator introduces noise that is included in the output signal. This noise shows up as variations in the amplitude and phase of the sine wave. The output of a real oscillator then becomes

$$V(t) = (A + N(t)) \sin(\omega_0 t + \phi(t))\,,$$

where $N(t)$ is the amplitude noise and $\phi(t)$ is the phase noise. The phase noise can be imagined as variations in the zero crossings of the sine wave. The frequency fluctuates as well, but since phase and frequency modulation are equivalent, it is necessary to consider only the phase noise. The phase noise also is the quantity that is directly measured. The amplitude noise is usually unimportant because the output of the oscillator can be passed through a limiter effectively removing it. The phase noise, however, cannot be removed simply. It results in frequency modulation of the oscillator producing sidebands that broaden the oscillator frequency spectrum from a

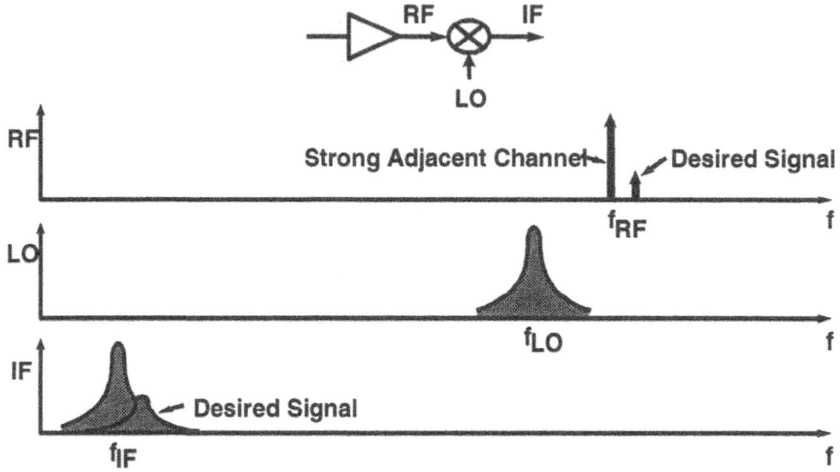

Figure 25. Schematic depiction of the consequences of finite phase noise for the case of two signals closely spaced in frequency.

delta function in the ideal case into one of finite width and shape for the real oscillator. This situation is shown schematically in Fig. 24.

It is the broadening of the spectrum that produces system degradation. As shown in Fig. 25, when two received signals are closely spaced in frequency, a larger signal can interfere with the reception of the smaller one. Such situations can happen in a communication system in adjacent channels, when one transmitter is close to the receiver and the other is far away. One of the most important applications of very low noise oscillators, however, is in doppler radar systems that rely on the doppler shift to find and identify targets. The strong signal of Fig. 25, for instance, is often the radar return from ground clutter (zero doppler shift) and the small signal might be the return of an airplane flying close to the ground and at slow speed, with small doppler shift so as to be shielded from detection by the ground return mixing with the noise of the LO. Lower phase noise leads to better ability to see small, slow-moving targets using the radar.

Since the spectrum of the oscillator is of interest for system performance, $\phi(t)$ is usually characterized in the frequency domain by

$$S_\phi(f) = \int |\phi(t)|^2 e^{i2\pi ft}\, dt\,,$$

where $S_\phi(f)$ is the power spectral density of phase fluctuations, and f is the offset frequency from the carrier $f_0 = \omega_0/2\pi$. A further commonly used definition is $\mathcal{L}(f) = S_\phi(f)/2$, the power spectral density of phase fluctuations in a single sideband of the carrier. A typical phase-noise spectrum is shown in Fig. 26, showing the power-law dependence that is usual for the various regions of frequency offset. An ideal oscillator would yield a delta function centered at zero offset frequency, i.e. a delta function at f_0.

A typical feedback oscillator circuit is shown in Fig. 27. A narrowband resonator is used as the feedback and frequency-stabilizing element. This circuit

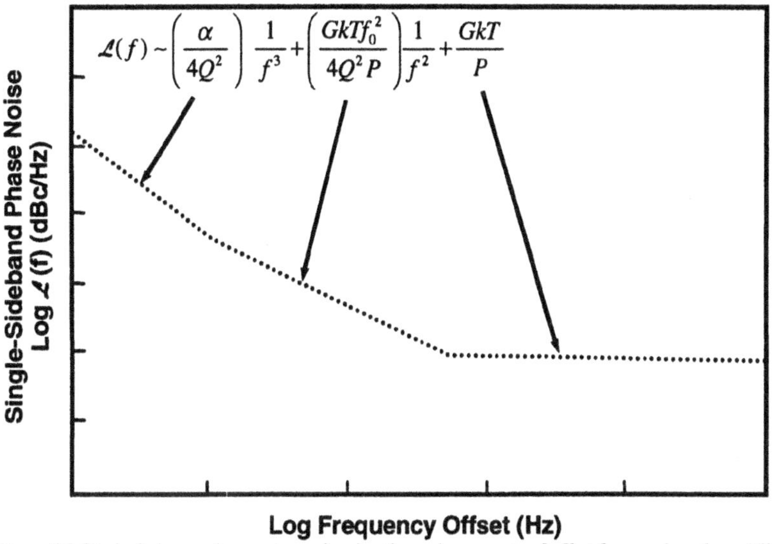

Figure 26. Typical phase-noise spectrum showing the various ranges of offset frequencies where different power-law relations apply.

oscillates when the gain of the amplifier is equal to the insertion losses of the other elements and the phase shift around the loop is equal to $2n\pi$, where n is an integer. The phase shifter shown adjusts the phase shift to $2n\pi$, and the coupler provides energy to the outside world. Small changes of f_0, within a small fraction of the bandwidth of the resonator, can be made by adjusting the phase shifter.

The phase noise of a feedback oscillator has been modeled by Leeson26 and is given by

$$\mathcal{L}(f) = 10\log\left[N^2\left(1+\frac{f_0^2}{4Q^2f^2}\right)\left(\frac{GFkT}{P}+\frac{\alpha}{f}\right)\right], \tag{7}$$

where N is the factor representing multiplication of the fundamental frequency of the oscillator, α is the flicker-noise ($1/f$) constant of the amplifier, Q is the resonator quality factor, G is the gain of the amplifier, F is the noise figure of the amplifier, P is the power in the loop, k is Boltzmann's constant, T is the temperature of the amplifier, f_0 is the oscillator center frequency and f is the offset frequency from the carrier. The flicker-noise constant of the amplifier is related to the rf $1/f$ noise of the amplifier, although not correlated to the base-band $1/f$ noise [27]. It must be empirically determined. Equation (7) has been shown to agree well with numerous experiments [28]. The expansion of the Leeson expression that follows shows explicitly the power-law behaviors of the different regions of the phase-noise spectrum that are shown in Fig. 26:

$$\mathcal{L}(f) = 10\log N^2\left[\frac{\alpha f_0^2}{4Q^2f^3}+\frac{GFkTf_0^2}{4Q^2Pf^2}+\frac{\alpha}{f}+\frac{GFkT}{P}\right].$$

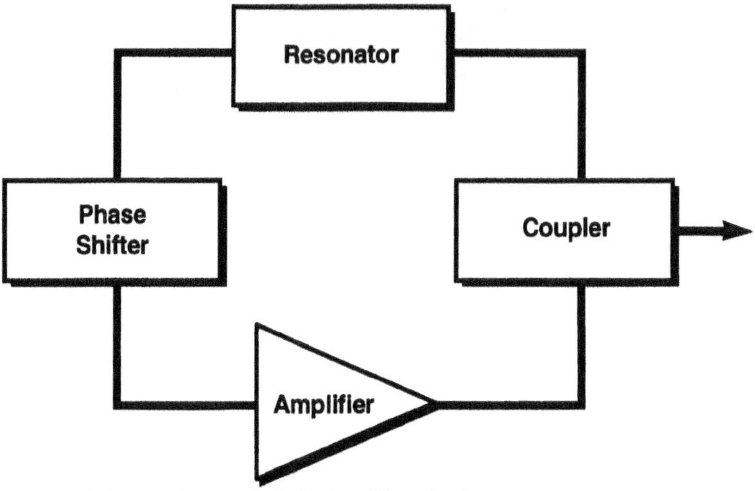

Figure 27. Schematic diagram of a feedback oscillator circuit.

This expansion of Eq. (7) shows that at small enough values of f, the first two terms will dominate.

The derivation of Eq. (7) shows that phase noise originates as thermal noise and flicker noise in the amplifier. This noise is then fed back through the resonator as shown in Fig. 27. The white phase noise at the output of the amplifier is converted to frequency noise in the resonator, because modulation of the phase shift in the feedback loop changes the frequency of oscillation. The frequency noise thus created has a $1/f^2$ offset-frequency dependence when measured as phase noise. For more detail of noise generation and modeling, see Ref. 28. It is clear, however, from Eq. (7) that a high-Q resonator will reduce the phase noise, especially in the low-offset-frequency region of the phase-noise spectrum close to the carrier. This dependence of $\mathcal{L}(f)$ on Q is the reason for interest in incorporating superconducting resonators in high-performance, very low noise oscillators.

There is an important qualification of Eq. (7) that must be kept in mind. The $1/Q^2$ reduction of phase noise in the flicker-noise-dominated part of the spectrum is only true if the sole source of the flicker noise is the amplifier. If the resonator itself is the source of the flicker noise, because of center frequency fluctuation or nonlinear processes mixing baseband $1/f$ noise [29] into the microwave frequencies, then it tends to *increase* proportional to Q^2, and as a result, there is no net improvement of the phase noise if the Q of the resonator is increased [28]. This preponderance of resonator $1/f$ noise has been shown to be true in some cases for thin films of Nb [30] and YBCO [31]. Thus, the benefits of the high-Q superconductors may be limited by fluctuations in the materials themselves. It is generally believed that film quality has an influence on $1/f$ noise, so higher-quality films may exhibit lower flicker noise. Additionally, vibration and other environmental factors such as temperature fluctuations will degrade phase-noise performance.

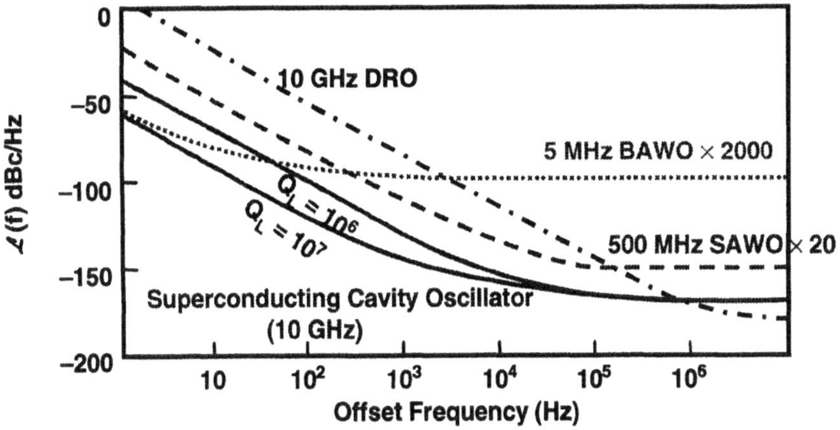

Figure 28. Projections of phase noise for superconducting cavities. After T. Parker [32].

Figure 28 shows projections of phase noise for two different values of resonator Q, compared with several other technologies [32]. These projections assume no flicker noise contribution from the resonator. In these cases, the advantage of a high-Q resonator over other technologies is clear.

Low-temperature superconductors forming a cavity have been used to produce resonators with Q factors greater than 10^{10} at microwave frequencies [25]. These high Qs, however, have been achieved only at very low temperatures. Much effort has been made to use high-T_c materials operating at temperatures at or just below 77 K. One fundamental problem in using high-T_c materials is the lack of bulk material with low microwave R_s to fabricate a cavity resonator exclusively from HTS material. Cavity resonators have the highest Q of any microwave component, but planar geometries utilizing thin films have been the only way to introduce the high-Q factors that are possible with HTS materials.

To date, the best results have been obtained with sapphire dielectric resonators that use superconducting ground planes as the shielding elements [33]. Figure 29 shows a schematic view of such a cavity [34]. This arrangement, for the TE_{011} mode of the cavity, uses superconducting films on the ends of a dielectric cylinder. Such resonators have been widely used for measurement of surface resistance of HTS films [35]. Sapphire is the usual choice for the dielectric because it has a very high dielectric Q. The total loaded Q of a resonator Q_l is given by

$$\frac{1}{Q_l} = \frac{1}{Q_s} + \frac{1}{Q_d} + \frac{1}{Q_r} + \frac{1}{Q_c},$$

where Q_s is the superconductor Q, Q_d is the dielectric $Q = \tan\delta$, Q_r is the radiation Q of the structure and Q_c is the circuit Q resulting from loading of the external circuit. For this geometry, radiation losses are small so $1/Q_r$ can be neglected. Loaded Q values of approximately $2\text{x}10^5$ at 60 K have been reported [34]. The Q is usually limited by the superconductor. Coupling to the resonator is adjusted by the coupling

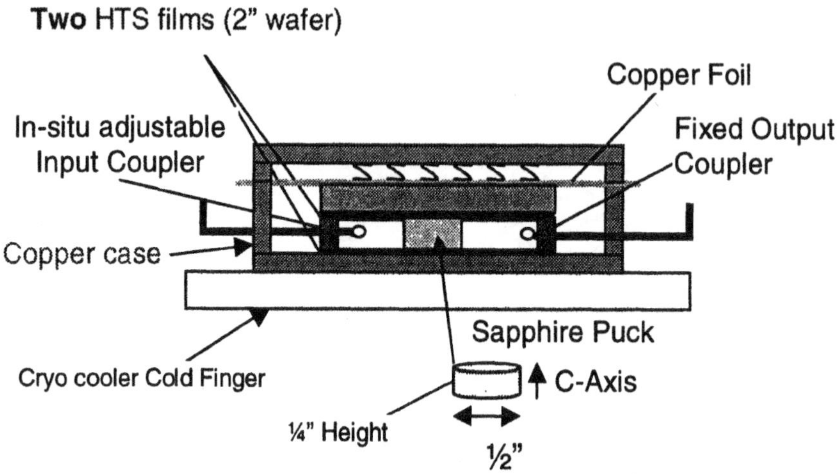

Figure 29. Typical sapphire loaded superconducting resonator that has been used in fabricating low noise oscillators and has been used for R_s measurement.

loops, which are located on the open sides of the resonator, to obtain a moderate value of insertion loss and to inhibit excessive loading of the *Q*.

Figure. 30 shows the results of measurements of phase noise in a high-*Q* HTS resonator [36]. These results are among the best reported to date. Also shown in Fig. 30 are results from the best competing room-temperature technology [37], which uses sapphire in a whispering-gallery mode. For a description of whispering-gallery resonators, which use high-order modes to obtain high *Q*, see Ref. 38. Superconducting low-noise oscillators do not yet offer better performance than room-temperature technology using sapphire whispering-gallery modes. It should be expected, however, that as superconducting material technology improves, better results will be obtained. Additionally, where low phase noise is needed (but not the very lowest), and where superconducting technology is already in use, then an oscillator stabilized with a superconducting resonator would be advantageous. It also is possible that planar superconducting resonators could be used with savings of size and weight.

5. Summary and Conclusions

There is a variety of potential applications of analog devices incorporating superconducting components. We have discussed several of the most promising here: delay lines, tapped delay lines, phase shifters, and low-noise oscillators. At this time, some of these seem to have advantages when compared with competing technology. Tapped delay lines in the form of chirp filters, for example, show great promise for applications in compressive receivers which perform real-time wideband spectrum analysis for signal-intercept systems.

Phase shifters currently are showing performance that merits serious consideration for inclusion in real systems. The issue of cooling is one problem that

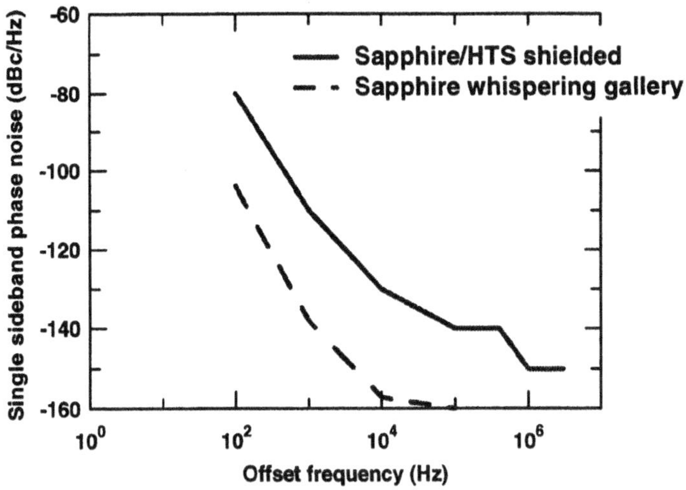

Figure. 30 Results of phase-noise measurements.

remains in using phase shifters in phased-array antennas. To gain the full benefit of the low loss obtainable with superconductors, phase shifters must be located very close to the antenna. Indeed, integration with the radiating elements would be the best method to gain the full benefits. This would require clever cryogenic design to minimize cooling requirements, since antenna elements would be colder than ambient and load the cooling system. As discussed, there are several implementations of phase shifters that have shown potential. It is not yet clear which will be preferable.

Applications of superconductors in low-noise oscillators continue to show great potential because of the extremely high Q values that are possible with superconductors, and have been demonstrated at low temperatures with low-T_c materials. However, to take full advantage of high-temperature superconductors in this application, better materials are needed, as well as better understanding of the losses and $1/f$ noise. This application perhaps awaits the development of bulk HTS with low surface resistance and low $1/f$ in order to fabricate microwave cavities entirely from HTS. At present, several existing technologies, which also continue to improve, perform adequately in applications that require ultra-low phase noise. Better low-noise oscillators including superconducting, cryogenic and conventional are under development.

6. Acknowledgments

This work was supported by DARPA. I would like to thank A. D. Smith of TRW for technical discussions of JJ phase shifters and for sending me the data used in Sec. 3.1. I thank W. G. Lyons of Lincoln Laboratory for discussions and for the figures used in Sec. 2.4. Also, I thank F. W. Van Keuls of NASA for sending me the figures used in Sec. 3.2. and G. C. Liang of Conductus for permission to use the figures and data in Sec. 2.2.

7. References

[1] Maxim Inc. May 1999 data sheet, unpublished.

[2] Oppenheim, A. V. and Shafer, R. W. (1975) *Digital Signal Processing,* Prentice-Hall, Englewood Cliffs, New Jersey.

[3] Kuchnir, M., McCarthy, J. D., and Pasquinelli, R. J. (1983) Superconducting delay line for stochastic cooling filters, *IEEE Trans. Nuclear Sci.* **30**, 3360-3362.

[4] Liang, G. C., Shih, C. F., Withers, R. S., Cole, B. F., and Johansson, M. E. (1996) Space-qualified superconductive digital instantaneous frequency-measurement system, *IEEE Trans. Microwave Theory Tech.* **44**, 1289-1299.

[5] Lyons, W. G., Arsenault, D. R., Anderson, A. C., Sollner, T. C. L. G., Murphy, P. G., Seaver, M. M., Boisvert, R. R., Slattery, R. L., and Ralston, R. W. (1996) High temperature superconductive wideband compressive receivers, *IEEE Trans. Microwave Theory Tech.* **44**, 1258-1278.
Lyons, W. G., Arsenault, D. R., Anderson, A. C., Sollner, T. C. L. G., Murphy, P. G., Seaver, M. M., Boisvert, R. R., Slattery, R. L., and Ralston, R. W. (1996) High-T_c superconductive wideband compressive receivers, *Linc. Lab. J.,* **9**, 33-67 .

[6] Matick, R. E. (1968) *Transmission Lines for Digital and Communication Networks,* McGraw-Hill, New York, Chapter 6.

[7] Biehl, M., Vogt, A., Herwig, R., Neuhaus, M., Crocoll, E., Lochschmeid. R., Scherer, T., Jutzi, W., Kratz, H., Berberich, P., and Kinder, H. (1995) A 4 bit instantaneous frequency meter at 10 GHz with coplanar YBCO delay lines, *IEEE Trans. Appl. Supercond.* **5**, 2279-2282.

[8] Kawecki, T. G., Golba, G. A., Price, G. E., Rose, V. S., and Meyers, W. (1996) The high temperature superconductivity space experiment (HTSSE-II) design, *IEEE Trans. Microwave Theory Tech.* **44**, 1198-1212.

[9] Pozar, M. (1993) *Microwave Engineering,* Addison-Wesley, Reading, Massachusetts.

[10] Lyons, W. G., Murphy, P. G., Boisvert, R. R., Anderson, A. C., Seaver, M. M., Ralston, R. W., and Sollner, T. C. L. G. (1995) Demonstration of a 40-ns $YBa_2Cu_3O_{7-\delta}$ chirp filter on 125-μm bonded-wafer $LaAlO_3$ substrates, *Solid State Research Report,* 1995:3, MIT Lincoln Laboratory DTIC#AD-A311816.

[11] Durand, D. J., Carpenter, J., Ladizinsky, E., Lee, L., Jackson, C. M., Silver, A., and Smith, A. D. (1992) The distributed Josephson inductance phase shifter, *IEEE Trans. Appl. Supercond.* **2**, 33-38.

[12] Van Keuls, F. W., Romanofsky, R. R., Bohman, D. Y., Winters, M. D., Miranda, F. A., Mueller, C. H., Treece, R. E., Rivkin, T. V., and Galt, D. (1997) ($YBa_2Cu_3O_{7-\delta}$,Au)/$SrTiO_3$/$LaAlO_3$ thin film conductor/ferroelectric coupled microstripline phase shifters for phased array applications, *Appl. Phys. Lett.* **71**, 3075-3077.

[13] Oates, D. E., Dionne, G. F., Temme, D. H., and Weiss, J. A. (1997) Superconductor ferrite phase shifters and circulators, *IEEE Trans. Appl. Supercond.* **7**, 2347-2350.

[14] Dionne, G. F., Oates, D. E., and Temme, D. H. (1995) YBCO/ferrite low-loss microwave phase shifter, *IEEE Trans. Appl. Supercond.* **5**, 2083-2086.

[15] Anlage, S. M., Snortland, H. J., and Beasley, M. R. (1989) A current controlled variable delay superconducting transmission line, *IEEE Trans. Magn.* **25**, 1388-1391.

[16] Hohenwarter, G. K. G., Martens, J. S., Beyer, J. B., Nordman, J. E., and McGinnis, D. P. (1989) Design of variable phase velocity kinetic inductance delay lines and their measured characteristics when fabricated by a simple Nb based process, *IEEE Trans. Magn.* **25**, 1100-1103.

[17] Luine, J. Smith, A. D. Sun, A. Dantsker, E., and Daly, K. (1999) Measurements of YBCO parametric amplifiers, *IEEE Trans. Appl. Supercond.* **9**, 3425-3428.

[18] J. Pond, proceedings of this NATO ASI.

[19] Galt, D., Price, C., Beall, J. A., and Ono, R. H. (1993) Characterization of tunable thin film microwave $YBa_2Cu_3O_{7-x}$/$SrTiO_3$ coplanar capacitor *Appl. Phys. Lett.* **62**, 3076-3079.

[20] Vendik, O. G., Hollmann, E. K., Kozyrev, A. B., and Prudan, A. M., (1999) Ferroelectric tuning of planar and bulk microwave devices *J. Superconductivity* **12**, 325-338.

[21] Rosenbaum, F. J. (1974) Integrated ferrimagnetic devices, in L. Young and H. Sobol, (eds.), *Advances in Microwaves* **8**, 203.

[22] Piqué, A., Harshavardhan, K. S., Moses, J., Mathur, M., Belohoubek, E., Venkatesan, T., Denlinger, E. J., Kalokitis, D., Fathy, A., Pendrick, V., Rajeswari, M., and Jiang, Wu (1995) Microwave compatible $YBa_2Cu_3O_{7-x}$ films on ferrimagnetic garnet substrates *Appl. Phys. Lett.* **67**, 1778-1780.

[23] Roome, G. T. (1969) Thin ferrites for integrated microwave devices, Technical Report AFAL-TR-69-149, Air Force Systems Command, Wright Patterson Air Force Base.

[24] Hansson, E. R. B., Aditya, S., and Larsson, M. A. (1981) Planar meanderline ferrite-dielectric phase shifter, *IEEE Trans. Microwave Theory Tech.* **29**, 209-211.

[25] Stein, S. R. and Turneaure, J. P. (1975) Superconducting-cavity stabilized oscillators with improved frequency stability, *Proc. IEEE*, **63**, 1249-1250 and Buchman, S., Turneaure, J. P., Lipa, J. A., Dong, M., Cumbermack, K. M., and Wang, S. A (1998) Superconducting microwave oscillator clock for use on the space station, *Proceedings of the 1998 IEEE International Frequency Control Symposium*, 534-539.

[26] Leeson, D. B. (1970) Short-term stable microwave sources, *Microwave J.* **13**, 59-69.

[27] Martinez, R. D., Oates, D. E., and Compton, R. C. (1994) Measurement and model for correlating of phase and baseband 1/f noise in an FET, *IEEE Trans. Microwave Theory Tech.* **42**, 2051-2055.

[28] Parker, T. (1987) Characteristics and sources of phase noise in stable oscillators, *Proceedings of the 1987 IEEE International Frequency Control Symposium*, 99-110.

[29] Surya, C., Israeloff, N. E., Widom, A., Seed, R., and Vittoria, C. (1995) Flicker noise in $YBa_2Cu_3O_{7-\delta}$ bicrystal grain boundary junctions in weak magnetic fields, *Appl. Phys. Lett.* **67**, 1307-1309.

[30] Oates, D. E., Anderson, A. C., and Shih B. S., (1989) Superconducting stripline resonators and high-T_c materials, *1989 IEEE MTT-S Digest*, 627-630.

[31] Oates, D. E. unpublished.

[32] Parker T. E. and Iafrate, G. J. (1988) High temperature superconductivity: novel concepts, fundamental issues, and frequency control aspects *Proceedings of the 1988 IEEE International Frequency Control Symposium*, 99-110.

[33] Ghosh, I. S., Schemion, D., and Klein, N. (1997) Low Phase Noise Microwave Oscillators Using Temperature Compensated High-Q HTS Shielded Dielectric Resonators, *Applied Superconductivity 1997: Proceedings of EUCAS 1997 Third European Conference on Applied Superconductivity*, Institute of Physics Publishing, Part 1, 611-614.

[34] Xin, H., Oates, D. E., Anderson, A. C., Slattery, R. L., Dresselhaus, G., and Dresselhaus, M. S. (1999) Comparison of power dependence of microwave surface resistance of unpatterned and patterned YBCO thin films, submitted to *IEEE Trans. Microwave Theory Tech.*

[35] Shen, Z. Y., Wilker, C., Pang, P., Holstein, W. L., Face, D., and Kountz, D. J. (1992) High-T_c superconductor-sapphire microwave resonator with extremely high Q values up to 90 K, *IEEE Trans. Microwave Theory Tech.* **40**, 2424-2431.

[36] Ghosh, I. S., Schemion, D., and Klein, N. (1997) Temperature compensated high-Q dielectric resonators for long term stable low phase noise oscillators *Proceedings of the 1997 IEEE International Frequency Control Symposium*, 1024-1029.

[37] Poseidon Scientific Instruments, Fremantle WA Australia, product brochure.

[38] Langham, C. D., Abbas, F., and J. C. Gallop, (1998) Observation of 'whispering gallery' modes in dielectric rod resonators, *1998 IEEE Conference on Precision Electromagnetic Measurements Digest*, 526-527.

HIGH TEMPERATURE SUPERCONDUCTING RADIO FREQUENCY COILS FOR NMR SPECTROSCOPY AND MAGNETIC RESONANCE IMAGING

STEVEN M. ANLAGE
Center for Superconductivity Research, and Materials Research Science and Engineering Center, Department of Physics, University of Maryland, College Park, Maryland 20742-4111 USA

Abstract. High-temperature superconductivity has a significant opportunity to improve the performance of nuclear magnetic resonance (NMR) spectroscopy and magnetic resonance imaging (MRI) systems. The low rf losses and low operating temperatures of superconducting coils allow them to improve the signal-to-noise ratio in applications where the system noise dominates that of the sample under study. These improvements translate into new capabilities and shorter measurement times for NMR and MRI systems. In this pedagogical paper, we discuss the implementation and impact of high-T_c superconducting thin films in rf coils for NMR and MRI applications.

1. Introduction

Nuclear Magnetic Resonance (NMR) has been employed for more than 50 years to study the physics and chemistry of atoms and molecules. It has developed into a highly refined and precise method for characterization and identification of complex molecules, as well as non-invasive imaging. Many generations of scientists have contributed sophisticated methods of using NMR to peer more and more deeply into molecular structure. High Temperature Superconductivity (HTS) comes into this mature field at a relatively late stage in its development. As such, HTS has not introduced any fundamentally new methodology for performing NMR spectroscopy or Magnetic Resonance Imaging (MRI). However, HTS presents the opportunity to significantly enhance existing measurement methods by improving the signal-to-noise ratio of the detection system. This has advantages in

H. Weinstock and M. Nisenoff (eds.), Microwave Superconductivity, 337–352.

certain applications of NMR spectroscopy or MRI for which the sample noise does not dominate the noise performance of the entire measurement system.

This paper presents a simple pedagogical discussion of the impact of HTS on NMR spectroscopy and MRI. It is written by an outsider; hence, the reader is cautioned to consult the literature for a more complete treatment of the subjects discussed here. A beginning student should consult Slichter's book for the theory of NMR [1], and any one of a number of practical introductions to NMR spectroscopy [2, 3]. There is also an excellent review paper on the use of HTS in NMR spectroscopy [4].

In this paper, we begin by reviewing the fundamentals of NMR, discuss some of the important uses of NMR in spectroscopy, and review the basic methodology of an NMR experiment. We then discuss the requirements for performing good NMR spectroscopy, setting the stage for the impact of HTS rf coils on the measurement process. The conditions under which HTS rf coils reduce system noise will be presented, followed by a discussion of the advantages and disadvantages of using HTS rf coils. We conclude by giving an example in which the HTS coil gives significant improvement in the signal-to-noise ratio in real NMR spectroscopy and MRI applications.

2. What is magnetic resonance?

Magnetic resonance occurs under a variety of conditions. First, we require a magnetized object, such as a ferromagnetic domain, or a collection of paramagnetic electron spins, or a set of nuclear spins. The magnetization can be represented by a vector μ. Next, a strong dc magnetic field $\mathbf{B}_0$ is applied to the system. In the classical picture, the magnetic field will exert a torque $\tau = \mu \times \mathbf{B}_0$ on the magnetization or spins, causing the magnetization vector to precess about the direction of the magnetic field at a rate proportional to the field strength, given by $\omega_{precession} = \gamma B_0$ where γ is the gyromagnetic ratio. The gyromagnetic ratio depends on the object carrying the magnetic moment. For electron spins it depends on the Landé g-factor, which includes spin and angular momentum degrees of freedom, and for nuclear spins γ depends on the particular nucleus studied. For example for a proton, $\gamma/2\pi$ = 42.5 MHz/T, while for a ^{13}C nucleus, $\gamma/2\pi$ = 10.5 MHz/T. The precessing spins create an oscillating magnetization in the plane perpendicular to the applied magnetic field. If an rf magnetic field, polarized in that plane, now is introduced, it can be resonantly absorbed if the rf frequency matches the precession frequency of the spins. This resonant absorption is magnetic resonance.

The quantum-mechanical description of magnetic resonance is as follows. The applied magnetic field gives rise to a Zeeman splitting between "up" and "down" spin states (i.e., spin aligned along and against the magnetic field direction). This comes about from the interaction of the magnetic moment μ with the applied field $\mathbf{B}_0$, with an energy $E = -m_I\gamma\hbar B_0$, where m_I is the component of the nuclear spin in the direction of $\mathbf{B}_0$, and $\hbar$ is Planck's constant divided by 2π. The Zeeman splitting energy between "up" and "down" states for a spin 1/2 is given by $\Delta E = \gamma\hbar B_0$ (the proton and the ^{13}C nuclei both have nuclear spin $I = 1/2$.) If an rf photon of energy $\hbar\omega_{rf} = \Delta E = \gamma\hbar B_0$ is present, and the polarization conditions are satisfied, there is resonant absorption. Both the classical and quantum-mechanical descriptions give the same condition for magnetic resonance.

As mentioned above, magnetic resonance occurs in many different systems: nuclear spins (NMR), electron spins (electron spin resonance and electron paramagnetic resonance), ferromagnetism (ferromagnetic resonance), and antiferromagnetism (antiferromagnetic resonance). In this paper we shall focus on NMR.

3. What does NMR tell you?

The precession rate of the nuclear spin is a sensitive function of the magnetic field experienced by the nucleus. Although the applied external field can be substantial (1 - 20 T), there are other magnetic fields at the site of the nucleus which can contribute to the precession rate. These fields come from electrons orbiting the nucleus, and their value will depend on the chemical structure of the atom or molecule into which the nucleus is incorporated. As a result, the nuclear spin precession rate will vary slightly depending upon the chemical bonding of the electrons around that nucleus. This variation in the precession rate, and therefore in the NMR frequency, is called the chemical shift [2, 3]. It gives important information about the molecule and can be used as a "fingerprint" of the chemical structure of the molecule, because in general, each nucleus experiences a slightly different chemical shift.

The degree of chemical shift depends on the gyromagnetic ratio of the nucleus involved. For protons, the chemical shift range is on the order of 10 parts per million (ppm). This means that the NMR frequency will vary from the nominal value by about 1 part in 10^5 due to chemical effects for a given applied magnetic field. For ^{13}C, the chemical shift range is about 200 ppm.

Other information which can be obtained from NMR relates to molecular dynamics. For instance, segments of molecules may be free to rotate, and nuclei in these segments give information about their rate of rotation. Other information comes from the way in which polarized nuclear spins become de-polarized through their interactions with the environment. There are two measurable relaxation time scales, T_1 and T_2. T_1 is the spin-lattice relaxation time and T_2 is the spin-spin relaxation time. Multi-dimensional NMR spectroscopy also can give information about which nuclei are neighbors in a complicated molecule, simplifying the structural determination and also providing unique information for molecular identification [2, 3]. All of this information contributes to understanding the molecules under study.

4. How is NMR done?

An NMR experiment begins by polarizing the nuclear spins with a strong static magnetic field. In modern NMR spectrometers and MRI systems, the magnetic field is created by a low-T_c superconducting magnet. State-of-the-art NMR spectrometers have static magnetic fields B_0 which are typically 7.1 to 18.8 T (proton precession frequency of 300 to 800 MHz), MRI magnetic fields are 0.5 to 1.5 T (proton precession frequency of 21 to 63 MHz), but low cost systems can be 0.1 T and clinical research systems can be 4.7 T. With this field applied, the nuclear spins are set into precession at a rate given by the field strength and the gyromagnetic ratio of the nucleus of interest, $\omega = \gamma B_0$. In NMR spectrometer systems the precession frequency is typically in the range of 100 to 800 MHz. In MRI systems, the precession frequencies are in the range of 10 to 100 MHz. Rf magnetic fields of the correct frequency and polarization now are introduced to manipulate the precessing spin system using rf drive coils. The rf fields are usually pulsed. These pulses are designed to perform basic quantum-mechanical manipulations of the spin system. The details of these pulse sequences can be quite intricate, and the reader is referred to the extensive and detailed literature on the subject [1, 2, 3]. A simple and useful rf pulse is the $\pi/2$ pulse. This causes the spins to rotate through an angle of 90° away from the applied magnetic field direction, causing the spins to lie in the plane perpendicular to the applied field. This produces a maximum transverse rf signal, which is easily detected by the pickup coils of the NMR spectroscopy or MRI system.

If there are several types of nuclei in the sample, then each nuclear spin precesses at a frequency depending on its gyromagnetic ratio and environment. Very often multiple-frequency rf pulses are used to simultaneously stimulate NMR in several different nuclear species. Once all of the spin manipulations have been made, the system is allowed to relax in a process called free induction decay (FID). The system uses pickup coils to collect the FID voltage signal as a function of time. This signal is then Fourier transformed to get the NMR spectrum of the system. The direct FID signal can give the relaxation time scales T_1 and T_2, while the FID spectrum gives the chemical shift and other information.

In good spectrometers with high signal-to-noise, the NMR spectrum is remarkably clean [2, 3]. In practice, the NMR spectrum linewidths are on the order of 1 Hz for NMR frequencies on the order of 500 MHz. This means that very small splittings, on the order of 0.01 ppm can be resolved, making NMR spectroscopy extraordinarily information rich.

One can get an idea of what an NMR spectrometer looks like from browsing the web sites of commercial manufacturers of spectrometers [5, 6]. The bulk of the instrument is made up of the low-T_c superconducting magnet and its associated dewar. The dewars are remarkably efficient and may require the addition of liquid nitrogen only once every few weeks, and liquid helium once every few months.

5. Requirements for Good NMR Spectroscopy

In NMR spectroscopy one often deals with small samples, sometimes smaller than 1 microgram. Hence, the most important consideration for an NMR spectrometer is maximizing the signal-to-noise ratio (S/N) of the system. The most direct way to maximize the S/N is to use the strongest static magnetic field, B_0, possible. It is found that the voltage S/N ratio scales as $S/N \sim \omega_{precession} M/\sqrt{R} \sim B_0^{7/4}$ [7, 8, 9]. One power of B_0 comes from the fact that the magnetization M of the (paramagnetic) sample increases linearly with B_0, and a second power of B_0 comes about because the rf voltage inductively picked up by the receive coil scales with $\omega_{precession}$, which is directly proportional to B_0. However, the (normal-metal) probe resistance R increases as $B_0^{1/2}$. In other words, as the precession frequency increases, the skin depth in the metallic coil decreases, enhancing the losses. Hence, the overall gain in S/N is a bit less than a power of 2. Because of this strong scaling of S/N with static magnetic field strength, and the fact that spectral features are more easily distinguished at high fields (greater dispersion), there are efforts to create stronger ($B_0 \geq 20$ T) superconducting magnets for use in NMR spectrometers. However, the great expense and engineering efforts of creating these more powerful magnets is prohibitive, so NMR spectroscopists are looking for other ways to improve system S/N.

HTS materials present a unique opportunity to improve S/N without great expense or replacement of the superconducting magnet. The low resistance and low operating temperature of HTS pickup coils significantly increase the S/N of the detection system. The fact that the HTS pickup coils are kept under cryogenic conditions also presents the opportunity to introduce other cooled circuit elements such as a preamplifier, matching network and transmit/receive switch. These considerations have made HTS pickup coils very attractive, in lieu of an increase in static magnetic field strength.

The improvement in S/N allows smaller samples to be studied. This in turn could reduce the cost of an NMR spectrometer because a superconducting magnet would not have to create a homogeneous field over a great volume.

Other considerations for good NMR spectroscopy include the need for good sensitivity in field and frequency to distinguish closely-spaced peaks in the FID spectrum, such as splitting coming from dipole-dipole interactions. Good sensitivity also requires extremely good magnetic field homogeneity. Each nucleus in the sample must see the same external magnetic field if the small ($\sim$ 10 ppb) chemical shifts are not to be smeared. This means that a 1:10^9 homogeneity of B_0 must be maintained over the volume of the sample (typically a cylinder about 5 mm in diameter and 15 mm tall). The magnetic field homogeneity is created by a series of shim coils (typically 28 or so) which excite magnetic fields with the symmetries of the spherical harmonics. One also needs good S/N and high dynamic range to see both large and small peaks in the NMR spectrum. Finally, an improved S/N also cuts down measurement time, a key consideration for MRI and for sophisticated multi-dimensional NMR spectroscopy techniques.

5.1. OVERVIEW OF AN NMR SPECTROSCOPY SYSTEM

There are three parts of an NMR spectroscopy system of interest to us. The first is the sample which contains the polarized spins producing free induction decay. The second is the probe, which is a tuned resonant rf coil probe for detecting the FID from the sample. The probe is a simple inductor-capacitor (LC) resonator circuit which is tuned to the NMR signals from specific nuclei, such as ^{1}H, ^{13}C, ^{19}F, etc. This is the part of the system which can be made with a large-area HTS thin film. Finally, we have the receiver system which mixes the detected FID signal down to lower frequencies. This part of the system can be improved with a cryogenic preamplifier and matching network.

The probe coil has a quality factor Q, and frequency selectivity given by $Q = \omega L/R$, where ω is the angular frequency of the NMR signal, L is the self inductance of the coil, and R is the equivalent resistance of the coil. Since $Q = \omega/\delta\omega$ is inversely proportional to the 3-dB bandwidth of the resonator, $\delta\omega$, it is a measure of the frequency selectivity of the coil. HTS coils have the advantage of a significantly smaller resistance R than their normal-metal counterparts with the same inductance. Hence, the quality factor of the receive coils is significantly higher for HTS coils, making them attractive for certain applications, as we shall see later.

5.2. NOISE IN NMR SYSTEMS

There are three sources of noise in the NMR spectrometer detection system of interest to us here. The first is thermal noise from conducting samples. One can think of the sample as a resistor at finite temperature (usually room temperature). For instance, biological samples are often in conducting fluids. There are thermally-induced fluctuation currents in this resistor which produce noise, called Johnson noise. Very little can be done about this source of noise if one wants to characterize the sample (or organism) at room temperature and in its natural state. A second source of noise is thermal noise from resistance of the FID voltage pickup coil. Again Johnson noise in the coil will add noise to any detected signal. A third source of noise is noise in the receiver system, including the preamplifier, the mixer and detector.

HTS is of benefit only when the second and/or third noise sources dominate the system noise.[8] In other words there is no benefit to using HTS pickup coils in NMR or MRI systems where the sample noise dominates the system noise.

5.3. HOW DOES HTS IMPROVE THINGS?

In the case where the probe and receiver dominate the noise of the NMR system, one has the following approximate scaling relation for the signal-to-noise ratio (S/N) of the system as the coil properties are varied [4]: S/N $\sim \sqrt{\frac{\eta Q}{T_p+T_A}}$, where η is the filling factor of the coil system (i.e. fraction of the rf energy created by the coil which is stored in the sample volume), T_p is the effective probe temperature and T_A is the receiver noise temperature. HTS rf pickup coils have higher Q (because of their lower surface resistance) [10], and they have a lower probe temperature T_p. These improvements are offset by a reduced filling factor η for the HTS coils. The reduced filling factor comes about because existing HTS coils are made on flat, rigid substrates, whereas optimized normal-metal coils are made on curved substrates which wrap around the sample. Table 1 summarizes the current state-of-the-art for the S/N improvement of HTS pickup coils [4].

In this table, the filling factor is normalized to that of the normal-metal probe. The filling factor is about a factor of 5 worse for the HTS coil because of present-day geometrical constraints in the deposition of the films. However, recent progress in ion-beam-assisted deposition of HTS thin films has created the opportunity to make HTS coils on flexible substrates. Thus improvements in the filling factor of HTS coils should soon be possible.

In Table 1 the quality factor, Q, of HTS coils is conservatively estimated to be about 10,000, versus about 250 for their normal-metal counterparts.

TABLE 1. Signal-to-noise comparison of normal-metal and HTS rf probes

Factor	Normal Probe	HTS Probe
η	1	0.2
Q	250	10,000
T_p	300	75
T_A	80	80
S/N	1	4.5

Higher quality factors in HTS coils have been demonstrated (e.g., Q = 70,000 at 7 T) [8, 11], and the Q appears to be limited by flux motion [8]. However, there are upper limits on the Q of the coil imposed by the NMR measurement process itself - see below.

The probe temperature comparison in this table is a bit unfair. The normal-metal probe is taken to be at room temperature, while the HTS coil is, of course, cooled below the HTS transition temperature. A more fair comparison would be between HTS coils and cooled normal-metal coils, as has been done for the case of MRI coils [8]. However, the amplifier noise temperature is taken to be the same for each system. Overall, an improvement of 4.5 in the S/N is expected simply by substituting an HTS pickup coil for a normal-metal coil. Such an improvement in S/N has been demonstrated in a commercial cryogenic product made by Conductus and marketed by Varian [4, 6, 12]. Similar improvements have been made with the commercial cryogenic probes made by Bruker [5].

6. HTS and MRI

The use of HTS pickup coils in magnetic resonance imaging (MRI) also is quite promising. The most familiar example of MRI, that of whole-body imaging, does not benefit from the introduction of HTS coils because the noise of the sample dominates the system noise. However, low-field MRI, with magnetic fields $B_0 \sim 0.1$ T and NMR frequencies on the scale of 5 - 20 MHz, can benefit from the use of HTS rf pickup coils. These systems are designed to look at small body parts or other small samples [13, 14, 15, 16]. They also are much less expensive than full-body MRI systems, and may find their way into other applications. However, in order to recover the small signals generated in low-field MRI, the S/N of the receiver system must be enhanced. HTS rf coils are a clear solution to this challenge. Here, HTS has an additional advantage because the rf coils are used only for receiving the

FID signal; hence, issues of nonlinearity in the material are less pressing.

In addition to standard MR imaging, there is the application of NMR microscopy, where MRI is performed over a small volume with high resolution [8, 10, 17]. In this case the sample (and therefore the signal) of interest is small. System noise can have a significant contribution from the receiver, and introduction of an HTS coil is clearly beneficial. A simple estimate shows that the coil S/N $\sim \sqrt{Q/T_p}$, where Q is the quality factor and T_p is the probe temperature [8]. One group has demonstrated a signal-to-noise improvement of a factor of 60 in an HTS coil at 10 K versus a Cu coil at room temperature [8]. This dramatically improves the capability of the NMR microscope.

Once again the main benefit of HTS is to improve the S/N. Improved S/N leads to shorter acquisition time for the image because the image acquisition time scales as $(S/N)^{-2}$ [8].

7. Constraints on HTS RF Coils

Although the above discussion makes it appear that HTS pickup coils should be clearly superior to their normal-metal counterparts, there are serious challenges to their introduction into NMR and MRI applications. Here, we discuss three significant constraints imposed on HTS coils in NMR spectrometer systems.

7.1. MAGNETIC FIELD PERTURBATION

First, the HTS coil must be introduced in such a way as to minimize the perturbation of the static magnetic field B_0 [4, 18]. As noted above, the nuclei in the sample must all experience the same external magnetic field to at least 1 part in 10^9. This is required to maintain NMR spectral line widths on the order of 1 Hz, or less, for high-quality spectroscopy. However, the introduction of a superconductor into this magnetic field is likely to alter its distribution since the superconductor will enter the critical state and create an inhomogeneous distribution of magnetic flux within itself. This in turn could ruin the homogeneity of the magnetic field in the sample.

The solution to this problem starts with use of thin-film superconductors for the rf coils. The thickness of these films is typically less than 1 μm. Next, an external field B_0 is applied in the plane of the film, leading to minimal disruption of the magnetic flux by the superconductor. In addition, the superconducting coils also must be kept relatively far away from the sample to minimize the inhomogeneity of the field at the sample. Unfortunately, this has the effect of decreasing the filling factor η of the coil, although this reduction is dictated to some extent by the cryogenic constraints as well. Finally, there also is the significant engineering challenge of keeping the superconducting rf coils aligned with respect to the direction of the field B_0 as the coils are cooled from room temperature [19]. If the coils tilt too much from the ideal orientation, the magnetic field homogeneity at the sample position would be lost. In addition, the HTS rf coils must not move during the pulsing of gradient coils used in spectroscopy, when significant forces can act on the currents flowing in the coil [19]. These pulses last for milliseconds, and can excite mechanical resonances of the coil assembly.

7.2. HIGH TRANSMITTER CURRENTS

The rf coils in NMR spectrometers often are used for both transmitting and receiving rf pulses. This is a challenge because it requires that the coils operate over a wide dynamic range, from high power during transmission to low-level signal recovery during receive mode. In transmit mode, a high-power pulse is sent to the sample. For instance, a $\pi/2$ rf pulse of H_{rf} = 1 mT for a duration of 10 μsec at a frequency of 400 MHz requires about 10 A of current to flow through the coil [4]. Superconductors have a limit to how much current they can support before dissipation sets in. This is defined by the critical current density, J_c. For this rf pulse, the film must have a critical current density $J_c \geq 10^6$ A/cm^2 [4]. Achieving such a large critical current density in a strong magnetic field ($B_0 \sim 20$ T) is not possible with conventional low-T_c superconductors. Only superconducting materials with high critical temperatures and high upper critical fields can operate in the environment of an NMR spectrometer. Hence, HTS thin-film superconductors are required. In addition, the critical-current constraint means that the films must be cooled significantly below their transition temperature to achieve $J_c \sim 1$ MA/cm^2. In practice this means that a superconductor such as $YBa_2Cu_3O_7$ with a T_c of about 90 K must be cooled below 77 K to operate under these conditions.

The HTS film should remain superconducting after transmitting such a pulse. In addition, its rf properties should remain linear [10]. This means that the Q and probe temperature T_p should remain constant during and after rf transmit pulses. This imposes another constraint on the HTS films. These constraints can be met through the use of epitaxial HTS thin films.

7.3. THE Q SHOULDN'T BE TOO HIGH!

In principle, the resistance of a superconductor at rf frequencies can be made arbitrarily small at low temperatures if it has an isotropic non-zero energy gap. This means that the Q of an rf coil made up of this superconductor can be made arbitrarily large, since $Q = \omega L/R$. However, larger Q is not necessarily good for an NMR spectrometer [10]. For an rf coil with a Q = 10,000, the probe has a bandwidth of 100 ppm at 400 MHz, limited by its 3-dB bandwidth. Recall that typical chemical shifts of ^{13}C are on the order of 200 ppm. Thus, a probe with a higher Q would not be able to measure all of the chemical shifts in a ^{13}C NMR spectrum. However, a larger Q may be of benefit for proton NMR, where the chemical shifts are on the order of 10 ppm.

Another problem with a high coil Q is the long ring-down time of the rf energy stored in the coil. The characteristic decay time for the circulating power in the coil is given by $\tau = Q/\omega_0$. For an NMR resonance at 400 MHz with a coil Q of 10,000, the decay time $\tau = 4$ μsec. This means that the coil will not be able to measure the free induction decay signal from the sample for some time after the drive pulse sequence is ended. Higher coil Q factors would increase this "dead time" further. In principle, after the pulse sequence ends, one can actively decrease the Q of the coil to dissipate the stored energy. However, another danger is that a high Q can decrease the characteristic time for energy transfer between the spin system and the pickup coil. This spin damping time scale $\tau_{damping} \sim 1/Q$, may become too short for accurate measurement of chemical shifts or of the relaxation time scales T_1 and T_2 [10].

In fact, the HTS materials do not have an isotropic energy gap. Hence they have intrinsic losses which cannot be eliminated, limiting the ultimate Q attainable from an HTS rf pickup coil [20]. Nevertheless, it is believed that higher Q factors can still be achieved at rf frequencies [8]. To summarize, if the NMR measurements requires significant bandwidth, the Q of the rf receive coils must be limited. Fortunately, a high-Q rf pickup coil can always have its Q reduced in a controlled manner, so that a variety of bandwidths can be achieved [10, 11].

8. HTS Probes for NMR Spectroscopy

We now are in a position to go into more detail about how HTS films are used in an NMR spectrometer.

8.1. BASIC PROBE CONFIGURATION

A basic NMR spectrometer consists of a room-temperature sample tube surrounded by rf coils, all of which are contained in the bore of a superconducting magnet. There are one or more self-resonant rf coils made of HTS films. Each coil is a simple lumped-element loop-gap LC resonator. The loop acts as the inductance and the gap acts as a capacitor. The geometry of the loop and the gap are adjusted lithographically to bring the resonant frequency, $\omega_0 = 1/\sqrt{LC}$, equal to the NMR frequency of interest for a given magnetic field B_0. A translating paddle can be used to fine tune the resonant frequency of the rf coil after it is installed in the probe [18, 21]. The NMR frequency depends on the nucleus of interest (e.g., ^{1}H, ^{13}C, ^{19}F, etc.) and the applied static magnetic field, B_0. Hence, probes are designed specifically with one or two nuclei and a certain magnetic field in mind [22]. Often the probes are used to excite simultaneously one nucleus of interest, in addition to deuterium. The deuterium signal (from a deuterated solvent) is used in a feed back circuit to control the magnetic field B_0. The intricacies of rf probe design for high-field and low-field applications are discussed in a number of papers [11, 16, 23, 24, 25, 26].

The films are typically grown on either sapphire, MgO or $LaAlO_3$ single-crystal substrates, which have low rf losses. In addition, the substrates must have good thermal conductivity because in use they may be cooled only along one edge [21]. These coils typically show quality factors $Q \sim 10{,}000$, even in a field of 9.4 T, as compared to normal metal coils which have Q on the order of several hundred to a few thousand.

8.2. CRYOGENICS AND PACKAGING

The HTS coils must be cooled to temperatures significantly below T_c to establish a large critical current density. This must be done with a cryogenic system which is independent of the neighboring superconducting magnet. Closed-cycle refrigerators, such as a Gifford-McMahon cryocooler, have been employed [4, 5, 11, 21]. There are several constraints on these cryocoolers. First, vibrations of the cold head must be minimized to maintain magnetic field homogeneity in the sample. This requires careful vibration isolation of the compressor from the cold head. Second, mK-level temperature control must be maintained because the rf properties and resonant frequency of the coil are temperature dependent. Third, the cryostat must be non-inductive, so that the rf pulses can pass through the cryostat walls without distortion. Finally, the cryostat also must accommodate a temperature-controlled sample holder which is kept near room temperature. Bruker has sold probes under the name CryoProbe^{TM} which uses coils held at 15 to 30 K and a preamplifier and transmit/receive switch at 60 to 90 K [5, 9].

8.3. PREAMPLIFIER

Once the signal is received by the HTS rf coils, it must be demodulated and detected. The first step is to preamplify the signal with a low-noise amplifier. Having cooled coils presents the opportunity to reduce the noise temperature of the preamplifier by cooling it as well [27, 28]. This will decrease the equivalent amplifier temperature T_A, and increase the system S/N. A GaAs field-effect transistor amplifier operating at 77 K has a noise temperature of just 20 K [4, 8]. The use of an all-77-K preamplifier and cables is expected to improve the system sensitivity by 75% [4]. The rf matching network also can be cooled, or even made superconducting [11]. Finally, the transmit/receive switch also can be kept at cryogenic temperature to minimize noise and losses [9].

9. Results with HTS Coils

9.1. NMR SPECTROSCOPY

There are several very convincing demonstrations that HTS provides significant S/N improvements in NMR spectroscopy and MRI applications. Bruker Instruments has demonstrated a factor of 4 improvement in S/N in the their 500-MHz and 600-MHz ^{1}H and ^{19}F cryogenic probes, as well as ^{13}C probes [5]. These improvements are along the lines of those estimated above. This translates to the ability to measure samples with one fourth the mass, or to measure in one sixteenth the time required with a conventional NMR probe. The measured NMR linewidths are almost as good as those measured by the conventional probe. These cryogenic probes are commercially available.

NMR spectroscopists use many sophisticated techniques to learn about the interactions of nuclear spins in complex molecules. One of these techniques is called Total Correlation Spectroscopy (TOCSY). This is basically a way of examining how spin polarization diffuses through a molecule due to nucleus-nucleus interactions. Hill has demonstrated that a Conductus HTS probe can significantly increase the S/N of a TOCSY spectrum - see Fig. 5 in Hill's paper [4].

Another spectroscopy method, called Heteronuclear Multiple Quantum Coherence (HMQC), excites two different nuclei simultaneously (e.g., ^{1}H and ^{13}C, this is an example of two-dimensional NMR) and establishes which of these nuclei are directly bonded to each other. This kind of information can be used as a fingerprint to uniquely identify a complex molecule. However, since the ^{13}C isotope is so rare (1.1% natural abundance), it takes considerable time (on the order of 12 hours) to acquire enough data to establish the HMQC spectrum. This is a situation for which HTS rf coils provide a clear advantage. With the improved S/N brought by HTS, and good long-term stability, the HMQC spectrum can be acquired in much less time. Bruker has a very clear dramatic demonstration of this improvement on their Web site for the cases of HSQC and HMBC two-dimensional NMR [5].

With the significant improvements in S/N brought by HTS coils, it may be possible to develop altogether new types of multi-dimensional NMR spectroscopy which would be impractical using conventional rf coils. Other companies thought to be investigating cryogenic and/or HTS rf coils for NMR spectroscopy include Dupont, IGC and Nalorac.

9.2. MRI

The improved signal-to-noise ratio of HTS pickup coils can significantly decrease image acquisition time in MRI systems. Penn, *et al.* have demonstrated a factor of 3 improvement in S/N using an HTS coil versus a silver coil, for both at 77 K in a 0.15-T low-field MRI system [29]. A factor of 10 improvement in S/N was demonstrated by Black, Early and Johnson in an HTS pickup coil versus a room-temperature copper coil in an NMR microscope with 30 μm resolution [10]. A factor of 5 improvement of S/N in MR imaging of a rat spine was found by Wosik, *et al.* [30]

MRI of sodium has been useful for examining cell integrity and has advantages over proton MRI for detecting certain kinds of brain disorders. However, sodium MRI suffers from lower S/N, making HTS pickup coils attractive for this application [24]. A very interesting market analysis of the impact of HTS on MRI has recently been published [31].

10. Conclusions

We have given a simple pedagogical overview of the impact of HTS on NMR spectroscopy and magnetic resonance imaging. Although HTS does not introduce any radically new measurement process to this field, it does make significant improvements to the signal-to-noise of the pickup system. This comes at a time when other improvements to signal-to-noise are becoming prohibitively expensive to pursue. Commercial products which employ cryogenic technology in the rf pickup coils are now available and have clearly established their superiority over normal-metal rf probes. Further improvements can be made to the HTS coils, most notably with films grown on flexible substrates for improved sample filling factor.

The author has benefited from fruitful discussions with Richard Withers of Bruker Instruments, Howard Hill of Varian, and Doran Smith of the Army Research Laboratory. Dr. Withers is also acknowledged for a critical reading of the manuscript. The author acknowledges the support of the NATO ASI on Microwave Superconductivity, as well as the Maryland Center for Superconductivity Research.

References

1. C. P. Slicther, *Principles of Magnetic Resonance*, 3rd Edition, (Springer-Verlag, Berlin, 1992).
2. A. E. Derome, *Modern NMR Techniques for Chemistry Research*, (Pergamon Press, Oxford, 1987).

3. Atta-ur-Rahman and M. I. Choudhary, *Solving Problems with NMR Spectroscopy*, (Academic Press, San Diego, 1996).
4. H. D. W. Hill, "Improved Sensitivity of NMR Spectroscopy Probes by use of High-Temperature Superconductive Detection Coils," IEEE Trans. Appl. Supercon. **7**, 3750-3755 (1997).
5. http://www.bruker.com/nmr/Products/Cryoprobes/index.html.
6. http://www.nmr.varian.com/.
7. D. I. Hoult and R. E. Richards, "The Signal-to-noise ratio of the nuclear magnetic resonance experiment," J. Magn. Res. **24**, 71-85 (1976).
8. R. D. Black, T. A. Early, P. B. Roemer, O. M. Mueller, A. Mogro-Campero, L. G. Turner, and G. A. Johnson, "A High-Temperature Superconducting Receiver Nuclear Magnetic Resonance Microscopy," Science **259**, 793-795 (1993).
9. R. S. Withers, private communication.
10. R. D. Black, T. A. Early, and G. A. Johnson, "Performance of a High-Temperature Superconducting Resonator for High-Field Imaging," J. Magn. Res. A **113**, 74-80 (1995).
11. R. S. Withers, G. -C. Liang, B. F. Cole, and M. Johansson, "Thin Film HTS Probe Coils for Magnetic-Resonance Imaging," IEEE Trans. Appl. Supercon. **3**, 2450-2453 (1993).
12. W. H. Wong, R. S. Withers, R. Nast, V. Kotsubo, M. E. Johansson, H. D. W. Hill, L. F. Fuks, K. A. Delin, B. Cole, A. Brooke, W. W. Brey, and W. A. Anderson, "HTS coils for high resolution nuclear magnetic resonance spectroscopy," Adv. in Cryo. Eng. **42**, 953-959 (1996).
13. M. Yap, R. D. Black, W. W. Brey, B. F. Cole, M. E. Johansson, G. A. Johnson, V. Y. Kotsubo, R. Nast, and R. S. Withers, "High Temperature Superconducting Probe for NMR Microscopy," Proc. Soc. Magn. Reson. **2**, 958 (1995).
14. G. van Heteren, T. W. James, and L. C. Bourne, "Thin Film HTS RF Coils for Low Field MRI," Magn. Reson. Med. **32**, 396-400 (1994).
15. M. Vester, F. Steinmeyer, B. Roas, G. Thummes, K. Klundt, "HTS surface coils with liquid nitrogen or pulse tube refrigeration," Proc. Soc. Magn. Reson. **3**, 1528 (1997).
16. Q. Y. Ma, "RF Applications of High Temperature Superconductors in MHz Range," IEEE Trans. Appl. Supercon. **9**, 3565-3568 (1999).
17. Z. H. Cho, C. B. Ahn, S. C. Juh, H. K. Lee, R. E. Jacobs, S. Lee, J. H. Li, and J. M. Jo, "Nuclear magnetic resonance microscopy with 4 μm resolution: Theoretical study and experimental results," Med. Phys. **15**, 815-824 (1988).
18. See for example US Patent No. 5,276,398, Superconducting magnetic resonance probe coil, R. S. Withers and G.-C. Liang, Jan 4, 1994; US Patent No. 5,565,778, Nuclear magnetic resonance probe coil, W. W. Brey, W. A. Anderson, W. H. Wong, L. F. Fuks, V. Y. Kotsubo, R. S. Withers, Oct 15, 1996; US Patent No. 5,585,723, Inductively coupled superconducting coil assembly, R. S. Withers, Dec. 17, 1996; US Patent No. 5,594,342, Nuclear magnetic resonance probe coil with enhanced current-carrying capability, W. W. Brey and R. S. Withers, Jan 14, 1997.
19. H. D. W. Hill, private communication.
20. Jian Mao, Steven M. Anlage, J. L. Peng, and R. L. Greene, "Consequences of d-wave Superconductivity for High Frequency Applications of Cuprate Superconductors," IEEE Trans. Appl. Supercon. **5**, 1997-2000 (1995).
21. V. Kotsubo and R. Nast, "Cryogenic system for a high-temperature superconductor NMR probe," Adv. in Cryo. Eng. **42** (1996).
22. J. R. Miller, Q. Y. Ma, E. S Yang, P. O. Alderson, I. K. Mun, J. Katz, D. W. Face, D. J. Kountz, "A Multiple Nuclei Superconducting Probe for MRI," Proc. Mag. Res. Medicine **1**, 172 (1997).
23. R. D. Black, P. B. Roemer, A. Mogro-Campero, L. G. Turner, and K. W. Rohling, "High temperature superconducting resonator for use in nuclear magnetic resonance microscopy," Appl. Phys. Lett. **62**, 771-773 (1993).

24. J. R. Miller, K. Zhang, Q. Y. Ma, I. K. Mun, K. J. Jung, J. Katz, D. W. Face, D. J. Kountz, "Superconducting Receiver Coils for Sodium Magnetic Resonance Imaging," IEEE Trans. Biomed. Eng. **43**, 1197-1199 (1996).
25. J. R. Miller, K. Zhang, Q. Y. Ma, I. K. Mun, K. J. Jung, J. Katz, D. W. Face, D. J. Kountz, "High Temperature Superconductor Receiver Coils for Sodium Imaging," Proc. Soc. Magn. Reson. **1**, 255 (1996).
26. J. R. Miller, S. E. Hurlston, Q. Y. Ma, D. W. Face, D. J. Kountz, J. R. MacFall, L. W. Hedlund, G. A. Johnson, "*In vivo* MR microscopy at 2 T using a superconducting probe," Proc. Mag. Res. Medicine **1**, 405 (1998).
27. P. Styles, N. F. Soffe, C. A. Scott, D. A. Cragg, F. Row, D. J. White, and P. C. J. White, "A High-Resolution NMR Probe in Which the Coil and Preamplifier Are Cooled with Liquid Helium," J. Magn. Res. **60**, 397-404 (1984).
28. D. R. Williams, W. Lum, and S. Weinreb, "L-band cryogenically cooled GaAs FET amplifier," Microwave Journal, p. 73, October 1980; M. W. Pospiezalski and S. Weinreb, "FETs and HEMTs at cryogenic temperatures - their properties and use in low-noise amplifiers," 1987 IEEE MTT-S Digest, p. 955; J. D. Gallego and M. W. Pospieszalski, "Design and performance of cryogenically-coolable, ultra low-noise, L-band amplifier," Natl. Radio Astronomy Observatory, Electronics Div. Internal Report No. 286, March 1990.
29. S. J. Penn, N. M. Alford, D. Bracanovic, A. A. Esmail, V. Scott, and T. W. Button, "Thick Film YBCO Receive Coils for Very Low Field MRI," IEEE Trans. Appl. Supercon. **9**, 3070-3073 (1999).
30. J. Wosik, K. Nesteruk, L. M. Xie, X. P. Zhang, P. Gierlowski, C. Jiao, J. H. Miller Jr., "Enhanced-resolution magnetic resonance imaging using high-T_c superconducting RF receiver coils," *Proceedings of the 10th Anniversary HTS Workshop on Physics, Materials and Applications*, Editors B. Batlogg, C. W. Chu, W. K. Chu, D. U. Gubser, and K. A. Muller, (World Scientific, Singapore, 1996), pp. 653-654.
31. D. G. Smith, P. S. Gerner, and C. P. Zuhoski, "A Study of the Applicability of High Temperature Superconducting RF Receive Coils to the MRI Industry," IEEE Trans. Appl. Supercon. **7**, 3710-3714 (1997).

HTS ANTENNAS

H.J. CHALOUPKA
University of Wuppertal
Fuhlrottstrasse 10/D 42097 Wuppertal, Germany

Abstract. The benefit of HTS technology to special antenna types is reviewed from a system point of view. Electrically-small antennas for low-noise and narrow-band receivers are identified as one candidate for HTS applications. Major benefits with regard to noise reduction and the integration of antenna and pre-select function are predicted for HTS multi-stage matching networks for electrically-small antennas. Decoupling networks for low-order superdirective multiport arrays with adaptive nulling capability are considered as other candidates, where HTS technology offers significant advantages over conventional solutions.

Table of contents:

H. Weinstock and M. Nisenoff (eds.), Microwave Superconductivity, 353–386.

1. Introduction

The low RF losses in epitaxially-grown HTS thin films have been sucessfully applied to various passive microwave components [1,2], and those HTS components have already been implemented into operational systems such as base transceiver stations for mobile communication. Whereas high-performance bandpass filters represent the majority of implemented passive HTS components, HTS applications to antennas and antenna networks have had much less attention. This is due to the fact that only a small class of special-purpose antennas can benefit from HTS technology.

Consequently, this review article on HTS applications to antenna circuits is focused on those cases where the low RF conductor losses in HTS materials provide capabilities which otherwise are not available. The main purpose of this chapter is to provide the reader with the engineering background required for an understanding of the benefits and limits of implementation of HTS technology to specific antenna types.

Payoff is obtained for a class of lower-frequency antennas at a size-limited platform (i.e., limited aperture space) which conventionally suffer from an unacceptably low efficiency associated with low sensitivity, low precision of direction finding, etc. Single electrically-small antennas, as well as moderately superdirective small arrays, are characterized by a reactive field energy stored in the antenna structure and the attached matching network which is much larger than the received or transmitted power integrated over a specified time period. Here HTS technology can be used to provide high efficiency and a low receiver noise figure if the inherent narrow frequency bandwidth of this type of antenna is consistent with system requirements. The low HTS surface resistance also provides access to novel types of bandpass antennas and frequency multiplexing arrays where the function of frequency selectivity becomes an integral part of the antenna circuit.

2. Specific technological problems associated with HTS antenna structures

Assuming that electrically small and/or superdirective antennas are required, but that conductor losses limit the system-relevant performance of these antennas and matching networks, implementation of HTS technology should be considered. However, the feasibility of HTS antenna structures is governed by a number of severe technological restrictions. In principle there are technological issues similar to those for other passive RF components such as filters, but antenna applications are affected by some specific additional problems.

Whereas in the case of non-radiating HTS components, e.g., filters, the component function is confined completely to the spatial region of the corresponding planar structure, the function of the antenna is to provide an interface between transceiver circuits (i.e., sensors) and the surrounding free space or subsurface structure, e.g., for subsurface remote sensing. Therefore, an "electromagnetically open" configuration must be employed. Hence, for operational temperatures of about 60 to 77 K, cooling of an antenna structure is more difficult than cooling of other passive RF devices. Part of the vacuum encapsulation of the antenna must consist of a "window" ("radome") which is transparent (with low losses) in the operational RF frequency band, but which minimizes the heat load due to thermal radiation - see Fig. 1.

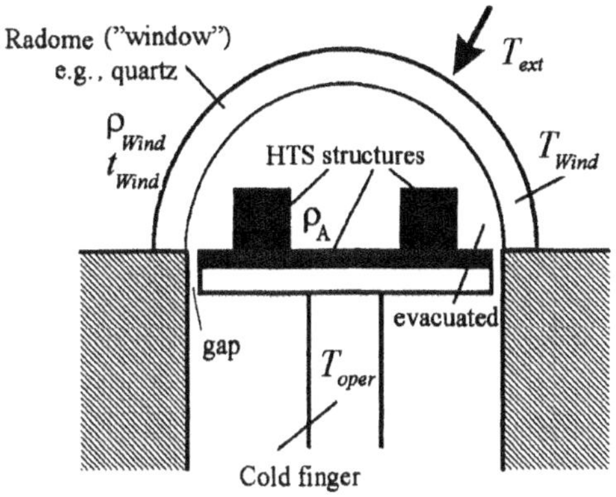

Figure 1. Schematic view of HTS antenna encapsulation. Parameters for heat-load considerations are ρ_{Wind} and t_{Wind} the thermal reflectance and transmittance of the window (with physical temperature T_{Wind}), respectively. The reflectance of the HTS structure at operational temperature T_{oper} is denoted by ρ_A.

Thermal radiation impinging from the inner surface of the window upon the superconductive structure depends on the sum of the thermal black-body radiation of the window (depending on the physical temperature T_{Wind} of the window) multiplied by its absorptance $(1-\rho_{Wind})(1-t_{Wind})$ and the external radiation (T_{ext}) multiplied by $(1-\rho_{Wind})$ t_{Wind} - see Fig. 1 for definitions. Since the window cannot be cooled significantly, both temperatures, T_{Wind} and T_{ext}, are approximately equal to room temperature T_0. In this case, the radiation incident at the superconductive structures becomes about 40 mW/cm^2 times $(1-\rho_{Wind})$, with ρ_{Wind} as the reflectance of the window. The thermal radiation absorbed by the antenna structure finally becomes proportional to $(1-\rho_{Wind})(1-\rho_A)$, with ρ_A as the effective reflectance of the cooled antenna structure - see Fig. 1. Consequently, reflectance of both the window and the superconductive structure must be chosen as high as possible. Quartz as a material for the window in the infrared regime provides a relatively high reflectance due to lattice resonances. The cooling problem can be relaxed if the radiating part of the antenna is comprised of a normal-conductor, and only the antenna network for matching, decoupling, etc. is superconducting. In this case, if thermal contact between antenna layer and network layer is avoided, the antenna layer can serve as a radiation shield.

In order to avoid a heat-conductive bridge between the cold section (with operational temperature T_{oper} on the order of 60 to 77 K) and the warm section, a "gap" (shown in Fig.1) with high thermal resistivity must be used. Suitable measures are required to avoid electromagnetic coupling between the antenna structure and the "cavity" formed by the cold finger and surrounding wall.

With regard to antenna properties, 3-dimensional structures with curved conducting surfaces would often be preferable, but a restriction to layered planar structures is imposed by the use of epitaxial HTS thin films. Polycrystalline HTS thick films could, in principle, be used to cover large curved surfaces. However, over the frequency and temperature range of interest (0.2 to 5 GHz, 60-77 K), the presence of grain boundaries with Josephson junction (JJ) behavior, results in a surface resistance two orders of magnitude higher than for thin films, and in an onset of unwanted nonlinear effects at very low field values. Furthermore, generation of excessive noise which exceeds the thermal (Johnson) noise can occur in granular HTS material.

Another design restriction results from the fact that all currently available compatible substrate materials, e.g., $LaAlO_3$, MgO and sapphire (with buffer layer), for deposition of YBCO and Tl-superconductor thin films possess relatively high permittivity (≥ 10). As will be shown in section 6, frequency-bandwidth requirements yield a lower limit for the substrate thickness of patch antennas. The use of relatively thick high-permittivity substrates renders the suppression of unwanted surface waves more difficult. One approach to overcome this problem is the employment of "mesa structures" where the thick high-permittivity material is restricted to the space below the antenna patches.

A further drawback is the limited power handling related to non-linearities at higher surface fields in conjunction with the "peaked" current-density distribution at the edges of thin planar conductors. Non-linearities occur if the RF surface magnetic field strength exceeds a certain temperature-dependent value on the order of 50 – 200 A/cm. Furthermore, due to the peaked current distribution, unwanted effects such as intermodulation and quality factor degradation can occur at a relatively low RF power level.

3. General properties of electrically-small and superdirective antennas

A general introduction to antenna theory and design, as well as to commonly-used terms and definitions, can be found in many books (e.g., [3]), and is therefore beyond the scope of this review article. Consequently, this chapter provides only those fundamentals which are in particularly relevant for understanding the impact of HTS technology on the class of electrically-small and superdirective antennas - see [4-8].

Antennas provide the transition between propagating electromagnetic modes of transmission lines and the propagating modes of free space. Through reciprocity, the behavior in the receive and transmit mode is closely related. In the transmit mode the far-field radiation pattern is characterized by the variations of the directivity function $D(\Theta,\Phi)$, with directional angles Θ and Φ - see Fig.2.

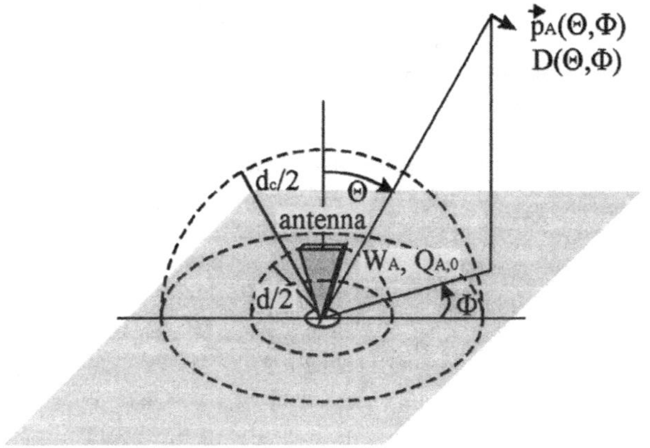

Figure 2.
Schematic view of antenna with coordinates for radiation pattern characterization, where d is the diameter of the antenna and d_c is the critical diameter- see Eq. 7.

The polarization $\vec{p}_A(\Theta,\Phi)$ (complex-valued vector with $\vec{p}_A \cdot \vec{p}_A{}^* = 1$ and with * denoting complex conjugate) represents the second important feature of the antenna. If in the receive mode a single plane wave with wavelength λ_0, radiation density S and polarization $\vec{p}_{inc}$ impinges upon the antenna in direction (Θ,Φ), the available power at the antenna port (with the assumption of a loss-free antenna) is given by

$$P_{rec,avail} = \frac{\lambda_0^2}{4\pi}\left|\vec{p}_A \cdot \vec{p}_{inc}\right|^2 D(\Theta,\Phi)S(\Theta,\Phi) \quad . \qquad (1)$$

The required "steepness" of the variations of D with Θ and Φ determines the required minimum size of the antenna. In order to give a more quantitative description of this steepness, $\sqrt{D(\Theta,\Phi)}\,\vec{p}_A$ can be represented as a sum of spherical wave functions with main mode numbers μ. In the transmit mode, if a time-harmonic wave is incident at the antenna port, spherical modes with complex-valued coefficients are excited. The set of these coefficients are characteristic for an antenna (or antenna port for multi-port antennas), and the radiation properties follow from a superposition of these modes. To discuss electrically-small and superdirective antennas, a measure of the contribution of the different spherical modes to the radiation characteristic must be introduced [7]. The totally radiated power P_{rad} due to the field orthogonality can be written as a sum of contributions $P_{\mu,\nu}$ from the different spherical modes. Furthermore, if all contributions which belong to the same main mode number μ are summed to give $P_{rad,\mu}$, we can introduce coefficients

$$\alpha_\mu = \frac{P_{rad,\mu}}{P_{rad}} \text{ with } \sum_{\mu=1}^{\infty}\alpha_\mu = 1 . \qquad (2)$$

The largest linear dimension is described by the diameter d of the smallest sphere enclosing the antenna structure - se Fig. 1. Whereas in the far-field region (with $r > 2d^2/\lambda_0$) electric and magnetic field components are in phase, reactive components predominate in the space which immediately surrounds the antenna. Energy storage in this "reactive near-field region" is characterized by the radiation quality factor Q_{rad}, which relates the maximum value W_A of the time-varying reactive field energy $W(t)$ to the radiated power P_{rad} via

$$Q_{rad} = \frac{\omega W_A}{P_{rad}} . \qquad (3)$$

The properties of electrically-small and superdirective antennas differ substantially from those of conventional antennas, mainly due to the relatively high radiation quality factor. Because of this fact, the definition and impact of the reactive field energy is considered first. An appropriate introduction to the term "reactive field energy" can be given by considering the radiation of a narrowband RF pulse with carrier frequency $f_0 = \omega_0/2\pi$ and time duration Δt. The time-dependent power possesses oscillations with frequency $2f_0$ around the slowly time-varying mean value $p_{rad}(t)$. The lower part of Fig. 3 depicts schematically this time-varying radiated power. The radiated energy W_{rad} is represented by the area under the curve $p_{rad}(t)$. The upper part of Fig. 3 illustrates the corresponding time-varying power $p_{in}(t)$ which is fed into the antenna port. In the time interval $0 < t < t_0$, power flows into the antenna port, so that a total energy W_+ - see Fig. 3 - is fed into the antenna. After $t = t_0$, power flows back to the generator, so that the totally-reflected energy is given by W_-. For the case of a lossless antenna, conservation of energy requires

$$W_{rad} + W_- = W_+ \tag{4}$$

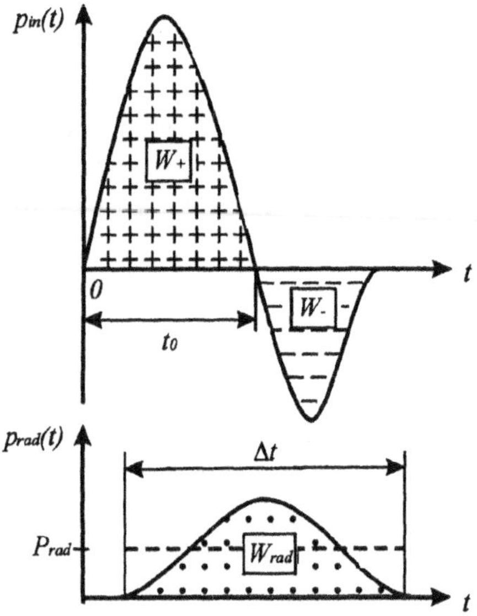

Figure 3.
Illustration of reactive power by means of the radiation of a narrow-band impulse of time duration Δt and carrier frequency f_0.
Upper part: Time-dependence of input power (mean value) with ingoing energy W_+ and reflected energy W_-.
Lower part: Time-dependence of radiated power with totally-radiated energy W_{rad}.

For the case illustrated in Fig. 3, the energy W_- is temporarily stored in the antenna nearfield and subsequently delivered back to the generator. This energy W_- is referred to as "reactive energy."

Assuming a radiated pulse with energy W_{rad} and constant spectral density within $f_0 - \Delta f/2 < f < f_0 + \Delta f/2$, one obtains [12]

$$\frac{1}{\Delta f}\int_{f_0-\Delta f/2}^{f_0+\Delta f/2} \frac{df}{1-|\Gamma(f)|^2} = \left\langle \frac{1}{1-|\Gamma(f)|^2} \right\rangle \geq \frac{W_+}{W_{rad}}, \tag{5a}$$

For the ratio of W_+ to W_{rad}, the approximation

$$\frac{W_+}{W_{rad}} \approx \sqrt{1+\left(\frac{Q_{rad}}{2\pi}\frac{\Delta f}{f_0}\right)^2} \tag{5b}$$

can be used [12]. Here, $\Gamma(f)$ denotes the input reflection coefficient of the antenna. Combining Eqs. (5a) and (5b) finally leads to a result for the upper limit of the fractional bandwidth:

$$\frac{\Delta f}{f_0} \leq \frac{2\pi}{Q_{rad}}\sqrt{\left\langle\frac{1}{1-|\Gamma(f)|^2}\right\rangle^2 - 1} \tag{6}$$

Eq. (6) provides a general estimation [12] for the achievable fractional bandwidth of power matching as a function of Q_{rad} and the in-band reflection. This issue will be discussed in greater detail in the next section.

The diameter d_c of the reactive near-field region depends on the radiation pattern. In the special case for which the radiation pattern is composed of only one spherical mode function with main mode number $\mu = \Lambda$,

$$d_c = \Lambda\frac{\lambda_0}{\pi} \tag{7}$$

holds, and the radiation quality factor is estimated via [7] to be

$$Q_{rad} \approx \left(\frac{\Lambda\lambda_0}{\pi d}\right)^{2\Lambda+1} = \left(\frac{d_c}{d}\right)^{2\Lambda+1}, \tag{8}$$

with d_c/d as the size-reduction factor. Eq. (8) clearly indicates that Q_{rad} sharply increases if (due to a limited platform size or due to required stealthiness) the antenna is chosen smaller than d_c. The steepness of this increase dramatically grows with the order Λ of the pattern. Antennas with a dipole-like radiation pattern correspond to $\Lambda = 1$ and are called electrically small if $d \ll d_c$. In this case Q_{rad} increases with the 3rd power of the size-reduction factor, so that, e.g., an electrically-small dipole of size $d = 0.03\ \lambda_0$ possesses $Q_{rad} > 1000$.

Antennas with a more complex radiation pattern require a significant contribution from higher spherical modes. In this case the radiation quality factor can be estimated by weighting the Q values of the individual spherical modes with the coefficients α_μ from Eq. 2:

$$Q_{rad} = \sum_{\mu=1}^{\infty} \alpha_\mu Q_{rad,\mu} = \sum_{\mu=1}^{\infty} \alpha_\mu \left(\frac{\mu\lambda_0}{\pi d}\right)^{2\mu+1} . \tag{9}$$

In a conventional antenna of diameter d the contribution to the radiation pattern of all modes with $\mu > \Lambda_c = \pi d/\lambda_0$ can be neglected , so that the radiation quality factor is small. If modes with $\Lambda \geq \mu > \Lambda_c = \pi d/\lambda_0$ contribute significantly to the radiation pattern, the antenna is called "superdirective." Here the mode with the highest mode number Λ dominates in Eq. 9 for the quality factor, so that Eq. 8 with Λ taken to be the highest significant mode number ("pattern order") can be used as an estimate. Superdirectivity can be employed to achieve (in comparison to a conventional antenna of the same size) a higher maximum directivity $D_{max} \approx \Lambda^2+2\Lambda$ or a higher number of radiation nulls. However, as will be pointed out below, a limit for the exploitation of superdirectivity is posed by bandwidth requirements.

Associated with the storage of field energy are losses in the conductors and dielectric materials of which the antenna is composed. These losses are characterized by the unloaded quality factor

$$Q_{A,0} = \frac{\omega W_A}{P_{diss,A}} \tag{10}$$

of the antenna structure. In the receive mode the available power at the antenna port is, with $P_{avail} = \eta_A P_{rec,avail}$, reduced by a factor

$$\eta_A = \frac{R_{rad}}{R_{rad} + R_{A,diss}} = \left[1 + \frac{Q_{rad}}{Q_{A,0}}\right]^{-1} , \tag{11}$$

which is called the radiation efficiency of the antenna - see Fig. 4.

In the final part of this section, the major conclusions about electrically-small and superdirective antennas are summarized for the receive mode.

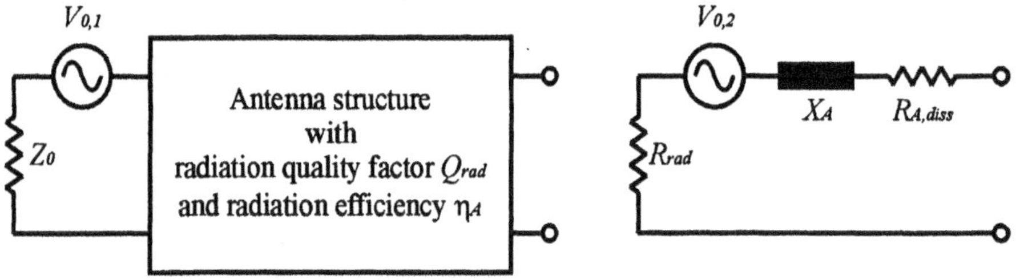

Figure 4. General representation of an antenna in the receive mode. Left: Representation of the antenna structure by means of a two-port fed by an equivalent voltage source. Right: Voltage source representation at the antenna port.

The left part of Fig. 4 shows a general representation of antennas in the receive mode. The equivalent voltage source, with voltage $V_{0,1}$ and internal resistance equal to the wave impedance of free space $Z_0 = 120\pi\,\Omega$, possesses available power according to Eq.(1). Hence, the voltage is given by

$$V_{0,1} = \sqrt{8Z_0 P_{rec,avail}}\,. \tag{12}$$

The two-port in the circuit on the left-hand side of Fig. 4 represents the storage of reactive power in the antenna structure associated with the delivery of power to a load, and is described by the radiation quality factor. Dissipative losses in the antenna structure, leading to a radiation efficiency $\eta_A < 1$ also are represented by the two-port.

The right-hand part of Fig. 4 shows an equivalent voltage source for the antenna port. The available power of the "driving voltage source" with $V_{0,2}$ and radiation resistance R_{rad} equals the available power of the voltage source in the left part of Fig. 4. Therefore,

$$V_{0,2} = \sqrt{8R_{rad} P_{rec,avail}} \tag{13}$$

holds. Radiation resistance R_{rad} as well as reactance X_A, result from the transformation of Z_0 to the antenna port by means of the two-port. Since in the case of electrically-small and superdirective antennas the "coupling to free space" is weak, the resulting value of R_{rad} becomes relatively small. $R_{A,diss}$ represents dissipative losses and can be on the same order as or even larger than R_{rad}, leading (according to Eq. (11)) to low values of the radiation efficiency η_A. The available power at the antenna port is lower (by the factor η_A) than the available power $P_{rec,avail}$ of the loss-free antenna. The reactive energy stored in X_A can be much higher than the power delivered per period (in time $1/f$) to the load due to the high radiation quality factor.

The special case of a monopole antenna over conductive ground may serve as an example for the circuit parameter of an electrically-small antenna. Part (a1) of Fig. 5 depicts the equivalent circuit for a monopole with height h smaller than a quarter of the free space wavelength. Inductance L_M and capacitance C are approximated via

$$L_M = \frac{1}{\sqrt{12\pi}}\frac{h}{c}Z_0 \quad \text{and} \quad C = \gamma \cdot \frac{h}{cZ_0}\,, \tag{14}$$

with c as the velocity of light and the constant γ between about 1 and 6, depending on the diameter-length ratio of the monopole. In the frequency region considered, $Z_0 >> \omega L_M$ holds. In this case the parallel circuit consisting of L_M and Z_0 can be transformed approximately into a series circuit consisting of L_M and a resistor (radiation resistance) which together with C forms a series resonant circuit with

$$R_{s,M} = \frac{\pi}{3}\left(\frac{h}{\lambda_0}\right)^2 Z_0 \text{ and } Q_{A,M} = (0.025 \div 0.125)\cdot\left(\frac{\lambda_0}{h}\right)^3 \qquad (15)$$

for the radiation resistance and the radiation quality factor, respectively. Furthermore, the reactance of the inductance can, in the frequency region of interest, be neglected in comparison to the reactance of the capacitance C, so that the simplified circuit (a2) in Fig. 5 results. The voltage in Fig. 5(a2) becomes

$$|V_M| = h\sqrt{2Z_0 S}\cdot \sin\Phi \qquad (16)$$

for a wave with power density S, incident at an angle Φ - see Fig. 2 - and with electric field polarized in parallel to the monopole.

In the lower part of Fig. 5 an analog representation for electrically-small loop antennas is shown - see Figs. 5(b1) and 5(b2). The radiation resistance and inductance of the loop are estimated via

$$R_{s,L} = \frac{8\pi^3}{3} Z_0\left(\frac{A}{\lambda_0^2}\right)^2 \text{ and } L = (0.4 \div 1.6)\cdot\frac{\sqrt{A}}{c} Z_0, \qquad (17)$$

with A denoting the area of the loop opening. The voltage becomes

$$|V_L| = 2\pi\frac{A}{\lambda_0}\sqrt{2Z_0 S}\ , \qquad (18)$$

if the wave is incident parallel and electric field vector perpendicular to the loop surface.

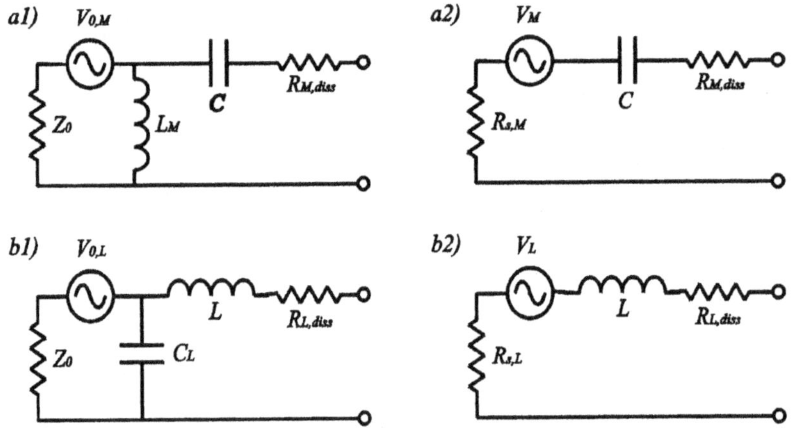

Figure 5. Equivalent circuits for electrically-small antennas. (a) Electrically-small monopole antenna: Full circuit in (a1) and simplified circuit in (a2). (b) Electrically-small loop antenna: Full circuit in (b1) and simplified circuit in (b2).

4. Matching of highly reactive antennas

4.1 POWER GAIN AND NOISE FIGURE OF A RECEIVER FRONT-END AS FUNCTION OF THE SOURCE IMPEDANCE. POWER AND NOISE MATCHING

Depending on the system function, the signal at the antenna port must be subjected to specific analog and/or digital signal processing in order to retrieve the required information. Consequently, in the receiver front-end the signal must be amplified to exceed a required minimum power level. However, amplification is always associated with a decrease in signal-to-noise ratio (SNR). Hence, the receiver must be optimized with respect to both a sufficient output power level and a low noise figure.

An analog linear amplifier is assumed to be at the output port matched to the load. The (frequency-dependent) input impedance of this amplifier is denoted by

$$Z_{in}(\omega) = R_{in}(\omega) + jX_{in}(\omega) = |Z_{in}(\omega)| \cdot \exp[j\Phi_{in}(\omega)] \,. \tag{19}$$

If a generator with available power P_{avail} and source impedance

$$Z_A(\omega) = R_A(\omega) + jX_A(\omega) = |Z_A(\omega)| \cdot \exp[j\Phi_A(\omega)] \tag{20}$$

is connected to the amplifier, the power P_{out} at the output of the amplifier is related to the available power of the generator via

$$\frac{P_{out}}{P_{avail}} = g \cdot \frac{4\cos\Phi_A \cdot \cos\Phi_{in}}{\dfrac{|Z_A|}{|Z_{in}|} + \dfrac{|Z_{in}|}{|Z_A|} + 2\cos(\Phi_A - \Phi_{in})} = g \cdot \left(1 - |\Gamma|^2\right). \tag{21}$$

with $|\Gamma|^2$ defined as the power reflection coefficient.

Maximum output power is achieved for power matching, i.e. if the output impedance of the generator is the conjugate of Z_{in}. In this special case the output power is given by

$$P_{out} = g \cdot P_{avail} \,,$$

with g as the available power gain of the amplifier. The expression $(1-|\Gamma|^2)$, introduced in Eq.(21), is referred to as the "power matching coefficient" and describes the reduction of the power gain P_{out}/P_{avail} relative to the available power gain. Power matching means that for a maximum power gain the source impedance is chosen to be $Z_A = Z_{in}^*$, which is equivalent to $|Z_A| = |Z_{in}|$ and $\Phi_A = -\Phi_{in}$.

In addition to power gain, the noise figure represents another important property of an amplifier. If the available spectral power density of the source noise is given by kT_{source}, with k as Boltzmann`s constant and T_{source} as the noise temperature of the source, the signal-to-noise degradation is given by

$$\frac{SNR_{out}}{SNR_{in}} = \frac{1}{(F-1)\frac{T_0}{T_{source}}+1}. \tag{22}$$

For the case that T_{source} equals $T_0 = 290$ K, the SNR is reduced by the noise figure F.

Like the power gain, the noise figure of the amplifier also depends on the complex-valued source impedance and becomes minimum for the "optimum source impedance" Z_{opt}. This is due to the fact that a noisy amplifier can formally be decribed by a noise voltage and a noise current source at the input port. Since these sources are partially correlated, their superposition leads to a minimum noise power at source impedance Z_{opt}. Noise matching exists, if for minimum noise figure F, the source impedance Z_A is chosen to be equal to Z_{opt}. It's worth noting that in general the optimum source impedance Z_{opt} of an amplifier differs from Z_{in}*, so that for a minimum noise figure, a different source impedance must be chosen than that for maximum power gain.

In the general case (i.e., no noise matching), the dependence of the noise figure on the source impedance $Z_A = R_A + \mathrm{j}\, X_A$ can be expressed as

$$F(Z_A) = F_{\min} + \frac{|Z_{opt} - Z_A|^2}{r \cdot R_A}, \tag{23}$$

with the minimum noise figure F_{min}, r and Z_{opt} as parameters of the amplifier, and Z_A and R_A as the antenna impedance and its real part, respectively.

These results are not restricted to amplifiers, but apply to all linear two-ports ("receiver two-ports"). In a practical receiver configuration, a cable section and a preselect bandpass filter often are inserted between the antenna port and the amplifier. Formally, the above relations can be applied if the cascade of cable, preselect filter and amplifier as a whole is considered as a "receiver two-port", and the two-port parameters Z_{in}, F_{min}, r and Z_{opt} of the "receiver two-port" are taken into account. However, for the case of a cascade of cable sections, filters, amplifiers, etc., the need for a modular design and absence of standing waves at the cable section, leads to the choice of a common real-valued characteristic impedance where all components in the cascade are matched. Usually, the characteristic impedance of standard cables (50 or 60 Ω) is used as a common reference. Consequently, the amplifier is noise- or power-matched to the real-valued cable impedance. Since the filter is also designed with respect to the common characteristic impedance, the antenna also must be matched to 50 or 60 Ω.

It is important to note that, especially for electrically-small and superdirective antennas, an interesting alternative consists of connecting the amplifier directly to the antenna port. In this case, the antenna impedance has to be matched to Z_{in} for maximum power gain or to Z_{opt} for a minimum noise figure.

4.2 MATCHING NETWORKS

4.2.1 General consideration

As was noted in section 3, the antenna impedance of electrically-small and superdirective antennas is dominated by its reactive part. Hence, a "very small" real part R_A is combined with a strongly frequency-dependent reactance with

$$\omega\left|\frac{d}{d\omega}X(\omega)\right| >> R_A \ . \tag{24}$$

In general, the impedance Z_A differs considerably from both the optimum impedance Z_{opt} for noise matching and from $Z_{in}{}^*$ for power gain maximization. It differs also significantly from typical real-valued characteristic impedances of cables on the order of 50 to 60 ohms. For that reason, a matching network (MN) is employed in order to transform the antenna impedance Z_A into the desired impedance Z_L , which may be equal to the optimum impedance for a minimum noise figure, or equal to the conjugate of the input impedance of the two-port connected to the antenna. For the sake of simplicity, in the following discussions, matching to a real-valued impedance Z_L is considered. Generalization to a complex-valued impedance is straightforward.

For some antenna structures the MN represents a two-port which can clearly be separated from the antenna structure, but in other cases the matching structure and radiating structure of an antenna form a non-separable unit. The impedance transformation is performed by a controlled storage of field energy W_N (maximum value of the time-varying energy) in the matching network. The ratio $\beta_{MN} = W_N/W_A$ is a measure of the matching expenditure. If the MN only serves to compensate for the reactive part of the antenna impedance at the center frequency, $\beta_{MN} = 1$ holds. For the cases that (a) a very low radiation resistance must be transformed to a higher value or (b) a multipole MN for achieving a higher frequency bandwidth - see section 4.2.3 - is employed, β_{MN} can become much higher than one.

The overall efficiency η, including antenna structure and MN, is given by

$$\eta = \left[1 + Q_{rad}\left(\frac{1}{Q_{A.0}} + \beta_{MN}\frac{1}{Q_N}\right)\right]^{-1} \tag{25}$$

where Q_N denotes the unloaded quality factor of the components which form the MN.

The signal delivered at the antenna port cannot directly be processed, but must be fed into a receiver or "sensor" for amplification, frequency conversion, analog-to-digital conversion, etc. The input impedance of this load is denoted by Z_L and the output impedance of the cascade consisting of antenna and MN by Z_{out} - see Fig. 6(a).

Highest power transfer requires $Z_{out} = Z_L{}^*$. Otherwise, a power mismatch with the reflection coefficient

$$\Gamma(\omega) = \frac{Z_{out}/Z_L - 1}{Z_{out}/Z_L + 1} \tag{26}$$

occurs, leading to a received power P_{rec} given by

$$\frac{P_{rec}}{P_{rec,avail}} = \left(1 - |\Gamma(\omega)|^2\right)\eta(\omega). \tag{27}$$

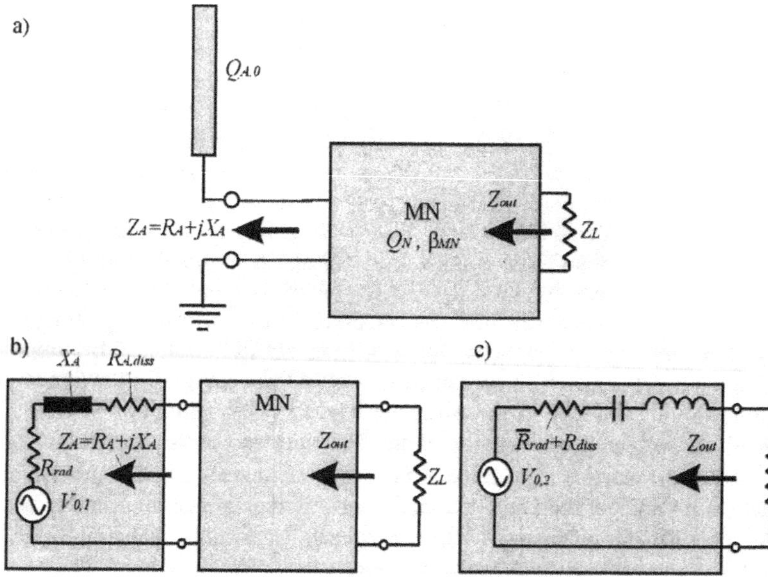

Figure 6. Characterization of matching networks: (a) configuration with antenna (only shown schematically), matching network (MN) and load; b) equivalent circuits; c) simplified circuit for one-pole configuration (single-section matching).

4.2.2 Single-section matching

For a matching circuit which, together with the antenna circuit, produces only one resonance in the operational frequency regime ("one-pole" MN or self-resonant antenna) the simplified circuit model in Fig. 6c can be used. Although highest power transfer at center frequency ω_0 occurs for $Z_L = \bar{R}_{rad} + R_{diss}$, we include the case of "overcoupling" by the introduction of an overcoupling coefficient, $\kappa \geq 1$, defined via

$$\kappa = Z_L / \left(\overline{R}_{rad} + R_{diss}\right) = \eta Z_L / \overline{R}_{rad} \quad . \tag{28}$$

In this case the frequency response of the mismatch factor becomes

$$1 - \left|\Gamma(\omega)\right|^2 = \frac{4\kappa}{(1+\kappa)^2 + \left[\eta Q\left(\omega/\omega_0 - \omega_0/\omega\right)\right]^2} . \tag{29}$$

We introduce a generalized frequency bandwidth $\Delta\omega$ by requiring the power transfer - see Eq. 27 - at the band edges, $\omega_0 \pm \Delta\omega/2$, to be 8/9 of the value at the center frequency. This definition is for the special case $\kappa = 1$, identical to the definition which requires that the voltage standing wave ratio (VSWR) is lower than 2 throughout the band. With this definition, the fractional bandwidth is given by

$$\frac{\Delta\omega}{\omega_0} = \frac{1+\kappa}{\sqrt{8}\,\eta Q} \quad . \tag{30}$$

Equations 25-30 clearly indicate the benefits as well as the limitations in the use of superconductivity in antenna and matching circuits. If Q_{rad} is very high (>1000), even with 3-dimensional antenna and matching structures having an unloaded quality factor ($Q_{A,0}$, Q_N) on the order of 2000, the efficiency η drops below 0.5. If MN and/or antenna are restricted to low-volume planar circuits with $Q_{A,0}$, $Q_N < 200$, even for moderate Q_{rad} values on the order of 100, efficiencies smaller than 0.5 result. With the use of superconductivity in the MN, and if necessary also in the antenna, the value of η can be increased to almost one. According to Eq. 30 (with $\kappa = 1$) a $1/\eta$ increased power transfer is obtained at the cost of a bandwidth reduction by the factor of η.

However, overcoupling with $\kappa > 1$ can be employed to control the bandwidth reduction at the expense of a reduced power transfer, but in such a way that there is still an improvement in comparison to the conventional case. If a conventional structure with a low efficiency $\eta = \eta_{nc}$ and a bandwidth $(\Delta\omega)_{nc}$ for $\kappa = 1$ is replaced with a superconducting structure with $\eta_{sc} \approx 1$, the bandwidth $(\Delta\omega)_{sc}$ can be chosen to be between $(\Delta\omega)_{nc}$ and $\eta_{nc}(\Delta\omega)_{nc}$ if κ is chosen between $(2/\eta_{nc})$-1 and 1. The corresponding improvement in power transfer is between (2- η_{nc}) and 1/ η_{nc}. For example, for the case of $\eta_{nc} = 0.1$, the bandwidth (with $\kappa = 19$) can remain unchanged with an improvement in power transfer by a factor of 1.9. If a bandwidth reduction is allowed, the power transfer becomes higher, with a maximum improvement of 10 in power transfer and a corresponding bandwidth reduction of 10.

For some system applications, e.g., in the case of strong interfering signals the narrow instantaneous bandwidth of electrically-small HTS antennas becomes a benefit. However, in the case of wideband operation, electronic tunability is required. Suitable tuning approaches which need to be further investigated are based on ferroelectric [13] or ferrimagnetic materials, semiconductor varactors or mechanical tuning via piezoelectric actuators.

4.2.3 Multiple section matching networks. Bandpass antennas

As pointed out in section 4.2.2 , for the case of a self-resonant antenna or an antenna with a single section (one-pole) MN, the fractional bandwidth for *VSWR* < 2 is limited to $1/\sqrt{2}\ Q_{rad}$. By using a multiple-section MN –see Fig. 7- a bandwidth improvement relative to the single-section version can be gained [9,10], but due to the stored reactive energy, this improvement factor is limited [12]. Based on Fano´s results [11], one obtains the approximation

$$\frac{\Delta\omega}{\omega_0} \approx \frac{1}{\sqrt{2}\,Q_{rad}} \frac{1.355 \cdot N^2}{1 + 0.355 \cdot N^2} \tag{31}$$

with N as the total number of poles, corresponding to N-1 additional matching circuits. Hence, the ultimate improvement factor for (theoretically) infinite circuits is 3.8; but with just $N = 4$, an improvement by a factor of 3.2 is achieved. The higher the value of N , the closer one comes to a bandpass response which cuts off sharply at the band edges.

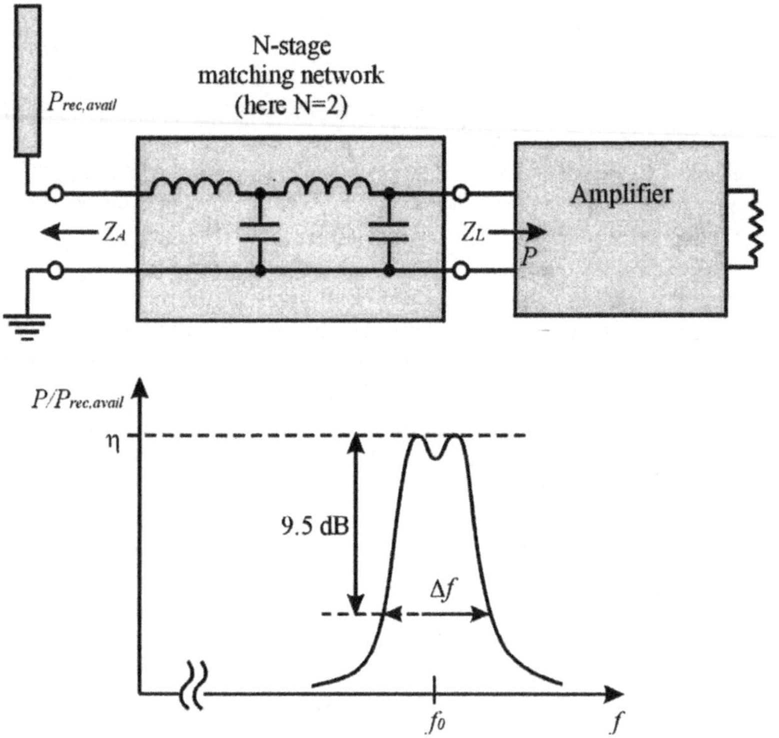

Figure 7. Multiple section matching.

Comparing this "bandpass antenna" with a bandpass filter, the common feature is that both are composed of coupled resonators. Whereas in the case of bandpass filters, both outer resonators are coupled via ports to transmission lines, in the case of a bandpass antenna, one outer resonator is directly coupled to free space and consists of (or includes) the antenna structure.

For Chebyshev response of a given order N and in-band $VSWR < 2$ (ripple 0.5 dB), fractional bandwidth and external coupling strength of the outer resonator (here characterized by Q_{rad}) have a fixed relationship to each other. Utilizing a novel concept for bandpass antennas by considering the required bandwidth and selecting an order N, one finds by means of filter synthesis the necessary radiation quality factor Q_{rad} (external quality factor) which then is realized by adjusting free geometric parameters of the chosen antenna structure. These bandpass antennas offer various advantages for particular system applications:

(a) As the antenna serves as an integral part of the preselect filter, strong out-of-band interfering signals are prevented from entering active elements (e.g., LNA and sensors) which are placed directly at the antenna port [8].

(b) Conventional antenna elements which cover the system bandwidth can be replaced with a set of small bandwidth antennas which perform the function of a frequency multiplexer. In the case of arrays, the small size and frequency selectivity allows bandpass subarrays for different frequency bands to be interlaced at a common platform, even if $\lambda_0/2$ element spacing is required. Different radiation patterns can be allocated to the different subarrays [8,17].

(c) For platforms with co-existing antenna systems for different frequency bands, mutual interaction represents a serious problem. In this case, bandpass antennas offer the advantage of a smaller interaction.

The benefit of HTS – at least in compact MNs - becomes apparent if one takes into account that the dissipative attenuation at the band edge is approximately

$$\frac{a}{\text{dB}} \approx 4.3 \frac{2+N^2}{3} \frac{Q_{rad}}{Q_0} . \tag{32}$$

If the efficiency is required to be no smaller than 8/9 (about 0.5 dB insertion loss)

$$Q_0 / Q_{rad} > 2.8\left(2 + N^2\right)$$

must hold, e.g., for 0.5 % bandwidth with $Q_{rad} = 460$ and $N = 4$, an unloaded quality factor of $Q_0 \geq 23{,}000$ is required.

4.3 APPROACHES FOR WIDEBAND OPERATION OF ELECTRICALLY-SMALL ANTENNAS

The conclusion drawn from sections 4.2.2 and 4.2.3 is that due to the highly reactive impedance of the antenna, power and noise matching is limited to a relatively small fractional bandwidth on the order of 2 % and less. Overcoupling - see Eq. 30 - can be used to achieve increased frequency bandwidth at the cost of reduced power transfer and an increased noise figure.

However, for applications with an operational bandwidth on the order of one octave or more, the power- or noise-matching concept introduced above becomes completely useless. In this case another approach for coupling between antenna and sensor (amplifier) must be chosen. This approach provides a large bandwidth at the cost of large power or noise mismatch.

The origin of narrow bandwidth in association with power or noise matching is related to the fact that for a highly reactive antenna, the phase Φ_A of the antenna impedance changes rapidly with frequency. Hence, the phase requirement of the matching condition can be met approximately only in a narrow frequency range. In order to overcome this problem, the ratio P_{out}/P_{avail} in Eq. 21 is maximized with respect to $|Z_A|$ only. From this, one obtains the optimum solution for

$$|Z_A| = |Z_{in}|. \tag{33}$$

If Eq. 33 is met, Eq. 21 becomes

$$1-|\Gamma|^2 = \frac{2\cos(\Phi_A)\cdot\cos(\Phi_{in})}{1+\cos(\Phi_A-\Phi_{in})} = 1-\frac{\sin\left(\frac{\Phi_A+\Phi_{in}}{2}\right)}{\cos\left(\frac{\Phi_A-\Phi_{in}}{2}\right)}. \tag{34}$$

For a load with a real impedance ($\Phi_{in}=0$), the matching condition becomes

$$Z_L = |Z_A| \approx |X_A| \quad . \tag{35}$$

Eq. (35) can be met approximately over a wide frequency range. For a highly reactive electrically-small antenna, one finds

$$\cos(\Phi_A) \approx 1/Q,$$

leading to

$$\frac{P}{P_{rec,avail}} = \frac{2}{Q}\cdot\eta = \frac{2}{Q_A} \tag{36}$$

for the power transfer. Important conclusions, contained in Eq. (36) are that

(i) wideband matching according to Eq. (35) is achieved at the price of a reduction of the transmitted power by the factor of $2/Q_A$. If, e.g., the radiation quality factor of the antenna is 200, the transmitted power is reduced by 20 dB in comparison to the case of power matching and $\eta = 1$.

(ii) the effect of dissipative losses in the antenna on the power transfer can (for matching according to Eq. (35)) be neglected if the dissipative component of the antenna impedance is much smaller than the reactive component. In this case, no benefit is achieved with superconductive antenna structures.

In [14] a SQUID amplifier is combined with an HTS loop antenna, in accordance with the matching concept described in this section.

5. Effect of increased efficiency on the signal-to-noise ratio of a receiver

Electromagnetic radiation incident at the antenna represents a superposition of "desired signals" (e.g., communication and navigation signals) and "unwanted signals" such as external noise sources. For a radiometer the noise under consideration represents the wanted, and background noise is the unwanted signal. The spectral power density of the external noise which is available at a (fictitious) lossless antenna can formally be described by an effective temperature of the radiation resistance (designated as the external antenna noise temperature T_A). It results from a summation of all noise sources seen by the antenna, weighted by the directivity of the antenna.

$$T_A = \frac{\iint D(\Theta,\Phi)\,T(\Theta,\Phi)\,d\Omega}{\iint D(\Theta,\Phi)\,d\Omega} = \frac{1}{4\pi}\iint D(\Theta,\Phi)\,T(\Theta,\Phi)\,d\Omega\,. \tag{37}$$

Below 100 MHz the external noise is primarily galactic noise and higher than 1000 K. For higher frequencies, galactic noise decreases, but atmospheric noise increases due to gaseous absorption. At about 1 GHz a minimum with combined galactic-atmospheric noise below 10 K occurs. For those portions of the antenna pattern that point toward the earth, a contribution on the order of 300 K results. In urban areas, man-made noise, e.g., from power lines (arcs and corona) and noise generated by vehicles, may dominate.

Thermal Johnson noise is added in the lossy antenna and matching structure. If this additional noise (as for the external noise) is referred to as an effective noise temperature of the radiation resistance, one obtains - see Fig. 8 -

$$T_{eff,A} = \frac{Q_{rad}}{Q_{A,0}} T_{A,phys} + \beta_{MN} \frac{Q_{rad}}{Q_N} T_{N,phys}\,, \tag{38}$$

with $T_{A,phys}$ and $T_{N,phys}$ as the physical temperatures of antenna and MN structure, respectively. Equation 38 shows that for a high Q_{rad} , the losses must be sufficiently small to assure that the thermal noise produced in the antenna is sufficiently lower than the physical temperature. Utilization of HTS instead of normal conductors at room temperature, reduces $T_{eff,A}$ in a two-fold manner, due to the reduced physical temperature and due to the higher Q factors, $Q_{A,0}$ and Q_N. However, cooling of a normal conducting MN would lead to a significant reduction of $T_{eff,A}$.

In order to fully understand the impact of an increased antenna efficiency on the SNR, the noise contribution of the active circuit with load impedance Z_L connected to the antenna port (e.g., preamplifier and sensor, hereafter denoted as "amplifier") must be considered - see the upper part of Fig. 8. The "interface" between the output of the unit composed of antenna and matching network (MN) on one side and the amplifier on the other side, is characterized by the reflection coefficient

$$\Gamma(\omega) = \frac{Z_{out}/Z_L^* - 1}{Z_{out}/Z_L + 1}.$$

As already explained in section 4.1 for the general case, the noise contribution of the amplifier becomes minimum for an amplifier-specific mismatch , described by the reflection factor - see Eq. 23 -

$$\Gamma_{opt} = \frac{Z_{opt}/Z_L^* - 1}{Z_{opt}/Z_L + 1}.$$

For the case of this "noise-matching," the noise contribution T_{min} is lower than for the case of power matching (T_{pm} for $\Gamma = 0$). If, e.g., the pre-amplifier can directly (without connecting cable) be implemented at the antenna port, the MN should be designed for noise-matching with $\Gamma = \Gamma_{opt}$, rather than for power-matching with $\Gamma = 0$ at the center frequency. The receiver noise temperature (due to internal thermal noise from antenna and amplifier noise) for arbitrary Γ values becomes

$$T_{rec} = T_{eff,A} + \frac{1}{\eta}\left[T_{\min} + \left(T_{pm} - T_{\min}\right)\frac{\left|\Gamma - \Gamma_{opt}\right|^2}{\left|\Gamma_{opt}\right|^2\left(1 - \left|\Gamma\right|^2\right)}\right]. \quad (39)$$

Equation 39 clearly indicates that with an efficiency of $\eta < 1$, as well as with $\Gamma \neq \Gamma_{opt}$, the amplifier noise contribution is increased.

In the case of a cable with wave impedance Z_{L0} and efficiency η_c in between the antenna port and internal noise-matched amplifier - see lower part of Fig. 8 - the amplifier noise becomes minimum for $\Gamma = 0$ (matched to Z_{L0}), and the receiver noise temperature is

$$T_{rec} = T_{eff,A} + \frac{1}{\eta \cdot \eta_c}\left[\left(1 - \eta_c\right)T_{cab,phys} + T_{Ampl,\min} + \frac{\left|\Gamma\right|^2}{1 - \left|\Gamma\right|^2}\left(\left(1 - \eta_c\right)T_{cab,phys} + \Delta T\right)\right]. \quad (40)$$

Here again, amplifier noise increases when the efficiency of the antenna and cable decreases. It is important to note that the total noise is given by the system noise temperature, which is the sum of T_A and T_{rec}. Increasing η and reducing the physical temperature by use of cooled HTS structures is therefore reasonable only if the antenna noise temperature (external noise) is "sufficiently low." This is not the case for frequencies far below 100 MHz (galactic noise), and at any frequency for the case of strong received man-made noise. From a noise point of view, an alternative bandwidth definition can be derived which requires $SNR < \gamma(SNR)_{max}$, with a properly chosen value of $\gamma > 1$.

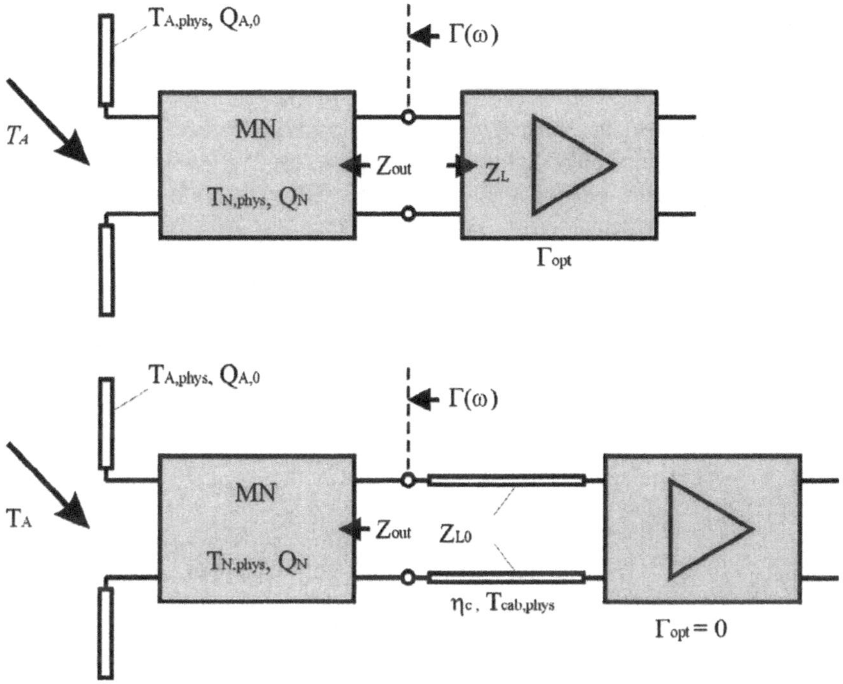

Figure 8. Impact on the receiver noise temperature of the reflection coefficient Γ between MN output and amplifier input. Upper part: Amplifier directly connected to antenna network. Lower part: Cable between antenna network and amplifier.

6. Examples for electrically-small HTS antennas and matching networks

Resonant half-wave dipoles possess at center frequency a real-valued impedance with $X_A \approx 0$ and $R_{rad} \approx 50 - 70\ \Omega$. Q_{rad} is on the order of 8 to 25 for fat and thin cylinders, respectively. The linear polarized field with a rotationally symmetric pattern has a maximum directivity of 1.5 (1.76 dB). Over large conductive platforms, $\lambda_0/4$ monopoles can be used with Q_{rad} similar to a half-wave dipole, but with twice as high maximum directivity. If due to lack of space or other system requirements, the accommodation of an $h = \lambda_0/4$ long cylinder (e.g., h = 375 mm for 200 MHz) is prohibited, one may use highly-capacitive, electrically-short monopoles ($h < 0.1\lambda_0$) with

$$R_{rad} / \Omega \approx 40\pi^2 (h / \lambda_0)^2 \text{ and } |X_A| / \Omega > 5\lambda_0 / h$$

- see Eqs. 14 and 15.

In contrast to the input impedance, the radiation pattern (and therefore the maximum directivity) is only slightly affected by the length reduction. If a real-valued input impedance on the order of 50 Ω is required, a MN with β_{MN} larger than 1 must be employed to compensate for the negative reactance and transform the very small radiation resistance (e.g., $X_A < -167\ \Omega$ and $R_{rad} \approx 0.35\ \Omega$, corresponding to $Q_{rad} \approx 477$ for 200 MHz and h = 45 mm). With an unloaded quality factor of the MN in the order of 200, the efficiency drops below 30%. Using HTS and a substrate material such as $LaAlO_3$ or sapphire with $\tan\delta < 2 \cdot 10^{-5}$ in the matching network, $Q_{MN} > 30{,}000$ and an efficiency, η, larger than 95 % can be achieved.

For the example in which a conventional MN with $Q_{MN} \approx 200$ has a bandwidth ($VSWR < 2$) of about 0.5 % , the price for increased efficiency in the HTS version is a reduced bandwidth of about 0.15%. If a larger bandwidth is required, overcoupling - see section 4.2.2 - or multiple-stage matching (discussed in section 4.2.3) can be used.

On the left-hand side of Fig. 9 an example is shown schematically for a monopole with a planar HTS matching network. Only the upper part consisting of a strip conductor with end-patch (top-loaded monopole) is located above the conducting surface of the platform. Whereas, at this "radiating part," conductors exist only on one side, the MN is realized as a stripline structure with a meander-line situated between two substrates, each with a conducting ground-plane on the backside. By a suitable choice of the length of the open-ended meander line and the location of the feed point, the matching requirements can be met. For another version of an electrically-small antenna, see [15].

Electrically-small loops are equivalent to magnetic dipoles. The right-hand side of Fig. 9 depicts an example based on a modification of the approach of Dinger et al. [6]. The half-loop is situated above the surface of the platform. The matching circuit below this surface is realized in stripline technology. A large patch at the end of the half-loop represents a short, and the single-stage MN consists of a capacitor. Whereas for the top-loaded monopole (left), the maximum conductor losses occur in the meander of the MN where the current has its maximum; in the half-loop (right) the radiating part of the antenna system dominates the losses. Hence, for the monopole, HTS in the MN would be sufficient, but for the half-loop, HTS is needed in the radiating part as well.

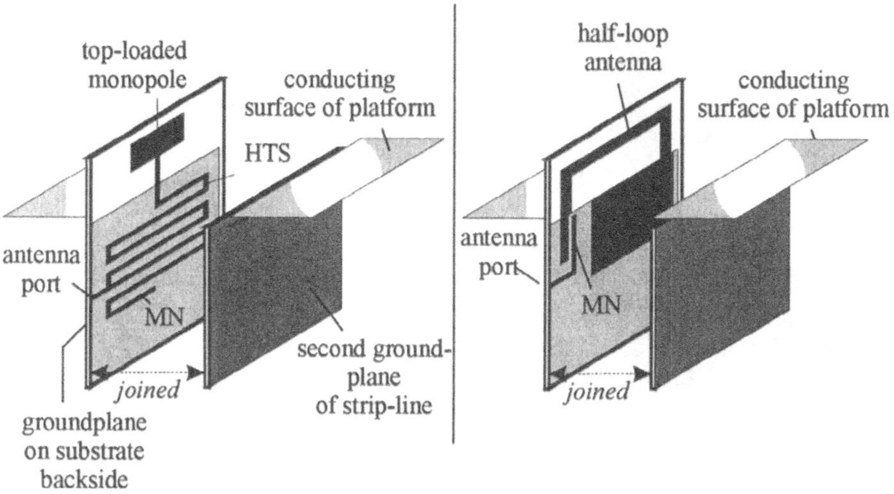

Figure 9. Examples of electrically-small HTS antennas with antenna structure and matching network (MN) separated from each other. Left: Top-loaded monopole. Right: Loop antenna [6].

Figure 10 depicts examples of "self-resonant" antennas where the MN is an integral part of the antenna. Figure 10a depicts an aperture-coupled microstrip patch antenna. If dimension a (square patch) corresponds to $0.5\lambda_0/\sqrt{\varepsilon_{r,eff}}$, resonance occurs at the operational frequency. The "patch-resonator" is coupled on one side to free space, with

$$Q_{rad} \approx 0.05\frac{\lambda_0^3}{V} \tag{41}$$

thus dependent on the volume $V = h\ a^2$ under the patch [8]. On the other side it is coupled through an aperture to the antenna port.

Aiming for a low Q_{rad}, conventionally a low effective substrate permittivity $\varepsilon_{r,eff}$ is chosen, leading to a maximum dimension a. Furthermore, for h, one uses the largest value with sufficiently low surface-wave excitation. If size reduction is of high priority, one ends up with a high $\varepsilon_{r,eff}$, and hence a low value for a at the expense of a narrower bandwidth. In addition to an increased $\varepsilon_{r,eff}$, a modified patch-shape is another means for size reduction. The "H-antenna" [16,17] represents a stepped-impedance resonator, and the meander antenna is a folded half-wave resonator. For all cases in Fig. 10, Q_{rad} increases with the 3rd power of the degree of miniaturization. Therefore, all considerations with respect to efficiency and bandwidth of antennas according to Fig. 9, also apply to the structures of Fig. 10.

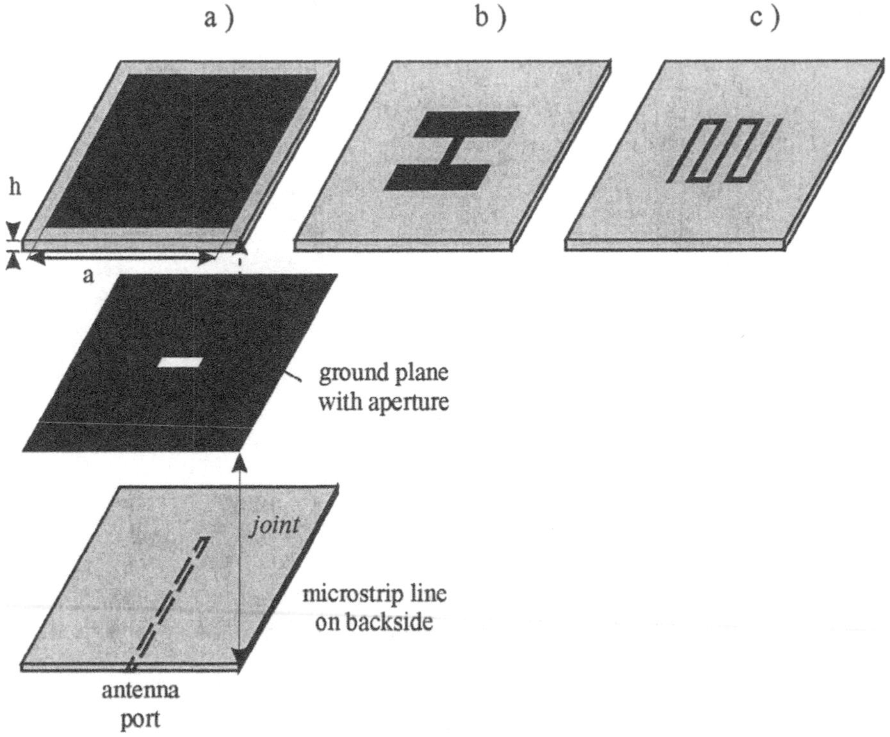

Figure 10. Examples of self-resonant electrically-small HTS antennas: (a) patch antenna on high-permittivity substrate; (b) H-antenna [16,17]; (c) meander antenna.

In section 4.2.3 the realization of antennas with multiple-section matching networks was discussed. For antennas as shown in Fig. 9, the single-stage networks must be replaced with multiple resonator structures (lumped element or distributed) similar to those in bandpass filters. For antennas as shown in Fig. 10, the antenna represents the first (radiating) resonator, and other (non-radiating) resonators can be added at the lower wafer (instead of a simple microstrip line as shown in Fig. 10).

7. HTS application to antenna arrays

7.1 SUPERDIRECTIVE MULTIPORT ARRAYS OF LOW ORDER

All antennas perform by means of the complex-valued radiation pattern and angular weighting ("spatial filtering") of the electromagnetic radiation with respect to amplitude and phase. The "steepness" of the angular changes of the radiation pattern can be characterized by the pattern order Λ introduced in section 3. This pattern order determines the required size of the antenna relative to the wavelength in free space. If a certain pattern order is realized with a size which is smaller than the "conventional size" according to Eq. 7, the antenna is called "superdirective."

As will be shown in this section, superdirective antennas only are practical for relatively low pattern order up to about $\Lambda = 3$. This section is therefore focused on arrays with a low-order radiation pattern and a directivity D_{max} - see section 3 for definition - smaller than about 15. Furthermore, the case is considered for which multiple simultaneous radiation patterns are required. The following system applications may serve as examples of the need for multiple radiation-pattern arrays:

(a) For direction finding, the incident electromagnetic waves must be received simultaneously via different radiation patterns. Comparison of the different signals received via different radiation patterns allows the direction(s) of incidence to be determined.

(b) Another application is for antennas that allow the adaptive rejection of disturbances based on their angular distribution of incidence. This approach is known as "adaptive nulling."

For adaptive forming of the radiation pattern, an array is needed which provides multiple (M) output ports with different linearly-independent radiation patterns. The output signals from these ports can be subjected to various forms of analog and digital signal processing. In particular, by means of a linear combination of these signals, an arbitrarily high number of different radiation patterns with M-1 degrees of freedom can be formed.

Since the maximally-realizable number of ports M equals the number of antenna elements (or $2M$ if 2 polarizations per element are utilized), the element spacing a - see Fig. 11 - should be chosen as small as possible to obtain the highest number of degrees of freedom for a given overall size d of the platform.

With $M > 1$ antenna elements, an (M-1) dimensional set of radiation patterns can be formed. For element spacing equal to or smaller than half a wavelength, a highest pattern order of

$$\Lambda_{max} \approx \sqrt{3M - 0.75} - 1.5 \tag{42}$$

belongs to a planar two-dimensional array with M elements.

For a conventional array, element spacing of $a \approx \lambda_0/2$ is used, leading to an array diameter of $d_{conv} \approx \Lambda\, \lambda_0/\pi$. If element spacing is chosen smaller than $\lambda_0/2$ and hence $d < d_{conv}$, the radiation quality factor for the highest pattern order rapidly increases. Therefore for a required frequency bandwidth, one sets a limit for the size reduction $d_{conv} - d$.

For a given bandwidth ($VSWR = 2$) and single-section(one-pole) matching the maximum size reduction is, via Eqs. 7, 8 and 30, approximated to be

$$\frac{d_{conv} - d}{\lambda_0} \le \frac{\Lambda}{3}\left[1 - \left(\sqrt{2}\,\frac{\Delta\omega}{\omega_0}\right)^{\frac{1}{2\Lambda+1}}\right]. \tag{43}$$

For a fractional bandwidth of 1%, an overall size reduction by about 0.38 λ_0 to 0.71 λ_0 for Λ between 2 and infinity results; for a bandwidth of 0.1%, the corresponding values are 0.49 λ_0 to 1.1 λ_0. The important conclusion is that in all practical cases the replacement of a conventional array by a superdirective array will be restricted to a size reduction of roughly $\lambda_0/2$. Consequently, a significant size reduction by a factor of about 2 is made possible only if the initial conventional array has a diameter of about one wavelength, corresponding to $M = 7$. The element spacing a can then be reduced from $\lambda_0/2$ to about $\lambda_0/4$, and the linear element size, from about $\lambda_0/3$ to $\lambda_0/6$. Hence, at a platform area that conventionally would allow for only 2 degrees of freedom per polarization, 6 degrees of freedom become possible - see Fig. 11. For direction finding applications a 2x2 planar array ($M = 4$) can be placed at a platform which conventionally encompasses only one element.

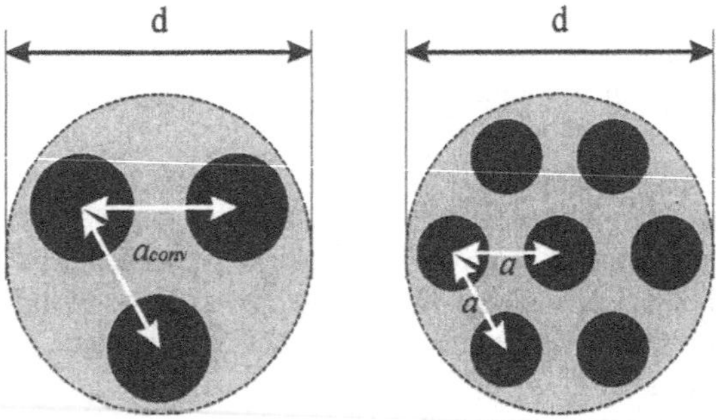

Figure 11. Enlarging the number M of antenna elements at a given platform by replacing a conventional array with a superdirective array. Circular patches symbolize antenna elements.

Since the following discussion on the properties of superdirective arrays relies on some fundamental facts about radiation patterns, these facts are first briefly summarized. If M antenna ports exist, M voltages V_1 to V_M are excited at the M loads if an electromagnetic field is incident on the array. Since the voltages are linearly related to the field strength of the incident field, the response of the antenna array to a single plane homogeneous wave (impinging upon the array in direction (Θ,Φ)) is sufficient to characterize the antenna. With respect to the coordinate system shown in Fig. 1, the electric field strength of the incident wave is given by

$$\vec{E} = \vec{p}\sqrt{2Z_0 S}\,\exp\left[-j\frac{2\pi}{\lambda_0}\vec{u}_i\cdot\vec{r}\right] \quad , \tag{44}$$

with S, $\vec{u}_i(\Theta,\Phi)$ and $\vec{p}$ as radiation density, unit vector in the direction of incidence and polarization, respectively. If Θ and Φ are varied, the voltages V_1 to V_M vary in amplitude and phase. This dependence defines the M radiation patterns $F_1(\Theta,\Phi)$ to $F_M(\Theta,\Phi)$, corresponding to the M different ports. In order to separate clearly the angular dependency from the absolute magnitude, functions normalized according to

$$\oint |F_\mu(\Theta,\Phi)|^2 d\Omega = 4\pi$$

are used as the radiation patterns. $|F_\mu(\Theta,\Phi)|^2$ is equivalent to the directivity function $D(\Theta,\Phi)$ introduced in section 3. For the (theoretical) case of a lossless antenna, the maximum available power at port μ is given by - see Eq. 1 -

$$P_{\mu,avail} = \frac{\lambda_0^2}{4\pi} |F_\mu(\Theta,\Phi)|^2 S(\Theta,\Phi). \tag{45}$$

For a lossless M-port array, the power P_μ delivered to load μ can be smaller than $P_{\mu,avail}$ due to mismatch and (cross-talk) coupling between the ports. This fact can most easily be understood for the transmission mode, where mismatch at port μ results in a partial reflection of the power incident at this port, and where cross-talk to port $\nu \neq \mu$ results in dissipation of power in the load of port ν. The ratio $P_\mu / P_{\mu,avail}$ can be described by the (generalized) matching factor

$$\Pi = P_\mu / P_{\mu,avail} \quad \text{with} \quad 0 \leq \Pi \leq 1 \quad . \tag{46}$$

A necessary condition for achieving $\Pi = 1$ is that the radiation patterns corresponding to the various ports are mutually orthogonal. With Ψ as the "angle" between pattern μ and ν defined via

$$\cos\Psi_{\mu\nu} = \frac{1}{4\pi} \left| \oint F_\mu(\Theta,\Phi) \cdot F_\nu^*(\Theta,\Phi) d\Omega \right|,$$

$\Psi_{\mu\nu} = \pi/2$ means orthogonality and $\Psi_{\mu\nu} = 0$ means "linear dependency."

Superdirective arrays possess substantially different behavior than conventional arrays. Furthermore, a specific antenna circuit design is required for realization. In the following, a lowest-order array, namely an array with two elements serves as an example in order to illustrate the basic ideas and problems associated with superdirective arrays. A generalization of these ideas to higher-order arrays is straightforward. The structure considered - see Fig. 12a - consists of two monopole antennas as elements, but the conclusions obtained are valid for all types of antenna elements.

In Fig. 12(a) the two antenna ports are denoted by 1-1′and 2-2′. In between the loads and the antenna ports, matching networks are inserted which transform the load impedance Z_L into (in general) a complex-valued input impedance Z_{in}. Fig. 12(b) depicts the equivalent circuit for the configuration. The two-port at the center represents the antenna structure between the ports 1-1′and 2-2′. The voltage sources in Fig. 12(b) represent the voltages induced in the elements 1 and 2, respectively.

A reduction of the element spacing a to values considerably smaller than $\lambda_0/2$ affects both the coupling between ports 1-1′ and 2-2′, and the difference between the voltages V_{01} and V_{02}.

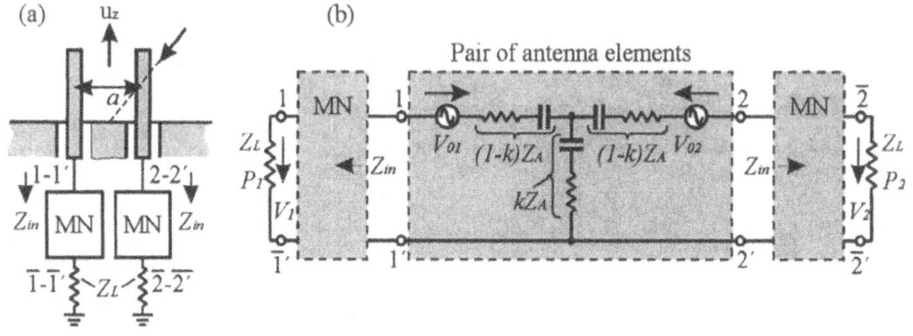

Figure 12. Example of a two-element array. (a) Structure with two coaxial monopole elements and a pair of matching networks (MN); (b) Equivalent circuit.

For the conventional case $a \approx \lambda_0/2$, the coupling (described in Fig. 12b by the complex-valued constant k) becomes negligible ($k \approx 0$). The shunted circuit becomes approximately a short, and one ends up with two independent voltage sources which can (via MNs) be separately matched to the loads. If the element spacing a is made much smaller than half a wavelength, the magnitude $|k|$ of the coupling constant rapidly increases, and in the (theoretical) limiting case $a = 0$, the series circuits in Fig. 12b become a short ($k = 1$).

The voltages V_{01} and V_{02} depend on the angle of incidence of the incoming electromagnetic field, which can be described by assigning radiation patterns to these voltages. For $a \approx \lambda_0/4$ the radiation pattern can be written as a linear combination of an "even radiation pattern" $F_e(\Theta,\Phi)$ and an "odd radiation pattern" $F_d(\Theta,\Phi)$, with

$$F_e(\Theta,\Phi)=\sqrt{3}\,(\vec{u}_z\cdot\vec{p}) \quad \text{and} \quad F_d(\Theta,\Phi)=\sqrt{7.5}\,\sin\Theta\cos\Phi\,(\vec{u}_z\cdot\vec{p}) . \qquad (47)$$

See Eq. (44) for the definition of the polarization unit vector $\vec{p}$ and the directional unit vector $\vec{e}_z$ parallel to the monopole axis. The even pattern represents the pattern of a (single) electric dipole with pattern order $\Lambda = 1$, whereas the odd pattern represents a quadrupole field with $\Lambda = 2$.

If the linear combination of $F_e(\Theta,\Phi)$ and $F_d(\Theta,\Phi)$ is written as

$$F(\Theta,\Phi)=\sqrt{1-\alpha^2}\,F_e(\Theta,\Phi)+\exp(j\varepsilon)\alpha\,F_d(\Theta,\Phi) \quad , \qquad (48)$$

the angular dependence of the voltages V_{01} and V_{02} in Fig. 12(b) is obtained from Eq. (48) via

$$\alpha_1 = \alpha_2 = \sqrt{\frac{2}{5}}\pi \frac{a}{\lambda_0} \tag{49}$$

and $\varepsilon_1 = -\varepsilon_2 = \pi/2$. The two patterns associated with the two voltage sources in Fig.12(b) are composed of even and odd patterns, and differ with respect to the sign of the odd part. As is obvious, the smaller the spacing between the two antenna elements, the smaller the odd part becomes in comparison to the even part and the more the two patterns resemble each other (i.e., angle Ψ according to Eq. 47 approaches zero).

Since the goal of the two-port array is to produce two different radiation patterns (preferably mutually orthogonal), the result of having two nearly linearly-dependent radiation patterns is contrary to this goal. This is exactly the situation in which the concept of superdirective arrays becomes useful.

As is shown in Fig. 12b, the load impedances Z_L can be transformed by means of matching networks (MNs) into arbitrary complex-valued impedances Z_{in}. Assuming equal loading for both antenna ports, one ends up with the voltages V_1 and V_2 at the loads Z_L (ports $\bar{1}-\bar{1}'$ and $\bar{2}-\bar{2}'$ in Fig. 12b), which depend on the choice of Z_{in}. The radiation patterns corresponding to V_1 and V_2 are (except for $Z_{in}\to\infty$) different from the radiation patterns corresponding to V_{01} and V_{02}. By the choice of appropriate Z_{in} values, one can realize mutually orthogonal patterns, in principle, even for extremely small spacing a. However, the configuration in Fig. 12 suffers from the fact that the matching factor Π - see Eq. 46 - also depends on Z_{in}. If, in the case of small spacing (yielding strong coupling between the ports), Z_{in} is chosen to produce an orthogonal pattern, the matching factor drops to very low values. This results in power delivered to the loads which is much smaller than the available power, and therefore strong signal-to-noise degradations occur. The value of Π can easily be determined by considering the transmission mode and relating the radiated power to the sum of radiated power and power dissipated in the source impedances due to cross-talk between and reflection at the ports.

The problem described above can be overcome by employing a matching-and-decoupling network (MDN) at the antenna ports instead of a pair of MNs as shown in Fig. 12. Figure 13 illustrates the basic idea of decoupling.

Instead of two separated two-ports (MN) as in Fig.12, a passive four-port (MDN) is inserted between the antenna ports $1-1'$ and $2-2'$ on one side and the loads ($\bar{1}-\bar{1}'$ and $\bar{2}-\bar{2}'$) on the other side. As can be shown, it is always possible to decouple the two antenna ports by means of this MDN. The result of decoupling and matching is shown in Fig. 13c. Both voltage sources are no longer coupled, and the voltages V_1 and V_2 correspond to mutually-orthogonal radiation patterns. In the case considered, the radiation pattern

$$F_{1,2} = \frac{1}{\sqrt{2}}\left[\sqrt{3} \pm \sqrt{7.5}\sin\Theta \cdot \cos\Phi\right](\vec{u}_z \cdot \vec{p}) \tag{50}$$

results. The enormous advantage of employing a matching-and-decoupling network is that now a matching factor $\Pi = 1$ results. This means that for the case of a lossless structure, the power available - see Eq. 45 - would be delivered to the loads.

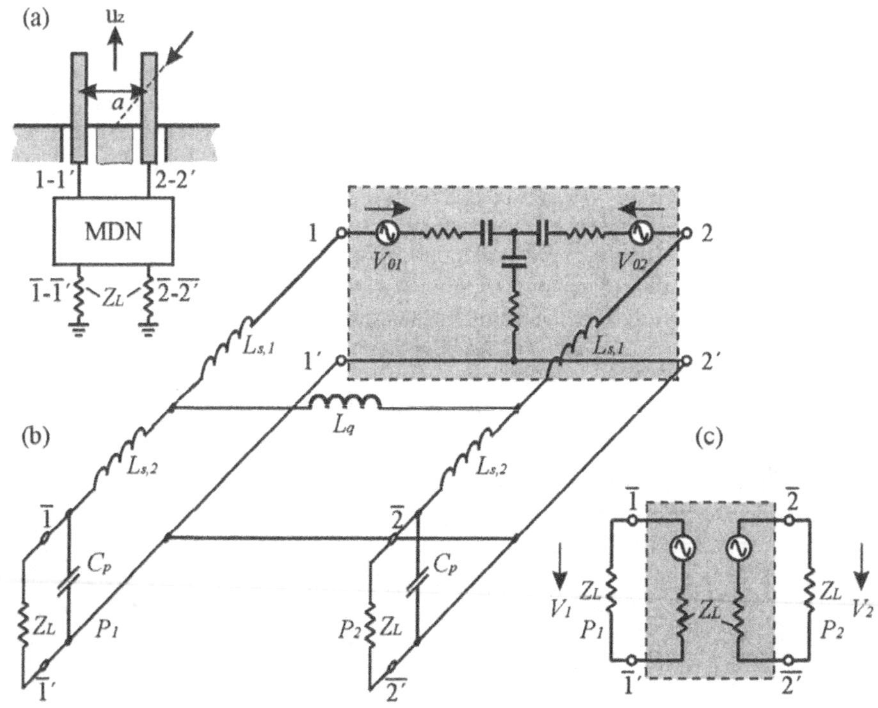

Figure 13. Two-element array with matching-and-decoupling network (MDN). (a) Schematic view of the structure. (b) Equivalent circuit for antenna and MDN. (c) Equivalent circuit for the decoupled and matched case.

A structure as shown in Fig. 13 represents a superdirective array. It is characterized by the fact that in contrast to conventional arrays, the entire array becomes a resonant structure due to the decoupling structure employed. An unloaded quality factor (radiation quality factor) can be assigned to this resonant structure. According to Eq. 8 this quality factor increases with reduced spacing a. Up to now, for ease of presentation, losses in the structures were assumed to be zero. Obviously, dissipation in the antenna and MDN structures leads to a reduction of the efficiency, and this degradation becomes stronger with an increase in the radiation quality factor. This fact provides the rationale for the strong interest in applying HTS technology in matching-and-decoupling networks.

Since the example in Figs. 12 and 13 was selected for ease of presentation, but does not represent a structure for HTS implementation, another example for a two-element array is depicted in Fig. 14. Furthermore, since the properties of a receive antenna are (via reciprocity) directly connected to its properties in the transmit mode, an alternative description in terms of the transmit operation is given.

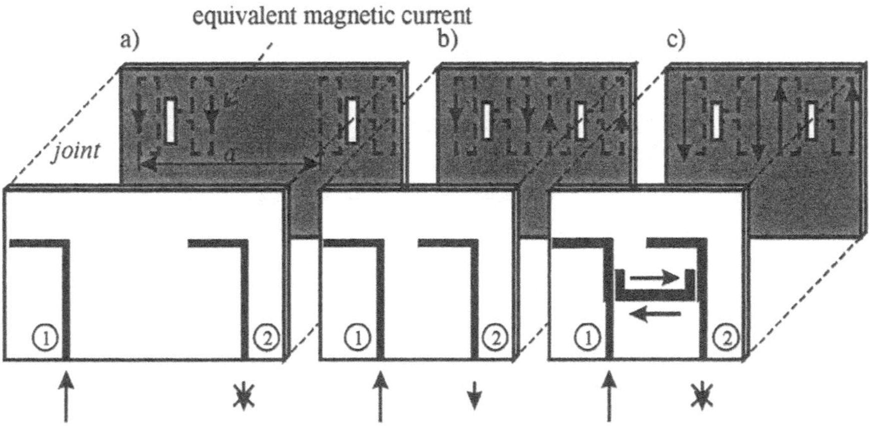

Figure 14. Example of a two-element HTS array with aperture-coupled H antennas as radiation elements. (a) Conventional half-wave spacing with negligible cross-talk. (b) Reduced spacing with considerably cross-talk between the ports. (c) Like (b), but with decoupling network.

In this example superconductive H antennas are chosen as radiation elements, with aperture coupling to microstrip lines - see Fig. 10. Fig. 14(a) depicts a pair of H antennas with separate antenna ports 1 and 2. If the element spacing a is on the order of $\lambda_0/2$, the radiation patterns owing to separately-excited antenna elements are approximately mutual orthogonal. If – as shown in Fig. 14(a) - a wave with power $P_{1,inc}$ is fed into port 1, practically only antenna element 1 is excited, and the crosstalk to port 2 can be neglected. However, if the element spacing is chosen considerably smaller than $\lambda_0/2$, as in Fig. 14(b), mutual coupling between the antenna elements must be taken into account. Since the radiation patterns of the individually-excited antenna elements can now no longer be mutually orthogonal, radiation coupling and additional capacitive or inductive coupling occur between the elements.

As can be seen from Fig. 7b, the wave incident at port 1 not only excites element 1, but also element 2. As a result of the power flow to the load at port 2, the matching coefficient of the antenna is reduced by a factor of $(1-|S_{21}|^2)$. To overcome this disadvantage, a decoupling network must be employed - see Fig. 14(c). The basic principle of this decoupling network is based on a "transmission path" between the antenna elements, which is compensated by a second path in the non-radiating antenna network. As seen from Fig. 14c, the additional path in the matching-and-decoupling network (MDN) forms (together with the transmission path between the elements) a type of ring resonator. If a wave is fed into port 1, a standing wave occurs at the second path in such a way that a voltage null is placed at the interconncction to port 2. Although both antenna elements are excited, port 2 is now decoupled from port 1. The corresponding radiation pattern of port 1 is produced by the superposition of contributions from both antenna elements with partially destructive interference. If the wave is fed into port 2, another radiation pattern is produced which is due to the decoupled ports mutually orthogonal to the radiation pattern of port 1. Theoretically, two orthogonal patterns are produced independent of the element spacing. It is noteworthy that, whereas in the conventional case only the antenna elements are resonant, in the superdirective array, the entire array including the network, is resonant, and different resonant modes are excited via different ports. The smaller the element spacing, the weaker is the coupling to free space and the ports. Hence, frequency bandwidth decreases, and sensitivity to losses increases. The benefit of HTS implementation lies in the fact that superdirective arrays of small order can be realized without a significant efficiency reduction, up to the limit given by the required frequency bandwidth.

These ideas, presented for the example of a two-element array, are now, as illustrated in Fig. 15, generalized to arrays of arbitrary order M. If waves are incident at the antenna ports, the power will not be completely radiated, but reflected or transmitted to other ports. The behavior of the multi-port antenna can be represented in general by a scattering matrix (Γ) with

$$\begin{bmatrix} b_1 \\ b_2 \\ \cdot \\ b_M \end{bmatrix} = \begin{bmatrix} \Gamma_{11} & \Gamma_{12} & \cdot & \Gamma_{1M} \\ \Gamma_{12} & \Gamma_{22} & \cdot & \cdot \\ \cdot & \cdot & \cdot & \cdot \\ \Gamma_{1M} & \cdot & \cdot & \Gamma_{MM} \end{bmatrix} \cdot \begin{bmatrix} a_1 \\ a_2 \\ \cdot \\ a_M \end{bmatrix} , \quad (b) = (\Gamma)(a), \qquad (51)$$

and relatively large off-diagonal elements are characteristic of the strong mutual coupling between the elements of a superdirective array. Since mutual coupling between the antenna ports reduces the available power, a matching and decoupling network (MDN) must be introduced between the element ports and the beam ports.

Within the MDN, transmission between the different ports is provided such that the cross-talk between the antennas is compensated. The result of this measure is that the individual beam ports are no longer related to the excitation of only one antenna element, but to the excitation of multiple antenna elements. If the beam ports are mutually decoupled, the corresponding radiation patterns are mutually orthogonal. The multiple-element excitations belonging to these radiation patterns are characterized by spatially oscillating equivalent currents, and the far-field pattern (even in the main beam direction) is the result of partially destructive interference. It can be shown that for simultaneous matching and decoupling, a network is required - see Fig. 15) - which (seen from the antenna elements) provides a scattering matrix

$$(a) = (\Gamma)_{MDN} \cdot (b), \text{ with } (\Gamma)_{MDN} = (\Gamma)^{*}.$$

With this MDN, the beam ports become decoupled, and the M different radiation patterns corresponding to the M different beam ports, are mutually orthogonal.

HTS technology provides a means to realize planar and compact matching and decoupling networks which carry a large amount of stored field energy without excessive losses.

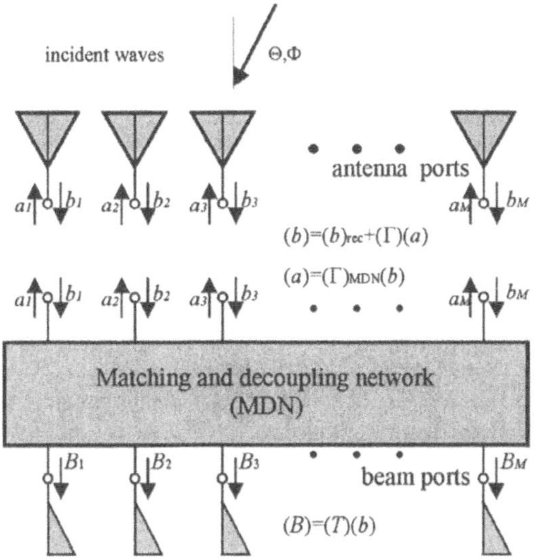

Figure 15. Schematic representation of a matching-and-decoupling network.

7.2 HTS FEED NETWORKS FOR CONVENTIONAL ARRAYS

Simultaneous beamforming of $M = 2^m$ beams for an array with M elements can be accomplished by means of an RF Butler matrix. This beamforming network requires $(m)2^{m-1}$ hybrid couplers and $(m+1)\ 2^m$ transmission lines for interconnection. If a 3-dimensional Butler matrix in coaxial or waveguide technology is to be replaced with a planar microstrip structure, crossovers become a problem. In microstrip technology each crossover can be replaced by a pair of 3-dB couplers. The number of required crossovers strongly increases with M, leading to 2, 16 and 88 crossovers for $M = 4$, 8 and 16, respectively. In order to realize a planar structure with such a large number of hybrid couplers within a compact volume, miniaturized hybrid couplers on high-permittivity substrates must be employed. Since these structures suffer from strong degradation due to dissipative losses, utilization of HTS can under certain circumstances be justified [8].

Highly-directive mm-wave arrays, e.g., with a linear size larger than 50 λ_0, and hence a half-power beamwidth of less than 1 degree, suffer from the disadvantage that with increased size, the increase in gain due to the narrowing of the beamwidth can be counterbalanced by the increase in dissipative insertion loss in the distribution network. Because of this fact, consideration has been given to the utilization of superconductive distribution networks for large arrays in the frequency range above about 60 GHz [18,19].

8. References

1. Lancester, M.J. (1997) *Passive microwave device applications of superconductors* , Cambridge University Press, Cambridge UK.
2. Chaloupka, H.J. (1999) Microwave applications of high-temperature superconductors, in Weinstock H. (ed.) *Applications of superconductivity*, Kluwer Academic Publishers.
3. Rudges, A.W. et al. (1983) *Handbook of antenna design*, Peter Peregrinus Ltd.
4. Walker, G.B. and Haden, C.R. (1969) Superconducting antennas, *J. Appl. Phys.* **40**, 2035-2039.
5. Hansen, R.C. (1990) Superconducting antennas, *Trans. IEEE Aerospace and Electr. Syst.* **26**, 345-354.
6. Dinger, R.J., Bowling, D.R. and Martin, A.M. (1991) A survey of possible passive antenna applications of high-temperature superconductors, *IEEE Trans. Microwave Theory Tech.* **39**, 1498-1507.
7. Chaloupka, H.J. (1992) High temperature superconductor antennas: Utilization of low RF losses and of nonlinear effects, *J. Superconduct.* **5**, 403-416.
8. Chaloupka, H.J. (1993) Superconducting multiport antenna arrays, *Microwave Opt. Technol. Lett.* **6**,737-744.
9. Hansen, R.C. (1998) *Phased array antennas*, John Wiley & Sons, Inc.
10. Hansen, R.C. (1981) Fundamental limitations in antennas, *Proc.IEEE* **69**, 170-182.
11. Fano, R.M. (1950) Theoretical limitations on the broadband matching of arbitrary impedances, *J. Franklin Institute* **249**, 57-83 and 139-154.
12. Chaloupka, H.J. (1992) On the frequency bandwidth of functionally small antennas, *Proc. URSI Int. Symp. Electrom. Theory (Sydney 1992),* 266-268.
13. Gevorgian, S.S. et al. (1997) HTS/Ferroelectric devices for microwave applications, *IEEE Trans. Appl. Superconduct.* **7**, 2458.
14. Luine, J. et al. (1999) Application of a DC SQUID array amplifier to an electrically small active antenna, *IEEE Trans Appl. Supercond.* **9**, 4141-4144.
15. Khamas, S.K. et al. (1988) High-Tc superconducting short dipole antenna *Electron. Lett.* **24**, 460-461.
16. Chaloupka, H.J. et al. (1991) Miniaturized high-temperature superconductor microstrip patch antenna, *IEEE Trans. Microwave Theory Tech.* **39**, 1513-1521.
17. Lancester, M.J.,Wang, H.Y. and Hong, J.-S. (1998) Thin-film HTS planar antennas, *IEEE Trans. Appl. Superconduct.* **8**, 168.
18. Newman, N. and Lyons, W.G. (1993) High-temperature superconducting microwave devices: Fundamental issues in Material, physics, and engineering, *J. Supercond.* **6**, 119-160.
19. Herd, J.S. et al. (1993) Experimental results on a scanned beam microstrip antenna array with proximity coupled YBCO feed network, *IEEE Trans Appl. Supercond.* **3**, 2840-2843.

HTS WIRELESS APPLICATIONS

BALAM A. WILLEMSEN
Superconductor Technologies Inc.
460-F Ward Dr., Santa Barbara, CA 93111,
United States of America

The application of HTS technology to passive microwave devices is finally coming of age. A brief overview of wireless telephony and why HTS filters may be beneficial in these systems will be presented in this chapter. This will be followed by a general overview of how HTS filters, come together with many of the individual technology areas discussed in other chapters of these proceedings to form turnkey systems which can be manufactured and sold to the wireless communications industry. Hundreds of these self-contained systems have been sold by the summer of 1999, and a brief discussion of recent market experience is also included here.

1. Introduction

Wireless forms of communication have become a ubiquitous part of life in the information age with the proliferation of wireless telephones (including cellular and satellite phones) and a variety of other devices such as pagers. No doubt this trend will continue driven by wireless access to the internet in the hand-held devices of the near future. By definition all of these systems are composed of elements which can transmit and/or receive radio signals. These elements can further be classified as base stations or handsets, depending on whether they are operated by the the provider or end-user of the service. We will focus on terrestrial systems where the provider's base stations are fixed and the handsets are mobile. (Satellite applications of HTS present many unique challenges and will be discussed in another chapter of these proceedings.) In this chapter I will focus on the applications of HTS in one such system, the US cellular telephone system.

The principal feature of HTS materials that we will use is their extremely low loss at microwave frequencies compared to conventional conductors and dielectrics. This allows for the development of a wide variety of passive microwave devices such as antennas, delay lines and filters. A good

H. Weinstock and M. Nisenoff (eds.), Microwave Superconductivity, 387–416.

overview of early efforts in this area is provided by Lancaster in [11] and more recent efforts are described in other chapters of this volume. Most commercial efforts have concentrated on filters, and the HTS technology developed is applicable to a wide range of frequencies and applications. At Superconductor Technologies Inc. (STI) we have developed high performance HTS filters spanning three orders of magnitude in frequency from $20 - 20,000$ MHz for both military and commercial applications. In this chapter we will focus only on the commercial communications band which roughly extends from $750 - 1,900$ MHz, with emphasis on the US cellular communications band around 850 MHz.

Filtering is an essential feature in wireless communications systems. In an increasingly dense electromagnetic environment, filtering allows the system to consider only the frequency band(s) which contains the desired signals, while disregarding all others. While filters are present in both base stations and handsets, at this time HTS filter systems can only be deployed in base stations. This limitation arises from the physical size and power requirements of the refrigeration system. Furthermore, due to the nonlinearity of the surface impedance of HTS materials (discussed in an earlier chapter of these proceedings), HTS filters are primarily deployed in the base station receivers, where the signal strengths are manageable.

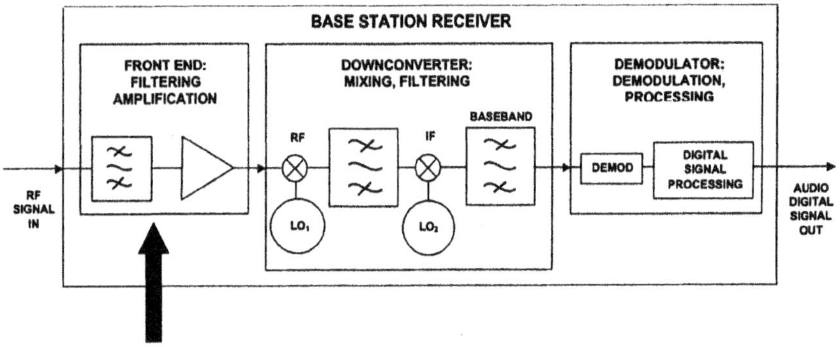

Figure 1. Block diagram of a typical receiver, such as might be present in a wireless base station. The arrow indicates the receiver front-end, where most HTS insertions are currently deployed.

The HTS filter system (consisting of one or more HTS filters and cryogenically cooled low noise amplifiers (LNAs), a cryocooler and enclosure) will be placed between the receive antenna and the first mixer within a typical base station (See figure 1) where it will be responsible for setting the overall noise characteristics of the receiver.

Serious efforts to manufacture and commercially deploy HTS filters in commercial wireless communications systems have been underway since about 1996. This is particularly apparent in the United States where three principal companies have emerged: Conductus Inc. (Conductus), Illinois Superconductor Corp. (ILSC) and STI. As of July 1st, 1999 these three companies employed approximately 300 people and had received hundreds of millions of US dollars in financing from government, as well as private and public sources. Even though most of the data and figures presented in this chapter will represent work done at STI, much of the information is relatively generic to all three US companies, as well as the parallel efforts by institutions in Europe and Japan.

There have been numerous overviews of the potential market for these devices. Three recent examples can be found in the proceedings of the 1998 Applied Superconductivity Conference. [1, 14, 17] Some of these surveys have been more optimistic than others, but all agree that if HTS is ever to achieve widespread success in electronics, it *must* succeed in wireless. I hope to convince you that there is no question that HTS *will* succeed in penetrating this market, it is only a matter of time. Hundreds of HTS filter systems had already been commercially deployed by the middle of 1999 and more should follow in 2000.

1.1. CELLULAR TELEPHONY

Cellular telephone systems derive their name from the topology of the network. A service provider will divide a desired coverage area into a number of sub-areas (or "cells") which have at their nucleus a base station. The base station must be able to detect and control multiple handsets within its coverage area and send/receive signals to/from those handsets. To be useful, the base station must be connected to the conventional, wire-line, phone system by some means. It must also be able to negotiate and transfer (hand-off) control of its handsets with adjacent cells as the handsets move from one cell to another. Figure 2 shows an over simplified situation, where the benefits of installing HTS filter systems to a service provider have clearly been established.

In this situation, coverage along this section of the road is provided by two adjacent cells. In this case, there is a gap in coverage (or dead zone) between the two cells. Because of this, a caller who crosses the boundary of a cell along this major thoroughfare while using their phone, will have his call dropped by the system before the call is completed. This often results in a needlessly irate customer, and contributes to "churn" or users switching to alternate providers. This coverage problem has been exacerbated by the move towards smaller, lighter handsets which are capable of lower uplink

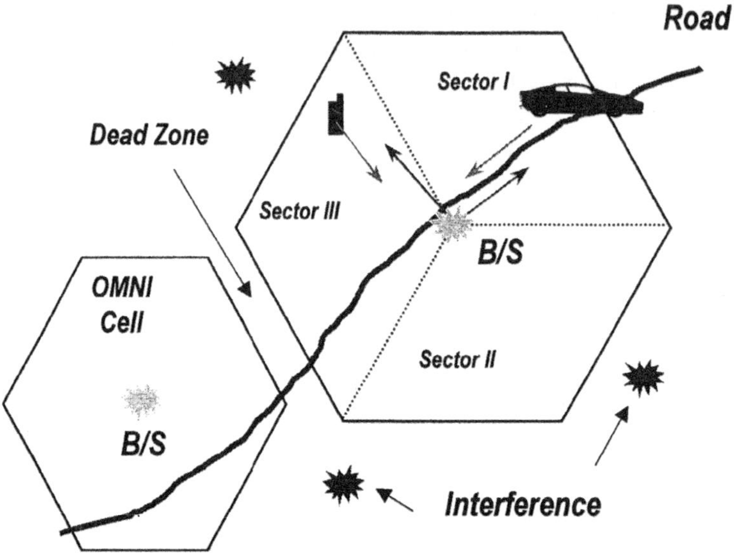

Figure 2. Schematic of a prototypical situation where HTS filters can be deployed: HTS filters deployed in this situation generally increase the range of the sites and fill in the dead zone between cells. Additionally the HTS filter system will eliminate many sources interference that could desensitize the receiver

powers and thus have shorter ranges than their predecessors.

As is also indicated in the figure, cells come in two basic flavors. In rural areas where there is generally little demand for service, providers tend to deploy larger omnidirectional cells. The base stations in these cells typically use omnidirectional antennas to cover a relatively large area. In urban and sub-urban areas where demand for service is higher, the cells are often split into sectors. Typically each cell is divided into three sectors, although some 6 and 12 sector sites are in use. Sectoring is an efficient way of increasing capacity by co-locating the equivalent capacity of three base stations. Sectoring also provides a certain amount of protection against interfering signals since the directional antennas will collect fewer signals outside their sector. Each sector or omnidirectional cell will typically have two RF antennas to provide "diversity", i.e. protection against multi-path interference as well as some redundancy.

1.2. WHY HTS FILTERS FOR CELLULAR TELEPHONY?

The use of high selectivity filters and low noise amplifiers (LNAs) to improve the performance of wireless base station receivers is not new. Conventional, room-temperature solutions are readily available from a multitude of vendors, including the Original Equipment Manufacturers (OEMs). These are the manufacturers of the base stations as a whole and include companies such as Motorola, Lucent, Ericsson and more recently Nortel. Other large multi-national companies such as Philips and Siemens are also beginning to enter the market. Many smaller companies are also offering tower mounted amplifiers (TMAs) which can be installed into existing base stations.

The existing cellular telephony service in the United States poses a number of unique problems that make it ideal for the deployment of HTS based systems. In the US, the American Mobile Phone System (AMPS) uses the frequency band from 824 to 849 MHz for signals sent from the handsets to their base stations. These signals are often referred to as uplink signals. Two providers, respectively denoted A and B, share this frequency band in every region. Due to historical reasons, the frequencies allocated to each provider are not contiguous within the cellular band.

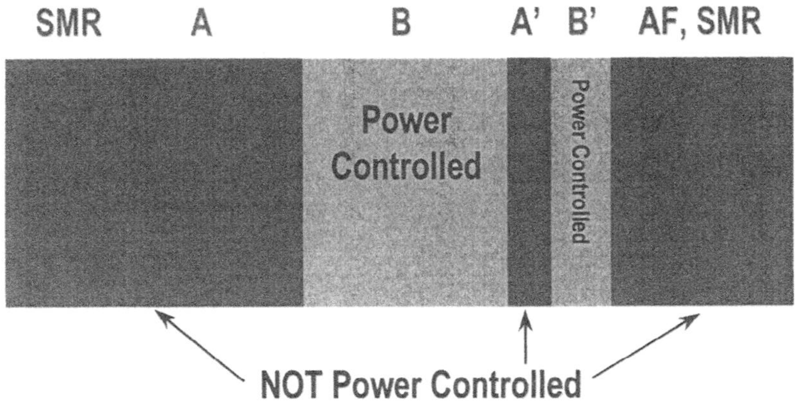

Figure 3. The AMPS/B situation in the US Frequency allocation and nearby interferers

While each provider is able to control the uplink power levels of the handsets on their system, they have no control over their competitor's signals which are then seen by the system as interfering signals. As shown in Figure 3 there are also other sources of interference which lie close to the cellular band. The most deleterious are the Airfone® (AF), pager and

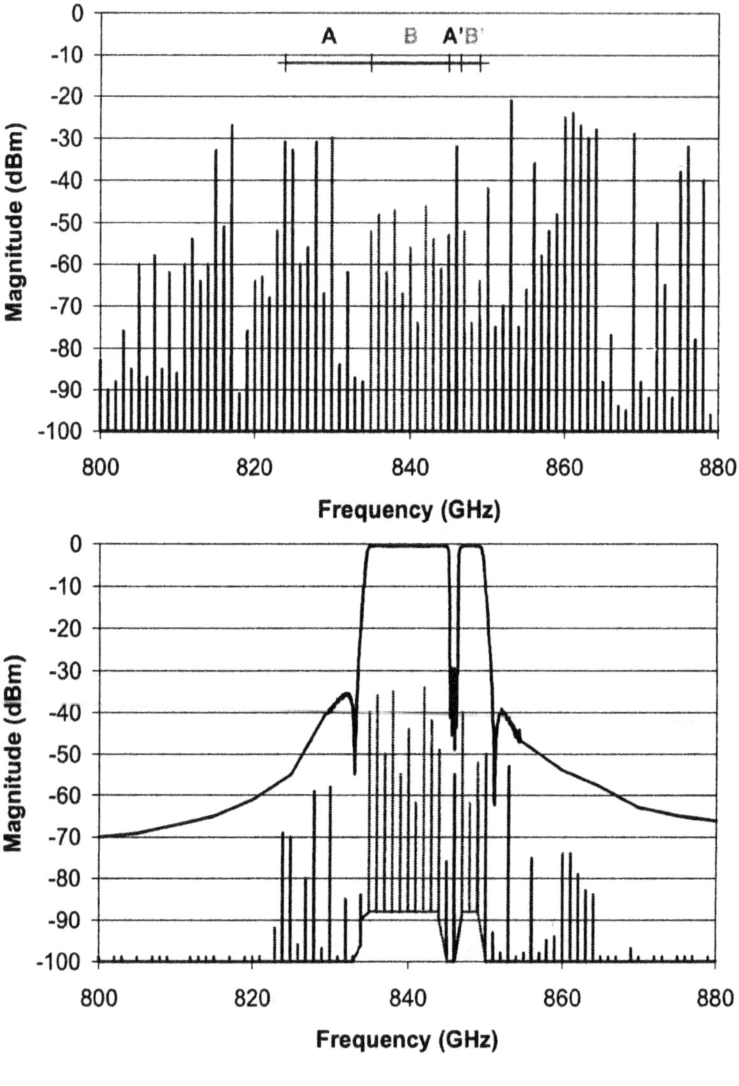

Figure 4. Spectral data before and after insertion of a SuperFilter® unit. The unit contains a pair of HTS filters designed for the AMPS/B frequency band filter and a cryogenic low-noise amplifier with +12 dB of gain. The non-contiguous AMPS/A and AMPS/B bands are also indicated in the figure.

other similar services operating in the Specialized Mobile Radio (SMR) bands that surround the cellular uplink band. All of these interfering signals can potentially be larger than those you are trying to detect and may thus desensitize the receiver. Furthermore, in many areas, the system currently in place was designed for high power car phones and handsets and thus, since mainly low-power handsets are currently in use, many coverage gaps of the type described in Figure 2 have developed. Sample spectral data before and after insertion of a SuperFilter® unit is shown in Figure 4.

The ideal receiver front end should be able to isolate only those signals of interest and pass them on to the subsequent stages of the receiver without alteration. These qualities can be quantified in terms of the selectivity and sensitivity of the front end. As was previously indicated, selectivity can be increased by the use of antennas, both simple directional antennas and "smart" antennas, as well as filters.

Receiver sensitivity can be harder to improve. There are two principal sources of noise which contribute to the overall receiver sensitivity: thermal noise and nonlinear effects in the mixers and amplifiers. Improving selectivity can have a significant impact on reducing nonlinearities as many of the large interfering signals will be significantly attenuated and thus have less opportunity to create undesired intermodulation products. Reducing thermal noise requires either lowering the device's operating temperature or reducing any resistive losses in the device. HTS based filter subsystems provide a natural solution in both areas (Figure 5).

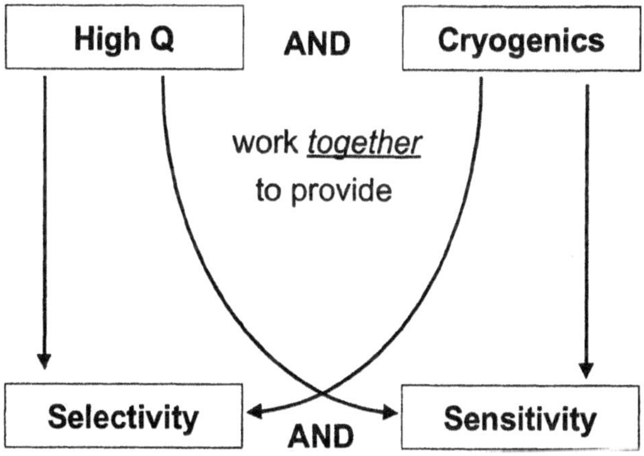

Figure 5. Unlike most competing technologies, HTS filter subsystems provide the ideal combination of selectivity and sensitivity in a small package

It is important to note that the cryogenic cooling that is required to enable HTS systems also provides significant advantages for more conventional, non-superconducting, elements. At room temperature it is well known that the contribution to noise figure arising from a passive element (such as a filter) is equal to its insertion loss. However, if the component is cooled, this rule of thumb no longer applies. This can easily be seen if we discuss noise temperature rather than noise figure. The effective noise temperature, T_e, of a passive device with insertion loss, L, at a physical temperature of T_p is: $T_e = (L - 1)\,T_p$, where T_e and T_p are expressed in absolute temperatures and L is expressed as a ratio. The noise figure, F, is calculated from the noise temperature as: $F = 1 + (T_e/290)$ and is normally expressed in dB as $N = 10\log_{10} F$. So, at room temperature ($T_p = 290$ K) the loss and noise figure are numerically equal. However, when the lossy passive device is cooled one can immediately see that the device will have a lower noise figure when $T_p < 290$ K.

For example, a 3 dB attenuator at liquid nitrogen temperature ($L = 2$, $T_p = 77$ K) will have a noise temperature of $T_e = (2 - 1)T_p = 77$ K but a noise figure of only $N = 1$ dB ($F = 1.26$). This represents an improvement of 2 dB over the noise figure of a comparable device operated at room temperature. In order to preserve benefit of this reduction in noise achieved by cooling, the passive cryogenic device needs to be followed by a low noise amplifier. For maximum benefit this LNA should have a noise figure as close to zero as possible, which can be achieved by cooling the LNA to the same temperature as the filter (i.e. 77 K).

2. Technology Overview

Many diverse technology areas, some of which have been discussed in depth in other chapters of these proceedings, must come together to form a useful turnkey system that can be marketed to the wireless industry. The key technologies which are required for HTS filter subsystems for wireless applications are: materials, RF/microwave engineering, cryopackaging, cryocooling and system control and integration. From an acceptance point of view, the ideal system should look no different from many of the other components which are already present in base stations. It should be compact, rack mountable, run using the base station's existing power infrastructure and present the appropriate RF input and output connectors.

Among the various companies which are pursuing wireless applications of HTS, STI is unique in that all of the key technology areas are developed and manufactured internally. Figure 6 shows an inside view of one of STI's commercial SuperFilter® systems with labels identifying each key technology area. Although we will focus on STI's approach and technology in the following sections, the key technology elements are common.

- Materials: HTS and other related materials.

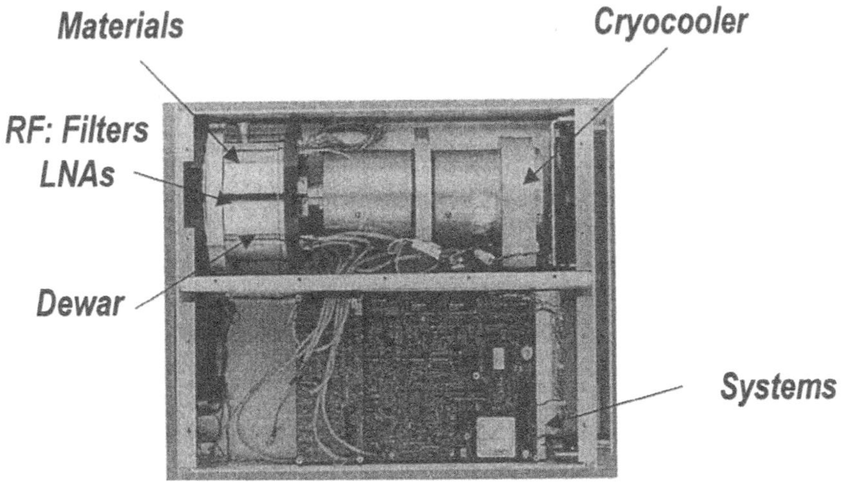

Figure 6. Inside view of an STI SuperFilter® system. The key technology areas are called out.

- RF circuitry: Filters and low noise amplifiers (LNAs).
- Dewar: Cryopackaging and vacuum encapsulation.
- Cryocooler: Long-life closed-cycle cryocooling.
- Systems: Cooler electronics, enclosures and interfaces.

2.1. MATERIALS

Two high temperature superconducting materials are most frequently used for microwave applications: $YBa_2Cu_3O_{7-\delta}$ (YBCO) and $Tl_2Ba_2CaCu_2O_{8+x}$ (TBCCO). Optimally these materials have critical transition temperatures of $T_c = 94$ K for YBCO and $T_c = 110$ K for TBCCO although in practice the T_c for materials used can be up to 10 degrees lower than optimum. YBCO has been used in the form of epitaxial thin films as well as textured bulk thick films, while the TBCCO efforts have been focussed on thin films alone.

Besides a high T_c a low surface resistance R_s at the frequency of interest that varies weakly with power is required for microwave applications. One also desires limited power dependence in the surface reactance X_s, as this has been linked to undesirable intermodulation distortion[6]. In the range of operating temperature from $60 - 77$ K, which is practical for filter subsystems, high quality YBCO and TBCCO films have similar properties.

Thick film technology (described elsewhere in these proceedings, and commercially used by ILSC) is simple and inexpensive to deposit. Furthermore, it can be used to coat three dimensional structures similar to those used in conventional filter design such as cylindrical cavities or split-ring resonators. Even though the material is textured to reduce structural defects, the quality of the materials (in terms of R_s, power handling and intermodulation) is less than their epitaxial counterparts. However the three dimensional nature and size of the resonators leads to low current densities which can often compensate for these deficiencies.

Epitaxial thin films are deposited as single crystals, and their quality far surpasses that of thick film material. Substrates with HTS films on both surfaces (commonly referred to as "double sided films") are extremely desirable as they can be used to form high Q resonators in traditional microwave geometries such as microstrip. This requires the use of compatible, low loss substrates. The two substrate materials used in thin film production today are $LaAlO_3$ ($\epsilon \simeq 25$, $\tan\delta \simeq 4 \times 10^{-5}$) and MgO ($\epsilon \simeq 10$, $\tan\delta < 1 \times 10^{-6}$).

Suitably buffered sapphire single crystals substrates have also been considered by many groups. Although sapphire is readily available in large substrates (greater than 3 inch diameter) it has yet to be shown that it provides any significant price or performance advantage over the alternatives when deployed in a production environment[13]. This may change if the larger die sizes enabled by larger area substrates are required for highly complex future sizes.

As is discussed in another chapter of these proceedings, various processes have been developed for HTS thin film deposition, but only two have been taken into production. At STI, we built our production capability around a proprietary method for depositing TBCCO on MgO. This process is an ex-situ process derived from pulsed laser deposition. STI has also developed an MOCVD process for YBCO deposition on MgO and has used traditional PLD to deposit both TBCCO and YBCO on $LaAlO_3$. All of the processes developed at STI yield high quality films, but only the TBCCO on MgO process allows for these films to be manufactured efficiently in volume. Other groups have refined the reactive co-evaporation technique originally developed by Kinder to produce YBCO films. This process is currently used by Conductus to produce films for their applications.

2.2. HISTORICAL PROGRESS AND THE FUTURE: HAMMOND'S LAWS

In 1998 we set out to counter an impression that seemed prevalent in the technical community; that the technology required to address HTS wireless applications was developing slowly. Two papers were published [5, 6] that demonstrated that the progress in the HTS materials and the microwave devices which can be built using them had in fact been dramatic. These

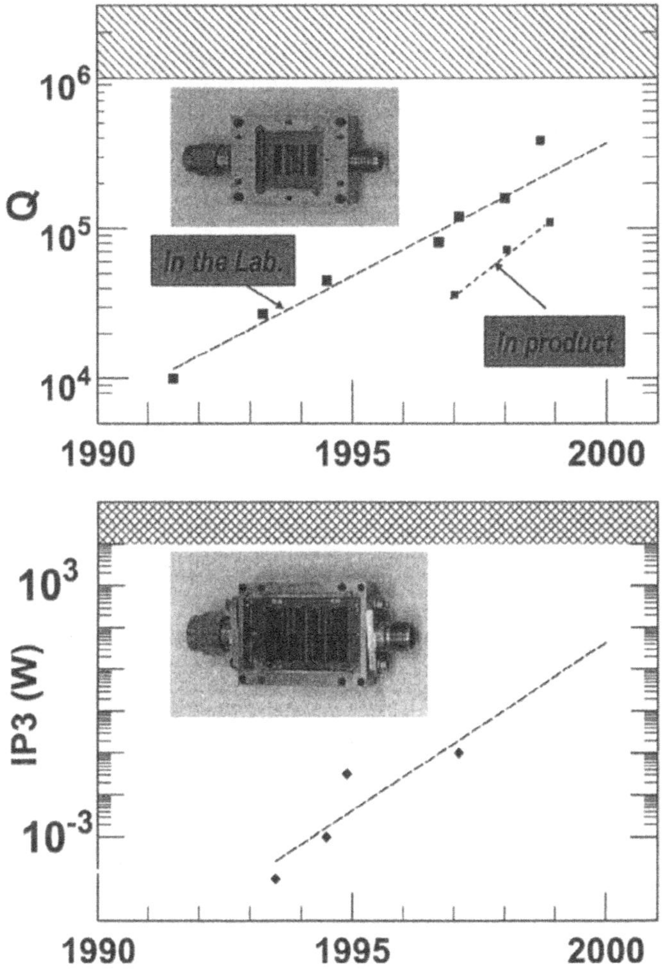

Figure 7. Historical performance of the quality factor of 850 MHz quasi-lumped resonators at STI (above) and the third-order intercept of intermodulation distortion products for 850 MHz band-pass filters. The ideal limits of performance are indicated by hatched regions.

two papers have become known as "Hammond's Laws" in direct analogy with Moore's Law in the semi-conductor world.

The first of these papers [5] focused on historical progress since STI first started to commercially develop filters for the US cellular system in 1991. Much resonator development must be carried out before useful filters can be realized, so a lot of data was available on lumped and quasi-lumped HTS resonators dating back to 1991. Only devices of similar size (approximately 1 cm$\times$1 cm) and which were measured at 850 MHz and 77 K were examined, and the data plotted versus time. (See Figure 7.)

This plot originally included only the results of the best devices demonstrated in the laboratory during any given period of time when significant improvements were made to either the HTS deposition process or the resonator design. In this version, we have added data points for the typical performance of devices that were produced in quantity since 1997. It is immediately apparent that production performance lags behind peak performance by a generation or more.

Besides resonator Q values, the historical nonlinear performance of band-pass filters designed to meet the similar specifications was examined. This was quantified in terms of the third order intermodulation product intercept (IP3). Peak IP3 of production filters was also found to increase exponentially from 1994 through 1997, and is also presented in Figure 7.

The ideal limits of performance that could be expected were examined separately [6], and it was determined that plenty of progress in both Q and IP3 are still possible. These limits assume that the nonlinearities in thin film material can approach those observed in high quality single crystal and that devices can be optimized to reduce the peak current levels. The headroom in Q has been demonstrated by various measurements of $Q > 400,000$ in the laboratory during 1999. This progress seems ahead of schedule according to the historical rate of improvement. Unlike Q, no such dramatic improvements in IP3 have been observed since 1997.

2.3. RF: FILTERS AND LOW NOISE AMPLIFIERS

The non-contiguous frequency bands allocated for cellular communications in the US (Figure 3) require complex filtering functions. A simple approach to generate the required functions is to use two filters for each band. While the A and A' bands can be addressed using two band-pass filters and a diplexer/duplexer pair, the B and B' bands are best addressed using a wide band-pass filter and a narrow band-reject (or notch) filter. We will focus on the B band system as it includes more diverse technology.

STI has focused on microstrip-like quasi-lumped technology for developing all commercial filters. The approaches and requirements for band-pass and band-reject filters are quite different, so let us examine them individually in more detail.

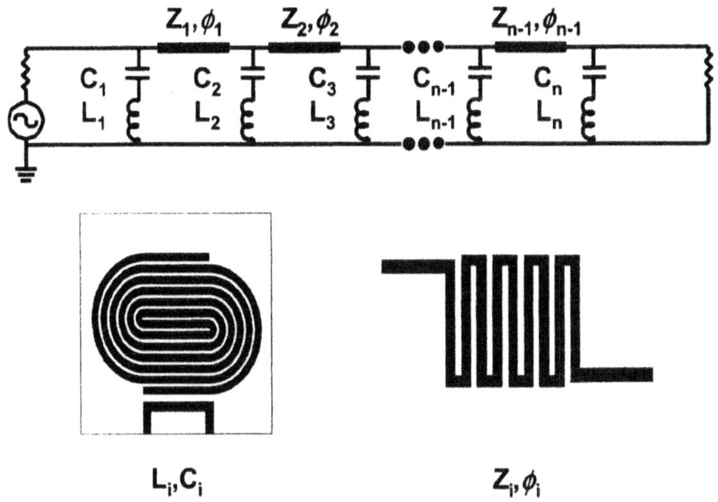

Figure 8. The idealized lumped band-reject filter we are trying to achieve, and the resonator and phasing elements used in the realization of such a filter.

2.3.1. *Band-reject filters*

STI's approach to compact band-reject filters is described in detail in a recent publication[19]. The desired quasi-elliptic filter shape can be represented as a simple ladder network of resonators (L_j, C_j) and phasing elements (Z_j, ϕ_j). (See Figure 8). Any coupling between the resonators that is not through the phasing elements is extremely deleterious to the filter shape.

The resonators can be any structure which can be generally be described as a series LC circuit shunted to ground. One such circuit is the quasi-lumped structure shown in the figure. It is quasi-lumped because it is smaller than $\lambda/10 \times \lambda/10$ in area, but unlike true lumped elements it does not have clearly identifiable Ls and Cs. The phasing elements consist of transmission lines, meandered to keep them compact.

Each resonator is compact (approximately 12 mm × 12 mm) and manufactured on individual chips. This allows for each resonator to be individually tested before it is assembled with the phasing manifold, allowing for greater filter yield. Additionally each resonator is housed within a metal box which provides the necessary RF isolation between resonators, and allows trimming of the resonant frequency by moving a dielectric rod through a hole in the top of the box. (See Figure 2.3.1). The box must also preserve resonator $Q > 60,000$ so that filter performance is unaffected.

Figure 9. Photograph of a finished band-reject filter

2.3.2. *Band-Pass filters*

STI's band-pass filter technology also consists of quasi-lumped quasi-elliptic filters and has been discussed in detail in recent publications[18, 7]. Like the band-reject filter the idealized filter can be represented as a ladder network of lumped inductors and capacitors. However, unlike the band-reject filter, couplings between the resonators are essential and must fully be understood to synthesize and fabricate filters. To more accurately characterize both the desired and parasitic couplings we replace the ladder network with a nodal model [7] which includes every possible coupling between elements. (See Figure 10))

Each element is considered to be a frequency variable susceptance b_{ij} between nodes. A rough layout of the filter elements is simulated using field solver software to determine b_{ij} at two extreme frequencies ω_1 and ω_2. Values for b_{ij} at intermediate frequencies are obtained by interpolation. The $b_{ij}(\omega)$ values are then used to generate the nodal model of the filter. If performance is not acceptable the nodal model can be optimized and the required changes folded back into the layout and field solver software. This process is far less time consuming than optimization using the field solver software directly and yields accurate results rapidly.

Once completed, the final circuit for a quasi-elliptic filter with 8 resonators and 2 transmission zeroes is presented in Figure 11. The filter

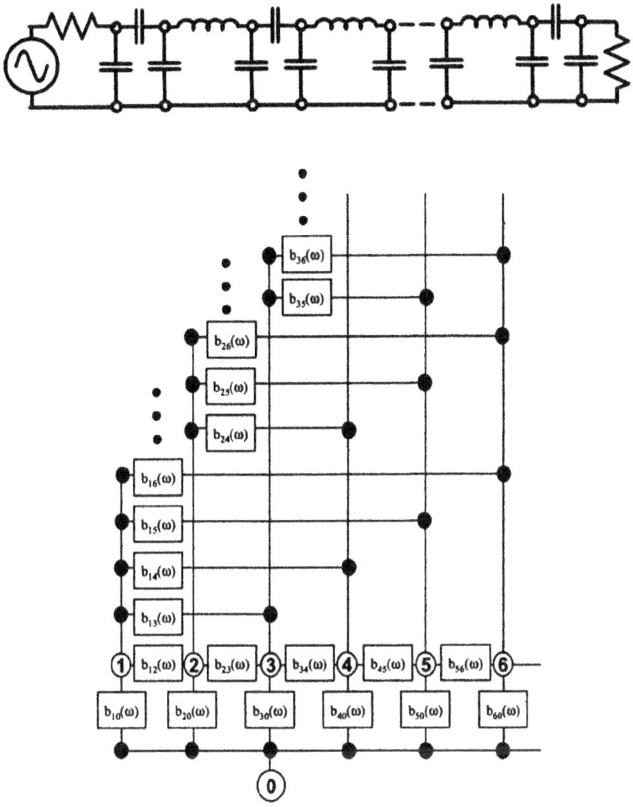

Figure 10. Traditional lumped element ladder network representing an idealized band-pass filter and a practical nodal representation of the same filter with frequency variable reactances including couplings beyond nearest neighbors.

return loss can be optimized by trimming the resonator frequencies using dielectric cylinders mounted in the housing similar to those used to trim the band-reject filters. As can also be seen, the fully optimized computer model describes the measured filter response very well.

2.3.3. *Filter/LNA subassembly*

For the B band product, the completed band-pass and band-reject filters are assembled into a single sub-assembly which we have termed a micro-enclosure. Each microenclosure also contains a low noise amplifier (LNA), designed for cryogenic operation. (See Figure 12).

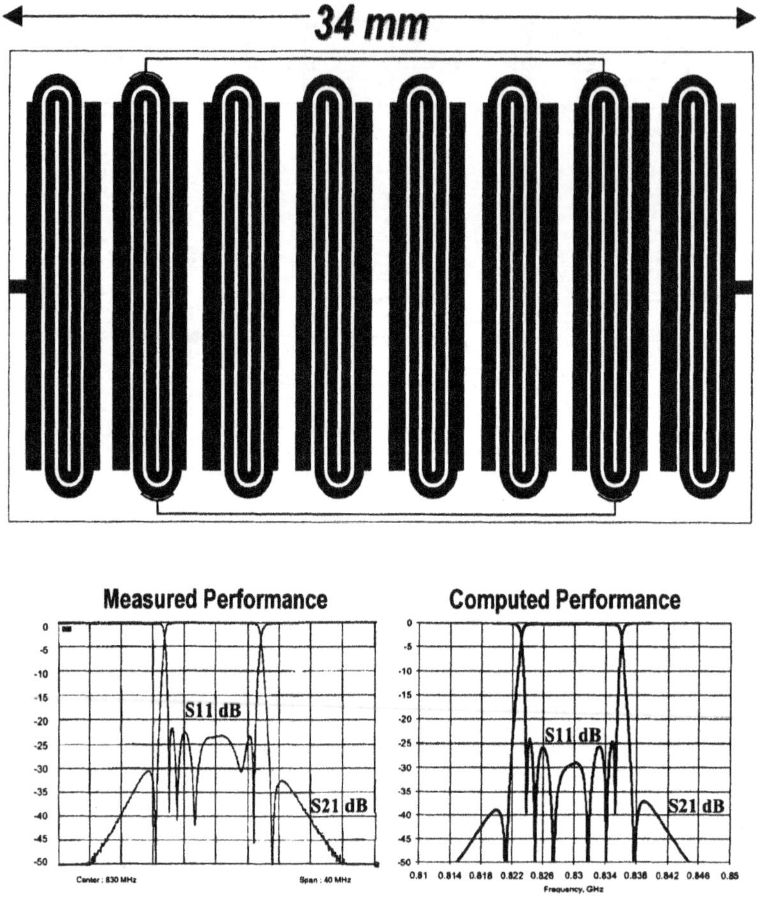

Figure 11. Band-Pass filter layout and performance compared to the model. The measured and computed insertion loss plots are on the same scale. The frequency scale is from 830 to 850 MHz and insertion loss from −50 to 0 dBm

2.4. PACKAGING AND SYSTEMS INTEGRATION

A finished filter/LNA microenclosure represents only a small fraction of a turnkey HTS system for delivery to the wireless industry. The microenclosures must be housed in a dewar that can be permanently evacuated so that it can be cooled and will stay cool. Once it has been populated with the appropriate combination of microenclosures, the dewar must be mated with a cryocooler and electronics must be provided to monitor and control the cryocooler. All of these elements must be designed to enable the

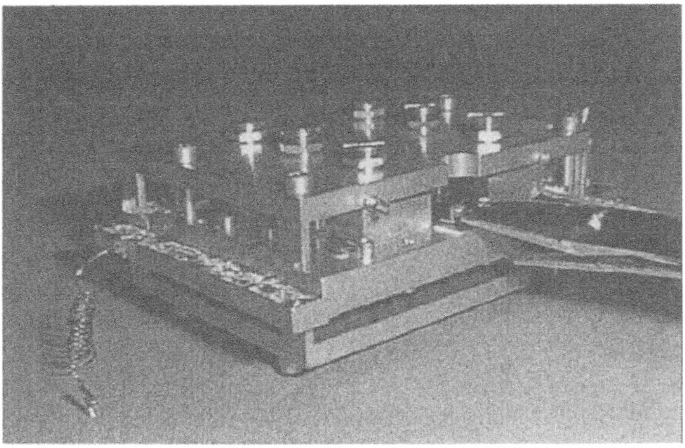

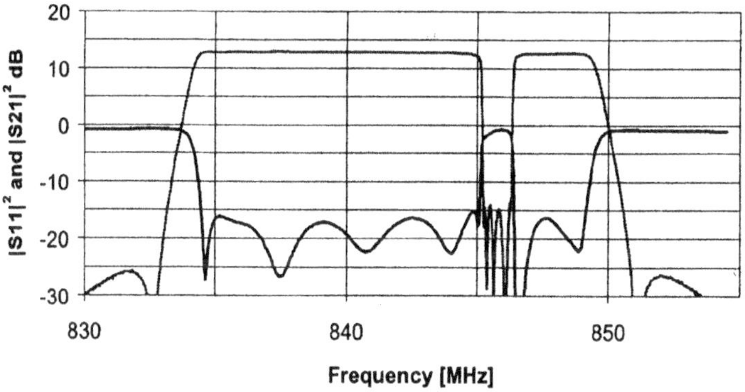

Figure 12. (above) Photograph of a completed B band filter/LNA subassembly (microenclosure) consisting of a band-pass filter, a band-reject filter and a cryogenic low noise amplifier (LNA). (below) Microwave performance of a completed B band microenclosure.

finished system to be installed into a base station. Many details of STI's standard commercial rack-mount platform are described in [16] and will be summarized here.

2.4.1. *Cryopackaging*

Cryopackaging techniques are adapted primarily from those developed for the infra-red detector industry with one major difference; wireless customers expect (and require) long, maintenance-free, lifetimes. Thus, one does not have the luxury of routine maintenance such as periodically evacuating dewars or recharging helium in coolers that might be acceptable in other areas. The dewar as well as all mechanical parts must be designed to last for more than 40,000 hours of operation, or approximately five years.

In order to attain long dewar life time, one must consider two principal factors, vacuum leaks and outgassing of the contents. Reducing the possibility of leakage requires that the dewar be permanently welded shut and also requires careful design of the RF and DC feedthroughs. In order to achieve long life all feedthroughs must have leak rates of less than 1×10^{-10} cc/sec He. This far exceeds levels of hermeticity typically defined for the electronics industry, and thus requires custom solutions.

Outgassing can be controlled by the use of appropriate materials and assembly processes within the dewar. The most volatile components can be further reduced by evacuating the dewar for an extended period of time at a moderately elevated temperature (typically days to weeks near 100^oC). Higher temperatures could accelerate this process, but may lead to degradation of the HTS properties.

Heat load is another critical consideration in the cryopackaging of HTS wireless systems as these losses determine the size, and thus cost, of the cryocooler. A principal source of heat load is radiation loss which can be reduced by minimizing the surface area of the heatsink as well as using high emissivity radiation shields. Reducing the size of the heatsink can also help reduce its thermal mass, which in turn will reduce the system's cooldown time. Although less of a concern for commercial systems than for military ones, cooldown times of less than an hour are desirable.

The heatsink used in STI's commercial platform is made of a single piece of OFHC copper. It has three bays each of which can each accommodate 2 filter/LNA microenclosures, allowing for up to 6 microenclosures per heatsink. The heatsink is directly attached to the cold finger that will ultimately provide the interface to the cryocooler. Attention must be paid to reducing the thermal impedance between the heatsink and microenclosure and/or cryocooler as this can also contribute to the heat load.

A pair of cryogenically compatible RF cables, or cryocables, are required to connect each microenclosure to the dewar wall. These cryocables introduce another significant source of heat load that is particular to RF applications. The ideal cryocable must present a large thermal impedance without adding significant RF loss, which would degrade the system noise figure. Commercially available 0.047 inch diameter stainless steel coaxial cable with reduced thickness outer jacket introduces 0.1 dB of insertion loss with 280 mW of head load for a 2 inch cable. STI's current proprietary cable cryocable introduces 0.2 dB of insertion loss with 100 mW of heat load for a similar length cable. The 12 cryocables required to fully populate a heatsink with microenclosures would then consume a significant fraction of the cooler's lift.

The dewar itself is modular in design, allowing for rework at any stage of assembly including after the dewar has been welded shut. A commercial dewar is shown at various stages of completion in Figure 13.

2.4.2. *Cryocooler*

The ideal cryocooler for our application should be capable of at least 4 W of lift at an operating temperature of 60 to 77 K at high ambient temperatures (60 to 70° C) with less than 150 W input. Furthermore it should be low-cost, small (weighing less than 5 kg) and have a long, maintenance-free life of more than 40,000 hours (approximately five years) of continuous operation.

The Gifford-McMahon (GM) cooler is attractive for this purpose as it is readily available from a number of commercial suppliers and is used throughout the semiconductor industry as a cryopump. In addition, it allows for separation between the cryocooler's compressor and its cold stage allowing for modular packaging. Conductus and ILSC have primarily used GM coolers in their products to date.

Early on, STI focussed on a design based on a linear free-piston, gas-bearing Stirling cycle cryocooler that was pioneered at Sunpower, Inc. An evolved version of this cryocooler is currently manufactured at STI for use in our wireless HTS products[12]. Over 150 of these coolers are installed in cellular telephone base stations across the US as of Summer 1999. Since they were first installed beginning in late 1997 these coolers have collectively logged more than 800,000 hours of total service, maintaining more than 99.9% overall system uptime.

Other Stirling cycle coolers have also been developed for this application and have been used for demonstration units both in Europe and Japan. Recently several Japanese prototypes have also used pulse tube cryocoolers.

2.4.3. *System Electronics*

All commercial SuperFilter® systems include a microprocessor-based controller. The main purpose of this controller is to maintain the temperature of the filters to within 0.1 K of $\approx$ 77 K under ambient temperature conditions from -10 to $+60^o$ C. This level of temperature control is necessary to keep the filters on frequency, as the temperature dependence of the superconductor penetration depth can lead to a sizable resonant frequency shift. For STI's planar filters, frequency shifts of 1 to 100 kHz/K have been observed for various resonator topologies. This temperature stability requirement can be relaxed by using a lower operating temperature (well below T_c) or by building a certain amount of temperature compensation in the filter.

Figure 13. Cryogenic technology elements: (above) STI's Stirling cycle cryocooler (below) a commercial STI dewar in various stages of assembly.

The STI controller makes use of a unique feature of the Stirling cycle cooler. The cooler lift is controlled by the amplitude of the drive, so the controller can provide maximum drive to the cooler during cooldown and reduce power consumption when the filter reaches steady-state operational conditions. While the maximum power requirement of a 6 channel SuperFilter® system is 200 W during initial cooldown, only 60 W or so are required under normal operating conditions, the equivalent of an incandescent light bulb.

The control electronics can also provide feedback on the system status including the ambient and cold stage temperatures. SuperFilter® systems provide an RS-232 serial interface which can be queried locally using a serial terminal, PC, or even monitored remotely via modem.

2.4.4. *System Enclosure*

Considerable engineering must go into the final packaging that contains the dewar, cooler and electronics. The principal role of this enclosure is to provide the necessary mechanical and electrical interfaces to the customer's system, e.g. to accept power at the +27 V which is standard in most base stations. The system enclosure must also remove any heat generated by the various components. This is most often done by means of forced air driven through the system using fans. As these fans are prone to failure, other passive approaches to heat rejection that do not require moving parts have been implemented and are highly desirable.

STI's commercial rack-mount system platform which can include up to 6 RF channels measures 7 inch × 19 inch × 22 inch (i.e. 1.7 cubic feet or approximately 0.03 m^3). This platform size has remained constant since late 1995 even though its contents have evolved. During this time, various government programs have necessitated the development of smaller variants of this platform. A program is currently underway to transfer many of these developments to the commercial product line and reduce the overall size of the system enclosure while preserving the functionality of its contents. Through this effort we expect that overall system size will be reduced by a factor of two by next year. Longer term predictions expect that a system with today's functionality can again be halved in size by 2002. It is also conceivable that at that time it would be preferable to preserve the overall system size while introducing additional functionality (i.e. additional filters or more complex filters). This evolutionary path is summarized in Figure 2.4.4.

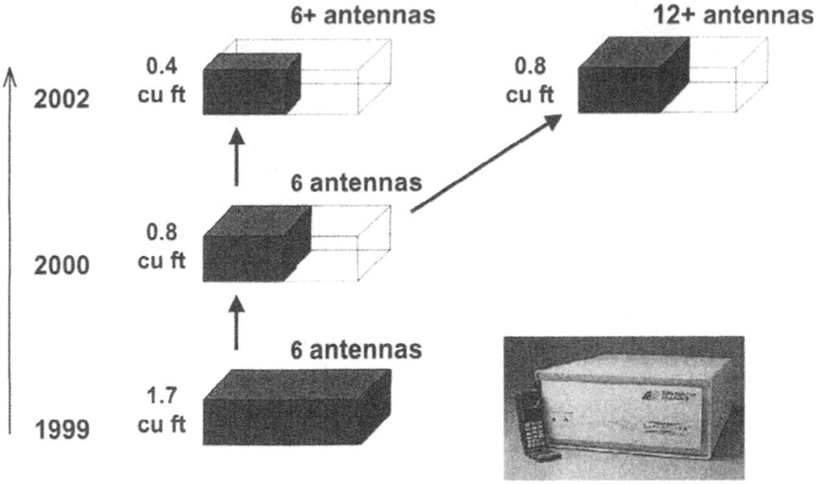

Figure 14. Size and functionality projections for wireless SuperFilter® platforms from the current 1999 model through 2002.

3. Market

The worldwide wireless telecommunications industry continues to grow faster than predicted. New systems are deployed in areas where coverage was not previously available and more consumers subscribe in areas where service is already in place. Whether sales of HTS filter subsystems will follow this trend in 2000 and beyond remains to be seen, and depends heavily on the percentage of base stations that will have HTS front ends.

3.1. US CELLULAR MARKET FOR HTS FILTERS

For HTS technology to gain international acceptance, it must first achieve significant penetration within the US market. This is the principal challenge for the three US HTS wireless companies for the year 2000. The US market is unique as it presents a more complex filtering problem arising from the non-contiguous frequency allocation and thus requires highly selective filtering. This makes it an ideal environment for the deployment of HTS filter subsystems.

Since 1996, the three US HTS wireless companies have primarily marketed their products directly to the end-user, the service provider, in the form of a retrofit kit to be installed in existing base stations. The US wireless industry is going through a consolidation phase and as of July 1999 only 7 companies control over 80 % of the base stations in the country (see Table 1). Thus, these 7 companies are the primary targets of the HTS wireless companies.

Top Seven Providers	Recent Acquisitions	Estimated Base Stations
SBC Wireless Inc.	(Ameritech, SNET, Comcast)	11,300
Bell Atlantic Mobile	(GTE Wireless)	11,000
AT & T Wireless	(Vanguard, LA Cellular)	9,500
AirTouch		7,900
BellSouth Cellular		4,600
Alltel		4,001
US Cellular		2,183
	Total Top Seven Base Stations	50,484
	Total Estimated US Base Stations	> 60,000

TABLE 1. Number of base stations in the United States as of June 1999 reflecting recent mergers and acquisitions (Source: RCR's Top 20 Cellular Carriers, March 1999)

These base stations primarily use the cellular frequency band around 850 MHz originally allocated for AMPS analog service, but most now also carry digital signals using various modulation types. Time-division modulation schemes such as GSM (Global System for Mobile, more commonly used outside the US) and TDMA (Time Division Multiple Access, a simple retrofit to AMPS) allow multiple users to share individual channels and allows for some low data-rate data transfer. CDMA (Code Division Multiple Access) is a newer spread-spectrum modulation scheme which spreads multiple users over a wider frequency band. In some areas all three digital modulation schemes, plus analog service, are present. All three digital modulation schemes also carry over to the higher frequency PCS band which is currently being deployed.

To date approximately 500 HTS filter systems have been deployed by the three US companies. This represents less than one percent of the total number of base stations in the US. While many of these systems have been sold for field trials, the vast majority remain permanently installed and in service. Exact numbers are hard to determine as system prices are not publically disclosed, but rough estimates can be made from published commercial revenue figures and are tabulated below. Note that at this early stage in the development of the market, the number of units sold may have more to do with time in the market than anything else, as ILSC and STI both established manufacturing capabilities before Conductus.

It should be noted that other US companies may also try to enter the market for HTS filter subsystems, once the market opportunity is es-

Company	Number of Units	Market Share
Illinois Superconductor	300	60%
Superconductor Technologies Inc.	150	30%
Conductus	50	10%

TABLE 2. Estimated number of HTS filter/LNA systems sold as of June 1999 by the US wireless HTS companies

tablished by the top three companies. Superconducting Core Technologies (SCT), once similar in size and product to Conductus, ILSC and STI, has re-emerged from bankruptcy with a new identity, Spectral Solutions Inc. (SSI). SSI employs approximately 10 people and has attended recent trade shows with demonstration units produced by SCT. DuPont Superconductivity and Intermagnetics General Corporation (IGC) have also worked in microwave cryoelectronics (using HTS and/or dielectric filters) and might potentially compete for market share if the opportunity arises.

3.1.1. *Europe*

The principal effort in Europe to commercialize HTS for wireless applications is the Superconducting Systems for Communications (SUCOMS) project which is part of the European Advanced Communications Technologies and Services program (ACTS) program. It involves a wide range of industrial and academic parters from across Europe including: GEC-Marconi, Leybold Vacuum, Thomson, the University of Birmingham and the University of Wuppertal. SUCOMS is an ambitious 3 year project which is scheduled to end in 1999.

The project completed its desired objectives for the first two years and has developed miniaturized lumped element filters[8, 9] and distributed power filters. The planned field trials of the equipment developed under this program had not yet begun as this manuscript was being prepared. Although originally scheduled for completion earlier, these were scheduled for the fall of 1999. Various other publications trace the progress of SUCOMS including [10, 4] from the 1998 Applied Superconductivity Conference and a report on the World-Wide-Web [3].

Various small companies have now formed in Europe that may pursue the HTS wireless market similar to their US cousins. CryoSystems Ltd. in the United Kingdom and CryoElectra GMBH in Germany, have grown out of SUCOMS and previous academic efforts. While they currently only employ a dozen or so people combined, these companies could easily grow given the right opportunity.

3.1.2. *Japan*

All three US HTS wireless companies have provided demonstration units for testing at various undisclosed locations in Japan. Carriers in Japan are looking at HTS not only for the existing Personal Hand Set (PHS) networks, but also for inclusion in a primary deployment of future third generation (3G) wireless networks. This may be a major opportunity for the the existing US or European companies, but they may face stiff competition from local Japanese companies.

Microwave superconductivity has traditionally not been the primary focus of HTS development efforts in Japan, unlike recent history in the US. In Japan, commercial development of HTS for wireless communications applications has been primarily centered in two multi-organization consortia: the so-called "Western Alliance", and Advanced Mobile Telecommunications Technologies (AMTEL).

The Western Alliance, was a loose consortium whose primary corporate members are Sumitomo, Matsushita and Kyocera. This consortium has focused on developing filters using thin film materials for both transmit and receive applications, using disk resonators for power handling and microstrip "hairpin" filters for the receive side. These have all been demonstrated in the lab, but no significant effort has been made to package these for field tests.

Efforts at AMTEL were funded for 6 years at a level of approximately 2.9 billion Yen (approximately 24 million US Dollars) with 70% of this funding coming from the Japanese government through the Japan Key Technology Center, the remaining funding equally from Alps Electric Co. and Denso Co. From 1994-1996 their efforts were focused largely on fundamental studies, but have since shifted to the development of demonstration systems. These efforts are expected to end by March 2000.

Like Conductus and STI, AMTEL has focused their efforts on thin film, planar HTS filters. However, their approach has primarily centered on "conventional" microstrip designs such as hairpin filters. Such filters have been demonstrated with up to 11 pole designs, rivaling the complexity of those in production in the US. AMTEL has constructed complete filter/LNA subsystems similar to those developed commercially in the US. However, these have so far only been tested in laboratory environments and cannot practically be deployed in the field due to their lack of advanced cryopackaging. These subsystems have been demonstrated with ≈ 0.5 dB noise figure.

It has been suggested that, as of 1998, these Japanese efforts were two years behind parallel efforts in the US[15]. While efforts in Japan for fundamental materials science and cryocooling as equal to or superior to those in the US, many of the other essential supporting technologies required to deploy HTS based cryo-subsystems are lacking.

Recently, Alps Electric Co. and Denso Co. announced the formation of a new company, Cryodevice Inc., which will perform technology development in high-performance systems for mobile communications. The primary goal of the new company will be to pursue the commercialization of components for next-generation communications systems based on the HTS filter system technology developed by AMTEL. Cryodevice has received 300 million Yen (approximately 2.5 million dollars) in funding from its corporate parents (70 % from Denso, the remaining 30 % from Alps). This level of funding is equivalent to one year of total funding for AMTEL, but it is significant to note that this is also equivalent to 3 years of AMTELs corporate funding, indicating continued support in this area. As of April 2000, Cryodevice will have 22 employees, and will be headed by M. Aoki (currently the Managing Director of AMTEL.)

3.2. VALUE PROPOSITION AND RECENT MARKET EXPERIENCE

In 1997 and 1998 most HTS filter systems were primarily sold to extend coverage in rural areas. While customers typically do not let it be known that they are buying HTS filter systems two customers did attach their names to early press releases: CellCom with Conductus and Carolina West with STI. Both of these are small rural analog service providers with less than 100 base stations that derive significant benefits from deploying HTS systems widely throughout their network.

Interest in using HTS to extend base station range in rural areas continued into 1999, and will likely continue well into the future. In addition, many field trials were underway to investigate other benefits HTS may bring in sub-urban and urban environments and using digital modulation schemes. In these environments the sensitivity enhancement and interference protection provided by HTS filter systems is likely to improve service quality, and can even potentially expand capacity in CDMA systems. These field trials are expected to continue into 2000 and motivate sales where HTS would be deployed into those situations.

The impact on overall system-level sensitivity of the insertion of an HTS filter subsystem has been evaluated through various laboratory and field tests. While individual results can vary widely depending on the particular conditions at each base station, system-level sensitivity improvements of 5 dB or more are commonly reported [2]. These improvements arise from the combination of high selectivity and high sensitivity (i.e. low noise figure) provided by the HTS filter/LNA subsystem, and cannot be achieved using competing technologies, such as TMAs, alone.

This system-level sensitivity improvement can be used for a variety of network-improvement purposes. Improving coverage quality is the most ob-

Type of system	AMPS	AMPS	GSM	CDMA
System-level sensitivity improvement (dB)	4.0	6.0	7.5	7.0

TABLE 3. Published HTS filter performance: A comparison of laboratory and drive test measurements for various modulation schemes[2]

vious application. This includes filling coverage holes at the edge of a cell or within buildings, etc. For digital systems, this amounts to keeping the bit-error-rate (BER) as low as possible. Various claims in press releases from the US companies imply that both coverage and traffic can be doubled with the use of HTS filter/LNA subsystems. Also, the number of dropped calls will be substantially reduced, by up to 40 % from recently reported results. Voice quality will also be improved, particularly for digital systems, where marginal BER can lead to completely unintelligible conversations.

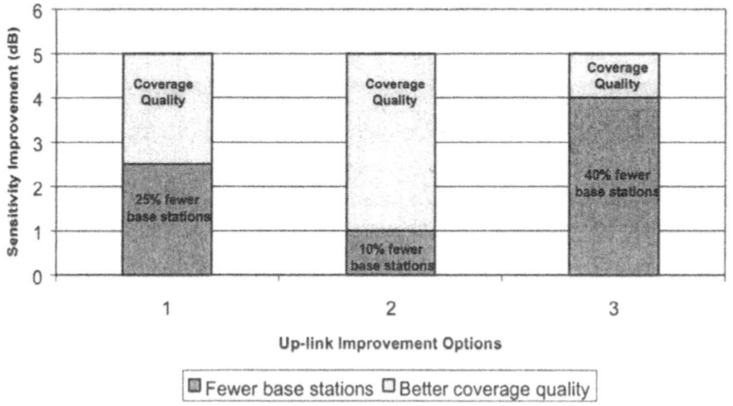

Figure 15. Options for using a 5 dB system-level receiver sensitivity improvement

The system-level sensitivity improvement can also be used to reduce the number of base stations required for a certain coverage level. As shown in Figure 15 the service provider can choose to dedicate most of the improvement they obtain by using HTS filter/LNA subsystems to reducing the number of base stations (Option 3), improving coverage quality (Option 2) or a balance of the two (Option 1).

Moving into 2000 we expect that service providers that have been evaluating HTS filter systems as retrofit kits for their existing base stations will begin to factor HTS into the deployment of new base stations and networks. The projected adoption of HTS filter systems is summarized in Fig. 16.

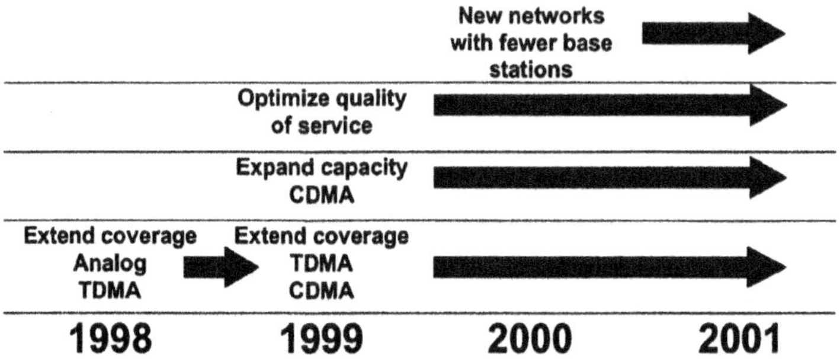

Figure 16. Projected adoption of HTS filter systems

This change was highlighted by the announcement by STI in the fall of 1999 of an agreement with US Cellular (the smallest of the seven major providers in Table 1). This agreement defines the terms under which US Cellular may purchase up to 500 SuperFilter® systems from STI over the next 5 years in addition to 50 units that were purchased earlier in the year. This is a milestone agreement as it indicates interest from one of the major providers in previously unheard of quantities. If all of the units covered by the agreement are ultimately delivered, HTS filter systems may ultimately be deployed in over well over 10% of their network.

4. Summary

Many diverse technology elements are required to make HTS filter/LNA subsystems practical. The obstacles to integrating these technologies have largely been overcome and at least three companies in the US are manufacturing turnkey systems that include HTS filters. Hundreds of these systems have been sold and permanently installed in base stations around the US and have been shown to substantially improve the coverage and quality of service where they are installed.

The critical parameter that remains to be determined is the percentage of base stations that will benefit from insertion of HTS. While some smaller, rural carriers have deployed HTS systems throughout their entire system, the larger carriers that control the vast majority of base stations have not.

With over 60,000 base stations in the US alone it is encouraging to see that volume sales to major providers are finally starting to materialize, and are expected to continue into 2000.

As results from ongoing field trials to determine the benefits HTS filter/LNA subsystems provide in urban and sub-urban conditions, particularly with digital modulation schemes, become available, the level of penetration that can be expected will become clearer. 10% penetration does not seem unreasonable from the sales agreement between STI and US Cellular, although higher levels of penetration are still likely.

We can also expect that the recent US experience will repeat itself on a global scale in the coming years. US companies will find markets abroad and foreign competitors will arise, particularly as new systems are deployed across the world. While HTS brings uncertain benefits to the high-data-rate, third generation (3G) wireless communications systems that are soon to be deployed in many areas, existing technology continues to be installed in Latin America and other areas of the world.

There are still many challenges ahead. Will HTS have a major impact on new 3G networks? Will this lead to a role for HTS in large-scale system deployments? Will volume sales really continue into 2000? And if so, will the manufacturing capabilities be able to keep up with demand? These are certainly interesting times for the budding HTS wireless industry!

5. Acknowledgments

The author would like to thank D. R. Chase, R. B. Hammond, G. L. Hey-Shipton, B. Isaacs, D. Jedamzik, S. K. Remillard, M. I. Salkola, and E. R. Soares for many useful discussions related to this work.

References

1. Braginski, A. I. (1999) Superconducting electronics coming to market. *IEEE Trans. Appl. Supercond.*, **9**, 2825–2836.
2. Bundy, S. C. (1998) Sensitivity improvements and associated benefits of tower-top amplifiers. *Microwave J.*, **41**, 88–99.
3. Gough, C. (1999) Superconducting electronics, a roadmap for Europe. Technical report, Superconductivity European Network of Excellence (SCENET). http://www.cm.ph.bham.ac.uk/reports/roadmap98/roadmap.html.
4. Greed, R. B.,Voyce, D.C., Jedamzik, D., Hong, J. S., Lancaster, M. J., Reppel, M., Chaloupka, H. J., Mage J.-C., Marcilhac, B., Mistry, R., Hf̈ner, H. U., Auger, G. and Rebernak, W. (1999) An HTS transceiver for third generation mobile communications. *IEEE Trans. Appl. Supercond.*, **9**, 4002–4005.
5. Hammond, R. B., Scalapino, D. J., Schrieffer, J. R., and Willemsen, B. A.. (1998) HTS wireless filters: Past, present and future performance. *Microwave J.*, **41**, 94–107.
6. Hammond, R. B., Soares, E. R., Willemsen, B. A., Dahm, T., Scalapino, D. J. and Schrieffer, J. R. (1998) Intrinsic limits on the Q and intermodulation of low-power

high temperature superconducting microstrip resonators. *J. Appl. Phys.*, **84**, 5662–5667.

7. Hey-Shipton, G. L. (1999) Efficient computer design of compact planar band-pass filters using electrically short multiple coupled lines. In *IEEE MTT-S Digest*, IEEE, Anaheim.
8. Hong, J. S., Lancaster, M. J., Greed, R. B., Voyce, D., Jedamzik, D., Holland, J. A., Chaloupka, H.J. and Mage, J. C. (1999) Thin film HTS passive microwave components for advanced communications systems. *IEEE Trans. Appl. Supercond.*, **9**, 3893–3896.
9. Hong, J.-S., Lancaster, M. J. and Mage, J. C.. (1999) Cross-coupled HTS microstrip open-loop resonator filter on LaO substrate. In *IEEE MTT-S Digest*, IEEE, Anaheim.
10. Jedamzik, D., Menolascino, R., Pizarosso, M. and Salas, B. (1999) Evaluation of HTS sub-systems for cellular base stations. *IEEE Trans. Appl. Supercond.*, **9**, 4022–4025.
11. Lancaster, M. J. (1997) *Passive Microwave Device Applications of High Temperature Superconductors.* Cambridge University Press, Cambridge.
12. Loung, V., O'Baid, A. and Harper, S. (1997) Path to low cost and high reliability Stirling cycle coolers, in R.G. Ross, Jr. (ed.), *Cryocoolers*, **9**, Plenum Press, New York, pp. 97-108.
13. Matijasevic, V. C. (1998) Reactive Evaporation of YBCO Films for Passive Microwave Applications, Presentation MBB-01, 1998 Applied Superconductivity Conference, Palm Desert.
14. Rowell, J. M. (1999) Recommended directions of research and development in superconducting electronics. *IEEE Trans. Appl. Supercond.*, **9**, 2837–2848.
15. Rowell, J. M., Beasley, M. R. and Ralston, R. W. (1998) WTEC panel report on electronic applications of superconductivity in Japan. Technical report, International Technology Research Institute. http://itri.loyola.edu/scel96/.
16. Scharen, M. J., Chase, D. R., Ho, A. M., O'Baid, A., Raihn, K. F. and Forse, R. J. (1997) Filter subsystem for wireless communications. *IEEE Trans. Appl. Supercond.*, **7**, 3744–3747.
17. Smith, D. G. and Jain, V. K. (1999) Superconducting filters for wireless communications: A reappraisal. *IEEE Trans. Appl. Supercond.*, **9**, 4010–4013.
18. Soares, E. R., Raihn, K. F., Davis, A. A., Alvarez, R. L., Marozick, P.J. and Hey-Shipton, G. L. (1999) HTS AMPS-A and AMPS-B filters for cellular receive base stations. *IEEE Trans. Appl. Supercond.*, **9**, 4018–4021.
19. Soares, E. R. (1999) Design and construction of high performance HTS pseudo-elliptic band-stop filters. In *IEEE MTT-S Digest*, IEEE, Anaheim.

HIGH TEMPERATURE SUPERCONDUCTOR MULTIPLEXERS FOR SATELLITE APPLICATIONS

RAAFAT R. MANSOUR
University of Waterloo
Waterloo, Ontario, Canada, N2L 3G1

1.0 Introduction

1.1 OVERVIEW OF SATELLITE COMMUNICATION SERVICES

The history of satellite communications began with the publication of "Extra Terrestrial Relays" by Arthur C. Clarke in the October 1945 issue of Wireless World [1]. In this article, Clarke suggested that three satellites in geo-stationary orbit could provide worldwide communications. The geo-stationary orbit is 35,860 km above the equator. Satellites in such an orbit appear to remain fixed in space in relation to a specific point on earth. Over the period 1945-1964 several experiments were carried out to demonstrate the concept of using the satellite as a repeater in the sky [2]-[3]. In 1964, eleven nations established the International Telecommunication Satellite Consortium (INTELSAT). The consortium launched INTELSAT I, known as "Early Bird" in April 1965. It had two transponders with 25 MHz bandwidth and 10 watt of effective radiated power each, and could handle 240 transmit-receive voice circuits at one time or one TV signal. Although, its design life was only 18 months, it lasted for four years. INTELSAT I was the first satellite to provide regular communication service, and so it began the commercial era of satellite communications

Since the launching of INTELSAT I, communication satellites have proliferated. Currently, there are hundreds of communication satellites in operations, and scores more are in the planning and construction stage. In just a few decades the applications of satellites have expanded to include communication, navigation, weather forecasting and defense applications. For communication satellites the expansion included the use of lower orbit satellites, the use of different frequency bands and the provision of various communication services.

Communication satellite services can be categorized into four main service areas: Fixed Satellite Services (FSS), which include the traditional voice trunking, remote telephony, direct to Cable TV and very small aperture terminals (VSAT's). Most of these services started in the 1970's with the exception of VSAT networks, which were introduced, in mid 1980. While voice trunking is currently being replaced by terrestrial

H. Weinstock and M. Nisenoff (eds.), Microwave Superconductivity, 417–446.

fiber optic networks, satellite services for remote telephony, VSATs and direct to cable TV services are still growing as driven by significant market demands.

Direct Broadcasting Services (DBS) or Direct To Home (DTH) services were introduced to the market place in a significant way initially in Europe and later on in North America and Asia. Although provision of these services has pushed innovation in the design of higher power satellites, the fundamental architecture of the DTH satellites remains the same as the FSS satellites, i.e. a bent-pipe repeater with no on-board processing capabilities.

The Mobile Satellite Services (MSS) and the interactive multi-media services are new emerging services. These services are used for paging, E-mail. telemedicine and Internet applications. Satellites for such services are candidates for more advanced multi-beam techniques and on-board processing. Figure 1 summarizes the services that are provided by communication satellites.

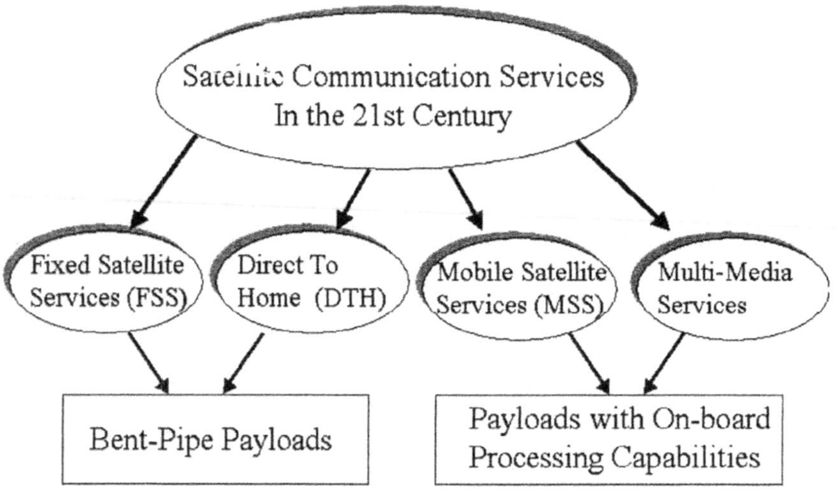

Figure 1. Satellite communication services in the 21st century.

Table 1 lists the frequency band designations in common use for satellite services. The higher frequency range in each of these bands is used for the up-link (earth to satellite) and the lower frequency range is used for the down-link (satellite to earth). The up-link and down-link frequency bands must be widely separated so that they do not interfere with one another. The L-band is used for mobile satellite services (MSS) and navigation systems. The C-band is used for fixed satellite services (FSS), while Ku-band is typically used for both FSS and DBS services. The X-band is dedicated for military applications. The majority of emerging multimedia satellites uses the Ka-band for broadband communications.

Table 1. Frequency band designations in common use for satellite services.

Band	Direction	Frequency (MHz)	Application	Satellite
L/S	Space-Earth	1525-1559	Mobile	Inmarsat III, MSAT
	Earth –Space	1626.5-1660.5		Inmarsat III
	Space-Earth	1980-2010		ICO
	Earth –Space	2170-2200		ICO
	Space-Earth	2483.5-2500		MSAT
	Earth –Space	1610.1626.5		Iridium
C	Space-Earth	3400-4200	FSS	Intelsat VIII, IX
	Earth –Space	5850-6650		Intelsat VIII, IX
Ku	Space-Earth	10950-11200	FSS	Intelsat VIII, IX
	Space-Earth	11450-11700		Intelsat VIII, IX
	Space-Earth	11700-12200		Intelsat VIII, IX
	Earth –Space	14000-14500		Intelsat VIII, IX
Ku	Space-Earth	11700-12200	DTH	
	Earth –Space	14000-14500		
Ka	Space-Earth	19700-20200	Multimedia	
	Earth –Space	29500-30000		
Q/V	Space-Earth	37500-40500	Multimedia	
	Earth –Space	47200-50200		

1.2 BENEFITS OF USING HTS TECHNOLOGY IN SATELLITE PAYLOADS

The mass, volume and power consumption of payload electronic equipment is a significant contributor to the overall cost of space systems. The High Temperature Superconductor (HTS) technology offers the potential of large reduction in mass and volume of electronic equipment, leading to significant cost reduction of satellite systems. Today's space segment represents a market of many billion of dollars for commercial and military applications and is growing. HTS technology has the potential of accelerating the development and implementation of new advanced satellite systems. It also represents the potential of performance enhancements of strategic defense communications systems.

The main features of HTS technology are very low RF loss due to the very low surface resistance, extremely low noise approaching the quantum limit due to operation at cryogenic temperatures and miniaturization through the use of planar thin-film structures on high dielectric constant substrates. The miniaturization is attributed to the low loss of the HTS films, which in turn allows bulky structures to be replaced with equivalent planar structures without sacrificing performance.

Mass reductions have a dramatic impact on the economics of a satellite program, since launch costs are related to satellite weight. Mass reductions can also be exploited to increase the capacity (by adding payload electronics) or extend operational life (by increasing station keeping fuel). As a result, market factors have constantly been pushing hardware suppliers to reduce mass and size of their products. Similarly, the reduction in noise figure can lead to improvement in the satellite receive gain to noise temperature (G/T) ratio which in turn would have a significant impact on the cost and size of satellite ground terminals. Table 2 provides an overall summary of the potential impact of HTS technology on satellite systems.

Table 2. Potential impact of HTS technology on satellite systems.

1. System and Component Mass Reduction • Launch cost reduction • Longer on-station time • Additional payload capacity
2. Improvement in Noise Figure and Receiver G/T • Effective increase in ground station equivalent isotropic radiated power (EIRP) • Reduction in antenna size • Possible cost impact to mobile services • Improved link margin against rain or fading
3. Reduction in subsystem assembly and integration • Potential cost reduction • Potential Improvement in delivery schedule
4. Prime Power Reduction • Mass reduction • Additional Capacity
5. Smaller satellite bus architectures due to equipment miniaturization

In communication satellites the practical constraints of non-linearity in high power amplifiers (HPA's) necessitate channelization of the frequency band into a number of RF channels, commonly referred to as transponders. Signals in each transponder are amplified by separate HPA's and then re-combined at the output for transmission via a common antenna. This division into individual transponders and the subsequent recombining is accomplished by employing the input and output multiplexing networks in the communication payload. These multiplexers comprise RF channel filters coupled via ferrite circulators (input multiplexers) or a waveguide manifold (output multiplexers). Therefore, communication satellites require a large number of microwave filters in the input and output multiplexing networks.

Figure 2 illustrates a simplified block diagram of a satellite payload. It consists of receive/transmit antenna, a beam forming network (BFN), input filter assembly (IFA), high gain receiver (RCV), input and output multiplexers, high power amplifiers (HPA) as well as several switch matrices for redundancy. The antennas, BFN network, receivers and HPA's are all wideband devices and therefore have little impact on the performance of the transponder. On the other hand, the amplitude and phase characteristics of the narrow-band filters in the multiplexing equipment play a significant role in shaping the overall performance of the transponder. As a consequence, the multiplexing equipment has been the focus of an extensive R&D over the past two decades leading to a series of innovations to optimize RF performance and minimizing size and mass. The emergence of the HTS technology has created another opportunity for more innovations in multiplexing equipment [4]-[12]. The following subsections describe the motivation of using the HTS technology in building input and output multiplexing networks for satellite payloads.

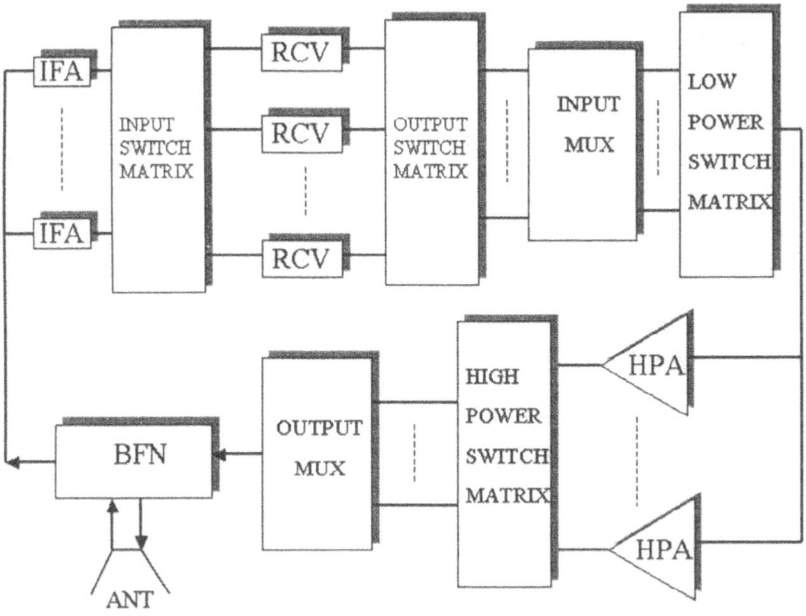

Figure 2. A simplified block diagram of a satellite payload.

2.0 C- band HTS Input Multiplexers

2.1 CONVENTIONAL TECHNOLOGY VERSUS HTS TECHNOLOGY FOR C-BAND INPUT MULITIPLEXERS

In communication satellites, the available frequency spectrum is a primary resource. In order to utilize the allocated frequency spectrum as effectively as possible, guard bands between transponders should be minimized and hence sharp cutoff filters are desirable. Furthermore the filters must have flat group delay and small gain slope to minimize the distortion and cross talk. The types of filter designs that have been employed over the past three decades for C-band input multiplexer applications are: i) high-order standard single-mode waveguide Chebyshev filters & equalizers (1971-1982), ii) dual-mode quasi-elliptic waveguide filters & equalizers (1978-1989), and dual-mode self-equalized & externally equalized quasi-elliptic dielectric resonator filters. (1983-present).

Conventional input multiplexers are built using the circulator-coupled approach shown in Figure 3. For input multiplexers, the absolute insertion loss - as long as it is less than a few dB - is not a constraint, since the Noise Figure is almost entirely governed by the front-end receiver section. Any insertion loss after the receiver contributes little to the overall Noise Figure. As a consequence, the relatively lossy circulator-coupled approach is employed as it provides a maximum flexibility, both in terms of realization as well as in the layout of the input multiplexer. However, it should be mentioned that the unloaded Q that must be achieved in channel filter is very important since it relates directly to the insertion loss variation and gain slope across the passband, which are two critical specifications for input multiplexers.

The first generation of C-band satellite input multiplexers were built with 24 input channel filters. In order to increase the communication capacity, there has been a push over the past decade to maximize frequency re-use and antenna spot beams resulting in a greater number of transponders. As an example, the C-band input multiplexer of the INTELSAT VII satellite (launched in 1991) has 44 input channel filters, while INTELSAT VIII satellite, which was lunched in 1996 had 60 channels.

Owing to the large number of channel filters, the mass and volume of multiplexing equipment constitute a considerable portion of the overall satellite payload. Any reduction in mass and volume can lead to greater communication capacity and/or increased lifetime, thus reducing the cost of a satellite channel. Figure 4 illustrates two types of superconductive filters: one uses the hybrid dielectric DR/HTS technology and the other is based on HTS thin film technology. In Section 2.3, we present the results of an input multiplexer built using hybrid DR/HTS filters while in Section 2.4, we present the results of a fully integrated 60-channel input multiplexer built using HTS thin film filters.

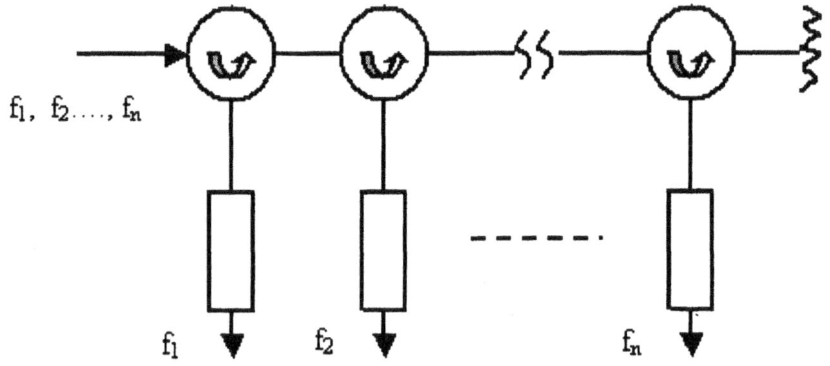

Figure 3. A circulator-coupled approach for input multiplexer applications.

Figure 4. Layout of two superconductor filters.

2.2 DESIGN CONSIDERATIONS FOR HTS C-BAND INPUT MULTIPLEXERS

The design and fabrication of superconductive filters differ in many respects from the conventional filters. As with all emerging technologies, there are a number of problems that must be overcome before commercial products can be designed. The following sections outline the three major design issues for the realization of HTS input multiplexers.

2.2.1 Defects in HTS films and substrates

The performance of highly selective narrow band filters is very susceptible to even minor defects in HTS films and substrates. It can be readily shown that a manufacturing defect or an inherent local defect with dimension in the range of 2 to 3 μm can cause about a 1 MHz deviation in the resonance frequency of a C-band HTS thin film resonator. The specifications of the present generation of satellite systems require that the center frequency of RF channels be held to within ± 300 kHz. As a consequence, a shift of 1 MHz in the filter center frequency would be unacceptable to the commercial satellite industry.

Similarly, small defects in the substrate in the form of non-uniformities, voids or twinning steps can also lead to substantial deviation in the resonance frequency of the filter elements. At the present time limited data is available on the variability of the dielectric constant from wafer to wafer or within a wafer. At C-band a difference in dielectric constant between 23.5 and 24 can cause a shift of 40 MHz, which is more than the whole bandwidth of the channel filter.

The quality, uniformity and reproducibility of HTS films and substrate will certainly improve in the future. Nevertheless, in view of the stringent specifications of satellite input multiplexers, we believe that successful design of HTS channel filters, that meet all the specification requirements, necessitates the use of tuning adjustment elements. It should be also noted that tuning elements are still being used in the design of conventional C-band input multiplexers.

2.2.2 Design tools for HTS thin film filters

The availability of accurate design tools is no longer an issue. Several commercial packages have become available recently for simulation and design of microstrip circuits. Although, these existing software packages for microstrip circuits do not take the physics of superconductivity into account, some of these packages provide reasonably accurate results for HTS thin film filters [13]. For HTS filters operating at low power, the impact of material defects and manufacturing tolerances on the filter performance far exceeds that caused by neglecting the physics of the superconducting materials.

The major problem with commercial EM simulators in the design of HTS thin film circuits is that they are very computation-intensive. The CPU time and memory space required to simulate and optimize a fully integrated HTS thin film multiplexer using present EM simulators far exceeds the capabilities of today's computer workstations.

Despite the fact that tuning elements would have to be used eventually to tune the channel filters performance to the required specifications, these CAD tools are considered a necessity. The role of the tuning elements should be restricted to "fine tuning". The task of designing HTS thin film filters without such advanced CAD tools would involve a great deal of trial and error efforts and more than likely would result in an inferior design.

2.2.3 Thermal stability of the HTS filters

Aside from mechanical defects and non-uniformities the resonance frequency of an HTS resonator is also affected by temperature variations. In conventional filters, temperature dependence is determined mainly by the thermal expansion coefficient of the resonant cavities and dielectric materials. For superconducting filters there is an additional effect that causes a strong thermal frequency shift. The inductance of a superconductor line has two parts: a geometric inductance (identical to that of a normal conducting lines) and a kinetic inductance whose value depends on the superconductivity penetration depth, which is a strong function of temperatures. At low temperatures, the kinetic inductance is rather small and relatively independent of temperatures. However, for temperatures close to the transition temperature, Tc, the penetration depth increases rapidly, leading to a strong increase in the kinetic inductance.

For operation at 77 K the frequency deviation for an HTS TBCCO filter operating at 4 GHz could be as large as 300 KHz per K. This frequency drift is considerably larger than that associated with the geometrical thermal expansion coefficient and may be difficult to compensate. HTS filters typically are tested and tuned in a laboratory environment before integration in the cryo-package. A temperature deviation could be generated from the following factors:

a) Temperature stability of the cryocooler (typically in the range of 0.1K for space cryo-coolers)
b) A temperature gradient across the package itself (typically depends on the cryo-package design)
c) A difference between the temperature in the laboratory, at which measurements were taken, and the actual operating temperature.

The designer must leave enough margins in the design to accommodate the temperature deviation generated from the above sources. Alternatively, the frequency stability could be improved by operating at temperatures further below the transition temperature, for example, near 60 K. This however, will have an impact on the achievable cooling power of the cryocooler.

2.3 INPUT MULTIPLEXERS EMPLOYING HYBRID DR/HTS

The current baseline design for C-band input channel filters is based on dielectric resonator technology. A conventional dielectric resonator filter consists of a number of dielectric loaded cavities operating either in dual-mode or single-mode configuration. The typical size for a C-band dielectric resonator cavity is 1.0"x1.0"x1.0", while the typical achievable Q is 10,000.

The size and mass of dielectric resonator filters can be reduced by using the concept of image-type resonators where a smaller size resonator is placed in contact with a conducting plate. A normal conducting plate will however significantly degrade the resonator Q. By replacing the conducting plate with a plate that is made from HTS thin film materials, considerably higher Q values can be achieved. Details of this filter configuration are given in [14]-[15].

Figure 5 illustrates the layout of a 4-channel C-band circulator-coupled multiplexer employing 8-pole hybrid dielectric/HTS filters with 1 % bandwidth. The overall size of the superconductive multiplexer including the mounting plate is 7.8 cm x 8.1 cm x 11.4 cm. The measured overall performance of the 4-channel superconductive multiplexer at 77 K is given in Figure 6. Channel #1 exhibits a minimum insertion loss of 0.26 dB. This overall loss is the sum of the losses encountered in the cryogenic circulator, the 0.085 stainless steel cable and the superconductive filter. Channels 2, 3 and 4 exhibit a relatively higher insertion loss due to insertion loss incurred during each trip through the channel dropping circulators.

A comparison between the superconductive multiplexer and a conventional 4-channel C-band input multiplexer, employing dual-mode dielectric resonator filters, is also depicted in Figure 5. The volume of the HTS multiplexer is less than one fifth the volume of the conventional dielectric resonator multiplexer.

The 4-channel superconductive multiplexer shown in Figure 2.4 was developed by COM DEV [15] for integration into the High Temperature Superconductive Space Experiment HTSSE II [5]. The objective of the HTSSE program was to demonstrate the functionality of advanced HTS devices and subsystems as well as advanced cryocoolers in space environment. The HTSSE package was launched on the Advanced Research and Global Observation Satellite in February 23, 1999. The data collected on orbit for this multiplexer indicate that the multiplexer survived launch conditions and about five years of ground storage without any noticeable degradation [16].

2.4 INPUT MULTIPLEXERS EMPLOYING HTS THIN FILM FILTERS

A consortium consisting of COM DEV International, Lockheed Martin Advanced Technology Center (LM-ATC), Lockheed Martin Communication and Power Center (LM-CPC), and DuPont, was formed in 1997 to develop and space-qualify integrated HTS subsystems for satellite payloads. The consortium was managed by NASA Glenn Research Center and was funded by the Defense Advanced Research Projects Agency (DARPA), Technology Reinvestment Program (TRP), Canadian Space Agency (CSA) and the Canadian Department of National Defense (DND).

Figure 5. A comparison between a 4-channel superconductive input multiplexer and a conventional 4-channel C-band dielectric resonator input multiplexer.

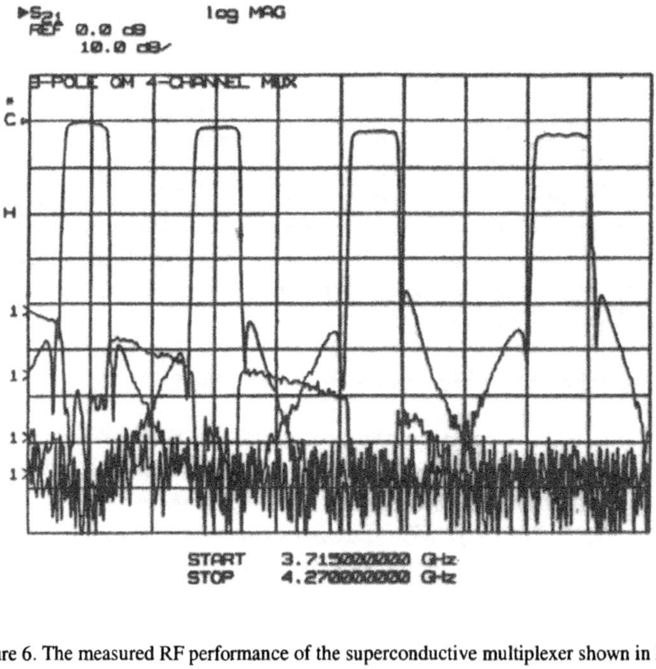

Figure 6. The measured RF performance of the superconductive multiplexer shown in Figure 5.

The objective of this program was to develop two space-qualifiable HTS satellite communication subsystems over a 3-year period. These two subsystems were intended to serve as a benchmark for satellite prime contractors to consider HTS technology as an alternative technology for their future space systems. One of the subsystems developed was a 60-channel input multiplexer that duplicates the requirements of the INTELSAT 8 program [9]. The following subsections provide design details of this multiplexer.

The input channel filters for the INTELSAT VIII program are required to comply with stringent passband characteristics. A 10-pole self-equalized quasi-elliptic HTS planar filter has been chosen to meet these requirements. An efficient way to realize the elliptic function response and the self-equalization is to use dual-mode resonators, which in turn allows coupling between non-adjacent resonators to be easily realized. Over the past years, several papers have been reported on the use of dual-mode patch resonators to design elliptic function HTS filters [13], [7]. However, one major concern with patch resonators is the difficulty in eliminating the undesired coupling between the various resonator elements. This, in turn, makes it difficult to realize filters with symmetrical frequency characteristics.

A quasi-dual mode lumped element resonator was used to build this filter [17]. The basic resonator is shown in Figure 7. Effectively, one can consider this structure as a dual-mode resonator where coupling between the two modes is controlled by adjusting the spacing G and the offset L. The advantages of the resonator configuration shown in Figure 7 is the ease of realizing and controlling the cross-coupling, which is necessary to realize quasi-elliptic and self-equalization functions. For example, Figure 8 illustrates an 8-pole filter realized using the resonator configuration shown in Figure 7. This filter consists of 4 dual-mode resonators. It can easily be shown that the filter layout permits easy realization of the cross-coupling M14 and M58. The sign of coupling, positive or negative, can be controlled by selecting the length of the coupling elements as well as the gaps: S1, S2, S3 and S4.

The dual-mode resonator shown in Figure 7 is used to construct the 10-pole self-equalized input channel filter for the 60-channel HTS multiplexer. Figure 9 illustrates a schematic diagram of the input channel filter. The filter is integrated with an input circulator and an output isolator. The circulator is part of the multiplexer function and used for channel dropping, while the isolator is inserted at the output to isolate the filter from any mismatch. The drop-in ferrite components were designed specifically for operation at cryogenic temperatures.

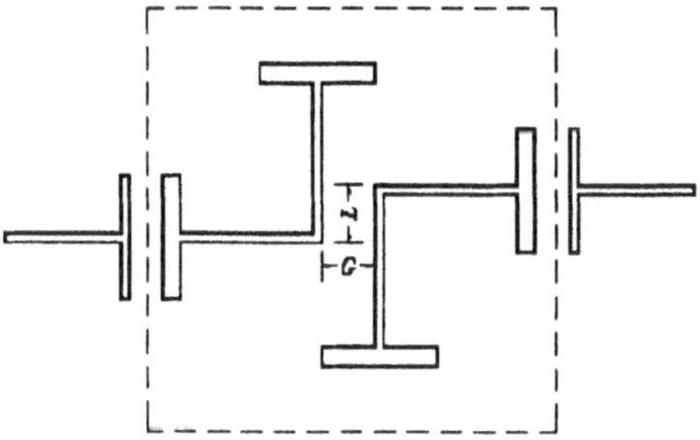

Figure 7. A dual-mode planar resonator.

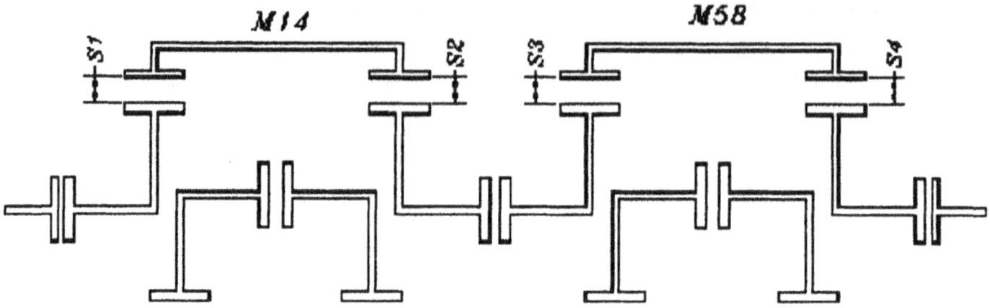

Figure 8. An 8-pole dual-mode planar filter.

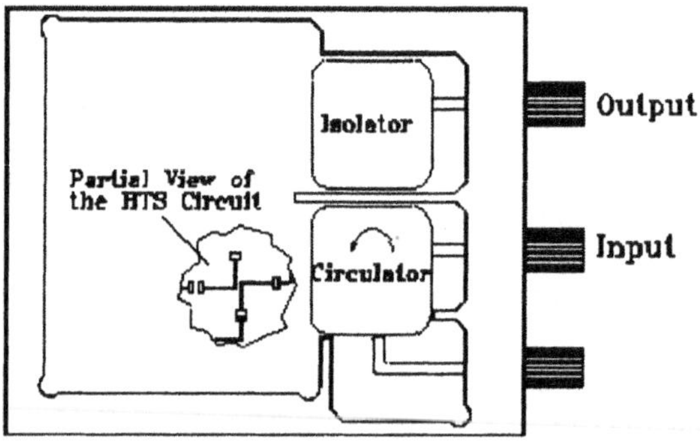

Figure 9. A schematic diagram of the input channel filter.

The measured isolation and group delay performance of a stand-alone filter with a channel dropping circulator and an output isolator is given in Figures 10 and 11. The typical specification requirements of the INTELSAT VIII program are also marked on these two figures. It can be seen that the filter exhibits symmetrical frequency characteristics. The measured performance meets the typical requirements of C-band input multiplexers. The RF test results of the integrated multiplexer are given in Section 2.4.2.

2.4.1 Cryogenic Packaging

The 60-channel HTS multiplexer is designed to duplicate the requirements of the INTELSAT VIII program [9]. The multiplexer has four operational channel filters and 56 dummy filters. The bandwidths of the four operational channels are 34 MHz, 41 MHz, 72 MHz and 112 MHz. These bandwidths were selected to correspond to all possible channel bandwidth requirements of the INTELSAT VIII program. The 56 dummy filters were inserted in the package to simulate the mass, size and heat load of an entire 60-channel system.

All four operational filters and 56 dummy filters were mounted in six stacks of ten filters each to an aluminum support structure. Figure 12 illustrates the filter support structure. This structure was machined from a single block of 6063 aluminum in order to increase its stiffness by eliminating bolted interfaces and decrease the thermal resistance through the filter stacks. The aluminum alloy 6063 was chosen for its high strength at room temperature and its high thermal conductivity at the filter operating temperature (77 K). The structure was designed to provide a mechanical resonant frequency of 100 Hz and provide a maximum temperature difference of 1 K along each 10-filter stack.

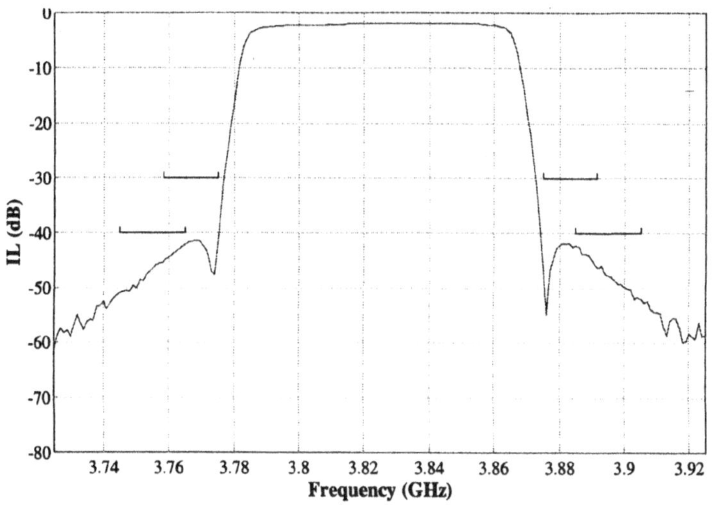

Figure 10. The measured isolation performance of a 10-pole self-equalized planar HTS filter. The horizontal lines represent the isolation specifications of the INTELSAT VIII program.

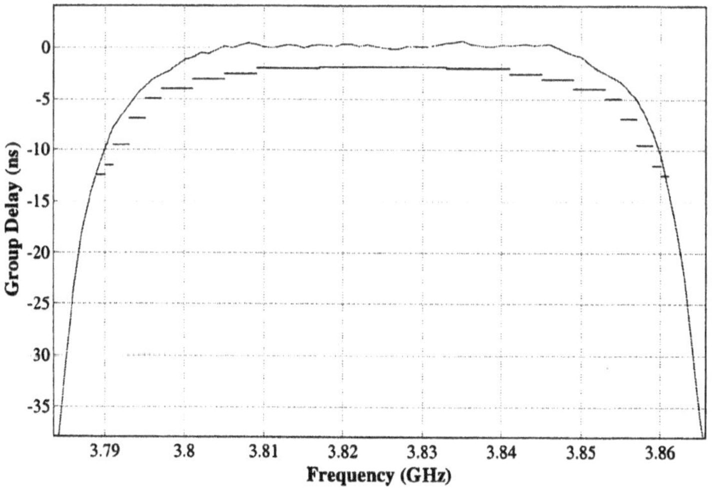

Figure 11. The measured group-delay performance of a 10-pole self-equalized planar HTS filter. The horizontal lines represent the group delay specifications of the INTELSAT VIII program.

Figure 12. Photograph of the 60 filters mounted to the aluminum support structure. Also shown is the fiberglass support tube in place and the cold coax connectors.

In the center of this aluminum structure, a beryllium shrink-fit plug was epoxied to the aluminum. This plug served as the thermal interface point between the cold head of the pulse tube cryocooler and the components to be cooled. A match-machined copper cup, connected to the cold tip of the cryocooler via copper flexible thermal braids, would slide over this plug when the system was at room temperature. The edges of the cup were flared and the tip of the plug rounded, allowing for a blind-hole interface. This shrink fit interface was designed to seize at approximately 200 K.

The package has 10 inputs and 60 outputs. A low temperature connector plate with 70 SMA connectors is securely mounted on the filter support structure. The purpose of the connector plate was to intercept parasitic heat leaks from the coax cables that connect to the warm environment; intercepting the heat at this point allows the aluminum structure to intercept the heat and minimizes the heat load to the filter packages. Copper coax cables with an outside diameter of 0.141 inches were used to connect the filters to the cold connectors and to provide the channel dropping function between filters.

The design of coaxial cables that connect the cryogenic package to the room temperature environment is a trade-off between heat leak and RF attenuation. An RF engineer, seeking to minimize the signal loss, will design a cable that is as short as possible, as large a diameter as possible, and is constructed of highly conductive materials (such as copper or silver). In contrast, the thermal engineer will strive to make the cable as long and thin as possible, constructing the cable from lossy, low thermal conductivity materials (such as stainless steel), in order to reduce the heat leak. For the input multiplexer system, the compromising solution was to use a very thin walled cable with a length that is based on the maximum allowable attenuation.

Very small diameter "lossy" coax cables were used to transition from the cold connectors to the warm connectors. Sixty 0.020-inch outer diameter stainless steel cables served as the output from each filter and ten 0.034-inch outer diameter stainless steel cables served as the inputs to each set of six filters. These small cables provided an acceptable RF loss while minimizing the parasitic heat leak into the cold package. In order to pass through the multilayer insulation (MLI) blanketing, the 70 cables were bundled into four groups, each bundle being about 1-inch long, thus minimizing the number of MLI penetrations.

The aluminum filter structure and filter packages were in turn mechanically supported by a G-10 fiberglass tube support system. One end of this support system was attached to a mounting plate that simulated a spacecraft interface, while the other end was attached to the inside of the aluminum filter support at approximately its center of mass. This support structure was engineered to survive the vibration environments and acceleration levels that are typical during the launch of a communication satellite.

A cylindrical aluminum shroud, approximately 0.030-inches thick, surrounded the cold packaging and provided a mounting surface for the MLI blanketing. This shroud was designed with vent holes at the corners to allow the contained gases to vent during pump down. The vent holes in the shroud's corners were aligned with the seams in the MLI blanketing. Thirty layers of MLI, each layer consisting of doubly aluminized mylar and silk netting, were applied over the entire cold packaged and formed a blanket thickness of approximately 0.8 inches thick. Figure 13 illustrates the 60-channel package before and after MLI blanketing.

The pulse tube cryocooler used for this package is a miniature pulse tube developed by Lockheed Martin-ATC [18]. The cold package was outfitted with two silicon diode temperature sensors and seven differential thermocouples (DTC). One diode was mounted at the bottom aluminum structure near the shrink-fit interface with the cryocooler and the other was mounted at the top of the structure near the cold connector plate; the difference between these two diodes gives the maximum temperature drop along the filter stacks. Three DTCs were mounted on filter packages at the top of the stack, approximately 120° apart from each other around the circumference. Two DTCs were mounted on filter packages at the bottom of the stack and one DTC was mounted near the beryllium shrink-fit plug. The remaining two DTCs were mounted near the diodes for reference.

2.4.2 HTS Multiplexer RF and Thermal Test Results

Figure 14 illustrates the RF performance of the four channel multiplexer at 77 K. As an integrated package the absolute loss of the individual channels is much higher. The overall absolute loss is the sum of the filter insertion loss, the loss of the input/output cables, the loss of the output isolator and the loss of the channel dropping circulators.

For laboratory testing of the 60-channel input multiplexer was inserted into a vacuum bell jar, as shown in Figure 15 and turbo-pumped at a slow, controlled rate to a 2 x 10^{-6} torr vacuum. Once the vacuum was achieved, liquid nitrogen was used to rapidly pre-cool the package to about 85K. The liquid nitrogen flowed through Teflon tubes, connected to a heat exchanger loop epoxied to the center of the aluminum filter support structure. Without the use of liquid nitrogen, a much longer cool down time would be needed with just the pulse tube cryocooler. After the temperature of the package had reached 85 K, the flow of liquid nitrogen was stopped, and the pulse tube cooler was turned on to continue the cool down.

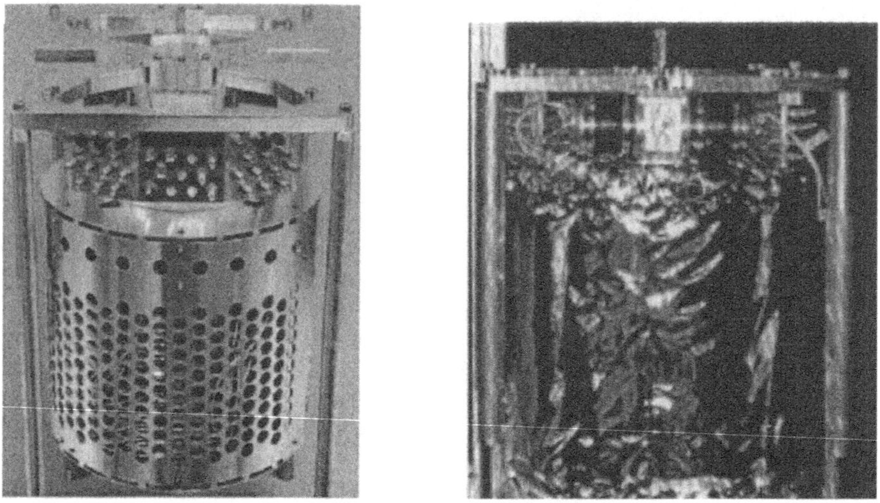

Figure 13. The HTS 60-channel multiplexer before and after MLI blanketing and cryocooler integration.

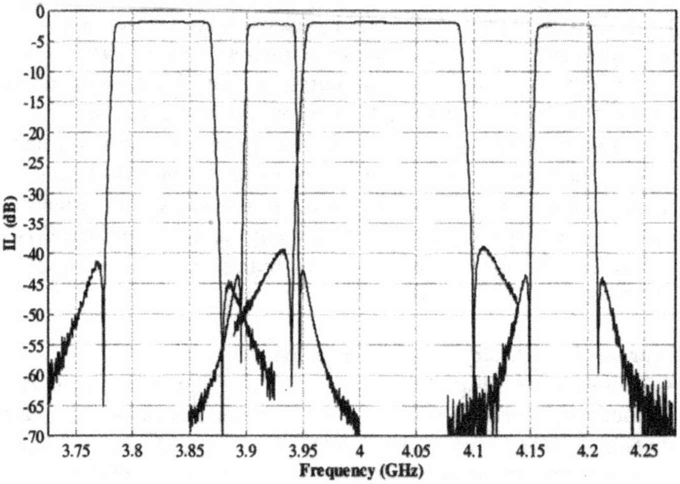

Figure 14. The measured RF performance of the individual four operational RF channels.

Two days were required to cool the package down from 85K to 77K. During the cool down, and for two weeks of continuous operation thereafter, the cooler showed no sign of performance degradation, easily maintaining the package temperature near 77K. The LN2 pre-cool is used for ground testing only. On orbit, using only the cryocooler, it would take approximately one week for the package temperature to reach 77 K.

Figure 15. Photograph of the 60-channel HTS input multiplexer and Integrated pulse tube cryocooler inside the vacuum bell jar for testing.

RF test data were taken at four temperatures. The temperatures at which this testing was performed are shown in Table 3. This Table also shows the temperature read by the bottom and top diodes. Figure 16 illustrates the measured RF performance of Channel 1 at 77.1 K, 82.3 K and 85.2 K respectively. It is observed that the filter performance drifts by 2 MHz over a temperature change of 7.1 K, i.e. a temperature drift of roughly 300 KHz/degree.

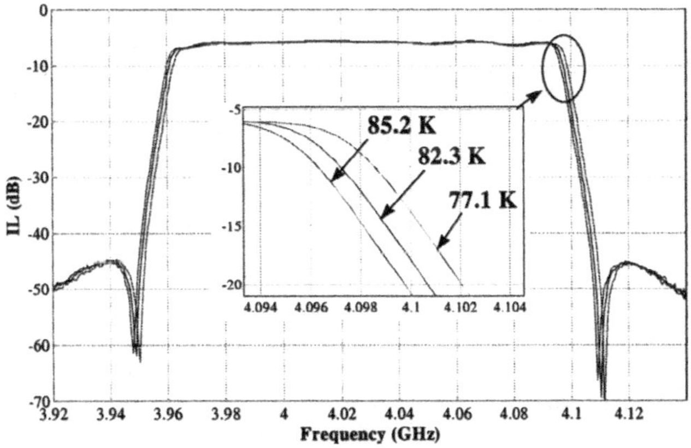

Figure 16. The measured RF performance of channel 1 at 77.1 K, 82.3 K and 85.2 K

The channel filters are designed with 0.6–1.2 MHz margin on both sides. i.e. allowing a minimum temperature margin of approximately 2K. It can be seen from Table 2.1 that the maximum temperature difference experienced along the filter stacks was less than 0.5 K. Considering the fact that the cryocooler temperature variation is ± 0.1 K, the overall temperature variation within the package is well within the maximum allowable temperature difference of 2 K.

Table 3. Summary of diode temperatures at points where RF data were taken.

Cold Tip Diode (K)	Bottom Diode (K)	Top Diode (K)
73.8	76.7	77.1
75.9	78.5	78.8
79.1	82.2	82.3
82.2	85.1	85.2

Table 4 shows a comparison of the predicted heat leak rate with the measured heat leak rate, as inferred from the cryocooler input power. The "design" heat leak rates are based on the 70 small diameter coax cables having a very low emissive (0.03) outer surface. Also, it was assumed in the design that the MLI blanket would be 1-inch thick and consist of 37 layers. As built, the 70 small coax cables had only their normal stainless steel outer surface (assumed emissivity of 0.2) and the MLI blanketing was only 0.8-inches thick with 30 layers.

Table 4. Summary of predicted and measured parasitic heat rates.

Component	Heat Leak rate (W)	
	Design	As-Built
Coax Cables	0.323	0.425
MLI/Insulation	0.246	0.314
Fiberglass supports	0.228	0.228
Total	0.797	0.967

2.4.3 Benefits of HTS technology for Input Multiplexers

The current technology for C-band input multiplexer is the dielectric resonator technology. This technology was used to build the C-band input multiplexers for the INTELSAT VIII program. Several spacecrafts were built under this program. The last one was lunched in 1996. COM DEV built all the C-band input multiplexing equipment for the INTELSAT VIII Program, which in turn was integrated into the spacecraft by Lockheed Martin.

Figure 17 illustrates a comparison between the 10-pole self-equalized HTS planar filter and the 10-pole self-equalized dielectric resonator filter similar to the one used in INTELSAT VIII. Table 5 summarizes the overall mass saving and size reduction. It can be seen that more than 50 % reduction in mass and 50% reduction in size can be achieved with the use of HTS technology. Future multimedia satellites are expected to have a larger number of RF channels (200-600 filters). As the number of channels increases the benefits of inserting the HTS technology become more substantial.

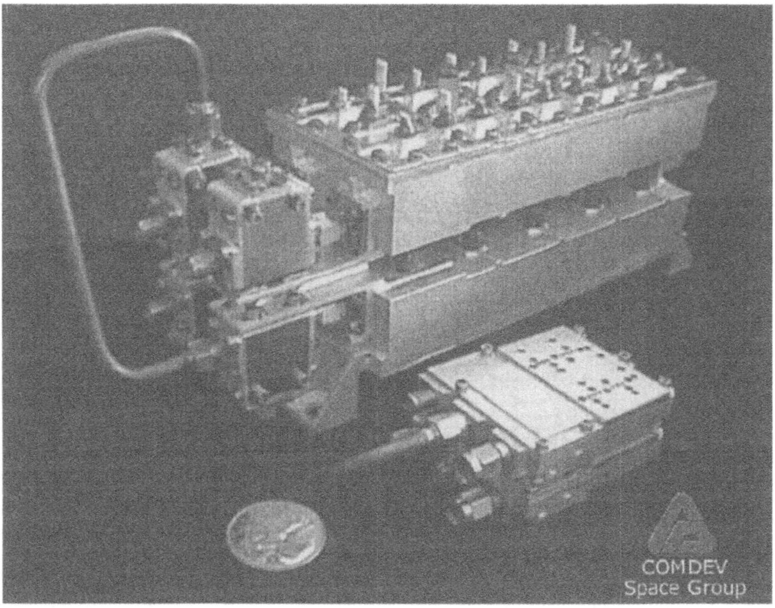

Figure 17. A comparison between a 10-pole self-equalized HTS planar filter and a 10-pole self-equalized dielectric resonator filter.

Table 5. Comparison between dielectric resonator technology and HTS technology for a 60-channel HTS input multiplexer.

Parameter	Mass	Size
Conventional Dielectric Resonator Technology Including Mounting Hardware	25 Kg	50×10^{-3} m^3
HTS technology including packaging, two cryo-coolers and two electronic-controllers	12.2 Kg	24×10^{-3} m^3
Percentage Saving	> 50 %	> 50 %

3.0 High Power HTS Multiplexers

3.1 HIGH POWER HTS FILTERS

The power handling capability of HTS filters is governed by the quality of the HTS materials as well as the filter geometry and its electrical characteristics. The parameters, which are related to filter characteristics include filter bandwidth, filter order, filter unloaded Q and operating temperature. In [19] theoretical and experimental results are presented to illustrate the effect of these parameters on the power handling capability of HTS filters. Table 6 illustrates a comparison between the power handling capability of HTS filters having six different filter configurations as shown in Figure 18.

Table 6. Maximum power handling capabilities of HTS filters having different configurations.

Filter Structure	Maximum Power (Testing in LN2)
4-pole hybrid DR/HTS filter (YBCO films, Figure 18a)	50 watts
4-pole hybrid DR/HTS filter (TBCCO films, Figure 18a)	50 watts
8-pole hybrid DR/HTS filter (TBCCO films, Figure 18b)	30 watts
4-pole patch resonator thin film filter (TBCCO films, Figure 18c)	6 watts
3-pole single-mode thin film filter (TBCCO films, Figure 18d) 50 Ohm impedance	0.7 watts
3-pole single-mode thin film filter (TBCCO films, Figure 18e) 12 Ohm impedance	2.5 watts
6-pole lumped element thin film filter (TBCCO films, Figure 18f)	0.12 watts

The data given in Table 6 indicate the importance of the structure geometry and its impact on the power handling capability of HTS filters. The results also show that hybrid DR/HTS filters have a much higher power handling capability than traditional HTS thin film filters. A detailed description of the configuration of hybrid DR/HTS filters is given in [14]. The hybrid DR/HTS resonator is the basic building block of these filters. It consists of a dual-mode dielectric resonator mounted an HTS wafer which acts as an image-plate. In contrast to HTS thin film resonators where there is a build-up of currents on the resonator edges, the current in hybrid DR/HTS resonators is distributed over a much larger area of the HTS films leading to a much higher power handling capability. Another advantage of hybrid DR/HTS filters over traditional HTS thin film filters is that no etching is required. Etching has been known to degrade the critical current density of HTS films. Hybrid DR/HTS filters have been proven to handle extremely high power levels in the range 50-100 watts [19].

For thin film filters the current density is a critical parameter in designing high power HTS thin-film filters. Using an HTS thin film resonator with a larger size is a natural choice to lower the current density, and hence increase the power handling capability of the filter.. However, the increase of resonator size has its limitations, i.e., the resonant mode along the resonator width direction may become too close to the desired resonator mode which in turn degrades the performance of the filter.

Over the past years, several papers [20]-[22], suggested the use of HTS patch-type resonators operating at TM_{01} modes. For resonator operating in TM_{01} modes, the magnetic fields are entirely confined between the resonator and ground plane and the current peaks, which typically exist at the edges of conventional resonators, are completely eliminated for resonators operating in TM_{01} modes. However, HTS filters built using TM_{010} patch resonators are considerably larger in size compared to conventional planar HTS thin film filter designs.

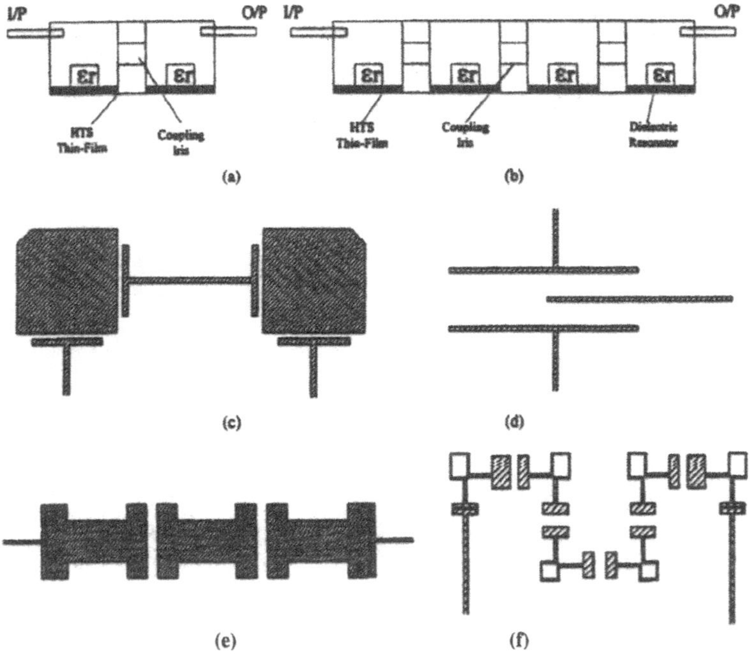

Figure 18. Six different filter configurations.

In [23] a high power thin film filter was proposed based on a novel split-resonator concept. Figure 19 illustrates the layout of the filter proposed in [23]. Each resonator consists of two (or more) parallel pieces separated by gaps. Full wave EM simulation has shown that the resonant frequency of the split resonator is very close to that of a similar solid resonator. Conventional microstrip lines have high current concentration close to the outer edges. However, by using the split-resonator concept shown in Figure 19, the current redistributes itself such that the current density at outer edges is reduced while that at the inner edges is increased [23].

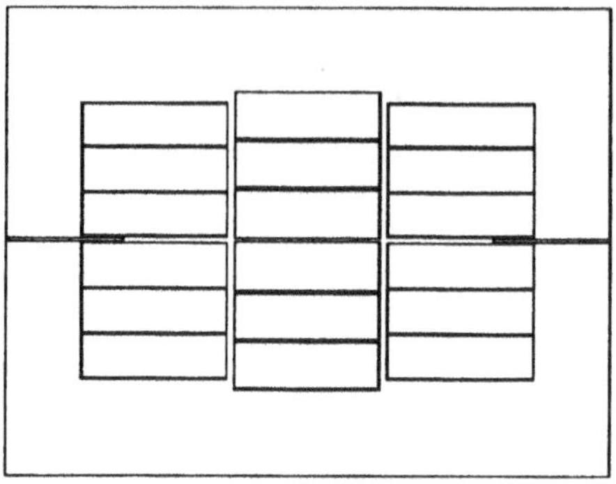

Figure 19. A schematic microstrip filter using split-resonator and inserted lines I/O coupling structures.

There is another important advantage of applying the split resonator concept. With the same total resonator width, the resonant frequency corresponding to the width of the resonator, i.e., the mode orthogonal to the desired operating mode, can be moved much higher. In other words, a much wider resonator can now be used without the potential interference of the unwanted orthogonal mode. In this filter gold films were deposited on the input/output lines. It should be mentioned that the gold films deposited on the input/output lines have a little impact on the Q of the resonators, which are completely made out of HTS films.

3.2 DESIGN CONSIDERATIONS OF HIGH POWER OUTPUT MULTIPLEXERS

The operating power of input and output multiplexers can vary from one satellite to another. Typically, input multiplexers operate at very low power levels (milliwatts), while output multiplexers operate at 10-50 watts. The type of multiplexing approach is another key parameter, which differentiates input and output multiplexers. While the circulator-coupled approach is widely used in input multiplexer applications it is seldom used in the design of output multiplexers due to its relatively poor insertion loss performance. In Section 2 the measured performance of a circulator-coupled low-power superconductive input multiplexer has been presented. In this Section we deal with the performance and design considerations of high power superconductive output multiplexers.

The number of channels on the manifold can range from 3 to 14, depending on the type of services to be provided by the satellite. A satellite payload can consist of more than one output multiplexer. Typically, the channel filters for C-band output multiplexers are either 4-pole or 5-pole 1% bandwidth quasi-elliptic filters with two transmission zeros. In contrast to the circulator-coupled approach, described in Section 2, where there is no interaction between channel filters, channel interaction is quite strong for the manifold-coupled approach. The channel filters must be properly designed to take into account the loading effects from adjacent channels.

For a 4-channel output multiplexer with a channel filter peak power of 50 watts, the instantaneous peak power due to the combined signal on the manifold can be as high as = $(4)^2 \times 50 = 800$ watts. In view of the extremely high peak power level the manifold must handle, it is not feasible to construct the manifold using currently available HTS materials. The manifold then must be built using conventional technology.

In superconducting output multiplexers, passive intermodulation (PIM) products could be generated from the inherent non-linearity of the HTS materials and/or the manifold. Even though the manifold is not made of HTS materials, any micro cracks or voids on the manifold can generate intermodulation products. For a typical communication satellite system, the level of any intermodulation products produced by the transmit portion of the satellite which fall within the satellite's receive band has to be at least 20 dB below the power of the received signal. This condition results in a carrier to passive intermodulation (C/I) requirement of anywhere from 140 to 200 dB depending upon the system gain and transmitter/receiver configuration. The feasibility of building high power hybrid DR/HTS filters with a third order intercept point of more than 100 dB has been demonstrated in [19].

Multipaction and/or ionization breakdown is another problem, which could limit the power handling capability of the superconducting high power multiplexer. Multipaction is an electron resonance phenomenon, which takes place in components operating in atmospheric pressure lower than 10^{-5} torr. At atmospheric pressure higher than 10^{-5} torr, multipaction cannot occur since ionization is the limiting breakdown condition. However, for satellite output multiplexers, the ionization is less common than multipaction breakdown as it requires partial pressure conditions, which are unlikely to occur in space.

Multipaction represents a possible payload failure mechanism for communication satellites since it can destroy RF components and/or significantly raise noise levels. If the superconductive channel filters are not properly designed, their power handling capability could in fact be limited by the multipaction breakdown rather than by the inherent power handling limitations of the HTS materials.

3.3 OUTPUT MULTIPLEXERS EMPLOYING HYBRID DR/HTS FILTERS

Figure 20 illustrates the layouts of a 4-channel superconductive C-band output multiplexer and a similar conventional waveguide output multiplexer. The HTS multiplexer employs the hybrid DR/HTS filter which is described in detail in [14]-[15]. For the conventional C-band multiplexer, both the channel filters and the manifold are built using the waveguide technology. For the superconductive multiplexer, the channel filters are hybrid DR/HTS filters while the manifold is realized using coaxial technology. The superconductive multiplexer occupies less than 5% of the volume of the waveguide multiplexer. The measured RF performance of this multiplexer is given in Figure 21. The multiplexer has been tested for high power; the channel filters were capable of handling 50 watts with no major performance degradation.

3.4 OUTPUT MULTIPLEXERS EMPLOYING PLANAR THIN FILM HTS FILTERS

Figure 22 illustrates the layout of a 4-channel HTS output multiplexer in comparison with that of a 4-channel conventional waveguide multiplexer. The measured RF performance of this HTS multiplexer is shown in Figure 23. Integration of planar filters on the coaxial manifold is more challenging than the integration of hybrid DR/HTS filters. An approach combing experimental testing as well as advanced modeling was used to design this multiplexer [24]. The planar filters shown in Figure 22 have a traditional patch design and were only capable of handling 7 watts. These filters were only used to demonstrate the feasibility of designing manifold-coupled multiplexers using planar HTS filters. With the use of the high power planar split-resonator filters similar to that shown in Figure 19, the power handling capability of the multiplexer can be extended to a much higher level.

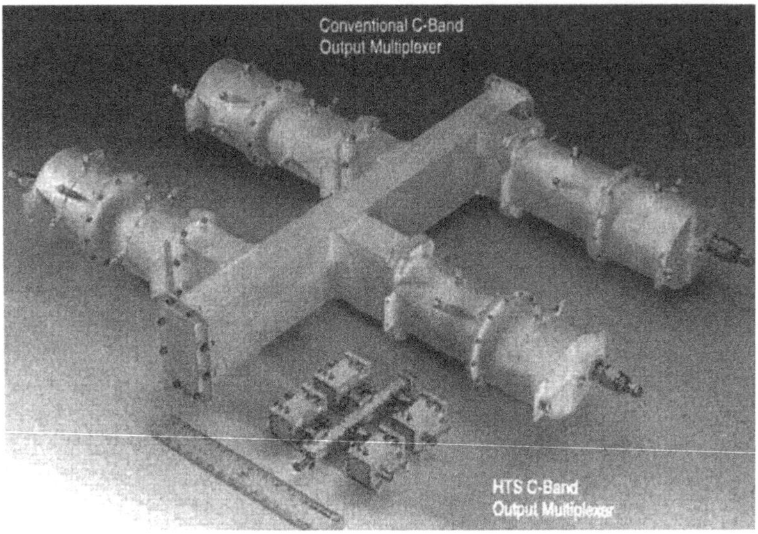

Figure 20. A comparison between a 4-channel superconductive manifold-coupled multiplexer and 4-channel C-band output multiplexer built using conventional waveguide technology.

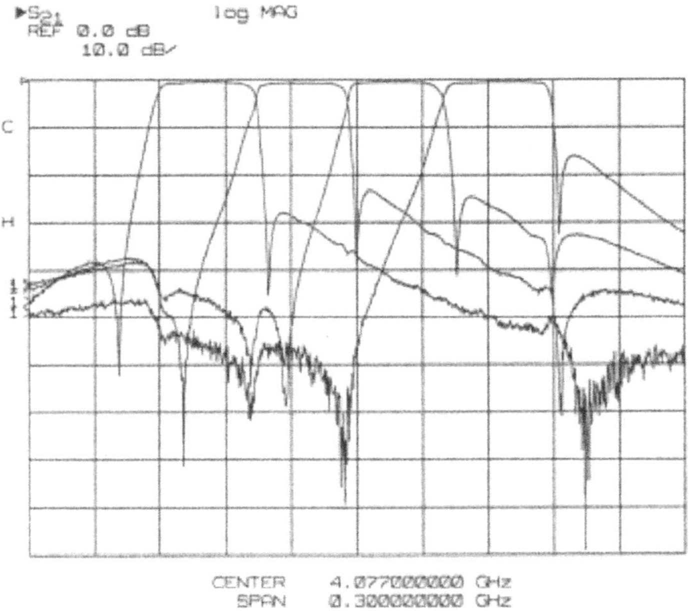

Figure 21. Measured results of the multiplexer shown in Figure 20.

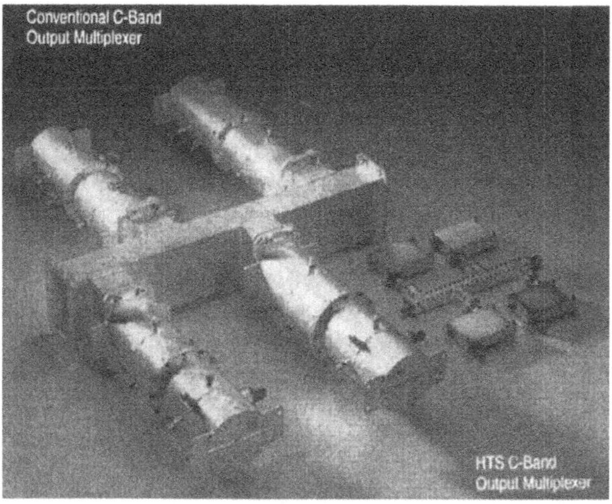

Figure 22. The layout of a 4-channel HTS output multiplexer employing planar HTS filters in comparison of a 4-channel waveguide output multiplexer

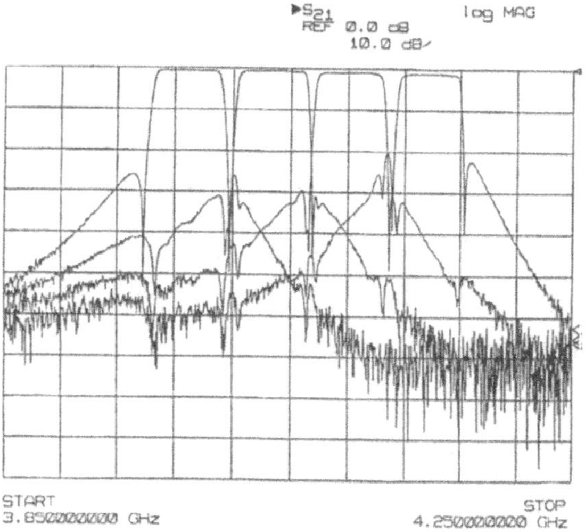

Figure 23. The measured RF performance of the HTS multiplexer shown in Figure 22

3.5 COOLING LOAD OF HTS OUTPUT MULTIPLEXERS

The potential advantages of using HTS technology in the design of output multiplexers are mass and volume reduction as well as insertion loss improvement. The insertion loss improvement can translate into improvement in satellite EIRP or reduction in the dc power required for the power amplifiers. However, the mass penalty of the cryocooler and associated electronic controller may overshadow any advantages gained with the use of HTS technology.

In contrast to input multiplexers, superconductive output multiplexers typically have a considerably higher heat load. Today's space-qualified cryocoolers that are capable of dealing with such heat loads are bulky and extremely heavy in mass. In this section, we present a detailed trade-off analysis for output multiplexers. The objective is to highlight the issues, which need to be addressed before HTS output multiplexers can be used for space applications.

The cooling load of superconductive high power multiplexers is mainly determined by:

1) The heat dissipation inside the channel filters
2) The heat load of the input/output cables
3) Heat leaks through support structure and radiation

In the case of output multiplexer assembly, the heat leaks through support structure and radiation through Multi-Layer Insulator (MLI) blanketing are typically small compared to the overall heat load. In this Section, we will focus on the heat load generated in the channel filters and input/output RF cables.

3.5.1 Heat Dissipation Inside the Channel Filters

In designing superconductive output multiplexers for satellite applications, the insertion loss of the HTS filter is a critical parameter. As the filter insertion loss increases more power will be dissipated in the form of heat, which in turn would increase the cooling demand to a level that may be significantly higher than what is available from current cooling technology. This concern is especially critical for multiplexers with a large number of channels.

There have been recent reports on the possibility of building HTS resonators with extremely high values of unloaded Q in the range of 50,000-100,000 at 4 GHz. Although such high Q values can be potentially achieved for single resonators, it is extremely difficult to achieve such high values for an integrated HTS filter that requires the use of tuning elements. The insertion loss of the filter is not only determined by the surface resistance of the HTS materials and the loss tangent of the dielectric substrate but also by the package configuration and the type of tuning elements used.

The use of tuning elements is considered a must to meet the stringent performance typically required for output multiplexer applications. It should be mentioned that although waveguide technology has been known for more than two decades, tuning screws are still being used in designing waveguide multiplexers. Therefore, it is important that one takes into consideration the impact of the tuning screws in estimating the Q of the filter, which in turn would have a considerable impact on the heat dissipation.

Conventional waveguide output filters typically operate over a wide temperature range (-20 to 100 C). The unloaded Q of the waveguide filter at 100 C could degrade considerably. A typical C-band 4-pole waveguide output filter with a 1% bandwidth can have an insertion loss as high as 0.25 dB at 100 C. The HTS filters described in Section 3.3 exhibited an insertion loss of 0.125 dB, which corresponds to an unloaded Q of 15,000.

3.5.2 Heat Dissipation of Input/Output Cables

For optimum use in HTS high power applications, the RF input/output cables must satisfy the following three conditions:

i) The RF cables must have low insertion loss to reduce heat dissipation
ii) The RF cables must provide low conduction heat load.

iii) The RF cables must be capable of handling high power.

It is difficult to find a commercially available RF cable that can meet all of the above three conditions. For example, having a cable with low conduction heat load requires the use of a small diameter cable. On the other hand, small diameter cables typically have a high insertion loss and may not be able to handle high RF power. For a 4GHz input power of 25 watts, a 15-cm long 0.085 Stainless Steel cable would have a total heat load of approximately 1.3 watts due to both heat dissipation and heat conduction. Such a heat load is considered high especially for systems, which use a large number of input/output RF cables. It is evident that there is a need to develop thermally isolated low loss RF transitions.

3.6 DC POWER REQUIREMENTS

The dc power required for the cryocooler is proportional to the cooling load, which the cryocooler has to handle. The typical cryocooler efficiency, for a cold temperature of 77 K and a rejection temperature of 300 K, is 15:1(assuming a high power cryocooler). This means that 15 watts of dc power is required at 300 K for each 1 watt of cooling load at 77K. The cryocooler dc power could be offset by the potential saving in the dc power generated in the following two areas:

i) Amplifier dc power: The typical channel insertion loss (filter and manifold) of conventional C-band waveguide multiplexers is 0.25 dB. The use of HTS technology can potentially achieve a saving of 0.1 to 0.2 dB in the insertion loss (based on the achievable HTS filter Q). Such saving in the insertion loss translates into less RF amplifier power, which in turn translates into less amplifier dc power.
ii) Heater dc power: Conventional waveguide output multiplexers use heaters in order to minimize the frequency drift with temperature. The objective of these heaters is to ensure that the multiplexer temperature is within the designed temperature range in the case when some of the channels are in the back off mode[3]. A dc power of 2 watts per channel is typically required for heaters on such systems. The use of HTS technology could eliminate the need to use these heaters, which represents another area of potential savings in mass and dc power.

A detailed trade-off analysis has been presented in [24] on the use of HTS technology for high power multiplexers for satellite systems. It was concluded that in view of the current cryocooler technology and spacecraft ambient environment for electronic hardware, the only clear advantage of using the HTS technology for a typical high power multiplexer (50-200 watt, 24 –36 channels) is size reduction. However, the size reduction advantage may be overshadowed by the mass and dc power penalties associated with the use of the cryocooler. For systems, which have large number of channels and operate at relatively low RF power (i.e 5-10 watts, more than 80 channels), in addition to size reduction, the use of HTS technology can also lead to mass reduction as well as saving in dc power. However, in order to achieve these advantages there is a need to develop low-loss thermally isolated RF transitions. The development of such transitions is as important as the development of high Q HTS filters. The key issue, however, for successful implementation of HTS output multiplexers for commercial applications, is the development of a highly efficient cryogenic environment in the spacecraft.

4.0 Conclusions

HTS technology offers the potential of large reduction in mass and volume of electronic equipment, leading to significant cost reduction for satellite systems. It could also provide performance discrimination not attainable with other technologies. The results given in this chapter demonstrate the feasibility of building C-band HTS input multiplexers with convincing advantages over current conventional dielectric resonator multiplexers. The advantages are significant reduction in size and mass. Even with the use of two cryocoolers for redundancy more than 50% reduction in mass and 50% reduction in size was achieved for a 60 channel C-band HTS multiplexer. For a program similar to INTELSAT VIII, such mass saving can add 5 to 6 months to the life of the spacecraft generating substantial revenue. Alternatively, it may be used to reduce the launch costs.

The mass of the two cryocoolers used for the 60-channel HTS multiplexer represents almost 45 % of the overall mass of the package. On the other hand, The HTS 60-channel multiplexer exhibited a heat load of less than 1 watt, which is much less than the 3-watt cooling capability of these cryocoolers. The same cryocoolers could be integrated with a much larger number of RF filter channels. Some of the multimedia satellite systems, which are currently under study, propose the use of 200-600 filters. The use of HTS technology for such systems can potentially lead to more than 80% reduction in mass in comparison with conventional dielectric resonator technology.

The results given in this Chapter also demonstrate the feasibility of designing a high temperature superconductive manifold-coupled contiguous–type multiplexer that is capable of handling the power requirements of conventional C-band satellite systems. In view of the current cryocooler technology, the trade-off analysis has shown that the use of HTS technology for output multiplexer applications leads to a significant size reduction. However, the size reduction may be overshadowed by the penalties associated with the use of cryocoolers, particularly for systems where the output power is in the range of 50-200 watts.

The tradeoffs considered for the use of high power HTS equipment also assume current spacecraft configuration and the available HTS materials. Any reduction in the temperature environment for HTS equipment via passive cooling, or other means, can provide a large reduction in mass and power required for active cryocoolers. The same advantage will result if new HTS materials can be found with higher T_c. Assuming such advances can provide an acceptable tradeoff for the required cryogenics, one can conceive of HTS-based communications payloads, including low power receive sections as well as high power output circuits.

In order to convince satellite prime contractors to consider HTS as an alternative technology for new systems, the following steps are needed: i) demonstration and validation of HTS equipment integrated with highly reliable cryocooler in operating environment, ii) demonstration of significant cost saving for overall systems costs including cost of cryocooler, and iii) demonstration of performance discriminators available only with HTS technology. These represent essential steps in bridging the transition from R&D to commercial viability of HTS technology.

The Naval Research Laboratory (NRL) HTSSE program has been a harbinger in addressing these commercialization imperatives. The program has provided a sharp focus in addressing all aspects of HTS technology especially those that pertain to performance in space environment. A variety of HTS devices, developed under HTSEE-II, were flown on the ARGOS satellite in February 1999, demonstrating the viability of HTS technology for space applications [16].

A joint (US/CANADA) program is currently underway involving COM DEV, Lockheed Martin and DuPont [9] to develop space qualifiable HTS communication subsystems for satellite applications. The focus of this program is on the design, integration and cryo-packaging of HTS subsystems. Another HTS program on space applications is currently underway in Europe [10]. Successful completion of these programs would be a big step forward in the continued development of HTS technology for space applications.

Acknowledgement

I would like to express my gratitude to Dr. Marty Nisenoff for reviewing the manuscript and for his constructive and helpful comments. I would like also to thank my colleagues at COM DEV Ltd., V. Dokas, S. Ye, B. Jolley, G. Thomson, S. Peik, W. C. Tang, and C. M. Kudsia for their support, help with analysis, simulation, design and testing. I would like to thank the management of COM DEV Ltd for their support and encouragement.

References

[1] Clarke A. C. (1945) Extra-terrestrial relays, Wireless World, vol. 51, No-10, 305.
[2] Cheung W. S. and Levien F. H. (1985) Microwaves made simple, Chapter 15, Artech House, Inc. Dedham, MA 02026.
[3] Roddy D. (1996) Satellite Communications, second edition, McGraw Hill, New York.
[4] Ritter J. C., Nisenoff M, Price G. and Wolf S. A (1991) High temperature superconductivity space experiment (HTSSE), IEEE Trans. Magn., vol. 27, pp. 2533-2537.
[5] Kawecki, T. G, Golba G. A, Price G. E. , Rose V. S. and Meyers W. J. (1996) The high temperature superconductivity space experiment (HTSSE-II) design, IEEE Trans. Microwave Theory Tech., vol. MTT-44, pp. 1198-1212.
[6] Mansour R. R., Ye S. Dokas V. Jolley B., Thomson G., Tang W. C. and Kudsia C. (1996) Design considerations of superconductive input multiplexers for satellite applications, IEEE Trans. Microwave Theory Tech., vol. MTT-44, pp. 1213-1228.
[7] Fiedziuszko S. J, Curtis J. A. and Holme S. C. (1996) Low-loss multiplexers with planar dual-mode HTS resonators, IEEE Trans. Microwave Theory Tech., vol. MTT-44, pp. 1248-1257.
[8] Klauda M. et al (2000) Superconductors and cryogenics fro future communication systems, IEEE, Trans Microwave Theory Tech., vol. MTT- 48, No. 7, 1227-1239.
[9] Mansour R. R. et al (2000) A 60 channel superconductive input multiplexer integrated with pulse-tube cryocoolers, IEEE, Trans Microwave Theory Tech., vol. MTT- 48, No. 7, 1171-1180.
[10] Soares E. R. et al, (2000) Applications of high temperature superconducting filters and cryoelectronics for satellite communications, IEEE, Trans Microwave Theory Tech., vol. MTT- 48, No. 7, 1190-1198.
[11] Romanofsky R. R. et al (2000) A cryogenic K-band ground terminal for NASA's direct-data-distribution space experiment, IEEE, Trans Microwave Theory Tech., vol. MTT- 48, No. 7, 1216-1220.
[12] Kwok R. S. et al (1997) Miniaturized HTS/dielectric multiplayer filters for satellite communications, IEEE Trans. Appl. Superconduct., vol. 7, pp. 3706 –3709.
[13] Mansour R. R. (1994) Design of superconductive multiplexers using single-mode and dual-mode filters, IEEE Trans. Microwave Theory Tech., vol. MTT-42, pp. 1411-1418.
[14] Mansour R. R. and Dokas V. (1996) Miniaturized superconducting dielectric resonator filters and method of operation thereof, US PATENT US05585331.

[15] Mansour R. R., Dokas V. Thomson G. Tang W. C. and Kudsia C. (1994) A C-band superconductive input multiplexer for communications satellites, IEEE Trans. Microwave Theory Tech., vol. MTT-42, pp. 2472-2479.
[16] Nisenoff M. (2001) On-orbit status of the high temperature superconductivity space experiment, IEEE Trans. Applied Superconductivity, to be published.
[17] Mansour R. R., Ye S., Peik S. , Dokas V. and Fitzpatrick B. Quasi dual-mode resonators (2000), IEEE Trans. Microwave Theory Tech., Vol. MTT-48, No. 12.
[18] Kutsuko V. et al, (1998) Development of pulse tube cryocoolers for HTS satellite communications in the 10th International Cryo-cooler Conference.
[19] Mansour R. R., Jolley B. , Ye S., Thomson F. S. and Dokas V. (1996) On the power handling capability of high temperature superconductive filters, IEEE Trans., Microwave Theory and Tech., vol. MTT-44, pp. 1322- 1338.
[20] Shen Z. Y. et al, (1996) Power handling capability improvement of high temperature superconducting microwave circuits, IEEE Trans Applied Superconductivity vol. 7. no..2, pp. .2446-2453.
[21] Muller G. et al, (1996) Double-sided YBCO films for planar high power filters, IEEE Trans. Applied Superconductivity vol. 7. no..2, pp. .1287-1290.
[22] De..Obaldia E. al. (1996) Thin film superconducting filters for PCS transmit applications, IEEE Trans Applied Superconductivity vol. 7. no..2, pp. .1911-1916.
[23] Ye S. and Mansour R. R. (1997) A novel split-resonator high power HTS planar filter, IEEE MTT-S Digest, pp. 299-302.
[24] Mansour R. R., Ye S., Dokas V., Jolley B., Tang W.C, Kudsia C. M. (2000) System integration issues of high power HTS output multiplexers, IEEE, Trans Microwave Theory Tech., vol. MTT- 48,1199-1208.

RF SUPERCONDUCTING CAVITIES FOR ACCELERATORS

RUGGERO VAGLIO
INFM - Dipartimento di Scienze Fisiche,
Universita' di Napoli "Federico II",
Piazzale Tecchio 80 I-80125 Napoli, Italy

Abstract. In this chapter the state of the art of rf superconducting cavities for particle accelerators is presented and discussed. The main concepts of rf cavity design and the main physical parameters are first introduced. Fabrication techniques and diagnostic measurements are then described for both bulk and thin film niobium coated cavities. The application of superconducting cavities to Electron Linear Colliders, Proton Linacs and Ion Accelerators as well as Free Electron Lasers is also briefly outlined. A large set of measurements on thin film niobium cavities is then analyzed in order to get insight in the origin of residual and nonlinear losses. Finally the possible role and preliminary results on alternative, high T_C compounds, are discussed in detail.

1. Introduction

The application of superconductivity to the world of resonant cavities for particle accelerators dates back more than twenty five years and has made possible results that could not have been realized with conventional techniques. At present, many superconducting accelerating structures are installed and fully operational in Electron Linear Colliders, Proton Linacs and Ion Accelerators. Superconducting cavities represent a very active research field and new ideas are continuously developed, as demonstrated by the recent application to Free Electron Lasers.

The extensive research effort has brought niobium radio frequency (rf) cavities close to their intrinsic limits and new and better materials are under test. Cavities are unique system to understand some of the important details of the physics of rf superconductivity, such as the origin of residual and nonlinear losses.

The aim of this chapter is to present a review of this fascinating subject, with special emphasis on cavity fabrication techniques, diagnostic measurements, interpretation of the niobium results and the role of innovative materials. The fundamentals of rf cavity design will also be briefly discussed while the theory of rf superconductivity will be presented elsewhere in this book.

H. Weinstock and M. Nisenoff (eds.), Microwave Superconductivity, 447–471.

2. Main cavity parameters, measurement techniques, need for superconductors, and intrinsic limits.

A pillbox metallic cavity can be seen as the evolution of a RLC circuit [1]. In such a structure, the electric field is maximum along the central axis of the cylinder and fails to zero at the lateral walls. An accelerating cavity has essentially the same shape but with a small hole on the cylinder axis to allow the passage of the charged particle beam which is to be accelerated by the electric field (Fig. 1). As will be discussed later, the curvature of the walls is designed to limit resonant secondary emission processes (multipacing).

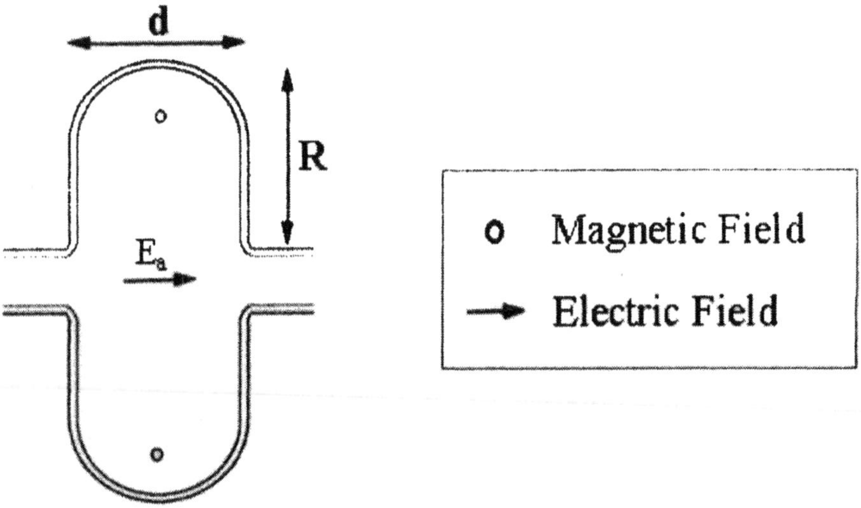

Fig. 1 Schematics of a pillow-box type resonant cavity. The accelerating electric field is along the axis of the cylinder. This cavity shape is adopted for high β (electron and proton) accelerators.

The fundamental resonant frequency is obtained for $\omega_o R/c = 2.4$ [1]. The surface currents that sustain the field produce losses that are proportional to the surface resistance of the metal. The cavity quality factor Q gives a measure of such losses and can be expressed as: $Q = \omega_0 W/P_d$, where W is the time averaged energy stored in the resonator and P_d is the power dissipated per cycle. Q can be measured in the "time domain" from the exponential decay of the energy stored in the cavity (Eq. 1), or in the "frequency domain" from the bandwidth of the frequency response of the resonator (Eq. 2):

$$W(t) = W_0 e^{-\frac{\omega_0}{Q}t} \tag{1}$$

$$Q = \omega_0 / \Delta\omega \tag{2}$$

The quality factor is related to the surface resistance by the relation :

$$Q = \frac{\mu_0 \omega_0 \int_V H^2 dV}{\int_S R_S H^2 dS} \tag{3}$$

If R_s is constant over the cavity walls we get the relation $Q = \Gamma/R_s$. In a very crude approximation, assuming a constant magnetic field, $\Gamma = \mu_0\omega_0 V/S$ where V and S are the cavity volume and surface respectively. For a cavity design as in Fig. 1 (electron cavity) with d = R we have V/S= R/4 and $\Gamma = 230\Omega$, independent of the cavity dimensions (and resonant frequency). For a normal metal $R_s \propto \sqrt{\omega}$ whereas for a superconductor $R_s \propto \omega^2$.

The second relevant parameter for rf cavities is the accelerating field E_a. The power dissipated in an accelerating structure per unit length can be written as $P_d/L \propto E_a^2\Gamma/Q\omega_0$. For a copper electron cavity operating at $f_0 = \omega_0/2\pi = 500$MHz and T=77K we get $R_s = 6$mΩ, Q = 4*10^4 and $P_d/L = 100$KW/m for $E_a = 2$MV/m. This explains why the maximum accelerating field that can be produced by a normal cavity is limited by the difficulties in cooling the structure. One solution could be to pulse the electric field to reduce the duty-cycle but this has severe technical drawbacks [2].

For the best superconducting element, Nb, we have $R_s = 50$nΩ at T = 4.2K and Q = 5*10^9 reducing the dissipated power by many orders of magnitude. In a simple approximation, the surface resistance of a superconductor can be written as $R_{BCS} + R_{res}$, where:

$$R_{\mathrm{BCS}} \propto \frac{1}{T}\exp\left(-\frac{\Delta}{k_B T}\right) \tag{4}$$

which goes to zero as the temperature is reduced [3], and R_{res} is a residual surface resistance related to defects, that can be, in principle, arbitrarily reduced. A typical measure of defects in a metal is the residual resistivity ratio RRR, defined as :

$$RRR = \rho(300K)/\rho_0 \tag{5}$$

where ρ(300 K) is the metal resistivity at 300K and ρ_O is its residual value at very low temperature. Since ρ(300K) depends only on phonon scattering, we have RRR $\propto 1/\rho_O \propto l \propto \lambda$, were l is the electronic mean free path and λ is the thermal conductivity. However, a definite relationship between R_{res} and RRR can not be established.

The accelerating field limit is set by the condition $H_{rf} \leq H_{sh}$ where H_{sh} is the "superheating" critical field [4] of the superconductor which is close to the thermodynamic critical field H_c ($H_{sh} = 0.75H_c$ for type II superconductors). For Nb, the value for H_{sh}= 200mT sets an intrinsic limit to the accelerating field at $E_a \leq 60$MV/m.

In principle Q should be independent of E_a, but, in practice, nonlinear losses tend to reduce Q at high fields ("Q-slope" problem).

3. Types and shapes of cavities, bulk niobium fabrication technologies and thin film technologies.

We have just seen that the cavity resonant frequency scales inversely with the cross section. The longitudinal shape is optimized for maximum energy exchange with the bunched particle beam.

First of all, the velocity of the particle when entering the resonator has to be synchronized with the frequency and phase of the rf fields. Secondly the transit time t* of the bunch has to be much shorter than half a period, so that the particle always sees a large and nearly constant accelerating field. Therefore we must have $t^* \leq T/2$ and $\omega_0 d \leq \pi\beta c$, where we define $\beta = v/c$ as the ratio of the particle velocity to the light velocity. For electrons, that can be accelerated close to c, $\beta \approx 1$ and this relation is satisfied for the cavity shape shown in Fig.1 with $d \approx R$. The cavity dimensions depend on frequency ranging from 350MHz (radiofrequency) to 3GHz (microwaves). A similar shape is used for high energy proton accelerators ($\beta \approx 0.5$) although the cavity is "squeezed" in the longitudinal direction to account for the lower β value.

Heavy ion cavities have $0.05 \leq \beta \leq 0.2$ and $5\text{cm} \leq d \leq 10\text{cm}$ and therefore must to operate in the frequency range between 50 and 150MHz. The ratio β determines the shape of the cavity. Various electron and proton accelerators use different numbers and sequences of cavities, all of which are essentially of the same type. On the other hand, complex shapes have to be designed for heavy ion accelerators to achieve low frequency and account for the required β value (Fig. 2) [5].

In practice, cavity cascades are often used. A typical accelerator operates with a series of four or more cavities in the same cryostat (accelerating module). Electron and proton cavities are generally directly immersed in a helium bath, whereas ion cavities require more complex refrigeration systems. The cavity cascade is operated in the so called "Π" mode. When the particle bunch is accelerated by the first cell, the accelerating field in the second cell is in opposition, but it is positive one half-cycle later, just when the beam arrives. The same holds for the other cells so that the beam is accelerated all along its path. This operational mode also stabilizes the bunch velocity.

As we will see later, due to the stringent high thermal conductivity requirements, niobium is the only superconducting material that can be considered for bulk cavity fabrication techniques. Historically e-beam welding has been the most successful way to produce electron cavities. In this case, two half cells are made by deep-drawing and the cavity was fabricated by subsequent internal (or external) e-beam welding at the equator. This method has been developed to a high degree of perfection, but problems related to micro-cracks and other welding defects are still unavoidable. More recently, seamless fabrication techniques are becoming competitive both in terms of cost and performances. A recent review of this subject can be found in ref. 6. The main techniques investigated are hydroforming, electroforming and spinning. Excellent results have been recently achieved by the last technique at LNL-INFN (Italy). In this method, a metal disc is plastically deformed by axial or radial motion of a tool acting on a work-piece clamped against a rotating chuck (Fig. 3). The method produces very few defects, is relatively inexpensive and is very promising for extensive use in electron/proton accelerators.

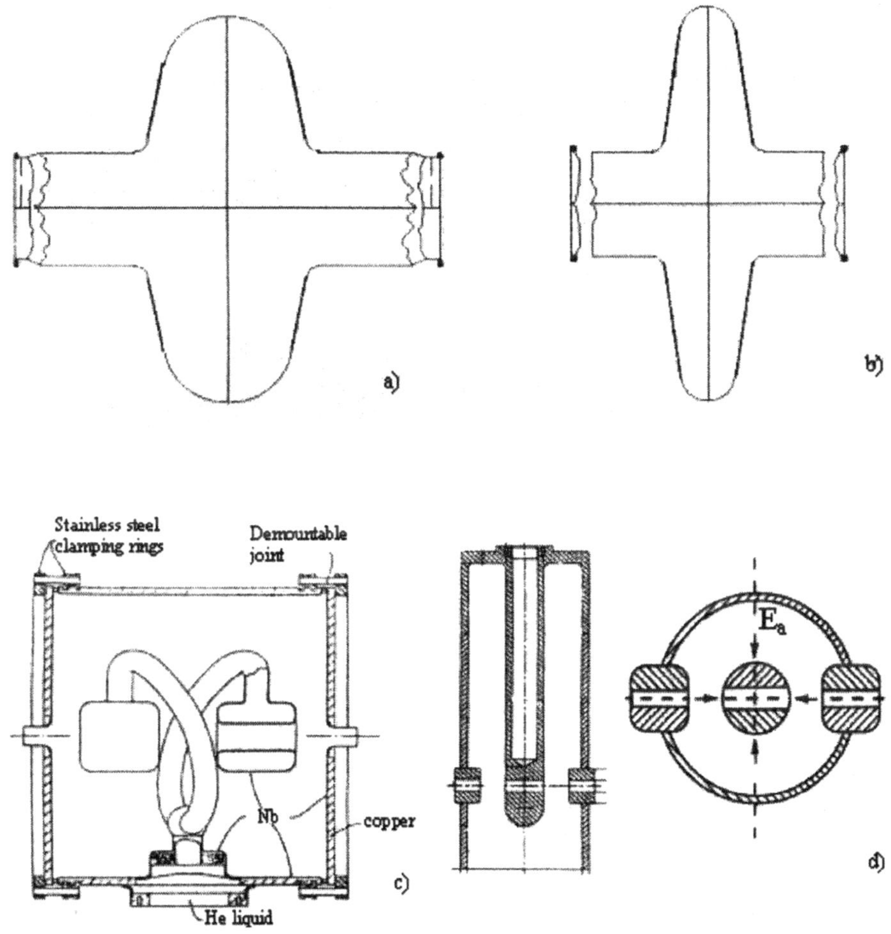

Fig.2 Different cavity shapes : a) $\beta = 1$ for electron cavity, b) $\beta = 0.5$ for proton cavity c) Split ring resonator for ion cavity, Argonne d) Quarter wave resonator for ion cavity (lateral and bottom view), Legnaro.

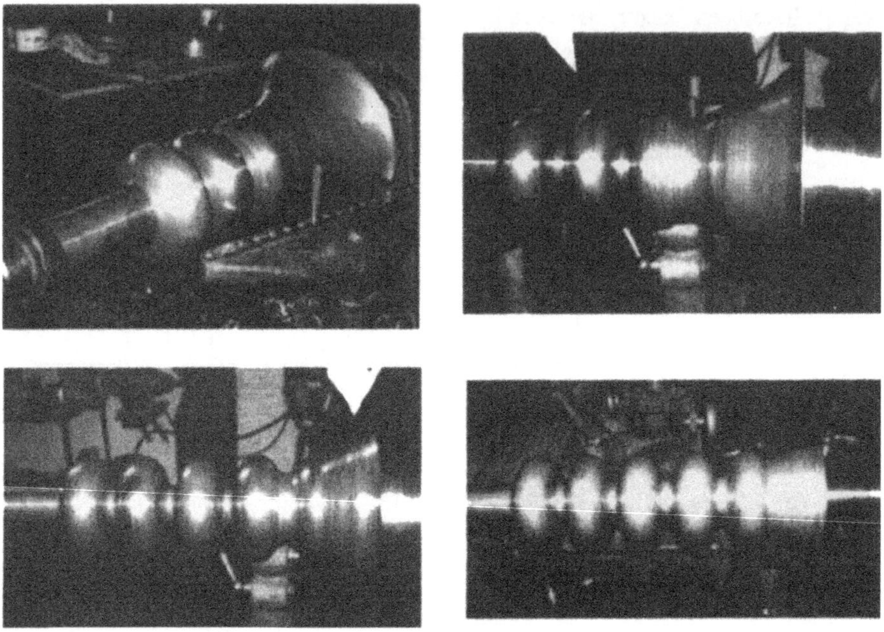

Fig. 3 Spinning of a 1.5 GHz multicell from a circular bank. From V. Palmieri [6]

In recent years, thin film sputtering deposition techniques have been developed based on the pioneering work done at CERN [7,8]. In this case, a copper cavity (obtained by one of the bulk techniques described above) is internally covered by diode or magnetron sputtering, with a thin superconducting film, (1-2 μm in thickness). This method has several relevant advantages:

- Strong reduction of thermal instabilities due to the high thermal conductivity of the copper substrates.
- Possible use of higher performance superconducting compounds
- Large material cost reduction

Other advantages are in the lower sensitivity to trapped dc magnetic fields [9] and higher Q values [10].

The cost saving is extremely relevant for large accelerators. At CERN, the Large Electron Positron collider LEP employs 288 4-cell 350MHz Nb coated copper cavities and the overall material cost reduction compared to bulk technology is estimated in 27.000.000 US$. The basic system for one-cell deposition processes is schematically shown in Fig. 4. Nb sputtered film coatings have been extensively studied and optimized [10,11]. A review of results obtained at CERN is presented in Sect. IV.

A comparison of bulk and film Nb cavity performances in terms of Q values as a function of accelerating field E_a is shown in Fig. 5.

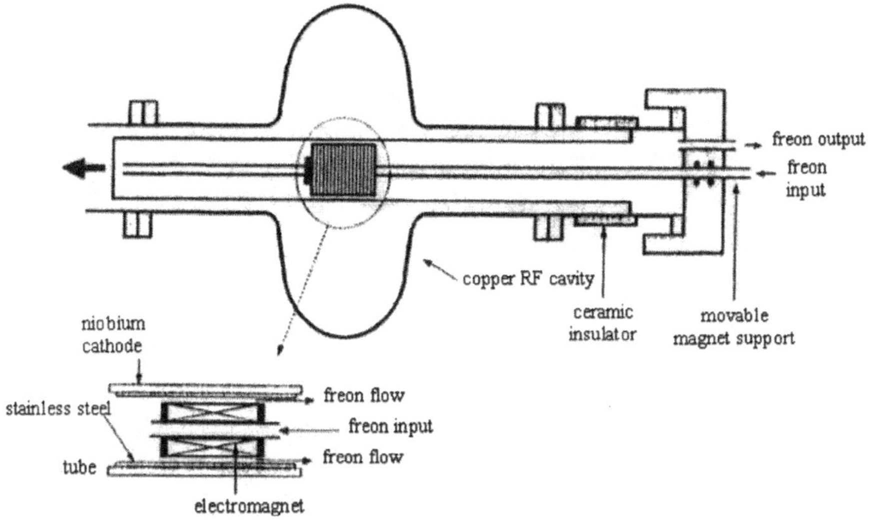

Fig. 4 Schematic view of the magnetron sputtering configuration for coating 1- cell 500 MHz cavities at CERN. From C. Benvenuti et al. [11]

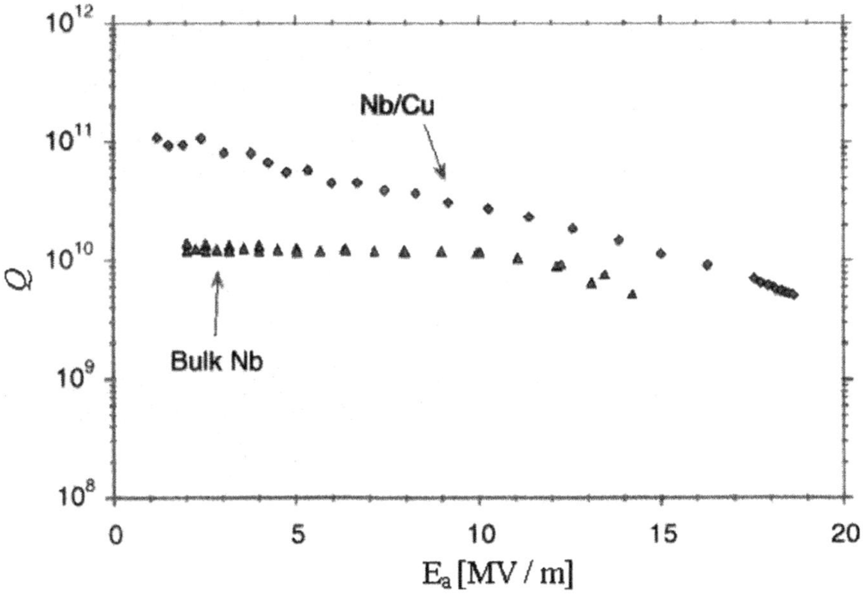

Fig. 5 Comparison of typical Nb film and bulk results for a 1.5 GHz single cell cavity at T = 1.7K. From C. Benvenuti et al. [11]

The main problem with thin film coating for accelerating cavities is the relatively high "Q-slope" whose possible origin and reduction strategies will be discussed in the next sections.

Thin film sputtering techniques have also been successfully used for quarter wave resonators [12]. Thick films techniques such as Pb electrodeposition and Nb plasma-spray or explosive bonding are also frequently used for ion cavities as an alternative to bulk techniques.

4. Experimental limits, in-situ and ex-situ diagnostics. Particle accelerators.

As seen in section I, the intrinsic limit for accelerating fields is related to the critical magnetic field and can be as high as 60MV/m. Effective experimental limits are in fact lower than this, reaching values only as high as 30-40MV/m in a few cases. This is related to field emission and thermal instabilities.

Field emission is strongly related to morphological defects (where the local electric field is high at sharp protrusions) and impurities on the cavity walls. The type of defects or impurities is strongly dependent on the specific fabrication process. Field emission at high fields is associated with additional heating and rapid Q degradation as well as to strong x-ray emission. Secondary electron emission can also occur and can lead to an avalanche breakdown if resonant conditions are established (multipacting). This latter effect can be limited by appropriate design of the cavity profile. The occurrence of field emission can be detected by Geiger counters or by x-ray photodiodes.

Field emission can be limited by chemical polishing of the cavity as well as other polishing techniques such as high pressure water rinsing and helium rf processing [13]. This technique is often used for thin film coated cavities.

Thermal instability can occur in ideal cavities due to the temperature rise of the internal cavity walls with respect to the external bath temperature. This can be written as:

$$\Delta T = q\left(\frac{d}{\lambda} + R_k\right) \tag{6}$$

where q is the heat flux, λ is the thermal conductivity of the cavity walls of thickness d and R_k is the Kapitza resistance [2]. A low value of λ or a high value of R_k can lead to a large ΔT and to an increase of the BCS surface resistance that depends exponentially on temperature (see Eq. 4).

Of course, an increased R_s would lead to an increase in the heat flux q and to thermal runaway. Defects that locally reduce R_s can magnify this phenomenon. Thermal maps are usually taken during cavity test by using rotating external resistors in contact with the cavity walls to identify the regions of temperature increase. As anticipated, due to the high thermal conductivity of copper, thermal instabilities are low in thin film cavities and could, in principle, be further reduced using high purity aluminum substrate cavities.

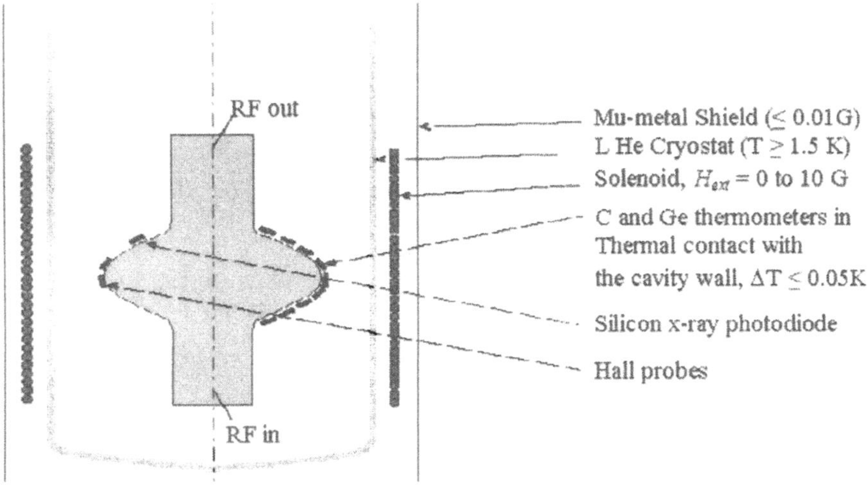

Fig. 6 Sketch of a typical "in situ" cavity diagnostic system. The sensors mounted on the rotating arm allow temperature mapping as well as x-ray and magnetic flux detection

For bulk cavities the only material with sufficiently high thermal conductivity is Nb. The only strategy to reduce thermal problems is to increase the RRR by high temperature purification [14]. Though expensive and time-consuming, the combined use of surface treatments and purification techniques can lead to very good results with accelerating fields up to 40MV/m for Nb bulk cavities.

As already mentioned, the quality factor Q is limited at $T \leq T_C/2$ by the intrinsic BCS value of R_s and, at lower temperatures, by the residual resistance. However trapped magnetic flux can have a detrimental effect on R_s degrading Q. Cavities have to be shielded from the external magnetic field during the cool-down. This can be achieved both by μ-metal shields or by compensating coils. The trapped magnetic flux can be detected by Hall probes. A typical cavity test arrangement is shown in Fig. 6.

The cavities after appropriate treatments and test are assembled inside the cryostat complete with rf coupling probes and fine tuning devices as an "accelerating module" to be installed in an accelerator. A large number of accelerators around the world are presently using superconducting bulk or film cavities. This is reported in Table I. Free Electron Lasers (FEL) driven by Superconducting Linacs are a rapidly growing field and represent a new frontier for cavity application. High electron intensity and stability are needed for FEL operation to provide the required quality of the optical beam [15].

Accelerators for electrons with superconducting structures {β=1}

LABORATORY	Operation	Frequency (MHz)	Active Length (planned) (m)	Average (planned) ($MV\ m^{-1}$)	Average current (mA)
Stanford HEPL, Recyclotron	1972	1300	6	2-3	0.5
University of Illinois,MUSL	1972-92	1300	6	2-3	0.01
CERN, SPS	1987	352	5.1	55	0.5
REK, TRISTAN	1988-94	508	48	30-47	14
Darmstadt, S-DALINAC	1990	2997	10.25	5	0.04
DESY, HERA	1991	500	19.2	2	35
CEBAF	1996	1497	169	2	0.4
CERN, LEP	1997	352	340 (492)	(6)	4

Accelerators for electrons with superconducting structures {β<1}

INSTITUTION (operation)	Type	Frequency (min-max, MHz)	Optimum velocity (min-max)	Total numbers of cavities
Argonne National Laboratory (1978)	Nb split-ring, Nb interdigital	48.5/145.5	0.023/0.160	74
Stony Brook (1983)	Pb-Nb splanning, Pb-C	150	0.055/0.100	42
Florida State University (1987)	Nb	97	0.065/0.100	14
University of Washington (1987)	Pb-Cu	148.9	0.10/ 0.20	37
CEN Saclay (1989)	Nb helix	81/160	0.085	50
JAERI (1991)	Nb quarter-wave		0.10	44
INFN Legnano (1995)	Pb-Cu quarter-wave, Nbquarter-wave, Nb/Cu quarter-wave		0.055-011	64
ANU Canberra (1997)	Pb/Cu split-ring	150	0.100	10

Existing SRF FELs

Existing	f (MHz)	E (MeV)	Q (pC)	I (mA)	λ (μm)	P (W)	ε_n (mm-mr)
Stanford	1300	20-40	13	0.02	3-100	0.6	8
Darmstadt	2997	30-50	6	0.06	2.5-10	4	<50
JAERI	499.8	13-23	200	0.04	20-80	10	>20

Table 1 . Accelerators presently employing superconducting technologies. From D. Proch [2] and C.L. Bohn and S. Benson [15].

5. Analysis of Nb thin film cavity data. Discussion of the origin of residual and non-linear losses.

RF cavities represent a unique system for a detailed study of the electromagnetic behavior of superconductors. In particular, the measure of the residual losses is unaffected by "spurious" effects such as radiation or dielectric losses that are present in other resonating structures.

Moreover, the fundamental role that superconducting cavities perform in particle accelerators provide extremely high quality and well-characterized set of "samples" for detailed physical investigations. This is the case of Nb thin film coated 1.5 GHz copper cavities recently fabricated and analyzed at CERN [16]. The deposition parameters are reported in Table.II .

P_{ar} [mbar]	V [V]	I [A]	T_s [°C]	d [μm]
1.5 10^{-3}	360	3	150	1.5

Tab.II Sputtering parameters for a 1.5 GHz Nb thin film coated copper cavity; V and I are the sputtering voltage and current respectively, T_S the substrate temperature, d the film thickness.

Different values of RRR were obtained using Ar/Ne sputtering gas mixtures. All films were 1.5μm thick with an average grain size a = 100 ± 50 nm. Spun copper cavities were primarily used as well as hydroformed and electroformed ones. The authors measured a large set of Q - E_a curves as a function of temperature and external dc magnetic field (field cooling). The data were converted in R_s – H_{rf} and analyzed using the empirical expression :

$$R_S = R_{BCS}(H_{rf}, T) + R_{res}(H_{rf}) + R_{fl}(H_{rf}, H_{ext}, T) \tag{7}$$

where R_{fl} represents the contribution to losses due to trapped dc magnetic flux. Eq. 7 assumes that there is complete independence of the different possible loss mechanisms and gives a reasonable description of the data. In Fig. 7 the rf field dependence of R_{BCS} and of the residual losses R_{res} are shown.

The temperature dependence of the BCS contribution could be accurately described by the standard theory with $2\Delta / k_B T_C = 4.11$ [3]. As shown in Fig. 8 the dependence of R_{BCS} on the mean path, exhibits a minimum for $l \approx \xi_0$ (RRR ≈ 12) was in excellent agreement with the theory [17]. This was the first experimental test of this effect which explains why films with values of RRR ≈ 10 have a lower surface resistance compared to high RRR bulk Nb.

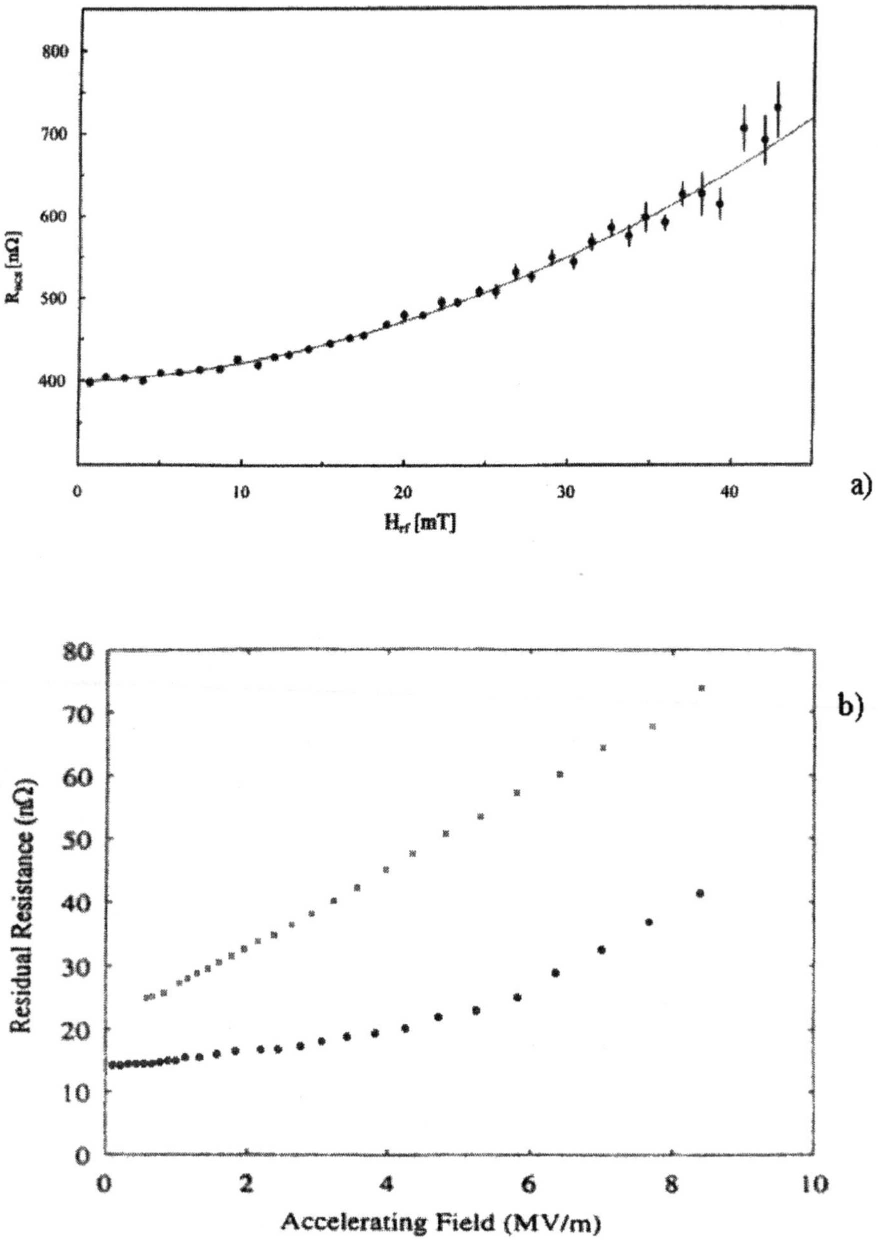

Fig. 7 The rf magnetic field dependence of R_{BCS} at T= 4.2K (a) from C. Benvenuti et al. [16] and of R_{res} (b). Squares correspond to n=1 behavior whereas dots to n= 2 in Eq. 9.

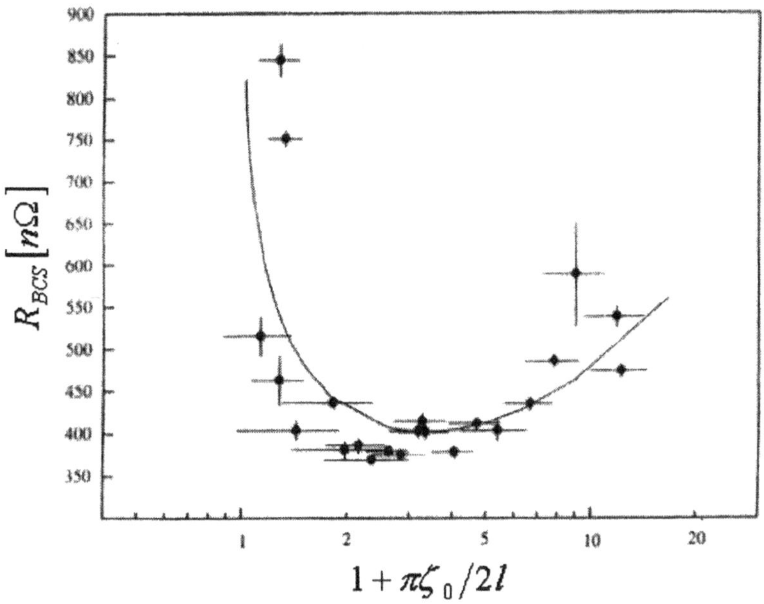

Fig. 8 Values of R_{BCS} at T = 4.2K as a function of the normalized mean free path. The continuos curve represents the BCS theory [3]. From C. Benvenuti et al. [16]

The field dependence of R_{fl} could be described by the following expression:

$$R_{fl} = \left(R_{fl}^0 + R'_{rf} H_{rf}\right) H_{ext} \tag{8}$$

As is clear from Fig. 7, R_{res} can exhibit different field behavior. This can be described by :

$$R_{res} = R_{res}^0 + R'_{res} H_{rf}^{n} \qquad (n = 1\text{-}2) \tag{9}$$

The field dependence of the BCS term was the same for all the measured samples as well as for a reference bulk cavity, and this was somewhat unexpected, since the "Q-slope" is generally attributed only to the residual contribution. A quantitative explanation of this effect has been given recently by I. Kulik and V. Palmieri [18]. Following Tinkham [3] they considered that in a current carrying state the energy gap decreases as $\Delta = \Delta_0 - p_f v_s$ with:

$$v_s = \frac{\hbar q}{m} = \frac{e}{\hbar}\int_0^t E(t)dt \qquad \Rightarrow \qquad \frac{e}{m} H_{rf} \lambda_L \tag{10}$$

The density of states is also modified. Within the framework of the BCS theory, the surface resistance depends exponentially on the gap (eq. 4) and its field dependence induces a change in R_s that can be easily computed, giving good agreement with the data.

The "bilinear" field dependence of R_{fl} is not surprising at all. This term has to be proportional to the trapped flux and hence to H_{ext} [2] and the flux flow losses are generally linear in the rf field [19]. The behavior of R_{res} is, instead, more difficult to understand. As is clear from Fig. 9, the absolute value of R^0_{res} can change by more than one order of magnitude for samples grown nominally under identical conditions. The only significant correlation is that better values are exhibited, on average, by the spun copper "substrate" cavities compared to hydroformed or electroformed cavities. In addition R^1_{res} exhibits a rather erratic behavior with no apparent correlation with R^0_{res}.

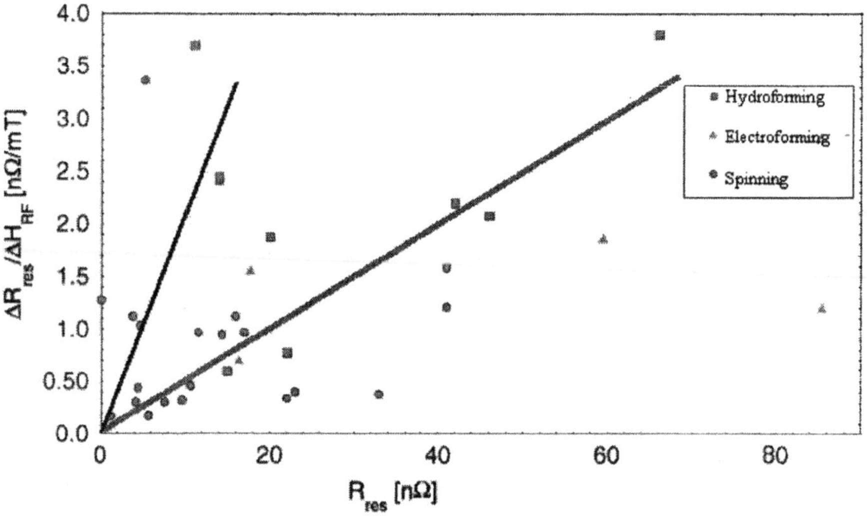

Fig.9 The R_{res} "slope" at 1.7K as a function of R_{res} in zero field. The slope is calculated by linear interpolation of the low field behavior. The continuos lines show possible data correlation. From C. Benvenuti et al. [16]

To gain insight into the origin of the residual, nonlinear losses, a quite general model has been proposed recently [20]. It can be shown that many types of losses (vortex penetration [21], weak link vortices [22,23], weakly coupled grains described by a nonlinear Josephson inductance [24,25] and others) can be described using the simple equivalent circuit for a superconductor shown in Fig. 10. Here σ_1-$i\sigma_2$ represents the

BCS conductance and α-iβ represents a conductance describing the losses due to "a series array" of weakened superconducting regions of various possible origins.

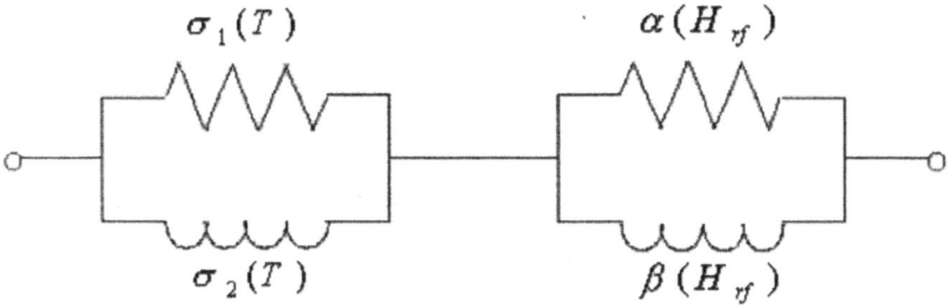

Fig.10 Equivalent circuit used to describe the electrodynamics of a non-ideal superconductor

Assuming $\sigma_1 << \sigma_2$ and $\alpha << \beta$, we are led to the following general expressions:

$$R_S = \frac{1}{2\lambda}\left[\frac{\sigma_1}{\sigma_2}\left(1+\frac{\alpha^2}{\beta^2}\right)+\frac{\alpha}{\beta^2}\right] \tag{11}$$

$$\lambda^2 = \frac{1}{\mu_0\omega}\left[\frac{1}{\sigma_2}+\frac{1}{\beta}\right] \tag{12}$$

We note that, due to the inductive character of the related circuit element, $\beta \propto 1/\omega$ whereas α can be assumed to be frequency independent. Furthermore, α and β can be assumed to be temperature independent (for $T < T_C/2$) and rf field dependent ($\alpha = \alpha(H_{rf})$, $\beta = \beta(H_{rf})$). With these assumptions, since σ_1 goes exponentially to zero with temperature, and σ_2 is temperature independent for $T < T_C/2$, from Eq. 11 we can identify the "residual" term :

$$R_0 = \frac{1}{2\lambda}\frac{\alpha}{\beta^2} \tag{13}$$

It is worth observing that with this model the empirical expression (eq. 7) $R_S = R_{BCS} + R_0$ (we are assuming $H_{ext} = 0$ and $R_{fl} = 0$) is no longer valid and we can write instead:

$$R_S = R_{BCS}(T)f(H_{rf})+R_0(H_{rf}) \text{ with } f(H)=1+\alpha^2(H_{rf})/\beta^2(H_{rf}) \tag{14}$$

However the residual losses are small, as is the case for the CERN Nb films, $f(H_{rf})$ represents only a small correction to R_{BCS}.

A theoretical model [24] considering weak-links at grain boundaries as the origin of the surface resistance of high T_C (and possibly low T_C [25]) superconductors has been extensively discussed elsewhere in this book. The model presented here (called "α,β") is equivalent to the weak-link one assuming:

$$\alpha = \frac{1}{R_J a} = \text{cost.} \qquad \beta = \frac{1}{\omega L_J a}$$

where R_J is the weak-links resistance and

$$L_J = \frac{\hbar}{2eI_C F(H_{rf})} \qquad F(H_{rf}) = \left[1 - \left(\frac{H_{rf}}{H_{cj}}\right)^2\right]^{1/2}$$

This gives:

$$R_{res} \cong R^0_{res}\left[1 + \left(\frac{H_{rf}}{H_{cJ}}\right)^2\right] \tag{15}$$

That is equivalent to (Eq. 8) with

$$\begin{cases} n = 2 \\ R'_{res} = R^0_{res} / H^2_{cJ} \end{cases}$$

Assuming vortices flow at grain boundaries would lead instead to n = 1. In any case the "α,β" model predicts a strong correlation between R^0_{res} and R^1_{res} which was not observed in the CERN experiment, although some possible correlation can be seen for the lower quality samples (see continuos lines in Fig.9). A key experiment to test the validity of the weak-link model for Nb would be to increase the film grain size a by increasing the film thickness d (considering the relation $a = Ad^{0.4}$). Unfortunately such experiment did not give a conclusive answer [26].

As we will see better in the next section, the data found for innovative materials can be easily interpreted in the frame of the "α ,β " model. In particular the data on NbTiN can be interpreted by assuming for α and β the expressions obtained from the weak-link model and, for Nb_3Sn, the expressions from the grain boundary vortex model. As a matter of fact, though different mechanisms seem to be responsible of the losses in different materials, in all cases the improvement of surface structural and morphological properties would lead to a significant reduction in loss.

6. Alternative materials to Nb : NbTiN, Nb_3Sn and others.

The possibility of using superconducting compounds with T_C higher than niobium (for example, MoRe, NbTa, NbN, Nb_3Sn) for the fabrication of high quality r.f. cavities for particle accelerators (or other applications) was already proposed in the mid seventies (see, as an example, [27-29]). However the poor quality of the superconducting surfaces produced fairly high losses, which, combined with quench problems related to the poor thermal conductivity, prevented any real success. As already mentioned, the development by the CERN group [7-8] of the innovative sputter-coating technique allows, in principle, the use of superconducting compounds with T_C higher than Nb, since the thermal conduction is provided, in this case, by the copper "substrate". Moreover, the improvement of surface treatments and analysis techniques has resulted in encouraging results using surface reaction techniques [30] .

In the last ten years advanced research was carried out in this field with fairly encouraging results, although many problems remain to be solved before superconducting compound cavities can be used in operating particle accelerators.

For a type II superconductor in the dirty limit, the relevant quantities for rf applications are (CGS units, $T < T_C/2$):

$$R_S(BCS) = A\sqrt{\rho_0}\omega^2 \exp(-1.76\eta T_C/T) \tag{17}$$

$$\lambda = B[\rho_0/hT_C]^{-1} \tag{18}$$

$$H_C = C\gamma^{1/2}\eta T_C\,,\ H_{sh} = 0.75 H_C \tag{19}$$

(where $A= 6*10^{-21}$, $B =10^{-2}$, $C= 2.4$, $\eta \geq 1$ is the strong-coupling correction, γ is the Sommerfeld constant and eq. 17 is equivalent to eq. 4)

These approximate expressions clearly show that for superconducting alloys and compounds, at a given operating temperature, the best r.f. performances (that is low R_S, low λ and high critical fields) are obtained for high T_C and low ρ_0 materials .

This can also be seen in Fig. 11. The R_S(BCS) v.s. T_C,ρ_0 normogram is computed from eq. 15. The data points represent typical values for different superconducting compounds. Based on Fig. 11, better results can be expected with A15 compounds (Nb_3Sn, V_3Si). Nitrides (such as NbN and (NbTi)N) have higher R_S (BCS) due to the higher value of ρ_0 while transition metal alloys, such as MoRe, do not have an advantage with respect to Nb since T_C is only slightly higher.

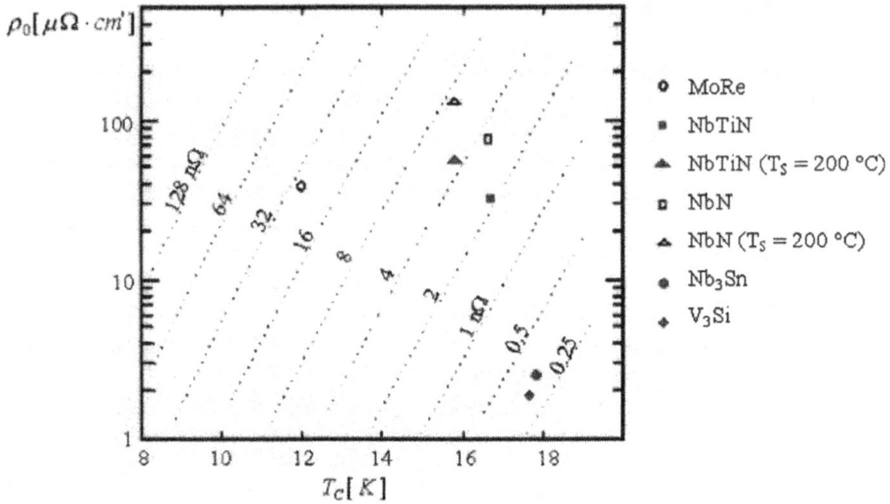

Fig. 11 Lines of equal R_{BCS} in the (ρ_0-T_c) plane atT=4.2K and 500MHz. (for Nb $R_s \approx 50n\Omega$)

Of course these ideas represent only a "first approximation" approach, since many other consideration come into play. First of all, as we know, the surface resistance for a real superconductor can differ significantly from the BCS prediction at low temperatures, due to the presence of additional (often nonlinear) losses caused by a number of possible sources (R_{res}) and these losses depend strongly on the degree of structural and morphological perfection of the surface.

Important parameters are also the low field emission coefficient as well as the low sensitivity to radiation damage. Moreover, depending on the specific preparation technique (sputtering or surface reaction) some materials can present additional problems. As an example, A15 films need to be deposited onto high temperatures (700-800 °C) substrates, and this can be a serious problem in the sputter-coating technique, due to the possible inter-diffusion with Cu.

In this contest, it is very difficult to consider high T_C materials for cavity applications since high T_C films that are free of strong weak link defects can only be obtained by high temperature deposition on single crystalline substrates and certainly not on complex metallic cavity surfaces.

On the basis of these considerations the first attempts to use alternative materials were based on the sputter technique to produce NbN coatings by reactive sputtering in N_2 atmosphere, since it was known that high T_C value could be realized for NbN even when the coating was deposited at relatively low temperature substrates (T_S = 200 °C). However as is clear from Fig. 11, the high normal state resistivity (due to tendency to have vacancies on the N sites and granularity) resulted in a relatively high value of the BCS surface resistance.

These problems can be partially overcome by the use of (NbTi)N alloys (close to 50% Ti percentage) that have a lower normal state resistivity and the same T_C as NbN [31]. The stronger metallic character of (NbTi)N with respect to NbN gives a technical advantage in the sputtering process [32] and therefore most of the work done on real cavities with the sputtering technique has been carried out with this material.

An interesting alternative is thin films of a high T_C material obtained by high temperature surface reaction of a previously deposited film or bulk "substrate". By this technique most of the work has been carried out on Nb_3Sn, due to the availability of extremely well characterized Nb "substrates" and the relatively easy Sn-vapor reaction technique [33].

The rf properties of innovative materials such as (NbTi)N and Nb_3Sn have been analyzed using small planar test samples deposited on dielectric substrates using the endplate replacement cavity techniques [33, 34] or the microstrip techniques [20,35]. The indications obtained by different methods were fairly consistent and indicated, as expected, a much lower surface resistance at 4.2 K with respect to Nb (see eq. 17) but a tendency toward higher nonlinear losses ("field slope"). In Fig. 12 and Fig. 13 results on NbTiN [34] and Nb_3Sn [20] are reported. The data can be interpreted in terms of the "α, β " model described in the previous section and the origin of nonlinear losses can be tentatively interpreted in terms of weak-links for (NbTi)N (n=2) or flux penetration at grain boundaries (n=1) for Nb_3Sn [20].

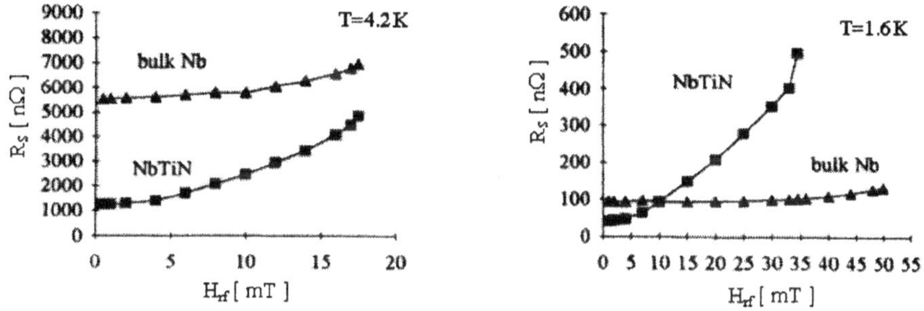

Fig. 12 The surface resistance of a $(Nb_{.55}Ti_{.45})N$ film on copper and bulk niobium at 4.2 and 1.6 K at 3.5 GHz. The rf field dependence has a parabolic behavior (n=2). From P. Bosland et al. [34]

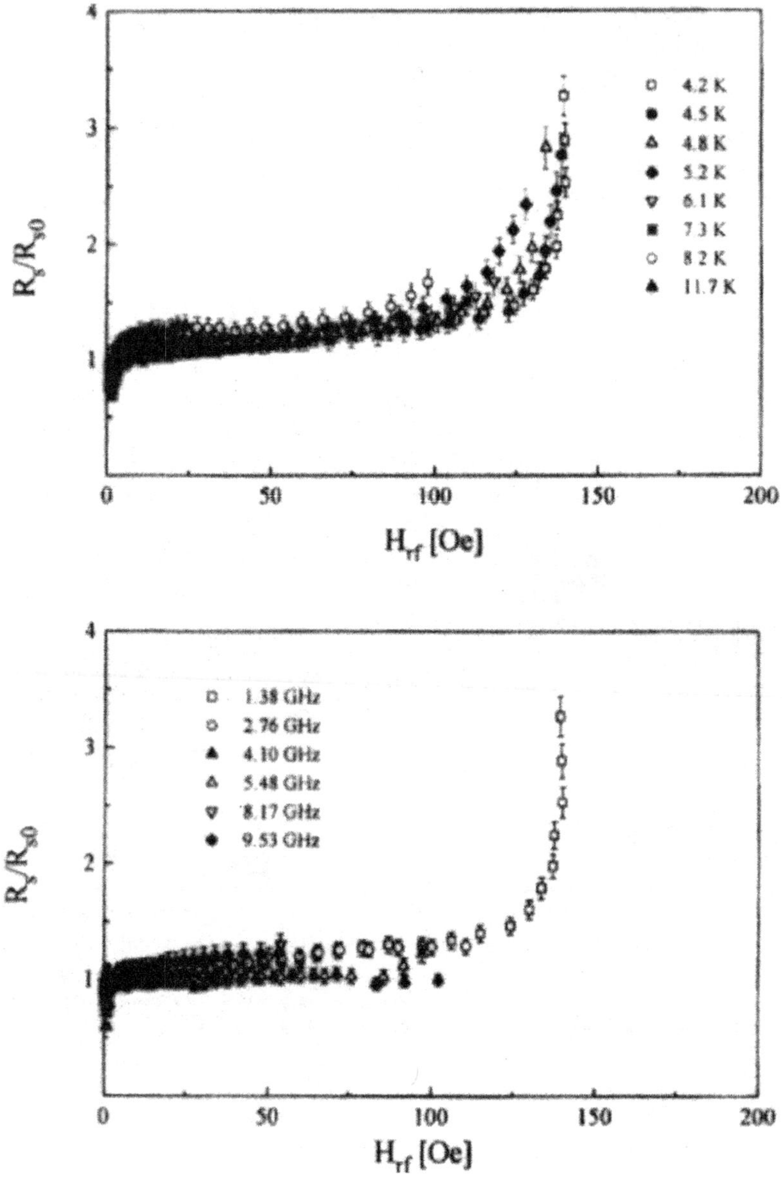

Fig. 13 The surface resistance of a Nb_3Sn film measured by microstrip resonator at different temperature and frequency . The rf field dependence has a linear behavior (n=1) in a large field region. From Andreone et al. [20]

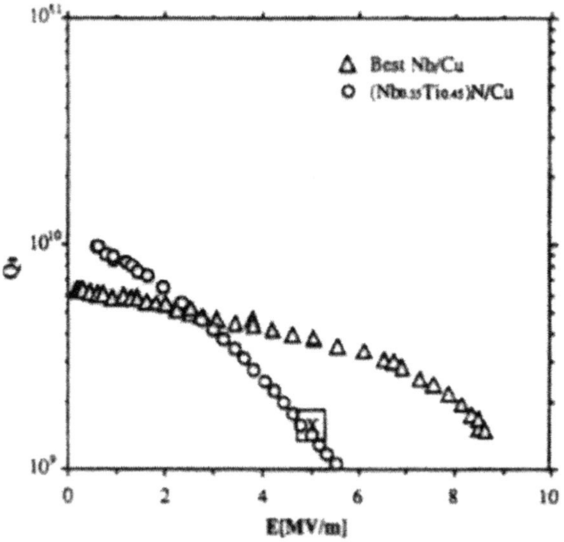

Fig. 14 Quality factor of a NbTiN cavity at T=4.2K and 500MHz. Nb bulk result is shown for comparison. The cross indicates the LEP specifications. From C. Benvenuti

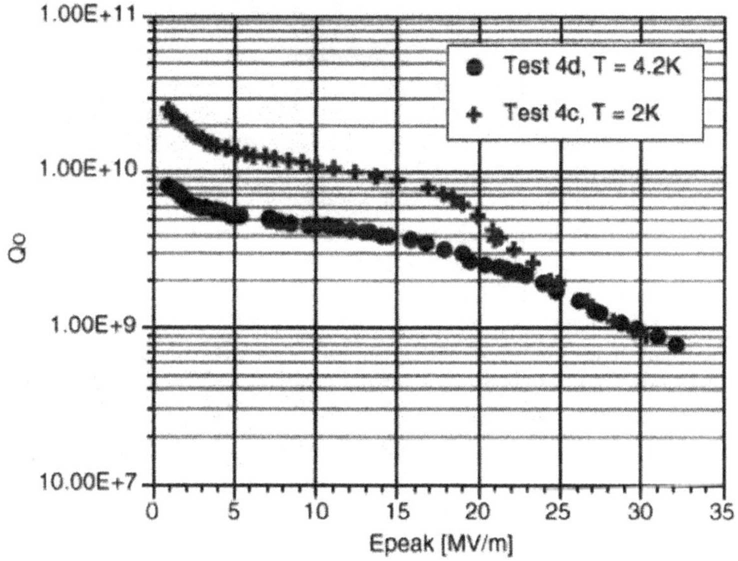

Fig. 15 Q vs E_a dependence for the best Nb_3Sn cavity produced by surface reaction technique. From P. Boccard et al. [36]

There has been a considerable effort at CERN to produce (NbTi)N sputtered thin film cavities. The sputtering system was the one used for the production of Nb films on copper (Fig. 4), but, in this case, the central section of the cathode was made of a $Nb_{.35}Ti_{.65}$ alloy and a N_2 gas inlet was added. The Cu cavity was held at a temperature near 200°C prior to the deposition. The determination of the optimum N_2 reactive sputtering partial pressure was achieved by placing test samples at the cavity equator. Values of T_C as high as 15.8K were obtained with RRR= 1.5. Unfortunately, due to the large Q-slope due to the weak-link induced nonlinear losses, the full cavity performances were superior to Nb only up to values of the accelerating fields of 5MV/m (Fig.14). Even more encouraging results were obtained by the surface reaction technique. High quality Nb_3Sn coatings were obtained at the University of Wuppertal [36] by Sn surface reaction at 1150°C of a standard high RRR Nb-bulk accelerating cavity.

The results showed the possibility of obtaining high electric field gradients (30MV/m) while preserving Q values higher than Nb at 4.2K (Fig.15).

7. Conclusions and perspectives

Superconducting resonant cavities play a fundamental role in many operating particle accelerators and allow the development of new ideas as shown by the increasing number of Free Electron Lasers operating with superconducting Linacs.

Bulk niobium cavities with rf properties approaching intrinsic limits can be made by advanced mechanical, chemical and heat treatments. This is a good approach and has produced important achievements, but this strategy is certainly close to its limits. Spinning technique is very promising and its optimization has to be pursued further.

Nb-film sputtering techniques on copper have also reached a high degree of development and successes have been obtained by this method. Optimization of 2-cathode process and the use of high quality spun copper "substrates" can lead to further improvements.

Innovative materials have been studied, but a much larger effort will be necessary to improve the sputter-coating techniques and/or the surface reaction techniques to achieve materials with a sufficient degree of surface perfection so that performance is not limited by spurious losses due to defects, grain boundary weak-links etc., and to avoid the "Q-slope" problem. More physics studies are required to fully understand the origin of residual and nonlinear losses in the different experimental situations. Encouraging results were obtained with NbTiN coatings and recently very promising results have been obtained with Nb_3Sn.

High T_C materials are at present not suitable for cavity applications, due to inherent difficulties in growing epitaxial films that are free of low critical current weak-links on complex metallic surfaces. But never say never!

In the light of these achievements and of the continuing progress in the surface coating and surface characterization techniques, it is certainly worthwhile to invest further in the development of innovative materials and techniques for the next generation of high performance accelerating cavities.

Acknowledgments

The author wishes to express his gratitude to V. Palmieri, C. Benvenuti, C. Attanasio, L. Maritato, A. Andreone and A. Cassinese with whom he shared an intense research season on the subjects presented here.

References

[1] Feynman R.P., (1965) "The Feynman lectures on Physics", Addison Wesley Publ. Comp., vol II.
[2] Proch D., (1998) "Superconducting cavities for accelerators", Rep. Prog. Phys. **61**, 431
[3] Tinkham M., (1996) "Introduction to Superconductivity", Mc Graw Hill, New York.
[4] Dolgert A.J. , Di Bartolo S.J. and Dorsey A.T. (1996) "Superheating fields of superconductors: Asymptotic analysis and numerical results", Phys. Rev. B **53**, 5650
[5] Palmieri V., (1991) "Superconducting resonant cavities", LNL- INFN Rep 051/91
[6] Palmieri V., (1998) "Seamless cavities: the most creative topic in RF Superconductivity" in "Proceedings of the Eight Workshop on R.F. Superconductivity" V. Palmieri and A.Lombardi Ed. LNL- INFN Rep 133/98, pp.553
[7] Benvenuti C., Circelli N. and Hauer M. (1984) "Niobium films for superconducting accelerating cavities", Appl.Phys.Lett. **45**, 583
[8] Benvenuti C., Circelli N. , Hauer M. and Weingarten W., (1985) "Superconducting 500 MHz accelerating copper cavities sputter-coated with niobium films", IEEE Trans. on Magn. **MAG-21**, 153
[9] Arnold-Mayer G. and Weingarten W., (1987) "Comparative measurements on niobium sheet and sputter coated cavities", IEEE Trans. on Magn., **MAG-23** ,1620
[10] Russo R.and Sgobba S., (1998) "Influence of coating temperature on Niobium Films" in "Proceedings of the Eight Workshop on R.F. Superconductivity", V. Palmieri and A.Lombardi Ed. LNL- INFN Rep 133/98, pp.890
[11] Benvenuti C., Calatroni S.,Campisi I.E., Darriulat P., Durand C., Peck M., Russo R. and Valente A.M., (1998) "Niobium sputter-coated copper resonators", Ibid. pp.1028
[12] Stark S., Palmieri V., Baldan L., Durand C., Kulik I.I., Preciso R., Stvanello F., Venturini W.and.Zandolin S., "Study of the role of the interface between niobium films and copper R.F. resonators", Ibid. pp.1050
[13] Saito K., Inoue H., Kako E., Fujino T., Noguchi S., Ono M.and Shishido T. (1998) "Superiority of Electropolishing over Chemical Polishing on High Gradients" Ibid. pp.795
[14] Kako E., Bolore M., Boudigou Y., Charrier J.P., Coadou B., Noguchi S., Ono M., Saito K.and Shishido T., (1998) "Cavity Performances in the 1.3. GHz Saclay/KEK Nb Cavities" Ibid. pp.491

[15] Bohn C.L. and Benson S.V. (1998) "Radiofrequency Superconductivity Applied to Free-Electron Lasers" Ibid. pp.107

[16] Benvenuti C., Calatroni S.,Campisi I.E., Darriulat P., Durand C., Peck M., Russo R. and Valente A.M. (1999) "Study of the surface resistance of superconducting niobium films at 1.5 GHz", PHYSICA C, **C-316**, 154

[17] Turneraure J.P., Halbritter J. and Schwettman H.A. (1991) "The surface impedance of superconductors and normal conductors: the Mattis-Bardeen Theory" Journal of Superconductivity, **4**, 341

[18] Kulik I.and Palmieri V. (1998) "Theory of Q-degradation and non-linear effect in Nb-coated superconducting cavities" in "Proceedings of the Eight Workshop on R.F. Superconductivity", V.Palmieri and A.Lombardi Ed. LNL- INFN Rep 133/98, pp.309

[19] Halbritter J., (1990) "RF residual losses, surface impedance, and granularity in superconducting cuprates", J. Appl. Phys.**68**, 6315

[20] Andreone A., Cassinese A., Di Chiara A., Iavarone M., Palomba F., Ruosi A. and Vaglio R.,(1997) "Non-linear microwave properties of Nb_3Sn sputtered superconducting films" J. Appl. Phys. **82**, 1736

[21] Clem J.R. and Coffey M. W., (1992) "Effects of Flux Flow, Flux Pinning, and Flux Creep upon the rf Surface Impedance of Type-II Superconductors", J. Supercond. **5**, 313

[22] Halbritter J., (1992) "On Intrinsic and Extrinsic Effects in the Surface Impedance of Cuprate Superconductors", J. Supercond. **5**, 331

[23] Portis A. M., (1992) "Electrodynamics of High-Temperature Superconductors", World Scientific Ed., Singapore

[24] Hylton T., Kapitulnik A., Beasley M. R., Carini J. P., Drabeck L., and Gruner G., (1988), "Weakly coupled grain model of high-frequency losses in high T_C superconducting thin films" Appl. Phys Lett. **53**, 343

[25] Attanasio C., Maritato L., and Vaglio R. (1991) "Residual Surface Resistance of Polycrystalline superconductors" Phys. Rev. B **43**, 6128

[26] Benvenuti C., Calatroni S., Orlandi G. and Scalambrin (Newport News, Virginia 1993) "Expected dependence of Nb-coated RF cavity performance on the characteristics of niobium" F. Workshop on R.F. Superconductivity, Ed. R.M. Sundelin , pp.718

[27] Yasaitis J.A. and Rose R.M. (1975) "Microwave surface resistance of a superconducting Mo-Re Alloy" IEEE Trans. on Magn. **MAG-11** , 434

[28] Agyeman K., Puffer I. M., Yasaitis J. A. and Rose R. M., (1977) "Superconducting $Mo_{0.75}Re_{0.25}$ cavities at X-band", IEEE Trans., **MAG-13**, 343

[29] Kniesel P., Stoltz O. and Halbritter J. (1979) "Measurements of Superconducting Nb_3Sn Cavities in the GHz range" IEEE Trans. on Magn. **MAG-15** , 21

[30] Dasbach D. , Mueller G., Peiniger M., Piel H. and Roth R.W. (1989) "Nb_3Sn Coating of High Purity Nb cavities" IEEE Trans. on Magn. **MAG-25** ,1862

[31] Di Leo R., Nigro A., Nobile G. and Vaglio R., (1990) "Niobium-titanium nitride thin films for superconducting R.F. accelerating cavities" Low Temp. Phys.**78**, 41

[32] Benvenuti C., Calatroni S., Hauer M., Minestrini M., Orlandi G. and Weingarten W., (DESY, Hamburg 1991) "NbTiN-NbTi coatings for superconducting accelerating cavities" in "Proceedings of the fifth Workshop on R.F. Superconductivity", D. Proch Ed.k, pp.518

[33] Perpeet M., Hein M.H., Mueller G., Piel H., Pouryamont J. and Diete W. J. (1997), "High-quality Nb_3Sn thin films on sapphire prepared by tin vapor diffusion" Appl. Phys. **82**, 9

[34] Bosland P., Cantacuzene S., Gobin J., Juillard M. and Martignac J. (Newport News, Virginia 1993) "NbTiN thin films for RF applications" J. Proc. of the 6th Workshop on R.F. Superconductivity, B. Bonin Ed., pp.1028

[35] Andreone A., Di Chiara A., Peluso G., Santoro M., Attanasio C., Maritato L. and Vaglio R., (1993) "Surface impedance measurements of superconducting NbTiN films by a ring microstrip technique" J. Appl. Phys. **73**, 4500

[36] Boccard P., Kneisel P., Mueller G., Pouryamount J.and Piel H. (1998) "Results from some tempearture mapping measurements on Nb_3Sn RF cavities" in "Proceedings of the Eight Workshop on R.F. Superconductivity",V Palmieri and A.Lombardi Ed. LNL- INFN Rep 133/98, pp.1094

THE MICROSTRIP DC SQUID AMPLIFIER

JOHN CLARKE, MICHAEL MÜCK, MARC-OLIVIER ANDRÉ,

Department of Physics
University of California, Berkeley
and
Materials Sciences Division
Lawrence Berkeley National Laboratory
Berkeley, California 94720

JOST GAIL and CHRISTOPH HEIDEN

Institut für Angewandte Physik
der Justus-Liebig-Universität Gießen,
35392 Gießen, Germany

Abstract. The fabrication, operation and performance of the dc Superconducting QUantum Interference Device (SQUID) are briefly described. The SQUID is used as a radiofrequency amplifier in a new mode in which the input signal is coupled between the square washer that forms the body of the SQUID and one end of the spiral coil overlaying the washer. The coil and the groundplane, together with the intervening dielectric layer, form a microstrip which resonates when its length is equal to one-half the signal wavelength. The resonance can be tuned from about 0.1 to 0.8 GHz by varying the length of the coil The magnetic flux generated in the microstrip is coupled into the SQUID. Positive feedback from the output of the SQUID to the microstrip enhances the gain and the quality factor. At 1.8 K and with a cooled semiconductor post-amplifier, a gain of 24.5 ± 0.5 dB and an intrinsic noise temperature of 0.19 ± 0.06 K have been achieved at 365 MHz. Cooled to 0.48 K in a ^{3}He refrigerator, a microstrip SQUID amplifier achieved a gain of 20 ± 0.5 dB and a noise temperature of 0.12 ± 0.10 K at 438 MHz. The resonant frequency can be reduced by about 40% by means of a varactor diode connected across the open end of the microstrip. Two microstrip SQUIDs have been operated in cascade to achieve an overall gain of 33.5 ± 1 dB at 386 MHz. The microstrip SQUID amplifier is intended for a second-generation axion detector.

H. Weinstock and M. Nisenoff (eds.), Microwave Superconductivity, 473–504.

Table of contents:

1. Introduction

Traditionally, most dc Superconducting QUantum interference Devices (SQUIDs) [1] are used as ultrasensitive detectors of magnetic flux at low frequencies, that is, below a few kilohertz, most notably for biomagnetic applications [2]. In these applications, one operates the SQUID in a flux-locked loop [3] to linearize the response of the SQUID, which is periodic in the applied flux with a period of one flux quantum, $\Phi_0 \equiv h/2e \approx 2.07 \times 10^{-15}$ Wb. For some applications – an example is low-frequency nuclear magnetic resonance (NMR) and nuclear quadrupole resonance (NQR) [4] – one requires higher bandwidth, a need that can be met with a flux-locked loop at frequencies up to about 5 MHz [5, 6]. DC SQUIDs have been used at much higher signal frequencies, however, in an open loop configuration in which the input signal is required to be less than $\Phi_0/2$. Amplifiers with gains approaching 20 dB and noise temperatures of about 1 K have been operated successfully at frequencies of around 100 MHz [7] and used as sensitive detectors of NMR and NQR [8]. At still higher frequencies, parasitic capacitance between the superconducting input coil and the body of the SQUID causes the gain to drop substantially [7]. As a result, SQUIDs have not been competitive as amplifiers at frequencies around 1 GHz where cooled high electron mobility transistors [HEMTs] offer substantial gain and noise temperatures of a few kelvin [9].

However, further improvement of the recently developed axion detector [10] depends critically on the existence of amplifiers in the 1 GHz frequency range with noise temperatures at least one order of magnitude lower than can be achieved with semiconductors. This need spurred the development of a new approach to SQUID amplifiers in which the input coil deposited on the SQUID body is used as a resonant microstrip, thus making a virtue of the parasitic capacitance [11]. With this approach, gains of well over 20 dB and noise temperatures as low as 0.1 K have been achieved [12]. Furthermore, the resonant frequency can be readily tuned over a range of nearly two [13].

This chapter outlines the principles, operation and performance of the microstrip SQUID amplifier. We begin in Sec. 2 with a summary of the principles of the dc SQUID, its practical realization with low-transition-temperature superconductors and its implementation as a radiofrequency amplifier. In Sec. 3, we describe the principles of the microstrip amplifier, and its operation at liquid ^{4}He temperatures. Section 4 is concerned with the reduction in noise temperature achieved at ^{3}He temperatures, and Sec. 5 describes how the amplifier can be tuned and how two SQUIDs in cascade can increase the gain. Section 6 gives a short description of the axion detector that motivated this work, and Sec. 7 contains some concluding remarks.

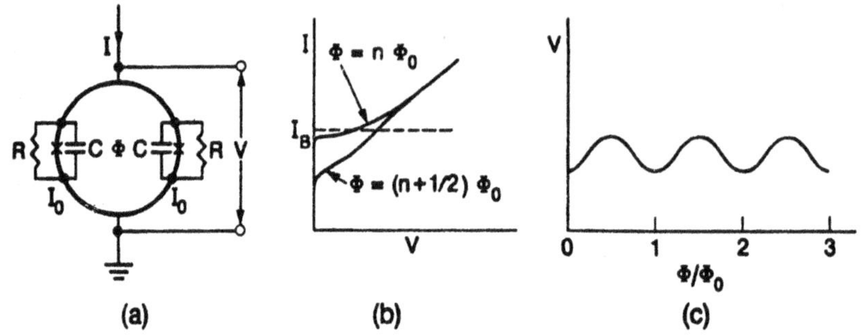

Figure 1. (a) The dc SQUID; (b) I-V characteristics; (c) V vs. Φ/Φ_0 at constant bias current I_B.

2. The dc SQUID

2.1 THEORETICAL CONSIDERATIONS

The essence of the dc SQUID is shown in Fig. 1(a). Two junctions are connected in parallel on a superconducting loop of inductance L. Each junction is in parallel with a resistance R, which may be an external shunt, and a capacitance C. According to the resistively-shunted-junction model [14, 15], the current-voltage (I-V) characteristics will be nonhysteretic provided that $\beta_c = 2\pi I_0 R^2 C/\Phi_0 \leq 1$, where I_0 is the critical current of each junction. The I-V characteristics are shown in Fig. 1(b) for $\Phi = n\Phi_0$ and $(n + 1/2)\Phi_0$, where Φ is the external flux applied to the loop and n is an integer. When the SQUID is biased with a constant current ($>2I_0$), the voltage across the SQUID oscillates with period Φ_0 as Φ is steadily increased, as indicated in Fig. 1(c). The bias current is adjusted to give the maximum voltage change, and the SQUID is operated on the steep part of the V-Φ curve where the flux-to-voltage transfer coefficient, $V_\Phi \equiv |(\partial V/\partial \Phi)_I|$, is a maximum. Thus, the SQUID produces an output voltage in response to a small input flux $\delta\Phi(<<\Phi_0)$, and is effectively a flux-to-voltage transducer.

To calculate the signal and noise properties of the SQUID one must solve numerically the equations relating the voltage across the SQUID to the bias current and the applied flux [16-21]. The noise arises from Nyquist current noise in the shunt resistors, which has a spectral density $4k_BT/R$. One finds that V_Φ peaks smoothly as a function of bias current and has a shallow maximum around an applied flux of $(2n + 1)\Phi_0/4$. By using the computer to explore the behavior of the SQUID over a wide range of parameters one finds the conditions for optimal performance [16, 20]. Noise plays a key role in determining these conditions, and enters in two ways. First, the Josephson coupling energy of each junction, $I_0\Phi_0/2\pi$, must be substantially greater than the thermal energy, k_BT. This constraint is conveniently written in terms of a dimensionless noise parameter as

$$\Gamma = 2\pi k_B T/I_0 \Phi_0 = I_{th}/I_0 << 1 , \tag{2.1}$$

where $I_{th} = 2\pi k_B T/\Phi_0 \approx 0.18\ \mu A$ at 4.2 K. Second, the magnetic energy per flux quantum, $\Phi_0^2/2L$, must also be substantially greater than $k_B T$. We can express this requirement as [20]

$$\Gamma\beta_L = 4\pi L k_B T/\Phi_0^2 = L/L_{th} << 1\ , \tag{2.2}$$

where

$$\beta_L = 2LI_0/\Phi_0\ . \tag{2.3}$$

The thermal inductance $L_{th} = \Phi_0^2/4\pi k_B T$ is equal to 5.9 nH at 4.2 K. The simulations show that optimum performance is achieved for $\beta_L \approx 1$, and that the noise level increases rapidly for $\Gamma\beta_L > 0.15$. From Eq. (2.2) we see that this constraint places an upper limit on L of about 900 pH at 4.2 K.

The computed values will, of course, depend on the chosen parameters, but for a typical SQUID at 4.2 K with $\beta_L = 1$ and $\Gamma = 0.05$, one finds the transfer function

$$V_\Phi \approx R/L \tag{2.4}$$

under optimum bias conditions. The corresponding white voltage noise across the SQUID has a spectral density

$$S_V(f) \approx 16 k_B T R\ , \tag{2.5}$$

approximately 8 times the Nyquist noise in a resistance R/2. The spectral density of the equivalent flux noise is thus

$$S_\Phi(f) = S_V(f)/V_\Phi^2 \approx 16 k_B T L^2/R\ . \tag{2.6}$$

It is often convenient in comparing SQUIDs to introduce the flux noise energy, which can be written in several forms:

$$\varepsilon(f) = S_\Phi(f)/2L \approx 9 k_B T L/R \tag{2.7a}$$

$$\approx 16 k_B T (LC/\beta_c)^{1/2} \quad (\beta_c \lesssim 1) \tag{2.7b}$$

$$\approx 4\Phi_0 k_B T/I_0 R\ . \tag{2.7c}$$

Equation (2.7c) makes it clear that, provided one can maintain the optimum values of the parameters, the noise energy scales inversely as the I_0R product of the Josephson junction.

For a representative SQUID at 4.2 K with L = 100 pH, $I_0 = 10\ \mu A$ and $R = 4\ \Omega$, one finds $V_\Phi = 80\ \mu V \Phi_0^{-1}$ and $\varepsilon(f) \approx 1.3 \times 10^{-32}\ JHz^{-1} \approx 125\ \hbar$. At 4.2 K, the performance of dc SQUIDs with a wide variety of parameters is generally in good agreement with the predictions.

2.2 PRACTICAL LOW-T_c DC SQUIDS

The most commonly used SQUIDs are made from photolithographically patterned thin films of Nb, following the design of Ketchen and Jaycox [22, 23], who introduced the idea of depositing a superconducting spiral coil over the square washer that forms the body of the SQUID. The coil is separated from the SQUID with an insulating layer and may be coupled to one of several different circuits, depending on the application. Design guidelines were given by Jaycox and Ketchen [23], who showed that a square washer (with no slit), with inner and outer edges d and w, has an inductance L (loop) = $1.25\mu_0 d$ in the limit w >> d. They gave the following expressions for the inductances of the SQUID, L, and the spiral coil, L_i, and for the mutual inductance, M_i, and coupling coefficient, α^2, between the spiral coil and the SQUID:

$$L = L\,(\text{loop}) + L_j, \tag{2.8}$$

$$L_i = n^2\,(L - L_j) + L_s, \tag{2.9}$$

$$M_i = n(L - L_j) \tag{2.10}$$

and

$$\alpha^2 = (1 - L_j / L) / [1 + L_s / n^2(L - L_j)]. \tag{2.11}$$

Here, L_j is the parasitic inductance associated with the junctions (including the slit), n is the number of turns of the input coil and L_s is the stripline inductance of this coil, which is much smaller than L_i for $n \gtrsim 20$. Measured parameters are generally in good agreement with these predictions. A particularly useful overview of SQUID design issues has been given by Cantor [24].

Figure 2 shows a typical square-washer SQUID [25] involving trilayer Nb/Al-Al_2O_3/Nb junctions [26]. The trilayer is deposited on a thermally-oxidized 4-inch Si wafer using dc sputtering. The trilayer is patterned and the junctions are defined by reactive ion etching. Before the resist is stripped, a light anodization is performed around the junctions to protect them from processing damage. The base electrode is then patterned and etched to form the SQUID washer. Subsequently, the Pd-shunt resistors are formed by e-beam evaporation, and the Nb input coil and counter-electrode are deposited. The insulation between layers is SiO_2, which is deposited by plasma-enhanced chemical vapor deposition. For the device shown in Fig. 2, the estimated SQUID inductance is L = 0.32 nH, and the measured shunt resistance R = 12 Ω.

Many authors have described square-washer SQUIDs of this kind: a small, representative selection of papers is listed in references [27-34]. A key issue in this configuration is the presence of parasitic capacitance between the input coil and the

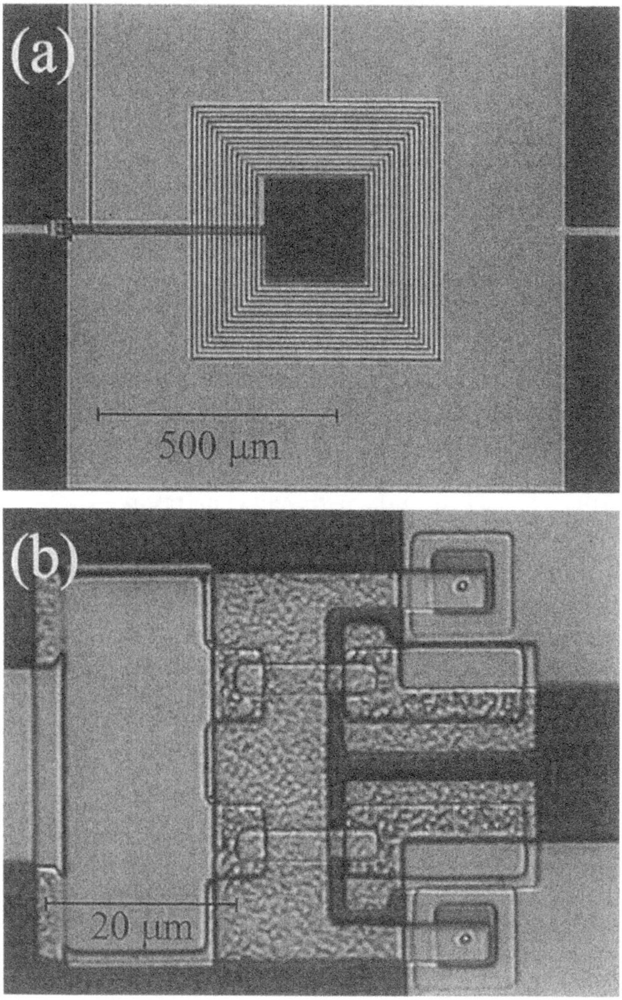

Figure 2. (a) Photograph of dc SQUID with 15-turn input coil. (b) Photograph of area to left of SQUID washer showing the two junctions and shunt resistors. (Courtesy Xiaofan Meng and Robert McDermott).

square washer that often results in a pronounced resonant structure on the I-V characteristic. This structure can be substantially reduced by means of damping resistors. However, the capacitance limits the frequency response of SQUIDs to at most 100-200 MHz [7].

In virtually all applications at frequencies up to a few megahertz, the SQUID is used in a flux-locked loop as a null detector of magnetic flux. Thus, if one applies a flux $\delta\Phi$ to the SQUID, the feedback flux generates a canceling flux $-\delta\Phi$ in the SQUID and an output voltage proportional to $\delta\Phi$. The flux-locked loop thus linearizes the response of the SQUID, enables one to track input signals of many flux quanta and provides a ready means of measuring the flux noise. Typical low-T_c SQUIDs have a noise of 1 to 2 $\mu\Phi_0\,\mathrm{Hz}^{-1/2}$ at frequencies down to about 1 Hz, a dynamic range of $\pm10\,\Phi_0$ to $\pm100\Phi_0$, a bandwidth of 20 to 50 kHz and a slew rate of 10^5 to $10^6\,\Phi_0\,\mathrm{sec}^1$.

A wide variety of readout schemes are also used, and have been comprehensively reviewed by Drung [5]. Drung gives a detailed discussion of the issues concerning dynamic range, bandwidth and slew rate.

2.3 THE SQUID AS A RADIO FREQUENCY AMPLIFIER

The dc SQUID has been configured and exploited as an extremely sensitive radio frequency amplifier at frequencies up to 100-200 MHz [7]. At these frequencies, the SQUID is operated open-loop with a flux bias near $(2n + 1)\Phi_0/4$ ($n = 0, \pm1, \pm2$ ..). To understand the theory for the noise temperature of this amplifier, we need to introduce the current noise $J_N(t)$ circulating in the SQUID loop [35], which has a spectral density

$$S_J(f) \approx 11k_BT/R \tag{2.12}$$

for $\beta = 1$, $\Gamma = 0.05$ and $\Phi = (2n + 1)\Phi_0/4$. Furthermore, the current noise is partially correlated with the voltage noise across the SQUID, with a cross-spectral density [35]

$$S_{VJ}(f) \approx 12k_BT. \tag{2.13}$$

This correlation arises because the current noise generates a flux noise, which, in turn, contributes to the total voltage noise across the junction, provided $V_\Phi \neq 0$.

One can make a tuned amplifier, for example, by connecting a suitable input circuit to the SQUID as shown in Fig. 3. In general, the presence of this circuit modifies the SQUID parameters and noise spectral densities. Provided that α^2 is small, however, one can neglect the mutual influences of the SQUID and input circuit on each other. The circuit can be optimized to find the lowest noise temperature T_N, which we define as follows. When the resistor R_i is at temperature T, the output noise voltage across the SQUID has a spectral density of the form $4k_B(T + T_N)A^2$, where A is the voltage gain. Thus, T_N is a convenient representation of the noise added by the amplifier for a particular source resistance R_i. For a given input coil inductance L_i, one can show that on resonance the minimum noise temperature is given by [7]

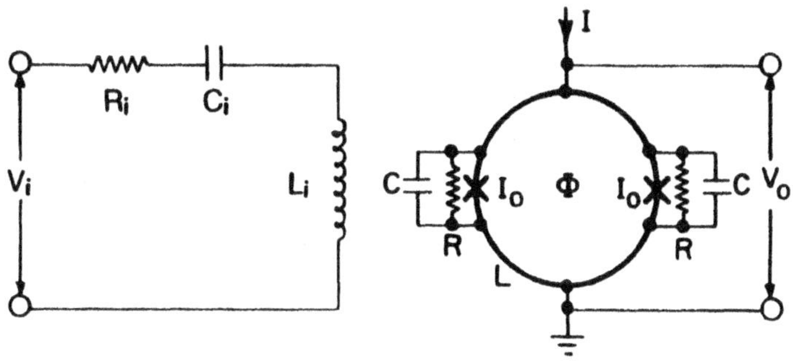

Figure 3. Tuned radiofrequency amplifier based on dc SQUID (from Ref. [7]).

$$T_N^{(res)}(\omega) \approx \frac{\omega}{2k_B V_\Phi}(S_V S_J)^{1/2} \tag{2.14}$$

for an optimized resistance

$$R_i^{(res)} \approx \alpha^2 \omega L_i L V_\Phi (S_J / S_V)^{1/2} \ . \tag{2.15}$$

Substituting for S_V and S_J, one can also write Eq. (2.14) in the form

$$T_N^{(res)}(\omega) \approx 7\omega T / V_\Phi \approx 0.7\omega\varepsilon(\omega)/k_B \ , \tag{2.16}$$

where we have used Eqs. (2.4) and (2.7a).

One can easily calculate the gain on resonance. For $\alpha^2 << 1$, an input voltage V_i produces an output voltage $V_0 = (V_i/R_i^{(res)})M_iV_\Phi$. The power gain is thus $G = (V_0^2/R_d) / (V_i^2 / R_i)$, where R_d is the dynamic output impedance of the SQUID. If we take $R_d \approx R$, we find

$$G \approx V_\Phi / \omega \ . \tag{2.17}$$

Hilbert and Clarke [7] made several radiofrequency amplifiers with both tuned and untuned inputs, flux biasing the SQUID near $\Phi = (2n + 1)\Phi_0/4$. There was no flux-locked loop. The measured parameters were in good agreement with predictions. For example, for an amplifier with $R \approx 8\ \Omega$, $L \approx 0.4$ nH, $L_i \approx 5.6$ nH, $M_i \approx 1$ nH and $V_\Phi \approx 3 \times 10^{10}$ sec^{-1} at 4.2 K, they found $G = 18.6 \pm 0.5$ dB and $T_N = 1.7 \pm 0.5$ K at 93 MHz. The predicted values were 17 dB and 1.1 K, respectively.

We emphasize that in this theory and these measurements one is concerned only with the noise temperature of the amplifier itself. Nyquist noise from the resistor adds a contribution which, in the example just given, exceeds the amplifier noise. Thus, the optimization procedure just outlined does not necessarily give the lowest system noise, and one would use a different procedure when the value of T_N in Eq.(2.14) is well below T.

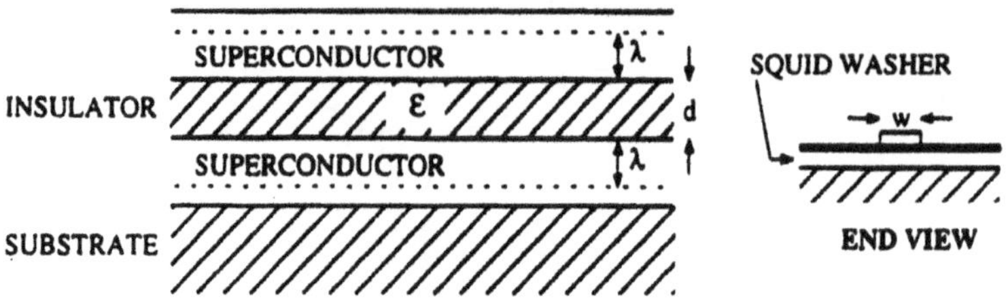

Figure 4. Side and end views of microstrip, showing two identical superconductors with penetration depth λ separated by an insulator of thickness d and dielectric constant ε.

3. The dc SQUID microstrip amplifier

3.1 THE MICROSTRIP

The microstrip consists of a superconducting strip of width w separated from an infinite superconducting sheet by an insulator with dielectric constant ε and thickness d (Fig. 4). We assume that the thicknesses of the two superconductors are much greater than the superconducting penetration depth λ, and that w >> d. The capacitance and inductance per unit length of the microstrip are given by [36]

$$C_0 = \varepsilon\, \varepsilon_0 w/d \text{ (F/m)} \qquad (3.1)$$

and

$$L_0 = (\mu_0 d/w)(1 + 2\lambda/d) \text{ (H/m)} . \qquad (3.2)$$

Here, $\varepsilon_0 = 8.85 \times 10^{-12}$ F/m and $\mu_0 = 4\pi \times 10^{-7}$ H/m are the permittivity and permeability of free space, and $c = 1/(\varepsilon_0\mu_0)^{1/2} = 3 \times 10^8$ m/s is the velocity of light in vacuum. The factor $(1 + 2\lambda/d)$ accounts for the penetration of the magnetic field into the (identical) superconductors. The velocity of an electromagnetic wave on the microstrip is thus given by

$$\bar{c} = c/[\varepsilon(1 + 2\lambda/d)]^{1/2} , \qquad (3.3)$$

and its characteristic impedance by

$$Z_0 = \left(\frac{L}{C}\right)^{1/2} = \frac{d}{w}\left[\frac{\mu_0(1 + 2\lambda/d)}{\varepsilon\varepsilon_0}\right]^{1/2} . \qquad (3.4)$$

The microstrip represents an electromagnetic resonator. For a microstrip of length ℓ and with its two ends either open or terminated with resistances greater than Z_0, the fundamental frequency occurs when ℓ is equal to a half-wavelength [36]. In this mode, the resonator is analogous to a parallel tuned circuit, and for a resistance $R_i > Z_0$ connected to one end, the quality factor of the resonance is given by [36]

$$Q = \pi R_i / 2Z_0 . \tag{3.5}$$

3.2 CONFIGURATION OF THE MICROSTRIP SQUID

The microstrip SQUID has the conventional square-washer configuration (Fig. 2). However, in contrast to the conventional input scheme in which the signal is connected to the two ends of the coil, the signal is instead coupled between one end of the coil and the square washer, which provides the groundplane for the microstrip. In our first experiments [11], the square washer had inner and outer dimensions of 0.2 x 0.2 mm^2 and 1 x 1 mm^2, and the input coil had $n = 31$ turns, a width $w = 5$ μm and a length $\ell =$ 71 mm. The SQUID loop and input coil were fabricated from niobium films and separated by a silicon film with a thickness $d \approx 0.4$ μm. The estimated inductance L of the SQUID loop was about 320 pH, leading to an estimated coupled inductance of the input coil $n^2 L \approx 300$ nH and a mutual inductance $M_i \approx 10$ nH. The critical current and shunt resistance per junction were typically 5 μA and 10 Ω, and the maximum value of V_Φ was about 60 μV Φ_0^{-1} . At 4.2 K, the white flux noise measured in a flux-locked loop at low frequencies was typically 2-3 $\mu\Phi_0$ $Hz^{-1/2}$. We note that virtually the entire length of the coil overlays the washer, which is at a uniform potential.

We estimate the parameters for the microstrip using Eqs. (3.1)-(3.5) for the linear case, although we emphasize that the presence of the hole and slit in the SQUID washer may have a profound effect; we shall return to this issue in due course. With this proviso, adopting the values $w = 5$ μm, $d = 0.4$ μm, $\varepsilon = 9$ and $\lambda = 0.15$ μm, from Eqs. (3.1) and (3.2) we find $C_0 \approx 1$ nF m^{-1} and $L_0 \approx 200$ nH m^{-1}. Thus, from Eq. (3.3) the velocity of propagation $\bar{c} \approx 0.25c$, and from Eq. (3.4) the characteristic impedance $Z_0 \approx$ 15 Ω. In this simplistic model, the fundamental resonance of the microstrip is at a frequency $\bar{c}/2\ell \approx 500$ MHz. When the microstrip is coupled to a 50-Ω signal source, from Eq. (3.5) we expect $Q \approx 5$, so that the -3 dB bandwidth is about 20% of the resonant frequency. In some devices, to increase the resonant frequency we progressively shortened the length of the coil by removing a portion of the Nb line using a photomask and reactive ion etching.

The SQUID chip was mounted on a printed circuit board, and each pad was wire-bonded to a 50-Ω trace patterned on the board. The grounds of the input and output SMA connectors were soldered to a ground plane on the reverse side of the board, and their center conductors were soldered to the traces coupled to the input of the microstrip and to the output from the SQUID, respectively. The assembly was rigidly mounted inside a box (made of lead foil) that eliminated fluctuations in the ambient magnetic field.

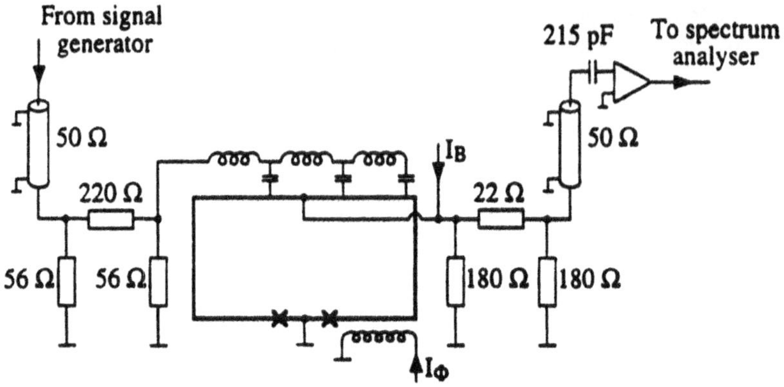

Figure 5. Circuit used to determine the gain of the microstrip SQUID, which is shown with counter electrode grounded. I_B is the current bias, and I_Φ provides the flux bias. The input and output π networks have attenuations of 20 dB and 5 dB, respectively.

3.3 GAIN OF THE MICROSTRIP SQUID

We used the circuit shown in Fig. 5 to measure the gain of our microstrip amplifier. In this schematic representation of the SQUID, the input coil over the square washer is shown as a distributed line. The current and the flux biases were supplied by batteries that could be floated relative to the system ground. The flux was generated by a small copper coil. The Hewlett-Packard 8620C sweep oscillator was coupled to the microstrip via a room-temperature 100-dB attenuator and a cold 20-dB attenuator that prevented noise produced by the generator from saturating the SQUID. The attenuator also presented an impedance of 50 Ω to both the input coaxial line and the microstrip. This impedance matching largely eliminated standing waves on the coaxial line. It also helped to minimize errors in the measured gain due to impedance mismatch. For the same reasons, for most of our measurements a cold 4-dB attenuator coupled the output of the SQUID to a room-temperature post-amplifier (MITEQ JS2-00100200-10-5P), which has a measured noise temperature of about 80 K. The gain of the system excluding the SQUID was calibrated by disconnecting the SQUID and connecting together the input and output attenuators. All measurements of the gain of the SQUID amplifier were referred to the baseline so obtained.

Since the washer SQUID is an asymmetric device – the two Josephson junctions are situated close together rather than on opposite sides of the SQUID loop – one can either ground the washer or ground the counter electrode close to the Josephson junctions. Since the washer acts as a groundplane for the input coil, one might be tempted to ground the washer. However, it is also possible to ground the counter electrode and have the washer at output potential. In the latter case, one can have feedback from the output voltage generated on the washer to the input coil, via the

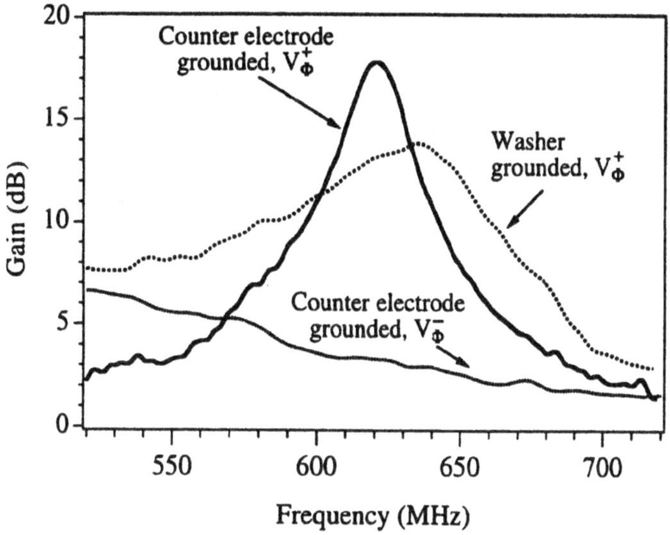

Figure 6. Gain vs. frequency for a microstrip SQUID with a 6-turn input coil, and with the signal applied to the innermost turn. Data are for V_Φ^+ and V_Φ^- with the counter electrode grounded, and for V_Φ^+ with the washer grounded.

capacitance between them. If the sign of V_Φ is such that the output voltage has the same sign as the input voltage, the feedback is positive; if the signs are opposite, the feedback is negative. We designate the flux-to-voltage transfer coefficients as V_Φ^+ and V_Φ^-, respectively.

The dramatic differences between the two cases, that is, for grounded washer or grounded counter electrode, are shown in Fig. 6, where we plot the measured gain as a function of frequency for a SQUID with a 6-turn input coil. The signal was connected between the innermost turn of the coil and ground. When we grounded the washer, we measured a gain of about 14 dB at 635 MHz, with a Q of about 10. On the other hand, when we grounded the counter electrode of the junctions and allowed the square washer to float at output potential, for V_Φ^+ the gain increased to 18 dB at 620 MHz and the Q increased to about 31. Conversely, for the other sign of the transfer function, V_Φ^- , the gain was greatly diminished with no evidence of a resonance. We attribute these effects to feedback from the output voltage of the SQUID to the input via the distributed capacitance between the coil and the washer. Evidently, positive feedback increases the gain compared with the case for the grounded washer for which there is no feedback, while negative feedback strongly suppresses the gain. We observe that positive feedback also enhances Q while negative feedback reduces it. In all subsequent measurements, we grounded the counter electrode of the SQUID and operated the amplifier with positive feedback.

To investigate the dependence of the resonant frequency on the length of the microstrip, we measured the gain for a device with a 31-turn coil, and then sequentially reduced the number of turns to 19, 11 and 6, counting from the inside turn. The results

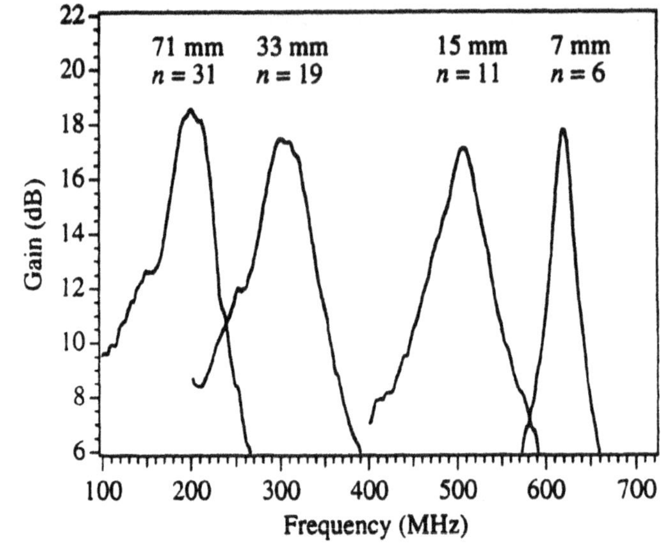

Figure 7. Gain vs. frequency for four coils on the same SQUID, with signal applied to the innermost turn. Data are for counter electrode grounded and V_Φ^+.

are shown in Fig. 7; in all cases the signal was applied to the innermost turn. We see that, as expected, the resonance shifted to progressively higher frequencies as the length of the microstrip was reduced, from 200 MHz for the 71-mm coil, to 620 MHz for the 7-mm coil. In addition, Q increased as the coil was shortened. However, the measured

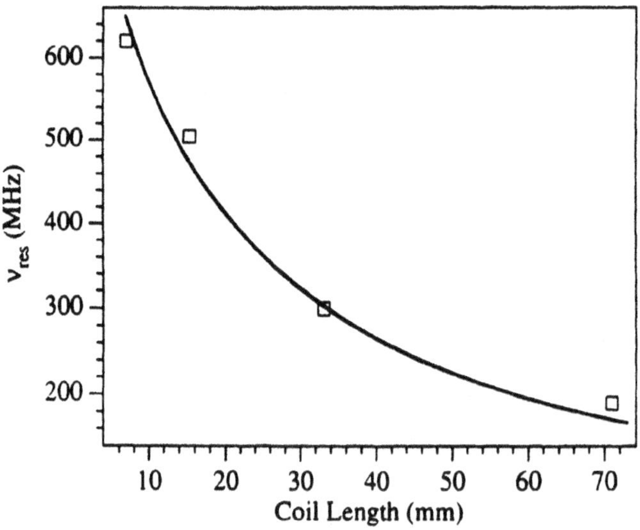

Figure 8. Resonant frequencies for the four lengths of microstrip shown in Fig. 7; fitted curve is $[1.49 \times 10^4 / (\ell + 16)]$ MHz, where ℓ is in millimeters.

resonant frequencies f_0 are significantly lower than we expected from our original model. In Fig. 8, we plot f_{res} vs. coil length ℓ together with the fitted curve

$$f_0 = \frac{1.49 \times 10^4 \text{ MHz}}{\ell + 16}, \tag{3.6}$$

where ℓ is in millimeters. The additive term, 16 mm, in the denominator arises from the wiring between the cold attenuator and the coil, which consists of five 4-mm-long wires in parallel bonded to the 3-mm-long niobium line on the SQUID. Since these leads are not over a ground plane, they increase the effective electrical length of the microstrip by more than their geometrical length, and decrease the resonant frequency. More surprisingly, the numerator is roughly a factor of three smaller than predicted from the parameters of the microstrip.

To investigate the low resonant frequency of the microstrip, we made a 195:1 scale model from a printed circuit board with a 31-turn input coil patterned on one side and a square washer containing a hole and a slit on the reverse side. The length of the coil was 12.25 m, its outer dimensions were 160 mm x 163 mm, and the turns were approximately 1-mm wide with a 1-mm spacing. The board was 0.8-mm thick and its dielectric constant was 5.5. We note that the thickness of the board was an order of magnitude greater than the correctly scaled value. As a result, the velocity of propagation on the line is likely to differ slightly from the prediction of Eq. (3.3), since the approximation $w \gg d$ is not strictly valid. We connected a sweep oscillator to one end of the line and a spectrum analyzer to the other. On resonance the output voltage was a maximum. When the hole and slit were covered with a copper sheet that was in electrical contact to the washer, the fundamental resonance occurred at 4.92 MHz, in reasonable agreement with the predictions of the simple model, namely 5.23 MHz. When we created a short circuit across the slit only, leaving the hole in the washer open, the resonant frequency was not significantly different. On the other hand, when we also uncovered the slit, the resonant frequencies dropped by a factor of about 2.6 to 1.88 MHz. We believe this reduction in the velocity on the microstrip arises from an increase in its effective inductance by a factor of $(2.6)^2 \approx 6.8$ due to screening currents that are forced to flow around the inner edge of the slit and washer, rather than on the surface of the ground plane as required by the simple model. This result is in quite good agreement with the reduced resonant frequency observed for the SQUID.

3.4 NOISE TEMPERATURE OF THE MICROSTRIP SQUID

We measured the noise temperature of several amplifiers. As a noise source, we used a 64-Ω resistor (51 Ω at room temperature) connected to the input of the microstrip via a length of stainless steel coaxial cable. The resistor was in thermal contact with a thermometer, an Allen-Bradley resistor, and 100 turns of manganin wire were wound around the combination. The assembly was potted in Stycast epoxy in a copper tube with a 5-mm internal diameter. By passing a current through the manganin wire, we could increase the temperature of the 63-Ω resistor to a measured value T. Thus, the

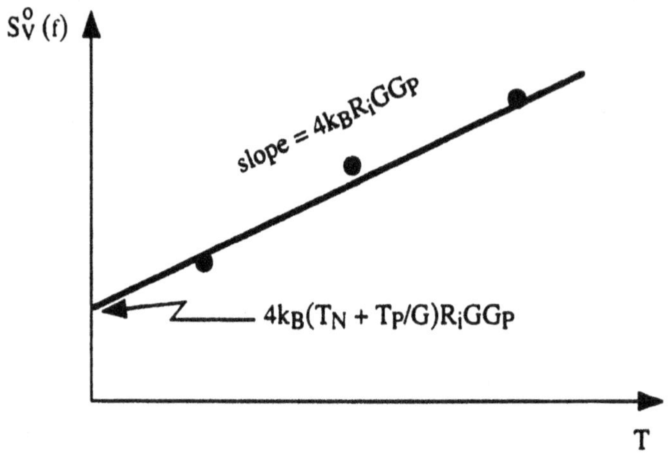

Figure 9. Measurement of noise temperature. The slope of the line through the points yields $4k_BR_iGG_p$ and the intercept of the line with T = 0 yields $T_N + T_{NP}/G$.

resistor provides a well-defined source of Nyquist noise power. The spectral density of the voltage noise at the output of the post-amplifier is given by

$$S_v^0(f) = 4k_B(T + T_N)R_iGG_P + 4k_BT_PR_iG_P\,, \tag{3.7}$$

where G and G_P are the (power) gains of the SQUID amplifier and post-amplifier, and T_N and T_P are the noise temperatures of the SQUID amplifier and post-amplifier. If one measures the output power for two or more values of resistor temperature, one can infer $T_N + T_P/G$, as shown schematically in Fig. 9. Measuring T_P separately using a similar hot/cold load method, one can deduce T_N. This discussion also makes clear that one requires $T_P/G \ll T_N$ to insure that the post-amplifier noise does not contribute significantly to the system noise temperature. To determine T_N, we measured the output with a spectrum analyzer, generally with the resistor at the bath temperature and at about 10 K.

Figure 10 shows the gain and noise temperature vs. frequency for a 31-turn SQUID; in this case there was no π-network between the SQUID and the post-amplifier. At the resonant frequency of 240 MHz, the gain is about 23 dB $\approx$ 200 and the system noise temperature is 0.8 ± 0.3 K. The contribution of the post-amplifier is about 80 K/200 $\approx$ 0.4 K, so that the intrinsic noise temperature of the SQUID microstrip amplifier is approximately 0.4 K. As the frequency moves away from resonance, the gain falls, and the system noise temperature, which becomes dominated by the post-amplifier, rises correspondingly.

While it is clear that the microstrip SQUID amplifier has a very low intrinsic noise temperature, it is also evident that the room-temperature post-amplifier makes a substantial contribution to the overall noise temperature. To ameliorate this problem, we built a single-stage amplifier using an HFET (Heterojunction Field Effect Transistor

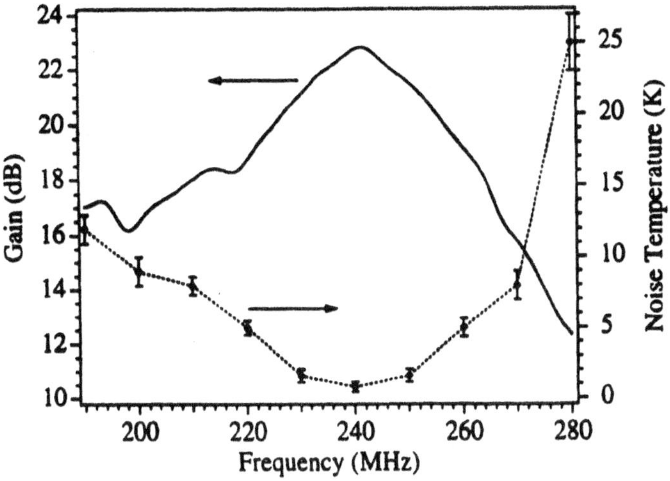

Figure 10. Gain and noise temperature vs. frequency for 31-turn SQUID.

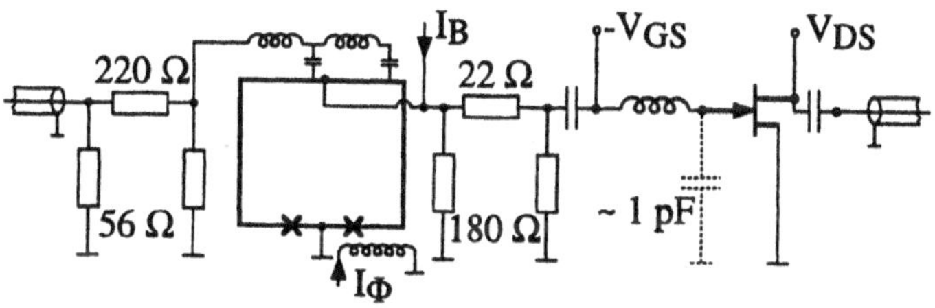

Figure 11. Configuration of microstrip SQUID amplifier and cooled HFET post-amplifier.

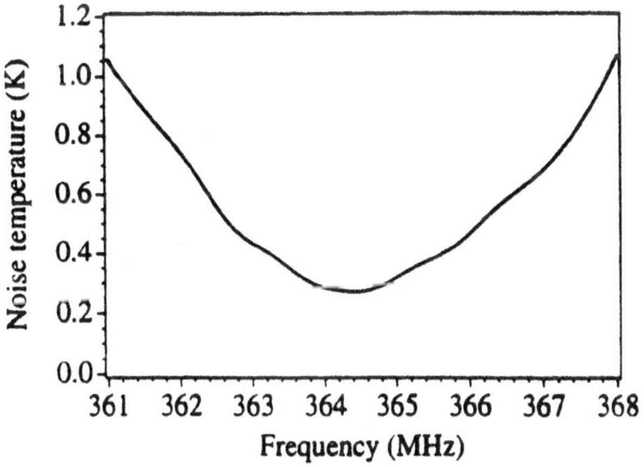

Figure 12. Noise temperature for a SQUID amplifier with 29-turn spiral input coil operated as $\lambda/2$ microstrip resonator. The device and its HFET post-amplifier were cooled to 1.8 K.

-- Fujitsu FHX 13LG), which we operated in the liquid helium bath. The configuration of the microstrip SQUID amplifier and the cooled post-amplifier are shown in Fig. 11. The output impedance of the 4-dB attenuator is boosted to the optimum value for the HFET by a resonant circuit which, unfortunately, limits the bandwidth over which a low noise temperature can be achieved to a few megahertz. We measured the gain and noise temperature of the microstrip SQUID in the usual way, and also measured the noise temperature of the post-amplifier, T_P . The results are listed in Table I for two amplifiers with resonant frequencies of 250 MHz and 365 MHz, and the system noise temperature is plotted vs. frequency for the latter device in Fig. 12.

TABLE I. Performance of microstrip SQUID amplifiers with a cooled post-amplifier at frequency f and bath temperature T. The power gain of the SQUID is G; G′ is the reduced power gain including the attenuator following the SQUID. The measured system and post-amplifier noise temperatures are T_S and T_P, and the inferred SQUID amplifier noise temperature is T_N.

f (MHz)	T (K)	G (dB)	G′ (dB)	T_S (K)	T_P (K)	T_P/G' (K)	$T_N = T_S - T_P/G'$ (K)
250	1.8	26.5 ± 0.5	22.5 ± 0.5	0.30 ± 0.05	10 ± 1.5	0.06 ± 0.01	0.24 ± 0.05
365	1.8	24.5 ± 0.5	20.5 ± 0.5	0.28 ± 0.06	10 ± 1.5	0.09 ± 0.02	0.19 ± 0.06

Table I lists the measurement frequency f, the bath temperature T, the gain G of the SQUID and the gain G' with the 4-dB loss of the output π-network subtracted, the system noise temperature T_S, the noise temperature T_P of the post-amplifier, the contribution T_P/G of the post-amplifier to the system noise temperature, and finally the intrinsic noise temperature T_N of the microstrip SQUID amplifier. We see that the intrinsic temperatures of the SQUID amplifiers operated at 1.8 K are low, 0.24 ± 0.05 K and 0.19 ± 0.06 K for the 250 MHz and 365 MHz amplifiers, respectively, but that the contribution of the post-amplifier, although small, is still non-negligible. As is evident from Fig. 12, the noise temperature increases rapidly once one moves away from resonance, because of the narrow-band response of the coupling circuit to the FET.

These experiments demonstrate that the microstrip SQUID amplifier is able to achieve gains well in excess of 20 dB and noise temperatures well below 1 K at frequencies from about 80 MHz up to about 800 MHz. These noise temperatures are significantly lower than one can achieve with cooled semiconductor amplifiers over the same frequency range. Since from Eq. (2.16) we expect that noise temperature should scale with bath temperature, in the next series of experiments we cooled the SQUIDs in a ^{3}He cryostat to see if still lower noise temperatures were achievable.

4. The SQUID microstrip amplifier below 1 K

In these experiments, we cooled the SQUID to 0.4 - 0.5 K using a charcoal-pumped, single-shot ^{3}He cryostat (RMC). The HFET post-amplifier (Fujitsu FHX 13 LG) was mounted on the 1-K reservoir; unfortunately, the power dissipated by the HFET, 34 mW, raised the temperature of the 1-K reservoir to about 1.8 K, degrading the performance of the refrigerator. In addition, we were no longer able to use an input load at various temperatures to determine T_N as before, because the small cooling power of the ^{3}He stage (about 10 μW) would not allow us to cool the noise source to the base temperature more than once. Instead, we measured the signal-to-noise ratio of the amplifier with a precisely calibrated power (-134 ± 0.5 dBm) applied to the input; to attenuate any room-temperature noise we added a second 20-dB attenuator at 1.8 K. Table II shows the results at two frequencies and two temperatures. The same SQUID

TABLE II. Performance of microstrip SQUID amplifiers with a cooled post-amplifier at frequency f and bath temperature T. The power gain of the SQUID is G; G′ is the reduced power gain including the attenuator following the SQUID. The measured system and post-amplifier noise temperatures are T_S and T_P, and the inferred SQUID amplifier noise temperature is T_N.

f (MHz)	T (K)	G (dB)	G′ (dB)	T_S (K)	T_P (K)	T_P/G' (K)	$T_N = T_S - T_P/G'$ (K)
89.6	4.2	24 ± 0.5	20 ± 0.5	0.90 ± 0.12	4.0 ± 0.5	0.04 ± 0.01	0.86 ± 0.12
89.6	0.39	24.5 ± 0.5	20.5 ± 0.5	0.10 ± 0.02	4.0 ± 0.5	0.04 ± 0.01	0.06 ± 0.02
438	4.2	20 ± 0.5	16 ± 0.5	1.40 ± 0.18	15 ± 2.0	0.38 ± 0.07	1.02 ± 0.19
438	0.48	20 ± 0.5	16 ± 0.5	0.50 ± 0.07	15 ± 2.0	0.38 ± 0.07	0.12 ± 0.10

was used throughout: the length of the coil was shortened after the 89.6-MHz measurements to increase the resonant frequency to 438 MHz. Because of the narrow bandwidth of the HFET post-amplifier, we made no attempt to measure at off-resonance frequencies. The intrinsic noise temperature of the SQUID amplifier at the lower temperatures was 0.06 ± 0.02 K and 0.12 ± 0.10 K at 89.6 MHz and 438 MHz, respectively. We note that the noise temperature at the higher frequency has a large error bar due to the dominant contribution of the post-amplifier. Nonetheless, within the error bars the noise temperature scales with the bath temperature at each frequency.

In the measurements described above, the signal source was resistive. Since the amplifier will ultimately be connected to the high-Q resonant cavity of the axion detector, to test the stability of the SQUID we also coupled it to a tuned circuit, as shown in Fig. 13, and measured the gain and noise temperature. The LC circuit consisted of a 1-pF capacitor and a 4-turn copper coil, about 4 mm in diameter, inductively coupled to the microstrip by means of a loop of wire. In a separate experiment, we connected the loop to a 50-Ω cable, and adjusted the distance between the coil and the loop to produce a 2-dB loss and reduce the Q by a factor of about two. A second loop, with a coupling loss of 10 dB to the tuned circuit, was connected to a signal generator via a cold 20-dB attenuator and a stainless-steel cable with a loss of 3 dB to enable us to determine the gain and Q of the system.

The measured transmitted power at 4.2 K is shown in Fig. 14(a) (upper trace). The Q is about 270 and the maximum gain is 22 dB. The gain is referred to the input coupling loop of the resonant circuit, and includes an added 12 dB to account for the

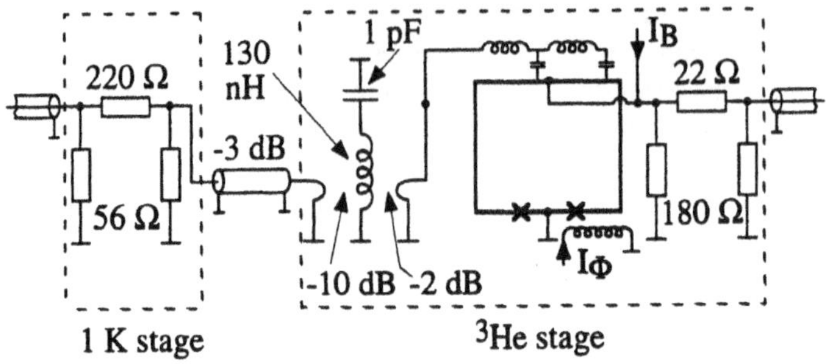

Figure 13. Configuration of circuit used to detect Nyquist noise in a resonant circuit inductively-coupled to the input of the microstrip amplifier.

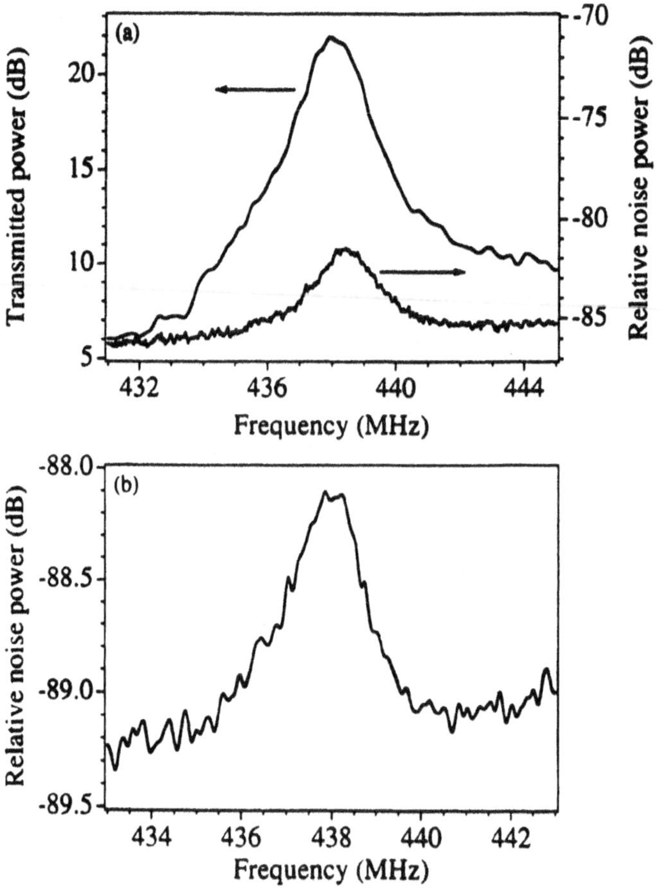

Figure 14. Microstrip SQUID amplifier coupled to the resonant circuit of Fig. 13: (a) Transmitted power (referred to loop that couples signal into resonant circuit) at 4.2 K, and noise power with no input signal; (b) noise power at 0.5 K with no input signal.

coupling losses. This result demonstrates that the microstrip SQUID remains stable when coupled to a resonant circuit.

Subsequently, we disconnected the signal generator and measured the noise generated by the resonant circuit, shown in the lower trace of Fig. 14(a). The measured peak is 4.7 dB above the (nearly white) noise at 432 MHz; to refer the peak to the input of the microstrip, we add 2 dB to account for the coupling losses to obtain a peak height of 6.7 dB. This peak contains contributions from the Nyquist noise of the resonant circuit and from the system noise of the SQUID amplifier. Although it is not entirely straightforward to separate these contributions, using the following *approximate* argument we show that this result is consistent with the measured noise temperature. On resonance, since the microstrip reduces the Q of the resonant circuit to roughly 1/2 of its unloaded value, the source impedance presented to the microstrip is roughly equal to the characteristic impedance of the microstrip. At frequencies well below resonance, the magnitude of the source impedance is approximately ωL_ℓ, where L_ℓ is the inductance of the coupling loop. Estimating $L_\ell \sim 10$ nH, we find $\omega L_\ell \sim 25\ \Omega$ at 400 MHz. In separate experiments, we found that the noise power of the SQUID was relatively unchanged when we varied the source impedance from 2 Ω to 330 Ω. Since $T_N << T$, to a first approximation we can ignore any variation in the noise power with source impedance. Thus, the total noise on-resonance can be characterized as a temperature $(T + T_N + T_P/G')$, while off-resonance the corresponding value is $(T_N + T_P/G')$. Using the peak height of 6.7 dB, we find T_N from the relation $10 \log [(T + T_N + T_P/G')/(T_N + T_P/G')] \approx 6.7$ or $T_N + T_P/G' = T_S \approx 1.1$ K. Given the approximations involved, this result is in reasonable agreement with the measured system noise temperature, 1.4 ± 0.18 K. It is clear that the predominant contribution to the noise peak in Fig. 14(a) is Nyquist noise in the resonant circuit.

We repeated the measurements at 0.48 K, and show the noise power in Fig. 14(b). The height of the peak above the noise at 434 MHz, taking into account the 2-dB coupling loss, is 3.1 dB. Making the same approximations as above, this result implies that $T_S \approx T \approx 0.5$ K, in good, but probably fortuitous, agreement with the measured value of 0.5 ± 0.07 K. In this case, the relatively small magnitude of the noise peak reflects the fact that the system noise is dominated by the post-amplifier.

These experiments demonstrate that the microstrip SQUID operating at 0.4 - 0.5 K can achieve an intrinsic noise temperature of about 0.1 K at frequencies up to 500 MHz. Clearly, to take advantage of this very low value, we have to reduce the noise temperature of the post-amplifier still further. It is possible that cooled amplifiers from NRAO [9] would overcome this problem, but unfortunately these are too large to fit into our existing ^{3}He system. A second SQUID as a post-amplifier offers an alternative solution, and might well have a higher bandwidth than we have been able to achieve with our cooled HFET.

5. Tuning and cascading

Despite the high gain and low intrinsic noise temperature that have been achieved for the microstrip amplifier, there are at least two important issues that need to be addressed for both the axion detector and for other potential applications. First, once one chooses the length of the input coil, the resonant frequency is fixed. Since the quality factor Q is typically in the range 10-30, the bandwidth over which the gain exceeds, say, 15 dB is relatively narrow, typically a few tens of megahertz. The resonant frequency of the axion-detector cavity, on the other hand, is swept over a wide range of frequencies to search for a broad range of axion mass, and it is thus necessary to provide an amplifier that either has an appropriately wide bandwidth or can be tuned. The second issue concerns the post-amplifier for the SQUID amplifier. In our experiments described in Secs. 3 and 4, we used a cooled HFET, but, at least at the higher frequencies, we found that the system noise temperature was dominated by the post-amplifier rather than by the intrinsic noise of the SQUID. Although this problem might be ameliorated by the use of a better post-amplifier, it would be required to have a low noise temperature (well below 10 K) over a broad range of frequencies. In this section we address both issues.

5.1 TUNING THE MICROSTRIP SQUID AMPLIFIER

The resonant frequency of a microstrip resonator of characteristic impedance Z_0 can be changed by terminating the (normally open) end with a reactive load, such as a capacitor, C. For a microstrip of length ℓ open at both ends (C = 0), the current undergoes a phase change of π on reflection at each end, leading to a fundamental resonance for which $\ell = \lambda/2$, where λ is the wavelength in the microstrip. If one increases C to infinity at one end of the line, there will be no phase change when the current is reflected, and the fundamental resonance will change, so that $\ell = \lambda/4$. For intermediate values of C, the phase shift ϕ on reflection at a frequency $f = \omega/2\pi$, is easily found from the reflection coefficient [36] $\rho = (Z_0 - 1/j\omega C)/(Z_0 + 1/j\omega C)$ using the relations

$$\phi = \tan^{-1}[2Z_0\omega C/(Z_0^2\,\omega^2 C^2 - 1)] + \pi, \qquad C < 1/Z_0\omega \qquad (5.1a)$$

$$\phi = \pi/2, \qquad C = 1/Z_0\omega \qquad (5.2a)$$

and

$$\phi = \tan^{-1}[2Z_0\omega C/(Z_0^2\,\omega^2 C^2 - 1)]\,. \qquad C > 1/Z_0\omega \qquad (5.3a)$$

Since the load is purely reactive, $|\rho| = 1$ for all C. We note that, for a given value of C, the value of ϕ depends on Z_0, which can thus be estimated from the dependence of the resonant frequency on C.

A convenient way of varying the capacitance is to use a varactor diode, in which the width, and hence the capacitance, of the depletion layer can be controlled by changing the reverse bias voltage. Since the capacitor must be adjacent to the SQUID, and thus at its operating temperature, we chose a GaAs diode [37] rather than the more commonly used Si diode. For $Z_0 = 50\ \Omega$ and $\omega/2\pi = 200$ MHz, a capacitance of about 20 pF enables one to reduce the resonant frequency by about 20%. The capacitance of the GaAs varactor can be varied from 1 pF to 10 pF by varying the bias voltage from

-15V to +0V; to obtain a larger range of capacitance, we used two diodes in parallel. Since these diodes have a diffusion voltage of about +1.1V, we were able to apply a forward voltage of +1V without decreasing the quality factor of the varactor significantly. In separate experiments, we confirmed that the voltage-capacitance relation was not significantly changed when we cooled the device to 4.2 K.

The configuration of the microstrip amplifier with the varactor diode is shown in Fig. 15. The gain for optimized current and flux biases for a SQUID with 31 turns is shown in Fig. 16 for 9 values of the capacitance of the two diodes. We see that the resonant frequency is progressively lowered, from 195 MHz to 117 MHz, as the capacitance is increased. The maximum gain is constant to within 1 dB over this range. In the absence of the varactor, the resonant frequency is about 200 MHz. We believe that the varactor increases the gain by increasing the degree of positive feedback; throughout these measurements the counter electrode of the SQUID was grounded.

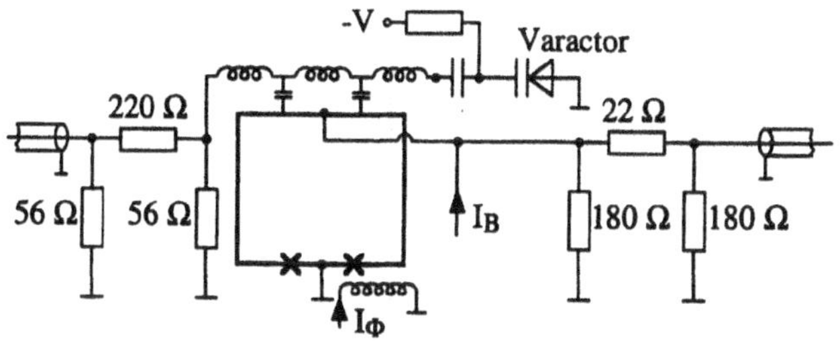

Figure 15. Configuration of SQUID microstrip amplifier showing a varactor diode connected to one end of the microstrip. The input and output π-network attenuators match the impedances to 50 Ω.

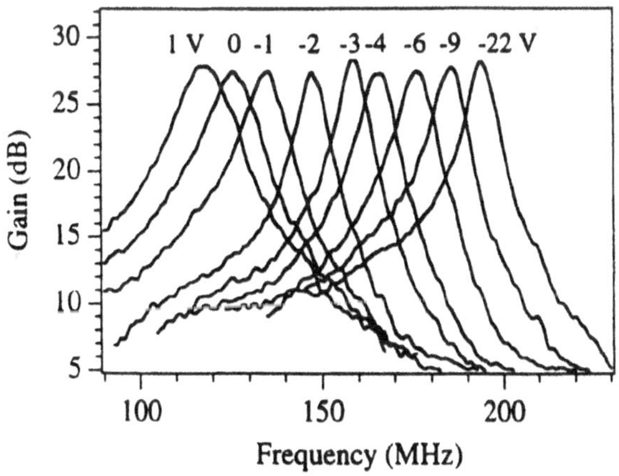

Figure 16. Gain vs. frequency for SQUID microstrip amplifier with 31 turns at 4.2 K for 9 values of reverse bias voltage applied to the varactor diode.

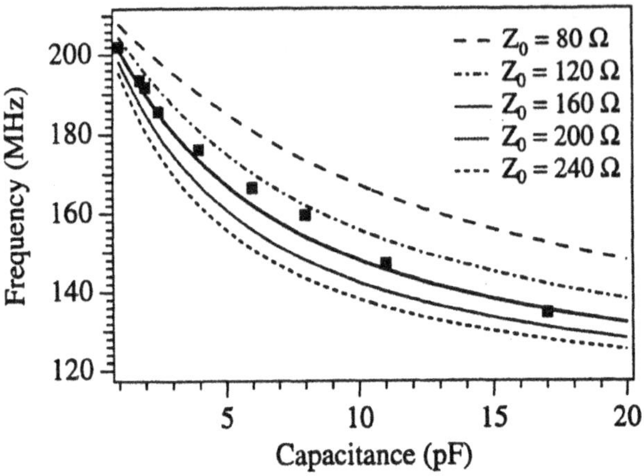

Figure 17. Calculated dependence of the resonant frequency f_0 of a microstrip with length 71 mm vs. varactor capacitance C [Eq. (5.2)] for five values of microstrip impedance Z_0. Squares are data obtained for the SQUID with 31 turns used to obtain the data of Fig. 16.

In Fig. 17, we plot the calculated resonant frequency,

$$f_0 = (\bar{c}\ /4\ell\ )(1 + \phi/\pi)\ , \tag{5.2}$$

vs. varactor capacitance for 5 values of $Z_0 = (L_0/C_0)^{1/2}$. Because of the uncertainty in the values of the capacitance at voltages of 0V and +1V, we have omitted the corresponding data in Fig. 17. The data indicate that the measured value of Z_0 for a particular SQUID (with 31 turns of total length 71 mm) is about 160 ± 40 Ω. Even though the errors due to parasitic inductance and uncertainties in the parameters of the microstrip and the varactor are substantial, it is clear that the measured value of Z_0 is much higher than that for a simple microstrip, about 15 Ω (Sec. 3.2). However, this result is not totally unexpected. As we noted in Sec. 3.3, the frequency of the fundamental mode is about 3 times lower than the simple model predicts, leading us to estimate a characteristic impedance of roughly 45 Ω. However, this value is still a factor of 3-4 below that deduced from Fig. 17. We suspect that positive feedback further enhances the effective microstrip impedance; further investigation is required to elucidate this issue.

We now turn to a brief discussion of possible noise sources associated with the varactor, of which there are at least three. The first is the Nyquist noise of the resistive component of the varactor, which, in the present case, can be represented by a series resistance of about 50 mΩ. Since the diode reactance $|1/j\omega C|$ is typically 100 Ω, the Nyquist noise current injected into the microstrip will be negligible. Second, the reversed-biased current I_r of the pn junction of the varactor generates a shot-noise current with an rms amplitude $(2eI_r\Delta f)^{1/2}$ in a bandwidth Δf. The manufacturer quotes an upper limit (near the breakdown voltage) on I_r of 10^{-8} A, leading to a rms shot-noise current of 60 $fAHz^{-1/2}$. Feeding into a 50-Ω resistance, this noise current would induce a voltage of 3 $pVHz^{-1/2}$, corresponding to a noise temperature of roughly 3 mK, which is negligible.

A third potential noise source is the bias supply of the varactor. Although it is straightforward to filter the voltage heavily at signal frequencies, noise at low frequencies will lead to fluctuations in the capacitance and thus variations in the resonant frequency of the microstrip. In turn, these variations will result in gain fluctuations, resulting in upconversion of the low-frequency noise, except at the maximum gain where $\partial G/\partial f = 0$. We can make a rough estimate of this effect as follows. Let us assume that the power supply produces a voltage noise at the varactor $\delta V_v = 50$ nV $Hz^{-1/2}$ in a bandwidth that extends from 0 to Δf. The resulting gain fluctuations will be given by

$$\delta G = \frac{\partial G}{\partial f} \frac{\partial f}{\partial C_V} \frac{\partial C_V}{\partial V_V} \delta V_V \ , \tag{5.3}$$

where C_v is the capacitance of the varactor diode at a voltage V_v. To estimate the magnitude, we assume $C_v = (10/V_v)$ pF, so that $\partial C_v/\partial V = (-C_v/V_v)$ pF/V = 0.4 pF/V at a representative capacitance of 2 pF (corresponding to $V_v = 5$ V). From Fig. 16, a change in V_v from 6 V to 4 V, corresponding to changing C_v from 1.67 pF to 2.5 pF, reduces the resonant frequency by about 10 MHz; thus, $\partial f/\partial C_v \approx 12$ MHz/pF for $C_v \approx$ 2 pF (Fig. 17). To estimate $\partial G/\partial f$, we note that for a gain of roughly 20 dB, the slope $\partial G/\partial f \approx 1$ dB/MHz or 1.3 MHz^{-1}). Putting together these values, we find $\delta G \approx 2 \times 10^7$ $Hz^{-1/2}$. Thus, if the bias voltage is heavily filtered down to (say) 100 Hz, one might hope to reduce the gain fluctuations to a few parts per million, a value that is negligible except for large input signals. We also reiterate that there should be no effect at all at the resonance when $\partial G/\partial f = 0$.

To investigate possible noise contributions from the varactor diode, we measured the noise temperature at 4.2 K and about 200 MHz with it in place and without it. We found $T_N = 0.9 \pm 0.2$ K in both cases, indicating that, as expected, the varactor has no discernible effect on the noise.

5.2 THE MICROSTRIP SQUID AS A POST-AMPLIFIER

In a preliminary attempt to replace a cooled semiconductor as the post-amplifier, we used a second microstrip SQUID as the post-amplifier. To prevent the two cascaded SQUIDs from interacting with each other, we found it necessary to separate them with a least 6 dB of attenuation. As an example, we were able to achieve a gain of 31 dB at 300 MHz with a π-network providing 8 dB of attenuation. Subsequently, in an attempt to reduce the coupling loss, we replaced the attenuator with a cooled circulator [38], which had forward loss of about 1-dB and a large reverse loss. We inserted one π network with a 1-dB loss to match the output of the first SQUID to the circulator, and a second π network very close to the input of the second SQUID to provide a termination of the resonant microstrip [Inset, Fig. 18(a)]. The measured gain is shown in

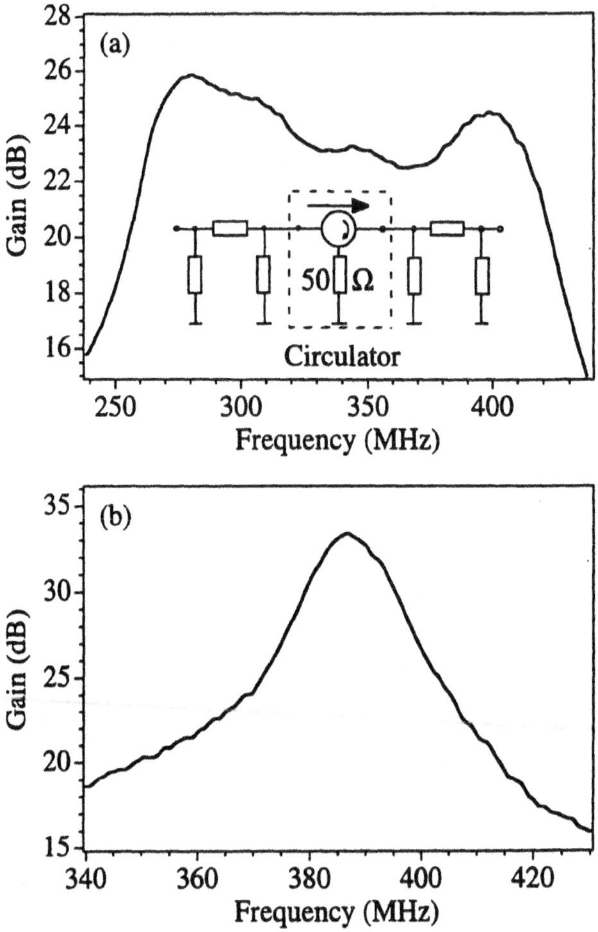

Figure 18. Measured gain of two cascaded microstrip SQUID amplifiers with (a) two different resonant frequencies, and (b) the resonant frequency of the second microstrip tuned with a varactor diode to coincide with that of the first. The coupling network between the two SQUIDs is shown inset in (a).

Fig. 18(a). The resonance frequencies of the first and second amplifiers were about 280 MHz and 400 MHz, respectively. We note that the gain was 22 dB or higher from about 260 to 420 MHz, with two peaks at about 26 dB and 24.5 dB.

Although the cascaded SQUIDs operated stably in this configuration, the first SQUID was apparently still loaded by the subsequent circuitry. When we inserted an additional 3 dB of attenuation between the first SQUID and the circulator, the resonant frequency increased to 386 MHz. Using a varactor diode to reduce the resonant frequency of the second SQUID to the same value, we achieved a gain of 33.5 ± 1 dB with a Q of 26, as shown in Fig. 18(b). However, since the total attenuation between the two SQUIDs amounted to 6 dB, in this case the circulator offered no advantage over a simple π network.

Even though more research is required, it appears likely that the use of a second microstrip would largely meet the need of a low-noise post-amplifier. Even with an 8-dB attenuator separating the two SQUIDs, one can expect an overall gain on resonance of at least 30 dB. Thus, if the noise temperature of the first SQUID were 0.1 K, a room-temperature amplifier with a noise temperature of 80 K would contribute 0.08 K to the input, giving a system noise temperature of 0.18 K. This would suffice for the axion detector. Nonetheless, a certain degree of care would be necessary to operate the two-SQUID amplifier: not only would one have to optimize two current and two flux biases, but in addition one would have to adjust two varactor diodes to achieve the desired resonant frequency.

6. The axion detector

There is overwhelming cosmological evidence that cold dark matter (CDM) is the predominant mass in the universe. The two leading contenders for CDM are the WIMP (Weakly Interacting Massive Particle), which supersymmetry theories predict to have a mass of 10-100 GeV, and the axion. The origins of this postulated particle lie in particle theory. The absence of any measurable electric dipole moment of the neutron places an upper limit 10^{11} smaller than the value expected from the Standard Model of particle physics. This result implies that CP conservation is vastly stronger than predicted. (Here, C is charge conjugation and P is parity inversion.) Peccei and Quinn [39] extended the standard model to suppress strong CP violation, and subsequently Weinberg [40] and Wilczek [41] independently showed that this leads to the existence of a neutral, spinless particle – the axion. There are good reasons to believe that the energy of the axion lies between 1 μeV and 1 meV (240 MHz and 240 GHz). To date, this particle has not been found.

In 1983 Sikivie [42] showed that in the presence of a high magnetic field the axion should undergo Primakoff conversion into a real photon, with energy equal to the mass of the axion, and a virtual photon. This prediction has led to the construction of axion detectors intended to search for the real photon; the latest and most sensitive detector is currently operating at Lawrence Livermore National Laboratory in a collaboration with the Massachusetts Institute of Technology [10]. In brief, the detector consists of a cavity 1-m long and 0.6 m in diameter cooled to about 1.5 K in a magnetic field of 8 T. The cavity can be tuned over the range 0.7-0.8 GHz. The output from the cavity is coupled into a cooled HEMT amplifier with a noise temperature of 1.7 K; thus the system noise temperature is about 3.2 K. The goal of the experiment is to look for a signal above the blackbody noise of the cavity that would signify the presence of the photon produced by the decay of an axion. Since the axion energy is unknown, one sweeps the frequency of the cavity by means of tuning rods.

The experiment has been running since February 1996 [10]. Over a limited range of frequencies, these experiments have demonstrated a power sensitivity below the KSVZ (Kim-Shifman-Vainshtein-Zakharov) limit [43], which is established by a particular axion model. A second model, designated DFSZ (Dine-Fischler-Srednicki-Zhitnitsky) [44], sets a limit an order of magnitude lower in power and is regarded as a more definitive goal. To achieve this limit on a realistic time-scale one requires a detector with a lower noise temperature. In turn, this implies a cavity at a lower temperature – say 0.3 K – and an amplifier with a lower noise temperature – preferably below 0.1 K. This need has been the driving force for the development of the microstrip SQUID amplifier. Provided funding is forthcoming, the cavity of the next generation axion detector will be cooled to 0.3 K (or possibly even lower, with the aid of a dilution refrigerator) and have a microstrip SQUID amplifier connected to it. The goal of the experiment would be to operate at the DFSZ limit over (say) one decade of frequency.

7. Concluding remarks

We have described a new approach to the operation of the dc SQUID in which the input signal is coupled between one end of the input coil and the square washer that forms the body of the SQUID. On resonance, when the half-wavelength of the signal is equal to the length of the microstrip, the magnetic field couples strongly to the SQUID, enabling one to achieve a power gain of 20-25 dB over frequencies that have varied over the range of 0.1 - 0.8 GHz. The intrinsic noise temperature of the amplifier is typically 1/4 of the bath temperature T and, within the measurement errors, scales with T as expected from Eq. (2.16). Thus, noise temperatures of around 0.1 K have been achieved for a bath temperature of about 0.5 K. The amplifier remains stable when an LC circuit – simulating a cavity – is coupled to its input; indeed, one can readily observe the Nyquist noise in this resonant circuit. The microstrip SQUID amplifier can be tuned by means of a pair of varactor diodes connected to the open end of the microstrip. A second such SQUID has been used as a post-amplifier for a first-stage SQUID, achieving a combined gain of about 33 dB and eliminating the need for a cooled HFET.

Needless to say, various issues and challenges remain. There are several aspects of the microstrip SQUID that clearly need further investigation. Although considerable insight has been gained from the large-scale model, we do not have an analytical understanding of the coupling of the inductances of the hole and slit in the SQUID washer into the microstrip, and of their influence on the microstrip inductance and the propagation velocity. Consequently, we do not have *a priori* predictions of the dependence of the resonant frequency on the SQUID and coil parameters. A detailed understanding of the positive feedback between the SQUID output and the distributed capacitance is also lacking, although it clearly plays a pivotal role in the gain, bandwidth and noise temperature. As a result, we do not have an adequate model for these figures of merit. Both of these issues deserve detailed computer modeling.

What are the experimental challenges? An exciting possibility is the expectation of quantum-limited performance. The intrinsic noise temperature of the SQUID amplifier, to within the measurement uncertainties, scales with T, and at 0.5 K and 438 MHz is measured to be 0.12 ± 0.10 K. Since hot-electron effects [45] remain unimportant at temperatures down to 0.1 K, a SQUID cooled to this temperature, with a second SQUID as a post-amplifier, has an excellent prospect of achieving the quantum limit $hf/k_B \ln 2 \approx 30$ mK. A second challenge, particularly pertinent to the axion detector, is the extension to higher frequencies. Could the microstrip amplifier be operated at, say, 10 GHz? If so, what noise temperature would be achievable? The geometry of the device would almost certainly have to be changed, from the square-washer format to, perhaps, an in-line configuration. One may hope that the design and testing of such a second-generation device will be undertaken in the not-too-distant future.

Acknowledgments

We thank Xiaofan Meng and Robert McDermott for supplying Fig. 2. We are indebted to C. Hagmann, Darin Kinion, L. Rosenberg and K. van Bibber for their ongoing encouragement, and to Xiaofan Meng for technical support in the SQUID fabrication. We acknowledge the use of the Microlab Facility of the Electrical Engineering and Computer Science Department, University of California, Berkeley. This work was supported by the Director, Office of Energy Research, Office of Basic Energy Sciences, Materials Sciences Division of the U.S. Department of Energy under contract number DE-AC03-76SF00098 and by the National Science Foundation under grant number FD96-00014.

This chapter is dedicated to the memory of Christoph Heiden.

References

[1] Jaklevic, R.C., Lambe, J., Silver, A.H. and Mercereau, J.E. (1964) Quantum interference effects in Josephson tunneling, *Phys. Rev Lett.* **12**, 159-160.

[2] For a review, see, for example, M. Hämäläinen, *et al.* (1993) Magnetoencephalography-theory, instrumentation, and applications to noninvasive studies of the working human brain, *Rev. Mod. Phys.* **65**, 413-497.

[3] Forgacs, P.L. and Warnick, A. (1967) Digital-analog magnetometer utilizing superconducting sensor, *Rev. Sci. Instrum.* **38**, 214-220.

[4] For a review, see Augustine, M., TonThat, D.M. and Clarke, J. (1998) SQUID Detected NMR and NQR, *Solid State Nucl. Magn. Reson.* **11**, 139-156.

[5] Drung, D. (1996) Advanced SQUID read-out electronics, in H. Weinstock (ed.), *SQUID Sensors: Fundamentals and Applications*, Kluwer Academic Publishers, Dordrecht, pp. 63-116.

[6] Koch, R.H., Rozen, J.R., Woltgens, P., Picunko, T., Goss, W.J., Gambrel, D., Lathrop, D., Wiegert, R. and Overway, D. (1996) High performance superconducting quantum interference device feedback electronics, *Rev. Sci. Instrum.* **67**, 2968-2976.

[7] Hilbert, C. and Clarke, J. (1985) DC SQUIDs as radiofrequency amplifiers, *J. Low Temp. Phys.* **61**, 263-280.

[8] Hilbert, C., Clarke, J., Sleator, T. and Hahn, E. L. (1985) Nuclear quadruple resonance detected at 30MHz with a dc superconducting quantum interference

device, *Appl. Phys. Lett.* **47**, 637-639. (See references therein for earlier work on NMR with SQUIDS).

[9] Bradley, R. F. (1999) Cryogenic, low-noise, balanced amplifiers for the 300-1200 MHz band using heterostructure field-effect transistors, *Nucl. Phys. B* (Proc. Suppl.) **72**, 137-144.

[10] Hagmann, C., Kinion, D., Stoeffl, W., van Bibber, K., Daw, E., Peng, H., Rosenberg, L.J., La Veigne, J., Sikivie, P., Sullivan, N. S., Tanner, D. B., Nezrick, F. Turner, M. S. . Moltz, D. M, Powell, J. and Golubev, N. A. (1998) Results from a high-sensitivity search for cosmic axions, *Phys. Rev. Lett.* **80**, 2043-2046.

[11] Mück, M. André, M-O., Clarke, J., Gail, J. and Heiden, C. (1998) Radio Frequency Amplifier Based on a Niobium dc Superconducting QUantum Interference Device with Microstrip Input Coupling, *Appl. Phys. Lett.* **72**, 2885-2887.

[12] André, M-O., Mück, M. Clarke, J., Gail, J. and Heiden, C. (1999) Microstrip DC SQUID Radio-frequency amplifier with tenth-kelvin noise temperature, *Appl. Phys. Lett.* **75**, 698-700.

[13] Mück, M. André, M-O., Clarke, J., Gail, J. and Heiden, C. (1999) The microstrip Superconducting QUantum Interference Device rf amplifier: tuning and cascading, *Appl. Phys. Lett.* **75**, 3545-3547.

[14] Stewart, W.C. (1968) Current-voltage characteristics of Josephson junctions, *Appl. Phys. Lett.* **12**, 277-280.

[15] McCumber, D.E. (1968) Effect of ac impedance on dc voltage-current characteristics of Josephson junctions, *J. Appl. Phys.* **39**, 3113-3118 .

[16] Tesche, C.D. and Clarke. J. (1977) DC SQUID: noise and optimization, *J. Low. Temp. Phys.* **27**, 301-331.

[17] Bruines, J.J.P., de Waal, V.J., and Mooij, J.E. (1982) Comment on dc SQUID noise and optimization, by Tesche and Clarke, *J. Low. Temp. Phys.* **46**, 383-386.

[18] De Waal, V.J., Schrijner, P., and Llurba, R. (1984) Simulation and optimization of a dc SQUID with finite capacitance, *J. Low. Temp. Phys.* **54**, 215-232.

[19] Ryhänen, T., Seppä, H., Ilmoniemi, R., and Knuutila, J. (1989) SQUID magnetometers for low-frequency applications, *J. Low Temp. Phys.* **76**, 287-386.

[20] Likharev, K.K. (1986) *Dynamics of Josephson Junctions and Circuits*, Gordon and Breach Science Publishers, New York.

[21] Koelle, D., Kleiner, R., Ludwig, F., Dantsker, E. and Clarke, J. (1999) High-Transition-Temperature Superconducting QUantum Interference Devices, *Rev. Mod. Phys.* **71**, 631-686.

[22] Ketchen, M.B., and Jaycox, J.M. (1982) Ultra-low noise tunnel junction dc SQUID with a tightly coupled planar input coil, *Appl. Phys. Lett.* **40**, 736-738.

[23] Jaycox, J.M. and Ketchen, M.B. (1981) Planar coupling scheme for ultra low noise dc SQUIDs, *IEEE Trans. Magn.*, **MAG-17**, 400-403.

[24] Cantor, R. (1996) DC SQUIDS: design, optimization and practical applications, in H. Weinstock (ed.), *SQUID Sensors: Fundamentals and Applications*, Kluwer Academic Publishers, Dordrecht, pp. 179-233.

[25] Xiaofan Meng and Robert McDermott, private communication.

[26] Gurvitch, M., Washington, M.A., and Huggins, H.A. (1983) High quality refactory Josephson tunnel junction utilizing thin aluminum layers, *Appl. Phys. Lett.* **42**, 472-474.

[27] De Waal, V.J., Klapwijk, T.M., and Van den Hamer, P. (1983) High performance dc SQUIDs with submicrometer niobium Josephson junctions, *J. Low. Temp. Phys.* **53**, 287-312 .

[28] Tesche, C.D., Brown, K.H., Callegari, A.C., Chen, M.M., Greiner, J.H., Jones, H.C., Ketchen, M.B., Kim, K.K., Kleinsasser, A.W., Notarys, H.A., Proto, G., Wang, R.H., and Yogi, T. (1985) Practical dc SQUIDs with extremely low l/f noise, *IEEE Trans. Magn.* **MAG-21**, 1032-1035.

[29] Pegrum, C.M., Hutson, D., Donaldson, G.B., and Tugwell, A. (1985) DC SQUIDs with planar input coils, *ibid*, 1036-1039.

[30] Noguchi, T., Ohkawa, N., and Hamanaka, K. (1985) Tunnel junction dc SQUID with a planar input coil, in H. D. Hahlbohm. and H. Lubbig (eds.), *SQUID 85 Superconducting Quantum Interference Devices and their Applications,* Walter de Gruyter, Berlin, pp. 761-766.

[31] Foglietti, V., Gallagher, W.J., Ketchen, M.B., Kleinsasser, A.W., Koch, R.H., and Sandstrom, R.L., (1989) Performance of dc SQUIDs with resistively shunted inductance, *Appl. Phys. Lett.* **55**, 1451-1453.

[32] Muhlfelder. B., Beall. J.A., Cromar, M.W., and Ono, R.H. (1986) Very low noise tightly coupled dc SQUID amplifiers, *Appl. Phys. Lett.* **49**, 1118-1120.

[33] Knuutila, J., Kajola, N., Seppä, H., Mutikainen, R., and Salmi. J. (1988) Design, optimization and construction of a dc SQUID with complete flux transformer circuits, *J. Low. Temp. Phys.* **71**, 369-392.

[34] Wellstood, F.C., Heiden. C., and Clarke. J. (1984) Integrated dc SQUID magnetometer with high slew rate, *Rev. Sci. Instrum.* **55**, 952-957.

[35] Tesche, C. D. and Clarke, J. (1979) DC SQUID: current noise, *J. Low Temp. Phys.* **37**, 397-403.

[36] Ramo, S, Whinnery, J.R. and van Duzer, T. (1965) *Fields and Waves in Communication Electronics*, J. Wiley & Sons, New York.

[37] M/A-COM MA 46H502, which has a hyperabrupt doping profile.

[38] PAMTECH UTE 1230-K

[39] Peccei, R. and Quinn, H. (1977) CP conservation in the presence of pseudoparticles, *Phys. Rev. Lett.* **38**, 1440-1443.

[40] Weinberg, S. (1978) A new light boson?, *Phys. Rev. Lett.* **40**, 223-226.

[41] Wilczek, F. (1978) Problem of strong P- and T-invariance in the presence of instantons, *Phys. Rev. Lett.* **40**, 279-282.

[42] Sikivie, P.(1983) Experimental tests of the 'invisible' axion, *Phys. Rev. Lett.* **51**, 1415-1417; (1985) Detection rates for 'invisible'-axion searches, *Phys. Rev. D* **32**, 2988-2991.

[43] Kim, J.E. (1979) Weak-interaction singlet and strong CP invariance, *Phys. Rev. Lett.* **43**, 103-107; Shifman, M.A., Vainshtein, A.I. and Zakharov, V.I. (1980) Can confinement ensure natural CP invariance of strong interactions?, *Nucl. Phys. B* **166**, 493-506.

[44] Dine, M., Fischler, W. and Srednicki, M., (1981) A simple solution to the strong CP problem with a harmless axion, *Phys. Lett. B* **104**, 199-202; Zhitnitsky, A. P.(1980) On the possible suppression of axion-hadron interactions, (1980) *Yad. Fiz.* **31**, 497 [*Sov. J. Nucl. Phys.* **31**, 260-263].

[45] Wellstood, F. C., Urbina, C. and Clarke J. (1994) Hot electron effects in metals, *Phys. Rev. B*, **49**, 5942-5955.

RADIO FREQUENCY SQUIDS AND THEIR APPLICATIONS

MICHAEL MÜCK
Institut für Angewandte Physik, Justus-Liebig-Universität Gießen, Heinrich-Buff-Ring 16, 35392 Gießen, Germany

BORIS CHESCA
Experimentalphysik II, Physikalisches Institut, Universität Tübingen, Auf der Morgenstelle 14, 72076 Tübingen, Germany

YI ZHANG
Institut für Schicht- und Ionentechnik, Forschungszentrum Jülich, 52425 Jülich, Germany

Abstract. Radio frequency superconducting quantum interference devices (rf SQUIDs) made of conventional superconductors had been very popular some fifteen to thirty years ago. Later on, they had been replaced by dc SQUIDs, which then offered higher sensitivity. However, the operation of rf SQUIDs at microwave bias frequencies has greatly improved their sensitivity, so that now rf SQUIDs can be fabricated which are comparable to the best dc SQUIDs in terms of field sensitivity, as well as energy resolution. The discovery of high temperature superconductors (HTS) has further increased interest in rf SQUIDs, mainly because the inductance of rf SQUIDs and thus their loop area, can be made much larger than that of dc SQUIDs before a significant reduction in signal voltage is observed. Today, rf SQUIDs are again used in numerous applications, such as biomagnetism, geophysical measurements and nondestructive evaluation.

1. Introduction

Since the time of their first demonstration in the 1960s, radio frequency superconducting quantum interference devices (rf SQUIDs) have been used as magnetic field sensors for applications requiring very high sensitivity. Later on, they were gradually replaced by dc SQUIDs, which then offered higher sensitivity. The discovery of High Temperature Superconductors (HTS) has again increased interest in rf SQUIDs. Most importantly, a sensitivity comparable to that of dc SQUIDs could immediately be achieved. Also, due to different geometry and readout, all rf SQUIDs offer interesting advantages when compared to dc SQUIDs.

The need for only one single weak link simplifies HTS SQUID fabrication. Additionally, all rf SQUIDs require no galvanic connection between SQUID and readout electronics. Due to the low impedance of a dc SQUID, a contact resistance of only a few ohms can significantly reduce its modulation voltage. As experiment and theory show, the

H. Weinstock and M. Nisenoff (eds.), Microwave Superconductivity, 505–540.

inductance of HTS rf SQUIDs, and thus their loop area, can be made much larger than that of dc SQUIDs before a significant reduction in signal voltage is observed. This leads to much higher field sensitivity of the bare SQUID, so that for a number of applications, a complex flux transformer is not needed. The absence of Josephson oscillations eliminates resonances in an integrated input coil, which, in dc SQUIDs can lead to a reduction in signal voltage and in sensitivity. When operated in a flux-locked loop, usually no 1/f noise is observed in rf SQUIDs down to frequencies below 1 Hz.

RF SQUIDs are often said to exhibit only mediocre sensitivity, which is true only when too low a bias frequency is used. When operating an rf SQUID at microwave frequencies, its intrinsic flux noise is of the same order as that of dc SQUIDs. In an unshielded environment the flux noise of all SQUIDs may increase drastically. This can be caused either by hopping of vortices trapped in the SQUID body, or by electromagnetic interference. Such effects can make the meaning of 'SQUID sensitivity' relative. Ultimately, therefore, the difference in performance between optimized dc and rf SQUID systems seems to be rather small.

In this paper, we shall review the basic operational principle of rf SQUIDs, as well as recent developments in rf SQUID theory and rf SQUID systems. Since the subject of this book is microwave superconductivity, we shall focus on the high-frequency aspects of rf SQUIDs and their readout electronics. In Sec. 2, we begin with a summary of the principles of the rf SQUID and of the effects of thermal fluctuations on rf SQUID operation. In Secs. 3 and 4, we describe the tank circuits and the electronics necessary to operate and read-out rf SQUIDs. Sec. 5 briefly reviews recent developments of rf SQUID systems and their applications.

2. Principles of rf SQUID Operation

2.1 THE BASICS OF OPERATION

Soon after the first demonstration of quantum interference devices, a consistent theory of their operation had been developed which was able to explain experimental results for low temperature superconductor (LTS) SQUIDs available at that time. A complete summary of this theory can be found in the excellent review articles by Silver and Zimmerman [2.1], Giffard, Webb and Wheatley [2.2], Jackel and Buhrman [2.3], Clarke [2.4] and in books by Likharev [2.5] and Gallop [2.6]. Recent work on rf SQUIDs has shown that in some cases this standard theoretical model needs further extension. Effects not included in early theories of the rf SQUID are low-frequency excess noise in rf SQUIDs and the influence of electromagnetic interference on rf SQUID operation. Also, the effect of SQUID inductance on the intrinsic noise of the SQUID recently has been investigated again, mainly because work on HTS rf SQUIDs has shown that these devices still function with much higher inductance than that for dc SQUIDs. We briefly summarize the most important results of this theory, but refer the reader to the above-mentioned papers for a more detailed review.

We limit the following description to the case for which the Josephson junction obeys the sinusoidal current-phase relationship and can be described by the resistively-shunted-junction (RSJ) model. We also assume that only small thermal fluctuations are present. The basic priciples are still valid, however, even for the case of a weak link with non-sinusoidal current-phase relation. The rf SQUID consists of a superconducting ring of

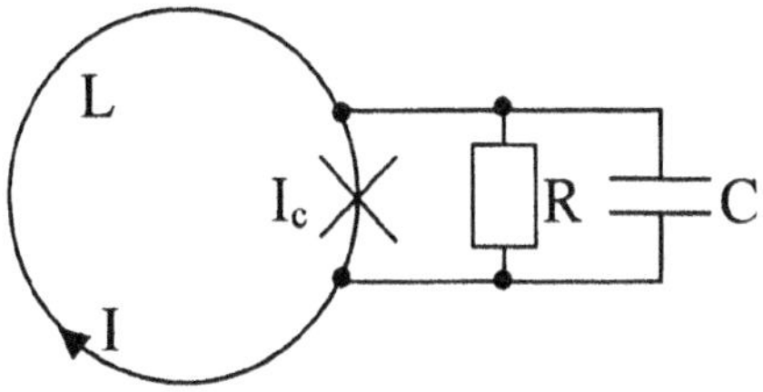

Fig. 1. Schematic diagram of an rf SQUID.

inductance L containing a single Josephson junction or weak link having a critical current I_C, a resistance R and capacitance C, as shown in Fig. 1. The resistance R is usually chosen to have a Stewart-McCumber damping parameter $ß_C = 2\pi I_C R^2 C/\Phi_0 < 1$, where $\Phi_0 = 2.0679 \cdot 10^{-15}$ Vs. This insures that the current-voltage (I-V) characteristic of the weak link is non-hysteretic. In that case, effects of C on the SQUID response can be neglected.

Flux quantisation requires that the phase difference $\Delta\varphi$ across the weak link is

$$\Delta\varphi = -2\pi\Phi/\Phi o \,. \tag{1}$$

The supercurrent I_S flowing through the weak link is connected to $\Delta\varphi$ by:

$$I_S = I_C \sin(\Delta\varphi) = I_C \sin(-2\pi\Phi/\Phi o) \,. \tag{2}$$

The total flux threading the SQUID ring is related to an external flux Φ_X by:

$$\Phi = \Phi_X + LI, \tag{3}$$

where I is the total ring current. In the limit where Φ_X is quasi-static,

$$\Phi = \Phi_X - LI_C \sin(-2\pi\Phi/\Phi o). \tag{4}$$

For the case of $ß_L = 2\pi LI_C/\Phi_0 < 1$, the $\Phi(\Phi_X)$ curve is monotonic and single valued — see Fig. 2. The weak link introduces an additional inductance into the superconducting ring:

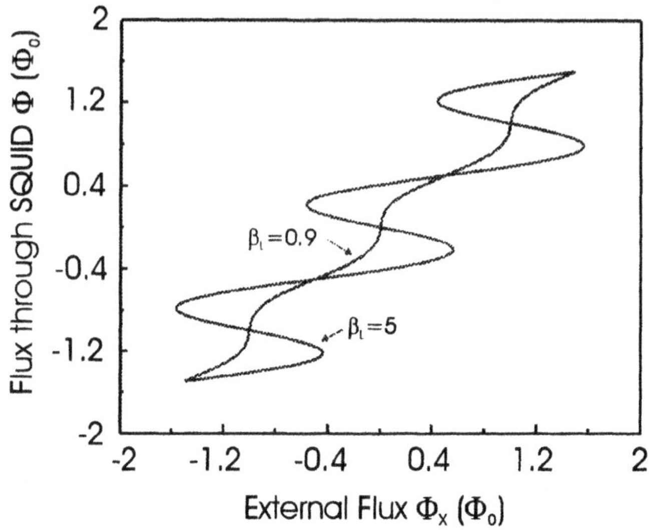

Fig. 2. Flux Φ threading the SQUID ring as a function of external flux Φ_X for $\beta_L = 0.9$ and 5.

$$L_J = \frac{\Phi_0}{2\pi I} \cdot arcsin\left(\frac{I}{I_C}\right) \tag{5}$$

This inductance undergoes a nonlinear periodic change with varying supercurrent — and thus flux — between $\Phi_0/2\pi I_c$ and $\Phi_0/4\pi I_c$. Measuring the flux-dependent inductance change can be accomplished by measuring the resonant frequency variation of a loosely coupled LC circuit [2.7]. This mode of operation of the rf SQUID is known as the "dispersive" or "inductive" mode.

For the case of $\beta_L > 1$, Eq. (4) describes a snake-like reentrant curve in the $\Phi(\Phi_X)$ plane — see Fig. 2. Kurkijärvi [2.8] has shown that segments with positive slope correspond to local minima of a potential

$$U(\Phi) = \int \{\frac{\dot{\Phi}}{R} + C\ddot{\Phi}\} d\Phi, \tag{6}$$

while those with negative slopes represent local potential maxima. Metastable states exist only on positive-slope segments and are separated from each other by potential barriers. With increasing flux, a SQUID will follow a segment with positive slope until, at a value $\Phi_C = n\Phi_0 + Li_C$, the supercurrent in the SQUID reaches the critical current of the weak link. The weak link momentarily switches to the resistive state and the SQUID undergoes a quantum transistion from the k = n to the k = n+1 state, i.e., the flux through the

SQUID changes by one flux quantum. For a system in which $\beta_C>1$, it may jump through several flux states before stabilizing. The time for this transition is given by the L/R time constant of the SQUID ring or by the quasiparticle relaxation time of the weak link $\tau = h/2eI_CR$ [2.9], depending on which is larger; h is Planck's constant. In the same way, as Φ is decreased, quantum transitions take place at $\Phi = n\Phi_0 - LI_C$. As the system undergoes quantum transitions from one state to another and back, a hysteresis loop is traversed along the $\Phi(\Phi_X)$ curve. The energy ΔE dissipated in traversing a hystersis loop is roughly $\Delta E \approx \Phi_0 I_C = LI_C^2$. For typical SQUIDs, ΔE is on the order of 10^{-21} J. Since hysteresis loops provide the flux dependent signal, this mode of operation is called the hysteretic mode.

In order to extract information about how much flux threads the SQUID, it can be

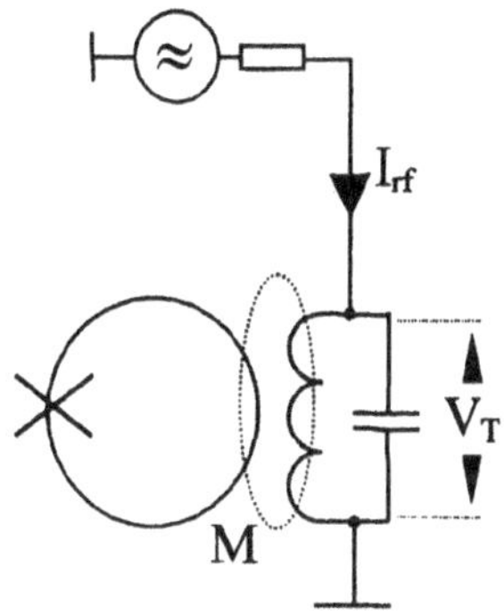

Fig. 3. rf SQUID inductively coupled to a tank circuit.

coupled inductively to a parallel tuned (resonant) circuit. Zimmerman and Silver [2.10] have investigated rf SQUIDs with a direct-coupled weak link, a method which is seldom used today. The tuned circuit, often called a tank circuit, is excited at its resonant frequency by an rf current with amplitude I_{rf} and the corresponding rf voltage across the tank circuit V_T is amplified and amplitude detected — see Fig. 3. If the SQUID is operated in the dispersive mode ($\beta_L < 1$), the flux-dependent change in SQUID inductance leads to a change in the tank circuit resonant frequency $\Delta\omega$, and thus to a change in the rf voltage V_T across the tank circuit. The rf angular frequency ω_{rf} is chosen such that it differs slightly from the resonant frequency of the tank circuit ω_0 in order to obtain maximum $\partial V_T/\partial\omega$. If $\omega_{rf} = \omega_0$, positive and negative flux changes lead to a reduction in V_T. If the flux changes periodically by, say $\Phi_0/2$, the case for which $\omega_{rf} = \omega_0$ leads to a period doubling, which can conveniently be used to identify operation in the dispersive mode.

In the hysteretic mode, provided that the supercurrent I induced by the rf flux $\Phi_{rf} = MQI_{rf}$ does not exceed the critical current I_C, V_T is a linear function of I_{rf}. Here, M denotes the mutual inductance between SQUID and tank circuit, and Q is its quality factor. As soon as the current in the SQUID ring exceeds I_c, the Josephson junction switches to the resistive state. The current in the ring then decays exponentially with a time constant L/R, until the Josephson junction switches back to the superconducting state. At higher I_{rf}, dissipation in the SQUID limits the tank circuit voltage V_T to a constant value, until at the positive and negative peak of every rf cycle, the current I in the SQUID exceeds I_c. A further increase in I_{rf} increases V_T linearly again up to a point, where during an rf cycle, a third quantum transition takes place — I_c is exceeded twice during the positive half of the rf cycle — which again limits V_T. The $V_T(I_{rf})$ characteristic thus looks like a staircase — see Fig. 4.

An applied external flux leads to an additional supercurrent in the SQUID, which adds to I_{rf} and leads to a shift towards lower V_T values for the step in the $V_T(I_{rf})$ characteristic — see Fig. 4. If the SQUID is biased at a constant I_{rf}, V_T is periodically changed by varying flux, with maximum values at $\Phi = n\Phi_0$ and minima at $\Phi=(2n-1)\Phi_0$. The flux-induced maximum voltage change, and thus the modulation amplitude, is [2.1]:

$$\Delta V_T = V_T(n\Phi_0) - V_T((2n-1)\Phi_0/2) = \omega\, L_T\, \Phi_0/2M\,, \qquad (7)$$

with L_T the inductance of the tank circuit coil, and ω the excitation frequency and resonant frequency of the tank circuit. It can be shown [2.11] that the energy transfer between SQUID and tank circuit — and thus the modulation voltage — is maximum when

$$k^2Q = \pi/4, \qquad (8)$$

where k is the coupling between SQUID and tank circuit.

The above model assumes that the times associated with the various superconducting processes in the SQUID are short compared to the period of an rf cycle. At high bias frequencies, this no longer is the case. Both the L/R time constant of the SQUID ring, as

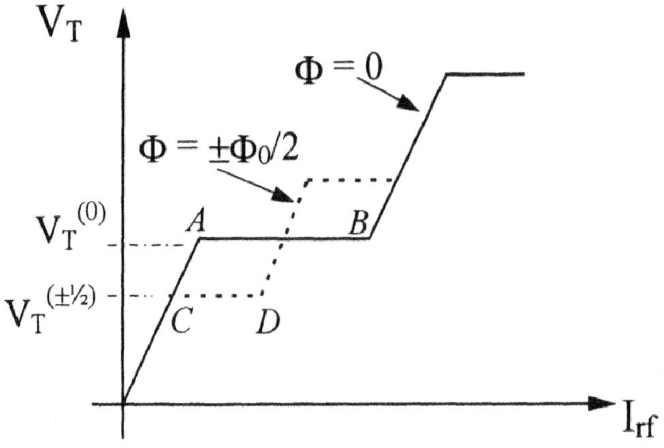

Fig. 4 Peak rf voltage across tank circuit V_T vs. rf current I_{rf} in the absence of thermal noise for flux $\Phi = \Phi_0$ (solid line) and $\Phi = \Phi_0/2$ (dashed line).

well as the quasiparticle relaxation time, can be on the order of the rf cycle period, if the SQUID is operated at microwave frequencies. Simulations carried out by Falco and Parker [2.12] and Jackel and Buhrman [2.13] have shown that even at bias frequencies above R/L, SQUID modulation can be observed. To obtain the highest possible modulation voltage, k^2Q should, however, be much larger than one.

Mercereau [2.14] introduced a SQUID model which explicitly contains a characteristic time associated with superconductive processes in the SQUID. He has taken into account the development of voltage pulses across the weak link during the quantum transition and the influence of the pulses on the modulation amplitude. He could show that the usual Bessel-type behaviour of the modulation amplitude with increasing rf current should no longer be observed at very high bias frequencies. Instead, a sine modulation of the Bessel behaviour is expected.

The first experimental evidence for such a behaviour was obtained by Rachford *et al.* [2.15] on rf SQUIDs operated at 9 GHz. This behaviour also could be seen on a few of our microwave SQUIDs [2.16].

For practical application to magnetometry, the equivalent flux noise of a SQUID system and thus its sensitivity to magnetic flux is the most important parameter. For the hysteretic SQUID, a theory of intrinsic noise had been developed in the early 1970s by Kurkijärvi [2.8, 2.17], and by Kurkijärvi and Webb [2.18]. Predictions of this theory have been proven experimentally by Jackel and Buhrman [2.3] among others.

Intrinsic noise in the rf SQUID originates from current fluctuations in the weak-link shunt resistor R. The ratio between SQUID response to the flux to be measured and to current fluctuations, determines its maximum achieveable sensitivity. The current fluctuations lead to a non-zero probability of making quantum transitions for flux smaller than Φ_C. If the external flux is increased slowly, a quantum transition will take place before the height of the barrier between two adjacent quantum states is reduced to below k_BT. This problem is similar to the case of a classical particle, with a mean energy k_BT, which tries to overcome a barrier with a height large compared to k_BT. The attempt frequency here is on the order of the SQUID ring frequency R/L. With the width of the distribution [2.17]

$$\Delta\Phi_C \propto LI_C(k_BT/\Phi_0 I_C)^{2/3}, \tag{9}$$

the equivalent flux noise becomes

$$S_\Phi^{1/2} \propto (1/\omega)^{1/2}\, LI_C^{1/3}\, (k_BT/\Phi_0)^{2/3}. \tag{10}$$

Jackel and Buhrman have shown that intrinsic noise leads to an increase in the tank circuit voltage with increasing bias current, and thus in a tilting of the steps in the $V_T(I_{rf})$ characteristic of the tank circuit, which, in the absence of noise, are horizontal. The tilt angle α is related to the SQUID noise by [2.3]:

$$S_\Phi^{1/2} = (2/\omega)^{1/2} \cdot \alpha\Phi_0/(1-\alpha). \tag{11}$$

Additionally, the step tilting reduces the modulation voltage by a factor of $(1-\alpha)$.

There are two additional noise sources in rf SQUID systems, which, in most cases, make a larger contribution to system noise than the intrinsic SQUID noise: the tank-circuit noise and the preamplifier noise. The tank circuit resistance R_T imposes a thermal noise current through the tank circuit inductance L_T. This current causes the bias point of the SQUID to fluctuate, which in turn leads to a fluctuation of V_T, since $\partial V_T/\partial I_{rf}$ is non-zero ($\partial V_T/\partial I_{rf} \approx \alpha$). Jackel and Buhrman [2.3] have calculated this noise contribution to be:

$$S_\Phi^{(tc)} = 4\pi\alpha^2/(1-\alpha^2) \cdot k_B T L/\omega. \tag{12}$$

Since the SQUID incremental modulation voltage (also called the transfer function) $\partial V_T/\partial\Phi$ is relatively small, e.g., 10 $\mu V/\Phi_0$ for a 30-MHz rf SQUID, thermal noise in the preamplifier usually limits SQUID system noise. The equivalent flux noise of the preamplifier is:

$$S_\Phi^{1/2\ (pa)} = S_V^{1/2\ (pa)} \cdot \partial\Phi/\partial V_T = S_V^{1/2\ (pa)} \cdot M\,[\omega L_T (1-\alpha)]^{-1}. \tag{13}$$

For conventional room temperature preamplifiers, $S_V^{1/2\ (pa)}$ is on the order of 2nV/√Hz for high (≈ 10 kΩ) and 0.5 nV/√Hz for low (50 Ω) input-impedance amplifiers. The energy resolution of an rf SQUID thus becomes [2.3, 2.19]:

$$\varepsilon \approx \frac{1}{\omega_{rf}}\left(\frac{\pi\alpha^2 S_\Phi}{2L} + 2\pi\alpha k_B T_a^{eff}\right), \tag{14}$$

where T_a^{eff} is the effective noise temperature of the preamplifier (noise temperature of the preamplifier plus losses in the transmission line between SQUID and preamplifier). One notices that the energy resolution is directly proportional to $1/\omega_{rf}$ and $1/L$. The sensitivity of SQUIDs can thus be enhanced by operating them at very high frequencies, and by choosing a high inductance. Since, however, the flux noise increases as the inductance increases, L cannot be made arbitrarily large. This effect becomes especially important at high temperatures and will be discussed in detail in the next section.

2.2 EFFECTS OF THERMAL FLUCTUATIONS ON RF SQUID OPERATION

Thermal fluctuations not only reduce the sensitivity of SQUIDs, but also place a lower limit on the critical current of the Josephson junction and an upper limit on the SQUID inductance. The lowest critical current allowed for the Josephson junction is given by the constraint that the Josephson coupling energy $I_c\Phi_0/2\pi$ [2.20] be larger than the thermal energy k_BT. Thermal activation reduces the observed critical current, and therefore there is a minimum value for I_c for which the two sides of a Josephson junction remain coupled. Computer simulations [2.21] have shown that this condition can be written as $I_c\Phi_0/2\pi > 5\ k_BT$. For T = 4.2 K, I_c should be larger than 0.8 μA, and for T = 77 K, $I_c > 15\mu A$. A similar criterion is valid for the SQUID inductance, because noise currents generated by the shunt of the Josephson junction produce flux noise $\Phi_N = (Lk_BT)^{1/2}$ [2.8, 2.17] which must be smaller than about $\Phi_0/4$ in order to still maintain SQUID operation. For T = 4.2 K the maximum usable inductance therefore is about $L < \Phi_0^2/(16k_BT) \approx 4$ nH, and for T = 77 K an inductance $L \approx 200$ pH results. It is interesting to note that for the optimum condition $\beta_L \approx 1$, the critical current required for the maxumum allowable SQUID inductance (4 nH for T = 4.2 K) is just the minimum value for the Josephson coupling energy. Experiments by Falco and Parker [2.12] and later on by Zheng *et al.* [2.31] have shown, however, that even for lower critical currents and larger SQUID inductances than given above, rf SQUID operation could be observed. Their observations formed the basis for the development of a new analytical theory [2.27] which can describe rf SQUID operation also in the presence of large thermal fluctuations.

We now turn to a more detailed description of thermal fluctuations in rf SQUIDs operated at higher temperatures (T > 10 K). When thermal fluctuations dominate (which is the case at T = 77 K), the processes that take place in the interferometer should be considered to be of a Markoffian type, and for calculations of important SQUID parameters, the Fokker-Planck equation should be applied. As the general theory of Brownian motion shows [2.22], this equation is equivalent to the Langevin equation which describes the SQUID dynamics in the framework of the RSJ model. Such an approach has been used for an isolated Josephson junction — see [2.23-2.25] —, and for an isolated SQUID ring — see, e.g., [2.5] chapter 6. The first attempt to include the interaction of the rf SQUID with the tank circuit into the Markoffian-type formalism had already been made some time ago [2.26]. In that work, however, the Fokker-Planck equation had not been solved because of its complexity. Nevertheless, the equation was written exactly for the case of a real SQUID (i.e., a SQUID inductively coupled to a tank circuit) with $\beta_L \geq 1$. It was assumed that the product of junction capacitance times resistance was small but not negligible. Recently [2.27], although in a different manner, this problem has been reconsidered for $\beta_L < 3$ and successfully solved for the case in which the CR product is negligibly small. In practice, this is almost always the case, because for proper HTS SQUID operation, it is necessary that the Josephson junction be overdamped. Hence, the Fokker-Planck equation is reduced to the Smoluchowski equation — for details see [2.27]:

$$\frac{1}{\beta_L \omega_c}\frac{\partial \sigma}{\partial t} = \frac{\partial}{\partial \phi}\left[\frac{1}{U_0}\frac{\partial U}{\partial \phi}\sigma\right] + \Gamma\frac{\partial^2 \sigma}{\partial \phi^2}, \qquad (15)$$

where U/U_0 is the normalized rf SQUID potential, $\phi = 2\pi\Phi/\Phi_0$ is the normalized flux in the ring, $\Gamma = 2\pi k_B T/(\Phi_0 I_c)$ is the noise parameter, $\sigma(\phi,t)$ is the probability density of finding the interferometer loop in a state ϕ at time t, and ω_c = R/L is the cut-off frequency.

For the case of $\beta_L \ll 1$, an exact solution for the non-stationary Smoluchowski equation (15) has been found [2.27]. One of the parameters which determines $\sigma(\phi,t)$ is the so-called fluctuation-threshold inductance $L_F = [\Phi_0/(2\pi)]^2/(k_BT)$; at 77 K, $L_F \cong 100$ pH. After the probability density distribution $\sigma(\phi,t)$ has been found, the statistical average of any variable can be calculated easily. Of particular importance is the expression of the screening current flowing in the SQUID inductance L:

$$\overline{I_c \sin\phi + \frac{\Phi_0}{2\pi R}\frac{d\phi}{dt}} = I_c \int_{-\infty}^{\infty} \sigma \sin\phi \, d\phi + \frac{\Phi_0}{2\pi R}\frac{d}{dt}\left(\int_{-\infty}^{\infty} \sigma\phi \, d\phi\right) \tag{16}$$

This expression basically differs from the one that corresponds to 4 K. In particular, the screening supercurrent is suppressed by a factor of $\exp[-L/(2L_F)]$ at higher temperatures. When analyzing the interaction between the SQUID ring and the tank circuit, the screening current through L is of prime importance, because the coupling is inductive via induced currents. As a consequence, the inductive interaction is directly affected and new equations for the output signal, transfer function, flux noise spectral density, and energy sensitivity have to be derived for T = 77 K. We shall give only the result for the energy sensitivity in units of $k_B T L_F / R \cong (ohm/R)\times 10^{-31} J/Hz$ — for details and all the other characteristics see [2.27]:

$$\varepsilon = 3\Gamma^2\left[\left(1+\omega_R^2\right)\left(1+\frac{T_k}{T}\frac{1}{k^2 Q}\frac{1}{\omega_R}\right)\right]\frac{\exp(L/L_F)}{L/L_F}. \tag{17}$$

Here, k is the coupling between the tank circuit and the ring, Q is the tank circuit loaded quality factor, and T_k is the tank-circuit temperature. Thus, in contrast to the case for small thermal fluctuations (T ≈ 4 K), the energy resolution exhibits a minimum with respect to two parameters: the bias frequency and the SQUID inductance L.

Equation (17) is valid under the assumption that the hysteresis parameter is much smaller than unity — strictly speaking it is accurate for β_L values up to 0.3. Next, the case of the hysteresis parameter β_L in the range of practical interest, namely, $0.3 \le \beta_L < 3$, shall be considered in the approximation that the operating frequency is much less than the SQUID cut-off frequency $\omega_R \ll 1$ — the adiabatic mode of operation. In contrast to the case of small thermal fluctuations (T ≤ 4 K) for which rf SQUIDs with $\beta_L < 1$ remain non-hysteretic, all rf SQUIDs having $\beta_L < 3$ remain non-hysteretic in the presence of large thermal fluctuations. Proceeding as before, one finds the correction to Eq. (17) for the energy sensitivity ε in units of $k_B T L_F / R \cong (ohm/R)\times 10^{-31} J/Hz$:

$$\varepsilon = 3\Gamma^2\left(1+\frac{T_k}{T}\frac{1}{k^2 Q}\frac{1}{\omega_R}\right)\frac{\exp(L/L_F)}{L/L_F}\frac{1}{g^2}, \tag{18}$$

with $$g = \exp\left\{ -\frac{\beta/2}{\exp[3.5(\Gamma - 0.5)] + \beta} \right\}.$$

From Fig. 5, where Eq. (18) is represented, one finds the optimal value for the SQUID inductance: $L = L^*$ if $\Gamma \leq 0.335$; and $L = L_F = 100$ pH if $\Gamma > 0.335$, where L^* is a monotonically increasing function of the noise parameter and takes values from $L^* / L_F = 0.55$ for $\Gamma = 0.3$ to $L^* / L_F = 0.9$ for $\Gamma = 0.335$.

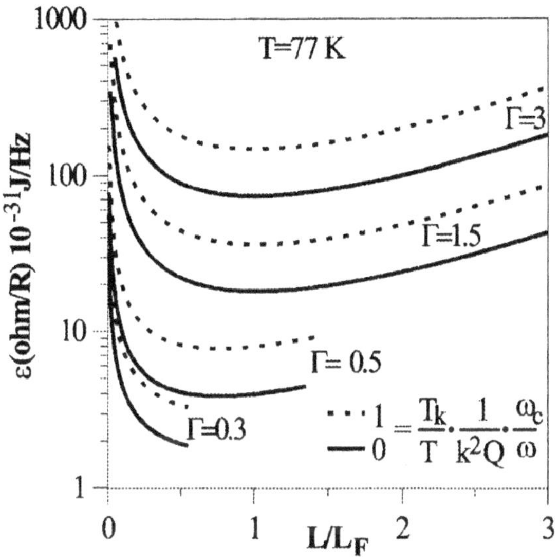

Fig. 5. RF SQUID energy sensitivity for several values of the noise parameter Γ as a function of the normalized inductance. Solid lines are for $(T_k / T)/(k^2 Q \omega_R) << 1$.

This theory provides an explanation of the well-known experimental fact that, in contrast to dc SQUIDs, rf SQUIDs preserve high energy resolution even at higher values for the inductance (up to about 300 pH) and the noise parameter Γ (up to about 2) — see [2.28-2.30].

In Table I, the calculated (using this theory) and experimentally-measured values of the energy sensitivity of some rf SQUIDs are shown — for further details on these HTS rf SQUIDs see Ref. [2.28]. All magnitudes marked with the superscript * were not directly measured, but indirectly estimated, since in practice it is difficult to measure the SQUID critical current in the presence of large thermal fluctuations. The first column for the theory estimates is for the ideal case when in Eq.(18) the term $(T / T_k)/(k^2 Q \omega_R) << 1$. In this case, for a given inductance, the optimum sensitivity is reached. The theoretical estimates in the second column of Table 1 are in good agreement with the measurements.

We note that the numbers given in the second column of Table 1 are for practical, non-zero values of the term $(T/T_k)/\left(k^2 Q\omega_R\right)$.

TABLE I. Calculated and experimental values of the energy sensitivity for five different rf SQUIDs.

	Josephson junction parameters			Tank circuit parameters			Derived SQUID parameters			Energy sensitivity (thermal noise only) $\varepsilon \times 10^{-31}$ (J/Hz)		
#	I_c (μA)	L (pH)	R^* (Ω)	k^*	Q^* loaded	ω/(2π) (MHz)	β^*	Γ^*	ω_R^*	Theory		Exper.
1	1.05	150	8	0.15	500	850	0.5	3	0.15	10.08	16	14.24
2	1.725	230	8	0.15	500	850	1.25	1.84	0.1	5.5	10.0	6.4
3	1.83	260	8	0.15	500	850	1.5	1.73	0.17	5.8	8.8	5.56
4	1.05	450	6	—	—	300	1.5	5	0.15	110	—	233
5	0.787	900	6	—	—	300	2.25	4	0.28	7202	—	7617

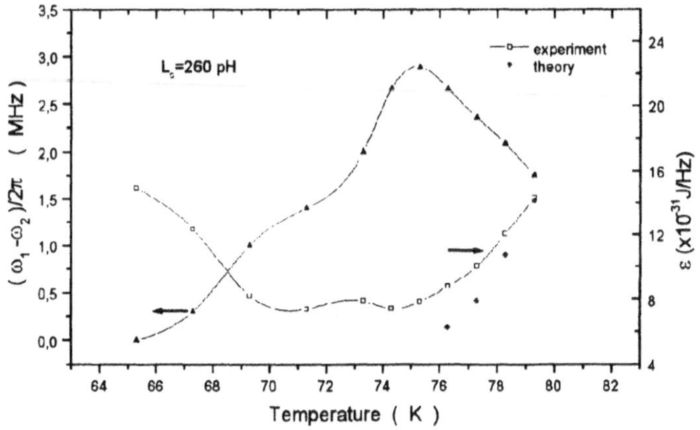

Fig. 6. Temperature dependence of resonance-frequency modulation by external flux and of the energy resolution. Lines are drawn to guide the eye.

In another study, Zheng et al. measured the amplitude-frequency characteristics (AFCs) of rf SQUIDs over a temperature range between 65 K and 79 K. Figure 6 shows the dependence of one SQUID's resonant frequency modulation (frequency difference at $n\Phi_0$ and $(n+1/2)\Phi_0$). It proves that in the temperature range investigated, that particular SQUID operated in the nonhysteretic mode. From such data, using theoretical expressions derived from the rf SQUID theory valid in the limit of high fluctuations [2.27], the basic SQUID parameters k^2, β_L and R could be determined. Table II shows the calculated parameters at four different temperatures in the range of validity, where $L/L_F > 1$ and $\Gamma >$

1. The values in the brackets were determined directly from measurements of I_c and R after destructively opening the SQUID loop. One sees that the agreement with values calculated from the AFC data is rather good. However, this good agreement may be somewhat fortuitous, because thermal fluctuations make the measurement of very low I_c values difficult. By making a linear extrapolation one can estimate that this SQUID remained nonhysteretic up to $\beta_L \approx 3.6$, which is consistent with the theoretical prediction.

Figure 6 also shows the temperature dependence of energy resolution. The open squares are experimental data points, after subtracting noise from the readout electronics.

TABLE II: Comparison of rf SQUID parameters calculated from AFC data [2.31], with direct measurement of I_c and R after opening the SQUID loop (data in brackets).

T (K)	k^2 ($\times 10^{-3}$)	β_L	I_c (μA)	R (Ω)	Γ
75.3		(2.19)	(2.8)	(10.5)	(1.2)
76.3	8.0	1.86	2.4	10.7	1.4
77.3	8.14	1.63 (1.84)	2.1 (2.3)	11.86 (10.8)	1.6 (1.4)
78.3	7.46	1.44	1.8	11.7	1.8
79.3	7.92	1.24 (1.37)	1.6 (1.7)	12.1 (10.9)	2.1 (1.9)

The solid circles are the theoretical values calculated from Eq. (18) using the data in Table II. One can see that experiment and theory show good agreement. Clearly, the optimum energy resolution is obtained in the dispersive regime, but for $\beta_L > 1$.

2.3. LOW FREQUENCY EXCESS NOISE IN RF SQUIDS

Certain applications of SQUIDs require high sensitivity at relatively low frequencies, often at 1 Hz or lower. However, excess noise, typically with a spectral density scaling inversely as the frequency f, can substantially reduce the sensitivity of SQUIDs at low frequencies. Although this 1/f noise has been known for many years in LTS SQUIDs, it has been particularly severe in HTS devices. In the case of the dc SQUID, the origins of 1/f noise in both LTS and HTS devices have been extensively studied, and at least two sources have been identified [2.32]. The first arises from fluctuations in the critical current I_C generated by the trapping of electrons [2.33] or by atomic reconfigurations in the barrier [2.34]; contributions from this source can be substantially reduced in the flux-modulated and flux-locked dc SQUID by one of several "double modulation" schemes — see, e.g., [2.32]. The second source of 1/f noise involves thermally-activated hopping of flux vortices between pinning sites in the body of the SQUID [2.35], and its magnitude depends strongly on the quality of the superconducting material. Noise from this source cannot be reduced by any bias modulation scheme.

Similar investigations had not been carried out for the case of rf SQUIDs until recently [2.36]. This may have been because the white noise of rf SQUIDs in most cases has been at a relatively high value due to preamplifier noise, so that the onset of 1/f noise occurs at much lower frequencies than for dc SQUIDs. Alternatively, it may be that rf SQUIDs are almost invariably operated in a flux-locked loop in which, as we shall see, the combination of rf bias and flux modulation eliminates 1/f noise due to critical-current fluctuations. This fact was recognized long ago by Giffard [2.37], who gave a correct, albeit concise, discussion of the theory.

2.3.1. Reduction of 1/f noise due to critical current fluctuations in the hysteretic mode

We consider the effect on tank circuit voltage V_T of a fluctuation δI_C in I_C at a frequency f $\ll \omega_{rf}$ for a SQUID operated in the hysteretic mode. Fluctuations in the SQUID critical current I_C lead to a fluctuation of the critical flux Φ_C, for which a quantum transition takes place. We note that due to white noise in the SQUID, Φ_C fluctuates in any case. However, fluctuations due to 1/f noise are slow compared to the frequency of the quantum transitions. An increase in I_C increases the value of Φ_C to $L(I_C + \delta I_C)$, and thus the transitions from the k = 0 state to the +1 and -1 states occur at the flux values $L(I_C + \delta I_C)$ and $-L(I_C + \delta I_C)$, respectively. As a result, V_T ($\Phi = 0$) is increased to

$$\tilde{V}_T(\Phi = 0) = \omega_{rf} L_T (\Phi_C + L\delta I_C) / M. \qquad (20)$$

We can see that 1/f noise in the critical current results in a 1/f noise component in the demodulated rf voltage.

However, when the SQUID is flux modulated and operated in a flux-locked loop, the effect of critical-current fluctuations is reduced. Consider first the effect of an applied quasistatic flux $\delta\Phi < \Phi_0/2$. The asymmetry introduced into the hysteresis loops causes the SQUID to make its transition from the k = 0 to the k = 1 state at a lower rf flux, reducing the voltage across the tank circuit to

$$V_T^{(\delta)} = \omega_{rf} L_T (\Phi_C - \delta\Phi) / M. \qquad (0 < \delta\Phi < \Phi_0/2) \qquad (21)$$

In the region *CD* of Fig. 4, the SQUID traverses only the k = +1 hysteresis loop. Similarly, if we now change the flux to $-\delta\Phi$ ($\delta\Phi < \Phi_0/2$), the voltage is

$$V_T^{(-\delta)} = \omega_{rf} L_T \left|(-\Phi_C + \delta\Phi)\right| / M. \quad (-\Phi_0/2 < \delta\Phi < 0). \qquad (22)$$

The SQUID now traverses only the k = -1 hysteresis loop; the modulus sign in Eq. (22) reflects the fact that the detection of the peak value of the rf voltage is insensitive to whether the transition occurs on a positive or negative peak of the rf current. Suppose now that I_C undergoes a slow fluctuation to a new value $I_C + \delta I_C$. The peak voltage across the tank circuit changes to

$$\tilde{V}_T^{(\delta)} = \omega_{rf} L_T (\Phi_C + L\delta I_C - \delta\Phi) / M \qquad (0 < \delta\Phi < \Phi_0/2) \qquad (23)$$

and

$$\tilde{V}_T^{(-\delta)} = \omega_{rf} L_T \left| -(\Phi_C + L\delta I_C) + \delta\Phi \right| / M. \qquad (-\Phi_0/2 < \delta\Phi < 0) \qquad (24)$$

We see that the effect of a positive fluctuation δI_C is to increase V_T uniformly for all values of applied flux. The effect on the demodulated voltage vs. flux is as follows: the demodulated output consists of a "rectified" sine wave with no Fourier components at a frequency ω_m, and a peak value $\omega_{rf}L_T\Phi_0/4M$ for critical currents of I_C and $I_C + \delta I_C$. When this signal is mixed with the fundamental frequency ω_m, the output of the mixer is zero, regardless of the voltage offset of the signal produced by the critical current fluctuations at frequencies $f << \omega_m$. In this way, the effect of low-frequency fluctuations in the critical current on the output of the flux-locked loop is substantially reduced. We note that any slow fluctuation of the rf driving current will be suppressed similarly.

All of the above discussion has neglected the effects of thermal noise, that is, white noise that adds a random component to the value of Φ_C at which each transition between quantum states occurs. The effect of this noise is to tilt the plateaus in Fig. 4, but it does not affect the mechanism for 1/f noise reduction. Furthermore, we have neglected the fluctuations in the resistance of the junction that invariably accompany fluctuations in critical current. Resistance fluctuations will, in turn, produce fluctuations in the quality factor Q of the tank circuit. Fortunately, the magnitude of V_T does not depend on Q, — at least for $k^2Q \approx 1$ — so that resistance fluctuations, to first order, will not give rise to noise at the output of the flux-locked loop.

2.3.2 1/f noise in SQUIDs operated in dispersive mode

We extend the discussion to the dispersive mode ($\beta_L < 1$) as well: a calculation [2.7] of the tank-circuit voltage V_T shows that in this case V_T also is proportional to δI_C. Here, however, the transfer function $\partial V_T/\partial\Phi$ is nonlinear and dependent on I_C. This not only leads to fluctuation of the static component of V_T, but also to amplitude fluctuation of the oscillating component generated by flux modulation. Since —an ideal — feedback loop is sensitive only to the phase of the flux modulation, amplitude fluctuations will not change the output voltage at the phase-sensitive detector. Noise produced by critical-current fluctuations will thus be suppressed as in the hysteretic mode.

2.4 RESPONSE OF RF SQUIDS TO HIGH FREQUENCY ELECTROMAGNETIC INTERFERENCE

It has long been known that SQUIDs are very sensitive to high frequency electromagnetic interference (EMI). EMI can severely degrade the performance of SQUID systems which are operated in an unshielded environment. Normal-conducting shields, placed around each individual SQUID, substantially reduce EMI, but at the expense of increased thermal magnetic noise [2.38], reduced bandwidth and a frequency-dependent phase shift due to eddy currents in the shield.

Recently, Ishikawa et al. [2.39] and Koch et al. [2.40] investigated the effects of EMI on dc SQUIDs, and Mück et al. [2.41], on rf SQUIDs. They found that EMI of sufficient intensity can reduce the modulation depth of SQUIDs, as well as induce low-frequency excess noise, especially if the EMI has an amplitude-modulated part.

The physical dimensions of SQUIDs in most cases are small compared to the wavelength of the EMI. Radiation applied to the SQUIDs can then be considered to be a pure rf flux bias [2.40], i.e., the SQUID senses only the magnetic component of the electromagnetic field. Indirect coupling can occur by pick-up of EMI in the tank circuit, and by the leads between SQUID and readout electronics. It is believed, however, that proper shielding of the transmission line between SQUID and readout, e.g., by using a semi-rigid coaxial cable, can minimize these effects. Besides, since the tank circuit is a resonant structure, substantial pick-up of the electrical EMI component will take place only at frequencies at or close to its resonant frequency.

For this reason, one primarily must consider effects of EMI magnetic flux Φ_{EMI} on SQUID operation. For simplicity, it is assumed that only the rf bias, the ac modulation flux Φ_{ac} used for a flux-locked loop, and single-frequency EMI flux are present. RF flux induced by EMI generates an additional screening current in the SQUID loop, which adds to that of the rf bias flux generated by the tank circuit. The tank-circuit voltage will then be the result of a superposition of the SQUID response to Φ_{ac} and the interference signal. EMI of "low" frequencies on the order of several hundred kHz probably will be sensed by the flux-locked loop, and thus will be present at the loop output. Higher frequencies will

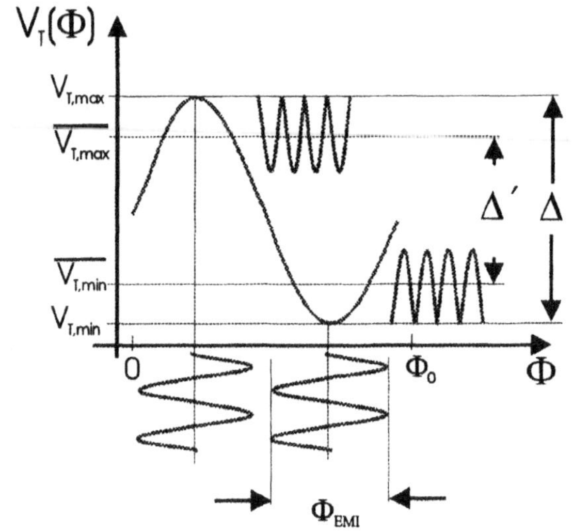

Fig. 7. Effects of EMI on the transfer function of an rf SQUID for $\Phi_{EMI} = \Phi_0/4$. Here, the modulation depth Δ of the undisturbed $V(\Phi)$ curve decreases to a lower value Δ', which is the difference between the averaged tank circuit voltages $\overline{V_{T,max}}$ and $\overline{V_{T,min}}$.

not appear at the detector output, since both the bandwidth of the detector circuit used in rf SQUID electronics and the bandwidth of the tank circuit, limit the frequency response of the system to a few MHz. Nevertheless, the SQUID itself will respond to much faster flux changes, since its response time is given only by the switching time of its Josephson junction and the L/R time constant of the SQUID ring, which — for conventional thin-film SQUIDs — is on the order of 50 ps.

The screening current induced in the SQUID ring by Φ_{EMI} will periodically change the flux threshold value Φ_C for which a quantum transition will occur [2.4]. This, in turn, leads to a periodic change in the rf voltage drop across the tank circuit. Due to the limited bandwidth of the tank circuit and readout electronics, however, sidebands generated by the amplitude modulation of the tank-circuit voltage are small. For EMI of sufficiently high frequency (> several MHz), the amplified and amplitude-demodulated tank-circuit voltage V_T, as measured at the detector, will not contain components of the EMI signal, but only will show its average value. No EMI is thus directly visible at the detector of the electronics, provided that the EMI is not amplitude modulated.

Due to the periodic nature of the rf SQUID transfer function, the shift in threshold value Φ_C generated by EMI will lead to a 'washing out' of the transfer function if Φ_{EMI} is sufficiently large. Suppose there is a static flux bias $\Phi_{dc} = 0$ (or $n\Phi_0$); the corresponding tank circuit voltage V_T has a value of $V_{T,max}$ — see Fig. 7. Φ_{EMI} modulates the bias point of the SQUID, leading to tank-circuit voltages of $V_T < V_{T,max}$ for the maxima and minima of Φ_{EMI}. EMI thus decreases the static $V_{T,max}$ and increases the static $V_{T,min}$. This leads to a reduction of the modulation depth $\Delta V = V_{T,max} - V_{T,min}$. Larger Φ_{EMI} can completely suppress flux modulation of the rf SQUID if the average tank-circuit voltages $\overline{V_T}$ are the same for a static flux of zero or $\Phi_0/2$. A static flux change will then no longer have an effect on the detected tank-circuit voltage. We note that Φ_{EMI} has the same effect on the static $V(\Phi)$ curve as broad-band noise generated by the tank circuit or the preamplifier. These effects already have been studied [2.12, 2.42]. When Φ_{EMI} is increased further ($\Phi_{EMI} > \Phi_0$), the modulation depth Δ oscillates between zero and a steadily decreasing maximum value. This is similar to the well-known oscillations in V_T caused by increasing the pump power; however, here it is due to the periodic nature of the $V(\Phi)$ curve. It should be noted that this effect can be used to measure precisely the field strength of high-frequency magnetic fields if the field-to-flux transfer coefficient of the SQUID is known.

Depending on the shape of the transfer function, the SQUID response of the ac flux stays sinusoidal with increasing Φ_{EMI} for a sinusoidal transfer function, or becomes somewhat flat-topped for a triangular transfer function. With increasing Φ_{EMI}, a dc offset is superimposed on Φ_{ac}, which can be attributed to a 'rectifying' of the EMI flux.

In summary, high-frequency electromagnetic interference leads to a reduction of the SQUID modulation voltage if the EMI flux is of the order of a flux quantum or larger. The only way to ameliorate this problem is by using a flux-locked loop with very high bandwidth (larger than ω_{rf}, which in most cases is impossible), or to prevent EMI flux from leaking into the SQUID by proper shielding.

3. Realization of the tank circuit

To extract flux information from a SQUID, it can be inductively coupled to a tank circuit. At lower bias frequencies, historically used in commercial rf SQUIDs, this usually consisted of a capacitor and an inductor, connected in parallel. For high bias frequencies, which one wants to use to minimize the intrinsic noise of the SQUID, the resonant frequency of the tank circuit must be increased, which requires a corresponding decrease in the capacitance and/or inductance of the tank circuit. However, because of stray and parasitic capacitance and inductance of the tank-circuit components, it is not practical to use discrete components at frequencies above 1 GHz. Several alternatives have been studied: for very high bias frequencies, a cavity [3.1], or a mismatched transmission line [3.2] can be used. Here, we shall concentrate only on practical planar designs.

For a planar SQUID, a planar transmission-line resonator (e.g., a half-wavelength ($\lambda/2$) microstrip resonator) is a suitable replacement for a discrete LC circuit — see Fig. 8. In order to obtain reasonable coupling between SQUID and resonator, the SQUID should be placed in the vicinity of a current anti-node. For simplicity, the SQUID also can be integrated into the resonator — see Fig. 8. Coupling the SQUID to the room-temperature electronics can be accomplished by a second microstrip on the substrate (coupling microstrip, Fig. 8), which runs in parallel to one end of the resonator, or by a small wire antenna.

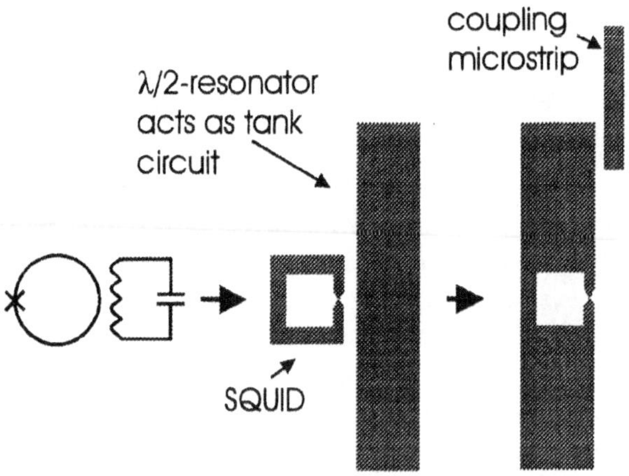

Fig. 8. Metamorphosis of a conventional washer-type rf SQUID with discrete tank circuit to a planar microwave SQUID integrated into a $\lambda/2$ microstrip resonator.

For a superconducting film to form a transmission-line resonator, dielectric material is required between the resonator and a metallic groundplane. The groundplane itself can be made of either normal or superconducting material. The dielectric layer can be formed by the substrate itself, and the groundplane can be formed by a piece of copper-clad printed-circuit material, which at the same time can serve as the sample holder. To avoid thermal magnetic noise from the normal-conducting groundplane, it should be very thin. Due to the groundplane, the SQUID inductance is lowered by a few percent.

An important requirement to be met by rf SQUIDs is that $k^2Q \approx 1$, where k is the coupling coefficient between resonator and SQUID, and Q is the quality factor of the resonator. The coupling between SQUID and resonator is determined by how much rf flux threads the SQUID ring if a given rf current flows in the resonator. Planar superconducting resonators can exhibit quality factors of several thousand to tens of thousand, if dielectric and radiation losses are small.

As examples of rf SQUIDs with planar transmission line resonators, we describe an LTS rf SQUID integrated into a half-wavelength microstrip resonator [3.3], and an HTS rf SQUID with coplanar resonator [3.4]. The layout for an LTS SQUID integrated into a $\lambda/2$ microstrip resonator is shown in Fig. 8. Such SQUIDs were made from thin niobium films sputtered onto 1-mm-thick sapphire substrates.

Microstrip SQUIDs made of niobium which had a resonator length of 20 mm exhibited a resonant frequency of about 3 GHz [3.3]. With a microstrip impedance of 80 Ω, unloaded Q's of up to 4500 were measured, reflecting mainly radiation losses and losses in the normal-conducting groundplane. Somewhat higher Q's can probably be obtained if the thickness of the substrate is reduced, thus reducing radiation losses. Loading by the readout electronics leads to values of Q between 1000 and 3000. The optimum SQUID signal voltage was obtained when the bias frequency equaled the microstrip resonant frequency. Signal voltages of up to 160 $\mu V/\Phi_0$ could be measured for tunnel-junction SQUIDs .

Noise measurements were carried out in flux-locked-loop mode with the readout system described in Sec. 4 and using a room-temperature preamplifier. Typical values for white flux noise of about 3×10^{-6} $\Phi_0/\sqrt{Hz}$ were obtained for tunnel-junction SQUIDs. Down to below 0.1 Hz no low frequency excess noise was observed in a flux-locked-loop mode. The measured values of the flux noise of about 3 $\mu\Phi_0/\sqrt{Hz}$ reflect the noise of the preamplifier of our readout electronics. Its input voltage noise is about 0.4 $nV/\sqrt{Hz}$, which together with a SQUID signal voltage of up to 160 $\mu V/\Phi_0$ results in a corresponding flux noise of about 2.5 $\mu\Phi_0/\sqrt{Hz}$. The uncoupled energy resolution of these SQUIDs is about 1.3×10^{-31} J/Hz for the best case.

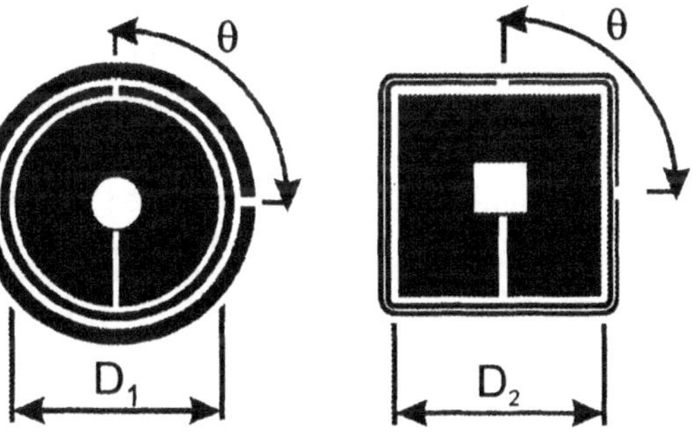

Fig. 9. Schematic of two flux-concentrator/coplanar-resonator layouts. θ denotes the mutual angular positions of the coplanar-resonator openings.

To obtain even lower flux noise and thus better energy resolution, a cooled preamplifier was used to reduce the amplifier noise. The intrinsic noise of the SQUID or the tank-circuit noise could then be investigated. Calculating the noise of these sources results in 6×10^{-7} $\Phi_0/\sqrt{Hz}$ for the tank circuit and 1×10^{-6} $\Phi_0/\sqrt{Hz}$ for the intrinsic

SQUID noise. With a two-stage amplifier in an open loop a flux noise of $9\times10^{-7}\ \Phi_0/\sqrt{\text{Hz}}$ was measured [3.5]. This corresponded to an energy resolution $\varepsilon = 2\times10^{-32}$ J/Hz = 30 h.

The same type of design (using a $\lambda/2$ resonator as tank circuit) was employed for (HTS) rf SQUIDs fabricated from epitaxial $YBa_2Cu_3O_7$ (YBCO) on $LaAlO_3$ substrates [3.6]. A λ resonator design also was tested [3.7], but the requirement $k^2Q > 1$ was difficult to fulfill with the layout employed.

In a different approach, modified microstrip hair-pin resonators have been used as tank circuits for washer SQUIDs with an rf bias frequency of about 850 MHz [3.8]. White flux noise about 20 $\mu\Phi_0/\sqrt{\text{Hz}}$ for $L_S = 150$ pH was measured. However, this white-noise level was attained only at frequencies above 100 kHz. It was concluded that the additional noise originated from the metallic ground plane. To eliminate this noise contribution, coplanar resonators were introduced to serve as the tank circuit [3.4]. Figure 9 shows two examples of the integration of a circular and a square coplanar resonator with a flux concentrator. All superconducting structures were patterned from 200-nm epitaxial YBCO on 1-mm-thick $LaAlO_3$ substrates. The inner hole of the concentrator was 1.5 mm in diameter. The mutual angular position θ of two microstrip resonator slits can be varied to adjust the resonant frequency. The width of the two microstrips was 100-200 μm each. The distance between the concentrator and the inner microstrip line, and between the microstriplines themselves, was about 100 μm. The quality factors Q_0, measured without electrical shielding, were in the range between 4,000 and 20,000 at 77 K.

For integration with coplanar resonators, washer-type SQUIDs with a diameter of 2.5 or 3.5 mm were used. A SQUID was inductively coupled to the coplanar resonator using a flip-chip configuration. The coupling was indirect, via a flux concentrator such as that shown in Fig. 9. The small size of the SQUID washers permits one to fabricate 4 (for 3.5-mm diameter) or 9 (for 2.5-mm diameter) of them on a 1-cm^2 substrate, thus enhancing the yield of these devices. The operation of the SQUIDs with the coplanar resonator has already been described in detail [3.9]. By changing the diameter of the matching loop or its distance to the resonator, the loaded quality factor Q is easily adjusted. Thus, the requirement $k^2Q > 1$ can be met without difficulty, since the unloaded quality factor Q_0 is sufficiently high. The value of k^2Q also influences the flux noise. Empirically, it was found that the noise is about minimum if this value ranged from 1 to 3, as determined from the loaded Q. With an rf bias frequency of 650 MHz and D_1 = 13.4 mm, the field resolution in the white-noise range was about 16 fT/$\sqrt{\text{Hz}}$. The SQUID parameter β_L was smaller than unity.

Yet another possibility to realize a tank circuit at higher frequencies is to use a dielectric resonator. A dielectric resonator usually consists of a cube or cylinder of dielectric material with high ε and low losses. It resembles a cavity resonator filled with dielectric, but with the outer conducting walls removed. The high ε keeps the electric field confined to the dielectric, but since the permeability μ of the dielectric is low, the magnetic field also penetrates the space outside the dielectric. Dielectric resonators are commercially available and usually used to stabilize oscillators at ultra-high frequencies, or are used in bandpass filters. They have a high Q of up to 10^4, especially when cooled to lower temperatures. Such resonators have been used first as tank circuits by Kornev et al. [3.10], who prepared a niobium rf SQUID directly onto a rutile cylinder of 3-mm diameter and 1-mm height. This rutile dielectric resonator had a resonant frequency of 10 GHz and was coupled to the readout electronics by placing it into a waveguide.

Recently, $SrTiO_3$ substrates have been used as dielectric resonators for HTS rf SQUIDs [3.11]. Multiple (100)-oriented $SrTiO_3$ substrates of $10\times10\times1$ mm^3 were stacked face-to-face to obtain the desired resonant frequency. Onto the top substrate, a 4×4-mm^2 washer-type rf SQUID was prepared. The dielectric resonator formed by these stacked substrates was connected to the readout by means of a wire antenna. With a stack of 6 substrates, a resonant frequency of 910 MHz and a Q of 6800 were obtained. The flux noise was measured in an open loop, and in the white-noise region a flux noise of $1\times10^{-5}\ \Phi_0/\sqrt{\mathrm{Hz}}$ was found. No measurements in a flux-locked loop were reported.

4. RF SQUID readouts

Especially in rf SQUID systems, the readout plays a key role in achieving the optimum performance of the system. Unlike dc SQUID systems, where the sensitivity of the entire system is mainly determined by intrinsic SQUID noise (and thus with simple electronics the optimum performance can still be achieved), in most cases rf SQUID system noise is limited by the readout. Therefore, it is of prime importance that the preamplifier noise, as well as the coupling between SQUID and tank circuit, and between tank circuit and readout, are perfectly optimized.

4.1 OPTIMIZATION FOR MAXIMUM SENSITIVITY

The sensitivity of an rf SQUID, in the absence of external noise, is given by the sum of intrinsic noise, tank-circuit noise and preamplifier noise. As already pointed out in Sec. 2, at low bias frequencies (e.g., 30 MHz) the system noise is limited by preamplifier noise. In order to reduce this noise contribution, the modulation voltage $V_m \propto \partial V_T/\partial\Phi$ should be maximized. Since $\partial V_T/\partial\Phi = (1\text{-}\alpha)\cdot\omega L_T/2M$, V_m can be maximized by using the highest possible tank-circuit inductance and the lowest posssible mutual inductance between SQUID and tank circuit, but without reducing k^2Q to below unity. To achieve a large L_T, the parasitic capacitance, e.g., of the transmission line to the room temperature electronics, must be made as small as possible. A rather ingenious way to achieve this is to use a cooled field-effect transistor [4.1], which is employed as an impedance match between the tank circuit and a coaxial cable, rather than as an amplifier. A large L_T and small C_T additionally lead to a high tank-circuit Q, which permits reduction of M and a decrease of tank-circuit noise. For LTS SQUIDs, the use of superconducting wire for the tank-circuit coil is advisable because of its intrinsic high Q and the absence of thermal magnetic noise. For HTS SQUIDs, either planar superconducting tank-circuit coils or the planar microstrip resonators (as discussed in the previous section) can be used.

4.2 COUPLING OF THE TANK CIRCUIT TO A TRANSMISSION LINE

For higher bias frequencies (e.g., 150 MHz), Eq. (6) no longer gives the attainable modulation voltage because of a phase shift between V_T and I_T. Instead, to obtain a realistic value for V_m, the available modulation power $P_m \approx 0.1\cdot LI_C^2\omega$ [4.2] should be used to calculate V_m. As already pointed out by Jackel *et al.* [4.3], with increasing bias frequency it becomes increasingly difficult to maintain a high tank-circuit impedance, which is desired for high sensitivity. This is mainly due to a relative increase in parasitic capacitances, and to a reduced input impedance Z_{in} of the preamplifier used. A realistic value for Z_{in} for FET amplifiers at 150 MHz is on the order of 5 kΩ, compared to an

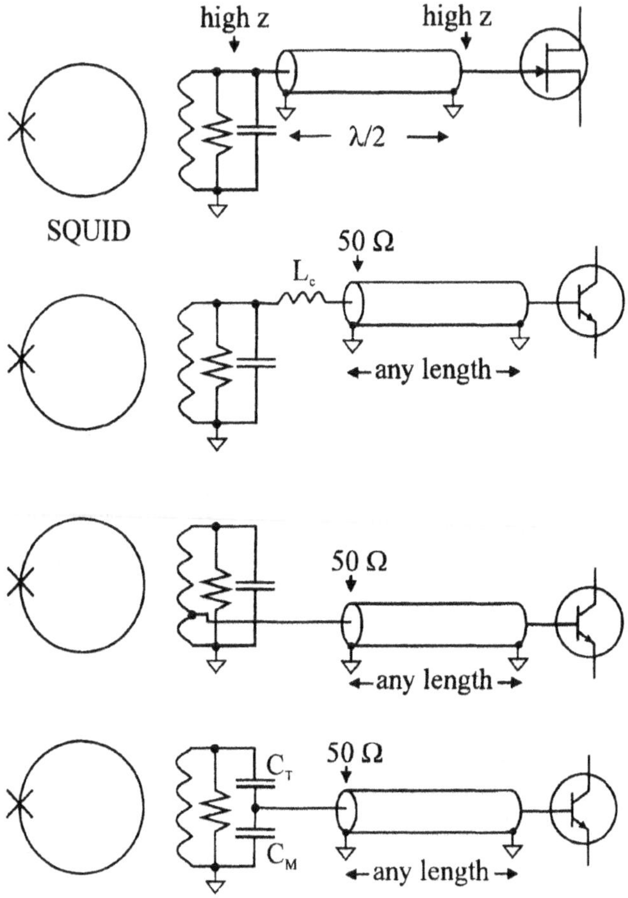

Fig. 10. Four different coupling schemes to obtain impedance matching between tank circuit and readout electronics

obtainable tank-circuit impedance of typically 30 kΩ. The low amplifier Z_{in} reduces V_m significantly when the amplifier is directly connected to the tank circuit. Additionally, an unmatched transmission line between tank circuit and amplifier introduces a high parasitic capacitance. For this reason, several different coupling schemes have been proposed in the literature.

A popular approach is the use of a low-impedance coaxial cable between tank circuit and amplifier — see Fig. 10a. In order to reduce its parasitic capacitance and the loading of the tank circuit, the coax is given an electrical length of $\lambda/2$. In this case, the high impedance of the tank circuit is transferred to a high-impedance amplifier. Since, however, in most cases the amplifier Z_{in} is smaller than that of the tank circuit, a somewhat shorter coax would lead to a better impedance match. We note that the use of an unmatched coaxial cable leads to a high standing-wave ratio along it. This leads to an additional loss in the coax, which increases with increasing intrinsic cable loss. This is usually relatively high, since in most cases, a thin coax is used to minimize thermal input. For a coax with 1-dB intrinsic loss, a total attenuation of 6 dB can result, which leads to a reduction in sensitivity by a factor of two! It seems advisable, therefore, to use coupling schemes which make use of a matched transmission line. A virtue of a matched line is that the condition $k^2Q=1$ guarantees matching of the amplifier to the tank circuit, so that all available signal power is transferred to the amplifier. Matching can be achieved by several different schemes: Robbes *et al.* [4.4] have used a coil in series with the transmission line (Fig. 10b), which can transform the high Z_{tc} to lower values. For optimum results, the inductance of the coil must be adjusted, which is difficult, since the coil is at low temperature. Jackel *et al.* [4.3] used a tapped tank-circuit coil — see Fig. 10c — which transforms the high tank-circuit impedance to a value of 50 Ω at the tap. Since it is difficult to tap the tank-circuit inductor used for a planar rf SQUID, a series connection of two individual inductors can be used, where only the high-inductance coil is coupled to the SQUID. It also is possible to use two separate coils, i.e., where there is a purely inductive coupling between the tank circuit and a second coil, which is connected to the coaxial cable [4.17]. The disadvantage of a tapped tank-circuit coil is that, with increasing bias frequency, stray inductances move the tap to higher inductance values, so that a 50-Ω impedance can no longer be obtained. For this reason, Duret *et al.* [4.5] developed a matching scheme using a capacitive voltage divider — see Fig. 10d. The capacitor C_T in conjunction with L_T determines the resonant frequency of the circuit, whereas C_M determines the output impedance. One could achieve a nearly perfect impedance match (return loss > 20 dB) by choosing the proper ratio between C_T and C_M [4.6]. It was found that the actual value of C_M is not very critical. A return loss of > 10 dB, which is sufficient, can be obtained for C_M varying by about 20%. For 150-MHz tank circuits with copper-wire coils, a ratio $C_M/C_T \approx 15$ gave optimum results. It should be noted that for quality factors of $Q > 200$ (e.g., when using a niobium coil with an LTS SQUID), matching by this technique becomes difficult. For tank circuits with superconducting coils, a resistor in parallel to the tank circuit can be used to reduce Q.

4.3 CONVENTIONAL READOUT ELECTRONICS

Figure 11 shows a diagram of all components needed to operate rf SQUIDs. An rf oscillator generates the bias frequency, whose amplitude can be controlled by an attenuator in a range of 20 to 30 dB. The attenuator is used to adjust the rf current in the SQUID so that it is operated at the first step of the $V_T(I_{rf})$ curve. An adjustment range of 20 dB allows for the use of SQUIDs with a spread in critical currents of a factor of ten. RF current is imposed in the SQUID tank circuit. The voltage drop across the tank circuit is amplified by a factor of several thousand and then amplitude demodulated, usually by an envelope detector. For flux-locked-loop operation, a lock-in amplifier is provided together with a low-frequency generator, which sweeps the SQUID field by a fraction of a flux quantum.

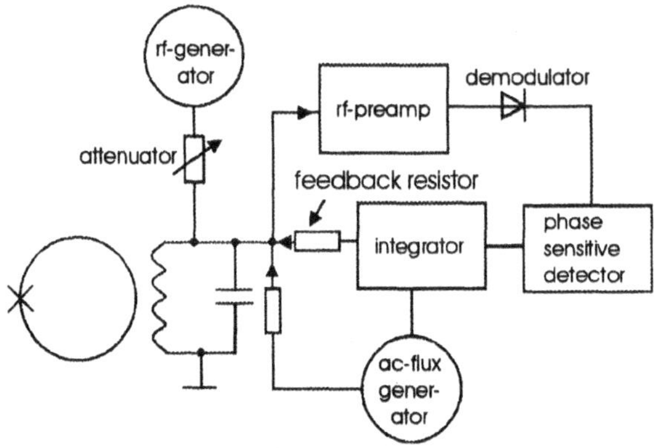

Fig. 11. Block diagram of a conventional rf SQUID readout electronics.

A readout system for SQUIDs operated at ultra-high frequencies (e.g., 3 GHz, [4.7]) is similar to conventional electronics used for 20-MHz SQUIDs. There are, however, a few details which distinguish it from a conventional configuration. If the tank circuit is in the form of a microstrip resonator, its resonant frequency is fixed. This requires tuning of the rf bias current generator. Microwave preamplifiers have a relatively wide bandwidth of hundreds of MHz. In order to decrease the bandwidth to the frequency range needed, a synchronous detector should be used instead of the usual envelope demodulator. The following lock-in amplifier can be a standard version such as that used in a 20-MHz system, although because of the high bias frequency, a much higher modulation frequency, and thus a higher slew rate, is possible.

It also is possible to operate an rf SQUID in a flux-locked-loop mode without ac modulation. Simacek et al. [4.16] and later others have measured the demodulated tank-circuit voltage and compared it to an adjustable dc voltage, so that the SQUID was always biased at the steepest part of its $\partial V_T/\partial\Phi$ transfer function.

4.4 CRYOGENIC PREAMPLIFIERS

The noise performance of rf SQUIDs can be improved by increasing their bias frequency. The intrinsic noise amplitude of the SQUID is proportional to $\omega^{-1/2}$, whereas the modulation voltage of the SQUID, V_m, is proportional to $\omega^{1/2}$. However, even at microwave frequencies, the signal voltage is so small that the noise of the entire SQUID system is still limited by the noise of the room-temperature preamplifier. In order to decrease this contribution to the overall noise, the preamplifier can be operated at cryogenic temperatures [4.8-4.11]. At a temperature of 4.2 K or less (used for LTS SQUIDs) and for microwave amplification, only III-V semiconductor devices (made of, e.g., GaAs or InP) are feasible at the moment. Additionally, they offer the advantage of a very low noise figure. Since a cooled preamplifier can be placed close to the SQUID, losses in the transmission line between SQUID and room-temperature electronics do not contribute to the system noise. This is particularly important at microwave frequencies, where thin coaxial cables can have considerable loss.

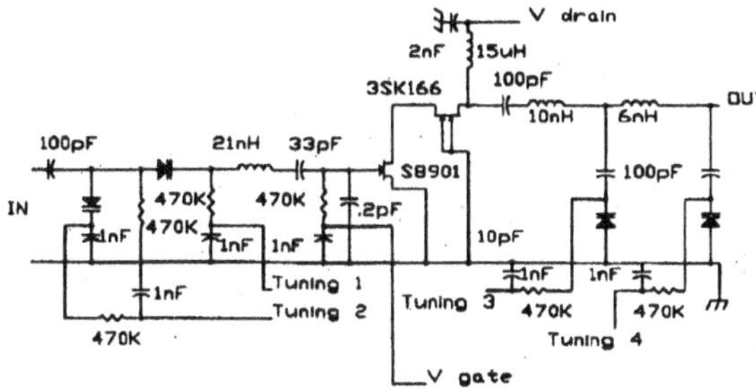

Fig. 12. Circuit diagram of a two-stage cryogenic preamplifier for use in rf SQUID systems with tunable input matching [from 4.12].

Mezzena et al. [4.12, 4.13] have developed a cryogenic preamplifier for LTS rf SQUIDs operated at 1.7 GHz and 3 GHz. In order to achieve sufficient amplifier gain and to have the possibility of noise-matching the amplifier, a two-stage version was used, where the input circuit could be tuned by two GaAs varactors — see Fig. 12.

The active elements of the preamplifier were arranged in a cascode configuration and included a microwave high-electron-mobility transistor (HEMT, Toshiba S8901) and a dual-gate MESFET (Sony 3SK166). The design was based on a cryogenic amplifier used in a 300-MHz SQUID system [4.11]. The cascode configuration was employed to reduce power dissipation, and thus, the temperature of the HEMT. The gain of the preamplifer was 15 dB with input and output loaded with 50 Ω. A careful determination of the amplifier noise temperature led to a value of about 5 K. With this low preamplifier noise, the noise of a SQUID system was dominated only by intrinsic SQUID noise.

4.5 MULTIPLEXED READOUTS FOR MULTICHANNEL SQUID SYSTEMS

With the growing number of channels in multichannel SQUID systems some efforts have been made to reduce their size and complexity. Using discrete readout electronics for each channel not only leads to a bulky and expensive system, but also to a troublesome alignment of each channel. In addition, the large number of wires between SQUIDs and the room-temperature part of the system — up to four in the case of dc SQUIDs — leads to increased heat input and accompanying increased coolant loss. The individual wires of each channel must be properly screened to prevent cross-talk between channels.

In principle some parts of the readout system, such as the amplifier and the phase detector, could be shared by all channels, if all the signals can be distinguished from each other. Nonlinearities in the amplifier and especially in the detector, however, can give rise to mutual mixing which makes the resulting output signals useless. Some interesting ideas have been published regarding this problem. Furukawa and Shirae [4.14] suggested a multichannel system where the sum of all channels modulates yet another SQUID, which then requires only one transmission line and amplifier for the readout. The response of the individual channels is separated by a number of phase detectors, but each SQUID still requires a discrete wire feeding its feedback coil.

Another solution to this problem is a serial readout of the channels [4.15] instead of a parallel one. Each SQUID is given a certain time window for operation during which all other SQUIDs are switched off. Under these circumstances, only one readout and one transmission line are needed (theoretically) with the addition of a logic circuit which controls the on-off switching of the SQUIDs. Using LTS rf SQUIDs biased at around 100 MHz, a three-channel multiplexed system could be realized. The SQUIDs were biased at sufficiently different frequencies (100, 120 and 140 MHz), and switching on and off of the SQUIDs was accomplished by switching the individual oscillators. The flux noise of the individual channels did not increase due to the multiplexed readout, and the slew rate of the system was sufficient to operate it in a low interference unshielded environment.

5. Recent developments in rf SQUID systems

The LTS rf SQUID has been widely used for many years because of its commercial availability and possibly because of easy set-up and operation. However, there was little new development until the advent of HTS created a renewal of interest.

Recent work on LTS rf SQUIDs has concentrated on improving SQUID sensitivity by increasing the bias frequency and/or decreasing preamplifier noise. A number of rf SQUID systems with bias frequencies above 1 GHz have been realized [5.1]. Their white flux noise and energy resolution are comparable to that of dc SQUIDs.

Since most rf SQUIDs suffer from low modulation voltage, preamplifier noise contributes to the overall noise to a substantial extent. Cooling the preamplifier to SQUID temperature decreases the thermal noise of the amplifier, as already discussed in Sec. 4. This work was pioneered by Gaerttner [4.8], who used a gallium-arsenide field-effect transistor (GaAs-FET) amplifier at 440 MHz, cooled to 4.2 K. The advent of low-noise high-electron-mobility transistors (HEMTs) has further increased the potential of this method. Cavalleri *et al.* [4.11] have operated a commercial 20-MHz rf SQUID at 300 MHz using a cryogenic HEMT amplifier. By this approach, they could reduce the noise of the SQUID system by a factor of ten, when compared to 20-MHz operation with a room-temperature amplifier.

Soon after the discovery of HTS, rf SQUID operation was demonstrated. A number of SQUID systems using bulk HTS samples [5.2-5.7] were realized. Here, either unstructured pieces of HTS, or one or two-hole SQUIDs were used. Weak links were formed by natural grain boundaries and critical current often was trimmed by some procedure to obtain the required critical current at 77 K. These procedures included heating [5.4], squeezing [5.6] or spark erosion [5.8] of a small bridge machined into the superconductor. The white noise level of these SQUIDs was in some cases comparable to that for LTS rf SQUIDs. However, pronounced low-frequency excess noise usually was observed. Nevertheless, due to the large hole diameter of up to several mm, relatively high field sensitivity was obtained, which was even sufficient for use in some biomagnetic measurements [5.5, 5.8].

Due to the multitude of grain boundaries in HTS bulk material, it is questionable whether the performance of bulk SQUIDs can be improved much further. However, since most of these devices were operated at 20 MHz, a slight decrease in white noise could be achieved with higher bias frequencies [5.9]. The low-frequency excess noise, which is attributed mainly to thermally-activated hopping of vortices in the bulk material, could probably be reduced by polarity reversal in the input circuit [5.10] if HTS wire can ever be used as an input coil. In that case, it seems likely that field sensitivities of 50 fT/√Hz could be achieved, which should be sufficient for a number of serious applications. The ease of fabrication may then make these devices competitive again with thin-film SQUIDs.

Low white noise and low 1/f noise are obtained for SQUIDs made of high quality, single crystalline HTS films. Josephson junctions can be fabricated utilizing natural grain boundaries, ion implantation [5.11-5.13], step-edge junctions [5.14,5.15] or bicrystal junctions [5.16]. It has been found that the white noise of these planar devices is not much lower than that of bulk SQUIDs. However, the crossover to 1/f noise can be reduced to frequencies below 1 Hz. Due to their small hole area, the field sensitivity in many cases is even less than that of bulk SQUIDs. But when making use of the flux-focussing effect of a relatively large SQUID washer or by using a separate flux focusser, a reasonable field sensitivity can be achieved, even without a flux transformer. To obtain sufficiently low flux noise, it is recommended that the bias frequency of the rf SQUID be as high as possible.

Washer-type rf SQUIDs have been made by a number of groups [5.13-5.18]. Common to all of them is — compared to dc SQUIDs — a relatively large loop area of up to 400 × 400 μm^2 and a low 1/f noise for frequencies above 1 Hz. Although in most cases the measured white noise is dominated by the preamplifier rather than the SQUID, the absolute value of the 1/f noise at a certain frequency is still lower than that of many thin-film dc SQUIDs. The value of the white noise is of the order of 10^{-4} $\Phi_0/\sqrt{Hz}$ for a 20-MHz bias frequency. Somewhat lower white noise can be achieved using a higher bias frequency [5.16-5.18]. A field sensitivity of a 150-MHz rf SQUID [5.17] of 90 fT/√Hz (@ 1 Hz) for a washer size of 10×10 mm^2, could be obtained, which is sufficient for many practical applications. Using a 4-cm-diameter flux focusser, a sensitivity of 24 fT/√Hz was obtained [5.19]. The lowest flux noise obtained so far for a magnetometer on a 10×10 mm^2 chip is approximately 30 fT/√Hz [5.20].

Even higher sensitivities of single-layer HTS SQUIDs should be obtained when working with bias frequencies above 1 GHz. Daly *et al.* [5.21] reported on the first microwave YBCO SQUIDs, which had been coupled to a 10-GHz TE_{011} cylindrical cavity and operated at temperatures up to 55 K. From the shape of their transfer function, one can estimate a flux noise of a few 10^{-5} $\Phi_0/\sqrt{Hz}$. Instead of using a cavity, one also can use a microstrip resonator as a tank circuit [3.6]. The flux-noise values obtained with these SQUIDs were an order of magnitude lower than those of SQUIDs operated at 150 MHz.

To futher enhance the field sensitivity of rf SQUIDs, flux transformers with a multi-turn input coil are desirable. For YBCO rf SQUIDs, attempts to use flux transformers remained unsuccessful until it was shown that the multi-turn input coil must be decoupled from the planar tank-circuit resonator to prevent damping of the latter. Recently, the feasibility of an HTS rf SQUID magnetometer with such a planar multi-turn flux transformer was demonstrated [5.22]. On a 10 mm × 10 mm substrate, the best sensitivity of 12 $fT/Hz^{1/2}$ was obtained for the layout shown in Fig. 13 [5.23]. Here, a small double-loop SQUID gradiometer was used. The multi-turn input coil was coupled to one SQUID loop while the tank resonator was coupled to the other. The transformer and resonator were laid out on the common substrate and flip-chip aligned with the SQUID gradiometer fabricated on a separate substrate. The 1/f noise for this structure still was found to be rather high.

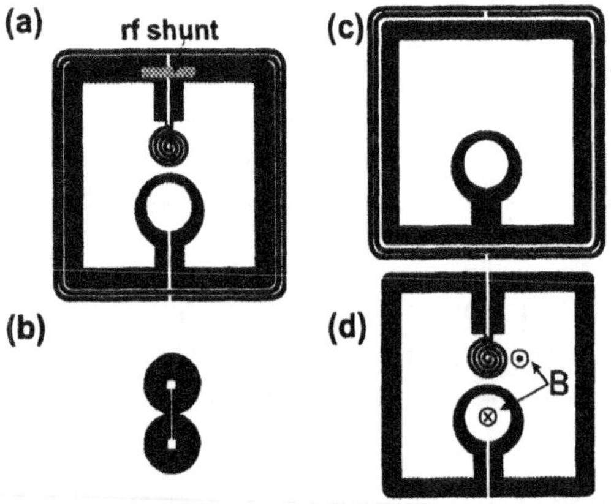

Fig. 13. Layout of a small double-loop SQUID gradiometer (b) with (spiral) multi-turn input coil (a) shunted by a metal strip, which is coupled to one SQUID loop while the (single turn) tank resonator is coupled to the other (from [5.23]). (c) shows equivalent rf and (d) equivalent dc circuit.

In an unshielded environment, one uses gradiometers of first or second order to reduce interference. The conventional (LTS) solution is to use wire-wound pick-up coils which are bonded by some procedure to the thin-film input coil of the SQUID. Unfortunalely, no low-noise HTS wire is available which can be used to wind the pick-up coils and the input coil of the SQUID. A superconductive connection between an HTS wire and an HTS thin-film input coil certainly is a very difficult task. Fabricating the gradiometer entirely using thin-film technology also is difficult because of the required baseline length of several cm. Additionally, one usually wants to measure the vertical ($\partial B/\partial z$) component of the magnetic field, which is not possible with a planar gradiometer.

Tinchev [5.24] has developed a thin-film rf SQUID gradiometer with a baseline length of 1.4 mm. The SQUID was operated at 20 MHz and the flux noise was about 1×10^{-4} $\Phi_0/\sqrt{Hz}$ in a flux-locked loop. Due to the relatively short baseline, this gradiometer must be close to the sample to be measured. Planar rf gradiometers with longer baseline, up to 5 mm, also were reported [5.25].

An alternative to a wire-wound gradiometer is an electronic gradiometer, where two or more magnetometers are separated by the required baseline length, and their output signals are subtracted. Subtracting the signals can be done by analogue as well as by digital (software) techniques. The requirements for slew rate and stability of the individual (magnetometer) channels of such a gradiometer are more severe than for a system with a gradiometric pick-up coil because such a pick-up coil reduces the response to homogeneous fields by a factor of 1000 or more.

To demonstrate the feasibility of these HTS rf SQUID gradiometers, biomagnetic measurements were performed inside a magnetically-shielded room, as well as in an unshielded environment. Magnetocardiograms of diagnostic value, as well as magneto-encephalograms, were measured [5.26, 5.27].

Tavrin *et al.* [5.28] and Borgmann *et al.* [5.29] developed first and second-order HTS SQUID gradiometers with mechanical-balance tuning, using washer-type thin-film rf SQUIDs operated at above 100 MHz. The SQUIDs were arranged so that the $\partial B/\partial z$ component of the magnetic field could be measured using a baseline length of 6 cm. Additional flux focussers were used to enhance the field sensitivity of the SQUIDs. A slew rate of 4×10^4 Φ_0/s and an unbalance of about 1×10^{-3} between the two channels was sufficient to operate the system in an unshielded, but relatively "quiet" laboratory. Magnetocardiograms (MCGs) were obtained which showed structure in the cardiac signal. To improve the signal-to-noise ratio, they used a bandwidth of about 40 Hz. By using a second-order gradiometer consisting of three rf SQUIDs integrated with coplanar resonators [3.4], it was possible to increase the measurement bandwidth to above 130 Hz [5.23]. They measured MCGs of a number of persons in real time and with a bandwidth of about 130 Hz. A signal-to-noise ratio of about 25 was obtained in the best cases.

Foley *et al.* [5.30] have described geophysical measurements using HTS rf SQUIDs. They showed that rf SQUID magnetometers could even be used for aeromagnetic mineral exploration from a moving platform.

RF SQUIDs also have been applied to nondestructive testing (NDE) of aircraft parts [5.18, 5.31, 5.32] and of pre-stressed concrete bridge structures [5.33]. In the latter case, usually dc measurements of the remanent magnetization of the steel reinforcement are made. Since these measurements must be performed in an unshielded environment, it is questionable whether the high sensitivity of a SQUID can be fully utilized, and thus it is possible that other types of magnetic sensors such as flux gates can be used equally well. To our knowledge, no data on a comparison between measurements of this kind using SQUIDs and conventional field sensors have been published for the case of bridge structure testing.

A field which has been widely explored is nondestructive testing of aircraft parts using eddy-current techniques. In order to achieve sufficient eddy-current density deep inside the sample to be tested, relatively low eddy-current frequencies (e.g., 400 Hz for a flaw at a 15-mm depth) must be used. Since the field sensitivity of induction coils which are used in conventional eddy-current NDE equipment scales as $1/\omega$, the high field sensitivity of SQUIDs at low frequencies makes these devices ideally suited for evaluation of relatively thick conductive test objects, such as layered stacks of aluminium sheets or aircraft wheels. In contrast to direct injection of dc or ac currents into the sample to be tested, the induction of eddy currents by an external excitation coil generates a spatially well-confined current distribution. Since no electrical connections to the sample are required, eddy-current NDE is more practical to use than direct current injection. However, the choice of the optimum measurement parameters, such as the excitation frequency, is more critical. Furthermore, calculation of the current distribution in the sample from the measured field distribution tends to be more difficult than in the case of direct-current injection. In order to minimize the excitation field at the location of the SQUID, a gradiometric excitation coil (a so-called double-D coil) is usually used. By carefully moving this coil below the SQUID to find its optimum position, the cross talk between excitation coil and SQUID can be kept lower than 0.1%. The field amplitude at the surface of the sample then can be as high as 5 mT.

Kreutzbruck et al. [5.32] performed measurements to compare their microwave SQUID NDE system with a commercial eddy-current NDE system (Elotest B1, [5.34]), which uses induction coils as field sensors. An induction coil without a ferrite core (PLA-44) was used. The test sample was an aluminium plate with an artificial crack 40-mm long, 0.15-mm wide and 1.2-mm deep, which was covered by another 13-mm thick aluminium plate. The crack was not detectable with the conventional system when a detection bandwidth of 30 Hz was used. In a bandwidth of 2 Hz, the crack could be detected. However, the signal-to-noise ratio was still less than 3. With the microwave SQUID system and a bandwidth of 50Hz, the crack could be detected with a signal-to-noise ratio of about 300.

NDE measurements of real samples, e.g., along a curved aircraft wing, require three-dimensional movement or tilting of the SQUIDs. This is not possible when using a conventional dewar for liquid coolants. Although special dewars are available for this purpose, a cryocooler is a promising competitor to liquid coolants in this case.

6. Concluding remarks

The rf SQUID continues to be an attractive magnetic field sensor for a number of applications. Operation of rf SQUIDs at very high bias frequencies is commonplace these days, and this results in substantial improvement in sensitivity compared to commercial 20-MHz rf SQUIDs. Although the readout electronics seem to be more complex than those used for dc SQUIDs, for commercial production on a larger scale, differences in the number and price of the electronic components used should be negligible. The absence of a galvanic connection between SQUID and readout electronics simplifies manufacture of a large number of devices and prevents ageing effects. In principle it is even possible to separate SQUID and tank circuit by a thin insulating wall, so that, for example, the SQUID could be situated in a vacuum space, whereas the tank circuit and the wires leading to the room-temperature electronics could be at atmospheric pressure. The possibliy large inductance of HTS rf SQUIDs, and thus the loop area, leads to much higher field sensitivity of the bare SQUID, so that for a number of applications, a complex flux transformer is not needed. Low-frequency excess noise caused by fluctuations in the critical current of the Josephson junctions is suppressed by conventional ac flux-modulated readout, whereas for dc SQUIDs, electronics with additional bias reversal must be used, which limits the bandwidth of the electronics. It should not be hidden, however, that there are a few applications where rf SQUIDs cannot be used, or are clearly inferior to dc SQUIDs. These applications include high-frequency amplifiers — see the chapter by John Clarke *et al.* in this book — and SQUID microscopes, where SQUIDs with very small loop area must be used and brought close to the sample under test.

To conclude, we have tried to give an overview of the current status of rf SQUIDs. Naturally, such an overview can never be totally comprehensive and there are many other SQUID designs and applications which we did not mention. We think it worthwhile, therefore, to refer the reader to the papers cited in this work and elsewhere, in order to obtain a more complete overview of rf SQUIDs and their applications.

7. Acknowledgments

We thank Alex Braginski and Harold Weinstock for critical reading of this paper and advice. Helpful discussions with John Clarke, Giorgio Fontana, Christoph Heiden, Renato Mezzena, Steffano Vitale, Shiguang Wang and James E. Zimmerman are gratefully acknowledged.

8. References

[2.1] Silver, A.H. and Zimmerman, J.E. (1967) Quantum states and transitions in weakly connected superconducting rings, *Phys.Rev.* **157**, 317-341.

[2.2] Giffard, R.P. Webb, R.A. and Wheatly, J.C. (1972) Principles and methods of low frequency electric and magnetic measurements using an rf-biased point contact supercondcuting device, *J. Low Temp.Phys.* **6**, 533-611.

[2.3] Jackel, L.D. and Buhrman, R.A. (1975) Noise in the rf SQUID, *J. Low Temp.Phys.* **19**, 201-249.

[2.4] Clarke, J. (1977) Superconducting quantum interference devices for low frequency measurements, in B.B. Schwartz and S. Foner, (eds.) *Superconductor Applications - SQUIDs and Machines*, Plenum Publishing Corp., New York, pp. 67-124.

[2.5] Likharev, K.K. (1986) *Dynamics of Josephson Junctions and Circuits* Gordon and Breach, New York.

[2.6] Gallop, J.C. (1991) *SQUIDs, the Josephson Effects and Superconducting Electronics*, Adam Hilger, Bristol.

[2.7] Hansma, P.K. (1973) Superconducting single-junction interferometers with small critical currents, *J.Appl.Phys.* **44**, 4191-4194.

[2.8] Kurkijärvi, J. (1973) Noise in the superconducting flux detector, *J.Appl.Phys.* **44**, 3729-3733.

[2.9] Eliashberg, G.M. (1970) Film superconductor stimulated by a high-frequency field, *JETP Lett.* **11**, 114-115.

[2.10] Zimmerman, J.E. and Silver, A.H. (1966) Macroscopic quantum interference effects through superconducting point contacts, *Phys.Rev.* **141**, 367-375.

[2.11] Pascal, D. and Sauzade, M. (1974) Experimental determination of optimum operating conditions of a superconducting magnetometer, *J.Appl.Phys.* **45**, 3085-3090.

[2.12] Falco, C.M. and Parker, W.H. (1975) Operating characteristics of thin-film rf-biased SQUIDs, *J.Appl.Phys.* **46**, 3238-3243.

[2.13] Buhrman, R.A. and Jackel, L.D. (1977) Performance factors in rf SQUIDs — high frequency limit, *IEEE Trans.Mag.* **MAG-13**, 879-882.

[2.14] Mercerau, J.E. (1970) Superconducting magnetometers, *Rev.Phys.Appl.* **5**, 13-20.

[2.15] Rachford, F.J., Huang, C.Y., Nisenoff, M. and Wolf, S.A. (1975) Observation of sine modulated bessel behaviour in microwave biased SQUIDs, *IEEE Trans.Mag.* **MAG-11**, 870-872.

[2.16] Mück, M. (1995) Developments and characterisation of low-noise rf SQUIDs, *Habilitation Thesis*, Justus-Liebig-Universität Giessen.

[2.17] Kurkijärvi, J. (1972) Intrinsic fluctuations in a superconducting ring closed with a Josephson junction, *Phys.Rev.B* **6**, 832-835.

[2.18] Kurkijärvi, J. and Webb, W.W. (1972) Thermal fluctuation noise in a superconducting detector, in *Proc. ASC Annapolis, IEEE Pub. No. 72CHO 682-5-TABSC*, IEEE, New York, pp. 581-586.

[2.19] Giffard, R.P., Gallop, J.C. and Petely, B.N. (1976) Applications of the Josephson effects, *Prog.Quant.Electron.* **4**, 301-402.

[2.20] Josephson, B.D. (1965) Supercurrents through barriers, *Adv.Phys.* **14**, 419-451.

[2.21] Clarke, J. and Koch, R.H. (1988) The impact of high-temperature superconductivity on SQUIDs, *Science* **242**, 217-223.

[2.22] Kramers, H.A. (1940) *Physica* **7**, 284.

[2.23] Goldman, A. M. (1970) Lifetimes of persistent currents in superconducting loops interrupted by Josephson junctions, *J. Low Temp.Phys.* **3**, 55.

[2.24] Fulton, T.A. and Dunkleberger, L.N. (1974) Lifetime of the zero-voltage state in Josephson tunnel junctions, *Phys. Rev.* **B9**, 4760-4768.

[2.25] Jackel, L.D., Webb, W.W., Lukens, J.E. and Pei, S.S. (1974) Measurement of the probability distribution of thermally excited fluxoid quantum transitions in a superconducting ring closed by a Josephson junction, *Phys. Rev.* **B9**, 115-118.

[2.26] Chesca, B. (1994) On the theoretical study of rf SQUID operation taking into account the noise influence, *J.Low Temp.Phys.* **94**, 515-538.

[2.27] Chesca, B. (1998) Theory of rf SQUIDs operating in the presence of large thermal fluctuations, *J. Low Temp.Phys.* **110**, 963-1002.

[2.28] Zhang, Y., Zander, W., Schubert, J., Rüders, F., Soltner, H., Banzet, W., Wolters, N., Zeng, X.H. and Braginski, A.I. (1997) Operation of high-sensitivity rf SQUID magnetometers with superconducting coplanar resonators at 77K, *Appl. Phys. Lett.* **71**, 704-706.

[2.29] Il'ichev, E.V., Zakosarenko, V.M., Ijsselsteijn, R.P.J. and Schultze, V. (1997) Current-phase relation of high-T_c bicrystal and step-edge Josephson junctions, *J. Low Temp.Phys.* **106**, 503-514.

[2.30] Zhang, Y., Mück, M., Herrmann, K., Schubert, J., Zander, W., Braginski, A.I., and Heiden, C. (1993) Sensitive rf SQUIDs and magnetometers operating at 77 K, *IEEE Trans.Appl.Supercond.* **AS-3**, 2465-2468.

[2.31] Zeng, X.H., Zhang, Y., Chesca, B., Barthel, K., Greenberg, Ya.S., and Braginski, A. I. (submitted) Experimental study of the amplitude-frequency characteristics of HTS rf SQUIDs, *J.Appl.Phys.*

[2.32] Koch, R.H., Clarke, J., Goubau, W.M., Martinis, J.M., Pegrum, C.M. and van Harlingen, D.J. (1983) Flicker (1/f) noise in tunnel junction dc SQUIDs, *J. Low Temp.Phys.* **51**, 207-224.

[2.33] Rogers, C.T. and Buhrman, R.A. (1984) Composition of 1/f noise in metal-insulator metal tunnel junctions, *Phys.Rev.Lett.* **53**, 1272-1275.

[2.34] Moeckly, B.H. and Buhrman, R.A. (1993) Electromigration studies of the role of oxygen defects in $YBa_2Cu_3O_{7-\delta}$ grain boundary weak links, *IEEE Trans.Appl.Supercond.* **AS-3**, 2038-42.

[2.35] Ferrari, M.J., Johnson, M., Wellstood, F.C., Clarke, J., Rosenthal, P.A., Hammond, R.H. and Beasley, M.R. (1988) Magnetic flux noise in thin film rings of $YBa_2Cu_3O_{7-\delta}$, *Appl.Phys.Lett.* **53**, 695-697.

[2.36] Mück, M., Heiden, C. and Clarke, J. (1994) Investigation and reduction of low frequency excess noise in rf SQUIDs, *J.Appl.Phys.* **75**, 4588-4592.

[2.37] Giffard, R.P. (1980) in H.D.Hahlbohm and H.Lübbig (eds.) *SQUID'80 'Superconducting Quantum Interference Devices and Their Applications'*, W. de Gruyter, Berlin pp. 447-480.

[2.38] Vant-Hull, L., Simpkins, R.A. and Harding, J.T. (1967) *Phys.Lett.* **A24**, 736.

[2.39] Ishikawa, N., Nagata, K., Sato, H., Kasai, N. and Kiryu, S. (1993) Effect of rf interference on characteristics of dc SQUID system, *IEEE Trans.Appl.Supercond.* **AS-3**, 1910-1913.

[2.40] Koch, R.H., Foglietti, V., Rozen, J.R., Stawiasz, K.G., Ketchen, M.B., Lathrop, D.K., Sun, J.Z. and Gallagher, W.J. (1994) Radio Frequency Interference in dc SQUIDs, *Appl.Phys.Lett.* **65**, 100-102.

[2.41] Mück, M., Dechert, J., Gail, J., von Kreutzbruck, M., Schöne, S. and Weidel, R. (1995) Response of Radio Frequency SQUIDs to Electromagnetic Interference, *Rev.Sci.Instrum.* **66**, 4690-4694.

[2.42] Brunet-Brunol, D., Pascal, D. and Duret, D. (1979) Effect of noise on the performance of rf SQUID magnetometers, *J.Appl.Phys.* **50**, 521-529.

[3.1] Kamper, R.A. and Simmonds, M.B. (1972) Broadband superconducting quantum magnetometer, *Appl.Phys.Lett.* **20**, 270-272.

[3.2] Hollenhorst, J.N. and Giffard, R.P (1979) High sensitivity microwave SQUID, *IEEE Trans.Mag.* **MAG-15**, 474-477.

[3.3] Mück, M. and Heiden, C. (1992) Planar microwave biased rf SQUIDS, *Appl.Phys.* **A54**, 475-479.

[3.4] Zhang, Y., Zander, W., Schubert, J., Rüders, F., Soltner, H., Banzet, M., Wolters, N., Zeng, X.H. and Braginski, A.I. (1998) HTS rf SQUIDs with coplanar resonator, *Applied Superconductivity* **AS-6**, 385-388.

[3.5] Fontana, G., Mezzena, R., Vitale, S., Cerdonio, M., Mück, M., Hallmanns, G. and Heiden, C. (1993) RF SQUIDs for application in a gravitational wave detector, *Proceedings of the Workshop on Cryogenic Gravitational Wave Antennae*, Legnaro, Italy (unpublished).

[3.6] Zhang, Y., Mück, M., Bode, M., Herrmann, K., Schubert, J., Zander, W., Braginski, A.I. and Heiden, C. (1992) Microwave SQUIDs with integrated YBCO resonator, *Appl. Phys. Lett.* **60**, 2303-2305.

[3.7] Zhang, Y., Mück, M., Braginski, A.I. and Töpfer, H. (1994) High sensitivity microwave rf SQUID operating at 77 K, *Supercond. Sci. Techn.* *7*, 269-272.

[3.8] Zhang, Y., Soltner, H., Wolters, N., Zander, W., Schubert, J., Banzet, M. and Braginski, A.I. (1997) HTS rf SQUIDS with fully integrated planar tank circuits, *IEEE Trans. Appl. Supercond.* **AS-7** , 2870-2873.

[3.9] Zhang, Y., Zander, W., Schubert, J., Rüders, F., Soltner, H., Banzet, M., Wolters, N.,. Zeng, X.H. and Braginski, A.I. (1997) Operation of high-sensitivity rf SQUIDs with superconducting coplanar resonator, *Appl. Phys. Lett.* ***71***, 704-706.

[3.10] Kornev, V.K., Likharev, K.K., Snigirev, O.V., Soldatov, Ye.S. and Khanin, V.V. (1980) Microwave SQUID with a high-Q dielectric resonator, *Radio Engineering and Electronic Physics* **25**, 2647-2655 (in Russian).

[3.11] Zhang, Y., Gottschlich, M., Soltner, H., Sodtke, E., Schubert, J., Zander W. and Braginski, A.I. (1995) Operation of high temperature rf SQUID magnetometers using dielectric $SrTiO_3$ resonators, *Appl. Phys. Lett.* **67**, 3183-3185.

[4.1] Goodkind, J.M. and Stolfa, D.L. (1970) The superconducting magnetic flux detector, *Rev.Sci.Instrum.* **41**, 799-807.

[4.2] Zimmerman, J.E. (1971) Recent developments in superconducting devices, *J.Appl.Phys.* **42**, 30-37.

[4.3] Jackel, L.D., Stoeckenius, J.W.H. and Epworth, R.W. (1978) Analysis and operation of UHF SQUID with tapped tank circuit, *J.Appl.Phys.* **49**, 471-472.

[4.4] Robbes, D., Dubuc, C., Lepaisant, J., Bloyet D. and Duret, D. (1985) Impedance matching of an rf SQUID to a standard transmission line, *IEEE Trans.Mag.* **MAG-21**, 892-895.

[4.5] Duret, D., Bernard, P. and Zenatti, D. (1975) A UHF superconducting magnetometer utilizing a new thin film sensor, *Rev.Sci.Instrum.* **46**, 474-480.

[4.6] Zhang, Y., Mück, M., Herrmann, K., Schubert, J., Zander, W., Braginski, A.I. and Heiden, C. (1992) Low noise $YBa_2Cu_3O_7$ rf SQUID magnetometer, *Appl.Phys.Lett.* **60**, 645-647.

[4.7] Mück, M. (1992) A readout electronics for 3 GHz rf SQUIDs, *Rev.Sci.Instrum.* **63**, 2268-2273.

[4.8] Gaerttner, M.R. (1974) in *Int.Mag.Conf. Toronto, Publ.No. 74 CHO 852-4-MAG* IEEE, New York.

[4.9] Long, A.P., Prance, R.J., Clark, T.D., Mutton, J.E., Potts, M.W., Widom, A. and Goodall, F. (1980) Quantum limited ac-biased SQUID magnetometer, in H.D.Hahlbohm and H.Lübbig (eds.) *SQUID'80 'Superconducting Quantum Interference Devices and their Applications'*, W. de Gruyter, Berlin, pp. 207-211.

[4.10] Ahola, H., Ehnholm, G.J., Östman, P. and Rantala, B. (1979) Cryogenic GaAs-FET amplifiers for SQUIDs, *J.Low.Temp.Phys.* **35**, 313-328.

[4.11] Cavalleri, A., Cerdonio, M., Fontana, G., Jung, G., Machietta, R., Mezzena, R., Vitale S. and Zendri, J.P. (1992) UHF thin film rf SQUID magnetometer with a cryogenic preamplifier employing a high-electron mobility transistor, *Rev.Sci.Instrum.* **63**, 5403-5407.

[4.12] Mück, M., Hallmanns, G., Heiden, C., Cerdonio, M., Fontana, G., Mezzena, R. and Vitale, S. (1992) Planar microwave biased rf SQUIDs with a cryogenic preamplifier, *Appl.Phys.Lett.* **61**, 1231-1233.

[4.13] Fontana, G., Mezzena, R., Vitale, S., Cerdonio, M., Mück, M., Hallmanns, G. and Heiden, C. (1993) Improved sensitivity of planar microwave biased rf SQUIDs using a cryogenic HEMT amplifier, *IEEE Trans.Appl.Supercond.* **AS-3**, 1820-1823.

[4.14] Furukawa, H. and Shirae, K. (1989) Multichannel dc SQUID system, *J.J.Appl.Phys.* **28**, L456-458.

[4.15] Mück, M. (1991) A three channel SQUID system with multiplexed readout, *IEEE Trans.Mag.* **MAG-27**, 2986-2989.

[4.16] Simacek, I., Zrubec, V. and Skakala, M. (1993) rf SQUID magnetometer with simplified circuitry, *Rev.Sci.Instrum.* **64**, 2401-2402.

[4.17] He, D.F., Zeng, X.H., Krause, H.-J., Soltner, H., Rüders, F. and Zhang, Y. (1998), Radio Frequency SQUIDs operating at 77 K with 1 GHz lumped-element tank circuits. *Appl.Phys.Lett.* **72**, 969-971.

[5.1] See e.g., Mück, M. (1993) Progress in rf SQUIDS, *IEEE Trans.Appl.Supercond.* **AS-3**, 2003-2010, and references therein.

[5.2] Colclough, M.S., Gough, C.E., Keene, M., Muirhead, C.M., Thomas, N., Abell, J.S. and Sutton, S. (1987) Radio frequency SQUID operation using a ceramic high-temperature superconductor, *Nature* **328**, 47-48.

[5.3] Zimmerman, J.E., Beall, J.A., Cromar, M.W. and Ono, R.H. (1987) Operation of a YBCuO rf SQUID at 81 K, *Appl.Phys.Lett* **51**, 617-619.

[5.4] Tinchev, S.S. and Hinken, J.H. (1989) 77K rf SQUID with 1/f noise only below 5 Hz, *Electron.Lett.* **25**, 1579-1580.

[5.5] Likhachev, A.G., Polushkin, V.N., Uchaikin, S.V. and Vasiliev, B.V. (1990) Magnetocardiometer based on a single-hole high-Tc SQUID, *Supercond.Sci.Technol.* **3**, 148-151.

[5.6] Zhang, Y., Xu, Y.H. and Heiden, C. (1990) Bulk BiPbSrCaCuO rf SQUIDs operating up to 101 K, *Appl.Phys.Lett.* **56**, 1597-1599.

[5.7] Chandra, R., Gupta, A.K. and Khare, N. (1993) Preparation of superconducting thin films of Bi-Sr-Ca-Cu-O by electron beam evaporation and a study of rf SQUID behaviour at liquid nitrogen temperatures, *Supercond.Sci.Technol.* **6**, 135-140.

[5.8] Tavrin, Y. private communication.

[5.9] Shnyrkov, V.I., Timofeev, V., Khvostov, S. and Tsoi, G. (1991) UHF high-Tc SQUID, *Mod.Phys.Lett.* **B5**, 1281-1286.

[5.10] Cunningham, C.E., Park, G.S., Cabrera, B. and Huber, M.E. (1993) Noise reduction in low frequency SQUID measurements with laser-driven switching *Appl.Phys.Lett* **63**, 1152-1154.

[5.11] Cui, G., Zhang, Y., Herrmann, K., Buchal. C., Schubert, J., Zander, W., Braginski, A.I. and Heiden, C. (1991) Ion-beam implanted YBCO weak links, *Supercond.Sci.Technol.* **4**, 130-133.

[5.12] Zani, M.Z., Luine, J.A., Lee, G.S., Murduck, J.M., Lanis, R.J.M.J., Davidheiser, R.A. and Eaton, L.R. (1991) Focused ion beam high-Tc superconducting SQUIDs, *IEEE Trans.Mag.* **MAG-27**, 2557-2560.

[5.13] Daly, K.P., Dozier, W.D., Burch, J.F., Coons, S.B., Hu, R., Platt, C.E. and Simon, R.W. (1991) Substrate step-edge YBCO rf SQUIDs, *Appl.Phys.Lett.* **58**, 543-545.

[5.14] Tinchev, S.S. (1990) Investigations of rf SQUIDs made from epitaxial YBCO films, *Supercond.Sci.Technol.* **3**, 500-502.

[5.15] Sloggett, J., Dart, D.L., Foley, C.P., Binks, R.A., Savvides, N. and Katsaros, A. (1994) Noise in thin-film $YBa_2Cu_3O_7$ step-edge junction SQUIDs, *Supercond.Sci.Technol.* **7**, 260-264.

[5.16] Wang, S.G., Peking University, priv.comm.

[5.17] Zhang, Y., Mück, M., Herrmann, K., Schubert, J., Zander, W., Braginski, A.I. and Heiden, C. (1992) Low noise $YBa_2Cu_3O_7$ rf SQUID magnetometer, Appl.Phys.Lett. **60**, 645-647.

[5.18] Bowman, R.W., MacFarlane, J.C., Cochran, A., Kirk, K.J., Pegrum, C.M. and Donaldson, G.B. (1993) SQUID systems for non-destructive testing by ac field mapping, *IEEE Trans.Appl.Supercond.* **AS-3**, 1926-1929.

[5.19] Zhang, Y., Krüger, U., Kutzner, R., Wördenweber, R., Schubert, J., Zander, W., Sodtke, E., Braginski, A.I. and Strupp, M. (1994) Single layer YBCO rf SQUID magnetometer with direct-coupled pick-up coil and flip-chip flux transformers, *Appl.Phys.Lett.* **65**, 3380-3382.

[5.20] Zhang, Y., Bousack, H., Bick, M., Montsch, A., Soltner, H., Wolf, W. and Braginski, A. I. (submitted) Magnetocardiography Using HTS rf SQUIDs with Coplanar Resonators, *IEEE Trans.Appl.Supercond.*

[5.21] Daly, K.P., Burch, J.F., Coons, S.B. and Hu, R. (1991) YBCO step-edge rf SQUID biased at 10 GHz, *IEEE Trans.Mag.* **MAG-27**, 3066-3069.

[5.22] Zhang, Y., Yi, H.R., Schubert, J., Zander, W., Banzet, M., and Braginski, A.I. (1998), A design of planar multi-turn flux transformers for radio frequency SQUID magnetometers, *Appl. Phys. Lett.* **72**, 2029-2031.

[5.23] Zhang, Y, Schubert, J., Zander, W., Yi, H.R., and Bousack, H. (1999) Operation of rf SQUID magnetometers with a multi-turn flux transformer integrated with a superconducting labyrinth resonator, *IEEE Trans. Appl. Supercond.* ***AS-9***, 3396-3400.

[5.24] Tinchev, S.S. (1994) YBCO thin film gradiometer, in H.C.Freyhardt (ed.) *Proceedings of the 1. European Conference on Applied Superconductivity (EUCAS)* Göttingen, pp. 1435-1437.

[5.25] Zhang, Y., Soltner, H., Krause, H.-J., Sodtke, E., Zander, W., Schubert, J., Grüneklee, M., Lomparski, D., Banzet, M., Bousack, H., and Braginski, A.I. (1997) Planar HTS gradiometer with large baseline, *IEEE Trans. Appl. Supercond.* **AS-7**, 2866-2869.

[5.26] Zeng, X.H., Soltner, H., Selbig, D., Bode, M., Bick, M., Ruders, F., Schubert, J., Zander, W., Banzet, M., Zhang, Y., Bousack, H., and Braginski, A.I. (1998) A high-temperature rf SQUID system for magnetocardiography. *Meas. Sci. Technol.* **9**, 1-9.

[5.27] Zhang, Y., Tavrin, Y., Mück, M., Braginski, A.I., Heiden, C., T.Elbert, T. and Hampson, S. (1993) Magnetoencephalography with high-T_c rf SQUIDs, *Physiol.Meas.* **14**, 113.

[5.28] Tavrin, Y., Zhang, Y., Mück, M., Braginski, A.I. and Heiden, C. (1993) Operation of a high-Tc rf SQUID gradiometer in a unshielded environment, *Appl. Phys. Lett.* **62**, 1824-1826.

[5.29] Borgmann, J., David, P., Ockenfuss, G., Otto, R., Schubert, J., Zander, W. and Braginski, A.I. (1997) Electronic high-temperature radio frequency superconducting quantum interference device gradiometers for unshielded environment, *Rev. Sci. Instrum.* **68**, 2730-2734.

[5.30] Foley, C.P. and Leslie, K.E. (1998) Potential use of high-T_c SQUIDs for airborne exploration, *Exploration Geophysics* **29**, 30-34; Foley, C.P., Leslie, K.E., Binks, R., Lewis, C., Murray, W., Sloggett, G.J., Lam, S., Sankrithyan, B., Savvides, N., Katzaros, A., Kuller, K.-H., Mitchell, E.F., Pollock, J., Lee, J., Dart, D.L., Barrow, R.R., Asten, M., Maddever, A., Panjkovic, G., Downey, M., Hoffman, C. and Turner, R. (1999) Field trials using HTS SQUID magnetometers for ground-based and airborne geophysical applications. *IEEE Trans. Appl. Supercond.* **AS-9**, 3786-3792.

[5.31] Hohmann, R., Maus, M., Lomparski, D., Grüneklee, M., Zhang, Y., Krause, H.J., Bousack, H. and Braginski, A.I. (1999) Aircraft wheel testing with machine-cooled HTS gradiometer system, *Trans.Appl.Supercond.* **AS-9**, 3801-3804.

[5.32] von Kreutzbruck, M, Mück, M. and Heiden, C. (1997) Experiments on Eddy Current NDE with HTS rf SQUIDs, *IEEE Trans.Appl.Supercond.* **AS-7**, 3279-3282.

[5.33] Sawade, G., Krause, H.-J., and Gampe, U. (1997). Non Destructive Examination of Prestressed Tendons by the Magnetic Stray Field Method, in M. T. Forde (ed.) *Proceedings of the 7th international conference on Structural Faults and Repair – 97*, Vol. I, Engineering Technical Press, Edinburgh, pp. 401-406.

[5.34] The Elotest B1 is manufactured by ESR-Rohmann GmbH, Frankenthal, Germany.

MILLIMETER AND SUB-MM WAVE JOSEPHSON FLUX-FLOW DEVICES

J. MYGIND
Department of Physics,
Technical University of Denmark, Lyngby, Denmark

and

N.F. PEDERSEN
Department of Electric Power Engineering,
Technical University of Denmark, Lyngby, Denmark

1. Introduction

Josephson fluxon devices have been reviewed by several authors in recent years [1-5]. In the context of this volume on microwave superconductivity we will concentrate on the use of Josephson vortices (fluxons) in flux-flow millimeter and sub-mm wave generators, which is a field that has made significant progress in recent years. All such flux-flow-based generators require long Josephson junctions (LJJs).

In Section 2 we first introduce the simple one-dimensional perturbed sine-Gordon model with boundary conditions as a model for the current-biased LJJ. The soliton solution which behaves as a quasiparticle with mass, energy and momentum is identified with the moving Josephson vortex (fluxon) containing one quantum of magnetic flux.

After a brief discussion of power and linewidth of the simple *short Josephson oscillator* we turn to the low-damping *Josephson Resonant Soliton Oscillator* (RSO). Here one or more fluxons oscillate, emitting radiation when being reflected at the ends of the LJJ. The oscillation frequency is essentially related to the geometrical length of the LJJ. In the dc current-voltage curve (IVC) this leads to the so-called Zero-Field Steps (ZFS) and, with an applied magnetic field, to Fiske steps (FS).

In the Josephson *Flux-Flow Oscillator* (FFO) the dynamics becomes more complex, and a chain of mutually interacting fluxons propagates through the junction and emits radiation when colliding with one of the junction ends. In the IVC, nonlinear mixing with the geometrical Fiske resonances generates a so-called Flux-Flow Step (FFS) for which the voltage is proportional to the applied magnetic field. The frequency of the emitted radiation can be tuned over a wide frequency range by either the applied dc bias current and the magnetic field.

The *Cherenkov Flux-Flow Oscillator* (CFFO) is a clever extension of the FFO in which an external "slow-wave" transmission line structure integrated along the junction is utilized to generate and extract Cherenkov radiation from each of the moving fluxons in the chain. This will increase the available power relative to the FFO. As with the FFO the frequency of the emitted radiation can be widely tuned by the dc bias current and magnetic field. The CFFO is still in its infancy. Here we concentrate on the annular geometry which is particularly simple, but still shows rich and highly non-linear

H. Weinstock and M. Nisenoff (eds.), Microwave Superconductivity, 541–580.

dynamics where the fluxons couple to (and self-excite) internal modes of the system. Also coupling to external structures and signals is discussed.

Finally in Section 2 we briefly review some significant experimental flux-flow oscillator results obtained with high-T_c samples.

Section 3 discusses the important topic of flux-flow devices based on stacked Josephson junctions. Vertical stacking and mutual locking is a straightforward way to increase emitted power. An important topic is the coupling between adjacent layers. Many of the dynamical features in the JSO, FFO, and CFFO can be seen as analogous to stacked systems of LJJs. Stacked junctions also appear to be useful for modeling flux-flow-related properties of natural crystals of high-T_c materials.

Related systems of discrete one- and two-dimensional arrays of small Josephson junctions, which in some sense also may be termed as microwave flux-flow generators, recently have been reviewed [6] and hence will not be considered here.

Since the early 1960s high-power mm and sub-mm wavelength (from 100 GHz to several THz) Traveling Wave Tubes, Backward Wave Oscillators, and FIR lasers have been developed mainly for military radar, tracking, surveillance and communication. Generally, as shown in Fig. 1, the emitted power decreases strongly at higher frequencies.

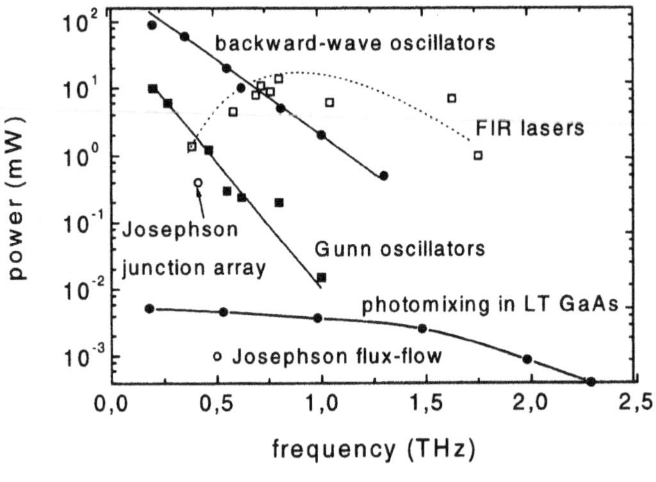

Figure 1. Emitted power versus frequency of mm and sub-mm wave sources (from [6]).

Currently there is an increasing demand for mm and sub-mm oscillators serving as local oscillators (LOs) in sensitive *heterodyne* receivers used in a variety of applications such as radio astronomy, and air- and space-borne spectroscopy for atmospheric research and environmental monitoring. The main constraints here are low weight, low power consumption, reliability, wide tunability and narrow spectral linewidth. Available oscillators are complex and expensive devices that do not meet these criteria. For spectral radio-astronomy applications, aside from the noise temperature and antenna beam pattern, the frequency resolution of the receiver, which is determined by both the instant linewidth of the local oscillator and its long-time stability, should be much less than 1 ppm of the center frequency.

In the heterodyne receiver, the LO must supply enough power to properly operate the mixing element which in most modern radio telescopes usually is an antenna-coupled, sub-micron sized highly nonlinear junction. In the mm and sub-mm wavelength range presently the best mixers are the superconductor-insulator-superconductor (SIS) mixer, the Josephson mixer, the superconducting hot-electron bolometric (HEB) mixer [7], and the novel impurity-controlled hot-electron transition-edge sensor (TES) [8]. Due to their small size, usually less than a microwatt of power transmitted to the mixer on chip is sufficient.

It is very attractive to integrate a LO and a mixer/antenna structure on a single cooled chip which can be mounted onto the back of a hemispherical lens. Thus, they can be matched and inserted directly into a parabolic antenna. An integrated receiver front-end may include the necessary superconducting stripline transmission lines, filters and eventually an IF amplifier, as well as a fast Rapid Single Flux Quantum (RSFQ) A/D converter for fast signal conditioning. Below we shall see that at least two types of flux-flow-based Josephson oscillators are ideally suited for this purpose. When phase-locked to a standard frequency reference (10–100 MHz, external or internal depending on the desired spectral purity and long-time stability) they fulfill all the above-mentioned criteria. Even the phase-locked-loop (PLL) circuit and active components may be integrated on the few-mm^2-area chip. We believe that such all-superconducting receivers, which can be mass-produced, cooled, and operated very economically, will have great impact as sensors for molecular spectroscopy, as well as in other fields of research (chemical, medical, biological, etc.).

For completeness it should be mentioned that *direct detection* with sub-micron TES superconducting bolometers operated at millikelvin temperatures recently has shown high sensitivity and may provide reasonably good spectral resolution [8].

2. Long Josephson junctions

2.1 THE SINE-GORDON SYSTEM; GEOMETRY AND BOUNDARIES

A Josephson tunnel junction consists of two superconductors separated by a thin tunnel barrier. The basic Josephson equations [1-5]

$$I = I_c \sin\phi \tag{1}$$

$$V = \left(\frac{\Phi_0}{2\pi}\right)\frac{\partial\phi}{\partial t} \tag{2}$$

describe the relations between the supercurrent, I, of Cooper pairs tunneling through the barrier and the voltage, V, across the junction. I_c is the critical current given by the materials and the barrier thickness, t. ϕ is the phase difference between the macroscopic wavefunctions describing the condensate of Cooper pairs in the two superconductors. $\Phi_0 = h/(2e) \approx 2*10^{-15}$ Wb is the flux quantum. The symbol h is Planck's constant and e is the electron charge. In a typical Nb/AlOx/Nb tri-layer junction an Al_2O_3 insulating barrier, t = 1-2 nm, is formed between two Nb thin films evaporated on a dielectric substrate. The Nb films usually are much thicker than the London penetration depth, $\lambda_L \approx 50$ nm. Here we shall restrict ourselves to long quasi-one-dimensional junctions with either overlap or annular geometry as shown in Fig. 2. The length, L, and width, W, satisfy $L >> \lambda_J$ and $W << \lambda_J$, where $\lambda_J = \sqrt{\Phi_0/(2\pi\mu_0 J_c d)}$ is the Josephson penetration depth, typically $\lambda_J \approx 10$ μm. J_c is the maximum critical current density and $d = (2\lambda_L + t)$ is the effective magnetic thickness of the junction.

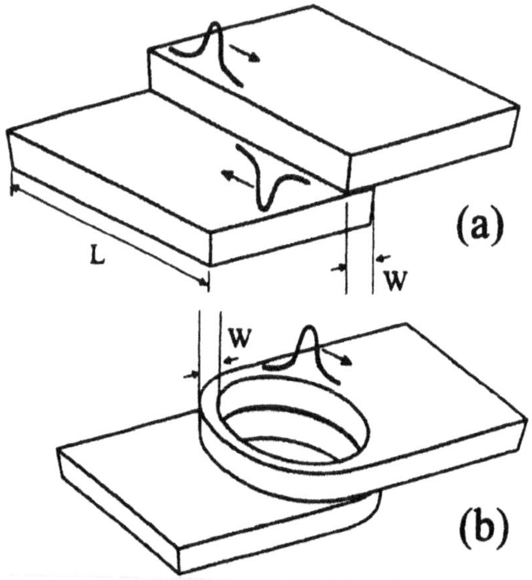

Figure 2. Geometries of quasi-one-dimensional Josephson junctions with (a) linear overlap and (b) annular overlap. Dimensions are not to scale. (From [9]).

If a homogeneous external magnetic field is applied parallel to the films in the W direction in Fig. 2a, self-generated screening currents will flow in a surface layer of thickness λ_L in the films in order to exclude the magnetic flux from the interior of the superconductors (due to the Meissner effect). For the same reason, the field only will penetrate a distance λ_J into the barrier from both ends of the junction. The rest of the junction appears diamagnetic until the applied field exceeds the critical field, $H_{c1} = \Phi_0/\{2\pi\,\mu_0 d\,\lambda_J\} = J_c\lambda_J$. For $H \geq H_{c1}$ the boundary conditions become unstable and magnetic flux enters the junction from both ends in the form of single fluxons (Josephson vortices). As a direct consequence of flux quantization and Cooper pairing each fluxon contains a single flux quantum, Φ_0. It is the dynamics of such fluxons driven by a dc bias current in the films that is the basis of the various flux-flow oscillators discussed below.

If we assume the somewhat unrealistic situation that the junction is completely decoupled from its electromagnetic environment except for the uniformly-distributed dc bias current, I_{dc}, supplied by an ideal current source, we can model the junction by the perturbed sine-Gordon equation in normalized form [10] (where the x-axis is along the junction)

$$\phi_{xx} - \phi_{tt} - \sin\phi = \alpha\phi_t - \beta\phi_{xxt} - \gamma_{dc} \tag{3}$$

subject to the boundary conditions

$$\phi_x(0,t) = \phi_x(L,t) = \Gamma . \tag{4}$$

In this equation time, t, is normalized to the inverse maximum plasma frequency ϖ_0^{-1}, space x is normalized to λ_J, and the sub-index refers to differentiation with respect to t and x. The parameter α describes the quasiparticle shunt loss $\alpha = 1/\sqrt{\beta_c}$, where β_c is the McCumber parameter, β is a dissipative term associated with the surface impedance of the superconducting films, $\gamma_{dc} = I_{dc}/I_c$ is the bias current normalized to the critical current, and $\Gamma = H/H_{c1}$ is the normalized applied magnetic field.

The LJJ which is so well described by the sine-Gordon equation is one of the simplest highly nonlinear systems with both time and space dependence. It can be modeled by a series of coupled simple pendula and is one of the best examples of the transition from the linear approximations used in the past to more realistic nonlinear processes [11]. Moreover the LJJ is one of the fastest experimental systems that operate typically over the wide frequency range from a few GHz to several THz. Thus, a life cycle takes less than a picosecond, meaning that ordinary dc measurements give the experimenter a very long time average value of its behavior. What we have learned in physics and mathematics from studying the sine-Gordon system may be used in a wide spectrum of other nonlinear systems, e.g., within biology, medicine, chemistry, and economics.

The solution to the unperturbed sine-Gordon equation, Eq. (3) with no drive or dissipation, i.e., zero r.h.s., is the *soliton*

$$\phi_\pm = 4\tan^{-1}\left[\exp\{\pm\frac{x-ut-x_0}{\sqrt{1-u^2}}\}\right], \tag{5}$$

where the + sign indicates a soliton and the – sign indicates an antisoliton. The function ϕ_+ is a smooth 2π step stretching from $x = \pm\infty$ moving with velocity u along the junction. For $t = 0$ it is located at $x = x_0$. The velocity is normalized to the Swihart velocity, $\bar{c} = \lambda_J \varpi_0$, the phase velocity of electromagnetic waves propagating in the junction. Due to the penetration of the magnetic field into the Nb films, $\bar{c}$ in a standard Nb/AlOx/Nb tunnel junction is only 2-3% of the velocity of light in free space.

According to Eq. (5) the width of a step is $\sqrt{1-u^2}$, and thus decreases to zero when u approaches 1. Notice that Eq. (5), which only is strictly valid for an infinitely long junction, has other solutions, e.g., soliton-antisoliton, breather and plasma oscillations.

In an LJJ, as modeled by the perturbed sine-Gordon equation (Eqs. (3) and (4)), the soliton (Eq. (5)) is a good approximation, and as shown in Ref. [10] the velocity u of the soliton can be determined from a power balance between the energy supplied by the bias and the energy dissipated in the two loss terms. Neglecting for convenience the β-term, we get

$$u = \frac{1}{\sqrt{1+(\frac{4\alpha}{\pi\gamma_{dc}})^2}} . \tag{6}$$

According to Eq. (2) ϕ_t is proportional to the junction voltage. It is easy to show that the local magnetic field is given by ϕ_x. As mentioned, due to self-field effects, magnetic flux only can enter the interior of the long tunnel junction in the form of Josephson vortices (fluxons), each containing one magnetic flux quantum, Φ_0. The fluxon is well modeled by the soliton. From Eq. (6) we see that $u \propto \gamma_{dc}$ for small bias currents while u asymptotically approaches unity for large values of γ_{dc}/α. The fluxon therefore, behaves like a relativistic particle and, as mentioned, its width decreases at high velocities due to Lorentz contraction.

2.2 POWER AND LINEWIDTH OF JOSEPHSON OSCILLATORS.

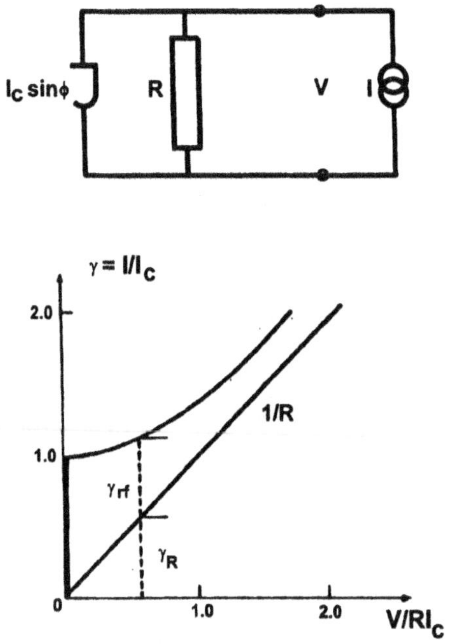

Figure 3. Current-biased resistively-shunted-junction (RSJ) model and the corresponding IVC valid for a short Josephson junction with negligible capacitance.

The Josephson junction inherently is an oscillator converting dc power into ac power with a frequency given by Eq. (2). Usually higher harmonics also are excited. We first consider the simplest case, the short non-hysteretic, current-biased Josephson junction ($L, W << \lambda_J$, no spatial dependence, $\alpha >> 1$, negligible capacitance), which can be modeled by the resistively-shunted-junction model (RSJ model). The equivalent circuit diagram and the corresponding hyperbolic IVC is shown in Fig. 3 [12]. Also drawn is a line with slope *1/R*. If the oscillator junction is at a given voltage, *V*, the total power supplied by the bias current γ_{dc} is $\propto (\gamma_R + \gamma_{rf})$. Since the power dissipated as heat in *R* is $\propto \gamma_R$, it is obvious that the maximum power the oscillator can deliver as rf radiation is given by γ_{rf}. Indeed, all power related to nonlinear Josephson processes is $\propto \gamma_{rf}$. Usually, depending on matching, frequency, up/down conversion, harmonic content, etc. only a small fraction of the power may be transferred to an external load impedance. From the IVC in Fig. 3 it is evident that nonlinear processes are dominant for small values of V, while dissipative processes dominate at higher voltages. These facts often are overlooked. For flux-flow oscillators with more complex IVCs these basic power arguments hold and are also valid for the case where the LJJ is biased on one of the step structures, e.g., ZFS, FS and FFS.

As indicated, power delivered to the environment (such as radiation to free space or ac currents and voltages in an external circuit) may be calculated by self-consistently solving the equations of motion for the total system. The general procedure is outlined in Ref. [12], where also a few simple analytically-solvable systems are analyzed. Experimentally, spectral analysis of emitted high frequency signals is the best tool for investigation of oscillator power, linewidth, intermittent behavior and chaotic motion. Also since the frequency of the emitted radiation from Josephson systems can be measured very accurately, determination of voltage via Eq. (2) is usually more precise and doesn't involve extra contact pads.

Dissipation in the total system determines the linewidth, $\Delta\nu$, of emitted radiation. As with the emitted power, theoretical prediction is possible only in very limited cases. For the RSJ model with no connection to the environment other than the dc current supply, and assuming only white noise fluctuations in R, one finds the FWHP (**F**ull **W**idth **H**alf **P**ower) linewidth of internal Josephson radiation [12] to be

$$\Delta\nu = \pi\left(\frac{1}{\Phi_0}\right)^2 R_d^2\, S_I'(\nu), \tag{7}$$

where R_d is the dynamic resistance, i.e., the slope of the IVC at the bias point, and $S_I'(\nu)$ is the current fluctuation power spectrum. For small fluctuations the current fluctuation spectrum consists of two contributions [13], one at $\nu = 0$ and one at $\nu = \nu_J$, where ν_J is the (Josephson) frequency corresponding to the voltage (Eq. (2)) at which the junction oscillates, so

$$S_I'(0) = S_I(0) + \frac{1}{2\gamma_{dc}^2} S_I(\nu = \nu_J)\,. \tag{8}$$

Actually the junction converts small current fluctuations to small voltage fluctuations with $S_V(\nu) = R_d^2\, S_I'(\nu)$, and via Eq. (2), to frequency fluctuations. The general form of the current fluctuations is given by [14]

$$S_I(\nu) = \frac{2h\nu}{R}\coth\frac{h\nu}{2kT}, \tag{9}$$

where T is the physical temperature and k is Boltzmann's constant. Eq. (9) includes both thermal and shot noise. For low voltages and high temperatures, $kT >> eV \approx h\nu$, Eq. (9) reduces to

$$S_I(\nu) = \frac{4kT}{R}, \tag{10}$$

which is the standard Nyquist expression for thermal noise.

If we insert Eq. (10) into Eqs. (7,8) we get a final expression for the linewidth of a short Josephson oscillator in the thermal low-voltage noise limit, $kT >> h\nu,\ eV$

$$\Delta\nu = \frac{4\pi kT}{\Phi_0^2} \frac{R_d^2}{R}\left(1 + \frac{1}{2\gamma_{dc}^2}\right). \tag{11}$$

Often one tries to correct for the omission of the 1/f spectral noise (and other noise sources) by replacing the physical temperature, T, with an effective temperature, T_{eff}.

The linewidth given in Eq. (11) relates only to internal dissipative processes due to the quasiparticles in the uncoupled junction. If it is weakly coupled to the environment, e.g., via a broadband stripline to a receiver, we may assume that the detected radiation has the same linewidth.

When the junction is strongly coupled to the environment, Eq. (11) must be modified to include also current fluctuations in the Cooper-pair current. This may seem surprising since the supercurrent is a result of the coherence between the macroscopic wavefunctions in the two superconducting electrodes. This process involves no dissipation and thus no fluctuations. Only when there is an interaction between the Cooper-pair condensate and a dissipative environment, e.g., by coupling to the photons in a lossy cavity [14], lossy waveguide modes, plasmons in a lossy junction, etc., the dissipation leads to fluctuations in the supercurrent (and the quasiparticle-pair interference current). The complexity significantly increases when we consider (below) one- and two-dimensional LJJs coupled to the environment via various spatial boundaries of the sine-Gordon system.

2.3 THE JOSEPHSON RESONANT SOLITON OSCILLATOR (RSO)

The Josephson resonant soliton oscillator (RSO) is the simplest flux-flow device. It consists of an open-ended 1-D LJJ forming a resonant section of a low-damping single flux quantum (SFQ) transmission line. A fluxon is driven by the bias current, and reflected at the two end boundaries, oscillates back and forth as an antisoliton and a soliton, respectively. This system has been investigated for more than 25 years. In the low-damping limit the sine-Gordon equation with boundaries as given in Eqs. (3,4) and the associated particle picture is a well documented model. The dynamics of the RSO is now understood and experimentally verified - see, e.g., [4,5,9,15] - and the properties relevant for this review may be summarized as follows.

These fluxon dynamics manifest themselves in the IVC as the so-called *Zero-Field Steps* (ZFSn) with asymptotic voltages

$$V_{ZFSn} = n\Phi_0\bar{c}/L\,, \tag{12}$$

where n is an integer. This follows from Eq. (2) if we insert the total $\Delta\phi = 4\pi$ phaseshift experienced in the two soliton-antisoliton reflections per oscillation period, $\Delta t = 2L/u \cong 2L/\bar{c}$. The integer n refers to n solitons and n antisolitons passing a given point in the junction per period. The power-balance velocity, u, approaches $\bar{c}$ at the top of the ZFS, which experimentally appears as a nearly constant voltage step. From Eq. (2) this also means that the oscillation frequency is practically constant and related to the length of the junction. The fundamental frequency of the signal emitted from one end of the junction is given by the modified relation (compared to Eq. (2))

$$\nu = \tfrac{1}{2}\frac{1}{\Phi_0}V_{ZFS1}\,, \tag{13}$$

where the factor $1/\Phi_0$ corresponds to 483 GHz/mV.

Experimental observations of the emitted fundamental-frequency radiation also when biased at the n'th ZFS [16] show that n fluxons can move together as a group. Early numerical simulations showed that the bunched multi-soliton state is stabilized by the surface damping (ϕ_{xxt}-term in Eq. (3)). Physically, a fast-moving fluxon leaves behind a "tail" of oscillating (plasma) waves which may act as a moving trap for other fluxons and thus form a multi-fluxon state. This picture has been verified in the annular junction [17]. The observed fine-splitting of the ZFS's [9,18,19] can be explained as a breaking of the equidistant fluxon chain into different bunched configurations, e.g., the ZFS4 with states (0+4), (1+3), and (2+2). For radiation at the m'th harmonic measured on the n'th step, a prefactor (m/n) is needed on the r.h.s. of the modified relation in Eq. (13) [20].

The other prominent structure in the IVC is the *Fiske Step* (FS), observed when a dc magnetic field is applied to the junction. In a *short* junction ϕ_x is proportional to the applied field and a Fiske resonance is observed when the spatial period coincides with the spatial period of the n'th resonant electromagnetic mode in the junction. The resonance is a highly nonlinear process that also generates self-induced Fiske steps (FSn) in the dc IVC. Due to the large electromagnetic impedance mismatch at the end boundaries, the quality factor, Q, of the Fiske resonance is given only by the internal shunt and surface losses of the junction. For high Q values the resonances are very sharp, and correspondingly, the Fiske steps have nearly constant voltages. In the IVC they appear as spikes, since we can only observe the low-voltage side of the resonance curve using an ideal current supply.

In the *long* junction the processes are more complex but it has been shown that Fiske steps appear as solutions to Eqs. (3,4) [21]. Physically this may be understood [16] in the fluxon model where a fluxon driven down the junction by the bias current collides with the ($x=L$) boundary and is annihilated. This process generates plasma waves (or breathers) that with sufficiently low damping return through the junction and at the ($x=0$) boundary trigger a new fluxon to enter the junction. This maintains the synchronization, but gives only a 2π phase shift per period. The FSn thus appear in the IVC at voltages

$$V_{FSn} = n\Phi_0\bar{c}/(2L) = (1/2)\,V_{ZFSn} \tag{14}$$

in the limit where both the plasma wave and the soliton attain the maximum (Swihart) velocity at the top of the step. Radiation from the Fiske steps follows the standard Josephson relation, Eq. (2). From Eqs. (13, 14) we see that even-numbered FSn's (n = 2, 4, ..) have nearly the same limiting voltages as the ZFSn's (n = 1, 2, ..). The voltage (or frequency) separation between successive ZFS's and FS's decreases inversely with junction length. The step height also decreases as L increases.

Originally, Kulik [22] used a superposition of modes to explain the FS phenomenon in the LJJ. The Fiske resonances are very important to the understanding of the FFO and the CFFO. In order to maintain the soliton bouncing at the ends of the LJJ, the collisions must be nearly elastic, i.e., the radiation loss at the junction ends must be small. This means that the available power is very low, usually attaining its maximum at the top of the step, as argued in connection with Fig. 3. In the other limit - if the end boundaries easily absorb the soliton - both the ZFS and the FS resonances disappear. This has been observed, e.g., in so-called window junctions where a smooth step in the SiO insulation seems to provide a "matching" condition.

Usually some radiation emission may be detected when the junction is biased at ZFS's and FS's, and the spectral linewidth, $\Delta\nu$, has been thoroughly investigated in the LJJ both experimentally and theoretically [23,24,20]. By adding white noise thermal fluctuations to the fluxon model in the low damping limit we find the following expression for the FWHP linewidth at the m'th harmonic of the fundamental frequency with the junction biased at ZFS1 [23]:

$$\Delta\nu_{ZFS1} = m^2 \frac{\pi kT}{\Phi_0^2} \frac{R_d^2}{V/I} . \tag{15}$$

Here R_d is the dynamic resistance at the bias point (associated with the momentum of the particle), and V/I is the "bias" resistance (associated with the kinetic energy of the particle). In comparing this relation to Eq. (11), a factor of 4 is missing due to the modified Josephson relation (Eq. (13)). Also, the γ-dependent factor is absent.

A large experimental and theoretical effort also has been devoted to fluxon dynamics in quasi-one-dimensional LJJs with annular geometry [25,26,27,28]. Due to flux quantization, solitons already trapped in the hole of the structure cannot escape. Localized e-beam or laser light heating, however, can be utilized to "drag" fluxons into or out of the structure [29]. The system simulates an infinite line without boundary conditions for the fluxons, and so is very useful for the study of fluxon-fluxon interactions, bunching of fluxons, and various damping mechanisms. Plasma waves, breathers and electromagnetic radiation, of course, resonate if they satisfy the 2π periodicity. The topology seems to prevent direct radiation emission.

In summary, the RSO is not a very practical oscillator. It may attain very small R_d values and thus small linewidths (a few kHz at 10 GHz) when biased on a ZFS or a FS [23]. Unfortunately, it has a very limited operational frequency since the step structures disappear well below voltages corresponding to about 100 GHz. The available power, therefore, is very low. About 90 nW at 75 GHz is the maximum that has been transferred to a tightly-coupled SIS junction [24]. Since the soliton assembles a δ-function, the harmonic content of the power emitted during its collision at the end boundary is very high. Last, but not least, its frequency cannot be tuned since it is linked to the physical length of the junction. The main feasible application of the RSO is as an on-chip, phase-locked clock oscillator.

2.4 THE JOSEPHSON FLUX-FLOW OSCILLATOR (FFO)

2.4.1 Basic properties of the FFO

The FFO is a quasi-one-dimensional LJJ in which an applied dc magnetic field, H_{appl}, exceeding H_{c1} introduces a tightly-packed chain of fluxons which (in a unidirectional flow) is driven through the junction by the bias current, I_B [30,31]. The mutual repulsion between the fluxons tends to keep the fluxons equidistant except at the two ends of the junction where generation and annihilation take place. Usually a current, I_{cl}, in an integrated overlaying control line generates the dc magnetic field applied to the FFO. The velocity and density of the fluxons and thus the power and frequency of the emitted mm-wave signal, may be tuned easily over a wide range by either of the two external parameters.

The main flux-flow resonance (with no external signals and no connection to the environment except via I_B and H_{appl}) is characterized by excitations which travel on top of the fast-rotating fluxon chain. The more tightly packed the chain becomes with increasing H_{appl}, the more linear the system becomes. It is therefore natural [32] to assume a solution to Eq. (3) of the form

$$\phi(x,t) = \varpi t + \Gamma x + \Psi(x,t) + \theta_1, \tag{16}$$

where $\psi(x,t)$ denotes a small ($\psi << 1$) modulation around the rotating background. An instructive picture of the background phase is a cork screw with (normalized) pitch Γ, that rotates with (normalized) angular frequency ϖ. θ_1 is an arbitrary phase constant. Inserting Eq. (16) into Eq. (3) and linearizing around $\Psi_0 = \varpi t + \Gamma x + \theta_1$, one gets an equation for $\psi(x,t)$ whose solution may be expressed as a Fourier series in space. The coefficients can be found by standard power balance. In the low-damping limit, noninteracting particle picture, one is able to calculate the IVC for the FFO [32].

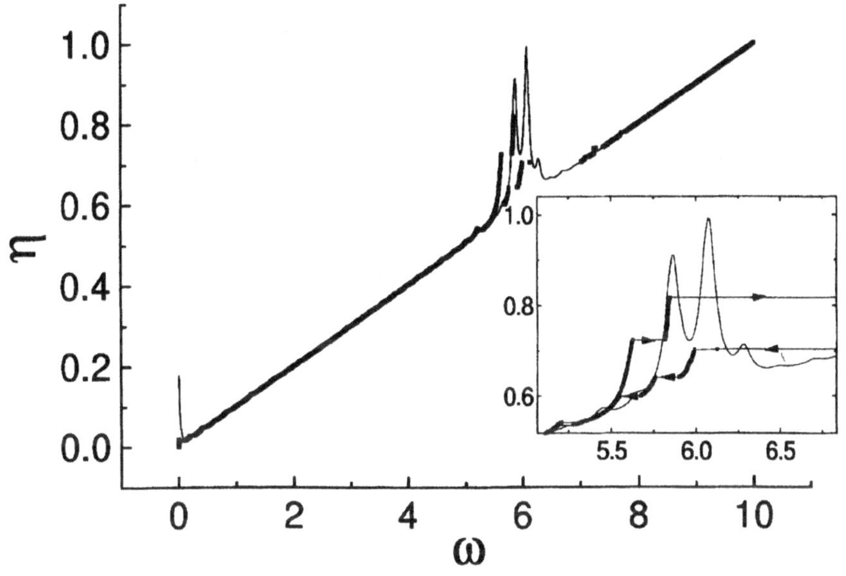

Figure 4. IVC for FFO with normalized length $L = 15$, damping $\alpha = 0.1$, and magnetic field $\Gamma = 6$ calculated numerically (thick line) and analytically (thin line) at the resonance condition where Γ equals the normalized frequency ϖ. Eq. (2) relates ϖ to the real voltage V, and $\eta = I/I_c$ is the normalized dc bias current (normalized quantities are defined in text). The FFS consists of an infinite series of Fiske steps, FSn, with fixed voltage separation and modulated by a fast Fraunhofer factor that strongly enhances only a few Fiske resonances around $\varpi = \Gamma$. All switching is horizontal due to the assumed ideal current bias source. Since the voltage separation between FS is $\propto 1/L$, the FSn merge as the junction gets longer, reaching the Eck peak for $L\rightarrow\infty$ even in the low-damping limit. For higher damping each FSn broadens, and the Fraunhofer pattern merges into a single Flux-Flow Step, FFS. (From [33]).

Fig. 4 shows a comparison between the analytical and numerical results obtained in Ref. [32]. The FFS consists of an infinite series of Fiske steps, FSn, equidistant in voltage - see Eq. (14) - with separation $\Delta V_{FSn} = \Phi_0 \bar{c}/(2L)$, and modulated by a fast Fraunhofer factor that strongly enhances only a few Fiske resonances around $\varpi = \Gamma$. In the infinite junction limit for $L\rightarrow\infty$, even in the low-damping limit it ends with a single peak (the Eck step) in agreement with the result given by Eck *et al.* [30]. The voltage of the Eck step varies linearly with the applied magnetic field.

At higher damping each of the Fiske resonances broadens, the FS*n*'s gradually begin to overlap, and the structure transforms into a single curve, the so-called Flux-Flow Step (FFS). In a real LJJ the dc voltage of the FFS depends linearly on the applied magnetic field, H_{appl}, and is approximately given by

$$V_{FFS} = d\,\bar{c}(\mu_0 H_{appl}), \qquad (17)$$

where $d = 2\lambda_L + t$ is the effective magnetic thickness of the junction. It is assumed that the fluxon chain (nearly) moves with the Swihart velocity.

It is somewhat misleading to call the low-voltage side of the FFS the Eck step. The term "velocity matching (VM) step" was used in the early literature, since the fluxon chain at the flux-flow resonance moves with a velocity that nearly reaches the Swihart velocity $\bar{c}$.

Recently, using the same theoretical approach, a uniformly-distributed rf current has been applied in order to study injection phase-locking of the FFO [33]. The rf current is homogeneously supplied from an ideal current source, and hence can be included as an additional term $\gamma_{ac}\sin(\Omega t + \theta_0)$ on the r.h.s. in Eq. (3). Injection locking of the FFO via the magnetic end-boundary conditions, Eq. (4), has been realized both experimentally and theoretically, see Section 2.4.3.

The FFO can deliver much higher power than the RSO. About a microwatt at 400 GHz was first reported [31]. Later, other designs [34,35,36] increased both the frequency and the power with a simulated maximum of 20-40 μW at 100 GHz – 1.2 THz in a NbN LJJ [37]. In contrast to the high harmonic content in the output from an RSO, the arrival of the tightly packed and nearly equidistant fluxon chain on the junction end ($x=L$) makes the output of an FFO nearly sinusoidal [38]. As also shown by Zhang [38], output power strongly depends on load impedance.

Emitted power from the FFO may be detected either with a remote sensitive microwave receiver, or as first suggested by Nagatsuma *et al.* [31], be coupled to a SIS detector placed on the same chip as the FFO, or as first shown by Dubash *et al.* [39] by frequency division in an RSFQ-TFF circuit. All three schemes enable linewidth measurements to be made with high spectral resolution.

The idea of using an FFO as a local oscillator (LO) for an SIS mixer to form an on-chip integrated heterodyne receiver was proposed in 1993 by Koshelets *et al.* [35]. Since then numerous improved chip designs of receiver components, see [40,41,42,43,44,45] and references therein, have been made. This, together with external phase-locking of the FFO [46,47] discussed below, has evolved the all-superconducting mm- and sub-mm-wave integrated receiver into a practical device for a variety of applications as mentioned in the introduction. Below we shall concentrate on the linewidth, tunability and phase-locking of the FFO.

2.4.2 dc IVC and linewidth

Preliminary FFO spectral linewidth measurements have demonstrated encouraging values (130 kHz at 70 GHz [29], about 1 MHz at 140 GHz [48] and 2.1 MHz at 320 GHz [34]). A linewidth as low as 750 kHz at 280 GHz was obtained when beating two free-running FFOs in an SIS mixer and by harmonic injection locking (n = 4, 5, 6, 7) of one of the FFOs to an external 70 GHz reference oscillator [36,49].

A typical Nb/AlOx/Nb test layout [50,51] for linewidth and power measurements is shown in Fig. 5. The chip contains an overlap FFO junction, microstrip transformers, dc-blocks and coupling circuit, and an SIS mixer. The FFO length, L, is ≈500 μm and the width, W, is about 3 μm. The critical-current density, j_c, is in the range 2 - 8 kA/cm^2, giving $\lambda_J \approx 8 - 4$ μm. The SIS mixer measures the power of the radiation emitted from the right end of the FFO. The power is transmitted to the SIS mixer via a microstrip transformer, a high-pass filter, and two dc blocks. As discussed below, the SIS junction is used also as a heterodyne harmonic mixer for linewidth measurements. The bottom Nb electrode forms a control line in which a current, I_{cl}, generates the applied magnetic field that threads the junction uniformly by penetrating a resistive film (hatched in Fig. 5) evaporated parallel to the FFO junction. The dc bias current flows through the resistive film (from "+" to "-" in Fig. 5). Not fully shown to the left in Fig. 5 is a fin-line antenna inserted in a rectangular 60-90 GHz waveguide from which an external microwave signal can be supplied for harmonic injection locking of the FFO [36,49].

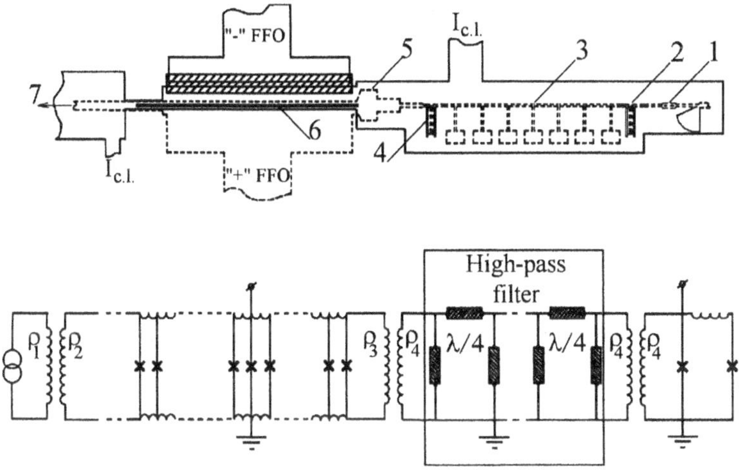

Figure 5. Chip layout and simplified equivalent diagram of the test circuit for FFO linewidth measurements. "1" - SIS mixer; "2", "4" – dc-blocks; "3" - high-pass filter; "5" - Chebyshev impedance matching transformer; "6" - FFO; "7" - connection to the input transformer and fin-line antenna. In the equivalent diagram ρ_i: ρ_j indicate impedance transformers. An array of discrete short junctions connected in parallel (marked by ×) models a long Josephson junction.

Typical IVCs of an FFO recorded at different magnetic fields (produced by I_{cl}) are shown in Fig. 6. It should be noted that the FFO is shaped with an "unbiased overlap extension" (left end of the FFO where the fluxons enter the LJJ - see Fig. 5) used to increase the slope of the FFS [31]. Consequently, the FFO has a strongly asymmetric magnetic pattern $I_c^{max}(H_{appl})$, where the maxima occur symmetrically at non-zero external magnetic fields.

In Fig. 6 Zero-Field Steps (ZFS's) are observed in the IVC of the FFO at very small magnetic fields. The so-called Displaced Linear Slope (DLS) is also observed at low magnetic fields. With increasing H_{appl} - see Fig. 6 at $V < 400$ μV - the DLS branch shifts almost linearly with H_{appl} toward higher voltages. With the FFO biased in the DLS regime, we found a "smearing" of the IVC of the SIS detector at $V \approx 0$ and $V \approx V_g$ (instead of the usual distinct superposition of quasiparticle and Shapiro steps). It means, most probably, that the FFO, when biased on the DLS, is in a chaotic state and operates as a wide-band noise source. Such an extremely broad radiation linewidth was first reported for an FFO biased at the DLS [52], where it also was shown by numerical simulations that the FFO dynamics at the DLS is characterized by irregular fluxon oscillations that resemble a chaotic state. Qualitatively, the complicated fluxon dynamics can be attributed to excitation of internal oscillation modes in a "soft" fluxon chain at weak magnetic fields.

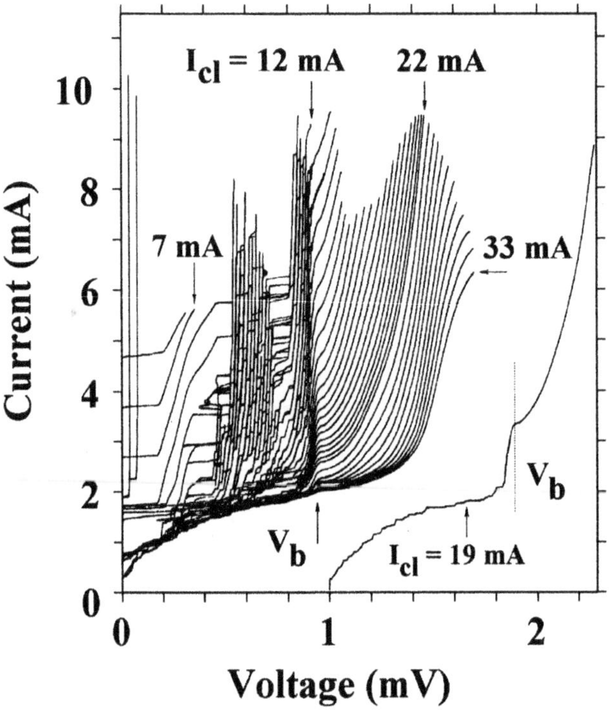

Figure 6. Typical IVCs of a test FFO recorded at different applied magnetic fields supplied by a control-line current, I_{cl}. The increment ΔI_{cl} is 0.5 mA for 7 mA < I_{cl} < 22 mA, while ΔI_{cl} = 1 mA for 22 mA < I_{cl} < 33 mA. The bottom right curve, which is shifted by 1 mV, shows the definition of the "boundary" voltage $V_b \approx V_g/3$ for one of the IVCs, recorded at I_{cl} = 19 mA. The gap voltage $V_g \approx 2.8$ mV at 4.2 K.

With increasing magnetic field the DLS abruptly transforms into a series of resonant high-numbered Fiske Steps (FS) as clearly seen in Fig. 6. This splitting takes place up to a specific "boundary" voltage, V_b, where the FS's disappear and evolve into the FFS. For fixed bias current the voltage of the FFS is proportional to the applied magnetic field, Eq. (17). At the same voltage, V_b, the maximum (switching) current of the FFS abruptly increases. As also seen in Fig. 6 for $V > V_b$, the FFS becomes smooth, and with increasing magnetic field it persists up to the gap voltage, $V_g \approx 2.8$ mV. It should be noted that this "boundary" is typical for all investigated LJJs with high current density ($j_c > 1$ kA/cm^2). This feature does not depend significantly on the exact junction geometry and its dimensions or on the coupling to the external microwave circuits. In order to clarify the definition of V_b the IVC recorded with a control-line current of 19 mA is shown as the 1mV-off-set curve in Fig. 6.

The IVCs depicted in Fig. 6 clearly show the influence of damping on fluxon dynamics. For low normalized damping $\alpha l < \pi$ (with normalized length $l = L/\lambda_J$), the most prominent step structure is the discrete and nearly equidistant FSn with n as high as 45. The fact that the Fiske steps are much higher than seen in Fig. 4 reflects the fact that the damping here is much lower than $\alpha = 0.1$. In Fig. 6 for $V > V_b \approx 950\ \mu V$, there is a transition from the region with (superimposed) characteristic FS's into a region where the FFS becomes smooth and single valued. This is because the normalized damping exceeds the critical limit $\alpha l \geq \pi$. As discussed above [32], this transition reflects the interplay between the distance between successive FS's (given by the physical length L) and the frequency width of the Fiske resonance (given by the effective junction damping).

Experimentally one finds [51] that the ratio $V_g/V_b \approx 3$ is independent of temperature, and thus the "boundary" voltage on the FFO IVCs can be explained by Josephson Self-Coupling (JSC) [53,54] which basically is absorption of the *ac* Josephson radiation energy by quasiparticles. This leads to the well-known phenomenon of *photon-assisted tunnelling* (PAT). The JSC which can be derived [53,54] from Werthamer's equation results in current steps at voltages V determined by $eV + phf = 2\Delta = eV_g$, where hf is the photon energy, p is an integer and Δ is the superconductor gap energy. Taking into account the Josephson relation $hf = 2eV$, the voltage position of the JSC current bumps, V_{JSC}, may be expressed as $V_{JSC} = V_g/(2p + 1)$, which gives $V_{JSC} = V_g/3$ for $p = 1$. The inherent JSC damping and the corresponding large value of the dynamic resistance, R_d, is a serious obstacle to the application of the FFO in the frequency range corresponding to voltages $Vg/3 < V < Vg$ due to increased linewidth. As discussed below, this obstacle may be overcome in a CFFO.

In the first FFO linewidth measurements [29], radiation was detected by a fin-line antenna - see Fig. 5 - coupled to a standard 76 GHz rectangular waveguide receiver. Since receivers for higher frequencies were not available, the next step was to beat two FFOs in an SIS mixer and measure the lowest beat frequency directly with a microwave spectrum analyzer. Assuming that the two FFOs were uncorrelated, one could derive a linewidth of 2.1 MHz at 320 GHz [34], and 1 MHz at 140 GHz [48]. A linewidth as low as 750 kHz at 280 GHz was obtained with the same scheme, but with one of the two free-running FFOs locked by harmonic injection (n = 4, 5, 6, 7) from a 70 GHz Gunn oscillator [36,49].

The R^2_d dependence in Eq. (11) was verified over nearly one decade for large values of R_d by measuring $\Delta\nu$ on and along FS's similar to the ones seen in the IVC of Fig. 6 [36,49]. In order to fit the results to Eq. (11) (without the γ-dependent term) we had to use an effective temperature $T_{eff} = 32$ K. For low R_d values, $\Delta\nu$ seems to depend linearly on R_d. In any case, it is surprising that the FFO data can be fitted to Eq. (11) (still without the γ-dependent term) which is valid for a short Josephson junction. A physical argument is that the values for R_d and R, which are derived from the IVC, in fact, are the long-term mean values "found" by the FFO itself, so that the values already include all internal processes, boundaries, etc.

In 1996 a novel and more versatile method for linewidth measurements was devised [50] using the chip lay-out shown in Fig. 5. In principle, the method can be used with *any* superconducting oscillator operated in conjunction with an SIS mixer. Due to the strong nonlinearity of the SIS array, it is one of the most effective high-number harmonic mixers in which the FFO signal beats with, say, the 50-th harmonic of an applied reference signal (~ 10 GHz) fed from a high-quality synthesized microwave generator to the SIS mixer via a standard (existing) IF coaxial cable. The proposed method resembles the use of an external harmonic semiconductor mixer (traditionally pumped by a low-frequency LO) for mm-wave upgrade of a microwave spectrum analyzer.

Using this novel method it is possible to measure the FFO linewidth all along any FS*n* or FFS. With the 70 GHz injection method it was possible to measure only the linewidth in a narrow range (± the bandwidth of the IF amplifier) around each of the voltages that correspond to the harmonics of the 70 GHz pump. The most recent measurements of $\Delta\nu$ versus R^2_d using this novel method are shown in Fig. 7.

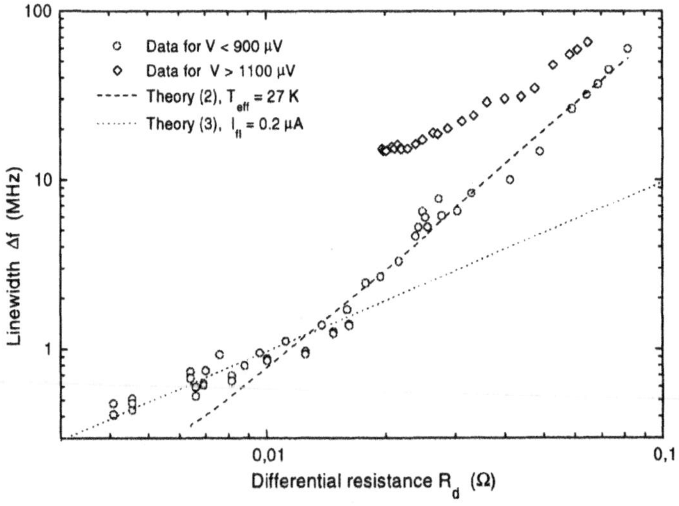

Figure 7. Dependence of the FFO linewidth on the differential resistance R_d. Data are recorded at $V < 900$ µV (circles) and $V > 1100$ µV (diamonds) - see Fig. 6. Theoretical dependence (Eq. 11) for the lumped (short) tunnel junction with T_{eff} = 27 K is shown by the dashed line; the dependence (Eq. 18) for a low-frequency noise current I_{lf} = 0.2 µA is indicated by the dotted line.

Eq. (11) (without the γ dependent term) describes a nonlinear superposition of thermal and shot noise. It should be noted that this relation does not take into account the spatial variation of the tunnel current along the FFO, the interactions of the moving fluxons and the influence of the external low-frequency interference. All these effects are believed to increase the FFO linewidth.

Fluctuations in the external magnetic field can be attributed to the differential tuning resistance of the control line $R_d^{CL} = \partial V_{FFO}/\partial I_{CL}$ for fixed dc bias current I_B. In the case of *external interference,* both the differential resistance $R_d^B = \partial V_{FFO}/\partial I_B$ and R_d^{CL} will "convert" low-frequency external noise currents, $I_{lf}^{(B,\ CL)}$, to frequency fluctuations following the same relations

$$\Delta\nu = (2/\Phi_0) I_{lf}^{(B,CL)} R_d^{(B,CL)} . \tag{18}$$

At the lowest R_d values in Fig. 7, external low-frequency noise (interference) seems to be dominant, with linear $\Delta\nu(R_d)$ dependence [51] indicating the presence of relatively strong external noise current. The dotted line in Fig. 7 is calculated using Eq. (18) with

$I_{lf} = 0.2$ µA. Of course, any external noise on the bias-current line and the control line should *a priori* be avoided. This, however, is a difficult experimental task.

According to Eqs. (11,18) the radiation linewidth may be further reduced by lowering the spectral noise density and/or the differential resistance at low frequencies, $\nu < \Delta\nu$. This can be done by appropriate shunting of the junction at low frequencies and/or suppression of low-frequency current fluctuations by an external phase-locked-loop (PLL) system with a bandwidth larger than $\Delta\nu$. The feedback can be applied through either of the two bias-current channels. A significant reduction of the FFO linewidth relative to the intrinsic FFO linewidth (determined by wide-band thermal fluctuations) is demonstrated experimentally by using an external electronic PLL as discussed in the next section.

In Fig. 7 for $V > 950$ µV in the "pure" flux-flow regime with high damping, the measured linewidth is much higher than that found in the low-damping "resonant" regime. Furthermore, the $\Delta\nu(R_d)$ dependence significantly deviates from Eq. (11), even at large R_d values, where the contribution from the low-frequency noise Eq. (18) should be small. This is not yet fully understood.

The first attempt to find theoretical expressions for the linewidth of the FFO [55], based on thermal fluctuations in the inter-fluxon distance on the propagating chain, gave far too large values for $\Delta\nu$. The theory was later corrected by Betenev and Kurin [56] and now seems to agree with the experimental linewidths given above in the limit of large R_d values. It also complies with the experimental linewidth given in Ref. [34]. Recently, Baryshev fitted the data given in Fig. 7 to a self-consistent phenomenological model [45,57] based on short-junction expressions, Eq. (11) and Eq. (18). In our opinion, however, there still exists no reliable and firmly-based theory for the linewidth of the FFO covering the entire range of parameters, especially including JSC and frequency-dependent surface damping at high frequencies. A straightforward attempt would be to accept frequency dependence in the damping terms, include a noise current fluctuation term on the r.h.s. of Eq. (3), and with the ansatz in Eq. (16), try to follow the linearization and power-balance approach used in Ref. [32].

2.4.3 *Phase-locking of the FFO*

The FFO may be injection phase-locked to external high-frequency sources via either the end boundaries [36,49,58,59,60] or a spatially homogeneous ac current applied (as is the dc bias current) to the long boundary [33]. Experimentally, locking via the "magnetic" end boundaries leads either to vertical (constant-voltage) satellite steps both on the FS*n* and on the FFS [36,58] or to symmetric sidebands with a replica of the FFS [59,60]. Even zero-crossing steps on the FFS have been observed [58]. Phase-locking by injection via the long boundary has not yet been seen in experiments, but simulations produce sidebands with replication of the FFS [33]. Also, harmonic injection phase-locking is possible via both boundary conditions [36,49]. The external Phase-Locked-Loop (PLL) discussed below is a much more advantageous method for controlling the FFO with respect to both tunability and simultaneous phase-locking. However, in order to apply the PLL we must reduce the broadening of the FFO linewidth so that it is considerably smaller than the snap-in range and effective bandwidth of the PLL.

A block diagram of the PLL set-up with linewidth measurements is shown in Fig. 8 [47]. The sub-mm-wave signal coming from the FFO is mixed in the SIS Josephson mixer with the n-th harmonic of the external synthesizer frequency, f_{SYN} (about 10 GHz). In order to prevent the external oscillator signal (as well as its lower harmonics) from reaching the FFO, a high-pass microstrip filter with a cut-off frequency of about 200 GHz is used.

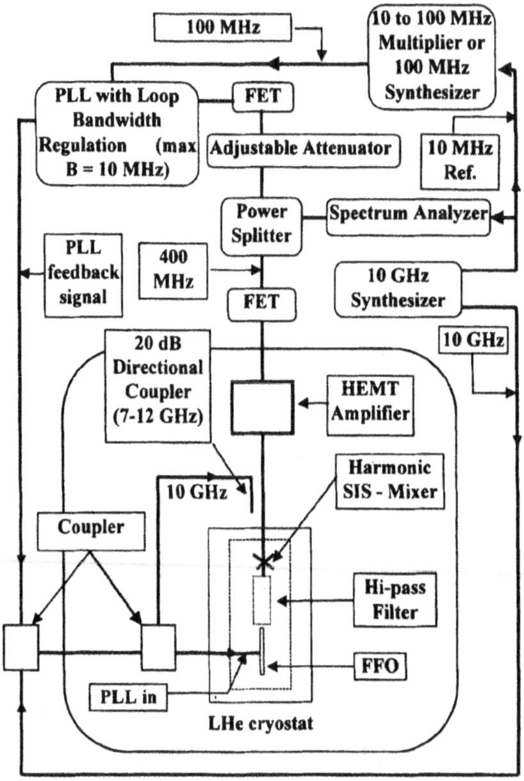

Figure 8. Block diagram of a PLL circuit and linewidth measurement set-up. The central components are the cryogenic chip with its FFO and SIS mixer, the cooled low-noise 400 MHz HEMT IF amplifier, and the 10 GHz synthesizer that also generates the 10 MHz reference signals for the spectrum analyzer and the phase detector in the PLL.

The intermediate-frequency (IF) signal with frequency, $f_{IF} = \pm(f_{FFO} - n\ f_{SYN})$ is amplified in a cooled amplifier with noise temperature $T_n \approx 20$ K and 27 dB gain. After additional room-temperature amplification, the signal enters the PLL system. A small fraction of the signal is applied via the directional coupler to a spectrum analyzer which also is phase-locked to the synthesizer by a common 10-MHz reference signal. By using this technique the down-converted FFO spectrum is measured - see Fig. 9. The recorded spectrum is the difference between the FFO signal and the n-th harmonic of the synthesizer, and thus the FFO phase noise is measured relative to the appropriate synthesizer harmonic.

In the PLL unit the signal frequency is divided by four, and in the Frequency-Phase Discriminator, it is compared with a 100-MHz reference signal, which also is phase-locked to the main 10-GHz synthesizer. The output signal, proportional to the phase difference, is returned via the Loop Bandwidth Regulator (with a maximum bandwidth of about 10 MHz) to the FFO current bias through the coaxial cable and the cold 50 Ω resistor mounted on the 4.2 K bias plate. The same coaxial cable entering the cryostat is used both for the 10-GHz synthesizer signal and the PLL control output. Couplers with microstrip filters are used to combine and split these signals.

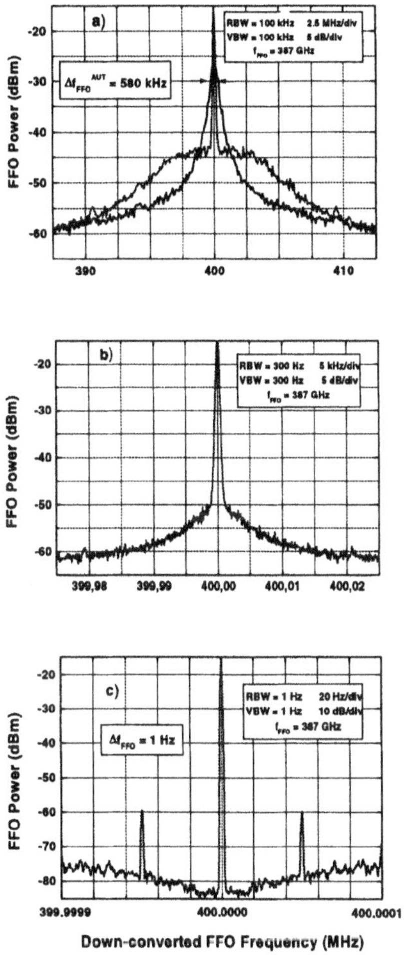

Figure 9. The down-converted IF power spectra of an FFO (f = 387 GHz) recorded with different frequency spans. The resolution (RBW) and viewing (VBW) bandwidth of the spectrum analyzer are shown in the insets to the right. (a, thick black curve). With $\Delta f_{AUT} > B_{PLL}$ only frequency locking is obtained, this yields a natural linewidth of 580 kHz. (a)-(b)-(c) Spectra recorded for $\Delta f_{AUT} < B_{PLL}$ with increasing spectral resolution clearly demonstrate phase-locking - see text. In c) two spurious 50-Hz modulation peaks are easily resolved with the 1-Hz resolution (minimum for the spectrum analyzer used).

In order to perform accurate linewidth measurements, the IF spectra must be averaged with a sufficiently small video bandwidth. The PLL system with a relatively low loop gain and narrow bandwidth setting (< 10 kHz) can be used for *frequency locking* of the FFO to the 10 GHz synthesizer in order to measure the linewidth, Δf_{AUT}, of the free-running FFO. In this case the spectral shape of the measured linewidth is rather wide but the average frequency remains stable.

The PLL system can considerably narrow the FFO linewidth if Δf_{AUT} is smaller than the PLL regulation bandwidth, B_{PLL}. For the opposite case, $\Delta f_{AUT} > B_{PLL}$, only frequency locking without a noticeable linewidth change is achieved. As seen in Fig. 9 the measured free-running linewidth of the FFO is 580 kHz at the chosen bias point. In the intermediate range where Δf_{AUT} is comparable to but smaller than B_{PLL} (2.5 MHz < Δf_{AUT} < 10 MHz), there is an increase of the FFO power at the central frequency while the FFO linewidth is reduced - measurements were done at FFO bias points with different values of R_d^B and R_d^{CL}. Full phase-locking takes place for Δf_{AUT} < 2.5 MHz. Fig. 9 shows typical IF power spectra of a phase-locked FFO measured at f_{FFO} = 387 GHz for different settings of the spectrum analyzer. An FFO linewidth as low as 1 Hz is presented in Fig. 9c. This value actually is determined by the limited resolution-bandwidth of the spectrum analyzer. It means that the FFO linewidth can be reduced below the value determined by fundamental shot noise and thermal fluctuations of the free-running tunnel junction.

A consequence of phase-locking is the appearance of a vertical step ($R_d^B = 0$) in the IVC at the voltage corresponding to the frequency f_{FFO} at which the FFO is locked. The position of this step also is insensitive to small changes in the control-line current, and accordingly also $R_d^{CL} = 0$. A voltage range as large as 1.5 μV, in which the FFO remained in phase-lock, has been experimentally measured. This corresponds to an effective PLL regulation band of about 750 MHz. The limits of the phase-lock range depend on the position of the operation point on the resonant Fiske step (FS), and it approximately equals the PLL snap-in range. It should be noted that *this step is not a harmonic Shapiro step*. First, it is shifted from the appropriate position by 0.8 μV (corresponding to a PLL input frequency of 400 MHz). Furthermore, the position of the vertical step can be tuned precisely by changing the reference signal. A reference signal in the frequency range 90 - 110 MHz can be applied from a second synthesizer phase-locked to the first one - see Fig. 8 - in steps of 0.1 Hz (which is the minimum increment of the synthesizer). This corresponds to a voltage precision of $2 \cdot 10^{-16}$ V.

It should be noted that phase-locking of the FFO at this time has been realized only on steep FS*n*'s where the free-running FFO linewidth is about 1 MHz due to small values of R_d^B. For operation at all FFO voltages including $V > V_b$, additional effort should be undertaken to decrease dynamic resistance and thus the initial FFO linewidth. An ultrawide-band PLL system with sufficiently low phase noise also is needed. This will enable continuous in-phase-lock tuning regions other than those with steep FS*n*'s. In this context the ongoing development of an on-chip integrated phase detector looks very promising. The PLL bandwidth in this case will not be limited by the electrical properties of the long interconnect cables. Also, a number of stability and noise problems related to room-temperature electronics may be avoided. The cryogenic phase detector, low noise amplifiers, etc. can be constructed using existing superconducting electronic components.

2.5 THE CHERENKOV FLUX-FLOW OSCILLATOR (CFFO)

Cherenkov radiation from moving fluxons was first suggested by Kivshar and Malomed [61]. Recently there has been increasing interest [62-70] for a diversity of reasons. First, the resonant radiation emission from nonlinear systems and structures is an interesting theoretical task in itself. Secondly, taking into account the effect of Cherenkov radiation, it is possible to explain a number of phenomena observed in experiments, e.g., microwave emission from HTSC ceramics [69], and fine structures in the current-voltage characteristics (IVC) of stacked annular Josephson junctions [70]. Finally, and rather important for applications, the tunable resonant interaction of Josephson fluxons with electromagnetic waves allows for a significant increase of the efficiency of Josephson oscillators based on fluxon dynamics. In addition to increased emitted mm- and sub-mm-wave power, the resonant nature of the Cherenkov effect simultaneously allows for significant reduction of the linewidth and the content of higher harmonics. Compared to the FFO and the RSO, where the power is emitted only from the end(s) of a one-dimensional long Josephson junction, the CFFO extracts energy from the entire fluxon chain as it propagates through the junction. The CFFO, therefore, may be considered as a new type of Josephson oscillator.

Cherenkov radiation exists when particle velocity exceeds (or equals) the phase velocity of the wave it generates. The type of radiation and its phase velocity depends on the surrounding medium. In air the phenomenon is observed when the speed of an aircraft exceeds the sound velocity. Fluxons in an LJJ cannot generate Cherenkov radiation since their velocity is always smaller than the phase velocity of its linear electromagnetic modes, $\bar{c}$, the Swihart velocity. However, as we shall see below, a "slow" wave structure can be attached to the LJJ or one can use stacked LJJs. The first clear evidence for Cherenkov radiation in an LJJ was observed in stacked annular junctions [69,70].

2.5.1 Coupling schemes

In the system under consideration an external transmission line is the medium which provides the "slow" waves that can be in resonance with the moving fluxons in order to satisfy the Cherenkov condition. There are several ways in which the waves in the external transmission line may appear slow relative to the fluxon. The simplest coupling scheme we can propose is to overlay the junction with a meander-shaped stripline that crosses (and couples) to the junction at regularly-spaced intersections (nodes). The wave generated in the stripline by a fluxon at a given node must travel a long distance in the meander path (and therefore is considerably delayed) before it again meets the fluxon at the next node. If the coupling between the junction and the stripline is weak, the propagating wave on the external transmission line obeys a simple wave equation containing a linear operator - see below.

The coupling scheme suggested above is an example of a general distributed system of Josephson junctions electromagnetically coupled to a waveguide. A distributed coupling scheme for an LJJ and the corresponding discrete electronic circuit model have been treated theoretically by Kurin and Yulin [71]. They show that the fluxons bunch in the decelerating phase of the field propagating in the waveguide in a process that closely mimics the way electrons bunch in traveling-wave and backward-wave tubes. If the waveguide system is periodic (e.g., contains resonant sections and thus appears dispersive) one may, for simplicity, treat the system in the long-wave limit assuming that its spatial period is much smaller than the wavelength of the excited wave in the waveguide. This avoids the additional difficulties with finding the necessary eigenfunctions and the dispersion relation. The general case, where the slow-wave also can be provided, but where all the coefficients in the equations describing the system depend on the space coordinate, can in principle be solved - see [71]. This, however, leads to very tedious calculations.

Recently Baryshev *et al.* [72] realized an alternative slow-wave structure consisting of a (main) transmission line (waveguide) with periodically-spaced resonant arms, each of which is end-coupled to the side of the LJJ - see Fig. 10. The delay of the excited wave in the arms makes the returning wave appear slow to the fluxons, while the excited waves in all of the side arms interfere constructively and propagate in the main line. The waveguide system is spatially periodic, but its wave equation can still be described by a linear operator. The dispersion characteristic of the linear waves exhibits a zone structure, and the required resonance condition can be achieved in a sufficiently high-numbered Brillouin zone (with resonance between a fluxon and a high spatial harmonic of the eigenfunction of the linearized operator).

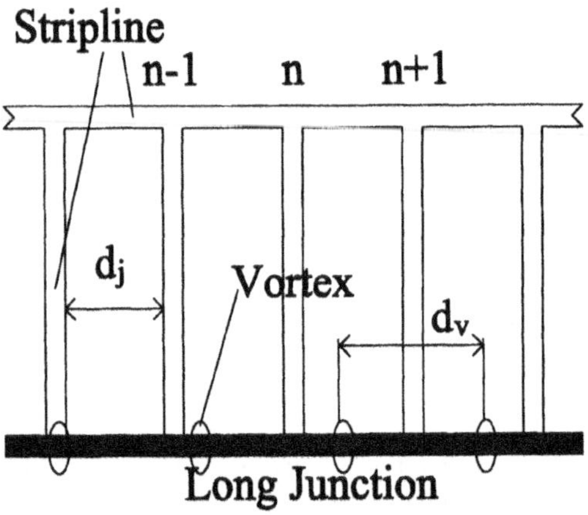

Figure 10. Scheme for coupling of a slow-wave transmission line to a long Josephson junction to form a Cherenkov Flux-Flow Oscillator (CFFO). (From [72]).

Greater detail of the CFFO working principle is as follows. The fluxons move along the LJJ, and radiation is generated and transmitted to the stripline stubs whenever fluxons scatter at the nodes between the LJJ and all the stripline ends. Let the fluxon scatter on the n-th node of the system at time $t = t_n$ where the radiation enters the stripline stub. The same fluxon scatters on the (n+1)-th node at time $t_{n+1} = t + \tau$, where $\tau = d_j / u$ is the delay time needed for the fluxon to travel in the LJJ from the n-th to the (n + 1)-th node, d_j is the junction length between successive nodes, and u is the fluxon velocity. The waves emitted due to fluxon scattering at the different nodes should be added in proper phase. In order to obtain an effective wave excitation, the following condition must be satisfied:

$$kd_l \pm \varpi(k)\tau = 2\pi l\,, \tag{19}$$

where k is the wave number, $\omega(k)$ is the dispersion of the waves in the stripline, and l is an arbitrary integer. The frequency of the radiation generated by a single fluxon can be found from Eq. (19).

For a moving fluxon chain, the radiation from all fluxons should be added in phase. This gives the additional condition, $k d_v = 2\pi m$, where d_v is the distance between fluxons, and m is an integer. This condition may be satisfied in practice by changing d_v by varying the magnetic field applied to the LJJ. First, it is possible to excite a large-amplitude wave if a large number of fluxons are scattered at the nodes and if the stripline has low dissipation. Second, the radiation amplitude can be controlled by adjusting fluxon density. It is assumed that the system has reached a stationary state where the moving fluxon chain maintains its periodicity as given by mutual fluxon repulsion, but it includes the effects of scattering at the nodes. Alternatively, one may assume that coupling to the stripline is so weak that it does not perturb the fluxon dynamics in the LJJ.

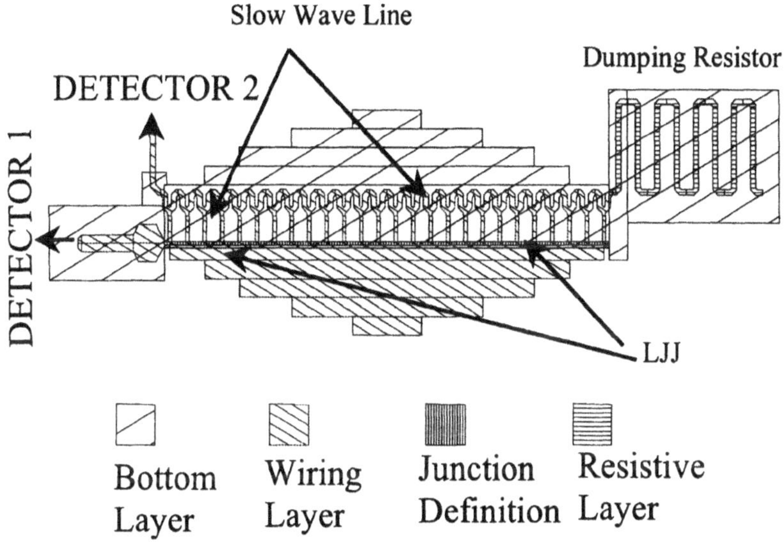

Figure 11. Practical lay-out of CFFO. (From [72].)

The layout of the experimental chip is shown in Fig. 11 [72]. The CFFO consists of a 510 μm long and 3.5 μm wide LJJ, and the sinusoidal "slow-wave" line with quarter-wave stubs ending at the LJJ. The critical-current density of the LJJ is ≈8 kA/cm^2. The distance between nodes is 20 μm. Quarter-wave stubs have been chosen in order to achieve a good impedance match between the 20-Ω impedance of the stripline and the low impedance (≈0.5 Ω) of the LJJ at 450 GHz. Additional delay in the slow line is provided by the sinusoidal meandering of the line. The external magnetic field is applied to the LJJ by a current in a control line integrated in the top electrode of the LJJ. A normal-metal stripline matching load is connected to the end of the slow-wave line to prevent reflections.

Two twin SIS-junction-array wide-band microwave detectors are attached to the experimental structure as shown in Fig. 11. Detector 2 is connected to the "slow-wave" line and detector 1 is connected to the LJJ with a microstrip circuit similar to that used for the FFO in Fig 5. Each detector has a band-pass filter and a dc block. The operating range of this circuit is 300-650 GHz. The detector design is described in detail in Ref. [45].

The power from both detectors could be measured simultaneously with the IVC of the LJJ. Each detector junction has a normal resistance of 14 Ω. As with the FFO, the microwave power is estimated from the increase of the SIS junction bias current with fixed bias voltage at the first photon-assisted-tunneling (PAT) step. This current is proportional to the absorbed *rf* power in the range of experimental parameters.

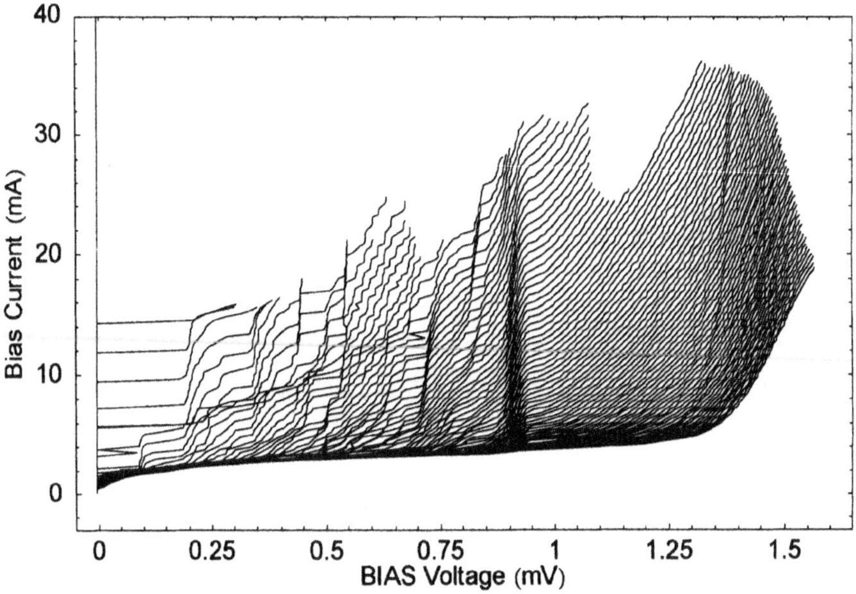

Figure 12. Set of IVCs of a CFFO recorded for incremented applied magnetic fields. Compare to Fig. 6 for the FFO. In the vicinity of 0.8 mV there is a set of resonances corresponding to the different modes of operation of the CFFO. Both the dynamic resistance, the emitted power, and the frequency can be changed with applied dc bias current and applied magnetic field. (From [72]).

The measured CFFO IVCs for increased applied magnetic field is shown in Fig. 12. It clearly resembles the usual FFO IVCs depicted in Fig. 6. The only difference is the set of resonances around 0.8 mV bias voltage at different bias currents and magnetic fields. These resonances correspond to the different modes of operation of the CFFO and confirm that the dynamic resistance, R_d, of the LJJ can be strongly reduced by the action of the slow-wave line. It is possible to tune the voltage of the resonances and thereby the oscillation frequency by adjusting the external magnetic field, while the position of the Fiske steps (FS*n*) remain fixed.

The emitted power collected in the slow-wave line is about 8 μW (detector 2). Actually, the power detected by SIS detector 1 directly from the FFO action in the LJJ is also about 8 μW, but we believe that the power in the slow-wave line may be increased considerably by proper design of the CFFO. Both forward- and backward-wave types of resonances have been observed by changing the direction of the flux flow [72]. This first CFFO design utilizes the resonant stub coupling between the LJJ and the transmission line collecting the radiation, but as mentioned above, alternative non-resonant slow-wave meander-type structures may be more useful.

2.5.2 Noise and linewidth of the CFFO

The influence of noise on the Cherenkov radiation effect has not yet been investigated experimentally. Theoretically, in the CFFO, where fluxons are moving in the periodic potential created by the accumulated radiated wave generated by the fluxons themselves, the problem becomes even more nonlinear, and currently the theory briefly described below is the only existing theory for the linewidth. As with the FFO, the first approach would be to fit the measured linewidth to the phenomenological equation (Eq. (11)) using measured (long-time average) values of R_d , V/I and T. No direct linewidth measurements have yet been made, but it is likely that one may obtain a reasonable fit using an effective temperature, T_{eff}.

In order to find analytical solutions for the CFFO with noise, we [73] have investigated the nonlinear dynamics of fluxons in general Josephson systems with dispersion and thermal fluctuations using the "quasiparticle" approach. We have restricted ourselves to the simplest case, that is for an annular system, i.e., a long annular junction coupled to an overlaid annular transmission line, e.g., a microstripline with the same radius and width as the LJJ. In the linear geometry, the interaction of fluxons with the LJJ boundaries, and collisions of fluxons and anti-fluxons; only can be dealt with using numerical simulations. We admit that the annular system may appear somewhat academic, but we can use it to demonstrate the most important features and effects. Nevertheless, the annular system can be realized in practice [27], and it is very important that the mm- or sub-mm-wave power generated by the Cherenkov process can be extracted using either the meander or the resonant antenna/waveguide systems suggested above. Both linear and annular geometry appear suitable for practical CFFO applications.

Due to flux quantization, the number, N, of fluxons trapped in an annular junction is constant, depending only on how the system was prepared. For a current-biased annular Josephson junction, the well-known sine-Gordon equation can be written [73]

$$\phi_{xx} - \phi_{tt} - \sin(\phi + \frac{2\pi Nx}{L}) = \alpha\phi_t - \beta\phi_{xxt} - \gamma_{dc} + (\xi\psi_{xx} + \kappa\psi_{tt}) + \zeta(x,t), \tag{20}$$

while the transmission line is described by the equation

$$\hat{D}\psi = -\xi\phi_{xx} + \kappa\phi_{tt}. \tag{21}$$

Here $\phi + \frac{2\pi Nx}{L}$ and $\zeta(x,t)$ are, respectively, the phase difference and the noise current. We ignore the surface-loss term and consider only white Gaussian noise. Also, we assume that the current density is constant and that the dc bias current is supplied uniformly along the junction. N is the number of fluxons trapped in the junction, and ξ and κ are dimensionless coefficients describing the coupling between the fluxons and the transmission line. Note that an annular junction can only be described locally by the differential equation for the Josephson phase difference for the case in which there is magnetic flux in the junction. The reason for this is that ϕ_J changes with $2\pi N$ when one follows a closed path along the junction. Hence, this variable is not differentiable at any point within the junction. Instead, we use in our equations the variable $\phi = \phi_J - \frac{2\pi Nx}{L}$, which is well defined in the system. The linear integro-differential operator $\hat{D}$ determines the dispersion properties of an external waveguide system that permits Cherenkov synchronization. These equations differ from the equations considered in [71] only by the noise current $\zeta(x,t)$ introduced on the right-hand side of Eq. (20).

For the special case where the frequency has an exponential correlation function, one can find an exact expression for the shape of the spectral line for arbitrary values of the modulation index, *m*. Experimental parameters show that a small modulation index, *m*<<1, is more realistic for practical CFFOs, and in this case, the radiation linewidth takes the form [73]

$$\Delta\nu_{CFFO} = \pi^2 S_\nu(0), \tag{22}$$

where $S_\nu(0)$ is the power spectrum of the frequency deviations. Still, for small values of the modulation index, it can be shown [73] that

$$\Delta\nu_{CFFO} \propto \frac{\alpha\delta}{A^2 k_r^2}. \tag{23}$$

Thus, the radiation linewidth is proportional to the damping coefficients of the wave, δ, and of the junction, α. It is inversely proportional to the emitted power, A^2, and the wave number, k_r^2, of the slow-wave at the Cherenkov resonance [73].

Inserting realistic values for the parameters, one finds that the linewidth may be reduced by factors as large as 10^5 compared to the corresponding "bare" FFO. Actually the reduction factor increases toward infinity with decreasing noise intensity. The linewidth has a minimum at a certain bias current which is slightly smaller than the current yielding the maximum amplitude of oscillation. Also, resonant behavior of both the mean amplitude and the linewidth as functions of bias current is found. In addition to the linewidth results given above, the results obtained enable an optimization of the main parameters (power, tunability and linewidth) of practical mm- and sub-mm-wave CFFOs.

We have neglected the fluxon-fluxon interaction, which is a valid assumption when the distance between neighboring fluxons is much greater than fluxon size. This condition probably will be fulfilled at the beginning of the Cherenkov radiation process. At a later time during the stationary resonant state, the interaction with the wave leads to the situation in which each fluxon is located near a minimum of the electromagnetic wave, so that the fluxon-fluxon distance doesn't decrease. The "slow" wave excited by the Cherenkov process in our description controls fluxon bunching. Inclusion of the fluxon-fluxon interaction, as well as putting the fluxons and the "slow" wave on equal footing, requires numerical simulation.

Analogous to the FFO, the CFFO can undoubtedly be phase-locked to an external reference oscillator. This, however, still needs experimental verification. The CFFO is particularly attractive, especially at frequencies above $\approx 1/3$ of the superconducting gap frequency, where the linewidth of a typical Nb/AlOx/Nb FFO increases due to higher damping caused by Josephson self-excitation of quasiparticles. Also, surface losses start to play a crucial role at such high frequencies. Finally, the CFFO concept appears very promising for oscillators fabricated with high-T_c superconductors.

2.6 HIGH-T_C FFO

A high-T_c $YBa_2Cu_3O_7$ grain boundary forms a transmission line for electromagnetic waves. The junction barrier acts as the dielectric medium with a capacitance per length given by the E field and charge distribution determined by the geometry of the grain boundary and the interconnecting grains. Similarly, the inductance per unit length is determined by the current distribution in the films and how flux distributes around (and in) the electrodes. The geometry resembles a co-planar waveguide [74]. Flux quantization must be satisfied also for the flux that penetrates into the grain boundary and into the film banks close to the junction. Following the Josephson equations it is therefore natural to describe the junction by the 1-D perturbed sine-Gordon model, Eq. (3) [75-78].

Here we discuss only the FFO aspect of flux flow in high-T_c materials as an instructive example for future applications. The large body of work on Josephson Vortex Flow Transistors (JVFTs), the Superconducting Current Injection Transistor (SCIT) and the Vortex Transistor, all is based on flux flow in LJJs, will not be considered – recent references to this work may be found in Refs. [79,80].

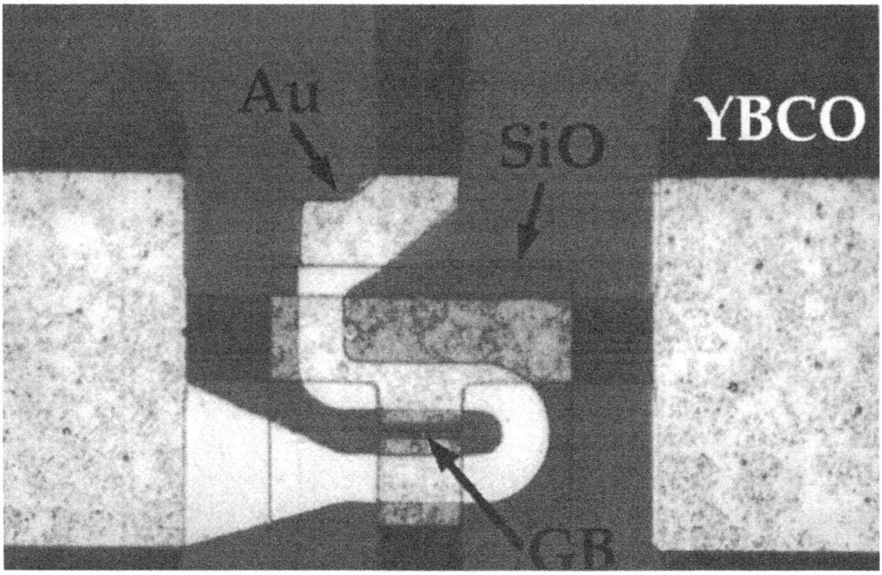

Figure 13. Photograph of YBCO grain boundary junction (GBJ) with length 20 μm A control current in an overlaying (SiO) insulated Au line generates the applied magnetic field. Layout doesn't contain the microwave antenna. Chip made by Z.G. Ivanov *et al.* [82].

Radiation emission from an HTS FFO has been observed only indirectly using the stimulated voltage response method devised by Divin *et al.* [81]. A weak microwave signal is applied to the junction, and from the shape of the ac-induced step in the IVC one may derive the linewidth using a Hilbert transformation. Generally the linewidth is several GHz wide as given by Eq. (11), reflecting the relatively large R_d values. We are presently producing a new chip design with a step-edge junction and an integrated fin-line antenna for coupling to a 60-90 GHz waveguide. Results with subharmonic generation and chaos [77,78] shown below have been obtained with a preliminary layout based on a $YBa_2Cu_3O_7$ grain-boundary junction (GBJ) with relatively low current density, as described in Ref. [76] - see Fig. 13.

The GBJs were fabricated [82] by growing a thin $YBa_2Cu_3O_7$ film (W = 200 nm, using laser ablation) on a grain boundary formed in an yttria-stabilized zirconia (YSZ) bicrystal substrate. The c-axis is oriented perpendicular to the substrate surface, and the a- and b-axes of the film were rotated by a 32° angle across the grain boundary. The Josephson junction is modeled as a very thin (thickness t) barrier sandwiched perpendicular to the substrate between the film banks. The junction area is $W * L$ (film thickness * patterned film width). A dc current, I_{mag}, in a 200-nm-thick overlying gold control loop - see Fig. 13 - produced a magnetic field in the plane of the junction.

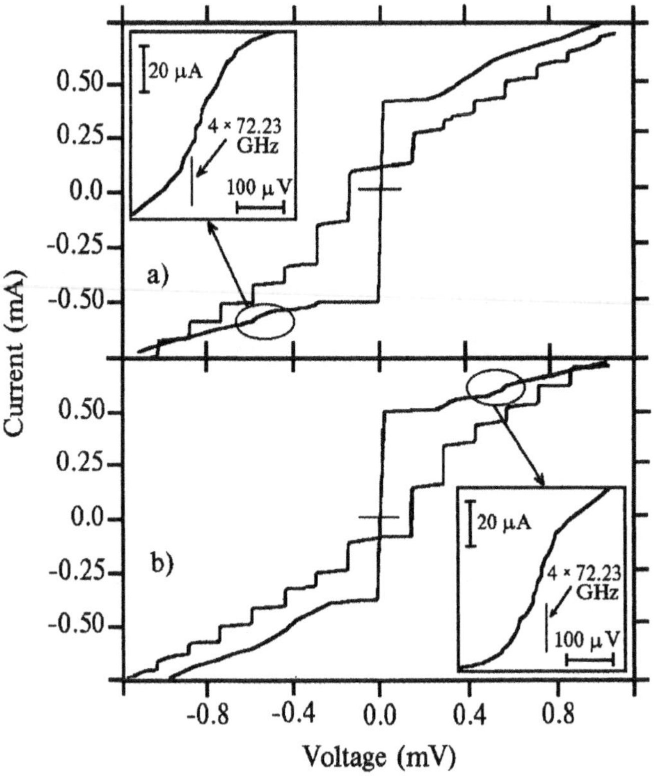

Figure 14. Pumped (72.23 GHz, att.= 2 dB) and unpumped (att.= 60 dB) IVCs of YBCO FFO at T = 4.2 K. The pump frequency is adjusted to the fourth subharmonic of the first Fiske resonance. A magnetic field must be applied in opposite directions: (a) I_{mag} = -0.21 μA and (b) I_{mag} = +0.15 μA in order to maximize the Fiske step structures FS-1 and FS+1, respectively. Unexplained fine-structure is superimposed on both FS-1 and FS+1 (blow-up in insets).

The behavior of these GBJs can be explained by the 1-D perturbed sine-Gordon equation, Eq. (3), valid for the low-T_c LJJs discussed above. This implies that the shunting capacitance, C, is distributed along the grain boundary [75,76]. The boundary conditions are as given by Eq. (4) for the overlap geometry. Space and time are normalized as usual to the Josephson penetration length, λ_J, and the inverse maximum plasma frequency, $1/\omega_{p0}$, respectively. With the shunt and surface damping described by α and β, as in Eq. (4), the total junction damping is amply described by the quality factor

$$Q = \frac{1}{(\frac{\varpi_J}{\varpi_{p0}})^2 \beta + \alpha}, \qquad (24)$$

where the (angular) frequency of the Josephson oscillation, $\varpi_J = (2\pi/\Phi_0) V_{dc}$, is related by (Eq. (2)) to the voltage, V_{dc}, at the bias point. Our junctions are overdamped with Q_{FS1} = 1.5 - 3.5 at the first Fiske step (FS1), and hence, the Fiske modes (FSn) appear as broad peaks superimposed on the nearly perfect RSJ-model IVC, shown in Fig. 14 as the unpumped curves. The Fiske-step voltages, V_{FSn}, inserted in Eq. (14) directly, give the Swihart velocity, $\bar{c}$, the propagation velocity of electromagnetic waves inside the junction. FSn with n as high as $n = 3$ have been seen [76] in these GBJs, and depending on frequency and temperature, the measured normalized Swihart velocity $\bar{c}/c_0$ = 0.02-0.04, where c_0 is the free-space light velocity.

By applying a microwave signal we have observed both high-numbered microwave-induced Shapiro steps (SSn) and chaos [77,78]. The frequency (f_a = 72.23 GHz) of the applied microwave signal was adjusted to equal the fourth subharmonic of the frequency, f_{FS1}, of the first Fiske resonance, $f_a = 1/4\, f_{FS1}$. Typical pumped curves (power is set by a room-temperature precision attenuator) are shown in Fig. 14. It should be noted that a small magnetic field must be applied in opposite directions: Fig. 14a, I_{mag} = -0.21 μA and Fig. 14b, I_{mag} = +0.15 μA in order to maximize the Fiske-step structures FS-1 and FS+1, respectively. The fine structure superimposed on the FSs is unexplained.

A typical magnetic interference pattern $I_c(I_{mag})$ at 4.2 K is shown in Fig. 15. Apart from a small off-set current, the pattern is symmetric and has the characteristic shape observed for a low-T_c long ($L/\lambda_J \approx 8$) overlap Josephson tunnel junction with few inhomogeneities. The inset shows the nearly linear $I_c(T)$ dependence. The critical temperature, T_c = (80±1)K.

The flux-flow step (FFS) is clearly identified on IVCs recorded with increased I_{mag}, and the linear dependence of its dc voltage given by Eq. (17) has been verified [76], giving

$$V_{FFS} = d\bar{c} I_{mag} F, \qquad (25)$$

where F is a demagnetization (or flux-focusing) factor, which can be derived independently from the $I_c(I_{mag})$ pattern in Fig. 15. If the dispersion is taken into account [75] the Swihart velocity found from these measurements is in experimental agreement with that derived for the Fiske steps discussed above. Actually, this enables us to number the Fiske steps unambiguously.

For the first time in long HTS junctions Shchukin *et al.* [77,78] observed chaos in YBCO grain-boundary FFO samples, as well as zero crossing microwave-induced constant-voltage steps of the type [83] which previously had been found only in hysteretic (underdamped, both long and short) low-T_c junctions. In the IVC, the steps cross the $I = 0$ axis curve, and they have the same voltage-frequency relation (Eq. (2)) as ordinary Shapiro steps. Work on HTS FFOs using step-edge samples is in progress.

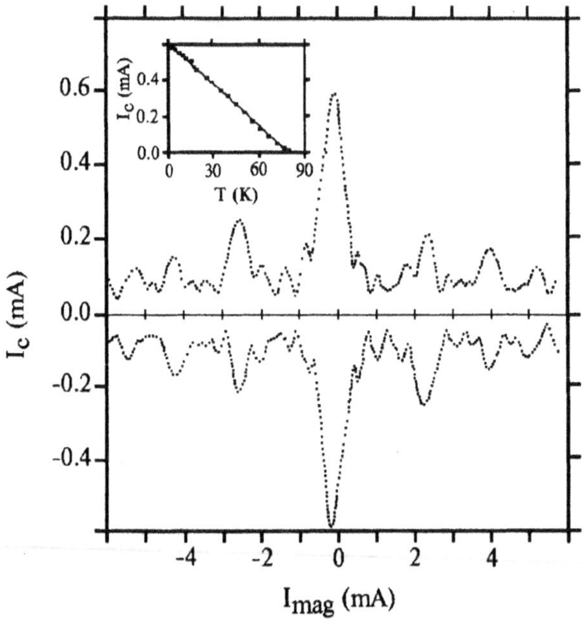

Figure 15. Typical magnetic interference pattern $I_c(I_{mag})$ at T = 4.2 K for YBCO FFO exhibiting the characteristic shape observed for a low-T$_c$ long ($L/\lambda_J \approx 8$) overlap Josephson tunnel junction with few inhomogeneities. The inset shows the nearly linear $I_c(T)$ dependence. The critical temperature, $T_c = (80\pm1)$K.

3. Stacked junctions

A new superconducting system that is described by special fluxon solutions is the long stacked Josephson junction which is sketched in Fig. 16. It consists of $N+1$ alternating superconducting and N isolating layers.

The equations for such a stacked junction system may be obtained from a generalization of the Josephson equations using the London and Maxwell equations [84,85].

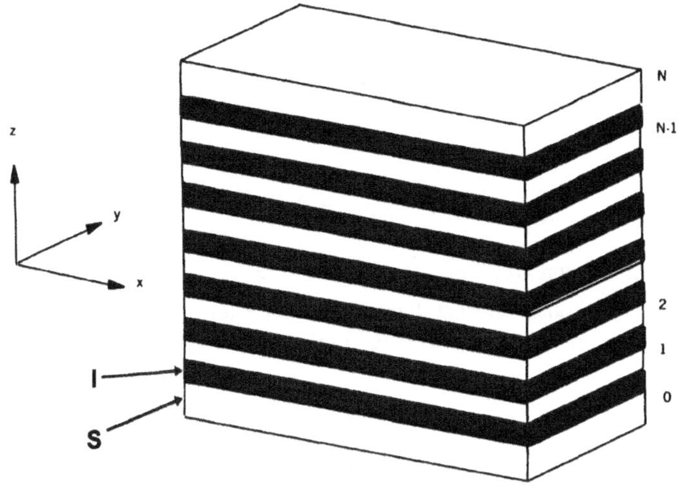

Figure 16. An example of a long Josephson stack. In a typical experiment the magnetic field H is applied in the plane of the tunnel barriers, and the bias current flows across them.

A simplified version of the stacked-junction equations may be written (in unnormalized units) as

$$\frac{\hbar}{2e\mu_0}\frac{\partial^2}{\partial x^2}\begin{pmatrix}\phi_{1,0}\\ \vdots \\ \\ \phi_{i,i-1}\\ \vdots \\ \\ \phi_{N,N-1}\end{pmatrix}=\begin{pmatrix} d'_{1,0} & s_1 & 0 & \cdots & & & \\ s_1 & d'_{2,1} & s_2 & 0 & \cdots & & \\ & & \ddots & & & & \\ & 0 & s_{i-1} & d'_{i,i-1} & s_i & 0 & \\ & & & & \ddots & & \\ & & & & & & s_{N-1} \\ & & & & 0 & s_{N-1} & d'_{N,N-1}\end{pmatrix}\begin{pmatrix}J^Z_{1,0}\\ \vdots \\ \\ J^Z_{i,i-1}\\ \vdots \\ \\ J^Z_{N,N-1}\end{pmatrix} \quad (26)$$

Here, $\phi_{i,i-1}$ is the phase difference between superconducting layers i and i-1. We define the effective magnetic thickness $d'_{i,i-1}$

$$d'_{i,i-1} = d_{i,i-1} + \lambda_i \coth\left(\frac{t_i}{\lambda_i}\right) + \lambda_{i-1}\coth\left(\frac{t_{i-1}}{\lambda_{i-1}}\right), \quad (27a)$$

and the parameter s_i, describing the coupling between layers, by

$$s_i = -\frac{\lambda_i}{\sinh(t_i / \lambda_i)}, \quad (27b)$$

where λ_i is the London penetration depth λ_L of the superconducting layer i, and t_i is the thickness of this layer. The isolating layer between superconducting layers i and i-1 has thickness $d_{i;i-1}$. Integer i runs in the interval $0 \leq i \leq N$.

From the usual Josephson-junction model, we now introduce the capacitive, resistive and Josephson currents between S layers i and i-1, and we define their total current density, $J^Z_{i,i-1}$, by

$$J^Z_{i,i-1} \equiv \frac{\hbar}{2e} C_{i,i-1} \frac{\partial^2 \phi_{i,i-1}}{\partial t^2} + \frac{\hbar}{2e} G_{i,i-1} \frac{\partial \phi_{i,i-1}}{\partial t} + J_{i,i-1} \,\mathrm{si}\; \phi_{i,i-1} \tag{28}$$

Here $C_{i,i-1}$ is the unit area capacitance, $G_{i,i-1}$, the unit area conductivity, and $J_{i,i-1}$, the dc maximum Josephson current density between S layers i and i-1.We note that an external bias current may be included in Eq. (26) if necessary, as described in Ref. [84].

3.1 FLUXON SOLUTIONS.

3.1.1 *A simple example; The coherent 2-fluxon mode*

The equations of the previous section have solutions in which fluxons in different layers move coherently (or are phase-locked). This is a very interesting type of motion that has been studied in planar systems [84,86].

For simplicity let us assume 3 superconducting layers with an evenly distributed bias current as in the overlap geometry. Also, we assume that the top and bottom layers have the same properties, and the two insulating layers have equal properties, i.e., N=3, $d_{1,0} = d_{2,1} \equiv d$, $t_2 = t_0$ (but may be different from t_1), $\lambda_2 = \lambda_0$ (but may be different from λ_1). The two tunnel junctions have the same properties described by C, G, and J. From Eq. (28) we obtain after some calculations - see Ref. [84],

$$\frac{\hbar}{2e\mu_0}\begin{pmatrix} \dfrac{\partial^2 \phi_{1,0}}{\partial x^2} \\ \dfrac{\partial^2 \phi_{2,1}}{\partial x^2} \end{pmatrix} = \begin{pmatrix} d' & s \\ s & d' \end{pmatrix} \begin{pmatrix} \dfrac{\hbar C}{2e}\dfrac{\partial^2 \phi_{1,0}}{\partial t^2} + \dfrac{\hbar G}{2e}\dfrac{\partial \phi_{1,0}}{\partial t} + J \sin\phi_{1,0} \\ \dfrac{\hbar C}{2e}\dfrac{\partial^2 \phi_{2,1}}{\partial t^2} + \dfrac{\hbar G}{2e}\dfrac{\partial \phi_{2,1}}{\partial t} + J \sin\phi_{2,1} \end{pmatrix} - I_B(d' + s)\begin{pmatrix} 1 \\ 1 \end{pmatrix}, \tag{29}$$

where

$$d' = d + \lambda_0 \coth\left(\frac{t_0}{\lambda_0}\right) + \lambda_1 \coth\left(\frac{t_1}{\lambda_1}\right) \tag{30a}$$

and

$$s = -\frac{\lambda_1}{\sinh(t_1/\lambda_1)}. \tag{30b}$$

Let us now look for a coherent mode with the property $\phi_{1,0}(x,t) = \phi_{2,1}(x,t) \equiv \phi(x,t)$. For this case Eq. (29) becomes

$$\frac{\hbar}{2e\mu_0}\frac{\partial^2 \phi}{\partial x^2} = (d' + s)\left(\frac{\hbar C}{2e}\frac{\partial^2 \phi}{\partial t^2} + \frac{\hbar G}{2e}\frac{\partial \phi}{\partial t} + J \sin\phi - I_B\right) \tag{31}$$

This equation is almost the same as the perturbed sine-Gordon equation (Eq. (3)) known from conventional long Josephson junctions [1]. The main difference is a change of the length scale due to the factor (d' + s) in Eq. (31). If we go through the normalizations, as for example done in Ref. [5], we find that the effective Josephson penetration depth $\lambda_J^{(2)}$ becomes

$$\lambda_J^{(2)} = \left(\frac{\hbar}{2e\mu_0 (d' + s)J} \right)^{\frac{1}{2}}, \tag{32}$$

and the velocity of light in the barrier is,

$$\bar{c}^{(2)} = \frac{1}{\sqrt{\varepsilon\mu_0}} \left(\frac{d}{d' + s} \right)^{\frac{1}{2}}. \tag{33}$$

Compared to the single junction soliton case, we note that

$$\frac{\lambda_J^{(2)}}{\lambda_J^{(1)}} = \frac{\bar{c}^{(2)}}{\bar{c}^{(1)}} = \sqrt{\frac{d'}{d' + s}}, \tag{34}$$

where index 1 refers to the single-junction soliton case. Since $s < 0$ (Eq. (30b)) we note that $\bar{c}^{(2)}$ is larger than $\bar{c}^{(1)}$, i.e., we may exceed the velocity of light in the single-junction case.

We also note that if we provide a bias such that we extract a bias current, I_{B1}, from the center layer with $I_{B1} = -2I_B$, the solutions discussed above become antisymmetric in the sense that now a soliton and an antisoliton move together in a coherent mode. The equation of motion, penetration depth and velocity of light can be obtained from Eqs. (31, 32 and 33) by substituting $-s$ for s. It should be noted that this coherent soliton - antisoliton state is not possible in conventional single long Josephson junctions.

Above we have concentrated on the coherent modes of the $N = 3$ system. Of course, modes other than the coherent ones are possible. An example will be shown below.

For $N > 3$ we can follow the same procedure as used above. A complication, of course, is the boundaries created by the top and bottom layers. For $N >> 1$ it may be justified to neglect effects from these two layers and assume that all the intervening layers are identical. In that case we may obtain coherent solutions of the same nature as for the $N = 3$ case discussed above. For the N-layer system ($N >> 1$) each of the N - 1 coherent solitons will then obey the equation

$$\frac{\hbar}{2e\mu_0} \frac{\partial^2 \phi}{\partial x^2} = (d' + 2s) \left(\frac{\hbar C}{2e} \frac{\partial^2 \phi}{\partial t^2} + \frac{\hbar G}{2e} \frac{\partial \phi}{\partial t} + J \sin \phi - I_B \right), \tag{35}$$

and the characteristic length and velocity are now

$$\lambda_J^{(N-1)} = \left(\frac{\hbar}{2e\mu_0 (d' + 2s)J} \right)^{\frac{1}{2}} \tag{36a}$$

$$\bar{c}^{(N-1)} = \frac{1}{\sqrt{\varepsilon\mu_0}} \left(\frac{d}{d' + 2s} \right)^{\frac{1}{2}}. \tag{36b}$$

3.1.2 The two modes of the two- fluxon case

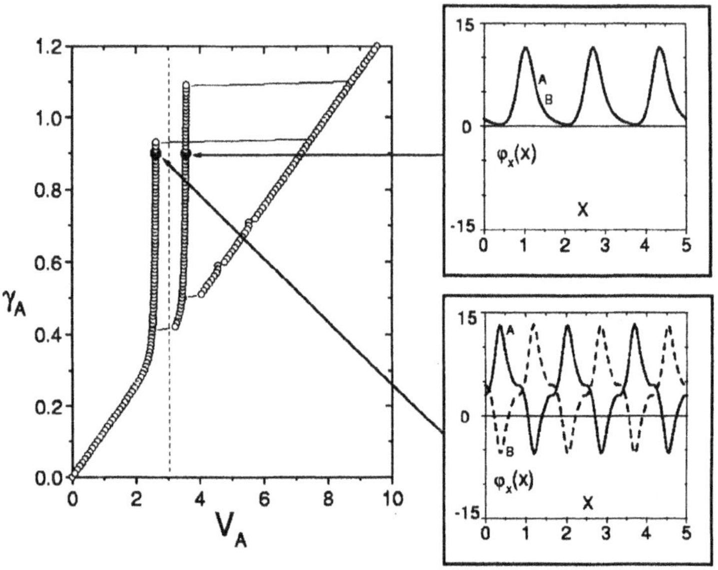

Figure 17. Numerical calculation of the IV curve for a twofold annular stack. Parameters are described in the text. The insets show the waveforms for the two branches corresponding to junctions A and B. (From [4]).

We have made numerical simulations corresponding to the two-fluxon coherent state described in Eq. (31). The normalizations were performed using the usual one-soliton case with the velocity of light given by $\bar{c}^{(1)}$ and the Josephson penetration depth $\lambda_J^{(1)}$. The parameters are: N = 3 (2 junctions), α=0.1, L=5 and the coupling parameter $S = -0.3$, where the loss parameter α is defined by $\alpha = G(\hbar/2eCJ)^{1/2}$ under normalization. Three fluxons are trapped in each of the annular junctions A and B , which are biased in series [4].

Fig. 17 shows the static IVC for the coherent 2-junction mode. Also shown is another mode found during the simulations. For this mode, the motion is symmetric in the sense that when the fluxon in junction (0;1) is at x=L, then the fluxon in junction (2;1) is at x=0. The waveforms of the in-phase and out-of-phase modes are shown in the insets. For the coherent mode in Fig. 17 the asymptotic value of the voltage is approximately 1.20 times higher that that of the single junction case (dashed line). This factor is in very good agreement with the theoretical expression $\bar{c}^{(2)}/\bar{c}^{(1)} = 1/\sqrt{1+S}$.

Here it should be noted that the two solitons are fully identical, i.e., there is no phaseshift between them, as is usually the case for coherent modes in other comparable systems [1].

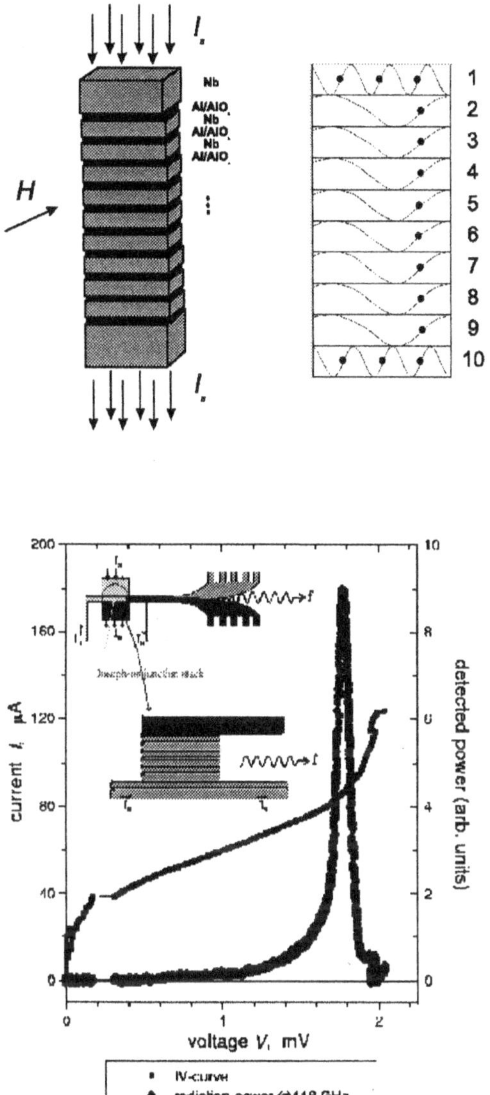

Figure 18. The junction model with some actual measurements on microwave radiation from low-T_c stacks. (Figure supplied by A.V. Ustinov).

In the simple analysis above we noted the existence of the two solutions for the two-stack, an in-phase solution and an out-of-phase solution. This picture exists not only for fluxons, but also for small-scale oscillations (Ref. [87]). This was demonstrated (Ref. [86]) very clearly by the experimental existence of two different series of Fiske steps corresponding to the in-phase and out-of-phase velocities. An analytic analysis of plasma oscillations in Josephson stacks (Ref. [87]) also has shown the existence of two different plasma frequencies in a two-stack, corresponding to in-phase and out-of-phase oscillations of the two layers.

This analysis (Ref. [87]) also treated the general behavior of oscillation modes for an N stack. It was shown here that there are N oscillation modes (N Swihart velocities) for an N stack. There is always an in-phase mode with all N junctions basically oscillating in phase. This mode has the highest Swihart velocity. There is also an out-of-phase mode where any two neighboring layers are oscillating out of phase. This mode has the lowest Swihart velocity. In addition, there are N-2 modes with intermediate oscillation patterns and velocities. When we consider flux-flow modes, we expect the same general pattern, although the general analytic solutions have not been obtained. However, for $N = \infty$, the equations for the in-phase mode becomes a sine-Gordon equation (Eq. (3)), but with another spatial scaling length. Thus for $N=\infty$, the flux-flow mode exists with coherent motion of fluxons in all the layers, and with a velocity (frequency) larger than for a single-layer junction. This, of course, can be utilized in a high-T_c single-crystal flux-flow amplifier or oscillator. Stacking has the further advantage that the impedance level becomes more compatible with that of ordinary microwave circuitry.

Ref. [88] gives a numerical calculation for a 10-junction stack. It shows that the in-phase flux-flow mode exists based on the 8 interior junctions. Only the top and bottom ones are not in phase with the rest. An example from this calculation is shown in Fig. 18, together with the actual measurement of microwave power [89].

Acknowledgments

This review is based on our fruitful collaboration with many former and present co-workers from a widespread network of institutes and institutions. Especially we thank A.A. Antonov, P. Barbara, A.M. Baryshev, P. Bodin, M. Cirillo, T. Claeson, G. Costabile, M. Darula, P.N. Dmitriev, T. Doderer, L.V. Filippenko, G.M. Fischer, H. Golstein, T. de Graauw, N. Grønbech-Jensen, G. Hallmanns, C. Heiden, R. Kleiner, H. Kohlstedt, V.P. Koshelets, V.V. Kurin, I.L. Lapytskaya, W. Luinge, M. Manscher, N. Martucciello, R. Monaco, P. Müller, M.A. Nordahn, V.A. Oboznov, A.L. Pankratov, R.D. Parmentier, S. Sakai, M. Salerno, M.R. Samuelsen, A. V. Shchukin, S.V. Shitov, A.V. Ustinov, V.L. Vaks, S.A. Vasenko, N. Whyborn, D. Winkler, A.V. Yulin and Y.M. Zhang. The work was supported in part by the Russian SSP "Superconductivity" project, INTAS project 97-1712, Russian Foundation for Basic Research (Project N~96-15-96718, Project N~97-02-16928 and Project N~99-02-17544), Ministry of High Education of Russian Federation (Project N~3877), Grant N~98-2-13 from the International Center for Advanced Studies in Nizhny Novgorod, the Nederlandse Organisatie voor Wetenschappelijk Onderzoek (NWO) grant, ESA TRP contract 11/653/95/NL/PB/SC, the Danish Research Academy, and the Danish Natural Science Foundation.

References

[1] N.F. Pedersen, "Solitons in Josephson Transmission lines". In 'SOLITONS', Modern Problems in Condensed Matter Sciences, Eds. A.A. Maradudin and V.H. Agranovich (North Holland) 1986, p. 469.
[2] N.F. Pedersen, "Fluxon Electronic Devices". IEEE Trans. on Magn. **MAG-27**, 3328 (1991).
[3] R.D. Parmentier, in *The New Superconducting Electronics*, Eds. H. Weinstock and R.W. Ralston (Kluwer, Dordrecht 1993) p. 221.
[4] N.F. Pedersen and A.V. Ustinov, "Fluxons in Josephson transmission lines: new developments". Supercond. Sci. Technol. **8**, 389 (1995).
[5] A.V. Ustinov, "Solitons in Josephson junctions", Physica D, **123**, 315 (1998).
[6] M. Darula, T. Doderer, and S. Beuven, "Millimetre and sub-mm wavelength radiation sources based on discrete Josephson arrays". Supercond. Sci. Technol. **12**, R1-R12 (1999).
[7] J. Schubert, A. Semenov, G. Gol'tsman, H.-W. Hübers, G. Schwaab, B. Voronov, and E. Gershenzon, "Noise Temperature of a NbN Hot-Electron Bolometric Mixer at Frequencies from 0.7-5.2 THz". Extended Abstracts ISEC'99, Berkeley, p. 527-529 (1999).
[8] B.S. Karasik, W.R. McGrath, H.G. LeDuc, and M.E. Gershenson, "A Hot-Electron Direct Detector for Radioastronomy". Extended Abstracts ISEC'99, Berkeley, p. 524-526 (1999).
[9] A.V. Ustinov, "Long Josephson Junctions and Stacks". *Lectures on Superconductivity in Networks and Mesoscopic Systems*. Eds. C. Giovannella and C. Lambert (AIP Conf. Proc. 427, NY, 1998), p. 31-55.
[10] D.W. McLaughlin and A.C. Scott, "Perturbation Analysis of Fluxon Dynamics", Phys. Rev. A, **18**, 1652 (1978).
[11] A. Scott, *Nonlinear Science, Emergence & Dynamics of Coherent Structures*, (Oxford University Press, 1999).
[12] K.K. Likharev. *Dynamics of Josephson Junctions and Circuits*. Gordon and Breach, New York (1986).
[13] A.N. Vystavkin, V.N. Gubankov, L.S. Kuzmin, K.K. Likharev, V.V. Migulin, and V.K. Semenov, "S-C-S junctions as nonlinear elements of microwave receiving devices", Rev. Phys. Appl. **9**, p. 79-109 (1974).
[14] A.J. Dahm, A. Denenstein, D.N. Langenberg, D. Rogovin, and D.J. Scalapino, "Linewidth of the radiation emitted by a Josephson junction", Phys. Rev. Lett. **22**, 1416 (1969); D. Rogovin and D.J. Scalapino, "Fluctuation phenomena in tunnel junctions", Ann. Phys. **86**, p. 1-90 (1974).
[15] T. Doderer, "Microscopic Imaging of Josephson Junction Dynamics", Journ. Mod. Phys. **11**, 1979 (1997).
[16] B. Dueholm, O.A. Levring, J. Mygind, N.F. Pedersen, O.H. Soerensen, and M. Cirillo, "Multisoliton Excitations in Long Josephson Junctions", Phys. Rev. Lett. **46**, 1299 (1981); see also B. Dueholm, E. Joergensen, O.A. Levring, J. Mygind, N.F. Pedersen, M.R. Samuelsen, O.H. Olsen, and M. Cirillo, "Dynamic Fluxon Model for Fiske Steps in Long Josephson Junctions", Phys. Rev. B **108**, 1303 (1981).
[17] A.V. Ustinov, T. Doderer, R.P. Huebener, N.F. Pedersen, B. Mayer, and V.A. Oboznov, "Dynamics of sine-Gordon solitons in the annular Josephson junction", Phys. Rev. Lett. **69,** 1815 (1992).
[18] J.B. Hansen and J. Mygind, "New Type of Zero-Field Steps in the IV-Curve of a Josephson Junction of Intermediate Length", Phys. Rev. B **32**, 178 (1985)
[19] R. Monaco, P. Barbara, and J. Mygind, " Fine structure on Zero-Field Steps in low-loss Josephson tunnel junctions", Phys. Rev. B **47**, 12292 (1993).
[20] J.B. Hansen, Yu. Ya. Divin, J. Mygind, and M. R. Samuelsen, " Measurements of Frequency and Linewidth of the Radiation Emitted from Overlap Josephson Junctions ", Eds. H. Hahlbohm and H. Lübbig, *SQUID'85 - Superconducting Quantum Interference Devices and their Applications*, pp. 489-494, (Walter de Gruyter & Co, Berlin, 1985).
[21] S.N. Erné, A. Ferrigno, and R.D. Parmentier, "Fluxon Propagation and Fiske Steps in Long Josephson Junctions", Phys. Rev. B **27**, 5440 (1983).
[22] I.O. Kulik, "Wave propagation in a Josephson tunnel junction in the presence of vortices and the electrodynamics of weak superconductivity", JETP. **24**, 1307 (1967).
[23] E. Joergensen, V.P. Koshelets, R. Monaco, J. Mygind, M.R. Samuelsen, and M. Salerno, "Thermal fluctuations in resonant motion of fluxons on a Josephson transmission line: Theory and experiment", Phys. Rev. Lett. **49**, 1093 (1982).
[24] M. Cirillo, I. Modena, P. Carelli, and V. Foglietti, "Millimeter wave generation by fluxon oscillations in a Josephson junction", J. Appl. Phys. **65**, 2376 (1989); and M. Cirillo, I. Modena, P. Carelli, G.S. Spagnolo, R. Leoni, and M. Pullano, "Experiments on tunable Josephson Millimeter-Wave Oscillators", IEEE Trans Magn. **27**, 3335 (1991).
[25] A. Davidson, B. Dueholm, B. Kryger, and N.F. Pedersen, "An experimental investigation of trapped sine-Gordon solitons", Phys. Rev. Lett. **55**, 2059 (1985).

[26] N. Grønbech-Jensen, P.S. Lomdahl, and M.R. Samuelsen, "Phase-locking of long annular Josephson junctions to an external rf magnetic field", Phys. Lett. A **154**, 14 (1991); and "Bifurcation and Chaos in a dc-driven Long Annular Josephson Junction", Phys. Rev. B **43**, 12779 (1991).

[27] N. Martucciello and R. Monaco, "Annular Josephson tunnel junctions in an external magnetic field: the statics", Phys. Rev. B **53**, 3471 (1996); and N. Martucciello, V.P. Koshelets, A. V. Shchukin, L.V. Filippenko, J. Mygind, and R. Monaco, "Fluxon Dynamics in Long Annular Josephson Tunnel Junctions", Phys. Rev. B **57**, 5444 (1998).

[28] A.V. Ustinov, B.A. Malomed, and N. Thyssen, "Soliton trapping in a periodic potential: experiment", Phys. Lett. A **233**, 239 (1997).

[29] A.V. Ustinov, T. Doderer, R.P. Huebener, J. Mygind, V.A. Oboznov, and N.F. Pedersen, "Multi-fluxon effects in long Josephson junctions", IEEE Trans. Appl. Supercond. **3**, 2287 (1993).

[30] R.E. Eck, D.J. Scalapino, and B.N. Taylor, "Self-detection of the ac Josephson current", Phys. Rev. Lett. **13**, 15 (1964); see also K. Yoshida, K. Hamasaki, and F. Irie, "Current Steps Induced by Vortex Motion in Long Josephson Tunnel Junctions", Jap. Appl. Phys. **18**, 373 (1979).

[31] T. Nagatsuma, K. Enpuku, F. Irie, and K. Yoshida, "Flux-Flow type Josephson oscillator for millimeter and submillimeter wave region", J. Appl. Phys. **54**, 3302 (1983); see also Pt. II, J. Appl. Phys. **56**, 3284 (1984); Pt. III, J. Appl. Phys. **58**, 441 (1985); Pt. IV, J. Appl. Phys. **63**, 1130 (1988).

[32] M. Cirillo, N. Grønbech-Jensen, M. Samuelsen, M. Salerno, and G.V. Rinati, "Fiske modes and Eck steps in long Josephson junctions: Theory and Experiments", Phys. Rev. B **58**, 12 377 (1998).

[33] M. Salerno and M.R. Samuelsen, "Phase-locking and flux-flow resonances in Josephson oscillators driven by homogeneous microwave fields", Phys. Rev. B **59**, 14653 (1999).

[34] Y.M. Zhang, D. Winkler, and T. Claeson, "Linewidth measurements of Josephson flux-flow oscillators in the band 280-330 GHz", Appl. Phys. Lett. **62**, 3193 (1993); see also Y.M. Zhang, D. Winkler, and T. Claeson, "Detection of mm and submm wave radiation from soliton and flux flow modes in a long Josephson junction", IEEE Trans. Appl. Supercond. **3**, 2520 (1993); and Proc. Fourth Int. Symp. on Space Terahertz Techn., Los Angeles, USA, pp. 485-499 (1993).

[35] V.P. Koshelets, A.V. Shchukin, S.V. Shitov, and L.V. Filippenko, "Superconducting Millimeter Wave Oscillators and SIS mixers integrated on a Chip.", IEEE Trans. Appl. Supercond. **3**, 2524 (1993).

[36] J. Mygind, V.P. Koshelets, A.V. Shchukin, S.V. Shitov, and I.L. Lapytskaya, "Properties of autonomous and injection locked flux flow oscillators", IEEE Trans. Appl. Supercond. **5**, 2951 (1995).

[37] S. Kiryu, S. Kohjiro, A. Shoji, S. Kodaira, J. Inatani, and S. Sato, "Flux-Flow-type Josephson oscillator consisting of epitaxial NbCN/MgO/NbCN junction", IEEE Trans. Appl. Supercond. **3**, 2528 (1993).

[38] Y. M. Zhang, "Dynamics and applications of long Josephson junctions", Ph.D. Dissertation, Chalmers University of Technology, Department of Physics (1993).

[39] N.B. Dubash, Y.M. Zhang, U. Ghoshal, and P.-F. Yuh, "Linewidth Measurements and Phase-locking of Josephson Oscillators Using RSFQ Circuits", IEEE Trans. Appl. Supercond. **7**, 3808 (1997); see also Y.M. Zhang, V. Borzenets, V.K. Kaplunenko, and N. B. Dubash, "Underdamped long Josephson junction coupled to overdamped single-flux-quantum circuits", Appl. Phys. Lett. **71**, 1863 (1997).

[40] V.K. Koshelets, S.V. Shitov, A.M. Baryshev, A.V. Shchukin, L.V. Filippenko, T. de Graauw, W. Luinge, H. van de Stadt, H. Schaeffer, and J. Mygind, "First implementation of the fully superconducting 500 GHz receiver with integrated flux flow oscillator", *Applied Superconductivity 1995, EUCAS'95*. ED. D. Dew-Hughes, Institute of Physics Conference Series, **148**, pp. 731-736 (IOP Publishing Ltd, Bristol and Philadelphia, 1995).

[41] V.P. Koshelets, S.V. Shitov, A.M. Baryshev, I.L. Lapitskaya, L.V. Filippenko, H. van de Stadt, J. Mess, H. Schaeffer, T. de Graauw, "Integrated sub-mm wave receivers", IEEE Trans. on Appl. Supercond., **5**, 3057 (1995).

[42] V.P. Koshelets, S.V. Shitov, L.V. Filippenko, A.M. Baryshev, H. Golstein, T. de Graauw, W. Luinge, H. Schaeffer, and H. van de Stadt, "First implementation of a superconducting integrated receiver at 450 GHz", Appl. Phys. Lett., **68**, 1273 (1996).

[43] V.P. Koshelets, S.V. Shitov, L.V. Filippenko, A.M. Baryshev, W. Luinge, H. Golstein, H. van de Stadt, J.-R. Gao, T. de Graauw, "An Integrated 500 GHz Receiver with Superconducting Local Oscillator", IEEE Trans. on Appl. Supercond. **7**, 3589 (1997).

[44] S.V. Shitov, A.B. Ermakov, L.V. Filippenko, V.P. Koshelets, A.B. Baryshev, W. Luinge, Jian-Rong Gao, "Superconducting Chip Receiver for Imaging Applications", IEEE Trans. on Appl. Supercond. **9**, 3773 (1999).

[45] V.P. Koshelets, S.V. Shitov, A.V. Shchukin, L.V. Filippenko, P.N. Dmitriev, V.L. Vaks, J. Mygind, A.B. Baryshev, W. Luinge, H. Golstein, "Flux Flow Oscillators for Sub-mm Wave Integrated Receivers", IEEE Trans. on Appl. Supercond. **9**, 4133 (1999).

[46] J. Mygind, V.P. Koshelets, S.V. Shitov, L.V. Filippenko, V.L. Vaks, A.B. Baryshev, W. Luinge, and N. Whyborn, "Phase-locking of 270-440 GHz Josephson Flux Flow Oscillators", *Extended abstracts International Superconductive Electronics Conference ISEC'99*, Berkeley, USA, Ed. T. Van Duzer, pp. 83-85 (1999).
[47] V.P. Koshelets, S.V. Shitov, L.V. Filippenko, V.L. Vaks, J. Mygind, A.B. Baryshev, W. Luinge, and N. Whyborn, "Phase-locked 270-440 GHz local oscillator based on flux flow in long Josephson junctions", Rev. Sci. Inst. (submitted April 1999).
[48] V.P. Koshelets, A.V. Shchukin, S.V. Shitov, L.V. Filippenko, G.M. Fischer, and J. Mygind, " Integrated Flux-Flow Oscillators for submillimeter wave receivers ", Physica B **194-196**, 81 (1994).
[49] V.P. Koshelets, A. Shchukin, I.L. Lapytskaya, and J. Mygind, "Spectral linewidth of autonomous and injection-locked flux flow oscillators", Phys. Rev. B **51**, 6536 (1995).
[50] V.P. Koshelets, S.V. Shitov, A.V. Shchukin, L.V. Filippenko, and J. Mygind, "Linewidth of submillimeter wave flux flow oscillators", Appl. Phys. Lett. **69**, 699 (1996); see also V.P. Koshelets, A.V. Shchukin, L.V. Filippenko, and J. Mygind," Linewidth of frequency locked flux flow oscillators for sub-mm wave receivers ", IEEE Trans. Appl. Supercond. **7**, 2905 (1997).
[51] V.P. Koshelets, S.V. Shitov, A.V. Shchukin, L.V. Filippenko, J. Mygind and A.V. Ustinov, "Self-pumping effects and radiation linewidth of FFO", Phys. Rev. B **56**, 5572 (1997).
[52] A.V. Ustinov, H. Kohlstedt, and P. Henne, "Giant radiation linewidth of multifluxon states in long Josephson junctions", Phys. Rev. Lett. **77**, 3617 (1996).
[53] N.R. Werthamer, " Nonlinear self-coupling of Josephson radiation in superconducting tunnel junctions", Phys. Rev. **147**, 255 (1966).
[54] L.-E. Hasselberg, M.T. Levinsen, and M.R. Samuelsen, " Theories of subharmonic gap structures in superconducting junctions", Phys. Rev. B **9**, 3757 (19974).
[55] A.A. Golubov, B.A. Malomed, and A.V. Ustinov, "Radiation linewidth of a long Josephson junction in the flux flow regime", Phys. Rev. B **54**, 3047 (1996).
[56] A.P. Betenev and V.V. Kurin,"Radiation spectrum of a long Josephson flux-flow oscillator", Phys. Rev. Lett. **56**, 7855 (1997).
[57] A.B. Baryshev, private communication (1999).
[58] A.V. Ustinov, J. Mygind, and V.A. Oboznov, "Phase-locked flux flow Josephson oscillator", Appl. Phys. **72**, 1203 (1992).
[59] A.V. Ustinov, J. Mygind, and N.F. Pedersen, "Millimeter-wave-induced fluxon pair creation in flux flow Josephson oscillators", Phys. Rev. B **46**, 578 (1992).
[60] M. Salerno and M.R. Samuelsen, "Josephson Flux-Flow Oscillators in Microwave Fields", preprint (1999).
[61] Yu.S. Kivshar and B.A. Malomed, "Dynamics of fluxons in a system of coupled Josephson junctions", Phys. Rev. B **37**, 9325 (1988); see also Rev. Mod. Phys. **61**, 763 (1989).
[62] A.M. Baryshev, A.V. Yulin, V.V. Kurin, V.P. Koshelets, S.V. Shitov, A.V. Shchukin, P.N. Dmitriev, and L.V. Filippenko, "Design and Fabrication of Cherenkov Flux-Flow Oscillator", IEEE Trans. Appl. Supercond., **9**, 3737 (1999).
[63] V.V. Kurin, A. Yulin, E. Goldobin, A. Klushin, H. Kohlstedt, M. Levitchev, and N. Thyssen, "Experimental Investigation of Cherenkov Flux-Flow Oscillators", IEEE Trans. Appl. Supercond., **9**, 3733 (1999).
[64] R.G. Mints and I.B. Snapiro, "Josephson-vortex Cherenkov radiation", Phys. Rev. B **52**, 9691, (1995).
[65] S.N. Artemenko and S.V. Remisov, Sov. Pis'ma v ZhETPh, **66**, vol. 12, 811, (1997) (in Russian).
[66] V.V. Kurin, A.V. Yulin, I.A. Shereshevsky, and N.K. Vdovicheva, "Cherenkov Radiation of Vortices in a Two-dimensional Annular Josephson Junction", Phys. Rev. Lett. **80**, 3372 (1998).
[67] J.G. Caputo, N. Flytzanis, N. Lazarides, V.V. Kurin, and A. Chiginev, "Dynamics of Fluxons in Narrow Window Junctions", submitted for Phys. Rev. B (1999).
[68] H.S.J. van der Zant, T.P. Orlando, S. Watanabe, and S.H. Strogatz, "Kink Propagation in a Highly Discrete System: Observation of Phase-locking to Linear Wave", Phys. Rev. Lett. **74**, 174 (1995); see also H.S.J. van der Zant and T.P. Orlando, "Eck peak in underdamped discrete superconducting vortex flow device", Journ. Appl. Phys. **76**, 7606 (1994), and A.E. Duwel, S. Watanabe, E. Triás, T.P. Orlando, H.S.J. van der Zant, and S.H. Strogatz, "Discreteness-induced resonances and ac voltage amplitudes in long one-dimensional Josephson arrays", Appl. Phys. **82**, 4661 (1997).
[69] G. Hechtfisher, R. Kleiner, A.V. Ustinov, and P. Mueller, "Non-Josephson Emission from Intrinsic Junctions in $Bi_2Sr_2CaCu_2O_{8+y}$ Cherenkov Radiation by Josephson Vortices", Phys. Rev. Lett. **79**, 1365 (1997).
[70] E. Goldobin, A. Wallraff, N. Thyssen, A.V. Ustinov, "Cherenkov radiation in coupled long Josephson junctions", Phys. Rev. B, **57**, 130 (1998).

[71] V.V. Kurin and A.V. Yulin, "Radiation of linear waves by solitons in a Josephson transmission line with dispersion", Phys. Rev. B **55**, 11659 (1997).
[72] A.M. Baryshev, A.V. Yulin, V.V. Kurin, V. P. Koshelets, P.N. Dmitriev, and L.V. Filippenko, "Forward and Backward Wave in Cherenkov Oscillators", ISEC'99 Extended Abstracts, pp. 418-420 (1999).
[73] A.A. Antonov, A.L. Pankratov, A.V. Yulin, and J. Mygind, "Influence of thermal fluctuations on Cherenkov radiation from fluxons in dissipative Josephson systems", Phys. Rev. B, **61**, 9809-19, 2000.
[74] R.G. Mints and I.B. Snapiro, "Electromagnetic waves in a Josephson junction in a thin film", Phys. Rev. B **51**, 3054, (1995).
[75] D. Winkler, Y.M. Zhang, P.Å. Nilsson, E.A. Stepantsov, and T. Claeson, "Electromagnetic properties at the grain boundary interface of a $YBa_2Cu_3O_{7-\delta}$ bicrystal Josephson junction", Phys. Rev. Lett. **72**, 1260 (1994).
[76] Y.M. Zhang, D. Winkler, G. Brorsson, and T. Claeson, "Parallel Plate Resonators in YBa2Cu3O7 Bicrystal Grain Boundaries", IEEE Trans. Appl. Supercond. **5**, 2200 (1995).
[77] A.V. Shchukin, Y.M. Zhang, D. Winkler, and J. Mygind, " Microwave-induced steps in long homogeneous YBCO grain boundary junctions", *Applied Superconductivity 1995, EUCAS'95*, Ed. D. Dew-Hughes, Institute of Physics Conference Series, **148**, 1327-1330 (IOP Publishing Ltd., Bristol and Philadelphia, 1995).
[78] A.V. Shchukin, J. Mygind, Y.M. Zhang, and D. Winkler, "Subharmonic generation and chaos in long YBCO grain boundary Josephson junctions", *Extended Abstracts of 5th International Superconducting Electronics Conference, ISEC'95, Nagoya, Japan*, pp. 507-509 (1995).
[79] T. Nguyen, G.A. Daniels, J.B. Beyer, and J.E. Nordman, "A Discrete HTS Josephson Flux-Flow Structure with Gain at 77 K", IEEE Trans. Appl. Supercond. **9**, 3945 (1999).
[80] M.A. Ketkar, J.B. Beyer, and J.E. Nordman, "Gain-Frequency characteristics of transistors based on flux flow in hysteretic Long Josephson Junctions (LJJ)", IEEE Trans. Appl. Supercond. **9**, 3949 (1999).
[81] Yu. Ya. Divin, O. Yu. Polyanskii, and A. Ya. Shulman, "Incoherent radiation spectroscopy by means of the Josephson effect", Sov. Techn. Phys. Lett. **6**, no 9, 454 (1980).
[82] Z.G. Ivanov, P.-Å. Nilsson, D. Winkler, J.A. Alarco, T. Claeson, E.A. Stepantsov, and A.Y. Tzalenchuk, " Properties of artificial grain boundary weak links grown on $Y\text{-}ZrO_2$ bicrystals", Supercond. Sci. Techn. **4**, 439 (1991).
[83] M.T. Levinsen, R.Y. Chiao, M.J. Feldman, and B.A. Tucker, "An inverse ac Josephson effect voltage standard", Appl. Phys. Lett. **31**, 776 (1977).
[84] S. Sakai, P. Bodin, and N.F. Pedersen, "Fluxons in thin film superconductor-insulator superlattices", J. Appl. Phys. **73**, 2411 (1993)
[85] R. Kleiner, P. Müller, H. Kohlstedt, N.F. Pedersen, and S. Sakai, "Dynamic behavior of Josephson-coupled layered structures", Phys. Rev. B. **50**, 3942 (1994)
[86] A.V. Ustinov, M. Cirillo, H. Kohlstedt, G. Hallmanns, C. Heiden, and N.F. Pedersen, "Coupled Fluxon Modes in Stacked Nb/Al/AlO_x/Nb Long Josephson Junctions", Phys. Rev. B. **48**, 10614 (1993).
[87] N.F. Pedersen and S. Sakai, "Josephson Plasma resonance in superconducting multilayers", Phys. Rev. B. **58**, 2820 (1998).
[88] A.V. Ustinov and S. Sakai, "Submillimeter-band high power generators using multilayered Josephson junctions", Appl. Phys. Lett. **73**, 686 (1998).
[89] N. Thyssen, H. Kohlstedt, and A. V. Ustinov, "Low-Tc tunnel junction stacks as models for intrinsic Josephson effect in high-Tc materials". Physica C, **293**, 264 (1997).

FLUX-FLOW OSCILLATORS AND PHENOMENON OF CHERENKOV RADIATION FROM FAST MOVING FLUXONS

EDWARD GOLDOBIN
Institute of Thin Film and ion Technology (ISI),
Research Center Jülich (FZJ),
52425, Jülich, Germany

Abstract.

We give a review of flux-flow oscillators based on low-temperature Josephson junctions and consider different methods for improvement. In particular we focus attention on the stacks of inductively-coupled long Josephson junctions and consider in detail a novel phenomenon which can be found in such systems — Cherenkov radiation of Josephson plasma waves caused by a fast moving Josephson vortex.

1. Introduction

Josephson devices attracted the attention of researchers immediately after discovery of the Josephson effect. Much research has been done since that time, and Josephson devices have proven to be extremely useful. We mention SQUIDs, high-frequency oscillators and mixers, digital circuits and voltage standards to name just a few.

In this work we focus on oscillator applications of Josephson junctions, *e.g.*, those which can be used for integrated receivers in radio-astronomy [1, 2]. The simplest oscillator can be built using a single Josephson junction, but its power, tunability and spectral characteristics leave much to be desired. Later it was proposed to use long Josephson junctions (LJJs), operating in the so-called flux-flow mode, as tunable oscillators for mm and sub-mm wavebands. The maximum frequency of such oscillators is limited by the gap voltage of the superconducting electrodes of the junction (1.4 mV for Nb-Al-AlO_x-Nb technology), and, at present, oscillators operating at frequencies up to 800 GHz have been reported. As an alternative,

H. Weinstock and M. Nisenoff (eds.), Microwave Superconductivity, 581–614.

it was proposed to use 1D arrays of Josephson junctions and other even more exotic devices such as Cherenkov oscillators.

As a next step for improving parameters of Josephson oscillators, it has been suggested to use stacks of coupled LJJs as coherently operating arrays of oscillators. The idea to use multi-layered devices is very fruitful in many fields of physics, *e.g.*, multi-layered magneto-resistive sensors. In addition, many materials which are of great interest today have a multi-layered structure on the atomic level. As an example, we mention high-T_c superconductors such as BSCCO or perovskites consisting of neighboring atomic layers with opposite magnetization direction. Thus, the study of artificially-prepared stacks also provides insight into the physics of natural multi-layered systems.

Multi-layered devices with N layers potentially can improve performance up to N times (sometimes even to N^2 times), while they do not occupy extra space on the chip due to the vertical integration.

In the following section we review the principles of operation of the oscillators based on long Josephson junctions, and conclude the section by showing the current performance of such oscillators. In section 3 we consider an interesting effect, namely the emission of electromagnetic waves by fast-moving fluxons, which arises in the stacks of LJJs. Possibilities for stimulation of such emission, and how to use it for oscillator applications are discussed.

2. Long Josephson Junction

2.1. BASIC EQUATIONS

Let us consider a single LJJ, which is a classical system for investigation of soliton dynamics and the Josephson effect. An LJJ consists of two superconducting electrodes separated by a barrier of thickness about ξ (coherence length) or less, as shown in Fig. 1. Usually, the barrier is made of an insulator, so that it forms a tunnel junction. Today, the most reliable way to produce such tunnel junctions is Nb-Al-AlO$_x$-Nb technology. We shall consider only a one-dimensional model of the LJJ, *i.e.*, the dynamics will take place in the x direction only.

The most common case of an LJJ will be considered here, *i.e.*, an LJJ with electrodes of finite thickness. That is because in stacked LJJs, which are interesting for our investigation, the thickness of at least one electrode is comparable with or less than the London penetration depth λ. The derivation of the differential equation describing the dynamics of the Josephson phase in an LJJ can be found in any Josephson-effect textbook — see, *e.g.*,

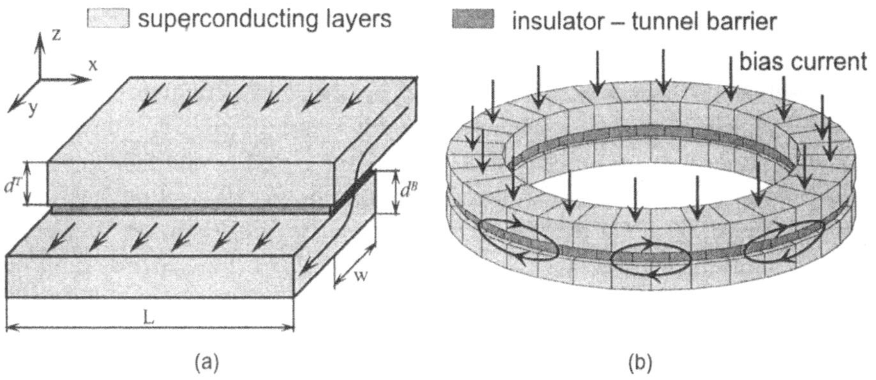

Figure 1. (a) Linear LJJ, (b) annular LJJ with 3 fluxons trapped in it. Arrows show the direction of current.

[3, 4, 5]. This equation is as follows:

$$\lambda_J^2 \phi_{xx} - \frac{1}{{\omega_p}^2}\phi_{tt} - \sin\phi = \frac{1}{\omega_c}\phi_t - \frac{j}{j_c} \tag{1}$$

This equation is called the perturbed sine-Gordon equation. If the r.h.s. of Eq. (1) is equal to zero, it is called the non-perturbed or simply the sine-Gordon equation. The quantity $\phi = \phi(x,t)$ is the Josephson phase, equal to the phase difference of the quantum mechanical wave functions of the top and bottom superconducting electrodes; λ_J is the Josephson penetration depth which describes the penetration of the magnetic field into the junction (in the direction of the x axis) and is given by

$$\lambda_J = \sqrt{\frac{\Phi_0}{2\pi}\frac{1}{\mu_0 j_c d'}}, \tag{2}$$

where

$$d' = d_I + \lambda^A \coth\frac{d^A}{\lambda^A} + \lambda^B \coth\frac{d^B}{\lambda^B} \tag{3}$$

is the so-called effective magnetic thickness of the junction, j_c is the junction critical-current density, and d_I is the thickness of the tunnel barrier. Here and below the parameters related to the top and bottom electrodes will be denoted by the subscripts A and B, respectively. Further, in Eq. (1)

$$\omega_p = \sqrt{\frac{2\pi}{\Phi_0}\frac{j_c}{C}} \quad , \tag{4}$$

where C is the capacitance of the junction per unit area ($\mathrm{F/m^2}$). Finally,

$$\omega_c = \frac{2\pi}{\Phi_0} R j_c \quad , \tag{5}$$

where R is the specific resistance ($\Omega \cdot \mathrm{m}^2$).

Eq. (1) is valid for an LJJ which can be described in the framework of the simple RSJ model. Nb-Al-AlO$_x$-Nb LJJs, most frequently used for experiments with fluxons, are most properly described by the microscopic tunneling model [3], which is very different from the RSJ model. Nevertheless, for the study of fluxon dynamics in an LJJ, the RSJ model is a good approximation of a real junction if, as a normal resistance R, one takes the subgap resistance R_J, defined as the resistance of the back part of the IVC at the bias point where the voltage drop across the junction is 2 mV (for Nb junctions), *i.e.*, $R_J = 2\,\mathrm{mV}/I(2\,\mathrm{mV})$. We also note that Eq. (1) works only for $\lambda_J \gg d' \sim \lambda$.

Eq. (1) assumes a one-dimensional model of an LJJ in which the Josephson phase ϕ depends only on one spatial coordinate x. This is valid if the width of the junction $w < \lambda_J$, *i.e.*, if the junction is relatively short in the y direction. The projection of the magnetic field H_x on the x axis must be much less than the first critical field for a small (short) JJ:

$$H_x \ll H_{c1}^{\mathrm{small}} = \frac{\Phi_0}{\Lambda w} \quad .$$

To solve Eq. (1) it is necessary to specify the boundary conditions which depend on the geometry of the junction. For a linear LJJ of length L, shown in Fig. 1(a), the boundary conditions can be written as follows:

$$\phi_x|_{x=0} = \frac{2\pi}{\Phi_0} H\Lambda \quad , \tag{6}$$

$$\phi_x|_{x=L} = \frac{2\pi}{\Phi_0} H\Lambda \quad , \tag{7}$$

where

$$\Lambda = d_I + \lambda^A \tanh\frac{d^A}{2\lambda^A} + \lambda^B \tanh\frac{d^B}{2\lambda^B} \tag{8}$$

is the effective depth of the magnetic field penetration into the superconducting electrodes of the LJJ.

To study fluxon dynamics, the annular geometry shown in Fig. 1(b) is of great interest. For such an LJJ the boundary conditions should require a continuous phase (modulo 2π) and phase derivative:

$$\begin{cases} \phi|_{x=0} &= \phi|_{x=L} \quad -2\pi N \\ \phi_x|_{x=0} &= \phi_x|_{x=L} \end{cases} \quad , \tag{9}$$

where L now denotes the average circumference of the ring, and N denotes the number of fluxons — see below — trapped in the junction.

For the sake of simplicity we introduce normalized units. The spatial coordinate x will be normalized to the Josephson penetration depth λ_J, and time will be normalized to the inverse plasma frequency $1/\omega_p$. Thus, Eq. (1) will appear as follows:

$$\phi_{\tilde{x}\tilde{x}} - \phi_{\tilde{t}\tilde{t}} - \sin\phi = \alpha\phi_{\tilde{t}} - \gamma \quad , \tag{10}$$

where the following new variables have been introduced:

$$\tilde{t} = \omega_p t; \qquad \tilde{x} = x/\lambda_J; \qquad \gamma = \frac{j}{j_c}; \tag{11}$$

$$\alpha = \frac{\omega_p}{\omega_c} = \frac{1}{\sqrt{\beta_c}} = \frac{1}{\sqrt{\frac{2\pi}{\Phi_0} j_c R^2 C}} \ . \tag{12}$$

The damping coefficient α describes the dissipation in the junction. As one can see from Eq. (12), it can be readily expressed in terms of the McCumber-Stewart parameter β_c, which characterizes the dynamic properties of the junction and, in particular, the presence of the hysteresis on the I–V curve (IVC). For investigation of ballistic soliton motion, an LJJ should have $\alpha \ll 1$. In typical tunnel (Nb-Al-AlO_x-Nb) LJJs, $\alpha < 0.01$ at $T = 4.2\,\mathrm{K}$.

After introduction of normalized units, the boundary conditions also change. For an LJJ of linear geometry the boundary conditions written in normalized units are:

$$\begin{cases} \phi_{\tilde{x}}|_{\tilde{x}=0} = & \tilde{H} \\ \phi_{\tilde{x}}|_{\tilde{x}=\ell} = & \tilde{H} \end{cases} \quad , \tag{13}$$

where

$$\tilde{H} = \frac{2\pi}{\Phi_0}\Lambda\lambda_J H = \frac{2H}{H_{c1}} \tag{14}$$

is a normalized magnetic field, and

$$H_{c1} = \frac{\Phi_0}{\pi\Lambda\lambda_J} = \sqrt{\frac{2\Phi_0 j_c}{\pi\lambda\mu_0}} \tag{15}$$

is the (so-called) first critical field of an LJJ, the physical sense of which will become clear later. Parameter $\ell = L/\lambda_J$ is a normalized length (circumference) for an LJJ.

For an annular LJJ the normalized boundary conditions become:

$$\begin{cases} \phi|_{\tilde{x}=0} & = & \phi|_{\tilde{x}=L} & -2\pi N \\ \phi_x|_{\tilde{x}=0} & = & \phi_x|_{\tilde{x}=L} & \end{cases} \ . \tag{16}$$

It is worth noting that the LJJ has a very simple mechanical analog, namely, a chain of bound hanging pendula shown in Fig. 2. This analogy is valid in the continuous limit, *i.e.*, when the distance between the pendula

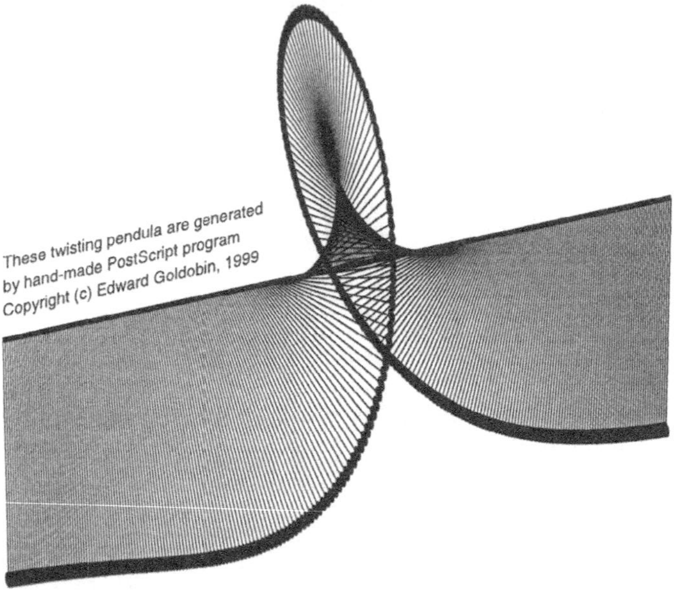

Figure 2. Mechanical analog of an LJJ — a system of bound hanging pendula. The system of pendula shown in this picture contains one soliton.

is infinitesimally small. The dynamics of such pendula is described by the same sine-Gordon equation (1), but the variables have different physical meaning: $\phi(x,t)$ is the angle of deviation from the vertical of the pendulum situated at point x at time t; $\gamma(x,t)$ is the torque acting on each pendulum; $\alpha(x,t)$ is a damping coefficient describing the friction on the axis of the pendulum.

2.2. SOME SOLUTIONS

The unperturbed sine-Gordon equation

$$\phi_{\tilde{x}\tilde{x}} - \phi_{\tilde{t}\tilde{t}} - \sin\phi = 0 \tag{17}$$

is a non-linear second-order PDE. It has many different solutions, but we restrict ourselves to the following physically interesting cases which are important for this work.

- **Josephson plasma wave:** $\phi = Ae^{i(k\tilde{x}-\omega\tilde{t})}$. If we suppose that the amplitude of the wave is small, *i.e.*, $A \ll 1$, the plasma wave is called a small amplitude wave or linear wave. In this case we can expand $\sin\phi$ in a Taylor series in the vicinity of $\phi = 0$. Substituting $\sin\phi \approx \phi$ into Eq. (17), we immediately get the dispersion relation for linear electromagnetic waves:

$$\omega^2(k) = k^2 + 1 \quad . \tag{18}$$

From here, the minimum phase velocity (Swihart velocity [6]) is

$$\bar{c}_0 = 1 \quad , \tag{19}$$

as expected when working in normalized units. $\bar{c}_0$ is also a maximum group velocity for electromagnetic waves. Recalling the normalization given by Eq. (11), it is possible to express c_0 in physical units:

$$c_0 = \frac{\text{length}}{\text{time}} = \lambda_J \omega_p = \sqrt{\frac{1}{\mu_0 C d'}} = c\sqrt{\frac{d_I}{\varepsilon d'}} \quad , \tag{20}$$

where $c = (\mu_0 \epsilon_0)^{-1/2}$ is the speed of light in vacuum. Thus, the dimensionless velocity which will be used below is normalized to the Swihart velocity.

- **Soliton.** It is easy to check that the function

$$\phi(\tilde{x}, \tilde{t}) = 4 \arctan\left[\exp\left(\frac{\tilde{x} - \tilde{u}\tilde{t} - \tilde{x}_0}{\sqrt{1 - \tilde{u}^2}}\right)\right] \tag{21}$$

satisfies Eq. (17). This solution represents a soliton (solitary wave), moving with velocity $\tilde{u} = u/c_0$. A soliton behaves as a relativistic particle which contracts according to the Lorentz law when its velocity approaches the Swihart velocity ($\bar{c}_0 = 1$), which, thus, plays the role of the velocity of light in an LJJ. Lorentz contraction of the soliton at high velocities follows from the Lorentz invariance of Eq. (17), which is easy to check by making the Lorentz transformation for x and t in Eq. (17). We stress that the soliton is essentially *a nonlinear wave* (since solitons do not exist in linear media). The fact that the maximum velocity of the soliton (21) is equal to the Swihart velocity (19, 20) (in principle) is just a coincidence! Later we shall see that this is not always the case. From a physical point of view a soliton carries a magnetic flux quantum, and therefore also is called a fluxon.
- **Chain of fluxons, the limit of high fluxon density.** If one imagines a junction with high fluxon density, the magnetic field in an LJJ is distributed almost uniformly, *i.e.*, $\phi \approx \tilde{H}(\tilde{x} - \tilde{u}\tilde{t})$, with a small correction which describes small oscillations of the field due to the presence of the chain $\delta\phi = A \sin\left[\tilde{H}(\tilde{x} - \tilde{u}\tilde{t})\right]$. Taking into account that $A \ll 1$, from Eq. (17) and using perturbation theory, it is easy to find

$$\phi = \tilde{H}(\tilde{x} - \tilde{u}\tilde{t}) - \frac{1}{\tilde{H}^2(1 - \tilde{u}^2)} \sin\left[\tilde{H}(\tilde{x} - \tilde{u}\tilde{t})\right] . \tag{22}$$

We see that the condition $A \ll 1$ is valid at rather high field H and not very high velocities of the fluxon chain u. In the relativistic limit $u \to 1$, the expression (22) can not be used! This is easy to understand, since at high velocities fluxons Lorentz contract, and at some velocity the fluxon width becomes smaller than the distance between the fluxons. The chain gets soft, and the single fluxon approximation (21) describes the situation much better than Eq. (22).

- **Chain of fluxons — finite fluxon density.** The most general case, embracing the two cases above, is for an arbitrary density of fluxons. In this case the solution of Eq. (17), which corresponds to the fluxon chain, can be written as [7]

$$\phi(\xi) = 2\mathrm{am}(\xi, k) + \pi \quad , \tag{23}$$

where $\xi = (\tilde{x} - \tilde{u}\tilde{t})/\sqrt{1-\tilde{u}^2}$, and $\mathrm{am}(\xi, k)$ is the Jacobi amplitude with modulus k, which defines the density of the fluxon chain. The distance between fluxon centers $\Delta\xi$ is related to k in the following way:

$$\Delta\xi = 2k\mathrm{K}(k) \quad , \tag{24}$$

where

$$\mathrm{K}(k) = \int_0^{\pi/2} \frac{d\theta}{\sqrt{1 - k^2 \sin^2(\theta)}} \tag{25}$$

is a complete elliptic integral of the first kind.

- **Resistive state** When the LJJ is in a resistive state, its phase rotates (roughly speaking) uniformly, $\phi(\tilde{t}) = \tilde{\omega}\tilde{t}$, where $\tilde{\omega} = \gamma/\alpha$. It is easy to check that such a solution satisfies Eq. (10) without the $\sin\phi$ term. The influence of the $\sin\phi$ term can be found by means of perturbation theory. Assuming $\phi = \tilde{\omega}\tilde{t} + A\sin(\tilde{\omega}\tilde{t})$ and $A \ll 1$, the phase is

$$\phi(\tilde{t}) = \omega\tilde{t} - \frac{1}{\tilde{\omega}^2}\sin(\tilde{\omega}\tilde{t}). \tag{26}$$

From here we see that $A = 1/\tilde{\omega}^2$ and is small at large values of $\tilde{\omega}$. Since in ordinary tunnel junctions $\alpha < 0.01$, then $\tilde{\omega} \gg 1$, and the condition $A \ll 1$ is valid always. If a magnetic field is applied, the expression (26) should be modified to become

$$\phi(\tilde{t}) = [\tilde{H}\tilde{x} + \tilde{\omega}\tilde{t}] - \frac{1}{\tilde{\omega}^2 - \tilde{H}^2}\sin(\tilde{H}\tilde{x} + \tilde{\omega}\tilde{t}). \tag{27}$$

Since usually $0 < \tilde{H} < 10$, then $\tilde{H}^2 \ll \tilde{\omega}^2$, and the condition $A = 1/(\omega^2 - \tilde{H}^2)$ is again always valid.

We note that in contrast to all previous solutions, we now consider the case of the perturbed sine-Gordon equation (10), rather than Eq. (17). Considering the R-state of an LJJ, the damping and the bias current are essential terms of Eq. (10), without which the formation of the R-state is impossible. Instead, the Josephson term of Eq. (10) is considered as a perturbation.

2.3. MOTION OF THE SINGLE FLUXON

First, we consider the motion of a fluxon in an infinite LJJ. The motion is described by the perturbed sine-Gordon equation (10). Usually the r.h.s. of Eq. (10) is small and, therefore, to a first approximation, the solution of Eq. (10) coincides with the solution of unperturbed Eq. (17), and the effect of the r.h.s. can be taken into account using a perturbation approach. The complete analysis was made for the first time in the classical work of McLaughlin and Scott [8]. The approach basically consists of abandoning the r.h.s. of Eq. (10) and introducing two new forces acting on the fluxon. The term $\alpha\phi_{\tilde{t}}$ describes the dissipation and is replaced by a corresponding friction force:

$$F_\alpha = \int_{-\infty}^{\infty} \alpha\phi_{\tilde{t}}\phi_{\tilde{x}}\, d\tilde{x}. \tag{28}$$

The term γ on the r.h.s. of Eq. (10) describes bias current that acts like a driving force:

$$F_\gamma = \int_{-\infty}^{\infty} \gamma\phi_{\tilde{x}}\, d\tilde{x}. \tag{29}$$

The driving force is sometimes also called a Lorentz force, because

$$\vec{F}_\gamma = \Phi_0[\vec{j} \times \frac{\vec{B}}{|B|}] \tag{30}$$

is similar to the force acting on the electron moving in a magnetic field $\vec{B}$ with a velocity $\vec{v}$,

$$\vec{F}_e = e[\vec{v} \times \vec{B}] \quad . \tag{31}$$

For a fluxon moving along the LJJ (in which α and γ are independent of coordinate x), from Eqs. (28) and (21), it is easy to obtain

$$F_\alpha = \frac{4\alpha u}{1-u^2}\int_{-\infty}^{\infty}\frac{d\xi}{\cosh^2(\xi)} = \frac{8\alpha u}{1-u^2} \quad , \tag{32}$$

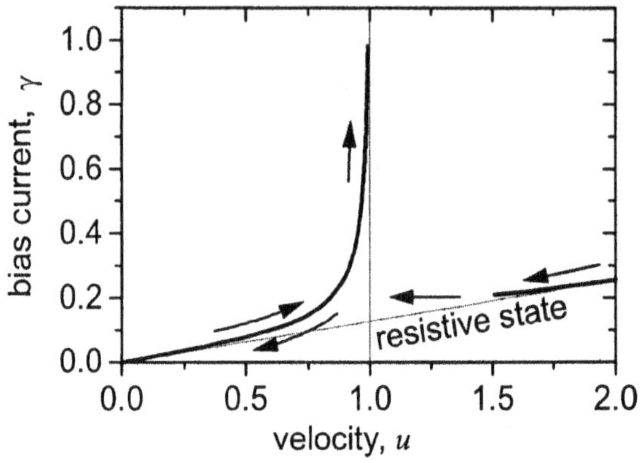

Figure 3. The dependence of the equilibrium fluxon velocity u (34) on the bias current γ for $\alpha = 0.1$. The results of numerical simulation are shown. The curve for $u < 1$ is well described by Eq. (34).

and from Eqs. (29) and (21) we derive

$$F_\gamma = \gamma\left[\phi(\infty) - \phi(-\infty)\right] = 2\pi\gamma \quad . \tag{33}$$

Writing a force balance equation (for the case of steady motion of the fluxon) $F_\gamma + F_\alpha = 0$, and substituting Eqs. (32) and (33) into it, we find the equilibrium velocity of the soliton:

$$u = \frac{1}{\sqrt{1 + \left(\dfrac{4\alpha}{\pi\gamma}\right)^2}} \quad . \tag{34}$$

The fluxon (21), moving with velocity u (34) is a very good approximation to the single soliton solution of Eq. (10). The important consequence of Eq. (34) is the fact that the velocity of a fluxon can't grow indefinitely with increasing bias current γ, but it does asymptotically approach $\bar{c}_0$. The soliton behaves as a relativistic particle which moves under the action of driving force F_γ and friction force F_α, and the velocity $\bar{c}_0$ plays the role of the velocity of light for our quasiparticle solitons. The dependence of the equilibrium velocity u (34) on the bias current γ is shown in Fig. 3. At small values of the bias current, the velocity is proportional to the driving force, as shown in Fig. 3 by the dashed line. This is equivalent to the motion of the particle in a viscous medium. When the velocity of the soliton approaches $\bar{c}_0$, the dependence $u(\gamma)$ deviates from the simple linear law and enters the relativistic region.

If one considers n fluxons in an annular LJJ [Fig. 1(b)] moving with velocity u (34), the average voltage across the junction is

$$V_n = \frac{\Delta\Phi}{\Delta t} = \frac{n\Phi_0}{L/u} = \Phi_0 \frac{u}{L} n. \tag{35}$$

One may see that the voltage V_n is proportional to the velocity of fluxons u. Therefore, below, instead of an IVC, $V(I)$, we sometimes will use normalized velocity *vs.* normalized current, $u(\gamma)$, still calling it IVC.

From Eqs. (34) and (35) one can see that for n fluxons moving along an LJJ, the IVC will look like one shown in Fig. 3 (in the case of a sparse fluxon chain), *i.e.*, one will observe a step with the asymptotic voltage $V_n^{fl} = \Phi_0 \bar{c}_0 n/L$ (for any fluxon chain density). These steps are called Fluxon Steps. The profile of the IVC $u(\gamma)$ for the finite fluxon density was derived in Ref. [9].

2.4. FLUX-FLOW

Now, let us consider a linear LJJ with no fluxons in it, in an applied magnetic field H and with bias current $I = 0$.

We must find a solution $\phi(x)$ of the sine-Gordon equation (17) satisfying the boundary conditions (13) at $x = 0$. It is easy to check that the single soliton solution (21) with $x_0 < 0$ satisfies this condition if the center of the fluxon x_0 is situated outside the LJJ. The position of the center is defined by the boundary condition (13), *i.e.*,

$$\frac{2}{\lambda_J \cosh \frac{x_0}{\lambda_J}} = \frac{2\pi}{\Phi_0} H\Lambda. \tag{36}$$

Sequential steps of the fluxon penetration into the LJJ are shown in Fig. 4. At a small value of field only the "tail" of a fluxon penetrates into the junction, as shown in Fig. 4(a). At a higher value of the magnetic field H, the center of the fluxon x_0 moves closer to the edge of the junction according to Eq. (36). At the value of the field

$$H = H_{\text{core}} = \frac{\Phi_0}{\pi \lambda_J \Lambda} \tag{37}$$

x_0 vanishes, and the fluxon can penetrate into the junction — see Fig. 4(b). In the same way, the second, third, *etc.* fluxons penetrate into the junction until they fill the whole LJJ with some density (approximately 1 fluxon per $\pi\lambda_J$ units of length) as shown in Fig. 4(c).

At $H > H_{c1}$ the fluxons freely penetrate into the LJJ from the edges and fill it with some density which depends on the value of the applied field

H. If now we apply a bias current γ to the LJJ, then, under the action of the Lorentz force (30), the chain of fluxons starts to move. Reaching the edge, the fluxons leave the LJJ (due to boundary conditions at $H > H_{c1}$), while new fluxons penetrate into the junction from the opposite edge. Such a mode of operation is called the **Flux-Flow** mode. In the flux-flow state the magnetic field on average completely penetrates into the LJJ in the form of fluxons. The voltage across the junction is equal to

$$V_{FF} = \frac{\Delta\Phi}{\Delta t} = \frac{H\Lambda L}{L/u} = H\Lambda u \quad . \tag{38}$$

As the bias current increases, the velocity of fluxons u approaches the velocity $\bar{c}_0$ (20) and, accordingly, a step with an asymptotic voltage $V_{FF} = H\Lambda\bar{c}_0$ appears on the IVC. This step is called the **Flux-Flow Step** (FFS). The interesting feature of the FFS is that the voltage of the step changes smoothly

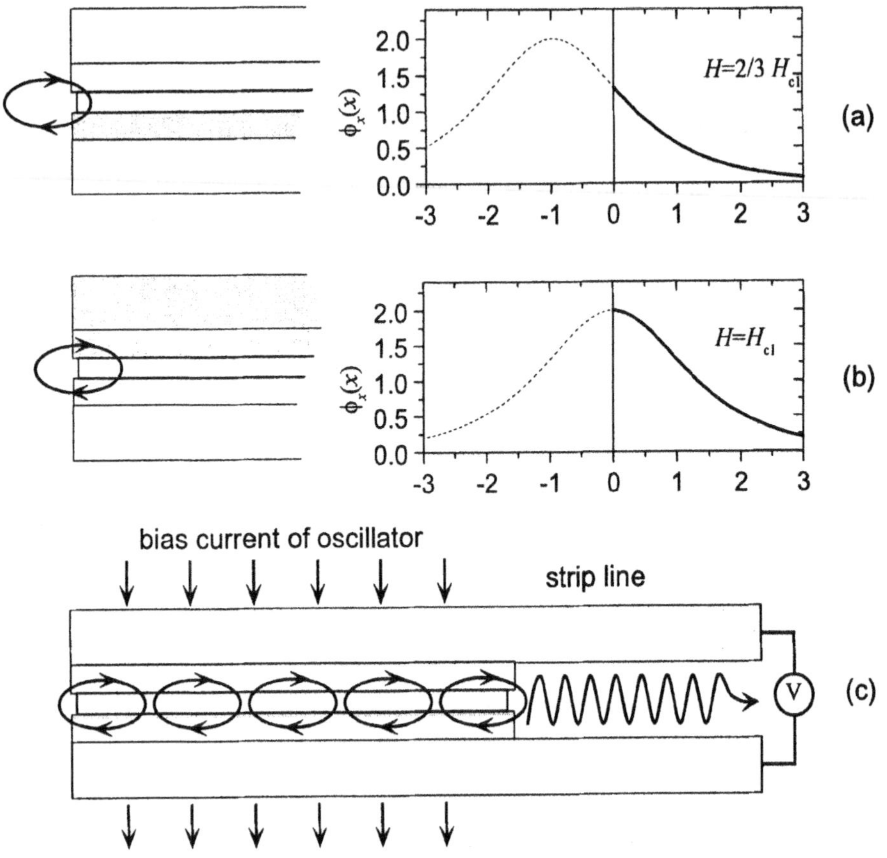

Figure 4. The penetration of a fluxon into a junction at different values of the magnetic field H. (a) $H = \frac{2}{3}H_{c1}$, (b) $H = H_{c1}$ (c) $H > H_{c1}$ and $I > 0$.

with magnetic field H. This makes an LJJ operating in the flux-flow mode a very attractive device for application as a tunable oscillator for mm and sub-mm wavebands [10, 1, 2]. Such oscillators are referred to as Flux-Flow Oscillators (FFOs).

Note that the magnetic flux quantum Φ_0 does not enter into the expression (38). In an LJJ with thin electrodes, the magnetic field generated by each fluxon spreads outside the junction boundaries in the z direction, and therefore the flux inside the junction, $-d^B < z < d^A$, corresponding to one fluxon, is equal to $\Phi_1 < \Phi_0$. The expressions for $\Phi_1(d^A, d^B)$ and $\phi(x, z)$ for different d^A and d^B are calculated in [11].

2.5. FISKE RESONANCES

Let us consider the important case of $\alpha\ell \lesssim 1$ and an LJJ in flux-flow mode. When one of the fluxons reaches the end of the LJJ and leaves it, due to interaction with the edge, a fraction of its energy is transformed into an electromagnetic wave which is emitted back into the junction. This electromagnetic wave propagates in the direction opposite to the direction of the fluxon motion. Its amplitude decays exponentially $A \propto e^{(-\alpha x)}$ over distances of the order of $1/\alpha$. The Lorentz force does not act on the electromagnetic (Josephson plasma) wave, since it does not carry magnetic flux, and the integral (29) is equal to zero. Thus, the pumping of energy into this wave by the bias current γ is absent. If the length of the LJJ ℓ is of the order of $1/\alpha$ or less, then the electromagnetic wave reaches the opposite edge of the LJJ, and, after reflection, interferes with itself, forming a standing wave. The frequency of oscillations of the Josephson phase in the standing wave is defined by the length of the LJJ, the (Swihart) velocity of the electromagnetic waves and mode number $\omega_n = 2Ln/\bar{c}_0$.

The fluxons moving along the LJJ interact with the standing electromagnetic wave. At some velocities of fluxon motion u_n, the oscillations of the Josephson phase caused by the fluxon motion occur in-phase with the oscillations of the Josephson phase caused by the standing wave. Thus, the velocities u_n are more favorable for fluxon motion than all other velocities. In addition, the pumping of energy into the standing electromagnetic wave, when the fluxon leaves the junction, takes place in phase with the wave. As a result, one can observe resonances in the IVC with asymptotic voltages

$$V_n^{FS} = \Phi_0 \frac{\bar{c}_0}{2L} n. \tag{39}$$

Consequently, in an LJJ with $\alpha\ell \lesssim 1$, the flux-flow voltage (measured by applying constant current through the junction) will not change smoothly as a function of the magnetic field H according to Eq. (38), but it will switch from one step given by Eq. (39) to another. These steps, caused by

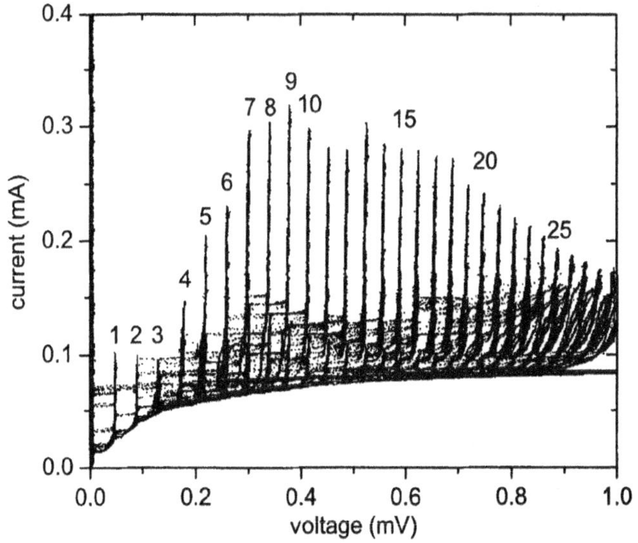

Figure 5. Fiske Steps on the IVC of an LJJ recorded over a wide range of magnetic field from 0 up to 35 Oe. The experiment was performed with a Nb-Al-AlO$_x$-Nb LJJ at $T = 4.2\,\mathrm{K}$. The resonance number is shown next to each step.

the geometric resonance of electromagnetic waves in the LJJ, are called Fiske Steps [12]. FSs obtained experimentally using a storage oscilloscope are shown in Fig. 5. For a given value of the magnetic field H, one usually can see only a few FSs — the ones with the voltage approximately equal to that in Eq. (38) with $u = \bar{c}_0$. The theory of Fiske steps for different cases was developed in Refs. [4, 13, 14, 15, 16].

Measuring the distance between FSs in an experiment, one can determine the Swihart velocity $\bar{c}_0$ from Eq. (39), and therefore, the capacitance of the junction per unit of area using Eq. (20).

2.6. REAL FFO

Due to the presence of the boundaries and the load at one of the edges of the LJJ in a real FFO of finite length, the motion of fluxons is quite different from that of the ideal fluxon chain described above. It was shown by Zhang *et al.*[17, 18] that the amplitude of the ac component of the voltage inside the junction is a function of x and behaves as shown schematically in Fig. 6. The bias current and magnetic field are applied as before. The fluxons enter from the left side, accelerate and exit at the right side. Depending on the boundary conditions and the load, the amplitude of $V_{ac}(L)$ can be different. Thus, to design a good working FFO one must perform a numerical simulation to choose the proper load-matching conditions for

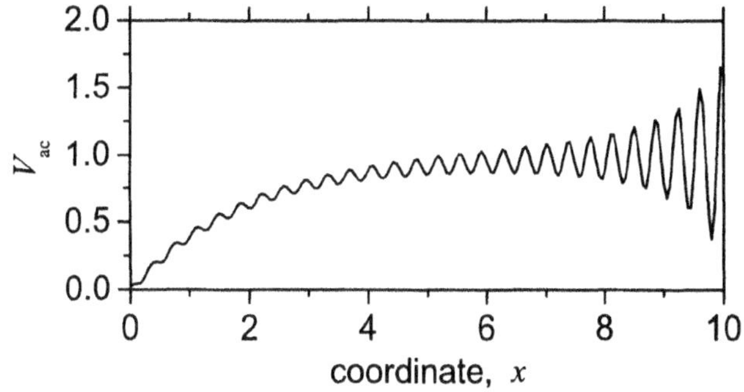

Figure 6. The schematically-shown dependence of the amplitude $V_{ac}(x)$. The fluxons enter from the left side, accelerate and exit at the right side. Depending on the boundary conditions and the load, the amplitude of the voltage oscillations at the right end $V_{ac}(L)$ can be different.

the maximum output power.

Fiske Steps, discussed above, prevent continuous tunability of an FFO by magnetic field as given by Eq. (38). Instead, the voltage, and therefore the operating frequency of the oscillator, have a set of discrete values, as can be seen in Fig. 5. Nevertheless, FSs have a finite slope which allows one to tune the frequency in a small region near the central frequency of a step. In a sufficiently long FFO, the distance between FSs can be rather small, and due to the final slope of the FSs, almost half of all frequencies can be covered (although not continuously). In spite of the difficulty of controlling such an oscillator — one has not only to change H, but sweep I back and forth to go to the proper step — Koshelets and co-authors [19] suggest that such an FFO may be used as a local oscillator in an integrated receiver which covers some frequency band continuously. This is possible if an intermediate-frequency amplifier has a bandwidth equal to or greater than a half of the voltage spacing between adjacent FSs. In the integrated receiver design reported in [19] the distance between FSs is 8 GHz, so that a 4-GHz IF amplifier will allow one to receive signals of any frequency over a wide band of several hundred GHz.

Another problem which arises is that a real FFO has a non-zero linewidth of emitted radiation. The broadening of the radiation linewidth can be associated with several factors:

- **Noise in the bias current supply and other extrinsic noise** are translated into variations of the voltage/frequency according to differential resistance R_d at the bias point of the IVC. In terms of reducing the linewidth caused by bias current noise, it is desirable to work on the most vertical part of the flux-flow step or at the FSs which are usually steeper than FFSs.

- **Self field** is the field generated by the bias current leads, and therefore, it is proportional to the bias current. It shifts the position of a FFS according to Eq. (38). In principle, making a proper chip design one can choose a negative self-field factor which will lead to a zero or even negative slope for the top part of a FFS. The object then is to make R_d vanish at the operating point of the FFO, and therefore effectively negate bias current noise as well as other sources of extrinsic noise.
- **Intrinsic noise** is caused by fluctuations in the resistors included in the equivalent circuit of the Josephson junctions. Other sources of intrinsic noise can be related to non-uniform critical current, the London penetration depth and tunnel-barrier thickness distribution. In general, if these contributions are small, they average to zero on a characteristic length scale equal to λ_J, *i.e.*, the characteristic fluxon size. There have been several attempts to derive an expression describing the broadening of the linewidth of a FFO due to thermal noise, which is a fundamental limit for all types of oscillators. In Refs. [20] and [21] the linewidth was derived for a chain of fluxons moving in an infinite LJJ for two different limits: dense and sparse fluxon chains. Over a wide range of magnetic field and velocity one may observe a crossover from one case to the other. Nevertheless, the values predicted by both theories are quite far removed from the experimentally-detected linewidths. The most probable reason for this is that the uniform fluxon chain approximation used in Refs. [20] and [21] is not valid near the edge, where fluxons leave the junction, as demonstrated in Fig. 6. At the moment we are not aware of any theory which takes into account the effect of the edge and load connected to the FFO and satisfactorily describes the experimentally-obtained linewidth values.

Nevertheless, the absence of a satisfactory theory for the FFO linewidth does not prevent experimentalists from improving power and linewidth of the oscillators. Tables 1 and 2 show some recent power and linewidth values reported in the literature.

TABLE 1. Power levels achieved by flux-flow oscillators.

Power	frequency	detection	Ref
1 pW	120 GHz	ac	[22]
430 nW/2JJ	320 GHz	on-chip	[23]
10 nW/2stack	410 GHz	ac	[24]
5 μW	440 GHz	on-chip	[25]

TABLE 2. Linewidths achieved by flux-flow oscillators.

Linewidth	frequency	detection	Ref
130 kHz	70 GHz		[26]
1 MHz	280 GHz		[27]
200 kHz	450 GHz		[25, 28]
2.1 MHz	320 GHz		[23]
1 Hz	270–440 GHz		[19]

2.7. IMPROVING THE FFO

Several approaches have been suggested to improve the performance of the FFO in terms of power and linewidth. We consider here several of them and discuss their advantages and disadvantages.

- **Phase-Locked Loop (PLL)**. The PLL is designed mainly to reduce the linewidth of the FFO. The idea is that the IF signal through the phase-locked feedback loop is added to the control-line (or bias-line) current, thus providing permanent compensation for the frequency deviation. Suggested and implemented by the group of Koshelets [29], an FFO in conjunction with a PLL shows record values of linewidth of 3.3 kHz at 310 GHz [29], 1 Hz at 270–440 GHz [19], and 3 Hz at 500 GHz [19]. We note that such a drastic decrease of the linewidth does not relax the requirements on the LJJ itself. The PLL has its own bandwidth and will reduce the FFO linewidth only for the case in which the FFO linewidth is less than the bandwidth of the PLL circuitry. Currently the PLL is implemented as an external feedback loop working at room temperature. Integration of the PLL circuitry onto the chip can provide narrow on-chip linewidth and eliminate the need for bulky room-temperature equipment.
- **Stacks of LJJs**. The obvious way to improve the power of any generator is to use N coherently-operating oscillators instead of one. To provide this coherence, LJJs must be coupled [30, 31, 32, 33, 34]. The simplest way to organize the coupling between the junctions topologically is to use a stack of coupled LJJs (CLJJs). If the thickness of the superconducting electrodes separating neighboring LJJs is such that $d < \lambda$, the interaction will take place by means of currents flowing along the common electrode of neighboring junctions. Vertical integration also minimizes wasted space on the chip of a receiver. This would allow one in the future to include extra functional blocks into an integrated receiver, *e.g.*, the block for high-speed digital signal processing

[35, 36, 37].

Some high-T_c superconductors (HTS) such as BSCCO have on the atomic level a layered structure which is very similar to a stack of intrinsic LJJs [38], and which can be described by the same equations. Thus, the interest in stacks is stimulated by the need to understand the processes taking place in HTS. In addition, a CLJJ is a very interesting subject in terms of non-linear physics. The non-linearity in combination with interaction between junctions results in a number of non-trivial physical phenomena [30, 39, 40]. The system of coupled sine-Gordon equations describing the dynamics of Josephson phases in CLJJs is not yet well understood. Therefore, the investigation of CLJJs is interesting from the non-linear physics point of view. The stacked LJJs are considered in detail below.

- **1D and 2D arrays of JJs**. A 1D array is a discrete analog of the LJJ. A 2D array is a discrete analog of a stack of LJJs. But one can go much beyond these simple analogies. Arrays can be used in the continuous limit in which from a fluxon motion point of view, they behave like an LJJ or a stack of LJJs when $a \ll \lambda_J$, where a is the cell size. Also, arrays can be used in the discrete limit where they exhibit dynamic behavior which is different in comparison with continuous systems. Even for the case when arrays are used in the continuous limit, they are much more flexible than multi-layered devices, since they allow one to put any active or passive element into horizontal and vertical branches, thus stimulating the in-phase locking of all array cells. In addition, one can change the geometry of the arrays from the classical rectangular form to the triangular form or even more exotic configurations. A set of publications concerning triangular arrays can be found in [41]. Recently obtained values of power ($\sim 45\,\mathrm{nW}$) from rectangular arrays and dc $\rightarrow$ac conversion efficiency (17%) are reported in [42]. As a disadvantage of arrays, it is worth noting that 2D arrays are fabricated in the plane of the chip and occupy considerable space. Therefore, it is problematic whether one can integrate them with other complex devices on the same chip.
- **TWT- and BWT-like devices**. Traveling Wave Tube (TWT) and Back Wave Tube (BWT) devices have been in use for decades and utilized for the generation of high-power electromagnetic waves by means of passing an electron beam through a dispersive medium. The same concept applies to Josephson devices. Fluxons, which play the role of the quasiparticles (electrons), move along an LJJ with a dispersive slow-wave system coupled to it. For a slow-wave system, one may use a stripline or related structures with stubs which look from above like a comb [43, 44]. Power will build in the slow-wave system and be emit-

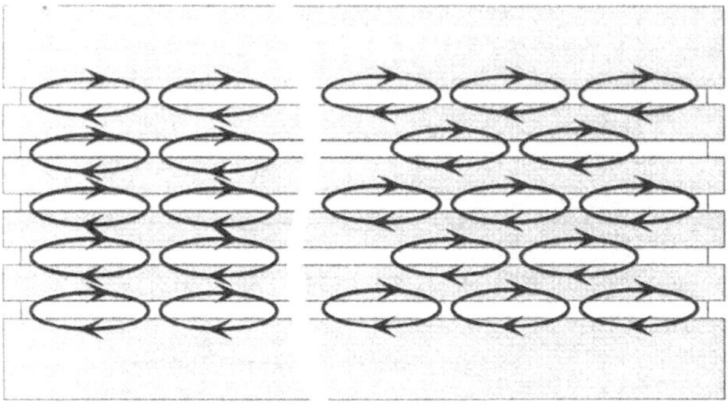

Figure 7. Different fluxon modes in stacks. While the in-phase mode shown on the left side of this figure is the best mode for oscillators, the mode shown on the right side is the most stable one.

ted to the load. It is expected to be a few times larger than for a conventional FFO.

It is important to note that such oscillators belong to a wider group of (so-called) Cherenkov oscillators — see the introduction and discussion of the Cherenkov effect in Josephson structures below. In Cherenkov oscillators, the fluxon-quasiparticle emits radiation which is used for practical purposes. Even a simple annular junction with a stripline attached to it can be considered as a primitive example of this class of systems, since periodic perturbation for the fluxon circulating around the junction introduced by the stripline effectively changes the dispersion relation for plasma waves and fluxons, resulting in the emission of Cherenkov radiation into the stripline. The emission of the radiation from such a system was reported in Ref. [45].

If one has a stack of N coupled junctions as shown in Fig. 7, a great number ($\gg N$) of fluxon states are possible. While the in-phase fluxon state shown on the left of Fig. 7 is the best mode for oscillators, the mode shown on the right is usually the most stable one.

Studies of the in-phase fluxon mode showed that in stacks with $N > 2$ coupled LJJs, the in-phase motion of fluxons is accompanied by the emission of electromagnetic radiation behind the fluxons as shown in Fig. 8. This radiation is known as Cherenkov radiation. The presence of the Cherenkov radiation (a) disturbs the in-phase motion of the next column of the in-phase fluxons, (b) results in additional loses and low R_d, *i.e.*, in large linewidth of the oscillator. Since this phenomenon has a major influence on the performance of the stack-based FFO, we study it in detail in the next section.

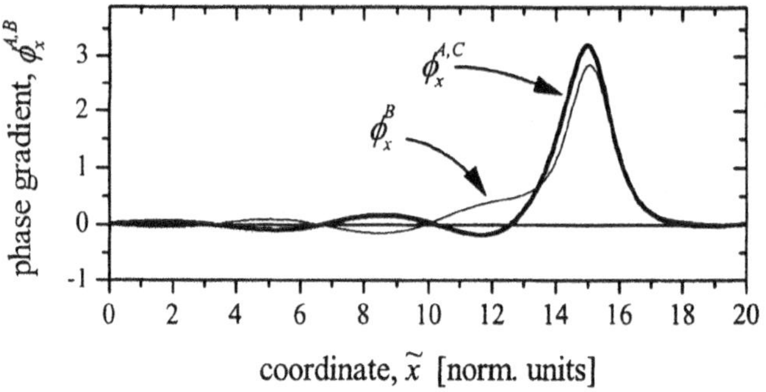

Figure 8. Cherenkov radiation of out-of-phase plasma waves generated by fluxons moving in the [1|1|1] state in an annular three-fold symmetric stack with $\gamma = 0.6$, $S = -0.5$, $\ell = 20$, $u = 1.71$ ($\bar{c}_+ = 1.84$).

3. Cherenkov radiation from fluxons

3.1. INTRODUCTION

From particle physics it is known that when a charged particle moves in a medium with velocity u which exceeds the velocity of light c_m in this medium, one can observe electromagnetic radiation emitted by the particle. This phenomenon has been known since the 1920s. It is called the Cherenkov effect, and the electromagnetic radiation emitted is known as Cherenkov radiation. Generally speaking, the phenomenon of Cherenkov radiation is not limited to a system of elementary particles. It may occur in any physical system where one can define a (quasi-)particle and an analog of the electromagnetic wave. If one considers Cherenkov radiation in long Josephson junctions and similar systems, a fluxon serves as a quasiparticle, and the Josephson plasma waves are the electromagnetic waves. From a mathematical point of view, a fluxon is a soliton with associated mass[1], velocity, momentum, energy, *etc.*, *i.e.*, it can be considered as a quasiparticle. Thus, we will not focus attention on its internal structure, *i.e.*, that it is actually a non-linear wave, but rather we focus on its particle-like characteristics. Josephson plasma waves are real electromagnetic waves with their own dispersion law.

[1]The mass of a fluxon is given by the expression

$$m = 4\left(\frac{\Phi_0}{2\pi}\right)^{\frac{3}{2}} \frac{\epsilon\epsilon_0 w}{d_I}\sqrt{\mu_0 j_c d'}$$

and is defined as a coefficient in front of the term $\propto u^2/2$ in the expansion of kinetic energy. For $j_c = 300\,\mathrm{A/cm^2}$, $w = 10\,\mu\mathrm{m}$, $d' = 300\,\mathrm{nm}$ and $d_I = 2\,\mathrm{nm}$, the value of mass will be $\sim 10^{-33}$ kg.

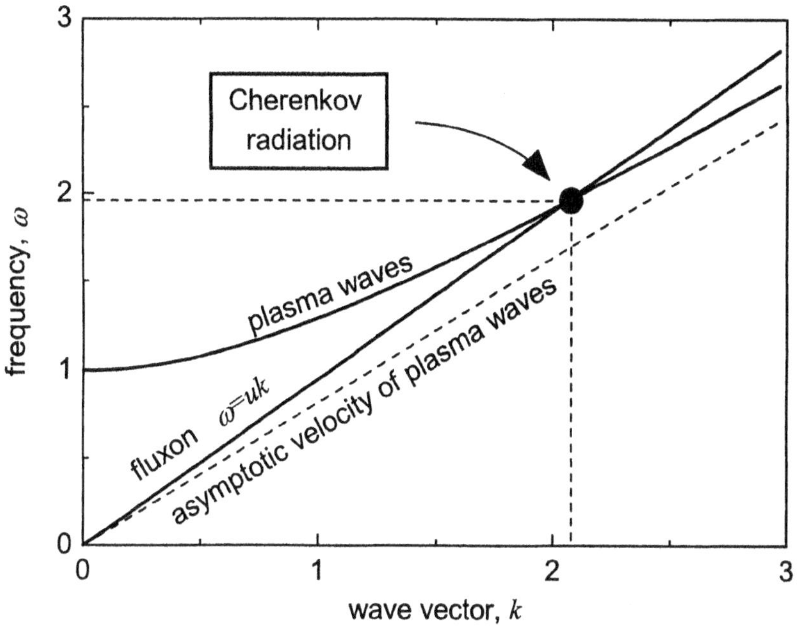

Figure 9. An example of a dispersion diagram where the straight line corresponds to a fluxon. If there is a crossing of the dispersion curve and a fluxon's straight line (as in this plot), the emission of Josephson plasma waves with corresponding k and ω (shown by dashed lines) occurs.

When Cherenkov radiation is emitted, the velocity of the (quasi-)particle is equal to the phase velocity of the emitted electromagnetic wave. This definition has two useful consequences. First, it relates the velocity of the particle to the dispersion characteristic of the medium. Second, it allows one to predict whether Cherenkov emission can take place in a particular system.

For an analysis of Cherenkov radiation it is convenient to use a dispersion diagram as shown in Fig. 9. The straight line corresponds to a fluxon. If the intersection of the dispersion curve and a fluxon's straight line takes place, the emission of Josephson plasma waves with corresponding k and ω occurs.

As a first example let us consider an LJJ. The dynamics of the Josephson phase is described by the sine-Gordon equation (17), which has, as explained above, fluxon and plasma-wave solutions. The fluxon velocity, according to Eq. (21) or (34), can have values ranging from 0 to 1 ($\bar{c}_0$ in physical units), while the possible range of the phase velocities of plasma waves — see Eq. (18) —

$$u_{\rm ph} = \frac{\omega(k)}{k} = \sqrt{1 + \frac{1}{k^2}} \quad , \tag{40}$$

is from 1 ($\bar{c}_0$) to infinity. Therefore, the criterion for Cherenkov radiation can never be satisfied, *i.e.*, the straight line corresponding to a fluxon in Fig. 9 always passes below the dashed line corresponding to the asymptotic velocity of plasma waves.

3.2. CHERENKOV RADIATION IN STACKS OF LJJS

As a second example we consider a stack of coupled long Josephson junctions, a geometry which has received considerable attention in recent years: first, due to progress achieved in fabrication of Nb-Al-AlO$_x$-Nb low-T_c tunnel junctions; and second, due to layered high-T_c superconductors that exhibit an intrinsic Josephson effect [38].

The behavior of Josephson phases in a system of coupled LJJs is described by a system of coupled sine-Gordon equations [46]. Since the dispersion relation for linear waves, as well as the maximum velocity of a fluxon, are influenced by mutual coupling between the junctions, one may find the conditions for which Cherenkov radiation appears in such a system.

For the sake of simplicity, let us first consider a simple system consisting of just two coupled (generally asymmetric) LJJs, which we denote as LJJ^A and LJJ^B. Such a system can be described by the two coupled perturbed sine-Gordon equations [47, 48, 49]:

$$\begin{cases} \dfrac{\phi^A_{\tilde{x}\tilde{x}}}{1-S^2} - \phi^A_{\tilde{t}\tilde{t}} - \sin\phi^A - \dfrac{S\phi^B_{\tilde{x}\tilde{x}}}{1-S^2} = \alpha^A\phi^A_{\tilde{t}} - \gamma^A \\ \dfrac{\phi^B_{\tilde{x}\tilde{x}}}{1-S^2} - \phi^B_{\tilde{t}\tilde{t}} - \dfrac{\sin\phi^B}{J} - \dfrac{S\phi^A_{\tilde{x}\tilde{x}}}{1-S^2} = \alpha^B\phi^B_{\tilde{t}} - \gamma^B \end{cases} , \tag{41}$$

where $\phi^{A,B}$ are the Josephson phases across LJJ ^{A,B}, respectively; S ($-1 < S < 0$) is a dimensionless coupling constant; $J = j_c^A/j_c^B$ is the ratio of critical currents; $\alpha^{A,B}$ and $\gamma^{A,B} = j^{A,B}/j_c^{A,B}$ are the damping coefficients and normalized bias currents, respectively. The effective magnetic thicknesses $d'^{A,B}$ of LJJ^A and LJJ ^{B} and the specific capacitances $C^{A,B}$ are assumed to be equal.

For analysis of linear modes we use Eq. (41) in the absence of perturbations. Considering small amplitude waves

$$\phi^{A,B} = A^{A,B}\sin(k\tilde{x} - \omega\tilde{t}) \quad , \quad A \ll 1 \quad , \tag{42}$$

in such a system and linearizing $\sin\phi^{A,B} \approx \phi^{A,B}$, we obtain the dispersion relation

$$\omega(k) = \sqrt{\frac{1+J}{2J} + \frac{k^2}{1-S^2} \pm \sqrt{\frac{(1-J)^2}{4J^2} + \frac{S^2k^4}{(1-S^2)^2}}} \quad . \tag{43}$$

It consists of two branches corresponding to two different modes of electromagnetic wave propagation in the system: an in-phase mode and an out-of-phase mode. From Eq. (43) the phase velocity $u_{\rm ph} = \omega(k)/k$ can be expressed as

$$u_{\rm ph} = \sqrt{\frac{1+J}{2Jk^2} + \frac{1}{1-S^2} \pm \sqrt{\frac{(1-J)^2}{4J^2k^4} + \frac{S^2}{(1-S^2)^2}}} \quad . \tag{44}$$

The phase velocities of the in-phase mode ["+" sign in Eqs. (43) and (44)] are in the range from $\bar{c}_+$ to ∞, and for the out-of-phase mode ["−" sign in Eqs. (43) and (44)] from $\bar{c}_-$ to ∞. The velocities $\bar{c}_\pm$ are given by the simple relation:

$$\bar{c}_\pm = \frac{1}{\sqrt{1 \pm S}} \quad . \tag{45}$$

Note, that $\bar{c}_\pm$ does not depend on J.

The variety of fluxon configurations is very rich even for 2 coupled LJJs. When discussing different fluxon configurations we will use the notation $[N|M]$, which means N fluxons located (trapped) in LJJ^A and M fluxons in LJJ^B ($N, M < 0$ describe anti-fluxons).

The [1|0] *state.* Consider now two coupled LJJs with $J \neq 1$ and some running solution $\phi^{A,B}(\tilde{x} - u\tilde{t})$. Introducing a new variable $\xi = \tilde{x} - u\tilde{t}$ and omitting the α and γ terms, the unperturbed Eqs. (41) can be rewritten as

$$\left\{ \begin{array}{lcl} \phi^A_{\xi\xi}\left(\dfrac{1}{1-S^2} - u^2\right) - \dfrac{S}{1-S^2}\phi^B_{\xi\xi} & = & \sin\phi^A \\ \phi^B_{\xi\xi}\left(\dfrac{1}{1-S^2} - u^2\right) - \dfrac{S}{1-S^2}\phi^A_{\xi\xi} & = & \dfrac{1}{J}\sin\phi^B \end{array} \right. \quad . \tag{46}$$

Now suppose that the solution $\phi^{A,B}$ moves with velocity $u = \bar{c}_-$ given by Eq. (45). Substituting $u = \bar{c}_-$ from Eq. (45) into Eq. (46), we obtain the following relation between ϕ^A and ϕ^B:

$$\sin\left(\phi^A\right) = \frac{1}{J}\sin\left(\phi^B\right) \quad , \tag{47}$$

After establishing Eq. (47), consider the [1|0] state with the only fluxon in LJJ^A, *i.e.*, ϕ^A grows from 0 to 2π as x changes from $-\infty$ to $+\infty$ and

ϕ^B changes with x, but $\phi^B(\pm\infty) = 0$. If $J = 1$, then from Eq. (47) it follows that $\phi^A = \phi^B + 2\pi k$. Thus, if ϕ^A grows from 0 to 2π, ϕ^B also grows in the same way, corresponding to the state [1|1] and contradicting the assumption that we are considering the state [1|0]. Therefore, in the state [1|0] with $J = 1$, the fluxon can not move with velocity $\bar{c}_-$ — it is the maximum asymptotic (unobtainable) velocity of the fluxon.

If $J < 1$ and ϕ^A grows from 0 to 2π as x changes from $-\infty$ to $+\infty$, ϕ^B will not change by 2π as x changes from $-\infty$ to $+\infty$. Imagine that ϕ^A grows from 0 and approaches $\pi/2$ at some point $x = x_0$, *i.e.*, $\sin(\phi^A)$ approaches 1. Then $\sin(\phi^B)$ will approach J $(J < 1)$ and ϕ^B will equal some value between 0 and $\pi/2$ (*e.g.*, $\pi/3$). At the next point $x = x_1 > x_0$ where ϕ^A is greater than $\pi/2$, $\sin(\phi^A)$ starts to decrease, so $\sin(\phi^B)$ becomes smaller than $1/\kappa$ and $\phi^B < \pi/3$, *i.e.*, ϕ^B will not change by 2π. This is an important point: the fact that ϕ^B does not twist means that there is no fluxon in LJJ^B, *i.e.*, the [1|0] state is possible at the velocity $u = \bar{c}_-$.

Mathematically (for a "good" differential equation), if a solution exists at point $u = \bar{c}_-$, it will exist also in the vicinity of this point, *i.e.*, also for values greater than $\bar{c}_-$. In fact, simulations show that for $J = 0.5$, $S = -0.5$ and $\alpha = 0.1$, the state [1|0] can survive up to about $\bar{c}_0$ for some parameters of the system. The region of velocities $\bar{c}_- < u < u_{\max}^{[1|0]}$ is a domain where Cherenkov radiation in the [1|0] state is excited by a moving fluxon. Of course, $u_{\max}^{[1|0]}$ is a function of parameters such as J, S and α.

For $J > 1$, a fluxon in the [1|0] state can not reach the velocity $\bar{c}_-$. Moreover, the maximum velocity of the fluxon in this case is smaller than $\bar{c}_-$.

The velocity of a fluxon (u) *vs.* the bias current (γ) for the three cases mentioned above is plotted in Fig. 10. This plot was obtained as a result of numerical simulation of a system with $S = -0.5$ and $\alpha = 0.2$. The Cherenkov radiation of plasma waves takes place on the curve $J = 0.5$ for $u > \bar{c}_-$.

If one places a fluxon in a system of two annular junctions (in a JJ with low j_c) of moderate circumference L ($\alpha L < 3$), the emitted Cherenkov waves can form a standing/running wave due to periodic boundary conditions. A fluxon moving along such a system interacts with plasma waves, resulting in resonances in the IVC. So, instead of the smooth curve shown in Fig. 10, one gets a curve as shown in Fig. 11. The curve shown in Fig. 13 was measured in an experiment with a (Nb-Al-AlO$_x$)$_2$-Nb stack and is very similar to the IVC which can be obtained from simulation. The details of the experiment and simulations can be found in Ref. [50].

The [1|±1] *state.* The state [1|−1], which has maximum velocity $\bar{c}_-$, can not excite any plasma wave mode. Only fluxons in the state [1|1] moving with velocity $\bar{c}_- < u < \bar{c}_+$ may, in principle, generate the out-of-phase mode

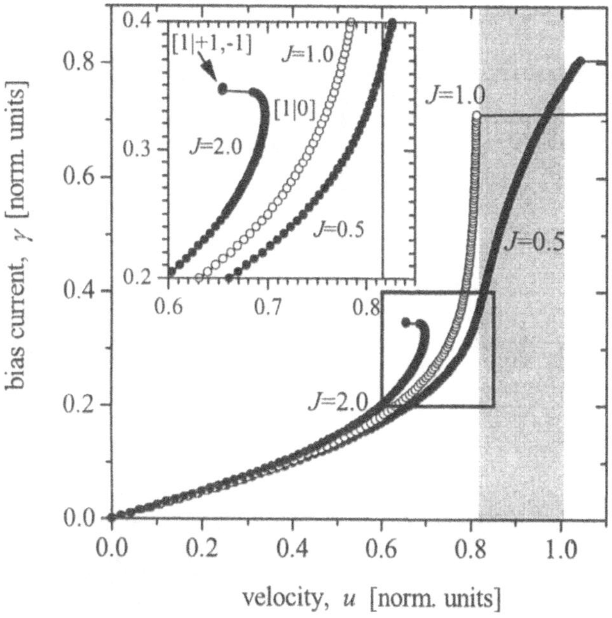

Figure 10. The family of IVCs of two inductively-coupled LJJs for different values of J in the [1|0] state. For $J < 1$, $u_{\max} > \bar{c}_-$; for $J > 1$, $u_{\max} < \bar{c}_-$.

of plasma waves. This may happen for two coupled LJJs with different parameters. In the symmetric case, Cherenkov radiation does not occur, since the solutions of the coupled Eqs. (46) are equal ($\phi^A = \phi^B$) and purely solitonic.

3.3. STACKS WITH $N > 2$ LAYERS

In the ideal case of high-T_c layered superconductors, the intrinsic LJJs have very similar critical-current densities. The system of equations which describes N inductively-coupled LJJs was derived in Ref. [46], and for the case of the Resistively Shunted Junction (RSJ) model and absence of perturbations (without α and γ terms, *i.e.*, quasiparticle and bias currents), it can be written as

$$\vec{\phi}_{xx} = \mathbf{S} \sin \vec{\phi} + \mathbf{S} \vec{\phi}_{tt} \quad , \tag{48}$$

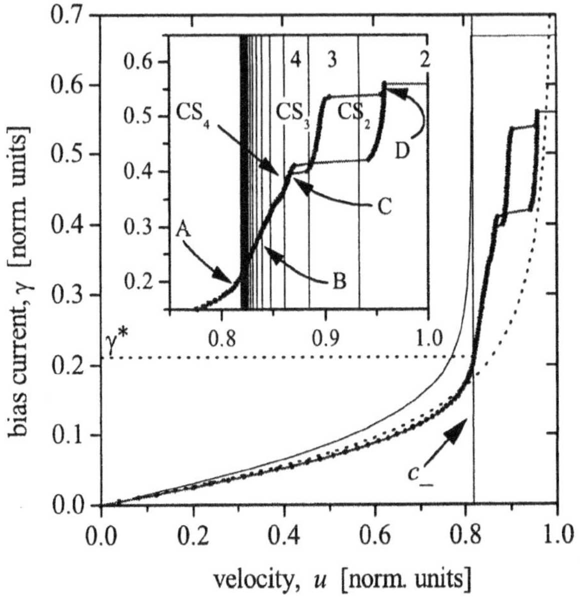

Figure 11. Calculated dependence of velocity u on bias current density γ (normalized IVC) for one soliton in a 2-junction ring for $L/\lambda_J = 7$, $\alpha = 0.1$, $J = 0.5$, $S = -0.5$. The inset shows in detail the top part of the IVC. The IVCs for different configurations are shown for comparison: dotted line corresponds to single annular junction ($S = 0$), thin line to the stack in the $[1|-1]$ state.

where

$$\mathsf{S} = \begin{pmatrix} 1 & S & 0 & 0 & 0 & 0 & 0 \\ S & 1 & S & 0 & 0 & 0 & 0 \\ 0 & \ddots & \ddots & \ddots & 0 & 0 & 0 \\ 0 & 0 & S & 1 & S & 0 & 0 \\ 0 & 0 & 0 & \ddots & \ddots & \ddots & 0 \\ 0 & 0 & 0 & 0 & S & 1 & S \\ 0 & 0 & 0 & 0 & 0 & S & 1 \end{pmatrix} \tag{49}$$

is the matrix which describes the interaction between the LJJs, $-0.5 < S \leq 0$ is a dimensionless coupling constant which for the N-fold case can be expressed as:

$$S = \frac{-\lambda}{d_I \sinh\dfrac{d}{\lambda} + 2\lambda\cosh\dfrac{d}{\lambda}}\;, \tag{50}$$

where d_I is the thickness of the insulator (tunnel barrier) between adjacent superconducting layers, d is the thickness of each superconducting layer and

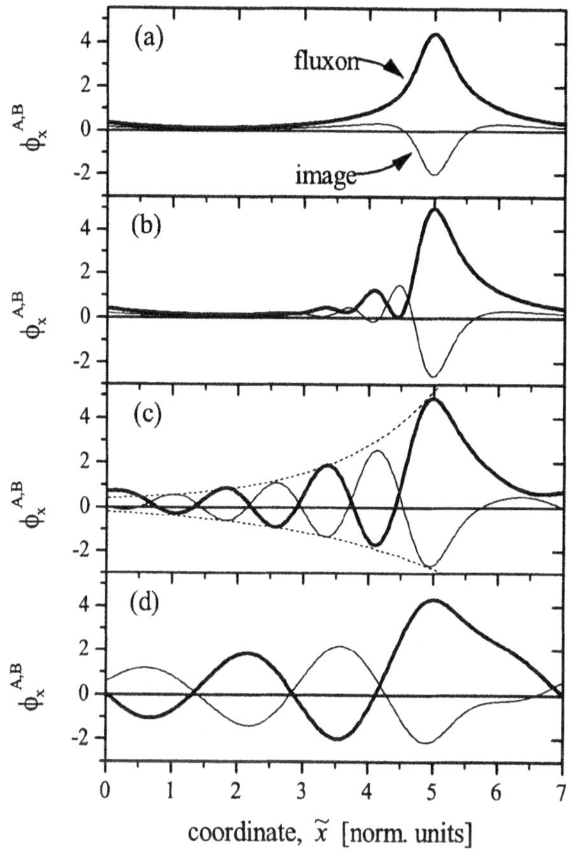

Figure 12. Profiles of $\phi_{\tilde{x}}^{A,B}(x)$ in both LJJs. (a), (b), (c), (d) correspond, respectively, to IVC points A, B, C, D shown in the inset of Fig. 11.

λ is the London penetration depth. In the limit of high-T_c superconductors, $d_I, d \ll \lambda$ and

$$S \approx -\frac{1}{2}\left(1 - \frac{d_I d}{\lambda^2}\right) \quad . \tag{51}$$

3.4. PLASMA WAVES

It has been derived that in an N-fold stack of LJJs there are N modes of plasma waves and, accordingly, N dispersion branches with N characteristic velocities. In the case of equal LJJ parameters, the expression for the n-th

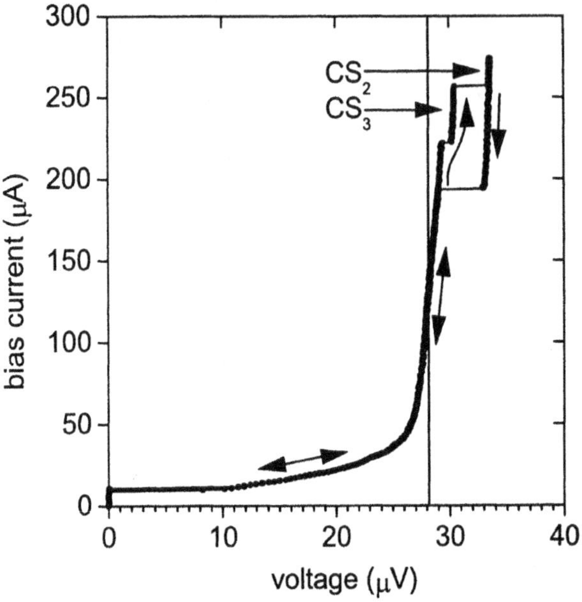

Figure 13. Experimental IVC of the double-layer annular junction in the [1|0] soliton configuration for $J \approx 0.5$. The Cherenkov resonances are marked $CS_{2,3}$.

characteristic velocity is simply [51, 47]:

$$\bar{c}_n^{(N)} = \frac{1}{\sqrt{1 + 2S\cos\left(\frac{\pi n}{N+1}\right)}}, \quad (n = 1 \dots N). \tag{52}$$

Note that, for given N, the values of velocities obey the following relation: $\bar{c}_N^{(N)} < \dots < \bar{c}_1^{(N)}$. As soon as a fluxon is accelerated to the velocity $u > \bar{c}_N^{(N)}$, *i.e.*, above the lowest Swihart velocity, Cherenkov emission of the slowest mode with characteristic velocity $\bar{c}_N^{(N)}$ may take place. If u exceeds several lowest Swihart velocities, the Cherenkov radiation may consist of a mixture of plasma waves of different modes with different k and ω as illustrated in Fig. 14. An example of a numerical simulation of a Cherenkov radiation wake for $N = 7$ is given in Ref. [52]

3.5. FLUXONS

For the case of two coupled LJJs, asymmetry is essential for Cherenkov emission to appear. An important question which arises for the case of an N-fold stack, and especially for layered high-T_c materials, is whether Cherenkov radiation may appear in a system where all junctions have the

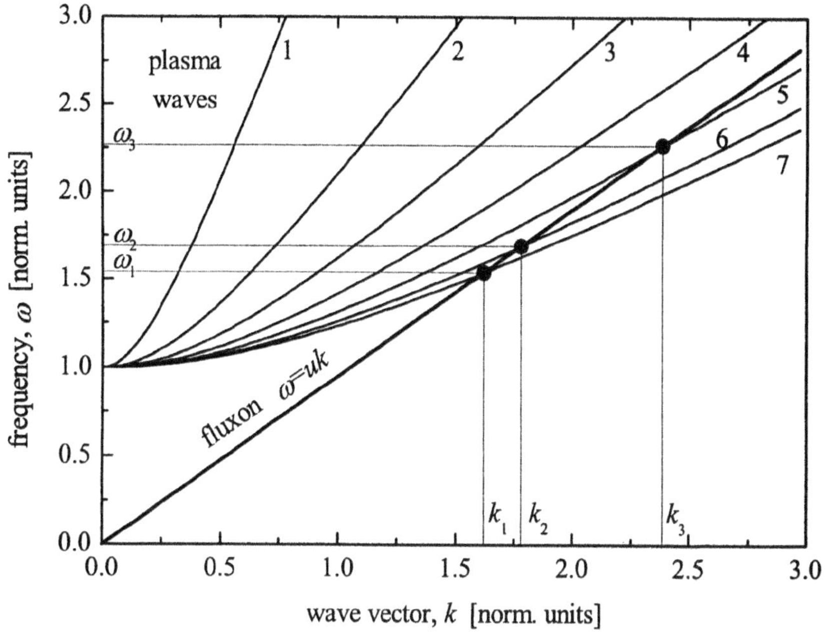

Figure 14. The dispersion curves of the 7-fold stack corresponding to 7 different modes of propagation of Josephson plasma waves.

same parameters. For the case $N \geq 3$ there is an additional decrease of the symmetry of the system in comparison to the case $N = 2$. For $N = 2$ and $J = 1$, both LJJs are in similar situations, and each of them interacts with its neighboring junction in the same way. But for $N = 3$, the middle LJJ^B has two neighbors, while LJJs^{A,C} have only one. Let us consider 3 coupled sine-Gordon equations and assume the [0|1|0] or [1|0|1] state with the symmetric solution $\phi^A = \phi^C$. In this case one can reduce 3 equations to two equations, one for ϕ^B and another for $\phi^A = \phi^C$:

$$\begin{cases} \frac{1}{1-2S^2}\phi_{xx}^{A,C} - \phi_{tt}^{A,C} - \sin\phi^{A,C} - \frac{S}{1-2S^2}\phi_{xx}^{B} = \gamma - \alpha\phi_t^{A,C} \\ \frac{1}{1-2S^2}\phi_{xx}^{B} - \phi_{tt}^{B} - \sin\phi^{B} - \frac{2S}{1-2S^2}\phi_{xx}^{A,C} = \gamma - \alpha\phi_t^{B} \end{cases} . \tag{53}$$

In Eq. (53) for ϕ^B, there is a factor of 2 in the coupling term. The problem of a 3-fold symmetric stack is reduced to the problem of a 2-fold asymmetric stack. Thus, asymmetry appears automatically when one goes from $N = 2$ to $N = 3$.

The analysis presented above for the case $J \neq 1$ can not be directly applied for the asymmetric coupling term as in Eq. (53). We simulated Eq. (53) for the case of [0|1|0] and [1|0|1] fluxon states, and found that the state $[0|1|0]$ is similar to the case $J < 1$, *i.e.*, the IVC bends to the right in the region $u > \bar{c}_3^{(3)}$, which results from the emission of Cherenkov radiation; the state $[1|0|1]$ is similar to the case $J > 1$, and results in an IVC with a region of back-bending at $u < \bar{c}_3^{(3)}$. Qualitatively, the IVCs for this case look like the IVCs shown in Fig. 10 for $J = 0.5$ and $J = 2.0$. Using this result, we predict that a fluxon moving in a layered high-T_c superconducting material produces Cherenkov radiation, and that its IVC will not have conventional single-junction asymptotic behavior.

Since N-fold stacks with $N > 2$ have lower symmetry, the $[1|\ldots|1]$ fluxon state will emit Cherenkov radiation even for the case of equal LJJ parameters. If the emission is very strong (for $N \gg 2$), the state $[1|\ldots|1]$ can not survive. We simulated a three-fold stack in a symmetric fluxon configuration $[1|1|1]$ and found that this state is still stable, emitting Cherenkov radiation for $u > \bar{c}_3^{(N)}$ as shown in Fig. 8.

In an N-fold stack of *linear geometry*, it was found rather difficult to stabilize the motion of a fluxon in a $[0|..|0|1|0|..|0]$ state in a ZFS-like mode. Due to the presence of images in the neighboring LJJs (relative to the LJJ with a fluxon), the reflection conditions become imperfect (since energy dissipates in the collision of images and anti-images at the boundary), and the fluxon tends to leave the junction. Ideal reflection conditions are realized in an uncoupled LJJ, *i.e.*, at $S = 0$. When $|S|$ and/or N increase, more energy dissipates during the reflection process. For a given N, at some critical value of S, the fluxon can not be reflected back and leaves the system. For layered high-T_c superconductors ($N \gg 2$), $|S|$ approaches its maximum theoretical value of 0.5. Thus, ZFS-like fluxon motion can most probably not be realized in the $[0|..|0|1|0|..|0]$ state in a high-T_c layered stack with open boundaries. Therefore, Cherenkov radiation from a *single* fluxon moving in a stack of intrinsic LJJs or in an N-fold stack ($N \gg 2$) can be studied only in an artificial annular structure.

In an experiment, one often deals with many fluxons moving in a flux-flow mode. Will Cherenkov radiation appear in this case or not? Assuming the $[0|\ldots|0|M|0|\ldots|0]$ state in the limit of a dense uniform fluxon chain and neglecting the junction edges, it is possible to show that Cherenkov radiation does not take place, since the maximum velocity of the chain in each mode is equal to $\bar{c}_n^{(N)}$. However, at high velocities when the fluxon chain becomes less dense, a crossover to the single fluxon behavior discussed above occurs. In this case, Cherenkov radiation results in a change of the top part of the IVC which bends to the right in comparison with the usual asymptotic behavior.

As a third example of a system in which Cherenkov radiation may take place, we consider a 1D array of JJs in the continuous limit, *i.e.*, with spacing $a \ll \lambda_J$. In this limit the dispersion relation for linear waves resembles one for a continuous system (18), but only for small k. For larger k it is necessary to recall that the dispersion relation for any periodic medium is periodic in k with period equal to π/a. The explicit expression for this dispersion relation is well known and can be written as:

$$\omega(k) = \sqrt{1 + \frac{4}{a^2}\sin^2\left(\frac{ak}{2}\right)} \quad . \tag{54}$$

In the continuous limit, $a \ll \lambda_J$, the shape of a fluxon to first approximation is the same as for a continuous system (21), and therefore, it has a maximum velocity equal to 1.

Since, the dispersion relation is periodic, the fluxon line corresponding to any non-zero velocity, $0 < u < 1$, will cross the dispersion relation (54) at some k and lead to Cherenkov radiation. Such an effect has been observed in simulations by many independent groups, *e.g.*, see Ref. [53].

4. Conclusion

The principles of operation of a flux-flow oscillator based on a long Josephson junction have been discussed. We showed that a Josephson flux-flow oscillator is a unique device in the 100 GHz–1 THz frequency range. While the theory of the FFO, arrays and stacks of LJJs has been developed over many years, there still is room for improvement, especially with regard to radiation linewidth. In spite of a lack of a complete theoretical description of these devices, there are many practical approaches to improve the (power and linewidth) performance of such devices. In the near future we may expect power levels of 10–100 μW from arrays of coupled, coherently-operating devices. The currently achieved linewidth of a single-junction FFO using a PLL is about 1 Hz.

We also introduced and discussed the consequences of Cherenkov radiation of plasma waves due to fast-moving fluxons. It is an important effect which must be taken into consideration when designing an FFO based on arrays and stacks. As a parasitic effect it can be suppressed [54] and even exploited [43] in the devices which are expected to have output power several times greater than for a conventional single LJJ FFO.

References

1. V. P. Koshelets, S. V. Shitov, A. M. Baryshev, I. L. Lapitskaya, L. V. Filippenko, H. van de Stadt, J. Mees, H. Schaeffer, T. de Graauw, *"Integrated sub-mm wave receivers"*, IEEE Trans. Appl. Supercond. **5**, 3057–3060 (1995).
2. V. P. Koshelets, S. V. Shitov, L. V. Filippenko, A. M. Baryshev, H. Golstein, T. de Graauw, W. Luinge, H. Schaeffer, H. van de Stadt, *"First implementation*

of a superconducting integrated receiver at 450 GHz" Appl. Phys. Lett., **68**, 1273–1275, (1996).

3. K. K. Likharev, *"Dynamics of Josephson junctions and circuits"*, Gordon and Breach Science Publishers, 1986.
4. A. Barone, G. Paterno, *"Physics and Applications of the Josephson effect"*, Wiley, New York, 1982.
5. V. V. Schmidt (Editors P. Müller and A. V. Ustinov), *"The Physics of Superconductors — Introduction to Fundamentals and Applications"* Springer (1997).
6. J. C. Swihart, *"Field Solution for a Thin-Film Superconducting Strip Transmission Line"*, J. Appl. Phys. **32**, 461–469 (1961).
7. I. O. Kulik, *"Wave propagation in a Josephson tunnel junction in the presence of vertices and the electrodynamics of weak superconductivity"*, Sov. Phys. J. Exp. Theor. Phys. **24**, 1307 (1967).
8. D. W. McLaughlin, A. C. Scott, *"Perturbation analysis of fluxon dynamics"*, Phys. Rev. A **18**, 1652-1680 (1978).
9. P. M. Marcus, Y. Imry, *"Steady Oscillatory States of a Finite Josephson Junction"*, Solid State Commun. **33**, 345 (1980).
10. T. Nagatsuma, K. Enpuku, F. Irie, and K. Yoshida, *"Flux-flow type Josephson oscillator for millimeter and submillimeter wave region"*, J. Appl. Phys. **54**, 3302 (1983); *"Flux-flow type Josephson oscillator for millimeter and submillimeter wave region. II. Modeling"*, J. Appl. Phys. **56**, 3284 (1984); *"Flux-flow type Josephson oscillator for millimeter and submillimeter wave region. III. Oscillation stability"*, J. Appl. Phys. **58**, 441 (1985); *"Flux-flow type Josephson oscillator for millimeter and submillimeter wave region. IV. Thin-film coupling"*, J. Appl. Phys. **63**, 1130 (1988).
11. E. Goldobin, Ph.D. Thesis (in Russian), Moscow (1997), available online http://www.geocities.com/Broadway/2442.
12. M. Fiske, *"Temperature and Magnetic Field Dependences of the Josephson Tunneling Current"* Rev. Mod. Phys., **36**, 221–222 (1964).
13. M. Cirillo, N. Grønbech-Jensen, M. R. Samuelsen, M. Salerno, G. V. Rinati, *"Fiske modes and Eck steps in long Josephson junctions: Theory and experiments"*, Phys. Rev. B **58**, 12377–12384 (1998).
14. I. O. Kulik, *"Theory of "steps" of voltage-current characteristic of the Josephson tunnel current"*, JETP Lett. **2**, 84–87 (1965).
15. I. O. Kulik, *"Theory of resonance phenomena in superconducting tunneling"*, Sov. Tech. Phys. **12**, 111–116 (1967).
16. Jhy-Jiun Chang, *"Fiske steps in Josephson junctions"*, Phys. Rev. B **28**, 1276 (1983).
17. Y. Zhang, Ph.D. thesis: *"Dynamics and Applications of Long Josephson Junctions"*, Chalmers University of Technology, Gothenburg, Sweden, 1993.
18. Y. Zhang, P. H. Wu, *"Numerical calculation of the height of velocity-matching step of flux-flow type Josephson oscillator"*, J. Appl. Phys. **68**, 4703–4709 (1990).
19. V. P. Koshelets, S. V. Shitov, L. V. Filippenko, V. L. Vaks, J. Mygind, A. B. Baryshev, W. Luinge, N. Whyborn, *"Phase locked 270–440 GHz local oscillator based on flux flow in long Josephson tunnel junctions"*, Rev. Sci. Instr. **71**, 289–293 (2000).
20. A. A. Golubov, B. A. Malomed, and A. V. Ustinov, *"Radiation linewidth of a long Josephson junction in the flux-flow regime"*, Phys. Rev. B **54**, 3047–3050 (1996).
21. A. P. Betenev and V. V. Kurin, *"Radiation spectrum of a long Josephson flux-flow oscillator"*, Phys. Rev. B **56**, 7855–7857 (1997).
22. A. V. Ustinov, H. Kohlstedt, and P. Henne, *"Giant Radiation Linewidth of Multifluxon States in Long Josephson Junctions"*, Phys. Rev. Lett. **77**, 3617–3620 (1996).
23. Y. M. Zhang, D. Winkler, and T. Claeson, *"Linewidth measurements of Josephson flux-flow oscillators in the band 280–330 GHz"*, Appl. Phys. Lett. **62**, 3195–3197 (1993).
24. S. V. Shitov, A. V. Ustinov, N. Iosad, and H. Kohlstedt, *"On-chip radiation detection*

from stacked Josephson flux-flow oscillators" J. Appl. Phys. **80**, 7134–7137 (1996).
25. V. P. Koshelets, S. V. Shitov, A. V. Shchukin, L. V. Filippenko, J. Mygind, A. V. Ustinov, *"Self-pumping effects and radiation linewidth of Josephson flux-flow oscillators"*, Phys. Rev. B **56**, 5572–5577 (1996).
26. A. V. Ustinov, T. Doderer, R. P. Huebener, J. Mygind, V. A. Oboznov, and N. F. Pedersen, *"Multi-fluxon effects in long Josephson junctions"*, IEEE Trans. Appl. Supercond. **3**, 2287–2294 (1993).
27. V. P. Koshelets, A. Shchukin, I. L. Lapytskaya, and J. Mygind, *"Spectral linewidth of autonomous and injection-locked flux-flow oscillators"*, Phys. Rev. B **51**, 6536–6541 (1995).
28. V. P. Koshelets, S. V. Shitov, A. V. Shchukin, L. V. Filippenko, and J. Mygind, *"Linewidth of submillimeter wave flux-flow oscillators"*, Appl. Phys. Lett. **69**, 699–701 (1996).
29. V. P. Koshelets, S. V. Shitov, A. V. Shchukin, L. V. Filippenko, P. N. Dmitriev, V. L. Vaks, J. Mygind, A. Baryshev, W. Luinge, H. Golstein, *"Flux Flow Oscillators for Sub-mm Wave Integrated Receivers"*, IEEE Trans. Appl. Supercond. **9**, 4133–4136 (1999).
30. M. B. Mineev, G. S. Mkrtchyan, and V. V. Shmidt, *"On Some Effects in a System of Two Interacting Josephson junctions"*, J. Low Temp. Phys. **45**, 497–505 (1981).
31. A. Petraglia, A. V. Ustinov, N. F. Pedersen, S. Sakai, *"Numerical study of fluxon dynamics in a system of two stacked Josephson junctions"*, J. Appl. Phys. **77**, 1171–1177 (1995).
32. A. V. Ustinov, H. Kohlstedt, and C. Heiden, *"Possible phase locking of vertically stacked Josephson flux-flow oscillators"*, Appl. Phys. Lett. **65**, 1457–1459 (1994).
33. N. Grønbech-Jensen, J. A. Blackburn, *"Hyperradiance from soliton oscillators synchronized by capacitive or inductive coupling"*, J. Appl. Phys. **74**, 6432 (1993).
34. T. Holst, J. Bindslev Hansen, N. Gønbech-Jensen, J. A. Blackburn, *"Phase locking between Josephson soliton oscillators"*, Phys. Rev. B **42**, 127–131 (1990).
35. E. Goldobin, P. G. Litskevitch, and V. P. Koshelets, *"Superconducting Digital Correlator for Integrated Sub-mm Receiver for Space Application"*, IOP Conf. Ser. (EUCAS'95 suppl.), 1697–1700, (1995).
36. E. Goldobin, P. G. Litskevitch, and V. P. Koshelets, *"Superconducting Digital Correlator for Integrated Sub-mm Receivers"*, presented at ISEC'95, Nagoya, Japan, September 18–21, 1995, report 7-20, 217–219 (1995).
37. A. V. Rylyakov, D. F. Schneider, Yu. A. Polyakov, *"A Fully Integrated 16-channel RSFQ Autocorellator Operating at 11 GHz"*, IEEE Trans. Appl. Supercond. **9**, 3623–3627 (1999).
38. R. Kleiner, F. Steinmeyer, G. Kunkel, and P. Müller, *"Intrinsic Josephson effects in $Bi_2Sr_2CaCu_2O_8$ single crystals"*, Phys. Rev. Lett. **68**, 2394–2397 (1992); R. Kleiner and P. Müller, *"Intrinsic Josephson effects in high-T_c superconductors"*, Phys. Rev. B **49**, 1327–1341 (1994).
39. A. F. Volkov, *"Solitons in Josephson superlattices"*, JEPT Lett. **45**, 376–379 (1987).
40. L. Bulaevskii, and J. R. Clem, *"Vortex lattice of highly anisotropic layered superconductors in strong, parallel magnetic fields"*, Phys. Rev. B **44**, 10234–10238 (1991).
41. P. Caputo, A. V. Ustinov, N. C. H. Lin, and S. P. Yukon, *"Radiation Emission from Triangular Arrays of Josephson Junctions"*, IEEE Trans. Appl. Supercond. **9**, 4538–4541 (1999); N. C. H. Lin, and S. P. Yukon, *"Maximizing Microwave Power From Triangular Josephson Junction Arrays"*, IEEE Trans. Appl. Supercond. **9**, 4320–4324 (1999).
42. P. Barbara, A. B. Cawthorne, S. V. Shitov, and C. J. Lobb, *"Stimulated Emission and Amplification in Josephson Junction Arrays"*, Phys. Rev. Lett. **82**, 1963–1966 (1999).
43. V. V. Kurin, and A. V. Yulin, *"Radiation of linear waves by solitons in a Josephson transmission line with dispersion"*, Phys. Rev. B **55**, 11659–11669 (1997).
44. A. Baryshev, A. V. Yulin, V. V. Kurin, V. P. Koshelets, S. V. Shitov, A. V. Shchukin,

P. N. Dmitriev, L. V. Filippenko, *"Design and Fabrication of Cherenkov Flux-Flow Oscillator"*, IEEE Trans. Appl. Supercond. **9**, 3737–3740 (1999).
45. V. Kurin, A. Yulin, E. Goldobin, A. Klushin, H. Kohlstedt, M. Levitchev, N. Thyssen, *"Experimental Investigation of Cherenkov Flux-Flow Oscillators"*, IEEE Trans. Appl. Supercond. **9**, 3733–3736 (1999).
46. S. Sakai, P. Bodin, and N. F. Pedersen. *"Fluxons in thin-film superconductor-insulator superlattices"*, J. Appl. Phys. **73**, 2411–2418 (1993).
47. S. Sakai, A. V. Ustinov, H. Kohlstedt, A. Petraglia, and N. F. Pedersen, *"Theory and experiment on electromagnetic-wave-propagation velocities in stacked superconducting tunnel structures"*, Phys. Rev. B **50**, 12905–12914 (1994).
48. E. Goldobin, A. Golubov, A. V. Ustinov, *"Two-fold stacks of long Josephson junctions with different parameters"*, Czech. J. Phys. **46** (LT-21 Suppl. S2), 663-664 (1996).
49. A. Wallraff, E. Goldobin, A. V. Ustinov, *"Numerical Analysis of the Coherent Radiation Emission by Two Stacked Josephson Flux-Flow Oscillators"*, J. Appl. Phys. **80**, 6523–6535 (1996).
50. E. Goldobin, A. Wallraff, N. Thyssen, and A. V. Ustinov, *"Cherenkov radiation in coupled long Josephson junctions"*, Phys. Rev. B **57**, 130–133 (1998).
51. R. Kleiner, *"Two-dimensional resonant modes in stacked Josephson junctions"*, Phys. Rev. B **50**, 6919–6922 (1994).
52. G. Hechtfischer, R. Kleiner, A. V. Ustinov and P. Müller, *"Non-Josephson Emission from Intrinsic Junctions in* $Bi_2Sr_2CaCu_2O_{8+y}$*: Cherenkov Radiation by Josephson Vortices"*, Phys. Rev. Lett. **79**, 1365–1368 (1997).
53. A. V. Ustinov, B. Malomed, and S. Sakai, *"Bunched fluxon states in one-dimensional Josephson-junction arrays"*, Phys. Rev. B **57**, 11691–11697 (1998).
54. A. V. Ustinov, and S. Sakai, *"Submillimeter-band high-power generation using multilayered Josephson junctions"*, Appl. Phys. Lett. **73**, 686–688 (1998).

INDEX

GPSR Compliance

The European Union's (EU) General Product Safety Regulation (GPSR) is a set of rules that requires consumer products to be safe and our obligations to ensure this.

If you have any concerns about our products, you can contact us on ProductSafety@springernature.com

In case Publisher is established outside the EU, the EU authorized representative is:

Springer Nature Customer Service Center GmbH
Europaplatz 3
69115 Heidelberg, Germany

Batch number: 10809436

Printed by Printforce, the Netherlands